Metal Contamination in Aquatic Environments: Science and Lateral Management

Metal contamination is one of the most ubiquitous, persistent and complex environmental issues, encompassing legacies of the past (e.g. abandoned mines) as well as impending, but poorly studied, threats (e.g. metallo-nanomaterials). Writing for students, risk assessors and environmental managers, Drs Luoma and Rainbow explain why controversies exist in managing metal contamination and highlight opportunities for policy solutions stemming from the latest advances in the field. They illustrate how the 'lateral' approach offers opportunities in both science and management, making the case that the advanced state of the science now allows bridging of traditional boundaries in the field (e.g. between field observations and laboratory toxicology). The book has a uniquely international and interdisciplinary perspective, integrating geochemistry, biology, ecology and toxicology, as well as policy and science. It explicitly shows how science ties into today's regulatory structure, identifying opportunities for more effective risk management in the future.

SAM LUOMA was a research scientist at the US Geological Survey for 34 years, before moving to the John Muir Institute of the Environment of the University of California at Davis. He was the first Lead Scientist for California's redesign of water management, gaining valuable experience in linking science and policy. He was a W.J. Fulbright Distinguished Scholar in 2004 and was awarded the US Presidential Rank Award for career accomplishments in 2005.

PHIL RAINBOW is Keeper of Zoology at The Natural History Museum in London, leading a staff of more than 100 working scientists in one of the premier scientific institutions in the world. He was awarded the annual Kenneth Mellanby Review Lecture Prize by the journal *Environmental Pollution* in 2002.

Metal Contamination in Aquatic Environments

Science and Lateral Management

Samuel N. Luoma
John Muir Institute of the Environment
University of California at Davis and
The Natural History Museum

Philip S. Rainbow
The Natural History Museum

Illustrations by Jeanne DiLeo

CAMBRIDGE UNIVERSITY PRESS

Cambridge, New York, Melbourne, Madrid, Cape Town, Singapore, São Paulo, Delhi

Cambridge University Press
The Edinburgh Building, Cambridge CB2 8RU, UK

Published in the United States of America by Cambridge University Press, New York

www.cambridge.org
Information on this title: www.cambridge.org/9780521860574

First published 2008

Printed in the United Kingdom at the University Press, Cambridge

A catalogue record for this publication is available from the British Library

Library of Congress Cataloging-in-Publication Data

Luoma, Samuel N.
  Metal contamination in aquatic environments : science and lateral management / Samuel N. Luoma and
Philip S. Rainbow.
    p.   cm.
  Includes bibliographical references and index.
  ISBN 978-0-521-86057-4 (hardback)
1. Metals–Environmental aspects. 2. Water–Pollution. 3. Water quality management. 4. Metals–Toxicology.
5. Contaminated sediments. 6. Trace elements in water. I. Rainbow, P. S. II. Title.

  TD427.M44L86 2008
  628.1′68–dc22

                                                                               2008022723

ISBN 978-0-521-86057-4 hardback

Contents

Preface   *page* xiii
Acknowledgements   xiv

1 Introduction   1
1.1 Why a book on metal contamination?   1
1.2 Metal issues are important   2
1.3 Understanding is growing but the
     science is complex   5
1.4 What are trace metals, and why
     should environmental managers
     know about them?   7
    1.4.1 Definitions   7
    1.4.2 Essentiality   9
    1.4.3 Toxicity   10
    *1.4.4 How are metals different from
          synthetic organic pollutants?*   11
1.5 Scope of this book   12
1.6 Conclusions   12

**2 Conceptual underpinnings: science and
   management**   13
2.1 Introduction   13
2.2 Framework for managing metal
     contamination   13
2.3 Approaches to managing metal
     contamination   15
    2.3.1 National-level assessments   15
    2.3.2 Site-specific assessments   16
    *2.3.3 Good chemical and good
          ecological status*   17
2.4 Assessing hazard and risk   18
    2.4.1 Ranking hazard or potential for risk   18
    2.4.2 Implementing risk assessment   19
    2.4.3 Risk management   20
2.5 Addressing uncertainty   20
    2.5.1 The Precautionary Principle   21

2.6 Adaptive management   24
2.7 Linkage of science and management   25
    *2.7.1 Connecting the scientific disciplines
          and their links to policy*   26
2.8 Integrating science and management   29
2.9 Conclusions   30

3 Historical and disciplinary context   32
3.1 Introduction   32
3.2 Disciplinary perspectives in metal
     science   32
3.3 Toxicity testing: applied metal
     ecotoxicology   34
    3.3.1 Priorities and benefits   34
    3.3.2 Limits   35
3.4 The environmental geochemistry
     perspective   36
    *3.4.1 Determination of metal
          concentrations*   36
    3.4.2 Metal cycling and metal reactions   37
    3.4.3 Limitations   39
3.5 Ecosystem and natural history
     perspective   39
3.6 Overlaps among core disciplines:
     detecting detrimental effects   42
3.7 Policy perspectives on the different
     disciplines   44
3.8 Conclusions   46

4 Sources and cycles of trace metals   47
4.1 Biogeochemical cycles of trace metals   47
    4.1.1 Natural cycles   48
    4.1.2 Human sources   50
    *4.1.3 Anthropogenic additions to
          biogeochemical cycles*   52
    4.1.4 Trends   55

4.2 Use of mass inputs to manage metal contamination   56
   4.2.1 *Non-point, diffuse sources of metals*   57
   4.2.2 *Anthropogenic sources and mass loading allocations*   59
   4.2.3 *Interaction of natural processes and mass inputs: primary, secondary and tertiary inputs of metals*   62
   4.2.4 *Direct determination of ecological risks from mass inputs*   62
4.3 Conclusions   64

5 Concentrations and speciation of metals in natural waters   67
5.1 Introduction   67
5.2 Loads and concentrations   68
   5.2.1 *What is dissolved metal?*   68
   5.2.2 *Determining dissolved metal concentrations*   70
   5.2.3 *Concentrations in natural waters and processes affecting them*   71
      5.2.3.1 Extreme contamination   71
      5.2.3.2 Dissolved metal concentrations in historically contaminated estuaries   71
      5.2.3.3 Dissolved metals in the oceans   73
      5.2.3.3.1 Vertical variability in concentrations   73
      5.2.3.3.2 Spatial and temporal variability   75
      5.2.3.3.3 Contaminated ocean waters   76
      5.2.3.4 Dissolved metal concentrations in rivers   77
      5.2.3.5 Dissolved metal concentrations in lakes   80
      5.2.3.6 Dissolved metal concentrations in estuaries and coastal waters   80
5.3 Speciation of trace metals   83
   5.3.1 *Oxidation state*   84
   5.3.2 *Speciation of redox-sensitive anionic metals*   84
   5.3.3 *Speciation of cationic metals*   85
   5.3.4 *Organometallic compounds*   87
   5.3.5 *Determining speciation in natural waters: models and analytical approaches*   89
5.4 Conclusions   91

6 Trace metals in suspended particulates and sediments: concentrations and geochemistry   93
6.1 Introduction   93

6.2 Units: concentrations   94
6.3 Metal concentrations in suspended particulate material and sediments   94
   6.3.1 *Decomposition and analysis*   95
   6.3.2 *Geological inputs: what is the natural background or regional baseline?*   96
   6.3.3 *Granulometric (particle size) biases*   98
   6.3.4 *Methods to improve comparability among data*   99
      6.3.4.1 Chemical extraction of anthropogenic metal   100
      6.3.4.2 Separating fine-grained sediment for analysis   101
      6.3.4.3 Normalisation   101
6.4 Spatial distribution of sediment contamination   104
   6.4.1 *Range and frequency of contamination in sediments*   105
   6.4.2 *Characteristics of distributions*   107
      6.4.2.1 Processes that determine metal distributions   108
      6.4.2.2 Diffusional flux of dissolved contamination   110
6.5 Sediment chemistry   110
   6.5.1 *Redox reactions affect metal partitioning*   111
   6.5.2 *The sediment column*   111
   6.5.3 *Metal form in oxidised sediments*   115
   6.5.4 *Differentiating reduced metal forms*   117
   6.5.5 *Effect of redox on metals that behave as anions*   119
6.6 Metal partitioning: the distribution coefficient K_d 120
   6.6.1 *Effect of water chemistry on K_d*   121
   6.6.2 *Pore waters*   122
6.7 Conclusions   123

7 Trace metal bioaccumulation   126
7.1 Bioaccumulation of trace metals   126
7.2 Processes governing bioaccumulation   128
   7.2.1 *Uptake processes*   129
      7.2.1.1 Processes at the membrane   129
      7.2.1.2 Uptake of dissolved forms   133
      7.2.1.2.1 Uptake rates and K_u   133
      7.2.1.2.2 External factors affecting dissolved uptake (physicochemistry)   134
      7.2.1.2.2.1 Chelation or complexation   135
      7.2.1.2.2.2 Major ion concentrations and salinity   136

7.2.1.2.2.3 Differences in pH   137
7.2.1.2.2.4 Other trace metals   138
7.2.1.2.3 *Internal (biological) factors affecting dissolved uptake (physiology)*   138
7.2.1.2.3.1 *Species specificity of* K_u 138
7.2.1.2.3.2 *Intraspecific variation*   140
7.2.1.3 *Dietary uptake processes*   141
7.2.1.3.1 *Feeding rate*   142
7.2.1.3.2 *Assimilation efficiency*   142
7.2.1.3.3 *Taxon specificity of assimilation efficiency*   143
7.2.1.3.4 *Range of variability: metal and concentration dependence*   144
7.2.1.3.5 *Digestive processes*   145
7.2.1.3.6 *Chemical nature of metal in the diet*   146
7.2.1.3.7 *Effects of excess metal*   150
7.2.1.3.8 *Exposure history*   150
7.2.2 *Physiological loss or excretion*   150
7.2.2.1 Efflux rate constant (K_e)   151
7.3 Accumulation   153
7.3.1 *Metabolic requirements for trace metals*   155
7.3.2 *Detoxified metal*   155
7.3.3 *Accumulation patterns*   156
7.4 Unifying the processes that determine bioaccumulation (biodynamic modelling)   159
7.4.1 *Trophic transfer of metals*   164
7.5 Conclusions   167

8 Biomonitors   169
8.1 Monitoring and biomonitors   169
8.2 Dissolved metal concentrations   170
8.3 Sediment concentrations   171
8.4 Biomonitors   172
8.4.1 *Requirements and uses of biomonitors*   172
8.4.2 *Cosmopolitan biomonitors of marine waters*   175
8.4.2.1 Solution   175
8.4.2.2 Solution and suspended particles   175
8.4.2.3 Solution and sediment   177
8.4.2.4 Herbivores and detritivores   185
8.4.3 *Tropical coastal biomonitors*   186
8.4.4 *Freshwater biomonitors*   187
8.4.5 *Interpretation of biomonitoring data*   193
8.4.5.1 Gut contents   193
8.4.5.2 Size, weight and age   194
8.4.5.3 Season   196
8.4.5.4 Moult cycle   197
8.4.5.5 Intrasample variability   197
8.4.5.6 Systematics   198

8.4.5.7 Intraspecific comparisons of absolute trace metal concentrations   199
8.4.5.8 Interspecific comparisons   200
8.4.6 *Translocation*   201
8.5 Conclusions   202

9 Manifestation of the toxic effects of trace metals: the biological perspective   204
9.1 Introduction   204
9.2 Toxicity and dose   204
9.3 Uptake, bioaccumulation and toxicity   206
9.4 Hierarchy of responses through levels of biological organisation   208
9.5 Biomarkers   209
9.5.1 *Molecular biology*   210
9.5.2 *Biochemistry*   211
9.5.2.1 Biomics   212
9.5.2.2 Metallothionein   212
9.5.2.3 General biochemical stress responses   213
9.5.3 *Cytology*   213
9.5.3.1 Micronucleus formation   214
9.5.3.2 Lysosomes   215
9.5.3.2.1 *Change in size and number*   215
9.5.3.2.2 *Destabilisation of the lysosomal membrane*   215
9.5.3.2.3 *Generation of lipofuscin and/or lipid peroxidation*   215
9.5.3.3 Phagocytosis   216
9.5.4 *Histopathology*   217
9.5.5 *Morphology*   217
9.5.6 *Physiology*   218
9.5.7 *Linking responses in suites of biomarkers to physiological change*   221
9.6 Responses that are integrative of lower-order effects   223
9.6.1 *Behaviour*   223
9.6.2 *Tolerance*   224
9.6.3 *Population*   226
9.6.4 *Community*   228
9.7 Conclusions   230

10 Toxicity testing   232
10.1 Introduction   232
10.2 History and goals of toxicity tests   233
10.3 Principles of toxicity testing   235
10.3.1 *Media*   237
10.3.2 *One species or many?*   238

10.3.3 *Endpoint and level of organisation*   241
    10.3.3.1 Measurement, assessment and performance   241
    10.3.3.2 Traditional lethality tests   243
    10.3.3.3 Sublethal responses   243
10.3.4 *Test species*   244
    10.3.4.1 Species differ widely in their susceptibility to metals   245
    10.3.4.2 Hardiness of test species might affect their representativeness   246
    10.3.4.3 At least some test organisms can only be tested in narrow range of conditions   246
    10.3.4.4 If sufficient data are available, statistical tests can help select the most sensitive species   246
    10.3.4.5 Some groups of species in different types of habitat appear to be highly sensitive to metals   247
10.3.5 *Life stage*   247
10.3.6 *Time of exposure*   250
10.3.7 *Environmental conditions (physical and chemical choices)*   252
10.3.8 *Exposure route*   253
10.4 Validation of extrapolations from toxicity testing   256
10.5 Conclusions   258

11 Manifestation of metal effects in nature   260
11.1 Introduction   260
11.2 Experimental evidence defines the feasibility of metal effects in natural waters   260
11.2.1 *Dietary exposure*   261
11.2.2 *Algal tests and speciation*   263
11.2.3 *Sublethal toxicity and population effects*   265
11.2.4 *Multiple sources of stress*   267
11.3 Establishing field-based effects   267
11.3.1 *Understanding exposure and natural variability*   268
    11.3.1.1 Distributions in time and space   269
    11.3.1.2 Temporal variability of exposure   269
11.3.2 *Fluctuations in biological properties*   272

11.3.3 *Statistical power*   274
11.3.4 *Multiple lines of evidence*   274
11.3.5 *Conceptual model of ecological change caused by metals*   276
11.4 Case studies for metal effects in natural waters   277
11.4.1 *Co-occurrence of contamination and metal effects in marine communities*   279
11.4.2 *Stream benthos: sensitive species, patterns of distribution and effects on ecosystem services*   282
    11.4.2.1 Toxicity tests   282
    11.4.2.2 Mechanisms   283
    11.4.2.3 Validation of toxicity testing with field observations   284
    11.4.2.4 Loss of ecosystem services   285
    11.4.2.5 Links to upper trophic levels   285
11.4.3 *Effects in lakes: elimination of species and cascades to upper trophic levels*   287
11.5 Indirect effects   290
11.6 Conclusions   291

12 Mining and metal contamination: science, controversies and policies   293
12.1 Introduction   293
12.2 Sources of metals: nature of mines and mining activities   296
12.3 Sources: primary contamination   298
12.3.1 *On-site disturbance*   298
12.3.2 *Smelting*   301
12.3.3 *Generation of acid mine drainage*   301
12.4 Secondary and tertiary contamination   303
12.4.1 *Contamination of rivers*   303
    12.4.1.1 Dispersal of metals in mine-impacted rivers   303
    12.4.1.2 Pre-mining dispersion of metals: what is background?   306
    12.4.1.3 Floodplains   307
    12.4.1.4 Accidental releases   308
12.4.2 *Estuaries and lakes as receiving waters*   308
12.4.3 *Atmospheric dispersal: smelter wastes*   309

12.5 Ecological risk: bioavailability,
   toxicity and ecological change   311
   12.5.1 Acid mine drainage   312
   12.5.2 Ecological risks from secondary
      particulate contamination:
      complicating factors   314
12.6 Signs of metal effects   316
   12.6.1 Ecological signs of risk
      in marine environments   316
   12.6.2 Ecological signs of risk
      in streams and rivers   318
12.7 Socio-economic consequences of
   ecological impacts   321
12.8 Recovery, remediation and
   rehabilitation   323
12.9 Conclusions   325

13 Selenium: dietary exposure, trophic
   transfer and food web effects   327
13.1 Introduction   327
13.2 Conceptual model   328
13.3 Sources: why is selenium
   an element of global concern?   329
13.4 Total mass loading   330
13.5 Water concentrations   333
13.6 Speciation   336
13.7 Phase transformation   337
   13.7.1 Plant uptake and
      transformation   337
   13.7.2 Adsorption   337
   13.7.3 Volatilisation   338
   13.7.4 Dissimilatory reduction   339
   13.7.5 Partition coefficient K_d 339
13.8 Bioaccumulation by consumer
   species   340
   13.8.1 Uptake   341
   13.8.2 Physiological loss rates   342
   13.8.3 Variability of bioaccumulation
      in natural waters   343
13.9 Trophic transfer   343
13.10 Toxicity   346
   13.10.1 Signs of Se toxicity   346
   13.10.2 Detecting toxicity:
      Se bioaccumulation in
      target organisms   347

   13.10.3 Differences in sensitivity   347
   13.10.4 Interaction of Se with other
      contaminants   349
13.11 Effects   350
13.12 Conclusions   351

14 Organometals: tributyl tin and
   methyl mercury   354
14.1 Introduction   354
14.2 Tributyl tin   355
   14.2.1 Sources and uses   355
   14.2.2 Concentrations and geochemistry   356
   14.2.3 Bioaccumulation   357
   14.2.4 Sublethal effects   358
   14.2.5 Toxicity testing   361
   14.2.6 Ecological observations from
      natural waters   362
   14.2.7 Risk management   362
   14.2.8 Recovery   365
   14.2.9 Was the partial ban adequate?   366
   14.2.10 Conclusions: tributyl tin   368
14.3 Methyl mercury   368
   14.3.1 Introduction   368
   14.3.2 Sources of mercury   369
      14.3.2.1 The global mercury cycle   370
      14.3.2.2 Historical mercury sources   371
      14.3.2.3 Modern mercury sources   371
   14.3.3 Mercury methylation   374
   14.3.4 Concentrations   375
   14.3.5 Bioaccumulation   377
      14.3.5.1 Invertebrate
         bioaccumulation   378
      14.3.5.2 Fish bioaccumulation   380
   14.3.6 Food web contamination   382
      14.3.6.1 Relationship of Hg
         loading to food web
         contamination   382
      14.3.6.2 Food web characteristics
         influence contamination   384
   14.3.7 Ecological effects   386
   14.3.8 Managing risks from
      methyl mercury   388
   14.3.9 Conclusions: methyl mercury   391

15 Hazard rankings and water quality
   guidelines   393
15.1 Introduction   393

15.2 Hazard identification   394
    15.2.1 Persistence   395
    15.2.2 Bioaccumulation and alternatives 396
    15.2.3 Toxicity   396
    15.2.4 Trophic transfer   398
15.3 Protocols: ambient criteria and
    risk assessment   399
    15.3.1 Ambient water quality criteria 399
    15.3.2 How are criteria derived? 401
    15.3.3 Ecological risk assessment 402
15.4 Uncertainty   403
    15.4.1 Geochemical uncertainties 403
        15.4.1.1 Total metal versus
                dissolved metal 403
        15.4.1.2 Hardness and water
                effects ratio   404
        15.4.1.3 Biotic Ligand Model (BLM) 405
    15.4.2 Biological uncertainties 409
        15.4.2.1 Acute-to-chronic ratios 409
        15.4.2.2 Most sensitive species 411
    15.4.3 Uncertainties deriving from the
        unique attributes of metals 413
15.5 Validating ambient water quality
    criteria   416
    15.5.1 How comparable are criteria
        from different jurisdictions? 416
    15.5.2 Are environmental metal
        standards congruent with
        expected metal concentrations
        in contaminated waters? 417
    15.5.3 When should corrections for
        geochemical conditions be applied? 419
    15.5.4 Were all data adequately considered? 420
15.6 Influence of assumptions on
    criteria derivation: cadmium   420
15.7 Conclusions   423

16 Sediment quality guidelines   425
16.1 Introduction   425
16.2 The setting   425
16.3 Toxicity   427
    16.3.1 Bioassays   427
    16.3.2 Complex sediment toxicity tests 429
    16.3.3 Geochemical uncertainties unique to
        sediment toxicity tests 432
        16.3.3.1 Sediment collection 432
    16.3.4 Appropriate uses of sediment
        bioassays   434

16.4 Metal concentrations in sediments:
    links to bioaccumulation
    and ecological risk   434
    16.4.1 How do metal concentrations in
        sediment relate to risk:
        bioaccumulation? 436
    16.4.2 Route of exposure 438
        16.4.2.1 Influence of sediment–
                water partitioning (K_d) 440
16.5 Metal bioavailability: variability in
    the relationship between
    concentration and toxicity   441
    16.5.1 Anoxic sediments 442
    16.5.2 Oxidised sediments 444
16.6 Physical processes   449
16.7 Management of sediment
    contamination   450
    16.7.1 Toxicity testing and sediment
        quality management 451
    16.7.2 Managing sediments on the basis
        of metal concentrations 452
    16.7.3 Guidelines defined relative to some
        reference concentration 452
    16.7.4 Empirically based guidelines 453
    16.7.5 Mechanistically based guidelines 455
    16.7.6 Lateral management: multiple
        lines of evidence 457
16.8 Conclusions   458

17 Harmonising approaches to managing
metal contamination: integrative and
weight-of-evidence approaches   460
17.1 Introduction   460
17.2 Integrative management
    approaches   462
    17.2.1 Dietary Metal Guidelines
        (DMG)   463
    17.2.2 Bioaccumulated Metal
        Guidelines (BMG) 467
        17.2.2.1 Inappropriate applications
                of a BMG   468
        17.2.2.2 Appropriate applications
                of a BMG   471
17.3 Wildlife criteria   476
17.4 Holistic or lateral approaches to
    managing metal contamination:
    weight of evidence   476

17.4.1 *Lateral management*   478
    17.4.1.1 Site-specific criteria
         developed from added
         lines of evidence   479
    17.4.1.2 Using weight of evidence
         from a water body to
         establish a regulation   480
    17.4.1.3 Schemes for weighting
         lines of evidence in
         holistic criteria   482
17.4.2 *Integrative risk assessment*   483
17.5 Prioritising   486
17.6 Conclusions   486

18 Conclusions: Science and policy   489
18.1 Introduction   489

18.2 State of knowledge   489
18.3 What is a science–policy interaction?   492
18.4 Developing a science dialogue
    useful to policy   494
18.4.1 *Goals*   494
18.4.2 *Trust*   496
18.4.3 *Scientific uncertainty*   497
18.5 The mechanics of a useful
    science–policy interaction   499
18.6 Constructing a science agenda   500
18.7 Conclusions   505

References   507
Index   556

Preface

This book provides a comprehensive background for those who are interested in the management of ecological risks from metal contamination in the aquatic environment. An important premise is that risk assessment and environmental standards will play an increasingly important role in the future in managing risks from metal contamination. Recent advances in scientific knowledge offer opportunities to improve existing approaches to risk management, and most effectively if a more lateral management approach can be initiated. Applications of ecotoxicology to risk assessment and risk management have not been yet fully successful at breaking down the boundaries between geochemistry, biology, ecology and toxicology; or penetrating the divide between reductionism and observational studies. Our purpose is to present the science of metal contamination from a perspective that integrates the different views of this complex field. We identify instances where boundaries impede risk management, and provide examples of how lateral interpretation can help the field move forward.

Metal contamination is one of the most ubiquitous, persistent and complex chemical contamination problems faced by human society. Yet, no comprehensive and coherent single analysis is presently available of metals science and how it applies to specific ecological risk management issues for metals. This book is such a synthesis. A premise is that practitioners must understand processes that determine the fate and effects of metals in nature, as well as they know basic toxicology or geochemistry. The first 11 chapters of the book present the fundamentals of the science that underlie metal policies and issues. The chapters are organised around a coherent conceptual model. Perspectives from geochemistry, biology/ecology and ecotoxicology are contrasted in the presentation of the science and analyses of issues. The next seven chapters more directly address the interaction between science and policy in specific circumstances. Some of the more difficult issues and some major approaches to risk management are discussed in individual chapters, as well as throughout the book. In the conclusions we address constructive dialogue between the scientific and the policy communities.

The controversies that surround specific metal management issues are not yet fully appreciated. We identify a number of those controversies and analyse their scientific underpinnings. We address the reasons for successful policies, or lack thereof, and suggest a way forward for many issues. A main goal is to help practitioners (risk assessors, environmental managers, stakeholder scientists) and students understand why such controversies exist, so they are prepared to develop policy solutions from the latest advances in the field.

The book builds insights from the respected careers, the broad experience and the rich publication records of the co-authors. A historical perspective will be a theme throughout. Interdisciplinary and international perspectives will be emphasised. The book is designed to be a text as well as a unique reference for the international community of risk assessors and scientists studying or consulting on metal issues.

Acknowledgements

Any work of this sort is the product, literally in our case, of decades of contributions in the form of ideas, shared works, discussions and collaborations with friends, colleagues, students, postdoctoral associates and family. The list of people we should acknowledge for such contributions is too long to include here; and surely we would miss someone crucial. Thank you to all those contributors. But there are those who must be thanked individually.

SNL wishes to acknowledge Geoff Bryan, who remains an inspiration; as well as ideas from and discussions with Nicholas Fisher, Johnnie Moore and Greg Cutter, all collaborators and co-authors for more than a decade. Decades of teamwork with Cynthia Brown, Daniel Cain, James Carter, Robin Bouse and Michelle Hornberger along with other past members of our research team are much appreciated, as are the recent contributions of Robin Stewart, Marie-Noele Croteau, B.G. Lee and David Buchwalter. SNL's participation in this work was partly supported by a W.J. Fulbright Distinguished Scholar Award, partly by the US Geological Survey, and would not have been possible without the full-time support of Elaine Dorward-King.

PSR wishes to acknowledge with gratitude collaborations with Cherif Abdennour, Graham Blackmore, Maria Caparis, Laurie Chan, Michael Depledge, Wojciech Fialkowski, Malika Galay Burgos, John Jennings, Malcolm Jones, Islay Marsden, Darren Martin, Geoff Moore, Dayanthi Nugegoda, Gabriel Nunez-Nogueira, Sian Pullen, Jason Weeks, Stephen White and Maciej Wolowicz in the many trace metal biology studies which have provided the background experience relevant to the writing of this book. Brian Smith in particular has provided stalwart research support over years, for which many thanks. PSR is especially indebted to Jean-Claude Amiard, Claude Amiard-Triquet, Dave Phillips and Wen Wang for many in-depth discussions over a long period that have formulated his thinking on concepts in trace metal biology. Very grateful thanks to them, and, of course, ultimately to my family for their support and forbearance.

1 • Introduction

1.1 WHY A BOOK ON METAL CONTAMINATION?

> The good news is that the environment continues to enjoy a high level of support among the public. Perhaps it is something from our biological origins; something we are hard-wired to yearn for. But greater resolve and creativity will be needed if we are to build upon the successes of the recent past.
>
> (paraphrased from Schnoor, 2004)

Prologue. The premise of this book is that cross-boundary and integrative (lateral) thinking will create new opportunities and lead the next generation of practices for managing risks from chemical contamination. The purpose of this book is to illustrate the use of lateral thinking in the application of a knowledge of science to one set of contamination issues: metal contamination.

Managing risks from toxic wastes will be an ongoing problem for human civilisation. We will not manage these problems efficiently and effectively until we better learn to cross boundaries in our thinking and practices. Diffuse disciplinary boundaries separate physical, biogeochemical, toxicological, biological, ecological and socio-political views of contamination issues. Boundaries also exist between process-orientated science and regulatory science; as they can exist between observational science and experimental science. Such boundaries create barriers to new ideas and to progress. When we begin to break down these barriers, we will begin to better incorporate all the knowledge and tools at our disposal into our

---

Box 1.1 Definitions

ecotoxicology: the field of study that integrates the ecological and toxicological effects of chemical pollutants on populations, communities and ecosystems with the fate (transport, transformation and breakdown) of such pollutants in the environment. (After Forbes and Forbes (1994) as *environmental toxicology*.)

fugacity: a measure of the tendency of a substance, often a fluid, to move from one phase to another, driven by a basic property. In organic chemical contamination, the tendency of the chemical to partition between octanol and water is fundamental to how it distributes among phases at equilibrium.

heavy metal: definitions may refer to metals with a specific gravity greater than 4 or even greater than 5. Nevertheless many authors would not describe the heavier metals in Groups 1 and 2 of the Periodic Table and the lanthanides as heavy metals although they fit this weight criterion, implying some expected chemical characteristics of the element concerned.

lateral thinking: a method of thinking concerned with changing concepts and perception, using reasoning that is not immediately obvious; exploring multiple possibilities and approaches instead of pursuing a single approach. (After Edward de Bono.)

ligand: an ion, a molecule or a molecular group that binds with a trace metal to form a larger complex.

metal: an element that has a lustrous appearance (high reflectivity for light), is a good conductor of electricity and heat, and usually enters chemical reactions as a positive ion (cation).

metalloid: an element with chemical properties intermediate between those of a metal and a non-metal, as defined by its location in the Periodic Table.

organometal: a combination of a metal and an organic entity (at least partly) covalently bonded. Usually the organic entity is a small alkyl group like a methyl or ethyl group. Examples are methyl mercury and tributyl tin.

trace metal: often used loosely without strict definition, although there is the implication that the metal is present in only 'trace' concentrations. Where defined, the concentration limit of 0.01% (also expressed as 100 parts per million (ppm), μg/g or mg/kg) (for example by dry weight in an organism) may be quoted, although this limit is often exceeded by the biological concentrations of many 'trace metals'. Here we use the term 'trace metals' to include metals and metalloids that share a common suite of chemical, biological and physical behaviours, all of which have the potential to be toxic to biota.

management practices, including risk assessment, risk management and precautionary practices. This approach is that of lateral thinking and, therefore, lateral management.

We choose metal contamination as an example because it is the most complex of today's contamination issues. Widespread uses of metals, the legacies of past contamination and new technologies continue to pose important ecological risks for aquatic environments across the Earth. Trace metals are very different from other pollutants. Our understanding of metal science is growing rapidly, but important aspects of metal behaviour are not yet fully known. The complexity of the subject and the growth of knowledge raise questions about the credibility of traditional management approaches. Demands on management are also changing and uncertainties that were less important in earlier eras are now leading to contentious debate and, in some cases, deadlock in decisions about managing risks.

Merging new scientific advances and new management approaches may be the solution to these challenges. But that will require that risk managers understand the state of a broad and complex science. The purpose of this book is to provide, in one place, a summary of the state of knowledge of trace metal science and how it applies to managing trace metal contamination. We use case studies to illustrate some of the more important risk management challenges. And we point toward opportunities to resolve some of the more difficult issues.

1.2   METAL ISSUES ARE IMPORTANT

Prologue. Humankind's historic legacy of environmental contamination includes instances of trace metal contamination with tragic ecological and human health consequences. Some of the easiest issues are moving toward resolution, but new and remaining issues present more difficult challenges.

The 1970s were a time when environmental awareness rose rapidly in the developed world, largely as a result of visual evidence of dramatically adverse ecological effects. In Lake Erie massive algal blooms annually resulted in malodorous die-offs of plants and animals, including decimation of fish populations. The River Thames was devoid of oxygen until the construction of sewage treatment facilities between 1850 and 1900, and devoid of most macroorganisms until treatment of water quality between 1950 and 1980 (Wood, 1982). In San Francisco Bay (USA) more than one oil spill occurred every day in the 1970s. There were 313 separate fish kills reported between 1963 and 1976, about 25 a year or two per month (Luoma and Cloern, 1982). The temperate zones of the Earth nearly lost some of our most magnificent wildlife during this period. Had scientific studies not shown that organochlorine pesticides were serious poisons, birds like the bald eagle, the brown pelican and the peregrine falcon would probably no longer inhabit our planet (Box 1.2).

Box 1.2 Bans on organochlorine pesticides, including DDT and PCBs

The history of organochlorine pesticides illustrate the necessity of crossing traditional scientific boundaries if an environmental problem is to be recognised and resolved. That history also shows that with these complex issues controversies can continue, even after the initial solutions seem to be found.

Testing first showed reproductive toxicity of organochlorines in wildlife in the 1950s but the potential for ecological risk was not fully appreciated. Uses in agriculture and for disease control expanded widely through the 1970s. Rachel Carson's now legendary observations suggested that wild bird disappearances were linked to such pesticides (Carson, 1962). In the 1960s field observations raised further alarms, describing crashing populations of birds of prey, including peregrine falcons, osprey, bald eagles and brown pelicans. For example, a National Audubon Society survey in 1963 showed only 417 bald eagle nests in the lower 48 states of the USA with an average of 0.6 birds produced per active nest (less than replacement). This was compared to what was once thought to be 100000 nesting pairs in the continental US (http://www.fws.endangered.fr95580.html). Bald Eagles were listed as endangered species under US law in 1967, brown pelicans in 1970 and peregrine falcon in 1972.

Thin eggshells were soon discovered to be a cause of the failure of young birds to hatch in some of the populations of concern, especially the fish-eating birds of prey (Hickey and Anderson, 1968; Ratcliffe, 1970). This observation was consistent with mechanistic studies which showed that DDT adversely affected calcium metabolism, and therefore limited calcium deposition in eggshells; an obvious explanation for the thin shells. Finally, sophisticated chemical analytical techniques became available that could detect DDT at the part per billion levels typical of nature. Environmental analyses verified that the organochlorines were ubiquitous in the environment and specifically concentrated in the birds whose populations were crashing (Risebrough et al., 1967). The chemicals reached such high concentrations in these predators because their concentrations were magnified at every level of aquatic food webs

(Woodwell et al., 1971). Each predator has a higher concentration of DDT than its prey so the worst effects occur at the highest trophic levels. Food web biomagnification was not (and usually cannot be) detected in the early testing.

The difficulty in detecting such effects experimentally was explained by the World Health Organization's (WHO) task group of experts on environmental criteria (WHO, 1989):

> DDT and its metabolites can lower the reproductive rate of birds by causing eggshell thinning (which leads to egg breakage) and by causing embryo deaths. However, different groups of birds vary greatly in their sensitivity to these chemicals; predatory birds are extremely sensitive and, in the wild, often show marked shell thinning, whilst gallinaceous birds (chickens) are relatively insensitive. Because of the difficulties of breeding birds of prey in captivity, most of the experimental work has been done with insensitive species (chickens), which have often shown little or no shell thinning. The few studies on more sensitive species have shown shell thinning at levels similar to those found in the wild.

DDT was banned in the US and Europe in the early 1970s. The evidence of effects on wildlife was unequivocal, stemming from a combination of laboratory experiments, field observations and mechanistic studies. The evidence of effects on human health was much less convincing at that time, and remains controversial.

A dramatic recovery of the predators after the ban was ultimate proof that diagnosis of the problem as chemical contamination was correct. Peregrine falcons were taken off the list of endangered US species in 1995. Brown pelicans were delisted on the Atlantic Coast and by California in 2006. In 1999 President Clinton proposed that the bald eagle be removed from the list; in 2007 that delisting moved forward.

The ban on organochlorines, in particular DDT, remains controversial, however. DDT is highly effective in killing mosquitoes that are vectors for malaria. Malaria is better controlled where DDT remains in use in the tropics than where it is banned. Advocates argue that the ban on DDT is therefore costing massive loss of human life. The uses of DDT responsible for spreading the chemical most widely were agricultural. Recognising that other uses may not be as great a threat to wildlife, the World Health Organization approved the indoor use of DDT for treatment of disease vectors in 2004; a compromise position between benefits to wildlife and risks to human health (of not using DDT). But an intense debate continues. Many of the lessons about the complexity of such an interplay of science and policy may be the same for trace metals as they were and are for DDT.

Tragic examples of unanticipated environmental effects also exist for trace metal contamination. The best known of the metal contamination catastrophes affected human health. Mercury poisoning was once termed Minamata disease. That is because at least 3000 people, including many newborn babies from small villages on Minamata Bay, Japan, suffered birth defects and terrible nervous system damage from inadvertently eating mercury-tainted shellfish and fish. Between 1932 and 1968, manufacture of fertilisers and petrochemicals released approximately 27 tons of mercury compounds into Minamata Bay. The methyl mercury that accumulated in the environment from these discharges was taken up by shellfish and fish that the local villagers depended upon for food. For decades, the company involved systematically paid off the families of villagers who suffered from an 'unclarified disease of the central nervous system'. The cause of the disease was first attributed to mercury in 1956 and again in 1965. Mercury discharges were halted in 1968. These were the first documented examples of human poisoning from an environmental pollutant in the food chain (Harada, 1995).

Japanese farmers were poisoned with cadmium, from 1940 to 1960, when cadmium-rich waters from a local mine were used to irrigate their rice. Thousands of men and women (20% of all women over 50 years of age in the most contaminated regions) suffered from kidney failure and severe osteoporosis, as a result of eating the rice. The disease was called itai-itai ('ouch-ouch').

A more modern example is occurring in Bangladesh and West Bengal in central Asia. In an ill-fated attempt to reduce cholera in people drinking from the Ganges River, the World Health Organization drilled millions of water wells, and encouraged the local populace to replace surface water with well water for drinking. Approximately 40% of the wells were contaminated with arsenic, mobilized from natural sources. Measured arsenic concentrations reach up to 1000 µg/L, far above the limit set for drinking water in Bangladesh (50 µg/L) or that recommended by the World Health Organization (10 µg/L). Consumption of this contaminated water has led to widespread death and disease, with millions of people affected (Nichson et al., 1998).

Although human health episodes raised consciousness of threats from metal contamination, ecological damage is probably more pervasive. Anecdotal evidence that metal pollutants might adversely affect plants, animals and ecosystems began to accumulate as the Industrial Revolution spread in the nineteenth century. For example, the deaths of deer caused by arsenic emissions from smelters were reported in Germany in the 1850s (Newman and Schreiber, 1988). Thousands of cattle, deer and horses were killed outright in the Deer Lodge Valley of Montana, USA when copper smelting first began in 1900 (Harkins and Swain, 1908). Food chain transfer of mercury also contributed to the near-demise of predatory birds (Chapter 14). Selenium contamination affects migratory birds over large areas of the western USA (Chapter 13), indicative of potential toxic effects in semi-arid irrigated lands across the world (Skorupa, 1998). Radioactive caesium from atmospheric tests of nuclear bombs by the USA and the Soviet Union was spread around the world in the 1950s, and ^{137}Cs in soils and sediments from that era is still easily detectable.

The closer we study, the more obvious it is that metals like copper, silver and zinc are potent toxins at the base of food webs, with documented effects on phytoplankton in lakes (Canada) and marine waters of the USA (Moffet et al., 1997; Sunda et al., 1990), as well as invertebrate populations in large areas of Rocky Mountain watershed, USA

(Clements *et al.*, 2000). Lead is both a human health issue and an ecological toxin. Tributyl tin caused well documented die-offs of bottom-dwelling invertebrates around the world until its ban from use as an antifoulant painted on the bottoms of small boats (Chapter 14). If not properly managed, tailings spills or chronic mobilisation from historic mining operations decimate river fish populations in otherwise pristine areas around the world (Chapter 12).

Decades of advances in science knowledge have aided progress with some important trace metal issues, especially in Europe, North America and Japan. Downward trends characterise some metal contamination in some of those locations. But science and policy face daunting new challenges and important questions in the future. Legacies of historic contamination remain. Abandoned mine sites and contaminated sediments are found throughout the world. Trends in some trace metal contaminants are not declining even where the strictest environmental protections are in place (selenium and mercury are two examples). New technologies, like nanotechnology, are introducing metals into the environment in forms with which we have little experience. Egregious contamination problems are reappearing where economies are growing rapidly. It is not yet clear that traditional management strategies will prevent us from repeating the old mistakes. The traditional regulatory frameworks must be re-examined to address such issues.

1.3   UNDERSTANDING IS GROWING BUT THE SCIENCE IS COMPLEX

Prologue. The risks from metals stem from their ubiquitous presence as contaminants wherever human activities impinge on aquatic environments. But there is controversy about how much of this metal contamination presents ecological risks. Those controversies can only be resolved by unraveling the complex effects resulting from metal exposures.

The challenges of managing future issues with trace metals will require:

(1) prioritizing understanding rather than simplicity in risk management;
(2) incorporating new advances in trace metal science, including differences between trace metals and organic contaminants;
(3) thinking laterally across the boundaries that prevent use of all the knowledge at our disposal; and
(4) reducing the inter-jurisdictional incoherence of management practices.

The public saw, from the first Earth Day in 1970, an immediate need for policies that protect against adverse effects of chemicals on the environment and human health. Policy makers interpreted the public desire to avoid repeating obvious disasters as an immediate need to clean up or eliminate problems, with or without fully integrated scientific knowledge. A political demand for immediacy led to a demand for 'simple measures' that could be applied to policy 'in a timely manner'. This political mandate had an immense influence on the historical development of understanding and risk management.

The interdisciplinary nature of chemical contamination issues has also been very influential in development of science and management. The science underlying the policies and management of metal issues is typically described as part of the scientific discipline of *ecotoxicology*, or *environmental toxicology*. The simplest definition of environmental toxicology (Forbes and Forbes, 1994) is 'the field of study which integrates the ecological and toxicological effects of chemical pollutants on populations, communities and ecosystems with the fate (transport, transformation, and breakdown) of such pollutants in the environment.'

Metal contamination issues require the skills of physicists, chemists, hydrologists, ecologists, biologists, physiologists and social scientists, in addition to toxicologists. They require experimentation and observational studies of natural waters. Applying that knowledge to risk management requires an entirely different skill set, including social and political sciences. Different approaches and different kinds of scientists characterise these different kinds of science.

Integrating work from different disciplines is always a challenge. Ecotoxicology is no exception. Depledge (1994b) wrote: 'it is only (since about 1990) that ecotoxicology has come to merit being regarded as a true science, rather than a collection of procedures'. One widely cited definition from Truhaut (1977) called ecotoxicology a sub-discipline of toxicology. Another (Moriarty, 1999) suggested that ecotoxicology should be a sub-discipline of ecology. But Moriarty noted that the only difference in practice between toxicology and ecotoxicology appears to be in the species selected for toxicity testing. Walker *et al.* (1996) stated that 'much early work . . . had little ecology or toxicology about it. It was concerned with the detection and determination of chemicals in the environment'. Many authors have lamented that the fields of toxicology and ecology are rarely associated (with one another) either academically or professionally. In discussing the future of ecotoxicology, Cairns (1993) opined: 'the field . . . must develop programs to fit a complex decision matrix that includes hazardous materials as an important part, but not the sole basis on which decisions are made'.

There are legitimate historical reasons for the inadequate integration of the past. The study of metal contamination is relatively young. The rapid growth of knowledge about ecological effects of metals really only started in the 1960s. In contrast, for example, modern biology built from a taxonomic system first published by Carl Linnaeus in 1735 in *Systema Naturae*. Charles Darwin's expedition aboard HMS *Beagle*, from which the theory of natural selection was developed, took place between 1831 and 1836. So biology has had 250 years to develop after the early descriptive studies, and ecology has had 150 years to digest the implications of its most fundamental theoretical advance. And these are young fields compared to physics and chemistry. The science that underlies our understanding of trace metal contamination has had 30 to 40 years to develop from its first solid roots. Integrated, interdisciplinary science can be expected to improve in the future with the growth of new knowledge. But that will require thinking laterally across the boundaries that separate the scientists and the sciences.

Another source of complexity is that the management approaches designed for organic contamination are often of marginal relevance for metals. For example, the number of organic contaminants that must be managed is huge; the number of metal products is much smaller. That makes simplified testing essential to cover the plethora of organic chemicals, but a hinder to understanding the complex circumstances that govern metal risks. The processes that govern the fate and effects of each are also quite different. The unique traits of metals are increasingly recognised, improving the likelihood of metal-specific management strategies.

New knowledge about trace metals offers opportunities to manage contamination more effectively. Credibility will increase as future policies better address local influences of geochemical processes; the implications of dietary exposure; biological differences in responses to metals; links between toxicity, population dynamics and community structure; and implications for human health. Data from natural waters are growing in sophistication, allowing the field to move beyond toxicology alone. Past policies include examples that remain justifiable (e.g. partial bans of lead and tributyl tin). But in other cases, controversy and/or policy deadlock are developing as new knowledge contradicts traditional approaches (some water quality standards, sediment quality criteria, guidelines for selenium management, guidelines for mercury management). More effective incorporation of new knowledge could improve management in each of the problematic areas.

Administrative incoherence adds to the challenge of managing metal contamination and can also threaten public trust. Water quality criteria (Chapter 15), for example, provide guidance about ecologically protective ambient concentrations for various metals. Jurisdictions typically must adopt water quality criteria that protect various uses for different kinds of water bodies, including protection of aquatic life. But water quality criteria, and the language used to describe them, can differ, sometimes widely, among jurisdictions; and can even differ among circumstances within one jurisdiction (Chapter 2). To the non-specialist this creates a confusing morass of laws and terminology, all with slightly different meanings. If regulatory systems appear to have their own language and their own art of enforcement then they are usually not easily penetrable by the stakeholder or the academic. Mastery of administrative process can become the emphasis, rather than mastery of the substance or intent of the law. Public trust in environmental protection depends upon continued investments in scientific studies, and transparent policies that increasingly assure that unexpected catastrophes will not occur again.

1.4   WHAT ARE TRACE METALS, AND WHY SHOULD ENVIRONMENTAL MANAGERS KNOW ABOUT THEM?

1.4.1   Definitions

Prologue. Heavy metals, metalloids and organometals have been previously defined as separate entities in formal chemical terms. Here we define all as trace metals, partly for convenience, but also because they all fall along a common continuum of geochemical, biological and hydrological behaviours.

Everyone knows what a metal is, but do we? We all understand iron (Fe), zinc (Zn) and lead (Pb), for example, to be metals, but what about selenium (Se) or arsenic (As)? What define a metal chemically are its physical and chemical properties. Crudely, chemists define metals as elements that have a lustrous appearance (high reflectivity for light), are good conductors of electricity and heat, and usually enter chemical reactions as positive ions (cations). What defines metal contamination, however, involves more than just chemistry. Each trace metal falls within a continuum of metal-specific bioavailability, biological and ecological behaviours that are just as important to the significance of contamination as is metal chemistry.

The Periodic Table (Fig. 1.1) lists all known elements, ultimately classifying elements on their atomic structure, so providing organisation to an understanding of their different physical and chemical characteristics. Of all the 90 elements occurring naturally on earth, most (67) can be considered as metals, of which all but one (Hg) is solid at room temperature (Williams and Fraústo da Silva, 1996). And yet, even chemists will differ in their opinions as to the exact limits of the definition of a metal. In a strict definition of a metal, its electrical conductivity decreases with increasing temperature. In the case of semiconductors, conductivity increases with increasing temperature. Some semiconductors (e.g. B, C, S, Si, Ge, Te, I) are generally accepted not to be metals, whilst others (As, Sb, Bi, Se) are often included as metals, although sometimes qualified as semimetals or metalloids (e.g. Walker *et al.*, 1996).

Different groupings of metals can be made according to context, and groupings overlap. So-called 'precious metals' include metals used in coinage and jewellery (Ag, Au, Pt, Pd, Ir), and 'noble metals' are those most resistant to high-temperature oxidation (Ag, Au, Pd, Pt, Rh). Contrasting with these are chemically reactive base metals – 'alkali metals' are those of Group 1 of the Periodic Table (Li, Na, K, Rb, Cs), and 'alkaline earth metals' are those of Group 2 (Be, Mg, Ca, Sr, Ba, Ra). 'Rare earth metals' are the lanthanides of atomic numbers 58 to 71. 'Transition metals' are the metals in the middle block of the Periodic Table (Fig. 1.1), from atomic numbers 21 to 30, 39 to 48, and 57 to 80, and include many of the metals of ecotoxicological interest in this book, such as Cu, Zn, Ag, Cd and Hg.

In the field of ecotoxicology two further groupings of metals are commonly used, 'heavy metals' and 'trace metals', and these groupings are similarly elusive to pin down. Historically the term 'heavy metal' has been used extensively to describe metals that are environmental pollutants. Definitions of heavy metals, where expressed, usually refer to metals with a specific gravity greater than 4 or even greater than 5. Such a definition would include the heavier Group 1 and Group 2 metals and the lanthanides, not usually referred to as 'heavy metals'. Thus it is apparent that implicit but undefined in the use of the term 'heavy metal' in ecotoxicology is some expected feature of the chemical characteristics of the element concerned. The second term 'trace metal' is similarly often used loosely without strict definition. The

Group

1	2	3	4	5	6	7	8	9	10	11	12	13	14	15	16	17	18
1 H																	2 He
3 Li	4 Be											5 B	6 C	7 N	8 O	9 F	10 Ne
11 Na	12 Mg											13 Al	14 Si	15 P	16 S	17 Cl	18 Ar
19 K	20 Ca	21 Sc	22 Ti	23 V	24 Cr	25 Mn	26 Fe	27 Co	28 Ni	29 Cu	39 Zn	31 Ga	32 Ge	33 As	34 Se	35 Br	36 Kr
37 Rb	38 Sr	39 Y	40 Zr	41 Nb	42 Mo	43 Tc	44 Ru	45 Rh	46 Pd	47 Ag	48 Cd	49 In	50 Sn	51 Sb	52 Te	53 I	54 Xe
55 Cs	56 Ba	57* La	72 Hf	73 Ta	74 W	75 Re	76 Os	77 Ir	78 Pt	79 Au	80 Hg	81 Ti	82 Pb	83 Bi	84 Po	85 At	86 Rn
87 Fr	88 Ra	89† Ac	104 Rf	105 Db	106 Sg	107 Bh	108 Hs	109 Mt									

* Lanthanides	58 Ce	59 Pr	60 Nd	61 Pm	62 Sm	63 Eu	64 Gd	65 Tb	66 Dy	67 Ho	68 Er	69 Tm	70 Yb	71 Lu
† Actinides	90 Th	91 Pa	92 U	93 Np	94 Pu	95 Am	96 Cm	97 Bk	98 Cf	99 Es	100 Fm	101 Md	102 No	103 Lr

Fig. 1.1 The Periodic Table of the elements with atomic numbers, and groups numbered according to the International Union of Pure and Applied Chemistry. (After Williams and Fraústo da Silva, 1996.)

description implies that the metal is present in only 'trace' concentrations. Where defined, the concentration limit of 0.01% (for example by dry weight in an organism) may be quoted. This concentration limit (also expressed as 100 parts per million (ppm), μg/g or mg/kg) does include most concentrations in organisms of the metals usually included under this term. However, there are some spectacular exceptions, such as the accumulated zinc concentrations in barnacles and oysters that typically exceed 10000 μg/g (see Chapter 7). Furthermore, for some authors the term 'trace metal' implies essentiality – that is a metal required in very small doses for the functioning of the metabolism of the organism (see below), but this is not the assumption of other authors who include both essential and non-essential metals under the definition of 'trace metal'.

In the search for objective definitions, Nieboer and Richardson (1980) proposed a chemical classification system based on the Lewis acid properties of metal ions, separating these into Class A, Class B or Borderline according to their degree of 'hardness' or 'softness' as acids and bases. Such a chemical classification depends ultimately on the affinity of metal ions for other elements. If a metal (M) binds to a ligand (L) as represented by the reversible equation

$$M + L = ML \quad \text{and} \quad K_{ML} = [ML]/[M][L]$$

then, the constant K_{ML} is a measure of the affinity of M for L. Using relative values of K_{ML} it is possible to construct preference tables of different metal ions for different ligands.

Class A metal ions are Lewis hard acids. These readily form cations (are 'more ionic') and have a ligand affinity order O > N > S, while Class B metal ions are Lewis soft acids, are 'more covalent' and have an affinity order S > N > O. Borderline metal ions have intermediate properties. This

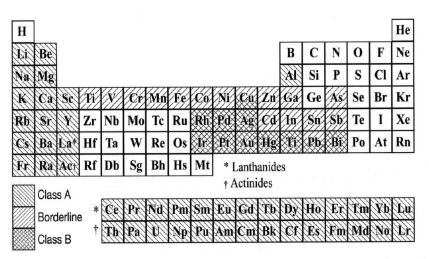

Fig. 1.2 The classification of metal ions into Class A, Borderline and Class B categories based upon Lewis acid properties. (After Nieboer and Richardson, 1980.)

classification of metal ions can be mapped onto the Periodic Table (Fig. 1.2). Class A ions include the ions of Group 1 and 2 metals (e.g. Na, Mg, K, Ca, sometimes referred to as 'major ions') and the lanthanides and actinides. Class B ions are those of some of the transition metals (e.g. Cu(I), Ag, Hg, Ti, Pt, Pd, Au) and Pb (Pb(IV)). Borderline metal ions also include ions of transition metals (V, Cr, Mn, Fe, Co, Ni, Cu(II), Zn, Cd), as well as those of Sn and Pb (Pb(II)), and the so-called metalloids As and Sb (Fig. 1.2).

Such a chemical classification of metal ions does have the great advantage of chemical objectivity, but it would be unrealistic to ignore the fact that the more general, undefined terms 'heavy metal' and 'trace metal' are still commonly used in the ecotoxicological literature. As a pragmatic compromise, throughout this book we will use the term 'trace metal'. We shall, however, define it, not on the basis of concentrations, but to refer to all metals, essential or not, with ions that fall into the Class B or Borderline categories defined by Nieboer and Richardson (1980), with the pragmatic additions of molybdenum (Mo) and selenium (Se). Another metal of ecotoxicological significance falls outside this definition of a 'trace metal' – that is aluminium (Al) which has ions with Class A properties. This example further typifies the general problem of fitting all metals of ecotoxicological significance into any objective chemical definition. Trace metals are listed in Table 1.1.

Table 1.1 *Trace metals and their abbreviations*

Antimony (Sb)	Manganese (Mn)
Arsenic (As)	Mercury (Hg)
Bismuth (Bi)	Molybdenum (Mo)
Cadmium (Cd)	Nickel (Ni)
Chromium (Cr)	Palladium (Pd)
Cobalt (Co)	Platinum (Pt)
Copper (Cu)	Rhodium (Rh)
Gallium (Ga)	Selenium (Se)
Gold (Au)	Silver (Ag)
Indium (In)	Tin (Sn)
Iridium (Ir)	Titanium (Ti)
Iron (Fe)	Vanadium (V)
Lead (Pb)	Zinc (Zn)

1.4.2 Essentiality

Prologue. Trace metals required for life processes in very small doses include Ti, V, Cr, Mn, Fe, Co, Ni, Cu, Zn, As, Mo, Sn and Sb; these are the 'essential' metals. Essential metals have been harnessed by evolution for use in metabolic pathways, exploiting their affinities for sulphur and nitrogen.

Some, but not all, trace metals are essential, required in very small doses for the functioning of the metabolism of the organism. This metabolic requirement historically would have been recognised by the appearance of deficiency symptoms, for example in livestock, in the absence or shortage of the metal concerned. More objectively today, essentiality is recognised by an identified biochemical role for a metal in a metabolic pathway. Metals may play an essential role in metalloproteins in which the metal is an integrated part required for the function of the protein, as in the case of zinc in the enzyme carbonic anhydrase, iron in the respiratory protein haemoglobin, or copper in another respiratory protein haemocyanin. Alternatively, metals may be more loosely bound in metal-activated proteins, as in the case of manganese and the enzyme aminopeptidase; the metal is still needed for the protein to function in metabolism but may be replaced by another trace metal.

Given the universality of most biochemical pathways amongst the biota of the Earth, it is not surprising that metals essential to one organism are essential to most others (especially for example among eukaryotic organisms). There are exceptions. Vanadium is essential for the nitrogen fixation biochemical pathway, unique to prokaryotes. Selenium is universally required in the enzyme glutathione peroxidase. But its additional role in anaerobic respiration in prokaryotes is absent from eukaryotes. Until relatively recently, cadmium was a classic example of a non-essential metal, but it is now known that particular oceanic phytoplankton have a carbonic anhydrase that depends on cadmium and not the usual trace metal zinc (Cullen *et al.*, 1999; Lane *et al.*, 2005).

Lists of essential trace metals therefore generally include Ti, V, Cr, Mn, Fe, Co, Ni, Cu, Zn, As, Mo, Sn and Sb. Non-essential trace metals usually include Pd, Ag, Cd (now qualified), Pt, Au, Hg and Pb.

It is the affinity of trace metals for ligands such as sulphur and nitrogen that means that trace metals are readily incorporated into proteins, the side chains of the many constituent amino acids of proteins containing these elements. The different redox properties of different trace metals, themselves affected by the surrounding amino acids within a particular protein, offer tremendous potential for natural selection to harness them in the evolution of metabolic pathways, including electron transfer and oxygen transport (see Lippard and Berg, 1994; Williams and Fraústo da Silva, 1996).

1.4.3   Toxicity

Prologue. The same high affinity for ligands such as sulphur and nitrogen that allowed the evolution of trace metals as essential components of metabolism makes all trace metals potentially toxic, binding to proteins or other molecules and preventing them from functioning in their normal metabolic role. Non-essential metals are not necessarily more toxic than essential metals.

On entry into an organism, a trace metal is available to bind to any nearby site or ligand for which it has an affinity. Adverse effects occur when one trace metal substitutes for another on a protein and blocks the catalytic active site of a metalloprotein. Alternatively a trace metal may bind elsewhere on a protein and exert a steric effect, distorting the geometry of the protein and preventing its normal metabolic action. Thus all trace metals are potentially toxic to life, including essential trace metals when present in excess.

The physiology of an organism may interact to excrete a metal or bind it to a particular molecule of high affinity from which the metal is unlikely to escape. Detoxification involves strong metal binding to proteins such as metallothioneins or to insoluble metalliferous granules (Mason and Jenkins, 1995). It follows therefore, that the manifestation of trace metal toxicity occurs when the rate of entry of a trace metal into an organism exceeds the combined rates of its excretion and detoxification by that organism (Rainbow, 2002) (see Chapter 7 for details).

It is important to remember that all trace metals are potentially toxic, whether essential or not, at a threshold of availability. Figure 1.3 compares the effect of increased availability of an essential

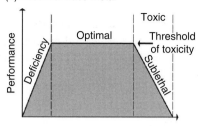

(a) Essential trace metal

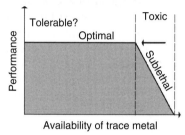

(b) Non-essential trace metal

Availability of trace metal

Fig. 1.3 The effect of increasing availability of (a) essential and (b) non-essential trace metals on the performance, e.g. growth, production, fecundity, survival, etc., of an organism.

and a non-essential metal on the performance (measured, for example, as growth, production, fecundity, survival, etc.) of an organism. The position of the onset of toxic effects along the trace metal availability axis (which might be measured for example as dissolved concentration) will vary greatly between organisms for the same trace metal, and between trace metals for the same organism. Furthermore, trace metals will not necessarily follow the same rank order of toxicities between organisms, depending on differences between the uptake rates, detoxification rates and excretion rates (see Chapter 7) of the different organisms compared. Nevertheless an approximate general order of toxicities of trace metals in inorganic form might be

Hg > Ag > Cu > Cd > Zn > Ni > Pb > Cr > Sn.

The rank order reiterates that non-essential trace metals are not necessarily more toxic than essential trace metals.

1.4.4   How are metals different from synthetic organic pollutants?

Prologue. The principles that govern environmental risks from trace metals are different in many ways from the principles that govern risks from organic chemicals.

As elements, metals are non-biodegradeable and cannot be broken down into less harmful components. Metals are also a natural component of the Earth's crust. Animal and plant life have thus evolved mechanisms to take advantage of their good traits and mitigate (to some degree) their bad traits. Among the adaptations to metals are mechanisms that aid detoxification. Detoxification requires trace metals to be bound within appropriate insoluble inorganic granules or in proteins for storage in a form unable to interact with important molecules of the metabolism (see Chapter 7). The specific characteristics of each trace metal affect their individual affinities for different ligands, helping to determine whether they are detoxified by being bound with sulphur, for example in a sulphur-rich protein or as an inorganic metal sulphide, or to the phosphate group in an insoluble metal phosphate. Furthermore, different organisms may make use of the different affinities of any one trace metal in the choice of detoxificatory binding available. For example, zinc, with ions of intermediate 'borderline' binding affinities, may be bound to sulphur in proteins and/or to phosphate in granules, the relative degree depending on the organism involved.

In contrast, organic compounds, be they naturally occurring polyaromatic hydrocarbons or synthetic organochlorines, are complicated molecules that often can be detoxified by enzymatic breakdown into less toxic (often more hydrophilic and less persistent) forms. Furthermore, such organic molecules tend to obey the same chemical rules of aquatic partitioning, fugacity modelling, quantitative structure–activity relationships (QSARs), etc. (Connell *et al.*, 1999). Aquatic partitioning means that the body concentrations of many such compounds will be predictable from a few relatively simple characteristics. No such general prediction can be made for accumulated concentrations of trace metals in organisms, any change in accumulated concentration being the result of more complex interactions between metal uptake, storage and excretion, themselves varying greatly between metals and particularly among organisms of different taxa (see Chapter 7).

1.5   SCOPE OF THIS BOOK

An important purpose of this book is to address the needs for better managing metal contamination. The scope is limited to trace metals and to the aquatic environment. The role of trace metals in human health is a fascinating subject of its own, but is included here only as a marginal consideration; that is largely an issue for a different volume.

We will also often use abbreviations throughout, spelling out the name of each trace metal when it is first used and when complete spelling is beneficial to the context. Chemical abbreviations are included in Table 1.1 as a reference for those unaccustomed to them.

In general the book is designed to help managers, risk assessors and scientists make use of the knowledge that has been accumulated about the fate and effects of trace metals in aquatic environments. Our emphasis is on the exciting new principles uncovered in the last decades, but we try to recognise where these have built from a longer legacy of work. In some cases the history of discovery is as important as the advances that built from that history. A complete understanding of administrative process is not our goal. We assume that managers and risk assessors will develop that understanding from management experience and training.

1.6   CONCLUSIONS

The basic hazards and risks of trace metals have been studied for decades. Laws and administrative process exist to manage metal contamination, but there is still disagreement about important issues. The scientific framework, upon which many of today's policies rest, is also fundamentally changing. Incorporating new science into policy requires understanding the principles of that science, its uncertainties and its trade-offs. The next generation of risk assessors and managers, therefore, must be sophisticated in connecting traditionally disparate disciplines and integrating the traditionally different worlds of science and policy.

The priority on simplicity and timeliness, inadequate integration among disciplines and a lack of balance in applying different disciplines are at the heart of our most serious (recognised and unrecognised) modern dilemmas about how to manage metal contamination, adding to the challenge of dealing with new knowledge and new issues. Some early contamination issues were readily visible and relatively easy to diagnose. Justifiable policies exist for most of those. The issues that are left are much more complex, and, in many cases, a great deal is at stake. History shows that incremental solutions to even the most complex problems are possible as new knowledge develops and we learn from experience. It also shows that creative solutions will be obstructed if priorities are askew.

Suggested reading

Cairns, J. Jr (1993). Environmental science and resource management in the 21st century: scientific perspective. *Environmental Toxicology and Chemistry*, **12**, 1321–1329.

2 • Conceptual underpinnings: science and management

2.1 INTRODUCTION

Prologue. This chapter introduces the framework for understanding metal contamination and its management. It also introduces different management approaches and discusses the linkage between policy tools and different aspects of metal science. A simple conceptual model is presented that is common to all disciplines and ties together science and policy approaches.

Trace metal contamination of aquatic environments occurs as a result of many human activities. Rational management of the ecological risks from contamination requires a coherent process that can identify contamination, recognise its implications and provide approaches that reverse or prevent effects adverse to nature and society. Management should become more rational as scientific understanding grows. The more information about degree of contamination, its sources and its site-specific damage, the more effective and efficient is risk management. The earliest approaches to managing chemical contamination were created in response to government mandates in the 1970s (e.g. the Clean Water Act in the USA). Various formulations of trace metals were among the most ubiquitous of the chemicals found in waste streams as a result of their widespread use in human activities. But little specific information existed about ecological risks from these trace metals.

2.2 FRAMEWORK FOR MANAGING METAL CONTAMINATION

Prologue. 'It will seldom if ever be possible, or justifiable, to eliminate risk of effects altogether since cost will be an important consideration and the essential requirement will be to strike a balance between the cost of reducing effects and the value of the benefit achieved in so doing' (Preston, 1979).

The original approaches to managing metal contamination issues evolved at a time when severe contamination occurred frequently and its effects were unambiguous. It was clear that justifiable legal means were needed to regulate the development, manufacture or release of chemicals that were at least potentially dangerous to humans and the environment.

Countries throughout the developed world followed a largely parallel history. Events in the USA were typical. After the first Earth Day on 1 May 1970, there was a burst of legislation directed at protecting the environment (e.g. Foran, 1993). For example, the Federal Water Pollution Control Act of 1972 (the Clean Water Act) was designed to clean the nation's waters. The Toxic Substances Control Act of 1976 mandated characterising the impacts of every chemical manufactured in the USA. The Ocean Dumping Act limited offshore dumping of wastes. The US Environmental Protection Agency was created to implement the legislation.

Scientific input was also needed to define the means and measures to implement the objectives of these laws. Preston (1979) succinctly described the consensus of the time on how to do so:

- Establish a criterion describing what dose of metal yields what response in organisms (the relationship between high concentrations and detrimental effects);
- Set primary protection standards for natural waters, based upon that relationship;

Box 2.1 Definitions

adaptive management: a systematic process for continually improving management policies and practices by learning from the outcomes of management actions.

bioavailability: describes a relative measure of that fraction of the total ambient metal that an organism actually takes up when encountering or processing environmental media, summated across all possible sources of metal, including water and food as appropriate.

Environmental Quality Standard: a value, generally defined by regulation, which specifies the maximum permissible concentration of a trace metal in an environmental sample. EQSs define acceptable limits for pollutants in water, designed to render a body of water suitable for its designated uses. Implemented as criteria, standards, objectives, directives, guidelines or trigger points.

Environmental Quality Objective: a non-enforceable goal, which specifies a target for environmental quality in a river, beach or industrial site. EQOs are generally not set by regulation (unlike Environmental Quality Standards), can be in the form of generally desirable objectives, and are usually not concrete quantitative measures.

exposure assessment: the degree of exposure that organisms experience in the specific circumstances from all potential media (e.g. food and water). In practice this involves estimation of emissions to environmental compartments, determination of the fate of such emissions in the environment, derivation of predicted environmental

concentrations, estimation of uptake by fauna (Floyd, 2006). See Chapter 16.

hazard: determination of the potential adverse effects that a substance has an inherent capacity to cause. Hazard is determined by a metal's intrinsic properties. Hazard identification criteria are thus independent of exposure conditions. Data that can be considered include chemical and physical properties, toxicity tests, field or epidemiological studies. Internationally consistent criteria are established to define the hazard of a chemical, although implementation can vary with different circumstances (Floyd, 2006). See Chapter 16.

Precautionary Principle: 'Where there are threats of serious or irreversible damage, lack of full scientific certainty shall not be used as a reason for postponing cost effective measure to prevent environmental degradation.' Rio Declaration.

risk assessment: a process of evaluating the likelihood and severity of an adverse effect(s)/event(s) occurring to humans or the environment following exposure under defined conditions to a trace metal. A risk assessment is comprised of hazard identification, hazard characterisation, exposure assessment and risk characterisation. A risk assessment must always include characterisation of attendant uncertainties.

risk (ecological): the probability of an adverse effect/ event occurring to humans or the environment following exposure to a trace metal, under defined conditions. The quantitative or semi-quantitative estimate of the probability of occurrence and severity of adverse effect(s)/event(s) in a given population under defined exposure conditions based on hazard identification, hazard characterisation and exposure assessment (Floyd, 2006).

- Identify which targets are to be protected (and how to protect them);
- Within the standards, take into account the characteristics of the particular receiving environment (the water body and its ecology) . . . and considerations of cost/risk and cost/benefit of regulating the discharge; and
- Derive standards to regulate metals in waste discharges based on the primary protection standards, targets, and local conditions.

Once the conceptual basis was set for primary protection standards, studies and data were necessary to support those standards. Concentrations and toxicity of effluents were some of the first data to be collected, to evaluate whether specific discharges were complying with the law. Beginning in the 1970s the National Pollutant Discharge Elimination System in the USA (NPDES) has issued permits to limit the pollutants discharged in effluents from specific facilities or point sources. The UK has a similar licensing system for effluents above a certain volume. The goal was to keep these discharges from severely polluting the water, making the water unsafe for public use, or unsuitable for aquatic life.

It was later recognised that data from natural waters were critical in order to justify the effluent controls. Environmental protection could be achieved by comparing toxicity data to environmental concentrations, and regulating to prevent toxic levels from being reached in the natural waters. For that, it was necessary to determine the maximum contaminant levels allowable. As issues were identified (contaminated waters, contaminated sediments) standards were set.

Early standards were based upon a combination of simplistic dose response studies and professional judgement (Cairns, 2003). Precautionary protections, not necessarily fully rooted in scientific evidence, were built into the regulatory rules with the goal of assuring environmental protection. That approach clearly worked in the first two decades after passage of legislation like the US Clean Water Act. Environmental benefits were visible, and progress was demonstrated in a few studies showing ecological recoveries (Box 3.10).

Today's needs are more challenging, more diverse and, potentially, technically more contentious than those in the past. Visibly obvious contamination (or effects from contamination) left no doubt that some action was better than none. Any attempts to manage contamination would result in progress. This situation may still exist in some parts of the world. But, in other places, trade-offs and uncertainties dominate the questions: How much clean-up is enough? Are ecological benefits resulting from investments in clean-up or reduced metal releases? What are the environmental and economic trade-offs of mitigating a specific risk? What are the uncertainties, and how do we make decisions in the face of uncertainty about effectiveness?

The suitability of the early standards is being questioned (e.g. Chapter 15). Some criteria appear to be overly protective of the environment; some appear to not protect the environment sufficiently. Some concepts in the original consensus, like taking environmental characteristics into account,

were not seriously incorporated into the early regulations. Today there is increasing demand for such considerations. The approaches to managing metals are in transition. We are confronted with a demand to address the 'balance' described by Preston (1979). To do that, new and more sophisticated approaches will be required.

2.3   APPROACHES TO MANAGING METAL CONTAMINATION

Prologue. Contaminants are managed at national and local scales, on the basis of both inherent hazard and local risk. Criteria, standards, objectives, guidelines and trigger points are the different science-based measures that regulatory agencies use to implement the intent of environmental legislation.

The scientific knowledge of metal issues and the general framework for managing risks are, today, applied in several ways. Firstly, regulatory bodies manage contamination at different scales.

2.3.1   National-level assessments

Prologue. Most governments set nationally consistent standards to protect natural waters from adverse effects from metals. As knowledge grew it became clear that determining the concentration of metals that can damage an aquatic environment over large scales (national criteria) is more technically complex than originally anticipated.

Most governments set nationally consistent environmental quality standards for metals (Box 2.2). Examples include *Maximum Contaminant Levels* (MCL) which are the highest concentration of a chemical allowed, usually in an effluent. *Continuous Concentration Criteria*, are the average concentrations allowed in a water body (USA terminology). *Ambient Water Quality Criteria* and *Sediment Contamination Guidelines* are other examples of chemical-specific national-level risk management criteria or guidelines (see Chapters 15 and 16). Australia and New Zealand take a slightly different approach, defining *trigger*

Box 2.2 Implementation of legislation: international criteria and standards

Around the world, criteria, standards, guidelines, objectives, directives and trigger points are the means by which higher-level jurisdictions (national governments or the EU Commission) interact with regional and local jurisdictions to set limits on the acceptable degree of ambient contamination. They provide a reference against which the state of water quality in a particular water body can be checked. Each approach dictates a different kind of action.

The terminologies are inconsistently used, unfortunately, among jurisdictions. In the USA, aquatic water quality criteria (USEPA, 2002b) are the governmental guidance that establishes the limits to alteration of water quality by human activities (Adams and Rowland, 2003). In the USA, federal criteria provide guidance for lower-level jurisdictions (e.g. states) who must adopt *water quality standards* that protect designated uses. The standards define the specific concentration of a trace metal in water that should not be exceeded to protect aquatic life. In British Columbia, Canada, *water quality objectives* are criteria enacted by the province, but they include consideration of the *national water quality guidelines*. The guidelines are not enforceable laws. Water quality objectives routinely provide policy direction for resource managers for the protection of water uses and guide the evaluation of

water quality, the issuing of permits, licences, etc. In Europe, Directives are the most common form of environmental law; they are flexible and contain differing requirements for different member states in the Union. European 'regulations' are more directly binding. In Australia and New Zealand, detailed risk analysis is required if concentrations exceed *trigger points*.

Exceedances of criteria are linked to further regulatory action such as analyses of mass loading on a watershed basis. In British Columbia exceedance can lead to basin-wide water quality studies. In the USA exceedance leads to Total Maximum Daily Load analyses at the watershed scale. The EU is initiating a slightly different approach, in which characterisation of all watersheds is mandated as the initial step (the Water Framework Directive). Exceedance of guidelines or criteria is one basis from which the watersheds are classified as to the quality of their water (from worst to best).

Implementation of the different methodologies can take several forms: (1) numerical values applied to all waters; (2) values modified to reflect site-specific conditions; (3) narratives that explain the goal rather than quantify it (e.g. no further deterioration); and (4) values derived using scientifically defensible methods other than those formally specified by the ruling jurisdiction. Some version or mix of all forms is used in most places, although broadly applied numerical values are the most common approach.

points as the concentration above which careful study of effects must begin. The European Union establishes broad narrative objectives and standards to meet these. Many national level methodologies are managed as absolute values above or below which specified actions are dictated. When metal concentrations exceed a specific value, reduction of the contamination is required.

Simple, single-value national standards contributed to eliminating egregious contamination, historically. The uncertainties were small compared to the problems, and protective standards were given the benefit of the doubt because scientific knowledge was minimal. Now such standards often generate controversy as we try to become more exact in specifying, in a national standard, what levels of contamination will cause what damage

in a specific setting. One part of the confusion stems from increasing knowledge of the influence of environmental conditions in determining risks. Because these conditions can vary widely across the geographical area, it is difficult to derive standards that are universally predictable of toxic conditions.

2.3.2   Site-specific assessments

Prologue. Site-specific objectives are set where scientific study suggests that a standard different from the national criterion is legitimate. In some jurisdictions these are initiated based upon the needs of the specific site. In others (e.g. the European Water Framework Directive) widespread watershed-by-watershed analyses are required.

The most detailed level of management is represented by site-specific assessments conducted to set policy for a particular location. These assessments can be initiated either because of suspicions that a national-level criterion is not sufficiently protective for a given water body, or because it is suspected that a water body may not be adversely affected even though metal concentrations exceed a national-level guideline. Examples where site specific assessments are used include local discharge permits and determining appropriate clean-up endpoints. Water discharge permits, for example, set a designated use for a water body and set criteria for individual chemicals that allow that use to be achieved, based upon local conditions.

National-level criteria are the starting point. Site-specific objectives require more detailed study, more data and a more thorough site-specific analysis of a problem (e.g. risk assessment) (Box 2.2). When, where and how to apply more advanced site-specific approaches depend upon the degree of the problem and, to some extent, how much is at stake (Chapter 15).

2.3.3   Good chemical and good ecological status

Prologue. In contrast to the chemical-by-chemical approach typical of traditional management efforts, some new programmes (e.g. Europe's Water Framework Directive) manage watersheds holistically, recognising contamination as but one of the possible human-induced pressures. Ecological status rather than chemical concentration is the measure used for management.

Watershed-scale evaluations of the status of water bodies are another approach to incorporating local variability in environmental conditions. One example is the Water Framework Directive of the European Union (e.g. Apitz *et al.*, 2006). Analogous efforts are underway in Australia and South Africa, for example. The EU's Water Framework Directive, which came into force in 2000, is an integrated approach to the protection, improvement and sustainable use of water bodies. It applies to rivers, lakes, estuaries, coastal waters and groundwater.

The Directive first sets environmental objectives (Box 2.1); which are broad narrative goals for the programme. Examples in the Water Framework Directive are:

- Achievement of good ecological status and good surface water chemical status by 2015.
- Prevention of deterioration from one status class to another.
- A progressive reduction in discharges of Priority Substances and a cessation of discharges of Priority Hazardous Substances.

Standards are set to prevent damage from contaminants like trace metals. But they are used as one of several benchmarks. Success is judged by the achievement of ecological goals, rather than meeting purely chemical standards set pollutant-by-pollutant. The ecological objectives are designed to protect and, where necessary, restore the structure and function of aquatic ecosystems and thereby safeguard the sustainable use of water resources.

The Water Framework Directive dictates that the current status of water bodies must be identified and classified as to condition (using a formal classification scheme). Measures must also be identified that will maintain and achieve the objectives. It also dictates ongoing monitoring to assure achievement of objectives. Implementation of the Directive will involve characterising the status of each watershed and water body. This means a detailed picture of ecological and chemical status. Reference conditions are defined to help judge ecological status for each type of surface water body, where reference conditions represent the condition of a water body in a relatively unimpacted state. A reference condition provides a calibration point against which other water bodies' quality can be measured. Pressures on ecological status are also identified, in particular the impacts of human activity. This is done to assess the chances of failing to meet environmental objectives.

Implementation of the Directive is just beginning (in relative terms). For each river basin a management planning system is envisioned that specifies environmental objectives, including ecological targets for surface waters. The management plan also must specify linkages, like those between surface and groundwater and water quantity and water quality; and must integrate management of wetlands, groundwater, rivers, canals, lakes, reservoirs, estuaries and other brackish waters, as well as coastal waters and the water needs of terrestrial ecosystems. The details of implementation vary among member states. The outcomes will be interesting to watch in the years ahead.

2.4   ASSESSING HAZARD AND RISK

Prologue. Hazard assessments and risk assessment are common components in the international strategy for managing metal contamination.

A structured approach is necessary to determine the probability of exposure to a hazard and the associated consequences. This structured approach is termed *risk assessment*. We introduce hazard assessments, risk assessments and setting of environmental quality standards (e.g. for water and sediment) here, so the terms are not mysterious through the rest of the book (see Box 2.2). In Chapter 15 we discuss the challenges of implementing these management approaches in the modern world.

2.4.1   Ranking hazard or potential for risk

Prologue. Hazard rankings define the potential to cause risks based upon properties of the contaminant such as persistence in the environment, tendency to bioaccumulate and toxicity.

At the broadest scale, most governments rank or categorise the many thousands of chemicals used by human societies on the basis of their *potential* to cause risk (this is termed designation of the *hazard* of a chemical). These are the chemicals most likely to pose an environmental threat. Attributes of the chemical are used as indicators of hazard. These include production volume, quantities released to the environment, mobility in the environment, persistence in the environment, tendency to bioaccumulate in organisms and toxicity. The categorisations and rankings set the priorities for acting to protect the environment from adverse effects of each chemical. Identification of the greatest

hazards assures that truly pressing environmental problems are the first to receive attention. The complications in applying the criteria to metals have led the US Environmental Protection Agency to begin developing a separate strategy for metals.

Many metal contaminants are on most priority lists as hazardous (e.g. As, Cd, Cr, Cu, Hg, Pb, Zn). A few are on lists in some jurisdictions and not in others (e.g. Ag, Se). Some are not typically considered as highly hazardous. The US Environmental Protection Agency classifies and prioritizes chemical hazards in several regulatory programmes, including the Toxics Release Inventory (TRI), the Hazardous Waste Minimization Prioritization Program (WMPT) and the New Chemicals Premanufacture Notification Program. Persistence, bioaccumulation and toxicity are the traditional means of ranking hazards for all chemicals.

Canada also has a systematic categorisation of hazardous substances on its Domestic Substances List (DSL). Environment Canada does not use persistence or bioaccumulation in ranking metal-containing substances; hazards are differentiated purely on the basis of inherent toxicity (Ball *et al.*, 2006).

Most recently, implementation of a new chemicals policy, called REACH, has begun in Europe: Registration, Evaluation and Authorization of Chemicals (Box 2.3). While implementation details of REACH will undoubtedly evolve as it develops, it includes a requirement for 'Authorisation' for use if the metal is classified as a hazard, based upon the mass produced and evaluation by risk assessment. The REACH approach is drawing global interest as a new concept in chemical regulation.

It is also likely that, in the longer term, 'new' hazards from use of metal products will be identified. These could require new data and new approaches. One example is the use of various metals in nanotechnology. Some of the new products could result in dispersion of potentially toxic forms of trace metals. There are also concerns that metallic nanoparticles can cross the membrane 'barriers' in organisms (including humans) and cause adverse responses in a manner unique to their properties. Although many uses of nanoparticles may be environmentally benign, very little is known about either hazard or risk.

Box 2.3 The new chemicals policy in Europe: REACH

REACH is a scheme whereby all of the estimated 30 000 substances placed on the market that are manufactured in (or imported into) the European Union in quantities of more than 1 tonne per year will need to be registered (Floyd, 2006). 'Substances of very high concern', those with the greatest hazard, will need to be authorised for use.

There are several somewhat new aspects to this programme, contributing to some level of controversy. The burden is shifted to the users of chemicals to prove that they should be authorised for use. Testing and risk assessments are the burden of industry not governments. Hazard rather than actual exposures and risks plays a much larger role in determining the regulatory fate of a chemical than has been the case in the past. Another concern is the work involved in registering, evaluating and authorising so many chemicals. Historically, such evaluations have been extremely difficult to complete.

Screening focuses on criteria designed to evaluate organic contaminants: persistence, bioaccumulation and toxicity (PBT). These categorisations are somewhat problematic for metals (Chapter 15). Metals are, of course, 'very persistent'. As elements, no change in the most basic inherent properties of metals is possible; so relative rankings of hazard on this basis are not feasible. Quantification of potential to bioaccumulate is controversial for metals, since nearly all metals bioaccumulate to some degree. Tendency for trophic transfer may be more informative than simple toxicity, but it is not yet widely accepted as a process relevant to metals by the regulatory community. Evaluation of toxicity is itself a complication in that data in the existing databases are fraught with uncertainties.

Nevertheless, most observers agree that it is likely that approaches can be agreed at an international level to resolve the problems with ranking metals and their effects. One advantage is that REACH moves a plethora of regulations into a single framework. Implementation could also reduce uncertainties associated with the use of chemicals and will lead to a more comprehensive data set on risks from chemicals use (Floyd, 2006). Such outcomes could facilitate the development of more reliable risk assessments, at the very least.

2.4.2 Implementing risk assessment

Prologue. Risks are defined by quantifying the threat from a hazardous substance in any specific circumstance. Risk assessment is the formal process for accomplishing that quantification.

Risk assessments for metals are formal approaches to provide qualitative and quantitative relationships between environmental exposures and effects of metals (Landis and Yu, 2004). USEPA (1998) describes risk assessment as a flexible process for organising and analysing data, information, assumptions and uncertainties in order to evaluate the likelihood of adverse ecological effects from environmental stressors (Dearfield *et al.*, 2005). Risk assessments can be used at any scale of management. Industry and the European Commission are conducting risk assessments of different metals and metal products in anticipation of the hazard classification that is entailed in REACH. Site-specific risk assessments are very common for specific metal problems (Chapter 15). Risk assessments also link to environmental quality standards, in that environmental conditions are often judged compared to such standards.

In the broadest sense, risk assessments are the tools for translating the research and findings of environmental science into public policy. The steps in a typical ecological risk assessment for metals include:

- Initial planning
  - Communicate among risk assessors, risk managers and interested parties
- Problem formulation (what will and will not be addressed)
  - Set goals
  - Select assessment endpoints
  - Define unacceptable risk levels
  - Prepare conceptual model that frames the problem
  - Develop analysis plan
- Analysis of stressor and effects
  - Evaluate exposure
  - Define relationship between exposure and effects

- Risk characterisation
  - Quantitatively estimate risk
  - Qualitatively describe risks
  - Communicate risks to interested parties

A great strength of risk assessment is the formalisation of the process to collect and analyse information relevant to the problem at hand. Risk assessments can also be controversial, however. The controversies probably stem more from the way risk assessment is practised than from the concept itself. The assessment of risk itself is only as strong as the clarity of the question and the relevance or quality of the data employed. For example, if an ecological risk assessment is conducted for a contaminated site, is the proper question 'What are the risks to organisms living at the site?' or is it 'What were the risks to the biological community that should occupy the site?' (Tannenbaum, 2005). The latter evaluates whether the contamination has eliminated some species (whose return would be the goal of risk management); the first question deals with survivors (the least likely organisms to be affected) if the contamination had been in place for some time.

An important criticism of risk assessments is that they are presented as the objective outcome of a scientific assessment. It is argued by some, however, that risk assessments also can include important, but obscure, assumptions and value judgements. Because risk assessments are predictive, they are inherently uncertain. At the simplest level, the uncertainty (e.g. about exposure) might result from lack of data on an existing situation. At the more complex level, there is almost always at least some uncertainty about the precise relationships between low levels of exposure to trace metals and ecological effects. Thus, at one extreme, there are those who question the use of risk assessment with uncertain data. At the other extreme, there are those who take little account of uncertainty in applying mandated government standards to quantify risk. In the end, a carefully conducted risk assessment is the most advanced tool for systematically evaluating a metal contamination issue, in a less than perfect world.

2.4.3 Risk management

Prologue. Risk management identifies strategies and tactics to reduce the likelihood of adverse ecological consequences from metal exposure.

Ecological risks from a contaminant are usually controlled in practice by reducing the probability of exposure. Examples include banning the chemical (e.g. some uses of tributyl tin (Chapter 14) and lead), substituting less hazardous for more hazardous chemicals, limiting release of the material to natural waters (water quality standards) and/or interrupting the pathways by which the substance reaches the water body where it could cause harm. For all materials, but especially new technologies such as nanotechnology, foresight of possible risks depends on a consideration of the life cycle of the material being produced. This involves understanding the processes and materials used in manufacture as well as the likely interactions between the product and individuals or the environment during its manufacture and useful life, and the methods used to dispose of it. A formal process also exists for understanding those processes, termed Life Cycle Analysis.

2.5 ADDRESSING UNCERTAINTY

Prologue. A valid risk assessment and an effective risk management approach are defined by how well and transparently scientific uncertainties are identified and addressed.

Sound policy recognises the inevitability of uncertainties. But making decisions in the face of uncertainty is one of the most difficult and one of the most common challenges in environmental management. A valid risk assessment characterises a circumstance, defines the uncertainties and identifies the choices that must be made. Explicit recognition of the types of uncertainties that affect the quantitative and qualitative evaluations of risk is an essential part of that process (Box 2.4).

A panel of scientific experts from the European Food Safety Authority (2006) recently presented

Box 2.4 Uncertainty in risk assessment

The most common error in recognising uncertainties is failure to recognise the different types of uncertainty. Uncertainties are usually grouped as follows (Floyd, 2006).

Knowledge uncertainty
Lack of basic knowledge about the mechanisms or interactions between different system components (we do not know what we do not know). This is the most insidious type of uncertainty because the effects of knowledge uncertainty cannot be quantitatively expressed. For example, we know that some types of metallic nanoparticles have unique properties that could lead to toxicity to aquatic animals; but there is essentially no knowledge about dose-response, whether or how that would occur; and/or which particles might represent the greatest threat. We cannot quantitatively express (percentage-wise) how far astray an expression of risk from such particles may be, or even whether that expression will be overprotective or underprotective of environmental resources.

Real-world uncertainty
It may be possible to characterise the range of 'natural' conditions, but there is often uncertainty in applying such general knowledge to specific situations. For example, we know concentrations of silver in uncontaminated ocean water, but we cannot extrapolate that general knowledge to a specific bay without site-specific data. We can predict that 50% of our laboratory animals will succumb to a given dose of a metal, but we cannot predict how that will manifest itself in a complex environment with many species.

Data uncertainty
This arises when knowledge is based on limited or incomplete sets of data, or data that may be subject to random variability. This type of uncertainty can be quantitatively expressed in terms of confidence limits.

Modelling uncertainty
The methods used to predict or forecast conditions beyond those observed always have some degree of uncertainty. These uncertainties can arise from a lack of knowledge, from the decisions made by analysts during the modelling process and from assumptions inherent within different models (models represent our best judgement).

the following suggestions for ways to improve risk assessments.

- Review and publish more explicitly the scientific basis of assumptions.
- Evaluate and describe how the various assumptions and refinements used in an assessment affected the overall conclusions.
- Increase interaction between scientists of the various disciplines, both when conducting individual risk assessments and when developing new, general guidance principles.
- Periodically review and update guidance documents to take account of advances in science.
- More fully document and justify approaches, assumptions and conclusions in individual risk assessments, especially refined or non-standard assessments.
- Develop effective ways of dealing with and communicating uncertainties.

2.5.1   The Precautionary Principle

Prologue. Application of the *Precautionary Principle* is a common approach to addressing uncertainty. A thorough impartial analysis of the available data, finding a tempered balance in applying precaution and acting before the media atmosphere becomes intense are critical ingredients to successful application of the Precautionary Principle.

The most widely used definition of the Precautionary Principle is from the Rio Declaration: 'Where there are threats of serious or irreversible damage, lack of full scientific certainty shall not be used as a reason for postponing cost effective measure to prevent environmental degradation.' This statement is interpreted in different ways, and as a result the Precautionary Principle is sometimes controversial.

Strong application of the Precautionary Principle was held up by the European Court of Justice in a case concerning the ban on the export of beef products:

> When there was such uncertainty regarding the risk to human health, the Community institutions were empowered [by the Precautionary Principle] to take protective measures without having to wait until the reality and seriousness of those risks became fully apparent.

Application of the principle by the UK's Royal Commission on Environmental Pollution typically associates it with taking pre-emptive action rather than waiting for proof of harm (Owens, 2006). For example, the 'Red List' of Substances is the UK's list of chemicals deemed to be most dangerous. The Royal Commission cited it as 'reasonable' to assume these chemicals 'guilty until proved innocent', and requiring toxicological testing in advance of marketing and monitoring for environmental impacts afterwards (Owens, 2006).

A second aspect of strong application of the principle is that it shifts the burden of proving risk from society (to prove something is risky before it can be eliminated from the market) to the producer of the chemical (to prove something is not risky before putting it on the market).

The Royal Commission also recommended that the Precautionary Principle be 'tempered' in application by the principle of proportionality: measures taken should be proportional to the potential threat. Acting ahead of evidence does not mean acting whatever the cost (Owens, 2006). The Royal Commission cited the elimination of lead additives from petrol (gasoline) as an example of tempered precaution. The phase-out of the additives occurred before the evidence of health effects on children was unequivocal (the evidence supporting such effects continues to grow). But the recommendation to phase out lead was facilitated by the finding that there was some evidence such effects were feasible and the partial ban could be accomplished at modest cost. The Royal Commission deemed the phase-out to be a rational response to the uncertainty.

Box 2.5 An example of tempered precautionary risk management is the Convention for the Protection of the Marine Environment of the North-East Atlantic (OSPAR Convention)

The OSPAR Convention led to an agreement to 'take all possible steps to prevent and eliminate pollution and to take the necessary measures to protect the maritime area against adverse effects of human activities so as to safeguard human health and to conserve marine ecosystems and, when practicable, restore marine areas which have been adversely affected'. In 1998 a directive from the European Union defined guiding principles from the OSPAR Convention for preventing pollution of the marine waters. Its wording balances strongly precautionary principles with a dedication to scientific study as a tool of implementation. Goals were to continuously reduce discharges, emissions and losses of hazardous substances, with the ultimate aim of achieving concentrations in the marine environment near background values for naturally occurring substances like metals. Means were also established to achieve those goals.

The guiding principles were:

(a) assessments made, and programmes and measures adopted . . . will involve the application of:
  (i) the precautionary principle;
  (ii) the polluter pays principle;
  (iii) best available techniques and best environmental practice, including, where appropriate, clean technology;
(b) in addition, the principle of substitution, i.e. the substitution of hazardous substances by less hazardous substances or preferably non-hazardous substances where such alternatives are available, is a means to reach this objective;
(c) emissions, discharges and losses of new hazardous substances shall be avoided, except where the use of these substances is justified by the application of the principle of substitution;
(d) in the work to achieve the objective, the scientific assessment of risks . . . is a tool for setting priorities and developing action programmes.

The Precautionary Principle becomes most controversial when data are scarce (Berry, 2006). Some argue that the absence of data is exactly the right time to be cautious about a perceived threat, if we are to avoid acting before it is too late. Others give examples where precaution in the absence of careful analysis (cited as an absence of data) exacted a high price in terms of the credibility of the parties involved. For example, a cost of the precautionary ban of all uses of DDT in 1972 appeared to be an increase in cases of malaria in tropical nations. In retrospect, a partial ban, for example a ban of agricultural uses, might have been more tempered and come with less cost to human health.

Politically popular or media-promoted causes can also build momentum for precautionary action. For example, in an intense, media-driven atmosphere it is not always clear if the risks are just a possibility to someone, or if there are credible studies to support such conjecture (Berry, 2006; we also refer readers to the *Brent Spar* incident: Löfstedt and Renn, 1997). Pressure develops for action if there is a suggestion of causative links to an environmental threat. Ultimately the public will lose trust in institutions that persistently overreact in such circumstances. But the public also loses trust when assured there is no threat, only to later find that a high-consequence threat existed, even if the risk to any individual was low (Löfstedt, 2005).

Underlying assumptions also characterise debates about precaution. For example, implicitly parties may have different definitions of environmental protection. When those assumptions are not explicitly stated, conflict can develop (Box 2.6).

Box 2.6 Example of underlying assumptions about environmental goals

Unstated assumptions about goals for environmental protection can add to controversy to applications of precaution. Assumed levels of environmental protection might include:

(1) Protect all species, in all places
It might be assumed that all species will be protected if protective limits are determined from the most sensitive species known. This approach seems the most precautionary in terms of protecting environmental values and therefore is an obvious choice. But it can be problematic (for example, if the protective concentration for a metal is less than the natural concentration of the metal in the water body). As more is learned about contamination and the complex ways its effects are manifested, it has become clear that a standard that is too precautionary ultimately is unenforceable.

(2) Avoid unnecessary regulation by implementing protective actions only when it is clearly demonstrated that some species will be affected
This is an attractive choice from an academic viewpoint, as a logical alternative to the precautionary approach. The basic assumption is that scientific knowledge is adequate to detect damage if it is occurring. Given the complexity of metal effects, however, there are very likely to be instances where effects cannot be conclusively demonstrated and are occurring. Thus this logically attractive approach can be the least cautious of the approaches for the environment, if important scientific uncertainties exist. If prematurely employed it can be extremely expensive in the long run if the result is unacceptable ecological damage.

(3) Protect some percentage of species
This is a statistical alternative that requires some knowledge of risks to at least some species. In practice it is a way to split the uncertainties of both of the above approaches, perhaps underprotecting in some instances and overprotecting in others. It splits the economic costs of both and it splits the ecological risks. It is the approach typically employed in many US regulatory decisions (protect 95% of species). What it does not take into account is that some species are consistently more vulnerable to specific metals than are others. An alternative question might be whether or not to protect specific vulnerable groups of species (there is no scientific basis for the choice of 95% but there is for identification of vulnerable groups).

There is no simple answer as to when or how to use precaution. The precautionary principle is difficult to characterise, just as risk is difficult to quantify. Risk managers must obtain the best possible data, and be explicit about how they are going to define precaution and uncertainties, recognising that inappropriate or ineffective action can be the cost of too much precaution, and degradation of the environment is the cost of too little. Political and media pressure can also create the perception that immediate decisions are necessary, when a more measured evaluation of the state of scientific knowledge might pay great benefits. Erratic, ill-informed risk management is the approach least likely to engender trust and result in efficient, effective environmental protection.

2.6   ADAPTIVE MANAGEMENT

Prologue. 'Reductionist science, i.e. the science of parts, provides bricks for the framework of environmental management, but not the strategic design of the edifice. Strategic design is needed for appropriate diagnosis and policy, and it has to emerge from a science of integration. A science of integration combines research and application, it is interdisciplinary. It faces the realization that knowledge of the system we deal with is always incomplete. It acknowledges that surprise is inevitable and that there will rarely be unanimity of agreement among peers – only an increasingly credible line of tested argument' (Holling, 1996).

Lateral management of risks from contamination must include adaptive management as a way to address the greatest uncertainties. Adaptive management is also a way to proceed experimentally when there is uncertainty about the outcomes of different proposals. It is the process of adjusting and refining management strategies over time as new information becomes available. Adaptive management can be viewed simply as learning while doing (Lee, 1999). Successes and mistakes are documented. Effectiveness is determined by monitoring outcomes. The result is continuous course correction. A more formalised view involves

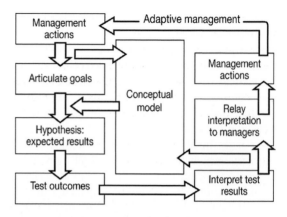

Fig. 2.1 An adaptive management scheme. Every management action is viewed as an experiment. The goals of the action should be explicit and a conceptual model must exist to explain the context of the action (e.g. implementation of a water quality standard). Each action generates hypotheses about the expected outcome. These hypotheses can be tested by formal, systematic observation or experimental design. Interpretation of the observations or experiments are passed on to management, and what is learned. The outcome is then assessed compared to the expectation, in a formally designed experiment. The interpretation of the assessment leads to advice to management about how to adjust the next action.

treating each management action as a more formal experiment that tests a new management approach against an alternative. Research results are used to refine knowledge about the environment and feedback to affect future management decisions (Fig. 2.1).

Some argue that adaptive management is simply good management. It involves trying something and then if it doesn't work, adapt and try something else. But adaptive management formalises this process. The advantage of an explicitly experimental approach is that it systematically tests assumptions in order to adapt and learn.

Philosophically, adaptive environmental management accepts the uncertainty that exists in the real world and adds flexibility where uncertainty is high. It is based upon the premise that managed ecosystems are complex and inherently unpredictable. Under ideal circumstances adaptive management is a substitute for postponing action until 'enough' is known. There is explicit acknowledgment that time and resources are too short to defer *some* action. Management can proceed despite uncertainty, but it is expected that uncertainties will be reduced by systematic observation of the outcomes of experience.

Like risk assessment, adaptive management has a formalised structure. It involves a six-stage process: problem assessment, experimental design, implementation, monitoring, evaluation and management adjustment (Murray and Marmorek, 2003) (Fig. 2.1). Adaptive management also has critics.

- If the formal process were applied everywhere all the time, it would encumber policy implementation. The lessons from experiments are rarely straightforward.
- Only in simple cases is the one large and formal adaptive experiment a feasible way to address a problem. Most complex problems require incremental advances on many fronts.
- The rigidity of the structure is interpreted in different ways by different users. Too much rigidity leads to controversy.
- Jurisdictions are hesitant to try (or to admit they are trying) management approaches that are not guaranteed to be successful. Alternatively, the resources at stake may be too precious to take risks with experimental management. Of course, proponents argue that most management involves risks anyway; they just are not acknowledged.

Trying alternative approaches in a systematic way and learning as uncertainties are addressed is the core of the adaptive management philosophy; and an approach that can be implemented. But there is yet little successful precedent for implementing formal adaptive management across a complex issue.

Adaptive management is much discussed in the resource management literature; it is less often applied to managing environmental risks from chemicals. At worst, experimental implementation of new management approaches on a small scale, and studying their effectiveness and outcomes, could be a valuable tool to defuse what are now just contentious debates. At best, adaptive management could become part of the solution for many of our most difficult contamination issues.

2.7   LINKAGE OF SCIENCE AND MANAGEMENT

Prologue. 'Heavy metals such as copper, zinc and lead are normal constituents of marine and estuarine environments. When additional quantities are introduced from industrial wastes or sewage they enter the biogeochemical cycle and, as a result of being potentially toxic, may interfere with the ecology of a particular environment. Within . . . organisms, the behaviour of heavy metals is described in terms of absorption, storage, excretion and regulation. At higher concentrations, the detrimental effects of heavy metals become apparent' (Bryan, 1971).

The general scientific issues with metal contamination were identified decades ago. Science and policy are linked at every step in the conceptual model that describes the processes that determine the fate and effects of trace metals.

The quote from the marine biologist G. W. Bryan, from the 1970s, signalled the onset of recognition that detrimental ecological effects in natural waters were possible from the metal releases of human activities. Many details were unknown in the 1970s about where metal contamination occurred, its trends, or its specific biological effects. Geochemical studies in the 1950s first evaluated what happens to metals introduced in high concentrations into the sea (e.g. Krauskopf, 1956; Schutz and Turekian, 1965). Around that time toxicologists documented that metals were toxic to aquatic organisms in the laboratory, at lower concentrations than they expected. Basic principles were developed: toxicity is what is of concern in the environment; and toxicity is a function of concentration. New knowledge was (and is) expected to aid effective management of chemical contamination in several ways (Cairns, 2003):

- Determine critical concentrations that determine ecological thresholds and breakpoints;
- Continually refine guidelines for implementing protection of the environment and achieve some socially acceptable 'balance' with cost;

- Develop ecological understanding to validate the legitimacy of standards and to assure that the goals of management were being met;
- Evaluate environmental change and thereby reduce the frequency and intensity of environmental surprises;
- Prioritise recognised problems, identify new problems, develop ways to remediate detrimental effects and create other novel solutions.

As appreciation grew for the complexity of metal behaviour, these simple demands turned into a myriad of complicated questions. As knowledge grew, the science of metal contamination contributed increasingly to assessing and managing risks. Nevertheless, the original goals remain a relevant guide to the general types of knowledge needed.

2.7.1   Connecting the scientific disciplines and their links to policy

Prologue. 'what regulatory authorities should be directing their attention to is the establishment . . . of a common approach to environmental assessments which will permit the uniform application of basic . . . standards, rather than uniform . . . standards (across very different localities), or uniform emission standards' (Preston, 1979).

The science that addresses metal contamination is, by necessity, multidisciplinary. Economics and engineering are necessary to determine the costs of reducing metal contamination. These were the original disciplines that dominated management. Environmental toxicology was next recognised as essential. Now we know that biology, ecology, geochemistry and geophysics are all necessary components in understanding and ameliorating metal contamination. The policy community needs knowledge from all those disciplines to develop and refine standards that effectively protect the environment and are otherwise justifiable. More importantly, effective policies will think laterally across these disciplinary boundaries, incorporating multiple tools into solutions.

The multiple disciplines that determine effects of metal contamination interact in a systematic way. A conceptual model is a graphical (Fig. 2.2) or verbal (Table 2.1) construct that systematically lays out important environmental processes into an understandable overview and provide a basis for 'the integrated application' of disciplines. The questions that frame the conceptual model for metal contamination guides both science and policy/management (Table 2.1).

The conceptual model starts with the sources of metals (point source, runoff or natural) (Fig. 2.3). Each emits a total mass or 'load' of a trace metal into the environment. The relative contribution from each source provides a primary basis for managing that contamination. Effluent standards, toxic release inventories, critical load inputs and total maximum daily loads (TMDL), are management tools that we will learn about later. They all consider the mass of metal that is released, in any form, from every source. Life cycle analyses can follow the mass of metals as they are mined from the Earth and passed through their many uses in human activities.

Loadings of a trace metal to the environment do not, alone determine risks from adverse effects. Metal concentrations, not metal mass, determine effects on organisms. Wastes are diluted by receiving waters, mixed by winds, adsorbed to particles, recycled within sediments and redistributed by water movement. These physical and chemical processes determine metal concentrations in water, sediment and organisms (e.g. Fig. 2.4). Elevated concentrations of trace metals are the first signs of risk from metal contamination. The physical location of concentrated deposits and the physical extent of the contamination must be known for effective management. Concentrations are usually managed via environmental quality standards; although more modern approaches may use ecological endpoints. Examples of standards include ambient water quality criteria or sediment quality criteria.

Trace metals each partition into different forms within water and particulate material, following reactions governed by the chemistry (and biology) of the water body. Metal forms in the environment modulate toxic effects by influencing *bioavailability*: a relative measure of that fraction of the total

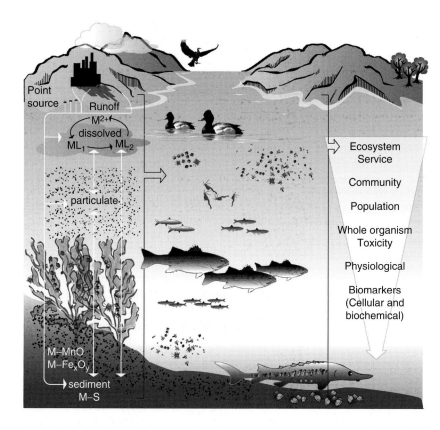

Fig. 2.2 A conceptual model in cartoon form describing the processes that determine the fate and influences of metals in a contaminated water body. The pieces of the model follow the questions in Table 2.1. In this example we show both a water column and a benthic food web through which metals can cycle and exert their influences. Abbreviations indicate metals (M) in the form of a divalent free metal ion (M^{2+}), bound to one of two ligands (ML_1, ML_2) in solution. Metals in sediments can be bound to oxides of manganese (M–MnO) or iron (M–Fe_xO_y), or bound to sulphide (M–S).

Table 2.1 *Questions guiding evaluations of metal contamination in nature (the conceptual model)*

- What are the sources of metal contamination and the magnitudes of those sources?
- What concentrations result when natural waters receive those wastes?
- How do the metal contaminants react in those waters to distribute themselves among different chemical forms?
- How do the chemical forms influence bioavailability of the metal(s) to organisms in the system?
- How much metal do different organisms accumulate internally; and how much of that accumulated metal is available to interact and damage living processes?
- How does each organism respond biochemically and physiologically to that threat of damage?
- How are populations of each species affected by different types of adverse effects on individuals (e.g. a small proportion of mortality, reproductive damage, slowed growth, behavioural change)?
- How do communities of species change when populations of some are damaged?
- How do ecosystem processes change as communities change?

ambient metal that an organism actually takes up when encountering or processing environmental media, summated across all possible sources of metal, including water and food as appropriate.

Environmental quality objectives often have the goal of 'taking into account the characteristics of the particular receiving environment' by incorporating consideration of bioavailability. Some standards use simple approaches to address aspects of bioavailability. The benefits of better addressing local conditions also carry with them some risks, however, that some of the precautionary protections

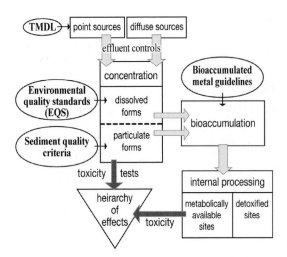

Fig. 2.3 Traditional regulatory approaches are linked to the different processes that determine the fate and effects of trace metal contamination. But individual management approaches skip links in the conceptual chain. Understanding the linked processes is critical to the next generation of management approaches.

for natural waters might be lost because of important uncertainties in the tools for defining bioavailability. Nevertheless consideration of bioavailability is an important next step in effectively managing contamination.

Bioavailability can also be assessed in nature by directly determining the concentration of metal bioaccumulated by particular organisms in a specific setting. Factors important in bioaccumulation as observed in natural waters include: (a) attributes specific to each metal, (b) attributes of the species, and (c) attributes of the environment. Management tools are developing that use biomonitors or bioaccumulation (dietary criteria, tissue-based criteria, wildlife criteria) (Chapter 17). Better integration of these with traditional environmental quality standards will be an important step towards a more lateral and integrated management approach.

Not all metal bioaccumulated within an organism is toxic. The organism's ability to detoxify metals after uptake is the next step in the conceptual model. Some recently enacted programmes monitor physiological stress indicators (termed biomarkers) as a sensitive way to identify the possibility

of adverse stresses on the organism (Chapter 9). As we begin to understand which species are most sensitive to metals, the presence or absence of stress indicators are a useful diagnostic for risk managers when evaluating the effectiveness of their actions.

If contamination is great enough to cause the rate of uptake to overwhelm the detoxification abilities of the organism then toxicity is the outcome. 'Toxicity', however, is an overly simplistic term for the many different ways that an organism can manifest detrimental responses to a metal. It can be expressed as physiological damage, such as inhibition of growth or reproduction. It can be expressed as death of sensitive young life stages. At the highest concentrations adults may die. Such events occur at different concentrations in different species, reflecting the sensitivity of the species.

Risk assessments generally use toxicity testing results to tie environmental concentrations to detrimental effects. Similarly, environmental quality standards are traditionally derived from direct toxicity testing on individual animals. Reliance on toxicity testing requires assumptions and precautionary correction factors that are increasingly a source of controversy (Tannenbaum, 2005) (Chapter 10). Validation of the effectiveness of such criteria in nature is difficult, however. Effectiveness will improve as ties to additional types of data (indicators of stress) are incorporated into risk management.

Just because individuals die or are ill, it does not mean the species will disappear. Extirpation of a species occurs when effects are sufficiently detrimental that the population cannot sustain itself. A toxin need not be lethal to adults to cause a population to disappear. Failure to recruit enough young to replace natural losses (for example, if reproduction fails) can cause the demise of a population just as readily as increased mortality. When a population disappears it affects the nature of the biological *community*; the assemblage of species in a locality. Complicated interactions within communities can disguise or balance the disappearance of some species, making the early stages of community simplification difficult to identify. Over time such effects can also accentuate the impacts of metals.

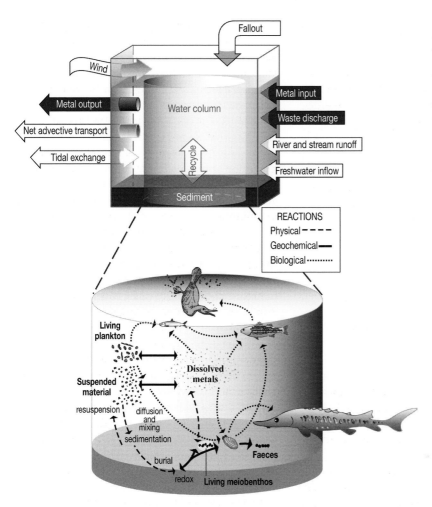

Fig. 2.4 Physical processes determine the concentration, fate and distribution of metals in a water body. In this figure the important physical processes in an estuary are depicted. Within the water column influenced by physical processes, a complex combination of geochemical and biological processes determine the effects of metal contamination.

Population and ecological responses can be used in retrospect to determine if metals are having (or have had) an effect. This arena offers great promise for both testing the validity of standards and overcoming some of the basic limitations of traditional standards. Ecological endpoints are the guide for success in the EU's new Water Framework Directive. A few environmental quality standards (Se in freshwaters) were derived from linking environmental concentrations and ecological effects of a trace metal (Chapter 13). But population and community considerations are not yet incorporated into the vast majority of metal-specific standards or standard-setting processes, even though that is what we are trying to protect. Nowhere is the fundamental difficulty of better linking eco- and ecotoxicology more evident (Cairns and Mount, 1990).

2.8 INTEGRATING SCIENCE AND MANAGEMENT

Prologue. The view of science through the lenses of environmental policy emphasises political, legal or administrative process as much as scientific data.

The policy community of risk assessors, government managers, non-government organisations (NGOs) and business people must view metal issues through the legal process of decision making and environmental management, creating and implementing the legal mandates to protect the environment. Policy and management, whether local, regional, national or international, require balancing politics, administrative process and science. Management has several roles:

- enforce existing policy, meeting deadlines established by politicians;
- anticipate new issues;
- identify those policies that are working and sustain them;
- revise policies that are not effective or can be improved by new knowledge; and
- find ways to incorporate new scientific knowledge into the process.

2.9   CONCLUSIONS

The legal mandates that drive management of metal contamination are derived from the processes that govern the fate and effects of metals. In the chapters ahead we will consider both processes and management. Existing management approaches rarely consider the full complexity of the conceptual model. For example, authorisation for the use of a metals (e.g. by REACH) is largely based upon mass used and toxicity as determined in controlled tests. But historic bans of specific metal uses (e.g. tributyl tin, Chapter 14; lead, Chapter 5) and identification of threats from the metals that pose the greatest risks (Chapters 13 and 14) have been based upon observations and analyses across processes that combine field evidence, process understanding and results from experimental testing. Therein lies the contradiction that typifies much of the management of metal contamination. That contradiction drives the need for a new generation of management tools.

Inventories and load allocation are used to identify and regulate sources of metals (Chapter 4). Used alone, this would be the most precautionary approach to managing contaminants. But loads are not the only factor driving effects; so managing on the basis of mass inputs alone is very difficult to justify. Organisms respond to concentrations in water (Chapter 5), concentrations partitioned into particulate material or sediments (Chapter 6), and concentrations bioaccumulated into their tissues (Chapter 7). Environmental quality standards employ these concentrations (Chapters 15 and 16). In theory, concentration-based standards for water and sediment should also be a precautionary approach to managing contamination, because bioavailability is not considered. But most of today's standards derive protective thresholds for aquatic environments from toxicity tests that involve a suite of trade-offs, some of which are overprotective of the environment and some of which are underprotective (Chapter 9). Existing standards are surprisingly variable across the globe. Judged against responses in nature, their validity varies widely.

New proposals for management incorporate consideration of the bioavailability of dissolved metals (e.g. biotic ligand modelling: Chapter 15). These standards better address some of the processes that govern adverse effects of metals, but leave others unaddressed (they are still largely based upon toxicity testing). As bioavailability is incorporated, some risks of mismanagement are reduced; but others might increase if processes are left unconsidered that could result in underprotection (Chapter 17). Bioaccumulated metal guidelines also consider metal bioavailability, albeit implicitly; but links to effects are complex. Finally, some modern management tools ('good chemistry, good ecology') use ecological response as the endpoint of concern, managing metal contamination within the context of other stressors. This is the ultimate measure of effect but, from these data alone, allocation of the sources of stress is not possible. Management of specific sources is therefore difficult.

Thus, the most cautious approaches to management emphasise the processes at the top of the conceptual model; but skip the most links and include the most guesswork. The most integrative measures are difficult to tie back to specific management actions. The good news is that the scientific uncertainties that characterise every choice for management tend to balance one another out when considered together in a systematic way.

Lateral, integrative management would balance the results from multiple tools in setting policies and making decisions, and use adaptive management experiments to evaluate the importance of uncertainties in new approaches. The challenge for the future is integration across the conceptual model, tying together the tools that were created to manage loads with those that manage concentrations, bioavailability and effects in nature (Chapter 17). An important part of that challenge is incorporation of validation studies that include bodies of evidence from natural waters (Chapter 11); including data that define exposures of organisms themselves (Chapters 8 and 17).

Suggested reading

Apitz, S.E., Elliott, M., Fountain, M. and Galloway, T.S. (2006). European environmental management: moving to an ecosystem approach. *Integrated Environmental Assessment and Management*, **2**, 80–85.

Bryan, G.W. (1971). The effects of heavy metals (other than mercury) on marine and estuarine organisms. *Proceedings of the Royal Society of London B*, **177**, 389–410.

Preston, A. (1979). Standards and environmental criteria: the practical application of the results of laboratory experiments and field trials to pollution control. *Philosophical Transactions of the Royal Society of London B*, **286**, 611–624.

3 • Historical and disciplinary context

3.1 INTRODUCTION

Different scientific perspectives can create different views of metal contamination. Toxicology, geochemistry, biological process studies and studies of natural history all contributed to what we now know. But history and approach shaped early findings and the impact each type of science had on policy. Different perspectives may explain why some policies exist as they do today.

3.2 DISCIPLINARY PERSPECTIVES IN METAL SCIENCE

Prologue. Studies of metal contamination began with different disciplines taking on different aspects of the problem with somewhat different goals. Alliances occur among applied ecotoxicology, geochemistry and the more holistic field and process-oriented perspectives. But differences in how metal contamination is viewed through the lenses of these different disciplines explain some of the contradictions and controversies in managing ecological risks.

A characteristic of science is the tendency toward specialisation. Depth of knowledge about a discipline leads to the greatest expertise; the world is too complicated for anyone to be expert in all disciplines. Unfortunately different types of specialists can view the same question differently, through the lens of their discipline. Generalists, such as policy makers, can take a different view yet.

The metal conceptual models shown in cartoon form in Chapter 2 clearly demonstrate the interdisciplinary nature of the science underlying our understanding of metal contamination. At least some of the controversy in managing metal issues stems from the challenges of integrating different, specialised disciplines. In the metal contamination field, scientific studies began with different disciplines taking on different aspects of the problem with somewhat different goals. These different views can be powerful tools for new discovery if integrated. However, if different scientists see the same issue through different lenses, and do not communicate well, contradictions can result over how science should be applied. Such circumstances lead to mistrust and confusion about application of the science.

Although alliances certainly have developed among specific groups in these fields, much in the different fields evolved separately. One perspective for trace metal science is rooted in toxicology and toxicity testing. This field, applied ecotoxicology, is focused on responding to regulatory demands, and many aspects of risk assessments. A second origin is the perspective of environmental geochemistry. Geochemists originally derived sophisticated chemical approaches capable of accurately detecting chemicals in minute quantities and interpreting their distributions and the chemical processes that determine their fate. The third perspective is more holistic and lies with field- and process-orientated biologists and ecologists, who study fate, biological implications and effects in natural systems. Their emphasis tends toward biochemical/physiological, biological or ecological observations and process-orientated experiments. This is a natural history perspective. The fourth perspective is that of the policy community, whose goal it is to combine science with policies to manage and protect the environment.

Box 3.1 Definitions

acute toxicity tests: toxicity to a test organism, expressed as the dissolved metal concentration that causes a percentage lethality (e.g. 50%) in a short exposure (traditionally 96 hours).

applied metal ecotoxicology: the development and application of knowledge of toxic metal science to meet legal mandates and support quantitative risk assessments.

bioaccumulation: accumulation of a metal in the tissues of an organism, resulting from uptake from all sources (e.g. water, diet).

bioassay: an assay of the response to a metal, using a biological system.

bioconcentration: concentrative accumulation in tissues resulting from exposure to the metal only from water (dissolved form).

biomagnification: accumulation of a chemical to a higher tissue concentration in a predator than occurred in its prey (trophic transfer is the uptake mechanism).

chronic tests: definitions vary widely. In practical terms often implemented as any toxicity studies lasting longer than 96 hours and with endpoints other than survival (De Schamphelaere and Janssen, 2004). Whereas Chapman (1989a) suggests that tests are only chronic if they extend over one full life cycle of an organism.

community: a group of populations of different species that interact in time and space.

dietary bioaccumulation: accumulation of a metal taken up from food ingested by the organism.

endpoint: the variable used to describe the response to the metal toxicant at or before the end of a test (e.g. death, growth rate over a given time, number of eggs laid).

geochemistry: the study of sources, reactions, transport, effects, and fates of chemical species in water, soil and air environments (Manahan, 1991).

lethal: causing death of an organism.

µg/g (micrograms per gram), mg/kg (milligrams per kilogram) or ppm (parts per million): units of measurement of metal concentration as metal mass (μg) per unit mass (g) of a substance. This order of magnitude is typical for sediment or tissue metal concentrations.

µg/L or ng/L: units of measurement of dissolved metal concentration as metal mass in μg (*micrograms*) or ng (*nanograms*) per *litre* (L) of water. These are units typical of natural waters. Dissolved concentrations are also expressed as parts per billion (*ppb*) or parts per trillion (*pptr*).

population: a local set of individuals of the same species.

sublethal: a response other than death (Chapman, 1989a). Effects determined experimentally which are of sufficient severity to carry with them a high probability of death or reduced reproductive capacity, which in the natural struggle for survival would have as detrimental an impact on the exposed population as prompt direct lethal effects.

trophic transfer: transfer of metal up a food chain from one trophic level to a higher one, for example to a predator from eating its prey.

uncertainty: the state of being unsure about an observation, prediction or anticipated outcome. Uncertainties develop from imperfections in the information available. The degree of uncertainty depends upon the specific circumstance: it can be great or inconsequential. Several types of uncertainty exist ranging from some that can be quantified statistically and some that cannot. 'Environmental science's (*sic*) ground state is uncertainty. And wisdom, for all parties, is defined by how one copes with it.' (Quote adapted from MedicineNet).

variability: the inherent heterogeneity in a variable of interest (Adams *et al.*, 1998).

A brief history of the development of some common perspectives illustrates the confluences and contradictions typical of this multidisciplinary field. Categorisations are inevitably oversimplifications, with exceptions and overlaps. But they can help us see how these different perspectives could develop. Throughout the book we will come back to these perspectives in discussing both the scientific fundamentals and the case studies.

3.3   TOXICITY TESTING: APPLIED METAL ECOTOXICOLOGY

there is no instrument devised by man that will directly measure toxicity. Toxicity tests are the only method available . . . toxicity testing may not be all we hoped for but the alternatives are unthinkable.

(Cairns, 1986)

3.3.1   Priorities and benefits

Prologue. 'In its earliest stages environmental toxicology depended on short-term laboratory tests with single species that were low in environmental realism, but with satisfactory replicability' (Cairns, 2003).

We will use the term *applied metal ecotoxicology* to describe the development and application of knowledge to metal contamination with the goal of meeting legal mandates and supporting quantitative risk assessments. This is science tied to policy mandates or decisions about policy (e.g. Box 3.2). The commonality in this perspective is a focus on studies that will directly affect the regulatory process.

The early priorities of contaminant science were driven by a demand for studies that could have an immediate impact on solving the most glaring problems. Assessments of discharges in the 1950s were carried out mostly by sanitary engineers and chemists. It was soon recognised that tests of toxicity could only be done with living material; hence the origins of biological testing. The

> **Box 3.2** Examples of some simple questions that drive applied metal ecotoxicology
>
> (1) Are metals toxic to aquatic life and at what concentrations?
> (2) How do we routinely estimate the toxic hazard of metals to aquatic life, such that timely appraisals are possible of new chemicals or new uses for chemicals?
> (3) Can biological tests be used to evaluate the potential for toxicity in waste discharges?
> (4) Since metals are natural components of all natural waters, at what concentrations do they become harmful? What are the highest concentrations that are safe?

developers of biological testing protocols sought tests that were standardised, predictive, simple, repeatable and inexpensive. If the protocols for doing this could be standardised, then they could be applied widely. It was also important that such tests be predictive (Cairns and Mount, 1990). Minamata disease, for example, was explained only after thousands of people had been affected. The effects of DDT on the reproduction of raptors was discovered only after populations of major species were in decline. The need was for tests that would predict these effects in advance, so that release of such chemicals could be regulated in advance.

The perception was that simple tests were needed to facilitate legal applications. The combination of simple and inexpensive would allow widespread use of predictive tests by the regulating and regulated community. Simplicity was necessary to test the thousands of chemicals whose environmental toxicity was unknown. Furthermore, it was perceived that the ecosystems receiving anthropogenic wastes were too difficult to understand in a timely manner, and too expensive to study directly (Cairns and Mount, 1990).

Science has a long tradition of using controlled laboratory testing to rigorously develop understanding. The approach to toxicity testing followed this paradigm. The goal was to control all other variables, and test increasingly higher metal concentrations until the test organism died. The concentration of chemical at which the animal died defined the maximum allowable concentration. To obtain repeatable results, rigorous control of testing conditions was necessary. The best results came from: testing for lethal concentrations; testing one species at a time; testing over a short time period (standardised at 96 hours); using rigorously standardised environmental conditions; testing only dissolved metal toxicity (diet was not contaminated).

A large body of data eventually developed from tests standardised to these criteria (Box 3.3). The results showed wide variability in toxicity among metals, but a repeatable ranking was possible. Thus the relative hazard of metals could be defined for test conditions. The differences in lethal concentration among test species were large, but results were repeatable. It was also clear that lethality changed greatly with environmental or test conditions.

Research extended traditional single species toxicity testing to much more sophisticated approaches. For example tests were developed for more sensitive endpoints than lethality. Longer duration tests were developed for some uses. A variety of life history stages was tested. The number of species that could be routinely tested was expanded. Test media were extended to sediments. Multi-species and *in situ* tests were also developed, although with less success from a regulatory perspective. Generalisations developed about effects of environmental conditions on toxicity.

The ability to get the same results over and over made toxicity testing well suited for legal applications. Part of the success of toxicity testing in influencing policy stems from pragmatism: *the tests can be interpreted to yield a single number to define toxicity*. Above this value, waters are judged to be toxic and below it they are not. The method facilitates rule making that is easy to understand, if not universally harmonised (different conditions can yield a different numbers). The simplicity of the outcomes makes the application to policy relatively straightforward.

3.3.2   Limits

Prologue. The regulations and assessments of risk based upon toxicity testing rest upon a methodology that carefully balances uncertainties that might lead to an underestimate of toxicity (time of exposure, surrogate species, chronic toxicity, and failure to account for dietary exposure) with methodologies that might lead to an overestimate of toxicity (worst-case chemical conditions, application factors, natural resilience).

Box 3.3 Early fundamental findings from toxicity testing

Toxicity testing showed that:

- metals are toxic, at what appeared to be low concentrations;
- the lethal concentration is repeatable if determined under the same conditions;
- the lethal concentration varies with test conditions;
- the lethal concentration varies among metals;
- the lethal concentration varies among species;
- correction factors and assumptions about resilience are necessary before applying test results to regulations.

> The comfortable margins of safety now used in water pollution regulation, which have helped us so much in the past, will probably disappear . . . aquatic toxicologists . . . will have to show marked improvement in the precision and accuracy of predictions in order to provide the defined response of aquatic communities without big margins of safety.
>
> (Cairns and Mount, 1990)

Although legally useful, there are influential uncertainties associated with predictions of toxicity from standardised testing. The most important are those associated with extrapolation from the laboratory to nature (Luoma, 1996) (Chapter 10).

- Geochemistry. Generally it is known that chemical conditions, like water chemistry, affect toxicity. The challenge is to extrapolate to the variety of environmental conditions that occur in nature.
- Biology.
  - Diversity of responses. There is an immense diversity of responses to metals among fauna and flora, but for practical reasons all species on Earth cannot be tested.
  - Exposure routes. Incorporating the role of dietary exposure into standardised toxicity assessments, in particular, is proving problematic.
  - Time of exposure. Exposures to metals are typically less than a full generation of the test organism and less than most exposure times in nature.
- Validation. It has proved difficult to validate whether toxicity test predictions are accurately reflecting toxicity in natural waters. This reflects the complexity of nature and the challenge of integrating observations and experiments.

The limits to toxicity testing were well recognised. In order to apply the results to policy making, systematic adjustments, or assumptions, were made in the choices of the concentration that would be protective of nature (Cairns and Mount, 1990):

(1) *Application factors*. These are somewhat arbitrary 'correction factors' applied to toxicity test data to take into account any underestimation of toxicity that resulted from the trade-offs.
(2) *Natural systems are resilient*. Metals are transformed, sequestered, volatilised and complexed in natural waters, reducing their bioavailability. Metals are also detoxified to some degree by organisms. It was assumed that these factors conferred a 'resilience of nature' (Cairns and Mount, 1990) that provided a buffer to counter aspects of the tests that might underestimate toxicity.
(3) *Chemical test conditions are set for worst possible case.* Standardised test conditions were established that maximised the likelihood of finding toxicity by minimising the factors that contribute to natural resilience, providing another margin

of safety (for example, using test water with a minimum capacity for complexing metals and reducing bioavailability).

As new science is incorporated into toxicity testing, these margins of safety are used less and less (Chapter 10), as predicted by Cairns and Mount (1990). The controversy lies in whether improved replication, precision and calibration to some of the conditions in nature equate to improved predictive accuracy for natural waters (Cairns, 1986).

3.4   THE ENVIRONMENTAL GEOCHEMISTRY PERSPECTIVE

Prologue. 'The decrease of three orders of magnitude in the accepted concentration of lead in sea-water over the past four decades is not a real effect but is an artifact of successive improvements in reduction and control of the level of contamination introduced during sampling, storage and analysis' (Bruland, 1983).

Geochemistry is the study of sources, reactions, transport, effects and fates of chemical species in water, soil and air environments (Manahan, 1991). Major challenges in metal geochemistry include (i) 'determination of the nature and quantity of specific metals/metalloids in natural waters'; and (ii) 'understanding the nature, cycling, reactions and transport of chemical species of these elements in the environment' (Manahan, 1991). The geochemical understanding built in the last four decades has been instrumental in the recognition of the environmental problems caused by metals, and will be even more important as we begin to address the complex problems of the future (Box 3.4).

3.4.1   Determination of metal concentrations

Prologue. Probably all data on trace element concentrations in natural waters from before the late 1970s, and many more recent data sets, are questionable; some are wrong by as much as three orders of magnitude (1000 times).

Box 3.4 General questions from geochemistry, applicable to metal contamination

The general questions that early geochemists addressed are still relevant today:

(1) How have humans changed the cycling and distribution of trace elements in the environment?

(2) What were natural, pre-industrial background concentrations of metals in nature, and what is the magnitude of the concentration change?

(3) How do the chemical and biogeochemical reactions of trace metals influence the fate and adverse effects that stem from metal contamination?

Instrumentation and associated methods for routine detection of relatively low concentrations of most trace metals were developed in the 1950s and 1960s. Atomic absorption spectrophotometers, the instrument of choice, were fixtures in laboratories interested in environmental chemistry by the 1970s. The instrument and methods seemed relatively simple, so data quickly began to proliferate on metal concentrations in effluents, waters, sediments and tissues from nature.

As so often happens, more careful applications of the procedures showed an unexpected bottleneck in the quality of the data. Even though the instruments could detect low metal levels, every step in the early protocols contributed metal to the sample (sample contamination). Laboratory dust, water films on washed glassware and the internal components of commonly used laboratory supplies all contained metals like lead, zinc, cadmium and copper, that would leach into the samples. The smaller the quantity of metal to be measured, the more influential was such contamination. The most serious contamination problems occur with concentrations in water (the medium used to test the toxicity of metals by the toxicologists). But metal analyses of all matrices (sediments, soils, biological tissues) always require careful use of contamination-free techniques.

The first accurate, contamination-free, measurements of lead concentrations in the ocean were made in 1976 to 1977 by Schaule and Patterson (1981). Accurate measurements of other metals in natural waters were reported soon thereafter, as Patterson's methodologies were quickly adopted, mostly by chemical oceanographers. By unequivocally demonstrating ubiquitous occurrence of elevated concentrations of industrial lead over large areas affected by human activities, Patterson (1994) cited 'irrefutable proof that humans have poisoned the earth's biosphere and themselves with excessive amounts of industrial lead'. Since that time, metal contamination has been documented in water bodies throughout the world. Monitoring these concentrations over decades has shown that environmental chemistry responds strongly to human actions. When human activities mobilise metal, concentrations increase. When metal releases are effectively regulated, concentrations decline. But the importance of accurate analytical work in describing those trends cannot be overemphasised (e.g. Box 3.5).

3.4.2   Metal cycling and metal reactions

The earliest geological and geochemical studies showed that trace metals are natural, ubiquitous components of the Earth's crust. They undergo natural cycling, between and through land, air and water. These cycles, termed biogeochemical cycles, are driven by chemical, physical and biological processes. Once metal concentrations could be determined reliably, it became possible to study both the natural cycles and human influences on those cycles. It quickly became clear that the primary effect of human activities is an acceleration of mass movements of metals into and between land, air and water, resulting in an increase in concentration wherever humans perturb the cycle. For example, metals are mined from enriched deposits, purified from their ores, and used in many of the products that humans now consider indispensable. An inevitable outcome of this widespread use is that metals are constituents of the waste streams that come from all human activities. Extraordinary measures have to be undertaken to remove them from those wastes.

Box 3.5 Trends or contamination?

An event during the early period of metal analysis illustrated the importance of reliable analyses. It is also indicative of how difficult it was for many in the environmental field to understand the importance of high-quality analyses. Results from the only nationwide, long-term water quality monitoring network in the USA (the US Geological Survey's NASQUAN program) were reported in the journal *Science* in 1987 (Smith *et al.*, 1987). One observation was a downward trend in lead concentrations in stream waters, presumably reflecting the ban on various uses of the metal. Other articles had also reported downward trends in Pb in lake water (from <5000 ng/L in 1965 to 20 ng/L in 1978: cited by Flegal and Coale, 1989). These observations attracted a great deal of attention, but geochemists were suspicious about the data (Windom *et al.*, 1991).

The sources of concern were the poor detection limits, and the exceptionally high historical and modern Pb concentrations. For example, the NASQUAN analyses could not detect Pb concentrations lower than 2000 ng/L. In checking the lake data, ultra-clean analyses showed that Pb concentrations in Lake Ontario and Lake Huron were actually <2 ng/L in open waters and 30 ng/L where contamination was most severe, in Hamilton Harbour (Flegal and Coale, 1989), differing greatly from the concentrations reported in the trend analyses. Both the Lake Huron trends reported for the early years and the NASQUAN trends turned out to be artifacts representing trends in the reliability of the analysis. It was suspected that the laboratories were actually analysing trends in Pb concentrations of laboratory dust. Reliable analytical studies published after the NASQUAN report did indeed show decreasing Pb concentrations in sediments from the Mississippi River and waters from the North Atlantic (Trefry *et al.*, 1985; Wu and Boyle, 1997), albeit at concentrations orders of magnitude below those reported in the early work. But at the time, the poor-quality analyses affected the credibility of the monitoring programmes, and confounded interpretations of whether laws to ban lead were actually working.

The mass movement of metals from human activities, through the natural cycles of the Earth, is now equal to or exceeds that mass involved in natural cycling on a global scale (Chapter 4).

Early in the history of the science of trace metals, geochemists also recognised that metal concentrations in natural waters were too low to be controlled by solubility (Krauskopf, 1956). Metal concentrations dissolved in water are controlled at very low concentrations by their tendency to be concentrated onto the surfaces of particles (scavenged). They are also taken up and utilised in biogeochemical reactions by microbes, and accumulated by plants and animals.

The outcome of these strong removal reactions is that metal concentrations on particles and in the tissues of living organism are typically 100 to 1000000 times higher than concentrations in water. The vast majority of the mass of metal in a contaminated environment is thus associated with sediments and suspended matter.

Strong biogeochemical removal reactions mean that inputs of metal contamination to a water body may result in only small changes in dissolved concentrations, even when there are large concentration changes in other media. Windom (1990) concluded that dissolved trace metal concentrations were insensitive measures of human perturbations in large rivers. But the author did suggest that 'suspended sediments . . . provide . . . a better basis for estimating anthropogenic influences'. This is because changes are magnified many times in the suspended media. So while toxicologists were developing tests for toxicity of dissolved metals, geochemists were discovering that this is not the aspect of the environment most affected by human inputs.

The low concentrations of metals in natural waters react with other constituents dissolved in water, forming a variety of ionic forms or 'species'. Each of these forms reacts differently with particles or with living organisms (Wangersky, 1986). Those metals that form cations (positively charged ions) associate with natural chemicals that occur in the water as anions (negatively charged ions). Dissolved metal carbonate complexes, metal hydroxide complexes, metal chloride complexes or complexes of metals with dissolved organic matter are much more common than free uncomplexed metal ions in solution. Sediments and suspended particles (including living materials) also have different components that bind metals: organic ligands, iron oxides, manganese oxides and sulphides are especially important. Thus, upon release to the aquatic environment, a metal is partitioned between solid and dissolved states. And within each phase, further partitioning to different forms or species occurs depending upon water chemistry.

Speciation controls distributions between dissolved and particulate metal, and affects the uptake of metals by living organisms from both food and water (bioavailability). An important attribute of metals, then, is that an organism is never exposed to a metal as a single chemical entity. Within food and solution, the organism is exposed to a variety of metal forms. Each form differs in its availability to the organism. So, the rigorously controlled environmental conditions of the standardised toxicity test were a minute representation of an almost infinite array of metal forms that could occur in nature. Generalisations about controlling processes were obviously necessary to link the two.

3.4.3   Limitations

Knowledge of the geochemistry of metals (Box 3.6) developed as a rigorous, mechanism-based science; the emphasis was on knowledge not legislative mandates. As a result, our knowledge of metal chemistry is probably more substantial than our knowledge of the biology or ecotoxicology of metals. But chemistry has an inherent limitation that will never be overcome. Concentrations of a chemical cannot be interpreted with regard to bioavailability or toxicity without an accompanying biological test or measurement. Chemical contamination alone is not sufficient justification for a policy decision, although it is a critical ingredient in justifiable policy. Only in the last decade has the understanding of chemical speciation progressed sufficiently that it is possible to develop quantitative and reliable biological interpretations

Box 3.6 Early fundamental findings from geochemistry

Geochemical studies showed that:

- determination of metal concentrations is repeatable and reliable, even at the very low concentrations typical of natural waters, but only if proper 'clean' procedures are followed;
- metals have natural cycles; the mass of metal released from human activities is equal to the mass participating naturally in those cycles;
- metal concentrations in natural waters are much lower than originally thought and substantially lower than usually used in experiments or standardised tests;
- human activities can greatly increase those concentrations, especially locally;
- increases in response to human activities are much greater in particulate materials like sediments than in water;
- metals react with the natural components of water and sediment;
- those reactions affect the fate and bioavailability of metals;
- collaboration with toxicologists and biologists is necessary to decipher the ecological significance of chemical observations.

of its implications. So, like toxicity testing, geochemistry cannot stand by itself as a measure of metal contamination. An interdisciplinary view of science creates the best justified policy.

3.5   ECOSYSTEM AND NATURAL HISTORY PERSPECTIVE

> The important issue in validation is the ability to predict the relationship between the response of the artificial laboratory system and the natural system.
>
> (Cairns, 1986)

Prologue. The most certain thing about the future is that there will be surprises. Observation of nature is necessary to detect surprises, identify their characteristics and develop preventative measures from that knowledge. Relating cause and effect in natural waters is difficult in the absence of a diverse body of work; nevertheless observation and field experiments address problems that cannot be addressed in toxicity testing or from geochemical determinations alone.

Natural history is the term we use to describe the more holistic view of metal contamination through the perspective of interacting physical, chemical, biological and/or ecological processes in natural waters. Few texts recognise this category as separate from the others. But recognising unique attributes of this perspective could be a key in understanding some of today's most difficult dilemmas.

The goal of studying metals from the holistic perspective is to understand phenomena that influence metal impacts in nature (e.g. Box 3.7).

The approaches to natural-history-based science are varied. Natural history intersects toxicology in experimental studies that identify the internal, biological modes of action whereby detrimental effects are expressed. Reliance on observations from nature is where natural history diverges most from the experiment-based toxicity testing perspective.

Box 3.7 Some questions addressed from the natural history perspective

(1) Are there examples of deleterious effects of metals on organisms in natural waters?
(2) What forms do those effects take?
(3) What are the early warnings of the potential for effects?
   • Concentrations in water and sediment?
   • Bioaccumulation?
   • Signs of stress in survivors?
   • Population changes in survivors?
   • Community change – what species are missing?
(4) What processes control effects in natural waters (predictive models)?
   • Bioavailability?
   • How different species ameliorate exposures to avoid effects (detoxification)?
   • Bioaccumulation and biomagnification?
   • Population and community resiliencies?

A basic premise has long dominated the thought of biologists and ecologists. A complete understanding of a system requires more than just understanding each part in isolation. The whole is, in fact, more than the sum of the parts. One must understand the interaction among the parts, termed the hierarchical (Odum, 1986) or holistic (Mayr, 1982) approach. But there is also a great tradition of experimentalism in science. Unfortunately there is often a tension between the reductionism of the experimentalist and the holism of nature's observers (see also Clements and Newman, 2002). There are those who deny the legitimacy of observing nature as a form of science, and those who suggest that view is arrogant. The legitimate approach lies somewhere in between, of course (Mayr, 1982):

> Science seeks to organize knowledge, provide explanations, and propose hypotheses that can be explored (or rejected). One approach is to subsume the vast diversity of phenomena into a smaller number of explanatory principles. The arrogance of the experimentalist lies in the presumption that experiments are the only approach to this goal. The alternative to experiment is observation (of nature). But, mere observation is not sufficient, and that is not what we mean by natural history. Scientists often go back and forth between a phase where they observe and describe phenomena to develop questions, followed by a phase of concept formulation (hypotheses about how things work) and experimental testing to evaluate the mechanisms or concepts in isolation.

Mayr (1982) also explained how to maximise the value of observation and natural history:

> Progress depends upon designing observations to answer *carefully posed questions*. One approach is the comparative method. . . . Observation of events at well chosen times and places can sometimes provide information as wholly sufficient for a conclusion as that which can be obtained from an experiment. The successful natural historian recognizes the many experiments nature is providing and takes advantage of opportunities to examine those

experiments. The answers may not always be unambiguous, but the insights are from the real system.

Mayr (1982) was a firm supporter of the long-term value of the natural history approach:

> Proof of the value of observation is deeply rooted in biology. Careful observations of nature resulted in the Linnaean hierarchy (the biological taxonomy of all life forms), natural selection, and other foundations of ecology. Observation in biology has probably produced more insights than all experiments combined.

Many of the most dramatic successes in reducing environmental damage stemmed first from observations of contamination problems in people or in nature. The 'haunting concern' (Waldichuck, 1979) of conservationists (and observers of nature) is that a metal pollutant may affect an organism in a most unexpected way. If so, important effects might be occurring that cannot be anticipated from standardised testing alone. The example of organochlorine chemical regulation (Chapter 1) illustrates how observations of a declining population of a species provided the first indication of a pollution problem. In another example, biologists Peter Gibbs and Geoff Bryan were shocked to find female snails with male genitalia on the rocky shores of southwest England in the 1980s (Chapter 14). Over decades of studying those habitats they had never seen such a phenomenon. They and others eventually identified the cause as widespread contamination from tributyl tin (TBT), an antifoulant used on the bottoms of the many watercraft in the area. Use of that chemical on small pleasure-craft was eventually banned.

The opposite concern is also expressed: do the safeguards of standardised testing go so far as to overprotect life from pollutant effects, thereby shifting resources and attention to the wrong problems in the environment? As the body of observations from nature grows in number and sophistication, it becomes clear that concerns about both overprotection and underprotection are legitimate.

Finally, some types of hazards from metals are extremely difficult (or impossible) to test fully in an experiment. For example bioavailability of metals is a subject that has long been studied because of its importance to metal effects. Geochemical methodologies that attempt to predict bioavailability in nature often require complex mathematical models with limited applications to natural conditions. On the other hand protocols for using organisms as biomonitors of metal exposure implicitly account for bioavailability. And practical protocols for collecting biomonitor data are available (Phillips and Rainbow, 1994). Similarly, methyl mercury (Chapter 14) and selenium (Chapter 13) are effectively transferred from prey to predator through food webs. Standard toxicity testing, alone, is not suited for studying large, long-lived organisms in the laboratory or for evaluating complex food webs. Experiments have difficulties studying effects that take a long time to develop or stem from the functional ecology of the species in question.

In Box 3.8, Larison et al. (2000) investigated such a circumstance in the field. In this case, a population of birds was living, by superficial appearances, undisturbed in an area with widely contaminated soils. Careful investigation showed that the population was struggling to survive because of metal-related, cumulative damage to the bone structure of adult birds. The damage occurred because the diet of this species was particularly contaminated with Cd. Extirpation of the population was likely, if it were not for immigration of juvenile birds from surrounding, less contaminated areas. Such observations can lead to predictions in other circumstances. But testing alone did not help anticipate the species most under threat. Nor has testing identified the specific diets that can lead to long-term damage or identified the dependence of effects upon long exposures. Key to perpetuation of the ptarmigan was the internal dispersal dynamics of the bird population; something that cannot be directly evaluated in a toxicity test. Field observations suggest such are the questions we must address in the future.

Most important, local regulatory decisions are best justified if the mechanisms underlying the

Box 3.8 Metal effects via food web transfer

Larison *et al.* (2000) investigated whether cadmium enrichment could explain the existence of fragile-bone disease noted in the herbivorous bird, white-tailed ptarmigan (*Lagopus leucurus*) in the Colorado ore-belt (an historically mined region in the western USA). They found that Cd was particularly magnified in the tissues of one type of plant common in the area: the willow, *Salix* sp. The digestive tracts of ptarmigan showed that willow was an important component of their diet. The result was high Cd concentrations in all ore-belt ptarmigan, compared to ptarmigan from outside the ore belt, especially in older birds. Cadmium accumulated in the birds until irreversible kidney damage occurred. Kidney failure affects the integrity of the skeleton, by affecting calcium balance. Calcium-deficient weak bones were found in the ptarmigan with the highest Cd exposures (Fig. 3.1). Larison *et al.* (2000) found that this resulted in reduced survival of adult ptarmigan. Females had higher Cd exposures than males and their survivorship was reduced more. The result was greatly reduced breeding and low ptarmigan densities in areas where Cd exposures were likely to be greatest.

 Thus subtle, but detrimental, effects were unambiguously demonstrated for an important species of bird, likely over a substantial area, due largely to their specific diet (willow). Larison *et al.* (2000) surmised that other herbivores that eat willow from contaminated areas could suffer similar chronic Cd poisoning. Toxicity testing alone did not, and could not have been expected to, anticipate these findings.

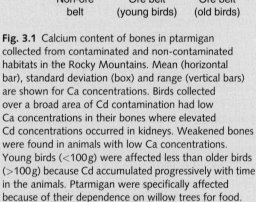

Fig. 3.1 Calcium content of bones in ptarmigan collected from contaminated and non-contaminated habitats in the Rocky Mountains. Mean (horizontal bar), standard deviation (box) and range (vertical bars) are shown for Ca concentrations. Birds collected over a broad area of Cd contamination had low Ca concentrations in their bones where elevated Cd concentrations occurred in kidneys. Weakened bones were found in animals with low Ca concentrations. Young birds (<100 g) were affected less than older birds (>100 g) because Cd accumulated progressively with time in the animals. Ptarmigan were specifically affected because of their dependence on willow trees for food. Willows accumulate more Cd than most plants in these contaminated areas. (From Larison *et al.*, 2000.)

problem are understood. Mechanistic understanding is rarely the goal of direct studies of toxicity. It is a fundamental goal of natural history studies. Box 3.9 lists findings from natural history studies.

3.6 OVERLAPS AMONG CORE DISCIPLINES: DETECTING DETRIMENTAL EFFECTS

 Theoretical ecologists have not been particularly helpful in providing end points for validating (toxicity) predictions.

(Cairns and Mount, 1990)

 Would we find any less ludicrous the proposition that studies of the response of isolated populations under laboratory conditions predict the ecological consequences of pollutant exposure on ecosystems? . . . this is what government regulations have been designed to do.

(Forbes and Forbes, 1994)

Like all categorisations, there are overlaps in interests and activities among the schools of thought described above. Natural history and environmental geochemistry overlap when it comes to the processes that control the fate or chemical speciation of metals in an environment, and how that

Box 3.9 Early fundamental findings from natural history

Natural history studies have shown:

- Metals have adverse effects in nature, but the concentrations at which they occur may not be easily comparable to toxicity test results.
- Manifestation of metal effects in nature is complicated by multiple exposure routes, unique characteristics of species and influences of environmental conditions.
- Observations of nature and experiments driven by those observations can uncover surprises (TBT) and define impacts that toxicity testing and geochemistry cannot predict (related to food webs).

- Metal exposures of biota in nature can be monitored using biomonitors and it is also possible to monitor sublethal responses of organisms to metals (biomarkers).
- It is challenging, although not impossible, to demonstrate unequivocally that metal exposure is the cause of an observed ecological effect. Complexity within all core disciplines, and natural variability, can confound simple approaches and individual studies. But a body of work can provide convincing evident from: (a) documentation of exposure, (b) knowledge of environmental conditions, (c) understanding of the species or community, (d) proper anticipation of the effect (from toxicological knowledge of the metal) and (e) study design that eliminates potential confounding factors.

influences biological responses. Natural history and applied ecotoxicology overlap in the study of physiological responses of organisms to metal contaminants. Effective overlaps need to involve more than the use of methods from one discipline by experts in another, however. It is in thinking across boundaries, and in the development of an integrated, interdisciplinary science that the greatest opportunities for more effective risk management lie (lateral management).

One obvious area where overlap should be important is validation of the predictions from traditional ecotoxicological testing that lead to environmental regulations. How well do such regulations protect nature but avoid unnecessary cost? How well did a risk assessment accurately assess risk? Evaluations of regulations and risk assessments are largely superficial to date. Policies to protect the environment ultimately are best justified by evidence that they are effective and were needed. But in reality few follow-up studies occur once a regulation is put in place or a risk assessment is completed.

By definition, validation studies for regulations must occur in nature for it is, after all, nature that we are trying to protect. Validation must be interdisciplinary, as well. It requires knowledge of sources, concentrations, forms, bioavailability, bioaccumulation and multiple aspects of effects

(Fig. 3.2). This means high-quality geochemistry, knowledge of the natural history of the species at risk and knowledge of the water body (among other things) in addition to high-quality toxicology. The body of work that results in a strong validation study should develop lines of evidence around the conceptual model that guides metal fate and effects (Fig. 3.2). This complex approach is necessary because nature is variable and manifestation of toxicity is often complex in all but the most extreme cases. Many factors affect the presence and absence of species (Chapter 11). Persistent study and a body of knowledge are necessary to overcome this challenge, along with a willingness to consider all lines of evidence. It traditionally has been difficult to identify, unambiguously, metal toxicity in nature (Bryan, 1984). This is probably not because adverse effects are not occurring, (e.g. Boxes 3.8, 3.9 and 3.10), but because our will to invest in the multiple lines of evidence and lateral thinking necessary to uncover the nature of any adverse effects is often insufficient. Despite the challenges, precedents now exist for how to address validation of regulations (Chapter 11).

One simple lesson we learn over and over from long-term observations is that as human releases of metals change, the environment responds (Box 3.10). The chemistry of the environment, the exposure of organisms and the response of

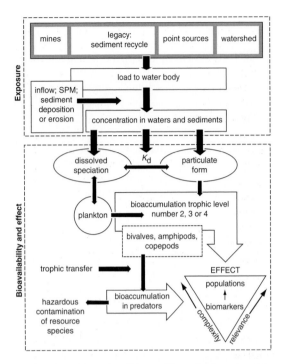

Fig. 3.2 A conceptual model to guide evaluation of metal exposure and effects. Such models provide a framework for systematic addressing of metal fate and effects. Traditionally, studies of the validity of regulations often attempted to directly compare ecological aspects of a community to toxicity test results. More successful approaches have addressed the linkages between exposure, bioavailability and effects in a more holistic way. SPM, suspended particulate matter.

organisms, populations and communities broadly change in response to human intervention. Although adverse changes are most frequently discussed, recovery from damage is also possible (Box 3.10).

3.7   POLICY PERSPECTIVES ON THE DIFFERENT DISCIPLINES

Prologue. Interdisciplinary science underlies the influences of metal contamination in natural waters and is fundamental to the approaches that we now use in managing that contamination. But the demands for simplicity and inexpensive approaches from the policy community also sometimes contradict rigorous application of geochemical and holistic ecosystem science.

The influence of the different disciplines on policy can be judged only subjectively and will continuously change. Toxicity testing seemed to have great influence on the early policies, although chemical analyses and observations of natural history were instrumental in identifying problems. The policies that were implemented on the basis of toxicity testing were instrumental in successfully reducing the most egregious cases of metal contamination. The question is whether that approach is adequate for the more complex modern issues.

The toxicity testing methodologies were the first choice for use by policy makers and regulators partly because alternative methods of managing contamination have not been accepted. Ambiguities are typically cited for every alternative. So, despite minimal validation, and serious trade-offs in their applications to nature, even the strongest critics of traditional approaches to establishing regulations have historically felt that there simply was no reasonable alternative to toxicity testing as a basis for metals policy (Cairns, 1986).

Geochemical aspects of the conceptual model are also instrumental in management of metals. Identifying mass movement of metals provides the conceptual basis for identifying the most important sources of metals in a basin, and determining how to reduce their loads to the environment (Total Maximum Daily Loads (TMDL) in US regulations). Toxic inventories of metal releases from specific industries (the mass of metal disposed of in the environment by an industry) are used all over the world to characterise inputs. Ambient water quality criteria and sediment quality criteria depend upon comparisons of concentrations in nature against a standard that represents toxicity. But at least some of the databases maintained by government agencies (e.g. the US Environmental Protection Agency's STORET database) still do not differentiate data on the basis of ultra-clean analyses. Until such oversights are corrected, the most basic aspects of the integration are questionable.

It is becoming more accepted that metal inputs and elevated metal concentrations, alone, are not a sufficient basis for regulatory action. Bioavailability is slowly becoming more important as a

Box 3.10 Detrimental effects of sediment contamination at a mudflat in South San Francisco Bay

Hornberger *et al.* (2000) conducted 23 years of monthly measurements on a mudflat 2 km from a domestic sewage outfall. Metal concentrations in surface sediment and clam tissues (*Macoma petalum*, as *M. balthica*), along with metal loads from the treatment works were studied. Two metals, copper and silver, were the major pollutants discharged from the treatment works. The study established a link between Cu and Ag discharge, bioaccumulation and detrimental effects throughout the mudflat.

Improvements in treatment of the wastes from this facility were progressively implemented during the 1980s as mandated by the Clean Water Act. In response, Ag concentrations declined from 106 µg/g to 4 µg/g; and copper from 287 µg/g to 24 µg/g in clam tissues during the 23 years of study. Concentrations of both metals in sediments also declined. Dissolved Ag was determined in the waters of South San Francisco Bay during the time of greatest contamination at 100 ng/L. In 2004 dissolved Ag concentrations were 6 ng/L.

Declining tissue metals were strongly correlated to declines in waste discharges of the metals (Fig. 3.3). Reproduction persistently failed in the clams in the mid-1970s to mid-1980s; the animals were not producing viable eggs and sperm. Reproduction recovered after metal contamination declined. The community of animals living in the mud also changed with recovery from the metal contamination. Animals that fed on the mud (deposit feeders) and species that laid their eggs in the mud reappeared or grew in abundance once the contamination subsided. Other potential explanations of the biological and ecological changes were considered: food availability, sediment chemistry, salinity, temperature. None of these changed unidirectionally over the 23 years. The causative link between metal contamination and reproductive failure in one species and reduced abundance of others was inferred because only metal concentrations were correlated with the change.

These effects occurred at dissolved Ag concentrations estimated to be 10-fold lower than the ambient water quality criteria (1500 ng/L). Although the comparison to existing criteria is indirect, it does point towards the need for further consideration of such criteria. (Ecological data are from Shouse, 2002; dissolved Ag data from Girvin *et al.*, 1975.)

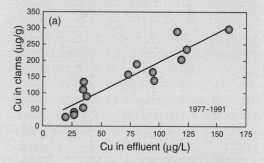

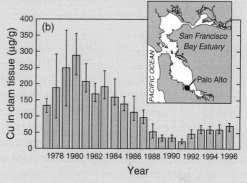

Fig. 3.3. Annual mean Cu concentrations in the tissues of the bivalve *Macoma petalum* (as *M. balthica*), a resident species on a mudflat 2 km from the outfall of the Palo Alto waste treatment facility, as a function of (a) dissolved Cu concentrations in the effluent of the treatment plant, and (b) time. Clams were collected eight to ten times per year, each year between 1977 and 1991. Changing bivalve concentrations closely followed changes in copper discharges to South San Francisco Bay as the treatment facility cleaned up its effluent (Hornberger *et al.*, 2000). Improved reproduction in *M. petalum* and return of species that laid their eggs in the mud accompanied the declining metal contamination in this species. The City of Palo Alto displayed their bivalve data to the public to demonstrate their improved environmental performance.

consideration (Fig. 3.2). Modern regulations consider at least some aspects of metal speciation in regulations and risk assessments, and at least simplistic consideration of the form of particulate metals. But protective regulations are based upon experiments with dissolved metal only. It is startling to realise that as applied ecotoxicologists were developing and expanding the use of standardised dissolved metal toxicity testing to include the complexities of speciation, geochemists were concluding that dissolved metal concentrations did not vary as widely in response to contamination as did concentrations in particulates and bioaccumulation by organisms.

The ecosystem approach has been historically difficult for managers and policy practitioners. Applied ecotoxicology has often been influenced by the argument that the ecosystem approach is simply too complex; and cause–effect too difficult to characterise. But knowledge is accumulating about many aquatic environments. These bodies of work provide numerous examples where the nature and occurrence of metal effects are unequivocal (Boxes 3.8 and 3.10; Chapter 11). As we proceed through this text, we will show how the science is approaching a point where the complexity and breadth of the geochemical, ecotoxicological and holistic paradigms can be reduced enough (through expanded experience and knowledge) to provide tools applicable to the requirements of the legal system for simplicity.

3.8   CONCLUSIONS

Untangling the complexities of natural environments is a slow process. Results can be ambiguous, especially from a legal perspective. Better understanding of all our major ecosystems, as well as generalisations about mechanisms and processes, are necessary for incorporation of local conditions into managing risks from metals. That goal is now realistic given the high-quality data and high-quality expertise that have developed in the last 30 years. What once seemed an impossible task is now an overt goal in some jurisdictions (see the European Union's Water Framework Directive). Vision at this scale is the source from which better management of metal contamination will grow.

The important differences between approaches to science must be recognised before investigators can successfully work across boundaries (Maciorowski, 1988). New understanding of metal contamination will expand as the perspectives of chemistry, applied ecotoxicology and natural history are integrated. This will bring opportunities for continually improving our management approaches, so long as the policy community is willing to think laterally as well.

Suggested reading

Bruland, K.W. (1983). Trace elements in sea-water. In *Chemical Oceanography*, vol. 8, ed. J.P. Riley and R. Chester. London: Academic Press, pp. 157–220.

Cairns, J. and Mount, D.I. (1990). Aquatic toxicology. *Environmental Science and Technology*, **24**, 154–161.

Hornberger, M.I., Luoma, S.N., Cain, D.J., Parchaso, F., Brown, C.L., Bouse, R.M., Wellise, C. and Thompson, J.K. (2000). Linkage of bioaccumulation and biological effects to changes in pollutant loads in South San Francisco Bay. *Environmental Science and Technology*, **34**, 2401–2409.

Larison, J.R., Likens G.E., Fitzpatrick, J.W. and Crock, J.G. (2000). Cadmium toxicity among wildlife in the Colorado Rocky Mountains. *Nature*, **406**, 181–183.

4 • Sources and cycles of trace metals

4.1 BIOGEOCHEMICAL CYCLES OF TRACE METALS

Prologue. Metals are a natural component of the Earth. They cycle between reservoirs in the environment, a process termed biogeochemical cycling.

Metals occur in the rocks of the Earth's crust and the soils that result from erosion of those rocks. They occur in streams, rivers and lakes in dissolved forms and associated with particulate material. They also occur in the atmosphere in particulates, aerosols and, in a few cases, as gases. In later chapters we will discuss concentrations (e.g. µg metal/g particulate) of metals typical of these different reservoirs (Chapters 5 and 6). But in this chapter we are interested in the masses (e.g. kg) of metals (the standing stock) that occur within and are exchanged between major reservoirs (e.g. kg/year). Load (in units of g or kg, for example) is another term for the mass of metal within a reservoir, and the term flux describes the rate of movement of the metal (e.g. moving from a human activity into the atmosphere). Fluxes and balances of mass movements allow us to quantify the contribution of human activities to the natural biogeochemical cycles of metals. Mass inputs also are useful in identifying and quantifying the relative

---

Box 4.1 Definitions

biogeochemical cycle: the biologically driven and/or geochemical interactions that exist between the atmosphere, hydrosphere, lithosphere and biosphere.

biosphere: the part of the Earth and its atmosphere in which living organisms exist or that is capable of supporting life.

budget or mass balance: the balance sheet of all inputs and outputs to a reservoir of metal. A steady state stock in a reservoir occurs when input rates equal output rates.

cycle: a system of two or more connected reservoirs of a substance, where a large part of the material is transferred through the system in a cyclic fashion.

flux: rate of movement of a metal between reservoirs (e.g. moving from sediment into the atmosphere); e.g. influx is the movement of a metal into an organism.

load: mass of metal (in units of g or kg, for example) within a reservoir (e.g. atmosphere, sediment, water, etc.).

reservoir: a reservoir (e.g. atmosphere, soil, sediment, water body) or group of like reservoirs that contains the material of interest.

stock (*or standing stock*): mass of material in a reservoir.

stratosphere: the region of the atmosphere above the troposphere and below the mesosphere.

Total Maximum Daily Load (**TMDL**): the mass of metal allocated to a given source measured as an input rate (loading rate) to the environment.

troposphere: the lowest layer of the atmosphere, containing about 95% of the mass of air in the Earth's atmosphere. The troposphere extends from the Earth's surface up to about 10 to 15 km.

importance of particular sources of metal contamination. The latter information is one approach used to manage risks from those sources.

4.1.1   Natural cycles

Prologue. Natural cycles involve the exchange of metals between the continents, the atmosphere and the oceans, as influenced by physical, geochemical and biological processes. Trace metals are released as rock is weathered. These metals are passed into soils and/or transported either with soil particles or in dissolved form, into rivers, estuaries and eventually to the deep oceans. Ecological risks from metal contamination occur when natural cycles are accelerated by human activities, resulting in concentrations that adversely affect ecological processes.

Environmental conditions on Earth are controlled by physical, chemical, biological and human interactions that transform and transport materials and energy. This 'Earth system' is characterised by complex responses and thresholds, with linkages between its various reservoirs (Jickells *et al.*, 2005). All metals have natural biogeochemical cycles that are important components of the Earth system.

The biogeochemical cycles of metals involve mobilisation and deposition within and among rocks and soils, rivers and streams, estuaries and oceans, as well as the atmosphere. Rates of movement (fluxes) between reservoirs characterise these biogeochemical cycles. The primary influence of humans is to accelerate these fluxes. In a global sense, this is merely a modification of what occurs naturally. But, in a local sense, it means the accumulation of metal concentrations that can far exceed what life forms have experienced during their evolution. If these concentrations reach sufficiently high levels, those life forms will be exterminated. The most egregious cases of accelerated fluxes (e.g. acid mine drainage), for example, can eliminate all life except some forms of bacteria. Progressively less disturbance of the natural fluxes eliminates progressively fewer forms of life.

The components of the Earth system are linked in complex ways. The atmosphere interacts with the oceans and the land masses in many ways that facilitate trace metal fluxes between the two. For example, volcanoes inject particulate metals high into the troposphere and even into the stratosphere (the layer of atmosphere above the troposphere). Particles in the stratosphere will be subject to global distribution. Particles in the troposphere are transported according to local and regional wind patterns. Nature's processes that contribute metals to the lower troposphere (the part of the atmosphere closest to the Earth's surface) include forest fires, windblown dust, vegetation, and sea salt sprays (Nriagu, 1979).

Metal fluxes into and out of the atmosphere may be restricted to a local or regional scale, if trace metals associated with larger particulates settle out of the atmosphere relatively rapidly (Salomons and Förstner, 1984). But metals can also be carried large distances under the appropriate circumstances. Dust from the Sahara Desert has deposited Fe, Mn, Al and associated trace metals across the Mediterranean, Atlantic and Caribbean Seas for millions of years (Box 4.2). Similar dust storms in the deserts of China result in detectable inputs to the Pacific Ocean. Riverine inputs move the greatest mass of metals from the land to the sea (Table 4.1). But those metals are quickly removed from the water column in coastal waters because of the tendency for trace metals to accumulate on the surfaces of particles that settle into the bottom sediments. Metals may also be taken up by living organisms in the surface waters (phytoplankton, in particular) and pass into food webs, recycle or settle to the bottom. In contrast, wind-driven dust carries particulate metal far offshore. This direct input is the primary source of metal inputs to the open ocean, in general. Some of the trace metals that are essential for life (e.g. Fe) are in sufficiently low concentrations in open ocean waters that they limit plant growth. It is the atmospheric input of Fe, for example, that supports growth of plants at the base of some oceanic food webs. In such areas, atmospheric inputs are crucial to ocean productivity.

Order-of-magnitude estimates of natural metal inputs to the atmosphere are available (Nriagu, 1979). These are derived from scattered local

Box 4.2 Dust storms and metal transport

It is now well established that winds can lift fine mineral particles from soils, and then carry those particles great distances around the planet. Regular wind patterns, like the trade winds of the subtropics, have typical dust plumes associated with them. But the most important inputs are from hyper-arid areas (deserts), such as the Sahara Desert. Hyper-arid lands occupy 0.9 billion hectares and dry lands occupy 5.2 billion hectares, one-third of global land area (Jickells *et al.*, 2005). Satellite data show that dust storms originating from these areas can cover huge areas of the world and that dust can be carried immense distances. Dust storms from the great deserts of Africa (the Sahara) and China (the Gobi) have particularly strong influences. For example, dust from

Northern Africa (the Sahara Desert) is the strongest dust source in the world (Prospero, 2001). Storms are frequent; therefore Africa exports dust much of the year, sending streams eastward across the Middle East and the Arabian Sea, north over the Mediterranean to Europe and west across the tropical Atlantic to North and South America. In ocean regions where satellites show clear air, the concentration of dust is typically negligible. But when a dust cloud advances into a region, particle concentrations can increase to hundreds of micrograms per cubic metre. Long-term variability in dust in the atmosphere of the Barbados Islands (in the Caribbean Sea in the Americas) is related to rainfall in Africa. The influence of African dust is seen in the atmosphere of Miami and elsewhere across the eastern USA. (After Prospero, 2001.)

Table 4.1 *Sources of Fe input to ocean waters. Riverine and glacial particulate Fe represent the largest supply to the edges of the ocean, but are efficiently trapped in these near-coastal areas, except where rivers discharge directly beyond the shelf. Hence phytoplankton depend upon dust transport as the dominant external input of Fe to the surface of the open ocean, mainly from the great deserts of the world. Thus the atmosphere seems like a small source of nutrients like Fe, but its inputs are crucial in the offshore open ocean because other sources do not reach those waters*

Fe source	Fe fluxes (10^{12} g/yr)
Fluvial (rivers) particulate	625 to 962
Fluvial dissolved	1.5
Glacial sediments	34 to 211
Atmospheric	16
Coastal erosion	8
Hydrothermal	14
Authigenic	5

Source: Jickells *et al.* (2005).

observations extrapolated globally, based upon broad assumptions about homogeneity of distributions. Nevertheless they provide a useful baseline of information about the relative importance of different undisturbed inputs (Table 4.2).

Knowledge of natural biogeochemical cycles is necessary to quantify anthropogenic inputs. The degree of disturbance is determined by comparing either emission rates, or fluxes in undisturbed environments with those in circumstances affected by humans. If it is necessary to extrapolate to the past to derive an undisturbed cycle, fluxes can be estimated by assuming that natural biogeochemical cycles were near steady state (inputs to major reservoirs matched outputs).

Human activities accelerate inputs to one or more reservoirs and thereby affect concentrations in all reservoirs (termed the multi-media nature of contamination). This is because the reservoirs are interlinked as described above. For example, if metals are emitted to the air in large quantities by human activities, that will result in higher concentrations in the atmosphere. But some of this increase also will be balanced by increased loss from the atmosphere as metals deposit in water, soil and sediments. Thus, accelerated inputs to the atmosphere accelerate inputs to land and the water and metal concentrations build up everywhere.

Mason *et al.* (1994) estimated pre-industrial fluxes of mercury by assuming the processes that governed Hg movement within the Earth system had not changed, although the magnitude of the fluxes had. They also assumed that inputs and

Table 4.2 *Natural emissions to the atmosphere of trace metals (thousand tons per year)*

Metal	Windblown dust	Forest fires	Volcanic particles	Vegetation	Sea salt sprays
Cadmium	0.25 (0.05–0.85)	0.01 (0.01–1.5)	0.5 (0.04–7.8)	0.2 (0.05–2.7)	0.002 (0.001–0.4)
Copper	12 (0.1–26)	0.3	3.6	2.5	0.08
Nickel	20	0.6	3.8	1.6	0.04
Lead	16	0.5	6.4	1.6	5
Zinc	25	2.1	7.0	9.4	10

Source: After Pacyna (1986); Nriagu and Pacyna (1988).

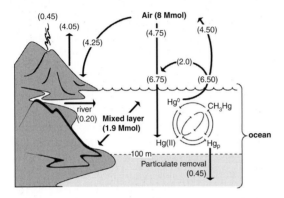

Fig. 4.1 The pre-industrial, undisturbed cycle of Hg. Bolded numbers represent standing stock of Hg in the reservoir (Mmol); unbolded numbers represent fluxes (Mmol/yr). (After Mason *et al.*, 1994 and Lindberg *et al.*, 2007.)

outputs balanced for each global reservoir before human inputs became significant (the global Hg cycle was at steady state). The major reservoirs that they considered were Hg in the atmosphere, Hg in terrestrial soils and Hg in the oceans (Fig. 4.1). Mercury is unusual in its large natural atmospheric reservoir which is dominated by a gaseous form of the metal, Hg^0. Atmospheric Hg exchanges rapidly with the land and surface water. The reservoir is relatively small and the fluxes are relatively large. Concentrations in the atmosphere and the surface of the ocean, for example, are detectably affected when the cycle is perturbed. In contrast, soils exchange Hg slowly relative to the large mass of Hg that resides there. Thus, it is a long time before changes in flux rates will affect soil concentrations. By knowing undisturbed concentrations in large reservoirs like soils,

undisturbed fluxes from them can be estimated from modern processes.

4.1.2   Human sources

Prologue. Metals are ubiquitous components of nearly all the products used by humans. The different sources of metal emissions to the biosphere by human activities are relatively well known, but differ among metals. Metals are emitted from concentrated sources like industrial establishments and from diffuse sources like street runoff. Soils and sediments are the predominant repository in the environment. The importance of the atmosphere differs among sources, as determined by metal chemistry and the form of source inputs.

Metals and their compounds are indispensable to modern society. Nearly every metal has some use in industry, commercial or personal applications, all of which generate wastes which must be discharged into the environment. Although most individual problems are local or regional, metals are a global contamination issue because of their long history of ubiquitous use wherever humans are active. The potential toxicity of all the metals being released annually into the environment by human activities exceeds the combined potential toxicity of all the radioactive and organic wastes, as measured by the quantity of water needed to dilute such wastes to the drinking water standards of the USA (Nriagu, 1988).

The principal anthropogenic sources of trace metals in the atmosphere are smelting of metallic

Table 4.3 *Anthropogenic emissions to the atmosphere of trace metals (thousand tons per year)*

Metal	Air	Water	Soil	Total
Cadmium	7.6	9.4	22	39
Copper	35	112	954	1101
Chromium	30	142	896	2169
Mercury	3.6	4.6	8.3	16.5
Zinc	132	226	1372	1730

Source: Nriagu (1988).

ores, industrial fabrication and commercial applications of metals, as well as burning of fossil fuels (Nriagu, 1988) (Table 4.3). Some inputs occur via large releases from individual activities (e.g. point sources like smelters). Discharges of animal wastes, irrigation drainage, food wastes, domestic and industrial waste waters, urban runoff and the dumping of sewage sludge are all contaminated by metals and can involve consolidation of wastes and release at a single point. More diffuse inputs also occur. For example, combustion of fuels from millions of motor vehicles and fleets of aircraft emit lead over large parts of the Earth. Lead emitted to the atmosphere is eventually deposited (sometimes over wide areas) on land or in natural waters. Once humans discharge Pb to the atmosphere, it becomes a source for the land and the oceans. Urban runoff is consolidated in some places and allowed to be diffuse in others. Many small point sources in the same geographical area can amount to a diffuse source on a regional scale (Box 4.3).

A comprehensive accounting must consider all stages in the 'life cycle' of a metal: mining and processing, fabrication of products, how those products are used and what happens to them when their useful life is exhausted (Gordon *et al.*, 2003). But the sources of individual metals are manifold, complex and widespread. For example, the US Environmental Protection Agency's National Emissions Inventory (NEI) contains data obtained by a variety of jurisdictions as part of EPA's hazardous air pollutant regulatory programmes. In 2002, the NEI estimated that there were some 13087

Box 4.3 A dispersed source of trace metal contamination: mining

Individual, concentrated sources of metal input are of great concern in determining human influences on metal cycles. But if the influences of many concentrated sources of metal input are combined on the scale of a region or continent, for example, the effect is the same as that of a diffuse input. Mining is an example of a concentrated input on a local scale and a diffuse input on a continental scale. The US Environmental Protection Agency cited more than 13 000 active mining operations in the USA in 1985 (Chapter 12), many of them relatively small (USEPA, 1985). By 2005 there were many fewer active mines in the USA, but an increased number of inactive, closed or abandoned operations. The period of intense mining activity left behind abandoned mines in 134 of the 387 National Parks. These include more than 3200 sites with about 10 000 mine openings, piles of tailings, and hazardous structures; and thousands of hectares of scarred lands (USEPA, 1985). Total waste accumulated by all active, inactive and abandoned mines in the USA between 1910 and 1982 was estimated at 50 billion metric tons; about fourfold more mass of waste per year accumulated than from all other industrial wastes combined. About 40% of the waste in 1982 was generated by copper mining, and 30% by phosphate mining (see Chapter 13 on selenium). Few accurate estimates of this type are available for areas on Earth or for global mining as a whole, however.

industrial, commercial or institutional sources in the USA that contain one or more processes that emit Pb to the atmosphere (Fig. 4.2). These were divided among 36 different categories of activities. The largest masses of Pb are released during processing of the metal (primary and secondary smelters) or in uses spread diffusely across the continent (combustion-related sources like boilers, motor vehicles or aircraft). Most of these sources emit less than 0.1 ton per year. But 1300 sources emitted greater than or equal to 0.1 ton per year, and these emitted 94% of the total Pb emissions in 2002 (94% of Pb point source emissions to the atmosphere were from 10% of the sources). Total

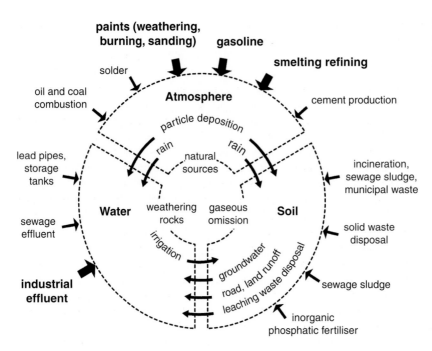

Fig. 4.2 Sources of Pb to different reservoirs of the Earth system. (After USEPA, 2002d.)

Pb emissions were 1640 tons per year (USEPA, 2002d). This all occurs on one subcontinent.

4.1.3   Anthropogenic additions to biogeochemical cycles

Prologue. Anthropogenic inputs accelerate natural fluxes for many of the most important metal contaminants. The mass movement of most metals from human activities is now equal to or exceeds the mass involved in natural cycling on a global scale. Estimates of specific outputs from human activities are not precise. They require assumptions and generalisations about different categories of activities, but they are sufficient to allow ranking of different types of activities.

The general principles that govern anthropogenic disturbances of biogeochemical cycles are understood but the data are only fragmentary. The challenge is not one of detecting metals, but of gathering information from a wide variety of heterogeneous imprecise sources (Gordon *et al.*, 2003). It is often necessary to assume that small studies provide representative information. Nevertheless, the information is sufficient to permit reasonably quantitative cycles to be developed at national, continental and global levels.

Even from fragmentary data, it is clear that the human inputs to the cycles of some trace elements can be immense compared to natural inputs (compare Table 4.2 to Table 4.3). For example, Pb of anthropogenic origin accounted for more than 95% of the Pb accumulated in both the troposphere and the surface waters of the North Atlantic Ocean in the 1980s. Erel and Patterson (1994) reported that 75% of the Pb in waters draining the soils of a remote canyon in California was of industrial origin in the late 1980s. The industrial Pb came from Pb aerosols, originating from automobile exhaust. The aerosols were carried through the atmosphere from urban areas and deposited on plant leaf surfaces in this canyon, far from human activities. The Pb was traced back to automobiles because of unique ratios of the different isotopes of Pb that characterise the Pb deposits (mines) from which the gasoline additive tetraethyl Pb was synthesised (the stable isotopes of Pb are ^{204}Pb, ^{206}Pb, ^{207}Pb, ^{208}Pb: Veron *et al.*, 1993).

Lead from motor vehicle exhaust was found in the atmosphere, remote surface waters and remote terrestrial sites during this era.

Mason *et al.* (1994) determined that undisturbed (pre-industrial) fluxes and reservoirs of Hg were about one-third of Hg fluxes of modern times (Fig. 4.3). Inventories suggest that total emissions of Hg to the atmosphere have increased about four to five times over the last century due to the activities of humans. But the mass of Hg in the atmosphere increased about three times, as a result of removal near many sources. The current atmosphere contains 2500 tons of Hg and an average concentration of 1.6 ng/m^3, compared to an estimated historical concentration of 0.5 ng/m^3. Mason *et al.* (1994) estimated that human activities in the last century have contributed 1700 tons of Hg to the atmosphere, 3600 tons to the surface ocean and 94700 tons to soils. Although soils cover less of the Earth than the ocean, they lose Hg more slowly and thus have accumulated a higher burden. Soils also have a higher natural burden of Hg than the other reservoirs, so human inputs have raised the inventory by only 15%. In contrast, concentrations in the surface waters of the ocean should have increased from 0.05 ng/L

to 0.15 ng/L as a result of human inputs, if Hg was dispersed evenly.

Yeats and Bewers (1987) estimated that Cd inputs to the ocean had been augmented by 60% due to human activities in the 1980s. The anthropogenic emissions of Cd were derived to be 7.3 million tons per year (Nriagu and Sprague, 1987). Of that, 2.0 million tons per year entered the ocean and 4.3 million tons per year were deposited on land. The higher proportion of land deposition reflects rapid removal of Cd from the atmosphere near inputs of air pollution. The atmosphere plays a small role in the Cd biogeochemical cycles compared to Hg (which has an abundant gaseous form) and Pb (which is emitted from combustion as a small aerosol) (Box 4.4). Input of Cd to the oceans is dominated by riverine sources (Yeats and Bewers, 1987). Rivers on the eastern coast of the USA carried 38% of the Cd consumed by the US in 1987 (4.3 million tons). Some of this was a natural component in the water and sediment, but nevertheless, a substantial proportion of the metal that was used eventually ended up as waste in rivers and streams and made its way into deposits in estuaries and the continental margins of the oceans.

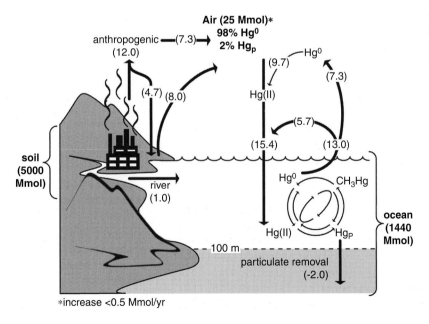

Fig. 4.3 Modern Hg biogeochemical cycle. (After Mason *et al.*, 1994 and Lindberg *et al.*, 2007.)

Box 4.4 Lead in the atmosphere

Riverine inputs and waste discharges are the largest input of Pb to the oceans on the basis of mass alone. But these sources are quickly scavenged, primarily by biological processes (uptake by plants) and deposited near the river mouths in coastal sediments. The mass input of Pb from the atmosphere is smaller than in the rivers. The particulate Pb emitted into the atmosphere by human activities occurs in a range of particle sizes. The most abundant particles are either large and settle out of the atmosphere quickly or are very small aerosols (<1 μm). The small aerosols are less susceptible to deposition as particles themselves (dry deposition) or to deposition by incorporation into clouds and rain; thus they are transported great distances. Larger-sized particulate Pb can also be transported some distance by remobilization back to the atmosphere via dust, resettlement and remobilization again.

Because aerosols can reach offshore sites in the oceans, atmospheric inputs dominate Pb input to the oceans, overall. Low biological productivity and the layering of oceanic water masses allow Pb to accumulate in surface waters and be transported regionally by oceanic circulation. For example, Pb originating in the subtropical North Atlantic (the area of the Sargasso Sea and the Gulf Stream off the eastern coast of the USA) was found to be the source of Pb in the eastern and central Atlantic waters (Veron *et al.*, 1993).

Thus, anthropogenic Pb and Hg are widely dispersed across the Earth partly due to the complex forms they take and exchange within the 'Earth system'. Anthropogenic Zn and Cd are dispersed through their ubiquitous uses, but are somewhat less widely dispersible. Anthropogenic Pb is detectable in the most remote regions on the Earth including the deep oceans and the Arctic/Antarctic ice sheets. Some of the best records of human influences on the Pb cycle are derived from cores into ancient glaciers and time series measurements in the open oceans (Boyle *et al.*, 1986).

If this increase were evenly distributed over the earth, average sediment concentrations of Cd would approximately double (from 0.2 to 0.3 μg/g to about 0.5 μg/g). Total (particulate plus dissolved) Cd concentrations in rivers would increase from between 90 and 120 ng/L to between 136 and 180 ng/L. Calculations like this are informative but do not accurately reflect metal distributions. Metals like Cd are not distributed evenly across the Earth's surface (Box 4.4). Increased concentrations occur locally wherever there are increased inputs. The result is that global concentrations increase in a complex way; as a mosaic or patchwork of hotspots, elevated contamination and unaffected areas. Broad increases occur near areas of human activity and sharp increases occur near the specific points of source inputs.

Amiard and Amiard-Triquet (1993) constructed a modern biogeochemical cycle for Zn (Fig. 4.4). Soils and sediments contain the greatest burdens of Zn among natural reservoirs: a mass 10^6 to 10^8 greater than found dissolved in water or in the atmosphere, respectively. Mass movements of Zn are much greater than are fluxes of Hg, because

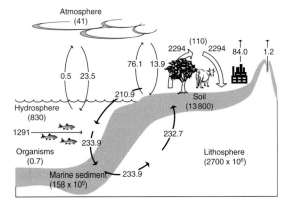

Fig. 4.4 Biogeochemical cycle of Zn. Mass is expressed as 10^{12} g Zn; fluxes are expressed as 10^{10} g Zn/year. (After Amiard and Amiard-Triquet, 1993.)

Zn is very abundant in the undisturbed Earth system. Like Hg, however, anthropogenic emissions (379 000 tons per year) are much greater than natural Zn emissions (43 700 tons per year: Gordon *et al.*, 2003).

Worldwide, anthropogenic emissions of Zn are dominated by industrial activities (e.g. metal production, iron and steel mills) as well as coal combustion, wood combustion and waste incineration (Amiard and Amiard-Triquet, 1993). Diffuse sources of Zn include the dust created as motor vehicle tyres break down during use. But the single greatest use of Zn is as a corrosion-resistant coating for ferrous metals (Gordon *et al.*, 2003); and ferrous metals are used prolifically, in nearly every human activity. Thus widespread uses of Zn globalise the mosaic of Zn contamination, which can be accentuated near mining and industrial point sources if not managed.

Because humans dominate the biogeochemical cycle, the fluxes of Zn among the reservoirs and between anthropogenic reservoirs and the environment now constitute a technological cycle (Gordon *et al.*, 2003). The source reservoir in the cycle is the combined ore bodies and their mines. The sink reservoir is the environment. Between origin at the source and deposition in the sink, the anthropogenic cycle includes four stages: processing, fabrication, use and waste management. Reservoirs include ingot processing facilities, the existing stock of Zn-rich products, and landfills where Zn-laden wastes are disposed of. Only 25 to 30% of the world's demand for Zn metal and Zn oxide is supplied by recycling (Gordon *et al.*, 2003). Thus growth in Zn use, as economies grow, requires growing inputs from the source reservoir and growing accumulation of Zn in both anthropogenic reservoirs and the biosphere sink. The risks in the biosphere can only be controlled by managing where the anthropogenic Zn reservoirs are accumulated. Zinc removed from waste streams and recycled or constrained in waste deposits provides less environmental risk (analogous to ore bodies) than when it is dispersed into the environment.

4.1.4   Trends

Prologue. Increasing usage of metals does not necessarily equate to increased dispersion in the environment. A ban on Pb in vehicle fuel and enforcement of clean air legislation were dramatically effective in reducing Pb inputs to the biosphere starting in the

1970s, whatever the trends in Pb production. But history also teaches us that permanently uncoupling metal production from metal emissions will require sustained and informed risk management.

Primary production of metals is increasing as economies grow, and new uses of metals arise. Trends in usage of Zn are typical of other metals (Fig. 4.5). Between 1930 and 1985, mine production of Al, Cr, Cu, Ni and Zn increased by 114-, 18-, 5-, 35- and 4-fold, respectively (Nriagu, 1988). Roughly half of all the Zn employed during human history was mobilised and put into use in the period from 1970 to 2000. Economic growth in the developing world (especially East and Central Asia) is undoubtedly further stimulating trends in metal usage in the first decade of the twenty-first century, although data are not yet available.

Although metal production is increasing, accelerated loadings to the biosphere are not necessarily inevitable. Trends in emissions are better measures of release to the environment than is production, as well as a way to judge the effectiveness of risk management strategies. For example, emissions of all air pollutants in the USA declined after passage of the Clean Air Act in the 1970s. Most pollutants declined to one-half their 1970 levels by 2005 (e.g. carbon monoxide, nitrogen oxides, sulphur dioxide); particulate emissions declined to about 16% of their 1970 value. For at

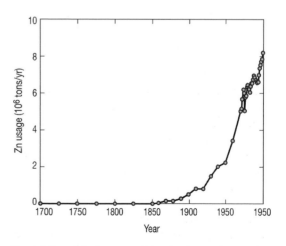

Fig. 4.5 Trends in Zn usage. (After Gordon *et al.*, 2003.)

Table 4.4 *Changes in usage (production from mines) of Cd, Cu and Zn between 1930 and 1985 (1000 metric tons per year)*

Metal	1930	1985
Cadmium	1.3	19
Copper	1611	8114
Lead	1696	3077
Zinc	1394	6042

Source: After Nriagu (1988).

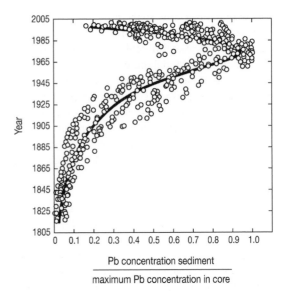

Fig. 4.6 Concentrations of Pb in sediments from sediment cores in Lake Michigan dated with isotope techniques. Trends reflect atmospheric uses of Pb. (After Yohn *et al.*, 2002.)

least some metals trends are the same: increasing usage but declining emissions.

For example, vehicle miles traveled doubled, and Pb production increased between 1970 and 1999 (Table 4.4). But Pb emissions were 1.5% of their 1970s levels in 1999 (http://www.epa.gov/airtrends/econ-emissions.html). Evidence suggesting Pb might have effects on brain development in children led to bans on Pb in gasoline beginning in the mid-1970s. Treatment of industrial emissions of Pb also became increasingly effective. Total Pb emissions declined from 74153 tons per year in 1980 to 1640 tons per year in 2002 in the USA. Emissions from the transportation sector fell from about 64000 tons per year to about 1200 tons per year in 10 years (1980 to 1990). Industrial emissions declined slowly between 1980 and 1995, but dramatically between 1995 and 2002.

Large and continuous increases or reductions in mass inputs of a pollutant to the biosphere are typically reflected in concentrations in water and sediments. Figure 4.6 shows Pb trends in a large number of dated sediment cores from the Great Lakes (Yohn *et al.*, 2002). Reflected in all sediments were the increased atmospheric inputs of Pb with growth in the use of leaded gasoline in the 1940s. Concentrations of Pb in sediments peaked in the 1970s but declined through the 1980s and 1990s after the ban on leaded gasoline was enforced. Oceanic concentrations of Pb showed similar trends (Chapter 5; Wu and Boyle, 1997).

Risk management can, thus, be effective in decoupling growth in economic activities from emissions and deposition in the biosphere. Such trends characterize a number of metal discharges in the historically developed world. But not all metals follow such trends, nor are such trends characteristic of all regions or continents, at least yet.

4.2   USE OF MASS INPUTS TO MANAGE METAL CONTAMINATION

Prologue. Ecological risks from metal contamination are traditionally identified from concentrations in each reservoir (air, soil, water, sediment: Chapters 15, 16 and 17). But one step in effective management of such risks is identification of sources and clarification of the magnitude of the mass inputs from each, compared to other sources. Many factors determine if such concentrations are the cause of risks; but the role of different sources, their influence on concentrations and the fluxes within the Earth system can all be useful in guiding effective risk management.

Historically, water quality was managed by programmes that limited pollutant concentrations in effluents and demanded use of best available technology to reduce such concentrations. Such programmes were not necessarily precise in linking contaminants with ecological risks. But they were successful in reducing contaminant inputs from specific industrial inputs and waste treatment facilities. As inputs from these point source inputs declined, it appeared that ecological risks were still occurring. Diffuse sources, underregulated point sources and/or the legacy of historical inputs trapped in sediments and soils appeared to all be potential sources.

4.2.1 Non-point, diffuse sources of metals

Prologue. Natural inputs of metals must be considered when defining the contributions of different loadings to metal concentrations in a water body, even if anthropogenic sources are known. Particulate metal (with exceptions) is the best measure of metal transport or flux from a diffuse source such as urban or natural runoff. But the concentration of suspended material is highly variable among environments and with time. Loads from major hydrological events must be carefully characterised to successfully quantify incoming loads of metal.

With the marked reduction in point source discharges to US waters, the predominant sources of metals in many watersheds are now attributed primarily to non-point sources. Natural sources contribute to these diffuse inputs (Box 4.5), along with surface runoff from impervious surfaces in urbanised and industrialised areas, groundwater seepage, remobilisation from contaminated sediments and atmospheric deposition. Erosion from land bared by human development can also be important. Distinguishing natural sources from anthropogenic sources is an important challenge for risk management. Even where substantial anthropogenic inputs are known, careful separation of the natural and anthropogenic is necessary (Box 4.5).

Assumptions and uncertainties can plague quantification of anthropogenic non-point source inputs to water bodies (Sanudo-Wilhelmy and Gill, 1999). The technology is available to make reasonable estimates, but is not always employed.

Good hydrological protocols are required for accurate estimates, in addition to ultra-clean sampling procedures and good geochemical analytical technique.

One difficulty is that the mass of metal inputs, with some exceptions, are dominated by particulate inputs. The metal mass (in kg) that is transported on suspended particulate material (SPM) is dependent upon suspended sediment concentrations themselves, as well as metal concentrations on the particulate material (measured in units of μg metal/g particulate). Thus a turbid river will transport a greater mass of metal in particulate form than a river with low suspended particulate concentrations, even if the concentration of metal on the particulate material is the same.

Horowitz (1991) presented a hypothetical calculation which illustrated that elements with the strongest preference for particle surfaces (e.g. Pb, Zn, Cr, Co, Cu) are typically >50% transported as particulate material even if suspended sediment concentrations are as low as 10 mg/L. On the other hand, elements that partition less to particulate material (e.g. Se, As and Sb) are transported primarily in the dissolved phase (on a mass basis), unless suspended material concentrations exceed 100 mg/L (quite a turbid river).

Suspended sediment in a water body is, itself, highly variable. Sediments are resuspended from the bottom by winds and tides and eroded into runoff during storms and floods (see Fig. 2.3). The complex nature of these events adds to the complexity of estimating the sources and fate of metal loadings within a water body. Runoff events, for example, are the largest source of metal transport in rivers. Rivers and streams typically transport more than half of their annual volume of water and mass of suspended sediments during spring floods. In a short pulse of high runoff, most of the metal mass discharged annually by a stream or river may be carried to its receiving waters. Thus, it is especially important to document discharge, suspended sediment concentrations and metal concentrations (particulate and dissolved) at a high frequency (continuously or, at a minimum, daily) during flood events. Few jurisdictions make such measurements.

Box 4.5 Inputs to San Francisco Bay

(a) Secondary natural Ni input to South San Francisco Bay

San Francisco Bay in general has elevated Ni concentrations in the bottom sediment (about 90 to 200 μg/g: Hornberger *et al.*, 1999) relative to average crustal concentrations (58 to 75 μg/g). Flux of Ni from sediment to water could be an important ecological stressor. The elevated dissolved Ni concentrations prompted examination of sources so that appropriate strategies could be designed to reduce inputs. It was known that computer hard drive producers around the South Bay used substantial quantities of Ni. It was also known that serpentine formations near the Bay provide a natural source of Ni that accumulates in bottom sediments.

It was not clear whether anthropogenic or natural sources of this metal were the cause of the contamination of the water column. Topping and Kuwabara (2003) showed that there was consistently a significant flux of Ni out of the sediments into the water column. This source averaged 42 ± 16 kg per day. Waste treatment plant inputs (including industrial sources) averaged 3 kg per day. The Ni loading associated with stormwater runoff was also large, at approximately 56 kg per day. These findings suggested that natural sources of Ni far exceed anthropogenic sources in terms of mass inputs. Sediment cores also showed that industrialisation of the Bay Area had not added to the Ni concentrations in Bay sediments.

(b) Diffuse inputs of Cr to San Francisco Bay

Sources of anthropogenic Cr to estuarine environments include chemical manufacturing, steel refining, and leather tanning. Chromium is also relatively abundant in the Earth's crust, resulting in substantial natural inputs from some types of geology. Abu Saba and Flegal (1995) quantified the relative magnitudes of natural and anthropogenic Cr fluxes to San Francisco Bay. Chromium concentrations and speciation were determined along the estuarine salinity gradient during the wet and dry seasons of 1992 to 1995. They showed that episodes of high inflow during the wet season transported Cr leached from alluvial sediments into the landward reach of the estuary. During periods of low inflow the principal dissolved Cr source appeared to be remobilisation from sediments, especially distant from the river mouths. Dissolved Cr fluxes from both sources were scavenged back onto particles; thus sediments served as both a source and a sink in the geochemical cycle of Cr.

Direct anthropogenic Cr inputs, which included a large steel mill in the landward estuary, were dwarfed by high-flow fluvial inputs but were comparable to low-flow dissolved inputs.

The importance of how metal inputs are determined is evidenced from uncontaminated watersheds. For example, >80% of the mass of suspended sediment and more than one-third of the annual inputs of dissolved Cu, Fe, Pb, Zn and DOC (dissolved organic carbon) from the Kuparuk and Sagavanirktok rivers in Alaska, USA are carried to the Beaufort Sea in the first 3 days and 12 days (respectively) of high river discharge during snowmelt (Rember and Trefrey, 2004). If the river were sampled only once a month (which is typical of many programmes), >80% of the metal input would be missed. Daily sampling during periods of potential high discharge is necessary to characterise such events with reasonable accuracy (Horowitz *et al.*, 2001). Alternatively, discharge and suspended sediments could be sampled at a more than daily frequency, with less frequent metal analysis.

Natural loads can be quantified from estimates of pre-disturbance concentrations of metals on particulate material (see Chapter 6), multiplied by potential suspended sediment loads. These are then subtracted from the modern mass of metal. The difference is the contribution from anthropogenic activities. This approach was used to determine metal mass input from the effluent of the Montreal (Canada) waste water treatment plant. The calculations showed that the sewage inputs contributed 60% of the Ag and 8–13% of the Cu, Zn, Mo, Cd and Bi transported by the St Lawrence River (Quebec, Canada) (Gobeil *et al.*, 2005). Sewage effluents delivered twice as much Cd and six times as much Cu as the atmosphere.

Waste streams, however contribute, at present, less than 3% of the Pb carried by the St Lawrence (Gobiel *et al.*, 2005). Atmospheric inputs are also small. On the other hand, the total Pb content in the river water is three times higher than the pristine level. Trace element fluxes are higher today than before the river was contaminated, but modern inputs are not the source. In this case, stable Pb isotope ratios (^{206}Pb/^{207}Pb, ^{206}Pb/^{208}Pb) show that high historic inputs were trapped in the sediments and are now a major source.

Urban runoff is another potential large source of metal contamination. Gammons *et al.* (2005) studied Cu loads, hourly, in a storm drain in Butte, Montana, USA (Chapter 12). The drain collects groundwater and surface runoff from an urban site affected by 130 years of atmospheric and direct inputs from mining, milling and smelting. Loads of dissolved and particulate Cu increased more than 100-fold during the first hour of the storm and remained elevated for hours after the storm. Loads of Cd and Zn increased during the storm, but decreased afterwards. Weekly or monthly sampling can entirely miss events like this, even though they provide more input than months of average input.

4.2.2 Anthropogenic sources and mass loading allocations

Prologue. Various schemes are being implemented to quantitatively manage mass inputs of metals to the environment. All necessitate a mass loading allocation, which is a formalised procedure to establish a (metal) pollutant budget for a watershed. The maximum acceptable concentration of metal is established for a water body, usually via ambient environmental quality standards. All loadings are quantified; then an allocation is granted to each such that the sum of them all would not result in a violation of the standards. The socio-political aspects of this complex process have been addressed more successfully than the scientific aspects in experience to date.

In order to deal with the combination of point source and diffuse inputs, limits must be determined for the metal concentrations that are allowed in natural waters (in the USA these are termed ambient water quality guidelines: Chapter 15).

In order to manage to that level of contamination, mass loadings from individual sources must be identified, quantified and regulated. In a general sense, knowledge of total incoming masses of metal to a water body can be used to characterise the significance of human influences compared to natural loadings, as shown above. But models (mass media models: Diamond, 1995) are necessary to demonstrate how natural processes combine with human inputs to create concentrations, distributions and fluxes of contamination in the environment.

Various schemes are used around the world to determine allowable loadings of trace metals to natural waters, where risks are managed. In Europe, two new proposals are being implemented to harmonise (align) management of water quality among member nations of the European Union (EU). REACH (registration, evaluation and authorisation of chemicals) is an approach for addressing specific chemicals (Chapter 2). It calls for collecting data on the overall mass of a metal product that is used in human activities. The potential hazard or toxicity of the metal formulation is determined. If the metal is produced or imported into the EU in quantities greater than 1 ton, then registering/authorisation are required. One element of the scientific data required by REACH is an estimate of the mass of chemical added to the technological cycle.

The European Water Framework Directive (Chapter 2) addresses cumulative ecological risks from all stressors for each watershed. Although this directive is in the process of implementation, and the details vary among member states, it is, in general, a watershed-by-watershed approach to risk assessment and risk management as compared to the chemical-by-chemical approach. Nevertheless, the loading data from REACH could ultimately be beneficial to evaluate potential loadings for individual watersheds via the Water Framework Directive.

The approach of the US Environmental Protection Agency is chemical specific. Where waters in the USA are not meeting appropriate water quality standards, the Federal Clean Water Act of the 1970s required that Total Maximum Daily Loads

(TMDL) be established for the pollutants of concern. Thus estimates of loadings from major sources to each watershed are required for each chemical deemed a threat to water quality. These rules were not enforced until the 1990s. Formal rules were not drafted until 2000; and their implementation was delayed even further by political controversy (NRC, 2001). The TMDL approach to water pollution reduction (Box 4.6) is however now a major effort. Some of the lessons from implementation of TMDL, REACH and the Water Framework Directive might be profitably merged as various nations try to cope with new ways to manage and regulate outputs from the technological cycle of different metals.

In theory, pollutants targeted for TMDL could also become commodities. If the cap on concentrations requires actual reductions in inputs, then different sources might trade with other sources to adjust individual allocations for input (e.g. sell their allocation, if, for example they find ways to reduce pollutant loads). Such 'cap and trade' schemes have worked in a few circumstances (although there is not extensive experience with metals). They fail, however, if caps on emissions are not strict enough to challenge the status quo.

Box 4.6 US Environmental Protection Agency TMDL scheme

USEPA defines a 'TMDL' as a calculation of the maximum amount of a pollutant that a water body can receive and still meet water quality standards, and an allocation of proportions of that load among sources. It requires each state to:

- develop a list of water bodies that do not meet water quality standards;
- document beneficial uses that are impaired;
- establish priority rankings of those waters;
- set maximum daily load limits for each contaminant of concern in that water body. This is a ceiling on the mass of metal that can be input.
- Allocate loads to each potential source. This is both a political and scientific exercise. Each source is limited to finite mass of metal input per unit time. Thus, data are required for every point and non-point source of (metal) contamination. The mass of metal inputs is determined by concentration in the inflowing water multiplied by inflow to the water body of concern:

g mass$_{input}$/d = g/L concentration $*$ L/d inflow.

Policy decisions then define the amount by which each point and non-point source must be reduced.
- Safety margins are required in total mass input to assure that the ambient concentration in the water body will be within limits.
- Treatment or restoration strategies must be developed to meet mass per day input limits.

Because the water body already exceeds standards, it is implicit that some corrective action will be required.
- Monitoring is also mandated to assure that both mass inputs per source and ambient targets for the water body are met.

There are many political nuances to this approach. The limits for each discharger or source of contamination are affected by what happens with other sources. Decisions must be made on a watershed basis, but with cooperation of institutions that have not historically cooperated. While top-down enforcement of such regulations is possible, this is an instance where a complex (and time-consuming) public process, involving interested parties and stakeholders is usually initiated to facilitate the allocation decisions. Cooperation and collaboration are essential, because, if one party in the watershed violates the input limits, all parties are affected. A second, unanticipated effect is that allocations tend to be most strict for those sources that are easiest to address. Thus point sources may be required to limit inputs, even though they are not a major source of loadings. Contamination from diffuse sources (urban runoff; agricultural wastes; exceedances linked to limited flows rather than concentrated inputs; etc.) is typically more difficult and expensive to reduce than inputs from discrete sources. Controlling point sources can at least give the appearance of action, even if controlling minor sources is unlikely to be effective environmentally.

Their effectiveness for circumstances where large concentrations of contaminant can be trapped in sediments is also unknown. Cap and trade schemes are being implemented for Hg in the USA, for example (Chapter 14).

The political challenges of deciding on waste load allocations are well known. Stakeholder engagement and the public process are the dominant characteristics of most TMDL undertaken through the first decade of the twenty-first century in the USA. The scientific challenges are less appreciated, however. These lie in the massive nature of the enterprise, the challenge of setting both ambient water quality standards that are commensurate with ecological risks, linking mass loadings to ultimate concentrations in the water body, and accurately determining mass loadings. Policies that short-change science to appear like action can end up being costly and ineffective (Box 4.6). The result is damage to public trust in environmental protection. Better understanding the scientific basis of the problem and addressing its difficult but quantitatively important aspects, must be an important part of future metal management.

A report by the US National Academies of Sciences (NRC, 2001) estimated that there were about 21 000 impaired river segments, lakes and estuaries in the USA, making up 300 000 river and shore miles, and 5 million lake acres. The number of TMDL necessary to satisfy the law would be greater than 40 000; according to one estimate about 20% of these would be for trace metals. In the state of Virginia alone, 220 TMDL were reported as completed between 1999 and 2004. They had 1410 additional studies slated for completion between 2004 and 2012–2016. The state further reports that their TMDL program has been working in six watersheds, and five have shown improvement in water quality. It is too early in the implementation process to determine if water quality is improving in the sixth watershed. The portion of the watersheds covered by the implementation plans is about 158 663 acres or 248 square miles of Virginia's landscape; a tiny proportion of the state (http://www.tmdls.net/Implementation/docs/report060605.pdf).

Virginia cites the transition from developing TMDL to actual water quality improvements as a challenge. In other words the socio-political process of creating a plan is challenging but usually manageable. The scientific process of implementing a plan that is demonstrably effective is more difficult. The Virginia characterisation is probably an honest statement about the status of these efforts in many localities. The National Research Council report (NRC, 2001) concludes that time limits and constraints on spending have resulted in a focus on 'administrative outcomes as measures of success' (plans completed, permits issued, money spent). The most important impediment is the 'pervasive lack of data' necessary to set water quality standards, define impairment, set caps on overall loadings, characterise loadings and provide ongoing assessments of success from the water body of interest. Similar challenges await the Water Framework Directive of the EU and its analogous approaches in places like Australia and South Africa.

To reduce costs, most states have tried to work with existing water quality standards and existing data on loadings. Thus, water quality standards are playing an increasingly important role in risk management for metals. Data on effluents are also available. But reliable data from diffuse sources are rare, especially data that separate natural and anthropogenic loadings. 'Programmatic issues' and 'limited budgets' were cited by the National Research Council (NRC, 2001) as reasons why jurisdictions rarely got to the point of quantitatively defining existing loads and reallocating them (as compared to just demanding that specific sources reduce their discharges). Investments in the detailed approaches necessary to achieve and demonstrate actual improvements in water quality (e.g. post-plan monitoring) were also rare. Thus the TMDL process is a rational idea, the plan is justifiable and the socio-political aspect of the process is healthy. But the scientific application has heretofore typically been inadequate. Lack of will to address complexities and/or lack of incentive to invest in the science seem to be important bottlenecks.

4.2.3 Interaction of natural processes and mass inputs: primary, secondary and tertiary inputs of metals

Management decisions concerning upstream water treatment should balance the results achieved by reducing loadings with the consequent remobilization of in-place pollution.

(Diamond, 1995)

Prologue. Consideration of historical sources is crucial when evaluating strategies to manage risks from metal contamination. Tightening controls on existing sources may not result in immediately detectable declines in concentrations in the water body if inputs from historical contamination are large.

Because metals are not broken down to simpler forms, all metal contamination activities create primary, secondary and tertiary contamination. Primary contamination is that directly generated by a human activity; these are the sources which TMDL are designed to manage. Secondary contamination is that transported away from the site of activity, by water or via the atmosphere. The complexity of primary source inputs and secondary redistribution make understanding difficult. For example, the mines at Bunker Hill (Coeur d'Alene, Idaho, USA: Chapter 12) produced 115 million tons of tailings, 60% of which are thought to have entered the river system. But because these metals were deposited across the floodplain after discharge, metals now move from the mining complex, to the river, and to the lake via:

- direct discharge of acid mine drainage to the river (until 1965);
- direct discharges from the smelters and concentrators (until 1974);
- discharges and seepage from the containment ponds;
- accidental spills or breakdowns in containment;
- airborne deposition when the smelter and mine were active, and remobilisation by wind from historically contaminated soils;
- remobilisation of historically contaminated bottom sediments from the river;

- overland flow eroding historically contaminated soils and their leachates into the river;
- reworking and remobilisation when river meanders cut into the floodplain.

Quantifying mass inputs from a suite of inter-related primary and secondary sources and allocating loading limits is complex.

Tertiary contamination is defined by remobilisation of secondary wastes deposited away from the area of source input, or remobilised after the sources are eliminated. For example, Pb was removed from gasoline in 1976 in the USA and direct emissions to the atmosphere declined dramatically. However during the times of high release, Pb deposited and accumulated to enriched concentrations in soils. The soils are constantly remobilised as dust, spreading the contaminant and reinjecting it into the atmosphere and surface soils. Concentrations of Pb in the Sargasso Sea off Bermuda reflected the use of Pb in gasoline, declining after the ban (Chapter 5). But 20 years after the ban was in place, Pb concentrations in the sea were still 10 to 20 times higher than in areas not affected by atmospheric inputs from the continents, probably reflecting continued remobilisation of Pb historically deposited on soils throughout the continent. Thus, even if improvements are implemented, the magnitude of the flux from historical deposits can be greater than known inputs from ongoing sources (Box 4.7).

4.2.4 Direct determination of ecological risks from mass inputs

Prologue. The simplest goal of critical load or mass media models is to calculate metal concentrations, distributions and risks that will result as elevated trace metal inputs from human activities first contaminate one environmental reservoir then equilibrate with other reservoirs.

A multi-media approach recognises connections, fluxes and the resulting concentrations in various environmental reservoirs. Critical load, 'unit world' or mass media models are examples of methods

Box 4.7 Legacy contamination in sediments

Mass media models illustrate the importance of processes within a reservoir of the Earth system (e.g. chemistry and transport within a lake ecosystem) and exchange between reservoirs. Diamond (1995) presented such a model for As releases from a mine in Ontario (Canada) upstream from Lake Moira. She first modelled conditions during mining conditions then compared conditions after mining ceased. In both cases, atmospheric inputs of As were irrelevant. During the mining era, massive As inputs occurred from the river draining from the mine into the lake: 37 tons of As per year entered the lake. From 11% to 55% of this As became associated with particulate material and settled to the sediment. Although 32 tons of As per year were exported downstream, the mass of As retained by the lake progressively increased during the mining era (inflows exceeded outflows). This was evidenced by a long-term increase from an estimated 0.2 µg/g As in pre-mining sediments (pre-1860) to a value of 360 µg/g As in the 1970s. Thus, during mining, the lake was a net sink for the contamination.

After mining ceased, As loadings from upstream were reduced (Fig. 4.7). But high concentrations of As in water were sustained by inputs from contaminated sediments and were exported downstream. Export of As from the lake was 1.63 times higher than imports after mine closure and remediation, in contrast to higher inputs than exports during mining.

A very slow, permanent burial of sediments occurred in the lake as cleaner incoming sediments deposited on top of the contaminated burial. But when winds and currents resuspended sediments in these shallow systems, contamination was mixed to the surface and exchanged between basins in the lake environment. Thus these physical processes continued the movement of As downstream, creating tertiary contamination.

Remobilisation and downstream export of As would occur in this lake as long as the in-place pollution remained. Sediment burial and reaction will permanently remove As from the lake in the very long term, but for the foreseeable future sediment–water exchange will maintain the lake as a source of As for the rest of the watershed (Diamond, 1995).

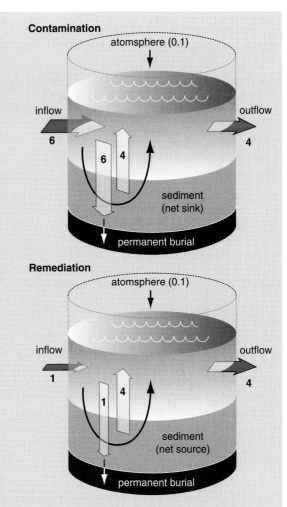

Fig. 4.7 Conceptual model of inflows, outflows, sediment deposition and sediment resuspension in Moira Lake, Ontario, Canada. Size of arrow is roughly proportional to flux. During the period of mining inputs, high mass inputs of As resulted in high As deposition in sediments (sediments were a net sink for As). After remediation the lake continued to contaminate the downstream river because of the flux of As out of sediments as contaminated sediments were resuspended by winds. (After Diamond, 1995.)

used to determine transfers (as emission rates) and thereby evaluate risks from specific sources of contamination.

For example, Hg loadings to the atmosphere have increased over the past centuries (Meili *et al.*, 2003) because of anthropogenic inputs (Chapter 14). Changing Hg inputs are of great concern where Hg sources from one jurisdiction can affect environmental contamination in another. Critical load models can be used to evaluate and limit such trans-boundary air pollution.

Based on data accumulated in Sweden over the past decades Meili *et al.* (2003) presented a model that used data from Hg in precipitation (atmospheric source) and linked it to Hg in large lake fish, organic soil layers and lake sediments. This is called an emission–exposure–effect relationship. Their goal was to determine risks from present and future loads of Hg input from the atmosphere (critical loads). They used historical data from lake sediment cores and the atmosphere to relate timing in changes in both, and among other, reservoirs. One conclusion was that the response of environmental Hg levels to changes in atmospheric Hg pollution is delayed by centuries, especially in soils. They suggested that current atmospheric concentrations of Hg could result in a continuation of increasing Hg contamination of fish, ultimately reaching levels beyond those recommended for human consumption. Meili *et al.* (2003) further concluded that if environmental Hg concentrations were to be kept below these critical limits, virtually no man-made atmospheric Hg emissions could be permitted.

Mass media models, also called unit world models, use a systematic mass balance approach to synthesise information on mass inputs, chemical characteristics of the water body and environmental properties (Arnot *et al.*, 2006). For example 'unit mass inputs' are calculated as concentrations in the source emissions times the mass of inflow in the receiving waters (Diamond, 1995; Mackay, 2001) (e.g. Fig. 4.7). Sediment concentrations might be specified or the ratio of concentrations in sediment/water might be used to define expected partitioning and hence concentrations in water

and sediment. Hydrodynamic modelling concepts are often employed to determine physical transport of contamination in water and sediment movement, and thereby predict distributions of metal and contributions of different sources.

Ultimately such models can include consideration of ecological risks directly. They use distribution of metal concentrations among environmental reservoirs, transport, reactivity, bioaccumulation into food webs and estimated concentrations at which toxicity occurs. A critical emission rate is calculated that results in accumulation of concentrations that pose an ecological risk. Finally, this estimated critical emission rate is compared with the estimated actual emission rate to assess the degree of existing problem or forecast risk.

4.3   CONCLUSIONS

Knowledge of biogeochemical cycles is essential to identifying the role humans play in changing inputs of metals to the biosphere. Human emissions typically overwhelm natural fluxes of metals between global reservoirs. For example, anthropogenic emissions of Pb, Cd, V and Zn exceed the fluxes from natural sources by 28-, 6-, 3- and 8-fold, respectively (Nriagu, 1990; Amiard and Amiard-Triquet, 1993). Industrial emissions of As, Cu, Hg, Ni and Sb were 100% to 200% of emissions from natural sources in 1988.

Although global biogeochemical cycles are of interest, they do not reflect how metal contamination is distributed. Metal contamination is a global-scale problem because humans use metals in everything we do. The waste products of metal use are spread widely wherever humans congregate. Distribution patterns of the contamination vary among metals. A portion of the atmospheric Hg contamination is distributed globally, because of the volatile form Hg can take. But most other metals do not have gaseous forms. Their contamination is distributed over the globe in complex mosaics (patchworks). Extreme contamination occurs in hotspots, usually near large sources if their wastes are not appropriately managed. A lesser degree of contamination is ubiquitous wherever water bodies border on urban or industrialised areas; physical or even biological processes may focus some of these wastes into additional hotspots. Rivers carry metal wastes to the continental margins. Winds and atmospheric circulations can carry aerosols and dust to the offshore oceans. For example, Pb aerosols were widely distributed from diffuse activities on the North American and European continents to the offshore oceans in the 1980s when motor vehicles burned Pb in their fuel. Silver contamination from activities on the Asian continent is now detectable in the western Pacific ocean (Chapter 5). The complexity of metal distributions adds important complexities to simple uses of loading data to manage contaminant inputs.

Human uses of metals are also growing extremely rapidly. But that growth need not necessarily result in greater metal contamination. Source control is the simplest approach to managing risks from metal contamination. Environmental regulations, governmental/domestic/industrial 'best practices' and recycling all can help uncouple growing uses of metals from waste release. Managing environmental risks from large sources of metal input was successfully implemented in North America, Europe and Japan after the 1980s using relatively rudimentary scientific knowledge of inputs and relatively simple, top-down command-and-control environmental regulations. Partial bans on some uses of metals, where risks were deemed unacceptable, were successful in reducing large-scale contamination for Pb. Mandates to use best management practices were actually the most effective tool in reducing contamination by other metals. The result was a decline in metal emissions from the easiest targets, and where effects were most obvious. These successes demonstrated that positive policies can result in positive responses in the environment.

Schemes to manage on a watershed basis (the Water Framework Directive in Europe) or allocate waste loadings source by source (TMDL process in the USA) represent the modern proposals for managing risks from metal contamination. Sociopolitical aspects of these approaches are being successfully implemented: goals, plans, stakeholder participation, expenditures of allocated funding. But the systematic collection of scientific data necessary to justify expenditures and make these plans work appears to be well behind the sociopolitical process. Monitoring accompanied by objective evaluation of what works and what does not is even more rare. Setting caps for allowable concentrations of metals in water bodies will require revisiting water quality standards, many of which appear to be seriously flawed (Chapter 15). Greater integration of different lines of knowledge is essential to more effective standards. Quantifying mass metal inputs from sources other than industries and domestic waste treatment facilities will require much better data than are presently available. Quantifying influences of historically deposited contamination also is necessary; but such data are only available from a few locations. Greater sophistication in the scientific input and greater knowledge of water bodies are necessary to improve on present conditions and prepare for future challenges.

One strategy for managing metal contamination is to demand ever greater reductions in input from what were historically (or seem intuitively to be) the major sources. But many of the traditional sources are now implementing strong source control. Where those point sources become quantitatively minor inputs of metal, continued restrictions will not only be expensive, but are likely to be ineffective in improving environmental conditions. The end result will be erosion of public trust in environmental regulation, ultimately. Such strategies may give the appearance of action, but they are no substitute for addressing diffuse sources and legacy sources.

Even the most rudimentary approaches for managing sources of metals present a challenge for newly growing economies like those in east and central Asia, at least initially. The social, environmental and economic costs of recovering from the environmental damage of the 100 to 200 years of industrialisation that preceded the recent environmental awareness in North America, Europe and Japan represent an ongoing lesson for growing economies. Minimising those same mistakes will minimise costs that are likely to be even greater in a more crowded world. The good news is that many leaders (public, private sector and governmental) seem to recognise the importance of quickly moving towards sustainable behaviour; the challenge is to make the investments, including investments in high-quality data, sophisticated models and growth of knowledge, that are necessary to implement that behaviour; and to stay the course when initial interest wanes.

Suggested reading

Diamond, M.L. (1995). Application of a mass balance model to assess in-place arsenic pollution. *Environmental Science and Technology*, **29**, 29–42.

Erel, Y. and Patterson, C.C. (1994). Leakage of industrial lead into the hydrocycle. *Geochimica et Cosmochimica Acta*, **58**, 3289–3296.

Nriagu, J.O. (1988). A silent epidemic of environmental metal poisoning? *Environmental Pollution*, **50**, 139–161.

Wu, J.F. and Boyle, E.A. (1997). Lead in the western North Atlantic Ocean: completed response to leaded gasoline phaseout. *Geochimica et Cosmochimica Acta*, **61**, 3279–3283.

5 • Concentrations and speciation of metals in natural waters

5.1 INTRODUCTION

Prologue. Important goals of this chapter are to provide a common perspective on metal concentrations in different types of waters, and to introduce the factors that influence concentrations and metal speciation. Perspective on metal concentrations and form in natural (contaminated and uncontaminated) waters is essential to assessing ecological risks.

Understanding the concentrations, forms and dynamics of dissolved trace metals is essential for the evaluation of metal effects, risk assessments and management decisions. Dissolved metal concentrations respond to human inputs and are one measure of metal influences on ecological processes. Metal concentrations in water (ambient dissolved metal) are one of the measures used by the regulatory community and risk assessors to determine compliance with regulations. Perspective on natural waters is critical to such decisions, but not always fully appreciated. For example, important questions include: What baseline should be used for comparison against the ambient concentrations of a potentially contaminated environment? What are the spatial and temporal patterns of variability in dissolved concentrations? What are the processes that influence dissolved metal concentrations and their variability, in oceans, estuaries and rivers? How do metal concentrations in natural waters compare to concentrations deemed toxic (e.g. in toxicity tests)? How reliable are speciation determinations and what processes influence speciation? The perspective obtained from these questions is necessary to avoid disconnects between risk management and actual risks from metal contamination.

Box 5.1 Definitions

colloidal metal: metal contained in a water sample that passes through a 0.45 μm filter but is retained by a 0.10 μm filter.

conservative: when a property is not removed from the water column during a change in chemical conditions (e.g. when freshwater meets seawater in an estuary), it is termed conservative. For example, in estuaries, conservative behaviour of dissolved metals is determined from a linear correlation with salinity.

dissolved metal: A metal in solution is dispersed homogeneously throughout a liquid (water) in a molecular form (Manahan, 1991). Dissolved metal is usually defined operationally as that which is contained in a water sample after passage through a 0.45 μm filter; although smaller filter sizes (e.g. 0.10 μm filters) are used in some applications.

fulvic acids: relatively soluble, low molecular weight, unstructured organic substances derived from the biological breakdown of organic matter, bearing no structural resemblance to the compounds from which they were derived.

humic substances (acids): high molecular weight organic substances of limited solubility derived from the biological breakdown of organic matter, that occur in solution or as colloids. Humic substances bear no structural resemblance to the compounds from which they were derived.

ligand: an ion or molecule which forms a complex in solution (or on a particle surface) with a metal ion.

particulate metal: metal associated with particles in suspension or deposited (e.g. in bed sediment) in a water body. Operationally suspended particulate metal is defined as that retained by a 0.45 μm filter after a water sample is filtered. Sedimentary and suspended particulate material will contain living biological material, (benthic microflora and meiofauna, or plankton), breakdown products from living organisms (detritus) and inorganic material.

5.2   LOADS AND CONCENTRATIONS

Prologue. Mass loadings differentiate sources of contamination; but concentrations in the receiving waters play a key role in determining ecological risk. Concentration is determined by the combined effect of mass loading and local hydrology.

Determination of the absolute mass of metal (measured in kg) entering from a source, or within a water body, can differentiate causes of contamination. But mass loadings do not provide a basis for understanding ecological significance. Concentration is the factor that determines biological exposure and dissolved concentrations are one component of that exposure. Thus, a regulatory tool may limit overall mass inputs to a certain level (e.g. Total Maximum Daily Load (TMDL): Chapter 4). But the target is a concentration in the receiving water that is protective against ecological risks.

Simplistically, load and concentration can be linked by:

$$C_w = I_{\text{all sources}} / V_{\text{all inflows}}$$

where C_w = concentration (weight per volume) in receiving waters; $I_{\text{all sources}}$ = composite input load (weight), and $V_{\text{all inflows}}$ = inflows (volume) from all sources.

Thus the metal concentration in a water body is determined by the influence of local hydrological processes combined with the mass of metal inputs.

As an example, let us assume that a river with low metal concentration is the receiving water for an industrial input with a constant but relatively high metal concentration. If the river has a typical seasonal cycle of high and low river flows, the metal concentration in solution in the river will fluctuate seasonally. It will be diluted during high flows and be more concentrated during low flows. Similarly, lower concentrations might be expected during the wettest years (higher annual discharge) and higher concentrations might be expected during drier years. This concept applies, but in different ways, to rivers, estuaries, coastal waters and lakes. Transformations to particulate material and physical factors such as residence times, tidal exchange and winds also will affect this simplistic example. But it does illustrate that loads alone are not sufficient to evaluate ecological risks from a metal source.

5.2.1   What is dissolved metal?

Prologue. Dissolved metal is defined operationally by filtration. Metals that pass through a 0.45 μm filter are defined as dissolved, although filter sizes of 0.2 μm, 0.3 μm and 0.4 μm are also occasionally used. Metal forms too large to pass a 0.45 μm filter are defined as particulate. Metals captured in the size range 0.10 μm to 0.45 μm are colloidal. Nanoparticles are defined as <100 nm. Total metal includes both the particulate and dissolved metal in a sample. Total concentrations are greatly affected by the mass of suspended particulate material in the sample.

Conventional knowledge tells us that trace metals exist in natural waters in dissolved (usually including colloidal) and particulate phases. In reality, different forms of metals exist across a continuum of sizes in all water bodies. Metals can be in true solution (e.g. the free ions or in small inorganic complexes, for example with hydroxide). They can be associated with nano-sized materials (large organo-complexes, aggregates, synthetic nanoparticles or colloids) or micro-sized particles (colloidal humates, bacteria). Or they can be associated with larger particulates in suspension or in the bed sediment.

'Dissolved' trace metal therefore needs precise definition, even if the definition is operational. In practice 'dissolved metal' is defined by filtration, being dependent on the pore size used to filter out particulate matter from the sample collected. In the case of oceanic systems, polycarbonate filters of either 0.3 or 0.4 μm pore size are commonly used to separate dissolved from particulate metal. Pore sizes of 0.2 μm have recently been used to ensure removal of even the smallest microorganisms (Donat and Bruland, 1995). The vast majority of publications, however, operationally define dissolved metal as that which passes a 0.45 μm filter. That pore size was chosen because it excludes nearly all bacteria. Colloidal metals are operationally defined as material that passes a 0.45 μm filter but is trapped on a 0.10 μm filter. Thus colloidal metal is usually included in the dissolved fraction, although some recent definitions of 'dissolved metal' mean colloid-free. Filtration with a 0.10 μm filter is infrequently used in routine sampling because of operational difficulties.

Sigg *et al.* (2000) found that the percentage of metal in the colloidal fraction was 11% for Cu and 5% to 6% for Zn and Ni (median) in the Thur River (Switzerland). But Dai and Martin (1995) found that about 50% of most metals was in the colloidal fraction in two Arctic rivers. In the Danube River (Eastern Europe), 40% of the Cu, 40% of the Cd and 60% of the Fe determined in the traditional dissolved phase was associated with colloidal material (Guieu *et al.*, 1998).

'Total' or 'total recoverable' metal concentrations are those in an unfiltered water sample. Metals associated with suspended particulate material will contribute to the metal concentrations determined in such a sample. Suspended metal has a great influence on the measured concentration because metal concentrations on suspended particulates are thousands of times higher, per unit mass, than metals in solution. A small mass of particulate material will contribute a large mass of metal to the sample (whether or not the suspended material is contaminated) (Fig. 5.1, Box 5.2).

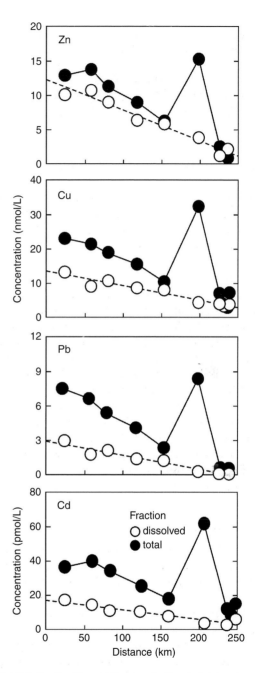

Fig. 5.1 Zn, Cu, Pb and Cd concentrations in total and dissolved fractions as a function of distance from the confluence of the Clutha River with the sea in New Zealand. The upstream peak in 'total' concentrations is the result of a high suspended load at that location; it has nothing to do with changes in metal concentration in either solution or per unit mass of particulate material. (After Kim *et al.*, 1996.)

Box 5.2 Effect of suspended material on 'total' metal concentrations

Let us consider a hypothetical, uncontaminated 1 L water sample that has 2 µg/L Zn in dissolved form (determined after filtration) and 0.100 g/L suspended material with 60 µg/g Zn. The Zn with particulate material would contribute 6 µg to the litre and the Zn from solution would contribute 2 µg. The filtered sample would measure 2 µg/L Zn, and the 'total' measurement would be 8 µg/L Zn. If another sample were collected 5 days later when the sediment load was different (say 0.010 g/L), the 'total' measurement would be 2.6 µg/L Zn. If a sample from the same water body was taken near the bottom and suspended sediment was 1 g/L, the 'total metal' would be 62 µg/L Zn.

In all these circumstances the metal chemistry is the same, the dissolved concentration is the same and the Zn on suspended material is the same; but 'total Zn' concentration varies 30-fold because of differences in suspended load. 'Total' water concentrations are usually biased upwards by the unremoved suspended material, and are difficult to interpret without knowledge of the suspended contribution.

Total metal concentrations in water (suspended and dissolved concentrations in a given volume of water) are the best choice to obtain estimates of the total mass (or load) of metal in a water body, for comparisons of mass loadings among water bodies or for mass balance calculations. However, total metal concentrations in water are not well suited for evaluating risk or for comparisons among water bodies or times. Despite that, total metal concentrations are sometime used to evaluate regulatory compliance because they are easier to determine (concentrations are not as low as dissolved concentrations); the trade-off is great uncertainty with regard to risk.

5.2.2 Determining dissolved metal concentrations

Prologue. There are now many laboratories that conduct reliable dissolved metal analyses, but concentrations also continue to be reported that reflect laboratory contamination. Summaries of existing values must be especially careful not to include the latter. Our understanding of metal concentrations in water has changed dramatically with the increasingly widespread use of ultra-clean methodologies.

For regulatory purposes (and many risk assessments), dissolved metal concentrations from a water body should be compared to laboratory-derived measures of toxicity to determine if organisms are at risk from metal toxicity. For example, in a site-specific risk assessment, a regulatory agency might define a Hazard Quotient for different localities from the ratio of 'site exposure levels' (dissolved metal concentrations) to 'levels believed to cause no or minimal effects' based upon the dissolved toxicity database for species relevant to a site (e.g. USEPA, 1999b). The limits of toxicity testing for defining risks to organisms are well known (e.g. Chapters 2 and 10; Luoma, 1996). Less appreciated are the challenges of identifying the ambient exposure in the water body.

Methodologies were commercially available to detect relatively low concentrations of metals in water as early as the 1950s. But, as noted in Chapter 3, many samples were contaminated by laboratory dust, water film on washed glassware, and the internal components of commonly used laboratory supplies. Plastic bottles and many filters are rich in Cd and Zn unless acid washed. Dust is rich in Zn, Cu and Pb. Metals from these sources leach into the samples during collection, handling, filtering, concentration steps or analysis. The measured concentration will be in error to the degree that the mass of metal from laboratory contamination is comparable to the mass in the original sample. Ultra-clean procedures are more important for metal concentrations in solution than for determinations of metals in bed sediments or animal tissues because concentrations in solution are so low. Even the smallest source of laboratory contamination can bias a metal determination upward in a water sample.

Ultra-clean methods are available to prevent contamination. Plastic bottles must be washed with acid, rinsed with ultra-clean distilled water, dried in a dust-free atmosphere and stored in dust-free bags. Water samples are collected with metal-free sampling gear (e.g. Kevlar drop lines, Teflon bottles, vinyl gloves), away from the engine exhaust of a boat, and cannot come in contact with any of the normal gear of the limnologist, hydrologist or oceanographer. Filtration must occur with acid-washed plastic equipment, within a dust-free apparatus. Nucleopore filters have predictable pore sizes and are often specified specifically as the brand most likely not to result in contamination. The need for these specialised, technically sophisticated techniques were (and sometimes remain) a major impediment for many laboratories. Low-quality data confounded understanding of metal concentrations in natural waters for decades, despite the availability of instrumentation that could detect those levels (Sanudo-Wilhelmy *et al.*, 2004).

To assure data quality, a publication must include citation of the ultra-clean approach used in sample collection and processing. Results from analyses of blanks must be reported. Analytical sensitivity must be low enough to detect the concentrations expected in uncontaminated waters. Inter-laboratory comparisons or reference standards must be analysed and results shown to demonstrate consistent outcomes (e.g. Flegal *et al.*, 2005).

The best test of a methodology is comparison to other reliable analyses. Ultra-clean methodologies show that Pb concentrations from the oceans are 1 to 5 ng/L (0.001 to 0.005 µg/L Pb), depending upon the degree of biological productivity (Schaule and Patterson, 1981). Large bodies of contaminated water have 20 times to perhaps 100 times those concentrations (for example Hamilton Harbour in Lake Ontario had 0.03 µg/L Pb in the 1980s: Box 3.5). Concentrations of 1 µg/L Pb, for example, might exist in untreated effluents, but not in natural water bodies (even if contaminated). Any values of such a magnitude are surely of suspicious analytical quality. A similar 'range of experience' can be derived for every metal from the existing literature.

5.2.3 Concentrations in natural waters and processes affecting them

5.2.3.1 Extreme contamination
Prologue. Most reports of dissolved metal concentrations, into the 1990s, included values reflecting what would appear to be extreme contamination by the standards of modern ultra-clean methodologies. Many of those values probably represented unreliable analyses, although some could have reflected reality. Nevertheless, reports of metal concentrations orders of magnitude higher than are now known to be typical of natural waters provided the first perspectives on the magnitude of dissolved metal contamination. This perspective may have influenced development of ecotoxicological tools and interpretations.

Surveys of dissolved metal concentrations in natural waters began, in earnest, in the 1970s and 1980s. Few of the earlier surveys included comprehensive use of ultra-clean techniques. As a result, there are uncertainties about the actual levels of metal concentrations during those years.

Table 5.1 shows data collected between 1978 and 1994 on British rivers severely affected by acid mine wastes. Low pH and proximity to the metal source contributes to the extremely high dissolved concentrations (see Chapter 12) although at least some laboratory contamination cannot be excluded. They are probably good examples of the worst that can be expected of metal contamination near an extreme source. The lowest values are probably the least reliable.

5.2.3.2 Dissolved metal concentrations in historically contaminated estuaries
There are only a few instances in the industrialised world where ultra-clean analyses of larger water bodies, like estuaries, can be compared to conditions before environmental regulations took force. Early uses of ultra-clean metal analyses along the Hudson River estuary (maximum concentrations are shown in Table 5.2) in 1974 and 1975 were compared to samples collected between 1995 and 1997 by Sanudo-Wilhelmy and Gill (1999). The median concentrations along the estuary had apparently declined 36% to 56% for

Table 5.1 *Dissolved concentrations (μg/L) of trace metals in two British rivers affected by mining contamination: the Carnon River, Cornwall leading into Restronguet Creek and the Afon Goch, Anglesey leading into Dulas Bay*

	Carnon River	Afon Goch Head	Entry to Dulas Bay
Copper			
Mean	593	19230	1420
Range	370–973	4480–59890	0–11430
Zinc			
Mean	12470	30030	1990
Range	3630–57230	15780–41910	170–5330
Manganese			
Mean	1777	10800	680
Range	955–3315	6140–49060	10–1640
Cadmium			
Mean	24.8		
Range	6.4–106		
Iron			
Mean	5024	193240	1360
Range	2670–10360	39280–259760	0–19190

Source: After Foster *et al.* (1978); Bryan and Gibbs (1983); Boult *et al.* (1994).

Cu, 55% to 89% for Cd, 53% to 85% for Ni, and 53% to 90% for Zn over a period of 23 years. Reliable (mostly ultra-clean) techniques were also used to establish the maximum dissolved trace metal concentrations typical of British estuaries known to be contaminated by distributed mine wastes or other industrial contamination (Table 5.2). Maximum dissolved concentrations of Cu, Cd, Ni and Zn were roughly similar between the Hudson and the British estuaries. Analyses of Hg and Pb were much more difficult to conduct and their reliability is less certain. The ranges of the Cu, Cd, Ni and Zn data are consistent with what we know about contaminated waters today; they establish the field's 'range of experience' for contaminated waters.

Data proliferated on dissolved metal concentrations between 1970 and 1990. Such data provided the perspective for ecotoxicologists developing toxicity tests. Unless reliable and unreliable data were differentiated (and they rarely were outside of an elite group of geochemists), it would have been easy to assume (and test for) exposures that were much higher than actually existed. Nevertheless such data represented the perspective of the time.

Table 5.2 *Dissolved concentrations (μg/L) of trace metals in industrial and mining-contaminated estuaries (highest concentrations reported in Bryan* et al.*, 1985; Law* et al.*, 1994; Sanudo-Wilhelmy and Gill, 1999*). **Receives acid mine drainage*

	Cu	Zn	Mn	Ni	Cd	Hg	Pb
Britain							
Tweed	3.1	1.9	13	0.81	0.033	0.0044	0.170
Tyne	1.2	22	120	2.8	0.130	0.0047	1.10
Wear	1.6	6.7	100	2.9	0.058	0.0045	0.410
Tees	10	14	120	1.0	0.056	0.012	0.820
Humber	4.9	15	59	6.3	0.220	0.019	0.610
Restronguet Creek**	176	20460	1513	18	38	–	4
Mersey	3.3	16	3.9	9.4	0.052	0.0021	0.880
Dee	1.4	2.4	5.5	0.87	0.018	0.0023	0.170
USA							
Hudson River estuary (maximum in 1974–75)*	3.5–4.5	10–17	–	4.5	0.25–0.30	–	–

5.2.3.3 Dissolved metals in the oceans

Prologue. Metal concentrations in the open oceans can be as low as anywhere on Earth. The diluting potential of the oceans is great because they are massive water bodies. Biogeochemical processes create vertical and spatial gradients independent of human influences. Physical and biogeochemical processes add substantial variability to concentrations in oceanic surface waters. Atmospheric deposition is the primary anthropogenic source of metals in the oceans. When human activities create contamination in the atmosphere (either in gaseous form as with Hg or in fine particulate materials coming off continents), anthropogenic contamination of open ocean waters is observed.

Because of biogeochemical processes, most trace metals occur at different dissolved concentrations at different depths in the ocean, and there are slightly different concentrations in different oceans (Bruland, 1983; Donat and Bruland, 1995). Understanding the causes of this variability is essential to interpreting metal concentrations in ocean waters.

5.2.3.3.1 Vertical variability in concentrations

Surface and deepwater concentrations of dissolved trace metals in the ocean, determined following reliable methodologies, are presented in Table 5.3.

The range between the lowest and highest values for many metals is high (>10-fold). That is partly because of the processes that affect concentration. Metals can be divided into three principal categories in the oceans: conservative, recycled and scavenged (Whitfield and Turner, 1987), with consequences for their concentration–depth distributions (Donat and Bruland, 1995).

Conservative trace metals interact weakly with particles, have long residence times (>10^5 years) relative to the mixing time of the oceans, and show a constant concentration relative to salinity and depth. Concentrations vary only as a result of water mixing (Bruland, 1983; Donat and Bruland, 1995). Metals in this category include the major metal ions Na^+, K^+ and the anion MoO_4^{2-} of the trace metal molybdenum.

Recycled (nutrient-type) trace metals have depth profiles exhibiting surface depletion and enrichment

Table 5.3 *Metal concentrations in the oceans*

	nmol/kg	µg/L = parts per billion (10^9)
Sodium	468 000 000	11 138 000
Magnesium	53 200 000	1 340 000
Calcium	10 300 000	427 000
Aluminium	0.3–40	0.01–1.1
Arsenic	15–25	1.1–1.8
Nickel	2–12	0.12–0.70
Zinc	0.05–9	0.003–0.61
Copper	0.5–6	0.03–0.39
Chromium	2–5	0.10–0.27
Manganese	0.08–3	0.004–0.17
Iron	0.02–2.5	0.001–0.14
Selenium	0.5–2.3	0.04–0.19
Cadmium	0.001–1.1	0.0001–0.13
Titanium	0.004–0.30	0.0002–0.014
Cobalt	0.004–0.30	0.0002–0.018
Lead	0.003–0.18	0.0006–0.038
Vanadium	0.023–0.036	0.0012–0.0018
Silver	0.0005–0.034	0.00006–0.004
Tin	0.001–0.02	0.0001–0.002
Mercury	0.0005–0.01	0.0001–0.002
Gold	0.00005–0.00015	0.00001–0.00003

Source: After Bruland (1983); Donat and Bruland (1995).

at depth as a result of interactions with biologically derived particulate material (Bruland, 1983; Donat and Bruland, 1995) (Box 5.3). Depletion in surface waters occurs by uptake by phytoplankton and/or adsorption onto particles created by organisms. Increases with depth occur as sinking detrital particles undergo microbial decomposition of their organic contents, and/or dissolution of their inorganic mineral phases (e.g. calcium carbonate) (Donat and Bruland, 1995). Residence times (10^3 to 10^5 years) are intermediate to those of conservative and scavenged trace metals.

Classic examples of recycled trace metals are cadmium and zinc (Bruland, 1983; Donat and Bruland, 1995). The vertical profile of Cd correlates closely with that of phosphate suggesting that Cd is cycled through organisms with the formation and subsequent decomposition of organic tissue rich in phosphate (Box 5.3). Zinc distribution

Box 5.3 Processes that determine recycling of trace metals in ocean surface waters

The correlations between some metals and major nutrients reflect the fate of the phytoplankton–metal association in oceanic surface waters (e.g. Lee and Fisher, 1994). Phytoplankton take up metals in surface waters in the presence of light, depleting metal concentrations near the surface. When phytoplankton are eaten by consumers (zooplankton), the metal is accumulated in soluble and insoluble stores in zooplankton tissues, or incorporated into faecal pellets (Lee and Fisher, 1994).

Unassimilated elements in faecal pellets or metals associated with the carcasses of dead organisms eventually are released into the water as they sink. The metals in the cell solution of organisms tend to recycle more readily than metal that is bound strongly to external or internal surfaces. Lee and Fisher (1994) showed that a large proportion of Cd, Se and Zn occurs in cell solution in phytoplankton and therefore these elements are rapidly released back into the ocean water as these forms of 'detritus' sink. Half the burden of Cd in faecal pellets, for example, is lost to solution in 0.4 to 2.6 days. That regeneration rate is very similar to the regeneration rate of some major nutrient elements like phosphorus. Cd, Se and Zn appear to be completely recycled before they reach the bottom of the ocean. In contrast a strongly particle-reactive element is rapidly exported into bottom waters by sinking faecal pellets before it can recycle. Americium (Am) and Pb are examples. One hundred days are necessary to lose half the burden of Am in faecal pellets.

The more easily recycled metals or nutrients accumulate toward the surface of the oceans; the more slowly recycled elements accumulate toward the bottom. For example, Si is a structural element in diatom valves and is recycled more slowly than phosphorus. Therefore the Si concentration maximum in the ocean is deeper in the water column than the P maximum. Metals that bind to tissues more strongly than Cd (e.g. Ag) recycle more like Si. Concentrations of specific metals and specific nutrients are correlated as determined by these detailed differences in recycling. The particle-reactive metals reach the bottom sediments before much is released. Dissolved concentrations do not build up in subsurface waters (nor are there correlations with nutrients that recycle).

correlates most closely with that of silicate, a common component of some common phytoplankton (diatoms) (Fig. 5.2) (Donat and Bruland, 1995). Other recycled trace metals are As (V) with a similar distribution to Cd (relatively shallow regeneration cycle leading to a mid-depth maximum as also observed for phosphate and nitrate), and Ni and Se with a combination of shallow and deep regeneration cycles (Bruland, 1983). Further recycled trace metals are Ag, Cr and V (Donat and Bruland, 1995).

Scavenged trace metals interact strongly with particles and have very short oceanic residence times ($<10^3$ years) (Donat and Bruland, 1995). The dissolved concentrations of these metals are maximum near their sources which include rivers, atmospheric dust, hydrothermal sources and bottom sediments. Concentrations decrease with distance away from the source, because little or no recycling occurs. Scavenged metals include Al, Am, Co, Mn and Pb (Donat and Bruland, 1995). Mn is scavenged through the water column on a timescale of 10 to 100 years and is appreciably regenerated only in low oxygen conditions; oceanic distributions are dominated by external input sources which can lead to maxima in surface waters or in deep water near hydrothermal zones, and regeneration in any oxygen-minimum zones (Bruland, 1983). Pb enters oceanic water by atmospheric input so concentrations are high in surface waters and decrease with depth (Bruland, 1983).

Hybrid distributions are shown by trace metals which include Cu and Fe. They are influenced by both recycling and scavenging processes (Donat and Bruland, 1995). Like recycled trace metals, Cu and Fe are often depleted in surface waters of high productivity and regenerated at depth. In less productive waters, or in areas with high atmospheric input, Cu and Fe can show surface maxima more typical of scavenged trace metals (Donat and Bruland, 1995). Instead of the rapid increase of dissolved concentration with depth shown by recycled trace metals, dissolved Cu concentrations increase only gradually with depth as a result of

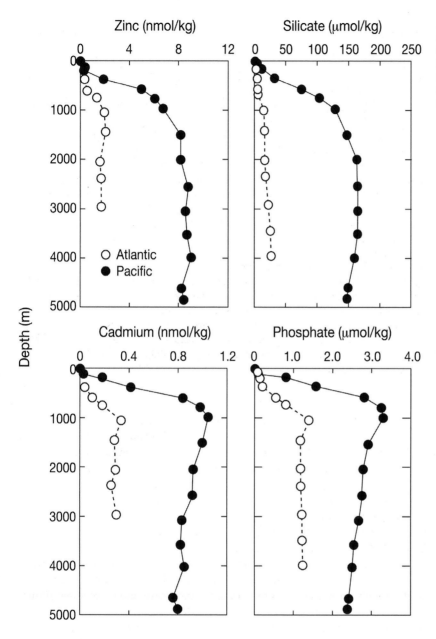

Fig. 5.2 Vertical dissolved concentration profiles of Zn, silicate, Cd and phosphate in the North Pacific and North Atlantic Oceans. (After Bruland, 1983; Donat and Bruland, 1995.)

the combined effects of regeneration and scavenging in deep water (Donat and Bruland, 1995). Regenerated Fe is scavenged relatively intensely at all depths, resulting in a short residence time (<100 years), dissolved Fe concentrations resulting from the balance between regeneration and scavenging (Donat and Bruland, 1995).

5.2.3.3.2 Spatial and temporal variability

Spatial and temporal variabilities of dissolved metal concentrations in ocean waters are driven by biogeochemical and physical processes. Depletion at the surface by phytoplankton uptake can reduce concentrations of some metals to very low values. Regeneration at depth causes subsurface oceanic waters to be enriched relative to surface waters by up to a 1000-fold for Cd and 180-fold for Zn (Donat and Bruland, 1995) (Table 5.3). For metals that are not taken up as efficiently by phytoplankton, variability in surface water concentrations is less extreme. For example, Pb concentrations vary about 40% on a weekly or monthly timescale in surface waters off Bermuda, because of difference in atmospheric inputs (Wu and Boyle, 1997). This relatively small variability is reflected in relatively small differences in Pb concentrations between water bodies or over time.

Movement of water bodies (e.g. advection) with different biogeochemical histories distributes metals in complex ways. For example, the elevated concentrations of the recycled trace metals in deep waters increase along the direction of the main advective flow of deep water in the world ocean, causing concentrations to be higher in deep North Pacific waters than in deep North Atlantic waters (Donat and Bruland, 1995).

Upwelling is an oceanic phenomenon whereby the complexities of major current movements cause waters from the subsurface to rise to the surface. Upwelling of nutrient- and metal-rich deep water is characteristic of the west coasts of the major continents. Offshore from San Francisco Bay, the enriched Cd and P concentrations that build up in the subsurface move to the surface

during upwelling periods. Dissolved Cd concentrations in surface coastal waters are about 0.03 µg/L during periods of low upwelling and increase to 0.08 µg/L when upwelling becomes intense (van Geen et al., 1992). The high values for Cd correlate with raised phosphate concentrations. The raised Cd concentrations resulting from natural Cd enrichment and translocation in the ocean are equal to anthropogenically enriched Cd concentrations in nearby San Francisco Bay (Table 5.4).

In fact, commercial culturing of mussels in British Columbia has been threatened with closure because the Cd concentrations in animals from pristine coastal waters exceed regulatory standards for food. The contamination is at least partly a result of upwelling rather than anthropogenic contamination. The Cd concentration does exceed standards for safe consumption by humans but the source is natural. This does not lower the health risk, but it does make management of that risk problematic. Mussel aquaculture is likely to have such problems wherever it is conducted in upwelled waters.

5.2.3.3.3   Contaminated ocean waters

In order to determine if human influences are contaminating the oceans, variability attributable to biogeochemical cycling must be understood, and sampling strategies must correspond to spatial

Table 5.4 *The range (highest and lowest) of metal concentrations (µg/L) observed in rivers, lakes and estuaries from remote and developed areas. The remote rivers are large rivers with unindustrialised watersheds (the Amazon in the 1980s and rivers from New Zealand and the Russian Arctic: Boyle* et al., *1982; Dai and Martin, 1995; Kim* et al., *1996). The ranges of values for remote areas are probably reasonable surrogates for undisturbed baseline concentrations expected for most river and estuarine waters of the world. Selected data are shown for rivers influenced by humans as reported for 19 east coast rivers (Windom* et al., *1991), in addition to the Mississippi, USA; St Lawrence, Canada; Rhone, France; and Ebro, Spain. These are compared to metal concentrations in the remote Kara Sea in the Arctic (Dai and Martin, 1995); selected estuaries/coastal waters (Table 5.5) and the Great Lakes, USA (Nriagu* et al., *1996)*

	Zinc	Copper	Nickel	Cadmium	Lead
Remote rivers	0.02–0.25	0.19–2.4	0.5–1.3	0.0006–0.0018	0.006–0.017
Remote coastal waters: Kara Sea		0.17–0.27	0.26–0.35	0.027–0.041	0.002–0.007
Great Lakes	0.02–0.38	0.7–1.06	0.47–1.00	0.0006–0.009	0.0004–0.032
Rivers influenced by humans	1.2–3.9	1.46–2.54	1.4–1.6	0.014–0.045	0.03–0.08
Developed estuaries and coastal areas	0.3–5	0.2–5	0.3–4.2	0.01–0.2	0.004–0.2

or temporal expectations of human influences. Even small changes in contamination of water bodies as massive as the oceans are quite important, but analytical chemistry must be extremely rigorous to detect changes. Change represents massive human inputs (to overcome the dilution power of these systems) and involves exposures of organisms probably not adapted to perturbations in metal concentration. They also usually signify that much more serious contamination is occurring in closer proximity to human activities.

Ranville and Flegal (2005) recently completed a survey of silver concentrations in the world oceans that illustrates the ability of human activities to contaminate even these massive water bodies. From earlier, but reliable, analyses they showed that Ag concentrations are naturally lower in the undisturbed surface waters of the Atlantic (0.25 pM or 0.027 ng/L Ag (0.000 027 µg/L Ag); cf. Table 5.3 quoting 1995 data summary) than in the undisturbed surface waters of the mid-Pacific (1 to 2 pM or 0.1 ng/L Ag). But in 2002, the highest Ag concentrations ever found in the open ocean were discovered off Asia (up to 12 pM or 1.3 ng/L Ag). The distribution of the contamination was consistent with an input of dust and industrial aerosols from the Asian mainland carried to sea by the prevailing westerly winds. The concentrations off Asia also had increased since a 1983 survey, suggesting the rapid urbanisation and industrialisation in Asia since then to be the cause. Water at mid-depths was even more contaminated, than the surface water, as would be expected from the recycling of Ag (Ag follows the nutrient Si). Ranville and Flegal (2005) concluded that the degree of anthropogenic contamination (50-fold difference between contaminated and uncontaminated surface waters) was greater for Ag than for any other metal.

The Ag contamination is reminiscent of what happened with Pb before it was tightly regulated in the developed world (Boyle *et al.*, 1986) (Box 5.4). In the case of Pb, detectable contamination of open ocean waters reflected much more serious regional and hotspots of contamination in onshore waters, estuaries, rivers and the terrestrial environment. A similar possibility is likely for Ag.

5.2.3.4 Dissolved metal concentrations in rivers

Prologue. Metal concentrations in rivers reflect the combined influences of weathering of local geology, floodplain and tributary inputs, human waste discharges, water chemistry and hydrology. Rivers tend to have higher dissolved concentrations of many metals than do the oceans to which they eventually discharge, except for some of the metals which are rare in the Earth's crust. The several influences on concentrations can cause temporal variability. But differences in dissolved metal concentrations among rivers are not as great as might be expected. Arguably, dissolved metal concentrations in rivers are not necessarily sensitive indicators of anthropogenic contamination.

The erosion of continents is a natural source of metals for rivers, estuaries and eventually coastal waters. Different geologies, rates of inputs from weathering and water chemistry add geographical variability to natural metal releases; human influences add additional variability. Data from pristine rivers provide the best comparison between rivers and oceans. Concentrations of Cu, Ni and Pb are higher in two pristine Arctic rivers than in the Kara Sea (Dai and Martin, 1995) (Table 5.4). Boyle (1979) compared metal concentrations in 17 rivers in the Amazon basin (Brazil) in the 1980s, to metal concentrations in the offshore ocean. Concentrations of Cu in the different rivers ranged from 0.4 to 2.8 µg/L; ocean water concentrations varied from 0.06 to 0.24 µg/L. Note the units for Cu are 1000 times higher (µg/L as compared to ng/L) than the previous units used for Pb and Ag, reflecting Cu's greater abundance in the Earth's crust.

Metals that occur in low concentrations on the undisturbed landscape, or undergo geochemical transformation between the river and the sea, are more likely to be found at comparable concentrations between pristine rivers and the sea. For example, Cd is a relatively rare element in the Earth's crust. Geochemical processes result in the release of dissolved Cd into estuaries, as rivers transit to the sea. As a result, Cd concentrations can be lower in pristine rivers than in the sea (Dai and Martin, 1995) (Table 5.4). Headwater streams can also have very low metal concentrations because of their small watersheds. Kim *et al.*

Box 5.4 Lead contamination of the oceans

Perhaps the first metal recognised as contaminating the oceans was Pb. Following its invention in the 1920s tetraethyl lead was (and in some places still is) added to fuel to improve performance of car engines. As fuel was consumed, Pb was released to the air, attached to fine particles or aerosols. A significant fraction (about 10%) was transported over long distances by the atmosphere and deposited into the ocean surface (Wu and Boyle, 1997). Removal of Pb from motor vehicle fuels was mandated by legislation in the 1970s in the USA and Germany, and in 1990 throughout the European Union. Because the USA was by far the largest consumer of Pb in fuel, global Pb emissions were at a maximum in 1970 and have declined since. The Atlantic Ocean receives Pb aerosols from the USA via westerly winds and from the trade winds which carry European aerosols. After it is deposited in the surface ocean, Pb is converted into soluble form, and later removed from the surface by uptake onto sinking particles of biological origin. The Pb concentration of surface water is expected to track Pb fluxes from the atmosphere into the ocean with a lag of about 2 years.

Determination of trends in Pb concentrations in the oceans was challenging. Dissolved Pb concentrations in seawater are low (1 to 5 ng/L Pb) and susceptible to contamination during sampling and analysis. 'Noise' from atmospheric and oceanic variability can also obscure long-term trends. For example, near-surface seawater Pb concentrations near Bermuda can change by up to 40% on timescales from a week to several months (Boyle et al., 1986) in response to changes in nearby atmospheric inputs. Boyle and his co-workers resolved these problems to produce two reliable time series of Pb concentrations in the oceans: one, a high-frequency set of samples from Bermuda,

combined with data from dated sections of corals; and one from surface water analyses from the Sargasso Sea in the North Atlantic (Boyle et al., 1986; Shen and Boyle, 1987) (Fig. 5.3). By analysing the Pb content of dated sections through corals, they found that Pb near Bermuda rose by about five times as a result of early industrialisation: from levels of about 6.3 ng/L Pb in 1883 to 17.8 ng/L Pb in the 1920s. Concentrations increased further to a peak of 31.5 ng/L Pb in the 1970s as a result of Pb use in fuels. It then declined to 8 to 14 ng/L Pb in the 1990s as Pb was removed from motor vehicle fuels. The remaining contamination in the 1990s appeared to represent inputs from modern industrial processes and dust from increased land use.

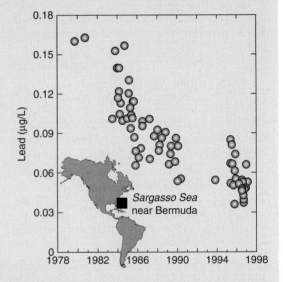

Fig. 5.3 Trends in lead concentrations in ocean waters off Bermuda as reported by Boyle and colleagues, between 1980 and 1996. (After Boyle et al., 1986.)

(1996) reported copper concentrations in the head-waters of a New Zealand stream (0.19 μg/L in the Clutha River) that were comparable to oceanic values.

Human activities affect dissolved metal concentrations in rivers draining developed landscapes, but the concentrations often differ less from natural concentrations than might be expected (Box 5.5).

Even though dissolved concentrations are not especially different among river systems, they can be variable with time within a river. The temporal variability is often underestimated as a source of uncertainty in defining 'site exposure levels'; and it is especially important when spatial differences are small. Box 5.6 illustrates some different time-scales of temporal variability and their relative importance.

Box 5.5 Response of dissolved metal concentrations in rivers to human perturbations

Windom *et al.* (1991) surveyed metals in 18 rivers in the eastern North America and compared the data to results from other major rivers of the world with large human populations in their watershed. Cu concentrations in East Coast rivers of the USA fell into the range found in the rivers of the Amazon (0.9 to 5.0 μg/L: Boyle *et al.*, 1982) and the Arctic (Dai and Martin, 1995). Cd and Zn were enriched in the east coast rivers, on average (0.011 and 1.2 μg/L respectively), and were highly variable (Cd, 0.001 to 0.050 μg/L; Zn, 0.05 to 3.92 μg/L). Elbaz-Poulichet *et al.* (1996) found similar results for the River Rhone, France. The surveys led Windom *et al.* (1991) to the conclusion that 'dissolved trace metal concentrations in (large) rivers provide inadequate bases from which to evaluate anthropogenic inputs'. Elbaz-Poulichet *et al.* (1996) concluded that 'the anthropogenic imprint . . . may exceed assumed unpolluted levels by a factor of 2 to 4'. The responsiveness to human contamination varies among metals. Metals that are relatively abundant in the Earth's crust (highest background concentrations) change least, in relative terms, in response to human inputs. Concentrations of elements that are relatively rare in the Earth respond to human influences with greater relative change (e.g. Cd, Ag). Internal processes (e.g. sequestration onto particulate material) are the primary factors reducing responsiveness to human perturbation.

Box 5.6 Variability of dissolved metal concentrations in mining-impacted streams

Variability in dissolved metal concentrations can occur on several timescales. These processes are especially accentuated in highly contaminated river systems; but are relevant for any system in which contaminated particulates have been deposited on the river's floodplain (see also Chapter 12). Some of the variability in dissolved concentrations is dynamically stable (e.g. predictable if controlling processes are known). For example, metal concentrations can vary diurnally (i.e. through day and night). Concentrations of metals like Cu, Zn and Mn show minimum values in the late afternoon and maximum values shortly after sunrise (Brick and Moore, 1996). Arsenic shows an opposite trend. Fuller and Davis (1989) attributed such cycles to variations in adsorption–desorption reactions at mineral surfaces as pH changed in waters of river bank soils (the riparian zone). A small pH increase during the day, is induced by the use of carbon dioxide in photosynthesis, which removes carbonate (carbonic acid) from solution causing pH to rise. High pH enhances adsorption of most metals but desorption or release of metalloids like As from particle surfaces. A decrease in pH at night, caused by an excess of respiration over photosynthesis (net release of carbon dioxide would create carbonic acid) appears to enhance Zn release (Jones *et al.*, 2004). In mine-contaminated Prickly Pear Creek, Montana, USA, Zn increased 300% at night, and As increased 33% during the day. Sampling waters only during the day would bias estimates of biological exposures and dissolved loads to the stream, accordingly.

Transport processes in the pore waters and groundwater adjacent to a stream are also variable. For example, metals leached from sediments by low pH groundwater in metal-contaminated, sulphide-rich floodplains eventually are flushed into streams during spring runoff (monthly scale variability) or after rainstorms in the summer (daily scale variability) (Nagorski *et al.*, 2003). The magnitude of such variability can be just as great on the scale of rainstorm events (days or hours) as it is seasonally.

River discharge is also an important source of variability, especially if waters of different quality mix. For example, pulses of increase in river flow with storms in the River Rhone, France appear to dilute trace element concentrations, suggesting that a constant input is diluted by cleaner runoff. But if high flow events carry a second source of metals (e.g. from a contaminated floodplain), high metal concentrations can increase with river discharge (Elbaz-Poulichet *et al.*, 1996). In the River

Rhone, variability was determined by the ratio of the standard deviation to the mean. The ratio was 23% to 34% for dissolved As, Cu, Ni and Pb, and 46% for Cd (Elbaz-Poulichet *et al.*, 1996).

5.2.3.5 Dissolved metal concentrations in lakes

Surveys of dissolved metal concentrations in lakes are not especially common in the literature. Atmospheric deposition is especially important because of the large surface areas of lakes. The internal processes that occur in large lakes show some similarities to the oceans, however. Nriagu *et al.* (1996) reported extensive surveys from the Great Lakes (USA, Canada): Lake Erie, Lake Ontario and Lake Superior. Although these water bodies are heavily industrialised on their shorelines, metal concentrations in 1993 more closely resembled remote rivers and the open ocean, than they did rivers, estuaries or coastal waters affected by human activities (Table 5.4). But, like the oceans, concentrations were greater near the shoreline than in deeper waters; and concentrations of typical recycled elements (Cd and Zn) increased with depth.

Many examples of contaminated lakes also exist. Air pollution from smelters in Sudbury and western Quebec heavily contaminated thousands of lakes, over hundreds of square kilometres, in the Canadian Shield until the 1980s (Yan *et al.*, 2004; see Chapters 11 and 12). For example, the mean Cd concentration in several pristine lakes surveyed in central Ontario was 0.010 µg/L Cd, similar to the Great Lakes (0.0006 to 0.009 µg/L Cd: Ball *et al.*, 2006). But Cd concentrations in lakes near enough to Sudbury to have been influenced by historic smelter aerosol discharges ranged from 0.05 to 0.58 µg/L Cd; >4 µg/L Cd was determined in one lake.

5.2.3.6 Dissolved metal concentrations in estuaries and coastal waters

Prologue. Dissolved metal concentrations in estuaries are influenced by processes that remove trace metals from solution and by the salinity gradient that characterises the mixing of a river with its oceanic receiving waters. Urbanised bays with extended residence times appear to be the most contaminated by human activities. Variability, overall, is about 10- to 20-fold

among locations that have been studied; but much remains to be learned about metal contamination in different water bodies.

The dissolved metal concentration in an estuary (and adjacent coastal waters) is influenced by the dissolved concentrations in the river(s) at its head, the ocean at its mouth, the geology of the local watershed and the presence of industrial or other anthropogenic activities along the estuary itself.

An estuary is defined as the water body where freshwater meets the sea. Thus the simplest estuaries have clear 'end members': the incoming river and the ocean waters that receive it. The area where these waters mix is characterised by a 'salt field', a gradient in NaCl created when the salty ocean waters mix with the freshwater from the river. Knowledge of the salt field is a fundamental prerequisite for understanding dissolved constituents like dissolved concentrations of metals. Because Na and Cl concentrations are not removed by biogeochemical processes or affected by human activities (other than freshwater inputs), dilution of oceanic salinity occurs in direct proportion to the percentage of freshwater mixed with the seawater. Salinity is thus said to be 'conservative'. Variability in salinity is a measure of the mixing of the two water bodies. Property–property relationships are simple correlations between salinity and another property, like dissolved metal concentration. They provide first order estimates of how mixing affects properties like dissolved metal concentration (Boyle *et al.*, 1977).

Several types of behaviour typify metal dynamics in estuaries (Fig. 5.4). Properties that correlate linearly with salinity are themselves 'conservative' constituents. It is assumed that they simply follow the mixing of salt and do not undergo reactions that change their concentrations in solution. (This, of course, assumes that removal and input reactions do not balance one another.) Positive deviations from a linear relationship with salinity, or a convex curve, indicates that 'internal inputs' of the metal are occurring within the estuary. For example, anthropogenic waste discharges, or release from sediments, might add dissolved metal to the estuarine waters, resulting in more metal in

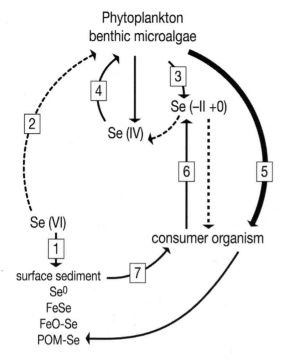

Phytoplankton
benthic microalgae

Fig. 5.4 The selenium cycle. The biogeochemical cycle of Se starts with selenate (Se(VI)) which is the most common form in oxidised waters (1). Selenate can be reduced by microbes to particulate forms in sediment or taken up (slowly) by plants (2). Plants (e.g. phytoplankton in aquatic systems) transform Se(VI) to organo-selenium (Se (–II+0)) and release selenite (Se(IV)) and organo-Se (3), which are reaccumulated by the plants (4) and therefore build up in a water body with long residence times. The kinetics of the back reaction to selenate are extremely slow (little is reconverted) (no arrow). Selenite and organo-Se in solution are bioavailable to animals (6) but are taken up very slowly; the predominant pathway to animals is ingestion of either sediments or plants (5, 7). Biogeochemical recycling of Se in the ocean (and many other water bodies) creates a predominance of organic selenide and selenite that would not be expected purely upon the basis of chemical thermodynamics.

solution than expected from freshwater–seawater mixing at a given point (Flegal *et al.*, 1991). Negative deviations from the gradient, or a concave mixing curve, indicate property 'removal' within the estuary. For example, scavenging of metals by colloid formation, adsorption onto precipitating iron oxides or complexation with organic ligands on the surfaces of particles are all well known in estuaries.

Trace metals can follow any of the three relationships depending upon their water chemistry, and the degree of adsorption, complexation or flocculation that occurs in each specific estuary (Fig. 5.4). But scavenging is particularly important for many metals. In general, it is assumed that estuaries act as traps for many of the trace metals transported by the rivers feeding into them. The bulk of the metal removed from solution by sequestration onto particulate matter is then held in the sediments.

Table 5.5 provides perspective on dissolved metal concentrations in estuaries from a global selection of data from the high salinity areas of these water bodies. These data are compared to concentrations in bays and coastal waters. Urbanised bays are characterised by some of the highest dissolved metal concentrations (away from point source of input). Their metal concentrations appear to represent cumulative effects of multiple human influences.

Biggs *et al.* (1989) classified urbanised water bodies like north San Francisco Bay (USA), and Hudson–Raritan Bay (USA) as very vulnerable to contamination from human activities. This was based upon the mass of waste input from urban activity, their volume, and the residence times of water in the system (how fast did they remove wastes?). In the 1980s San Francisco Bay had approximately 200 points of waste discharge and more than 50 streams that directly carried urban runoff into its waters. Residence times of months were characteristic of the Bay during low river inflows. Thus Biggs *et al.* (1989) classified the segment of the Bay with the most industry (Suisun Bay) as one of the most vulnerable water bodies in the USA. The Hudson–Raritan Bay system had similar characteristics and was similarly classified (Feng *et al.*, 1999). Table 5.5 shows dissolved metal concentrations typical of these systems.

The urbanised estuaries are compared to another type of contamination: regional contamination from many smaller sources. In this case we employ data from a number of locations off England and Wales. These water bodies are relatively simple estuaries with relatively short residence times. Some of the estuaries in this region of Britain are largely agricultural, some are affected by mining and others were

Table 5.5 *Higher of two dissolved concentrations (μg/L) of trace metals measured in subsurface seawater samples from estuary mouths and near coastal sites around England and Wales in 1991 and 1992 (after Law* et al., *1994) compared to the range of concentration in urbanised bays, estuaries and coastal water sites from the USA (after Flegal and Sanudo-Wilhelmy, 1993; Sanudo-Wilhelmy and Gill, 1999; Paulsen 1995); concentrations in many of these water bodies may have declined in recent years*

	Cu	Zn	Mn	Ni	Cd	Hg	Pb
(a) British estuary mouths							
Tweed	0.58	0.58	5.0	0.48	0.012	0.004 4	0.096
Tyne	0.43	1.2	2.1	0.36	0.018	0.004 2	0.086
Wear	0.46	1.3	2.5	0.51	0.025	0.003 4	0.230
Tees	1.6	2.6	13	0.3	0.020	0.004 5	0.096
Humber	1.8	5.1	6.8	1.6	0.076	0.001 4	0.024
Mersey	1.7	3.9	11	1.2	0.030	0.002 1	0.130
Dee	1.4	2.4	5.5	0.87	0.018	0.002 3	0.170
(b) Urbanised, semi-enclosed bays							
San Diego Bay (CA, USA)	0.89–2.60			0.35–0.94	0.006–0.19		0.03
San Francisco Bay (CA, USA)	1.4–4.64			1.2–4.2	0.06–0.16		0.018–0.064
Raritan, NY bays (New York, USA)	1.27–1.78			0.59–1.29	0.028–0.091		0.015–0.269
Hudson Estuary	1.14–2.6			0.41–1.17	~.011–0.13		0.021–0.166
(c) British coastal water							
Off Tyne	0.41	0.66	1.3	0.49	0.011	0.004 6	0.036
Off Tees	0.24	0.35	1.5	0.38	0.012	—	0.071
Humber	0.71	2.2	0.73	0.29	0.022	0.002 3	0.045
Wash	0.74	1.0	1.3	1.0	0.027	0.000 3	0.190
Outer Thames	0.83	0.92	0.9	0.9	0.032	0.005 0	0.073
Selsey Bill	0.48	1.0	1.1	0.47	0.018	0.006 3	0.024
Central English Channel	0.21	0.57	0.9	0.24	0.015	<0.000 2	0.032
Celtic Deep	0.36	0.27	0.27	0.3	0.018	0.002 4	0.024
Bristol Channel	1.0	2.1	0.9	0.71	0.081	0.000 3	0.097
Cardigan Bay	0.85	2.1	0.3	0.56	0.069	0.000 9	0.058
Liverpool Bay	1.5	1.8	7.8	0.87	0.035	—	0.170
Irish Sea	0.61	0.72	1.7	0.42	0.028	0.000 6	0.065
(d) Undisturbed coastal waters							
California coastal waters	0.10			0.27–0.32	0.003–0.08		0.004–0.013

heavily industrialised at the time of sampling (e.g. the Humber, Mersey, Cardiff and Bristol Channel). The data, together, typify regional variability in contamination among water bodies variably influenced by several hundred years of human activities.

All the coastal concentrations of dissolved metals exceed or are at the top end of the oceanic concentration ranges listed in Table 5.3, particularly when compared to ocean surface concentrations of the trace metals with nutrient-type distributions.

The higher concentrations in coastal water probably reflect the influences river inputs combined with widespread human activities. But it is notable that variability, overall, is relatively low among these very different environments. The difference between the lowest and highest values for each metal among all the estuaries and coastal data does not exceed 10- to 20-fold.

The dissolved metal concentrations from large bays and coastal waters are two to five times lower than the maximum values available from British and North American estuaries in the 1980s (Table 5.2; assuming the latter represent reliable analyses). An important question is how frequent, across the world, is the occurrence of the highest dissolved concentrations seen historically or in the urbanised estuaries? Surprisingly, we do not know. Sanudo-Wilhelmy *et al.* (2004) noted that, since 1975, 83 articles had been published on dissolved metal cycling in estuaries of North America. Seventy percent of those articles considered five estuaries: the Delaware, Chesapeake Bay, Galveston Bay, Narragansett Bay and San Francisco Bay. There were no data for about 50% of US estuaries. The state of knowledge is even worse on a global scale. Sanudo-Wilhelmy *et al.* (2004) concluded that quantifying how water quality is responding to human disturbance, and attempts to mitigate that disturbance, depend upon long-term, systematic studies of the biogeochemical cycling and fate of aquatic contaminants in the water column of a variety of estuaries and coastal water bodies. Until those studies are instituted, quantifying water quality problems is problematic.

Tables 5.2 and 5.5 also provide some perspective for risk assessments on both the range and variability expected from coastal waters and estuaries. Concentrations substantially higher than these should be viewed with suspicion; regulatory guidelines well outside these limits do not reflect the realities of nature.

5.3   SPECIATION OF TRACE METALS

Prologue. Metal speciation determines metal behaviour and ecological risks in natural waters. The ability to predict speciation, free ion concentrations and/or directly determine potentially reactive metal is growing rapidly. Precise predictions in site-specific situations remain subject to at least some uncertainties, however. If speciation is critical to the assessment of metal risks, multiple approaches are suggested to verify model outcomes.

The chemistry of natural waters greatly influences the behaviour of metals in these dilute solutions. The formula for a metal ion in aqueous

---

Box 5.7 Coordination chemistry

Complexes in solution (or on particle surfaces) are formed between a metal ion and a set of ligands. Ligands are ions or molecules which exist independently of the metal complex. Ligands are oppositely charged to metals and thus are electron donors. They form bonds with the electron-deficient metal centre. The widest range of complexes is formed with the transition metals. The denticity (from the Latin for teeth) of a ligand is the number of places in which it binds with the metal centre. Ammonia has one pair of electrons on the nitrogen atom, and it binds as a unidentate ligand. Some ligands may have large numbers of donor sites, such as the artificial organic molecule $EDTA^{4-}$, which has six donor sites. These are known as polydentate ligands. Equilibrium constants, also termed stability constants, define the strength of the metal–ligand complex and can be used in models to predict how metals will partition on particles or speciate in solution among ligands.

---

solution is usually written M^{n+}. But, in formal terms, that term represents the simplest hydrated metal cation. In fact, metals occur as a variety of 'species' in solution. It is the combination of those species that controls the states available for reaction (Wangersky, 1986). Speciation occurs as metal ions coordinate to water molecules and other chemical entities in the water, attaining a state of maximum stability (Box 5.7).

Investigation of metal speciation is very difficult, even under controlled conditions. The reactions are complex and the form that metals take in solution usually cannot be directly determined. Most descriptions of metal speciation are based upon models developed from first principles (which are well known). In natural waters, instrumental analyses can provide operational definitions of metal lability that can be used to infer some key interactions; but speciation is even more difficult to describe than in the laboratory, precisely because so many variables have to be taken into account.

We will summarise a few principles that are critical to understanding how metal speciation is related to ecological risks from metals. For more complete descriptions of the state of knowledge in aquatic geochemistry, however, some classic textbooks are available (Morel and Hering, 1993; Stumm and Morgan, 1996).

5.3.1 Oxidation state

Prologue. Forms of some trace metals (e.g. Fe, Mn, As, Cr, Se) are determined by the reduction–oxidation status of the water.

Trace elements that become fully hydrolysed in natural waters may exist in one of two or more oxidation states according to the redox potential (Box 5.8) of the medium. Examples of trace metals that exhibit redox changes in seawater include Fe [Fe(II)/Fe(III)], Mn [Mn(II)/Mn(IV)], Se [Se(IV)/Se(VI)], Cr [Cr(III)/Cr(VI)], As [As(V)/As(III)], and Cu [Cu(I)/Cu(II)] (Donat and Bruland, 1995). The higher of the two oxidation states is the thermodynamically stable form in normoxic (oxygenated and therefore oxidising) waters. Most natural waters, however, do not attain complete chemical equilibrium, particularly under the influence of light which can cause biochemical (e.g. by photosynthesis) or photochemical production of metals in oxidation states that are out of thermodynamic equilibrium with the environment (Donat and Bruland, 1995). In hypoxic (low oxygen) or anoxic (no oxygen) waters, such as interstitially in sediments rich in organic carbon, the lower oxidation state metal species will be promoted. For example, Mn(II) is the dominant form of dissolved Mn in such sediments.

5.3.2 Speciation of redox-sensitive anionic metals

Some fully hydrated trace metals do not exist dissolved in natural waters as cations but as anions. In seawater, these include As ($HAsO_4^{2-}$), Cr (CrO_4^{2-}), Mo (MoO_4^{2-}) and V (HVO_4^{2-}) (Bruland, 1983).

Arsenate ($HAsO_4^{2-}$) is the thermodynamically stable form of As in oxgenated seawater and the major arsenic species in the oceans, but reduction to arsenite ($HAsO_2$), and even methylation of arsenic in phytoplankton (see below), can occur within the photic surface zone (Bruland, 1983). The thermodynamically stable form of Cr in oxygenated

---

Box 5.8 Redox conditions

Oxidation-reduction (redox) reactions involve changes of oxidation state of reactants. In formal chemical terms these reactions can be visualised as the transfer of electrons from one species to another. Simplistically, Zn^{2+} may be removed from solution by an association with metallic Fe wherein it receives electrons from the metallic Fe by the equation:

$$Zn^{2+} + Fe \longrightarrow Zn + Fe^{2+}.$$

The electrons on both sides of the equation are equal but they change from Fe to Zn (or vice versa), with a concomitant change in the form of each.

The same reaction can be either an oxidation or a reduction reaction depending upon the direction of the arrow; these are called redox reactions. The predominant direction of the reaction is determined by environmental conditions (e.g. the presence of free oxygen). The species receiving electrons is reduced, that donating electrons is oxidised. Among their many manifestations, redox reactions determine the predominant form and mobility of many trace metals. A common redox reaction is that of Fe:

$$2Fe^{3+} + H_2 \longrightarrow 2Fe^{2+} + 2H^+.$$

Again, the charges on both sides of the equation are equal, but the form of Fe changes when conditions favour the 'reduced' form of the element (Fe^{2+}) over the oxidised form (Fe^{3+}).

Redox conditions determine if the reaction is dominated by oxidation or reduction and 'redox couples' are the potential end products of the reaction in one direction or another. In air-saturated waters the redox level is set by the oxygen/water redox couple. In anoxic waters it is set by major dissolved components, including Fe, Mn, and especially carbon and sulphur. Redox potential is an intensity parameter of overall redox reaction potential in the system derived from the several dominant redox reactions.

seawater is the chromate ion (CrO_4^{2-}) in which Cr is in the +6 oxidation state, although Cr(III) in complexed cation form [$Cr(OH)_2^+$ and $Cr(OH)_3^0$] will be present in anoxic conditions (Bruland, 1983). Molybdenum exists in oxygenated seawater in the +6 oxidation state as the oxyanion molybdate (MoO_4^{2-}); it is the most abundant transition metal in seawater and is atypical for trace metals because it has a conservative vertical depth distribution (Bruland, 1983).

Not all speciation, of anionic metals in particular, is governed by, or follows, thermodynamic expectations, however. Se can occur in the oxidation states − II, selenide; 0, elemental; IV, selenite; and VI, selenate (Fig. 5.4). In oxygenated surface waters, chemical thermodynamics predict that dissolved Se should almost entirely exist as selenate. In anoxic waters, thermodynamics predict that selenite and selenate should reduce to insoluble (particulate) elemental selenium, Se(0). Surprisingly, early speciation surveys showed that Se in the upper layers of the (oxidised) oceans is 80% organic selenide (Cutter and Bruland, 1984). A biogeochemical cycle appears to explain the unexpected aspects of Se speciation in the oceans (Cutter and Bruland, 1984) (Fig. 5.4):

(1) Se is predominantly released to, and is chemically retained as, selenate in natural waters.
(2) Selenate and the small amounts of selenite present are taken up by plants and/or microbes (e.g. phytoplankton in the ocean).
(3) Within the cells of the plants, the inorganic forms are reductively incorporated into proteins to form organic selenides; when the plant cells die or are consumed by animals, their detritus contains only organic Se.
(4) They release dissolved selenite and organic selenide upon cell death.
(5) Selenite and organo-Se remain in their original forms in solution because the rates of their reconversion to selenate occur very slowly (more than 100 years).
(6) In addition, consumer organisms eat the phytoplankton and
(7) release organic selenide in their carcasses, faecal pellets and fluids.

Because the rates of reconversion or breakdown of selenite and organic Se are slow they tend to accumulate in ocean waters.

5.3.3   Speciation of cationic metals

Prologue. Speciation of cationic metals is driven by the concentration of reactive ligands and the concentration of ions that compete with metals for ligands. The composition of seawater is relatively constant, with the exception of organic matter, so speciation is relatively predictable. Speciation is typically more variable in freshwater than seawater, because of greater variability in ligand concentrations.

A number of elements can produce free cationic forms in aqueous solution (Turner et al., 1981). The trace metals that can produce cations do not exist only as the free hydrated ion (e.g. M^{2+}), however. To achieve stability they tend to associate with ligands that occur in the water as anions. In natural waters there are five main inorganic, anionic ligands that compete for association with the cationic metals. These are fluoride (F^-), chloride (Cl^-), sulphate (SO_4^{2-}), hydroxide (OH^-) and carbonate (CO_3^{2-}). This distribution of a metal among ligands is termed metal speciation, and it plays a crucial role in the environmental chemistry of metals.

The concentrations of the different ligands vary widely among natural waters. Metal carbonate complexes, metal hydroxide complexes, metal chloride complexes, or complexes with dissolved organic matter are usually much more common than free ionic metal in all waters, but the relative distribution of a metal among all of them will differ with water chemistry.

In the absence of organic complexing agents, the theoretical inorganic speciation of trace metals in seawater at standard pH, temperature and pressure can be modelled from stability constants (e.g. Turner et al., 1981; Byrne et al., 1988). Table 5.6 lists the probable main species of trace metals in seawater in the absence of organic matter. The free hydrated divalent cation M^{2+} dominates the dissolved inorganic speciation of Zn (although representing

Table 5.6 *Predominant inorganic species of metals in seawater*

Metal		Probable main dissolved species
Ag	Silver	$AgCl_2^-$, $AgCl_4^{3-}$, $AgCl_3^{2-}$
Al	Aluminium	$Al(OH)_4^-$, $Al(OH)_3^0$
As	Arsenic	$HAsO_4^{2-}$
Ca	Calcium	Ca^{2+}
Cd	Cadmium	$CdCl_2^0$, $CdCl^+$, $CdCl_3^-$
Co	Cobalt	Co^{2+}, $CoCO_3^0$, $CoCl^+$
Cr	Chromium	CrO_4^{2-}, $NaCrO_4^-$
Cu	Copper	$CuCO_3^0$, $Cu(OH)^+$, Cu^{2+}
Fe	Iron	$Fe(OH)_3^0$, $Fe(OH)_2^+$
Hg	Mercury	$HgCl_4^{2-}$, $HgCl_3Br^{2-}$, $HgCl_3^-$
K	Potassium	K^+
Mn	Manganese	Mn^{2+}, $MnCl^+$
Na	Sodium	Na^+
Ni	Nickel	Ni^{2+}, $NiCO_3^0$, $NiCl^+$
Pb	Lead	$PbCO_3^0$, $Pb(CO_3)_2^{2-}$, $PbCl^+$
Se	Selenium	SeO_4^{2-}, SeO_3^{2-}, $HSeO_3^-$
V	Vanadium	HVO_4^{2-}, $H_2VO_4^-$, $NaHVO_4^-$
Zn	Zinc	Zn^{2+}, $Zn(OH)^+$, $ZnCO_3^0$, $ZnCl^+$

Source: After Bruland (1983); Donat and Bruland (1995).

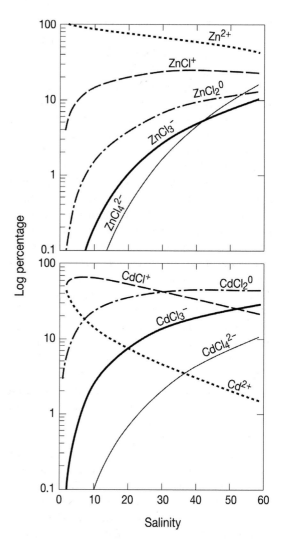

Fig. 5.5 The effect of changes in salinity on the inorganic speciation of dissolved Zn and Cd (after Rainbow *et al.*, 1993b). Full-strength seawater has a salinity of 33.

only 40% of dissolved Zn present: Mantoura *et al.*, 1978), and the first transition series metals Co, Mn and Ni (Donat and Bruland, 1995). Trace metals associated with hydroxyl groups include the trivalent trace metal ions of Al and Fe, although the inorganic speciation of Fe in seawater is complicated and not fully agreed (Donat and Bruland, 1995). Trace metals whose dissolved inorganic speciation is dominated by chloride complexation include Ag, Cd (only 2.5% present as free hydrated Cd^{2+}: Zirino and Yamamoto, 1972) and Hg. It follows that inorganic complexation, particularly by chloride, is reduced as seawater is diluted in an estuary. The modelled changes in Zn and Cd inorganic complexation with decreased salinity are shown in Fig. 5.5 (Rainbow *et al.*, 1993b).

Dissolved organic materials are also important metal-binding ligands in natural waters. Trace metals can be complexed by organic ligands such as the metabolites or decay products of prokaryotic plankton, eukaryotic phytoplankton or zooplankton and other animals, or humic substances (the unstructured aggregates of organic breakdown products). In coastal waters, the decay products of macrophytic benthic algae may also be a significant source of organic complexing agents. In deep ocean waters the recycled metabolites of living organisms are important. The differences in characteristics among these many types of organic material make the interactions of organic material with trace metals particularly difficult to generalise.

Organic complexation dominates the dissolved speciation of Cu and Zn in seawater, especially in surface ocean and coastal waters (Donat and Bruland, 1995). This is also the case for Cu in freshwaters. Organic complexation reduces the fraction of dissolved Cu in inorganic form in some surface ocean waters to less than 0.3% of which free hydrated Cu^{2+} is only 4% (0.012% overall: Donat and Bruland, 1995). While total dissolved Cu concentrations in the central northeast Pacific increase approximately threefold from surface to mid-depths, free hydrated Cu^{2+} concentrations vary about 2 000-fold. Complexation by organic ligands decreases with depth and causing free Cu^{2+} to increase from $10^{-13.1}$ M in shallow surface waters to about $10^{-9.9}$ M at 300 m (Coale and Bruland, 1990; Donat and Bruland, 1995). Similarly for Zn, 98.7% of dissolved Zn in central north Pacific waters above 200 m is organically complexed, and the free hydrated Zn^{2+} ion varies from $10^{-11.8}$ M at depths less than 200m to $10^{-8.6}$ M at 600 m, a 1400-fold increase (Bruland, 1989; Donat and Bruland, 1995).

Organic complexation may also be significant for other dissolved trace metals in seawater. For example, 70% of dissolved Cd is organically complexed at the surface in central North Pacific surface waters. Organic complexation decreases with depth where inorganic chloro-complexation of dissolved Cd dominates (Donat and Bruland, 1995). Similarly dissolved Pb may be complexed organically to a significant degree in ocean surface waters (Donat and Bruland, 1995).

Inorganic speciation, organic complexation and differences among cationic metals can all interact to create interesting differences in the behaviour of metals when chemical conditions in waters change. For example, organic complexation of dissolved trace metals will be enhanced wherever there is increased biological or other organic production. Metal speciation and concentration will also change with the changes in chemistry that occur when freshwater mixes with seawater in

estuaries (Fig. 5.6). Dissolved Cu was not removed from solution in the Rhone estuary, because strong organic complexes (colloidal or molecular organic moieties) prevented its removal by adsorption, complexation or flocculation. Pb is weakly enough associated with ligands in freshwater that it can be stripped from them and co-flocculated with Fe as river waters mix with the sea (Fig. 5.6). Cd increases in dissolved concentrations as salinity increases (as in the Rhone estuary) (Fig. 5.6), because it is desorbed from particle surfaces by the formation of strong chloro-complexes when salinity reaches 5 to 10.

Metal reactions follow the same principles in freshwater as in seawater, but speciation is more complex in freshwater. The concentration of potentially reactive ligands is much lower in freshwater as is the concentration of potential competitive ions. But range of variability in these concentrations is large among waters, making the range of possible interactions with trace metals greater. Hydrated forms (complexes with OH^-) of many metals are more common in freshwater than seawater. Chlorinated forms of metals are rare in freshwater but common in seawater (especially for Cd, Ag and Hg which form strong chloro-complexes). Reactions with organic ligands might be stronger in freshwater, because of less competition with calcium and magnesium ions (both are very abundant in seawater). The pH also varies widely in freshwater. Free ion activities can be greater in freshwater than seawater because of the fewer ligands overall.

5.3.4   Organometallic compounds

Prologue. Organisms can generate metals that are covalently bonded to an organic moiety. Humans also synthesise organometals, often as poisons or as a by-product of other reactions.

Dissolved trace metals in natural waters may also include organometallic compounds in which the trace metal is covalently bound to carbon. Examples in seawater include methyl (CH_3) forms

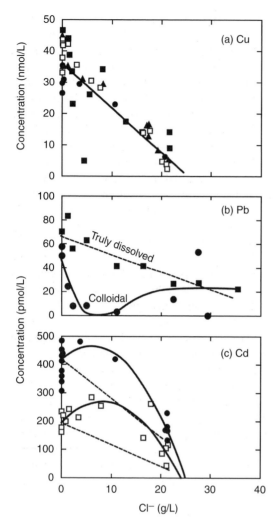

Fig. 5.6 Estuarine gradients for three metals, plotted in a property–property plot (metal concentration versus salinity) illustrate the three types of behaviour common in estuaries. The conservative Cu profiles and the Cd profile that indicates an internal addition of the metal are from the Rhone River, France (after Elbaz-Poulichet et al., 1996). The Pb profiles, indicating Pb removal, are from the Ob River, Russia (after Dai and Martin, 1995). Differences in speciation reactions (see Section 5.3) explain why the behaviour of these three metals is so different. A mid-estuarine input from an anthropogenic source (rather than a geochemical process) would also show a profile like Cd. The most notable aspect of these curves is that metal concentrations can vary widely along the mixing gradient of the estuary. Clearly, it is essential to understand where a sample was collected in the salinity gradient to interpret its significance.

of As, Hg, Sb, Se and Sn, ethyl (C_2H_5) forms of Pb and butyl (C_3H_7) forms of Sn (Donat and Bruland, 1995). Methylation of As to monomethylarsonate [$CH_3AsO(OH)O^-$] and (mostly) dimethylarsinate [CH_3AsOO^-] by phytoplankton occurs in the photic zone of the sea with subsequent release into solution (Bruland, 1983; Donat and Bruland, 1995).

The production of methyl mercury (CH_3Hg) from inorganic mercury in the aquatic environment occurs in water and sediments (particularly organically rich sediments) by bacterial action (Pelletier, 1995). Bacterial demethylation can also occur, for example in lake and estuarine sediments. The net amount of methyl mercury in an aquatic system is a result of the balance between methylation and demethylation processes (Pelletier, 1995).

Tin can also be methylated by microbial action in the environment leading to the presence of tetra-, tri- and dimethyl tins in natural waters and sediments (Maguire, 1991; Pelletier, 1995). Methyl tins in the aquatic environment may also originate from industrial activities given their use as stabilisers in polyvinylchloride (PVC) plastics (Pelletier, 1995), but it is another anthropogenic (synthetic) organometallic form of tin that is more notorious in aquatic environments – tributyl tin (TBT) (see Chapter 14).

Organolead compounds have also been introduced anthropogenically into aquatic environments, usually as alkyl lead compounds. For much of the twentieth century trimethyl and tetraethyl lead compounds were added to motor vehicle fuels to prevent pre-ignition (as an 'antiknock' agent). Although most of the alkyl lead compounds were degraded during combustion, the quantities released into the environment were eventually of such environmental concern that many countries have banned their use (see earlier). Anthropogenic alkyl leads also have industrial uses in alkylation processes, catalysis in polymer reactions, PVC stabilisation and biocides; these are still released into the environment (Pelletier, 1995). The final source of alkyl lead in the environment is by environmental methylation of inorganic lead. It is not certain whether the methylation is a biological or abiotic process (Pelletier, 1995).

5.3.5 Determining speciation in natural waters: models and analytical approaches

Prologue. Equilibrium models are the most used approach to estimating metal speciation in natural waters. The variable nature of organic matter adds uncertainty to most aquatic speciation models, however.

There are no analytical techniques suitable to detect unequivocally the concentrations of all the different forms of dissolved metal in natural waters, with a few exceptions (e.g. Se). This is important because speciation is so important in metal behaviour (and risk) and each metal exists in different forms in different waters.

One approach used to circumvent this difficulty is speciation modelling. Most speciation models are built from first principles governing metal behaviour. Three relatively well known processes control metal association with any specific ligand:

- abundance of the ligand;
- stability of metal binding to the ligand (affinity between the specific metal and the specific ligand);
- concentration of the metal.

In theory, one should be able to calculate the distribution of dissolved metal among different ligands, from knowledge of these three fundamental characteristics. Analytical techniques are available to determine metal and ligand concentrations. Experimental approaches are now well developed for determining the stability of metal binding to any specific ligand (Box 5.9). Large databases of such coefficients exist, determined under different conditions. Lists of the predominant forms of different metals in natural waters have been developed from such theoretical considerations (e.g. Table 5.6).

In simple solutions and in a few natural waters, equilibrium inorganic speciation of trace elements can be modelled effectively from a few limited field measurements. However, speciation based strictly on aqueous inorganic ligands often does not agree with behaviour in natural waters. The uncertainties occur because of the ubiquitous presence of both strong organic ligands and poorly characterised, complex, inorganic species. Recently discovered examples of the latter include meta-stable sulphide clusters and other sulphide species that are present in both oxic and suboxic environments.

Organic materials and sulphide clusters present analytical challenges (reliably characterising them in the water body) as well as modelling challenges. One of the problems is that organic materials occur in heterogeneous forms in natural waters. They can occur as identifiable, high molecular weight organic materials such as polysaccharides and proteins or simpler substances such as sugars,

---

Box 5.9 Complexation reactions control metal association with each species in solution

Metal association with a ligand can be expressed by a simple complexation reaction:

$$M_i^z + S_j \xrightleftharpoons{K_{M_iS_j}} M_iS_j$$

where M_i^z represents the metal ion i, S_j represents a site on the ligand j, and M_iS_j is the complex. $K_{M_iS_j}$ is the equilibrium or stability constant that drives the reaction. The equation may be resolved and rearranged to illustrate that three factors affect metal concentration associated with any specific ligand: intensity of binding, the concentration of metal available for binding and the number of binding sites as determined by the concentration of the ligand.

$$[M_iS_j] = K_{M_iS_j} * [M_i^z] * [S_j].$$

If this equation is solved for every species that metals might bind to, then the distribution of metal among all ligands (speciation) can be calculated. A key assumption is that there is time for all reactions to proceed to their culmination: the point at which forward and reverse reactions occur at equal rates so that the concentrations of the reactants and products do not change with time. Because this assumption is critical, these models are called equilibrium models.

amino acids and other small molecules. But most of the organic materials in a water body are usually loosely termed humic substances or humic acids. Humic substances have many definitions, but in general they are aggregates of breakdown products that have survived biological metabolism and bear no structural resemblance to the compounds from which they were derived. They are relatively high molecular weight substances of limited solubility that occur in solution or as colloids. More soluble, lower molecular weight, unstructured materials are also formed by secondary breakdown reactions. These are called fulvic acids.

The stability constants for the binding of at least some trace metals to organic materials can be extremely high (the same is apparently true of sulphide aggregates). But the binding is also complex. A continuum of binding strengths seems to characterise organic material, depending upon the nature of the organic material, the concentration and the metal-specific stability constants. This mix of complexities has long frustrated modellers. For example, one of the most commonly used models to estimate organic complexation is WHAM, or the Windermere Humic Aqueous Model (Tipping *et al.*, 1998). This model has been calibrated to multiple data sets of titrations of isolated organic matter with acid, base and a number of metals; it includes the most thorough analysis of metal–organic interactions available. Nevertheless WHAM V tends to overestimate Cu binding to DOM under natural conditions. The WHAM V-calculated Cu^{2+} activity under natural conditions is generally lower than the measured Cu^{2+} activity. For natural waters, Dwane and Tipping (1998) observed that the best match between WHAM-calculated and measured Cu^{2+} activity was obtained when 40% to 80% of the DOM was considered to be active fulvic acid and when the rest was considered to be inert for Cu complexation. But it is unlikely that this assumption applies to all natural waters. As a fallback they suggest that, if nothing is known *a priori* about the Cu complexation characteristics of the DOM, 50% active FA should be used as the input for the speciation calculations.

There are analytical methods to help bridge the uncertainties of speciation modelling in certain circumstances. The invariant inorganic composition of ocean waters makes speciation estimates more reliable than for freshwaters, although variations in organic matter are a challenge. Organic matter can be characterised for a given circumstances. Methods also exist to determine some metal forms directly. For example, ion selective electrodes can directly detect free ion forms of some metals in experimental solutions. But free ion concentrations in natural waters are usually far below the detection limits of most electrodes. Thus, in general, there is uncertainty in many estimates of free ion concentrations. Models remain the best tools available to understand metal speciation in nature, but they must be used carefully and tested against other lines of evidence. Model predictions of free ion concentration or activity, in particular, are most useful in well-known circumstances or as estimates of general conditions.

Operational methods that manipulate solution chemistry and/or infer form from metal behaviour are an alternative way to quantify 'reactive' metal concentrations. Some of the most effective of these determine a 'labile' form of metal from complex instrumental approaches (Box 5.10). 'Labile' forms undoubtedly include the free metal ion, but also include other forms that equilibrate rapidly with the free ion. These methods determine the proportion of metal in different operationally defined fractions. They have the disadvantage of being operational (showing tendencies rather than mechanistically exact forms) and they require great expertise to employ in the field. They have the great advantage of being sensitive to concentrations as low as those that occur in nature. They are also direct analytical measurements, rather than depending upon constructing an approximation from diverse input data of variable reliability. As a result of the advantages they are increasingly used in research studies and occasionally in a regulatory context.

Many studies exist where speciation models have played an important role in understanding metal speciation. But there are also examples of over-reliance on such models. As an example of the implications of uncertainties, Guthrie *et al.* (2005) studied speciation of Cd, Cu, Ni and Zn in freshwater lakes receiving smelter effluents in Rouyn-Noranda,

Box 5.10 Direct determination of labile metal forms in natural waters

Buck and Bruland (2005) employed an established competitive ligand exchange-adsorptive cathodic stripping voltammetry (CLE-ACSV) method with a competitive ligand (salicylaldoxime: SA) to determine labile metal concentrations in estuarine waters. This method is sufficiently sensitive to work in natural waters; a great advantages over the ion selective electrodes. They determined a 'comprehensive characterization of the spectrum of organically complexed Cu in San Francisco Bay: from the strongly complexed Cu to the weakly complexed Cu'. The organic complexes included a strongly binding ligand class, as well as intermediate and low binding strength ligands. From these they inferred ambient dissolved Cu speciation at sites throughout San Francisco Bay. More importantly they were able to show that the strong ligand class was extremely important in Cu speciation. Concentrations of Cu associated with the weak ligand class, the so-called reactive Cu, were far below levels that might cause toxicity to organisms in the Bay.

An alternative emerging technique is that of DGT (diffusive gradients in thin films) (Zhang and Davison, 2000). In DGT, metals are bound to a resin layer after passing through a well-defined diffusion layer. By controlling the pore size of the diffusive gel layer, conditions are created whereby small (inorganic) species are separated from larger complexes. The small species diffuse freely through all gels but larger fulvic acid and humic acid (organic) complexes diffuse less rapidly. It was thereby possible to quantify the inorganic and organic species separately. Tests show that the quantification agrees with predictions made using the WHAM speciation code. (The model was reliable because the components of the solution were well defined.) The authors then showed that they could determine the proportion of metal in different forms in a stream with high organic content; where modelling would be much more uncertain. They suggested that this direct speciation measurement may be preferable to modelling approaches which require diverse input data that are difficult to determine.

Quebec (Canada). They determined speciation using WHAM V (Tipping *et al.*, 1998). They compared model predictions of free ion concentrations to instrumental determinations of labile metal (similar to cathodic stripping voltammetry described in Box 5.10). The labile metal and free ion concentrations of Zn, Cd and Ni agreed in the two methods; but the model underestimated free ion Cu concentrations by 10 to100 times compared to the labile metal. The uncertainties in the model were probably the result of overestimating implications of organic matter complexation, as described earlier.

The ability to predict speciation, free ion concentrations and/or directly determine reactive metal is growing rapidly. Capabilities are vastly improved over those historically available. Precise predictions in site-specific situations remain subject to at least some uncertainties, however, and should be viewed as such. If speciation is critical to the assessment of metal risks, multiple approaches are suggested to verify model outcomes.

5.4   CONCLUSIONS

It is essential that risk assessors and ecotoxicologists are familiar with the concentrations and forms of metals typical of natural waters (contaminated and uncontaminated). Our understanding of dissolved metals has changed dramatically with the development of ultra-clean protocols for metal determination, speciation models and methods for directly inferring speciation. Many of the dissolved metal concentrations that were once thought to be characteristic of nature were found to be the results of laboratory contamination. In some cases actual metal concentrations in contaminated and uncontaminated water bodies are order(s) of magnitude lower than previously thought.

Understanding the level of contamination requires not only high-quality analytical technique, but appreciation of the natural processes that drive variability. A general characterisation of metal speciation can also be achieved from models and operational protocols for determining reactive metal. Precise determination of speciation is more difficult.

Differences between waters contaminated by human activities and unaffected waters are distinct if not particularly large compared to other environmental media (sediments, biomonitoring organisms). Human activities have generally raised dissolved metal concentrations across the Earth. Examples of contamination can be found from rivers and streams to the open oceans. Examples can also be found where waters are not contaminated, although these are typically remote and are increasingly rare. We now know that each metal covers a relatively predictable range of concentrations in pristine and human-influenced waters. That range provides a valuable perspective for risk assessment.

The differences between contaminated and uncontaminated waters are much less than was once thought. Thus characterising contamination requires some precision in analysis and interpretation. The relatively small range of concentration for most elements suggests that responses of dissolved metals to different human influences are relatively similar across the Earth, and/or that natural processes buffer the variability in dissolved concentrations caused by such influences.

Future management of metal issues will require appreciation of the ranges of metal concentrations that might be expected in nature and balancing those against the ranges of concentrations where adverse ecological effects actually occur. Measures of adverse effects must be as realistic as the analyses of the waters themselves for a valid comparison, however.

Suggested reading

Cutter, G.A. and Bruland, K.W. (1984). The marine biogeochemistry of selenium: a re-evaluation. *Limnology and Oceanography*, **29**, 1179–1192.

Lee, B.-G. and Fisher, N.S. (1994). Effects of sinking and zooplankton grazing on the release of elements from planktonic debris. *Marine Ecology Progress Series*, **110**, 271–281.

Nagorski, S.A., Moore, J.N., McKinnon, T.E. and Smith, D.B. (2003). Scale-dependent temporal variations in stream water geochemistry. *Environmental Science and Technology*, **37**, 859–864.

Ranville, M.A. and Flegal, A.R. (2005). Silver in the North Pacific Ocean. *Geochemistry, Geophysics, Geosystems*, **6**, 1–12.

Sanudo-Wilhelmy, S.A., Tovar-Sanchez, A., Fisher, N.S. and Flegal A.R. (2004). Examining dissolved toxic metals in U.S. estuaries. *Environmental Science and Technology*, **38**, 34A–38A.

6 • Trace metals in suspended particulates and sediments: concentrations and geochemistry

6.1   INTRODUCTION

Prologue. Perspective on metal concentrations and metal chemistry in particulate material is essential to evaluating ecological risks from metals. Sediments represent a concentrated mass of metal in potentially reactive form that can be biologically available, toxic and dispersed widely if not contained. Total metal concentration in sediment is not fully predictive of bioavailability or toxicity, however. Consideration of chemistry can improve assessment of risks.

A fundamental characteristic of all trace metals is their tendency to partition between aqueous (pore water, overlying water) and solid phases (sediment and suspended particulate matter). Understanding metals in the particulate phase is a requisite for any comprehensive evaluation of metal risks or rational metal management. There are several reasons for this (Horowitz, 1991):

- The concentrations of metals in particulate form are higher, by orders of magnitude, than dissolved concentrations in the water column.
- Sediments act, therefore, as a large and concentrated reservoir of potentially reactive metals in the water body.
- Metals continually equilibrate among sediments, pore waters, the water column and the biota, as determined by geochemical conditions in a water body.

Box 6.1 Definitions

bed sediment: a dense mixture of pore water and particles of many possible sizes and compositions. Components of the sediment include several types of inorganic materials, living organisms and detritus.

benthic flux: the movement of dissolved chemical species between the water column and the underlying sediment.

bioturbation: the stirring or mixing of sediment or soil by organisms, especially by burrowing, ingesting/egesting sediment or moving sediment into or out of tubes.

detritus: in ecological terms, detritus is organic debris formed by the decomposition of plants or animals. Detritus is a typical component of sediments and suspended material. Detritus and its accompanying microorganisms are a food source for some organisms.

overlying water: the water body overlying the bed sediment.

pore water: water filling the spaces between grains of sediment.

sediment cores: tubes driven deep into sediment from which progressively deposited layers of sediment are collected. Data from sediment cores help us understand the history of sediment deposition and sediment character in a water body. The core must be from a location where sediments were progressively deposited over time if historic processes are to be interpreted. The date of sediment deposition can be determined from isotope deposition or decay, and/or other markers.

solid phases: the inorganic and organic particulate components of suspended sediment, bed sediment and the associated living material.

- Dissolved concentrations result from a combination of metal inputs from external sources and their equilibration with local particulate material.
- Living organic material and biogenic organic detritus are part of the sediment. This organic component of sediment is the food source at the base of a major food web, and is a source of bioavailable metal.
- Even if inputs of metals cease, sediments preserve a record of historical inputs and remain a source of metal to the water column, organisms and, in some cases, habitats in distant locations.
- If physical and/or chemical conditions change, the form of metal in the sediments will change. What was once a small source of metal input can become a large one (or vice versa). What was once a metal form with limited exchange and bioavailability can become an exchangeable, bioavailable form (Fig. 6.1).

6.2   UNITS: CONCENTRATIONS

Prologue. The concentration of metal on particulate material is expressed as mass (μg) of metal per unit weight of particulate material, typically as μg metal/g particulate dry weight (dw). This concentration contributes to metal exposure, metal exchange and the risks that metals pose to ecological processes. Metal concentrations in suspended material are often expressed as mass of metal in the water column per unit volume, or μg metal/L water. This latter measure is useful for determining fluxes or mass balances, but it is not necessarily indicative of ecological risk.

Risk assessors and managers must be very careful of units when evaluating values for metals associated with particulate material. The concentration of particulate metal suspended in the water column is often reported as μg/L, mg/L or μm/L. Mass of metal per unit volume is useful for determination of fluxes and sources, as described in Chapter 5 (see Fig. 5.1 for an example), but it is the concentration of metal on the particulate material, expressed as μg metal/g particulate dry weight that usually contributes most to exposures and exchange in the environment, and therefore ecological risk. We

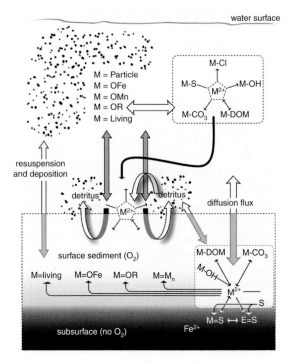

Fig. 6.1 A simplified conceptual model that describes some of the processes that influence the partitioning of metals within the aquatic environment. Particulate metal concentrations are influenced by concentration and speciation of dissolved metals, resuspension, transport and settling of sediment-bound metals, pore water fluxes into and out of the sediments, and burial of surface sediments which can ultimately be trapped in deep anoxic sediment layers, below the influence of disturbances.

will term the latter 'particulate metal concentration' or 'concentration on particulate material'.

6.3   METAL CONCENTRATIONS IN SUSPENDED PARTICULATE MATERIAL AND SEDIMENTS

Prologue. One consistent outcome of metal partitioning between dissolved and particulate phases is that the concentration of metal associated with particulate materials is much higher than the concentration in solution. It is important to understand the range of concentrations that we might expect for each metal in sediments, from natural to contaminated circumstances. But it is also critical to recognise factors that can confound the interpretation of metal concentration in sediments. For each confounding factor there are approaches to minimise those influences.

Metal can be released into an aquatic ecosystem in either dissolved or particulate forms. Whatever the original form, metals will repartition within the water body between various ligands in solution (speciation: Chapter 5) and ligands associated with particulate material (Fig. 6.1). The rate of partitioning is usually rapid (on timescales of ecological relevance) although it can be influenced by the original form of the metal. The outcome of the repartitioning will be determined by the metal and the geochemistry of the water body.

At pH near neutral (i.e. most natural water bodies), the partitioning of metals between solution and solid phases strongly favours particulate material, with concentrations in particles exceeding concentrations in solution by orders of magnitude (Fig. 6.2). As a result sediments and suspended

particulate material form the largest repository of metals in a water body.

From the broadest perspective, sediment metal concentrations can be a general indicator of the environmental contamination gradient to which all biological resources are exposed. But several factors must be considered when interpreting metal concentrations in sediments.

- analytical and digestion procedures;
- natural geological inputs;
- granulometric conditions (grain size or particle size in the sample);
- anthropogenic inputs;
- hydrological and hydrodynamic conditions.

6.3.1   Decomposition and analysis

Prologue. Methodologies for analyses of metals in sediments are well developed. Concentrations are usually higher than will be detectably influenced by minor laboratory contamination, but reproducible recoveries are more of a challenge. Total decomposition of sediments guarantees the most consistent and unambiguous recovery. Recoveries of metals from near total decomposition are biased slightly downward, but are nonetheless reproducible and useful for environmental surveillance. Analyses of suspended particulate material are extremely valuable, but have some of the same challenges as analysis of dissolved metal.

Analytical problems are much more manageable for metals in sediments than for dissolved metals. In particular, bias from laboratory contamination is rare, because concentrations in sediments are high. Precision or reproducibility should be verified by analysis of reference sediments, which are widely available. The method by which metals are removed from the sediment can influence concentration. Total decomposition methods (e.g. dissolution in concentrated nitric, perchloric and hydrofluoric acid) dissolve the entire matrix and should yield 100% recovery of all elements in the reference material.

'Near total' decomposition is also quite rigorous, if carried out over heat with vigorous refluxing

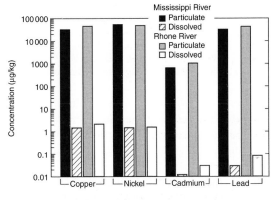

Fig. 6.2 A comparison of Cu, Ni, Cd and Pb concentrations on particulate matter and in solution in the Mississippi River and the Rhone River. Particulate concentrations are presented as μg metal/kg dry mass of particulate. kg is used to make the numerator and the denominator in comparable units. The strong preference of metals (by orders of magnitude) for association with particulate material is indicated. Typically (and elsewhere in this book) metal concentrations in sediments will be presented as μg/g dry weight particulate mass. (Data from Elbaz-Poulichet *et al.*, 1996.)

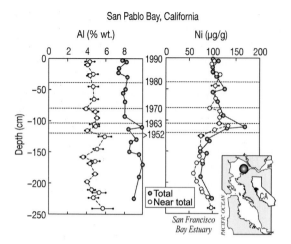

Fig. 6.3 A comparison of 'total' (hydrofluoric acid) and 'near total' (concentrated nitric acid) concentrations of Ni and Al in sediments with depth in the sediment column. The data are from a sediment core in San Francisco Bay. 'Near total' decomposition removes considerably less than 100% of the Al in the sediment, but the concentrations in the 'total' and 'near total' digests are correlated. 'Near total' decomposition removes nearly all the Ni; the downward bias is minor. (Data after Hornberger *et al.*, 2000.)

(Axtmann and Luoma, 1991) or in a combustion bomb. This approach removes all possible exchangeable metal in a sediment. Recoveries are consistently biased downward but are 80% to 95% of total metals for most metals (Fig. 6.3). Exceptions include Cr and V, where recoveries can be <50%. 'Near total' decomposition also leaves about 50% of the recalcitrant portions of the sediment undigested (e.g. Al and Fe) (Fig. 6.3). Reproducibility should be high for 'near total' decomposition, as should be the precision of the recovery from different materials. If the decompositions are done carefully, metal concentrations from the near total and the total decomposition approaches correlate closely (Fig. 6.3). Both methods are widely used. Near total analysis is practical because special laboratory facilities are less necessary and the acids used are less dangerous. But 'total' concentrations are the more exacting.

Partial extractions are also used to decompose selected fractions of metals in the sediment (e.g. 2 hour extraction with 1% hydrochloric acid). Recoveries are always biased considerably

downward in partial extractions. Rapid turn around of samples is an attraction of using this approach for surveillance of sediment contamination. For example, many samples can be quickly screened for anthropogenic contamination (if an uncontaminated comparison is available), from which a subset can be chosen for more detailed analysis (Hornberger *et al.*, 2000). However, inconsistencies result if the efficiency of partial extractions varies with environmental circumstances. Partial extractions can also be used operational screen for particular metal forms (see later discussion).

Data from suspended sediments can be extremely valuable, as this is a major food source for filter-feeding benthic and pelagic animals. But collection and analysis are more of a challenge, compared to bed sediments. Filtration of suspended particulate material typically gathers only a few milligrams of material. Analyses require microdigestion techniques and ultra-clean handling and analysis procedures. Methods are available to collect larger masses of suspended material. Continuous centrifugation is one but it is logistically challenging. Use of traps or other collection devices is a pragmatic approach, but raises uncertainties about unbiased sampling of all particle sizes.

6.3.2   Geological inputs: what is the natural background or regional baseline?

Prologue. A convincing interpretation of metal contamination in sediments requires defining natural background concentrations (concentrations deriving solely from local geology). Background must be subtracted from total metal concentration to determine the anthropogenic contribution. Sediment cores (analysing sediments deposited before human disturbances began) and/or analyses from uncontaminated locations of similar geology can be useful in identifying natural background concentrations. Use of global average geology is common in the literature. But local geology can differ widely from the global average, biasing interpretations either negatively or positively.

Mineral components of all natural particulates contain trace metals as part of their fundamental

Table 6.1 *Metal concentrations in major components of the Earth's crust, in descending order of metal abundance (as reported by Salomons and Förstner, 1984), compared to natural background of selected metals from San Francisco Bay. The San Francisco Bay data are from layers of sediment deposited before anthropogenic activity in the Bay area and from a nearby estuary (Tomales Bay) with similar geology but no industrial or urban development (data from Hornberger* et al., *2000). Units are μg metal/g sediment dw except where expressed otherwise*

Metal	Mean crust	Average shale	Sandstone	Limestone	San Francisco Bay
Aluminum (Al)	8.2%	8.0%	4.3%	0.7%	6.4–8.2%
Iron (Fe)	4.1%	4.7%	4.3%	1.7%	3.5–4.5%
Vanadium (V)	160	130	20	45	75–117
Manganese (Mn)	950	850	460	620	324–639
Chromium (Cr)	100	90	35	11	113–150
Nickel (Ni)	80	68	9	7	73–200
Zinc (Zn)	75	95	30	20	60–75
Copper (Cu)	50	45	30	5.1	16–25*
Cobalt (Co)	20	19	0.3	0.1	
Lead (Pb)	14	20	10	5.7	5.2 ± 0.7
Tin (Sn)	2.2	6.0	0.5	0.5	
Molybdenum (Mo)	1.5	2.6	0.2	0.2	
Arsenic (As)	1.5	13	1.0	1.0	
Antimony (Sb)	0.2	1.5	0.05	0.3	
Cadmium (Cd)	0.11	0.22	0.05	0.03	
Silver (Ag)	0.07	0.07	0.25	0.12	0.09 ± 0.02
Mercury (Hg)	0.05	0.18	0.29	0.16	0.06 ± 0.01

structure. But metal concentrations differ among the different components of the Earth's crust. Table 6.1 shows average concentrations among shale, sandstone and limestone (Salomons and Förstner, 1984). For every element, all three types of rock are different. The range between the highest and lowest concentrations in different rock types is threefold for metals like Pb, Ag and Hg, and greater than 10-fold for most metals. Carbonate-containing rock is not included in Table 6.1, but it is very poor in all trace metals. When carbonate is present in high concentrations, it should be treated as a metal diluent in sediment (see methods proposed by Horowitz, 1991).

Different geologies offer different mixes of these components. For example, average sedimentary rock is 74% shale, 15% limestone and 11% sandstone (Salomons and Förstner, 1984). But the most abundant sedimentary rock is argillaceous rock (the source of deep sea clays) which more closely resembles shale than it does the average mixture.

Determining natural background concentrations of metals is an important challenge because of the complexity and heterogeneity of natural inputs. A traditional approach is to use 'average geology', or, more commonly, average shale as an indicator of nature's contribution to sediment metal concentrations. Table 6.1 also compares average geological concentrations to more direct determinations of natural background concentrations in San Francisco Bay sediments. Concentrations of Fe and Al in the undisturbed Bay sediments were comparable to average shale. But average shale concentrations of Mn, Zn, Cu, Pb and Hg are higher than the range of natural background concentrations for these elements from the Bay. Use of the shale approach would overestimate natural background contributions to concentrations and thereby underestimate human contributions of these metals. On the other hand, transposing estimates of average geology onto San Francisco Bay would consistently underestimate natural background concentrations of Cr and Ni concentrations. This is because the watershed of the Bay contains substantial deposits of Ni- and Cr-rich rocks (e.g. serpentine).

Useful methods for determining site-specific natural background concentrations in sediments are:

- Use layers of sediment deposited before human disturbance, from sediment cores. This is the most direct approach if it is possible to obtain cores from sedimentary deposits in lakes, reservoirs, bays, estuaries or coastal waters. The cores must be deep enough to sample the pre-anthropogenic influence of the local geological inputs. Precise dating of cores using isotopes and other markers is necessary.
- Collect sediment data from locations within the same region where average geology is similar, and sediments are undisturbed by metal inputs from mining, industrialisation or urbanisation. Assuring a similar geology in the watersheds is important; and finding completely undisturbed watersheds is an increasing challenge.
- Mineralised regions that have been disturbed by mining or other human activities provide an especially difficult challenge in determining natural contributions (e.g. what is the natural versus human contribution to contamination in a naturally mineralised area?). In this case, neighbouring undisturbed watersheds are probably not similar in geology, but dispersion trains can be used to determine the area affected by the mineralised zone and thereby assess its influence on any specific location (Box 6.2).
- A more approximate approach is to establish a modern regional baseline from widespread sampling (see Section 6.3.4.3).

6.3.3   Granulometric (particle size) biases

Prologue. Grain size, or particle size, in a sediment will affect metal concentrations and confound comparisons of metal concentrations among places and times. Grain size biases should be removed if the goal of the study is to minimise ambiguities in interpreting degree of contamination, spatial distributions, trends or fluctuations. Elimination of granulometric biases

can also help avoid falsely concluding a lack of anthropogenic metal input in situations where contamination is diluted by a high proportion of sand. Contamination in fine-grained sediments probably also has the greatest biological implications.

The relationship between metal input and metal concentration in sediments is simple only if surface area per unit mass is not highly variable, and compositional differences in the sediments do not cause great differences in binding site numbers among different sediments. Increases in metal concentrations are directly indicative of anthropogenic metal inputs among sediments with similar granulometries and, to a lesser extent, from similar geologies and geochemistries. Across wide gradients in grain size, especially, the correlation between anthropogenic metal input and sediment metal concentration loses its robustness.

Grain size is a fundamental physical characteristic of sediments. Sediments are usually a mix of clay (<2 μm in diameter), silt (2 to 60 μm diameter) and/or sand (60 to 2000 μm diameter). The proportion of each size fraction depends upon inputs and differences in physical processes, including flow rates in rivers, currents in estuaries and the sea, wind disturbances, wave action, etc. Average grain size is a measure of the mix of the different particle sizes.

Grain size exercises an inordinate control over metal concentrations in sediment because metals primarily bind to the surfaces of particles. Finer-grained particles have more surface area per unit mass. The number of metal binding sites increases with more surface area. Numerous studies show that finer fractions within sediments and finer-textured sediments consistently have higher concentrations of iron oxides, manganese oxides, organic material and metals than do coarse materials. There are exceptions to this principle (Luoma, 1990), but those exceptions do not negate the well-documented observation that variability in grain size in a water body will bias comparisons of metal concentrations unless accounted for. The most important effect will be an overestimation of the frequency of low concentrations, because samples that are downwardly biased by high amounts of sand are included in the comparison (Fig. 6.4).

Box 6.2 What is the best measure of natural background from a mineralised deposit?

Controversies often arise over the causes of metal enrichment around mining sites, because mineral deposits are naturally enriched in metals. In fact, prospectors sometimes use 'natural enrichment' of metals in water, plants and sediment to find ore bodies. As the ore physically weathers, particles are eroded and enrich both the dissolved and particulate material in the runoff to stream systems. On the other hand, prospectors have long known that water concentrations decrease to background concentrations within hundreds of metres of unmined ore bodies, however; perhaps extending a few kilometres if the ore body is large (Helgen and Moore, 1996). This is partly because weathering and processes that erode or expose the deposit reach an equilibrium with processes such as dilution by unenriched tributaries, or reaction with rock adjacent to the ore body. Weathering eventually also reduces concentrations in the surficial contaminated soils over the thousands of years that unexploited deposits are exposed.

The size of the ore body and the concentration of a particular metal in the soils covering the ore body are important determinants of the dispersal of the particulate contamination away from ore bodies. As the area of watershed cumulatively increases downstream from the mineralised zone, concentration is diluted by watershed erosion and sediments from incoming streams. The degree of dilution is a function of the surface of the watershed, creating a predictable 'dispersion train' (Helgen and Moore, 1996). Helgen and Moore (1996) showed that dispersion trains in rivers or streams draining mined and unmined sites could be forecast from the area of contaminated surface soil, the concentration of contaminant over that area and the cumulative area of the watershed downstream from the deposit or mine.

Mining activities may not greatly change the concentration of metals at the site of the mineralised deposit. Their greatest effects, in the absence of careful management, are to expose the ore body to oxygen and spread the ore over a large area (increasing the area of contaminated surface). The result is mobilisation and dispersion of the contamination over a much wider area.

When they considered dispersion from unexploited mineral deposits, Helgen and Moore (1996) found trains of 0.4 to 2.4 times the area of the ore body (a few kilometres or river miles). Human activities vastly increased the footprint of contamination. The dispersion trains of uncontained mine wastes in the Clark Fork and Coeur d'Alene Rivers (USA), as examples, extend for hundreds of kilometres. Thus for mineralised deposits the question of 'background' for any given location can be solved by determining the land area affected by the original deposit and comparing that to the area of land affected by the mining. The dispersion train is what differs between natural and mined rivers. Estimate of the dispersion train allows determination of natural background concentrations downstream from the mineral deposit. Similar methods and logic might be applicable to other types of sources.

The most important benefits of establishing comparability among sediments of different grain size are reduction of variability and avoiding mis-diagnosis of a sediment as uncontaminated, when the contamination is really undetectable because it is diluted by some characteristic of the sediment. In essence, eliminating grain size effects is a scientifically valid precautionary tool. The existence of contamination, its trends and its spatial distribution can be irresolvable, or misinterpreted, if particle size biases are not removed from sediment data.

Knowledge of the particulate metal concentration as it exists *in situ* (bulk concentration) is valuable if total mass of metal in a given weight of sediment is of interest; for example for determining fluxes, mass balances or metal transport. Otherwise it can be misleading.

6.3.4   Methods to improve comparability among data

Prologue. Grain size and other biases add uncertainty to many of the metal concentration data from sediments in the published literature, in risk assessments and in monitoring programmes (e.g. USEPA, 1999b). Separation of grain sizes, normalisations or regression can improve comparability, avoid confusion and minimise contentious debate.

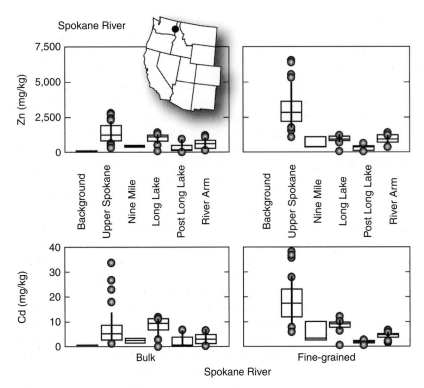

Fig. 6.4 A comparison of Cd and Zn concentrations in bulk (whole) sediment samples and sieved sediment samples (<63 μm) along the contamination gradient in the Spokane River, Washington (USA) (after Grosbois *et al.*, 2002). The dots represent the data distribution, the vertical bar represents the range, the box the standard deviation and the horizontal line the mean in this 'box and whisker' plot. Mean concentrations in the fine-grained sediments are significantly elevated closest to the source of contamination. The differences among stations are not statistically significant in the bulk sediments, primarily because some low concentrations are seen at the most contaminated site (caused by sand in the sediment).

Methods are available to improve comparability of metal concentrations in sediments that are heterogeneous in character (especially if grain size is variable). They include chemical extraction, physical separation of fine-grained materials, normalisation to particle size-sensitive natural components of sediments and statistical techniques. Each has advantages and disadvantages; but most offset the biases that can distort interpretations if sediments are variable in grain size. Careful studies suggest that a combination of methods might provide the most unambiguous interpretation of metal concentrations, relative to anthropogenic inputs (Rule, 1986).

6.3.4.1   Chemical extraction of anthropogenic metal

Prologue. Chemical extractions can be useful in characterising tendencies toward contamination in a sediment, but are not selective for 'anthropogenic metal'.

Chester and Hughes (1967) and Malo (1977) proposed that a combination of an acid and a reductant (acetic acid/hydroxylamine) might directly extract the adsorbed component on sediments, leaving behind the metals within the structure of the particles. The assumption of this approach is that when new metal is added by anthropogenic input, it is all in the absorbed phase; and that the natural background metal is primarily within the structure. More recent work shows that both assumptions are flawed. When metals are added to a particle, some proportion moves into the interior, gradually over time. So no surface extraction removes the entire anthropogenic component from a particle. In addition, adsorption to surface sites is probably continually renewed by metal cycling at the micro-scale in the absence of anthropogenic contamination, especially in estuaries and at complex redox interfaces in sediments. Thus uncontaminated oxidised sediments contain a substantial fraction of adsorbed metal that is removed by any low pH extractant.

6.3.4.2 Separating fine-grained sediment for analysis

Prologue. Analysis of only the <63 µm sediment fraction obtained from wet sieving is consistently recommended by experienced researchers as the most unambiguous way to reduce the biases introduced by the diluting effect of variable amounts of sands in different samples.

Granulometric techniques involve physical separation of materials of different grain size followed by analysis of one or more of the fractions. Several methods are available to separate or quantify the different grain size fractions in a sediment. These include the traditional dry sieving approach of the soil sciences, wet sieving to isolate a selected fraction for analysis, settling techniques, electronic techniques that count the number of particles that pass a certain size orifice (Horowitz, 1991) or geochemical techniques that directly determine surface area.

Basic characteristics of grain size association in a sediment can be determined from analysing metal concentrations in a variety of separated grain sizes (Brook and Moore, 1988). There are instances where metal concentrations do not increase in the smaller grain size fractions (Moore et al., 1989). But these are rare and usually associated with precipitation of iron or manganese oxides onto a very sandy sediment. Typically, metal concentrations increase strongly in the smallest grain sizes. The greatest disadvantage of determining the full distribution in every sample is the time- and labour-consuming effort. It is not practical for large numbers of samples. Concentrations of metal within each fraction are also not highly variable within a site, although the proportion of the different size fractions may vary among sediment samples from different microhabitats. Determination of the metal concentration on each particle size fraction from a few analyses will generally suffice for a water body. Then distributions can be estimated from grain size analysis alone.

If the goal is to establish comparability among a number of determinations of metal concentrations, then a widely recommended and pragmatic approach is to isolate one fraction for analysis; usually the silt-clay fraction of the sediments. This is typically carried out with a 63 µm non-metallic (e.g. Nytex) sieve (Luoma and Bryan, 1978; Salomons and Förstner, 1984). Then only the silt-clay fraction is digested and analysed.

Intuitively, it seems that the information obtained from a wet sieved sediment sample is incomplete, because it does not included all fractions present in the environment (Szefer, 2002). Furthermore, the fraction separated for analysis has an elevated metal level compared to the total. Nevertheless, its advantages outweigh its disadvantages. It increases sensitivity in detecting contamination (i.e. contamination anomalies are easily detected). The absolute concentrations in the <63 µm fraction are in unambiguous units (µg metal/g). Sieving is simple and thus suitable for large numbers of samples. Methodologies are available to conduct separations in the laboratory or in situ. Large data sets are available for metal concentrations in sieved sediments (Luoma and Bryan, 1981; Salomons and Förstner, 1984; Axtmann and Luoma, 1991; Grosbois et al., 2002). For example, the most extensive surveys of British estuaries (Bryan et al., 1980) and the US Geological Survey's National Water Quality Assessment, USA, use this approach.

An underlying source of controversy over analysis of a selected grain size is whether the concentration in situ or the concentration in the fine fraction is most influential in ecological risk. Metals in fine fractions correlate better than do bulk measurements with metal concentrations in biomonitors (see Chapter 8), especially if the sedimentary environment is heterogeneous (e.g. Cain et al., 1992). If the sedimentary environment is not heterogeneous, then the bias introduced by sieving is small anyway.

6.3.4.3 Normalisation

Prologue. Improved interpretations are obtained by normalising metal concentrations in sediments to percentage of a given grain size, or Al, Fe or organic carbon concentrations. Percentage grain size normalisations are the most imprecise. Normalisations that reflect both surface area and heterogeneous compositional influences (Fe and organic carbon) will confound rather than improve interpretation under

Box 6.3 Hypothetical normalisation and a real world example

Let us assume that we determine 100 µg/g Cu dw in fine-grained sediment #1 and 10 µg/g Cu dw in sandy sediment #2. In the absence of granulometric data we might conclude that sediment #1 was contaminated with Cu (if we assume a background like San Francisco Bay; Table 6.1) but sediment #2 was not. However, if we determined Al, an indicator of the abundance of clays, we might find sediment #1 had 5% Al and sediment #2 only 0.5% Al. If we normalise the Cu concentrations by the Al concentrations in the two sediments we obtain:

Sediment #1: Cu/Al = 100/5 = 20
Sediment #2: Cu/Al = 10/0.5 = 20.

The conclusion from the normalisation is that the sediments are receiving an equal input of Cu, but that the contamination in sediment #2 is being diluted by sand (few clays). The fraction of fine material in sediment #2 would probably show a similar Cu concentration to sediment #1, if that fraction was isolated. Organisms ingesting sediments at site #2, or living in the interstices, are probably exposed to as much Cu as organisms at site #1; but a bulk sediment analysis prevents detection of that exposure because of the presence of coarse-grained material.

Feng et al. (1999) noted that Fe and Al in the suspended sediments of the Hudson Estuary (New York, USA) co-varied widely among samples, with no clear pattern in the distribution of concentrations. They normalised metal concentrations to Fe concentrations, and patterns began to emerge (aided by correlations with natural tracers of upriver and lower estuary processes). Patterns for Ag/Fe indicated that urban sources from down the estuary, like waste water from the New York metropolitan area, were predominant. This is consistent with the known proclivity of Ag to occur in sewage. Patterns in Pb/Fe, Cu/Fe and Zn/Fe were more complex suggesting multiple sources, including the river itself. Upriver sources appeared to predominate for Cd, suggesting that an extreme hotspot of historical Cd contamination at an upriver munitions plant was still contributing contamination to the river. In the absence of careful characterisation of particle size effects and hydrology, the particulate metal concentrations in the estuary would not have been interpreted correctly.

some conditions. Normalisations to Al are the most common choice. Enrichment factors are calculated to simplify interpretation of normalisations. The most attractive approach, however, is regression of concentration against a conservative constituent (e.g. Al), if a large set of data is available.

Normalisation is brought about by dividing a metal concentration by the concentration of a property (in this case a chemical constituent) that is indicative of a factor that might consistently bias the metal concentration. Box 6.3 describes how a hypothetical normalisation to the concentrations of clay in the sediment helps correct for granulometric differences, and thereby improves comparability between metal concentrations in two sediments.

Common choices for normalisation are:

- percentage of a selected grain size fraction in the sediment;
- a constituent inert to binding metals (a diluent);
- a metal-reactive component of the sediment that is either surface area dependent or indicative of a surface area characteristic;
- a conservative constituent of sediment.

The simplest normalisation technique is constructing a ratio of metal concentrations in sediments to the percentage particles in the bulk sample less than a given particle size (Birch, 2003). This approach is effective in helping identify sources of contamination in some instances (Birch, 2003). In others it is imprecise (e.g. Moore et al., 1989). A basic problem is that grain size, as determined by nearly all techniques, is not a simple, constant characteristic in natural waters (e.g. Walling and Morehead, 1989). Sediments from natural water bodies are a complex of different substances, coated and aggregated in different ways (e.g. NRC, 2003). As a result grain size data can be difficult to reproduce, especially when the sediment is dominated by fine-grained material. The method for characterising grain size also influences the outcome. Physical sieving or settling is a pragmatic approach but results tend to be variable. Instrumentation is available for determining the proportion of particles that pass a certain size orifice, although a narrow range of orifice sizes limits the usefulness of this technique on natural sediments. Geochemical techniques are also available to determine surface area directly. These methods work best on simple surfaces or sands. They are of limited application in the complex fine-grained sediments from natural waters.

In some instances metal concentrations in sediments can be heterogeneously diluted by an inert component of the sediment. Comparisons among estuaries, for example, might encounter water bodies with more or less carbonate in the sediments. If the differences in carbonate concentrations are large, the metals in the sediments of the different estuaries are probably best compared after use of a normalisation to eliminate that bias (see Horowitz (1991) for methods).

Salomons and Förstner (1984) recommended normalisation of the metal concentration in each sample to a conservative constituent in the sediment. They suggested that, when combined with sieving, normalisation to an element like Al could provide powerful diagnostic data. The effectiveness of Al normalisation stems from its unambiguous relationship to the abundance of clays in the sediment. Clays consistently have a large surface area per unit mass and high concentrations of aluminosilicate minerals. If Al concentrations are variable among sediment samples, grain size has probably altered metal concentration data. Normalising to Al can remove those biases.

Normalisation to Al is subject to distorted interpretations in extremely sandy sediments, as a result of dividing by a number very close to zero (Feng et al., 1999), or in the case of any other geology where clays are rare (Loring, 1991). Otherwise this calculation is a useful way of accounting for biases from both historical background and the physical character of the sediment.

Metals in sediment are also sometimes normalised to iron or to organic carbon concentration (Feng et al., 1999; Schiff and Weisberg, 1999). Both coat particles so their concentrations can co-vary with particle surface area. Inputs of Fe are also affected by redox processes, however (see later discussion). Thus Fe is a consistent correction factor only if redox processes are relatively constant across all sampling locations (see later discussion). Total organic carbon (TOC) corrections have the same caveat. If inputs or the predominant nature of the organic materials vary, each can distort use of that normalisation.

The greatest disadvantage of normalisation is that the product is not a concentration; it is unitless ratio. Ratios are also more difficult to grasp conceptually because they can be affected by either the numerator or denominator (Horowitz, 1991). It is also difficult to evaluate risk from ratios.

To improve interpretability, anthropogenic input can be quantified by calculating an enrichment factor, relative to natural background (Rule, 1986). The enrichment factor (EF) for Al-normalised data, for example, is determined by

$$EF = (M_x/Al_x)/(M_c/Al_c)$$

where M_x is the metal concentration in sample x, Al_x is the Al concentration in sample x, M_c is the metal concentration in the undisturbed or crustal sediment (control), and Al_c is the Al concentration in the control or crustal sediment.

A variant of normalisation is to employ a regression of metal concentration against an indicator of a bias. Regression against a surface active component or an indicator of surface area, can establish the grain size-dependent baseline in the data set, for example Al for grain size. This is perhaps the most powerful tool to aid interpretation of metal concentrations in sediments.

The approach is to collect or identify a broad set of data, including a subset of the sediment data thought to be minimally affected by contamination. The baseline of the regression is then identified (the lowest quantile of the concentration data). The lowest quantile represents the lowest values of y at any given x value on the plot. The word quantile means the proportion of lowest values (e.g. the lowest 10% of Cu values at the value of 5% Al in Fig. 6.5). The lowest quantile represents the baseline in the data set (e.g. the

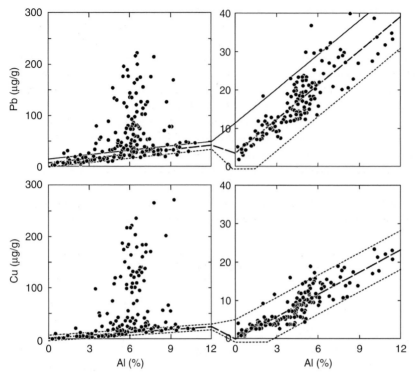

Fig. 6.5 Concentrations of Cu and Pb in coastal zone sediments (USA) regressed against Al concentrations. Sediments were collected by the National Status and Trends monitoring program. Reference conditions are established by a regression from the bottom quantile; the stations not expected to be enriched by human activities. The frequency of contaminated sites is evidenced by positive deviations from reference relationship. (After Hansen et al., 1996.)

regional baseline). A regression equation with confidence intervals can be determined for the chosen quantile (Cade and Noon, 2003) to quantify the baseline. The degree of contamination in each sample from the area can be characterised by calculating the predicted baseline Cu concentration at the Al concentration of a sample (Fig. 6.5) then subtracting the actual value from the predicted baseline.

Multivariate analysis can also be used to group metal–grain size relationships in sediments. For example, cluster analysis was combined with Al normalisation to identify two zones of metal enrichment in the St Lawrence River estuary (Coakley and Poulton, 1993); one influenced by circulation patterns and the other by local geology. R-mode factor analysis showed the influence of grain size on the geographical distribution of metal concentrations in Delaware Bay (USA) (e.g. Bopp and Biggs, 1981). Multivariate approaches are especially useful for demonstrating the concept

that particle size influences metal distributions in a water body and can be used to separate groups of sediments for further analysis.

6.4   SPATIAL DISTRIBUTION OF SEDIMENT CONTAMINATION

Prologue. Metal concentrations in sediments can vary over four orders of magnitude (10 000-fold) within and among water bodies.

Anthropogenic inputs are a primary cause of the variability of the metal concentrations in sediments, although natural factors can influence interpretations. Concentrations are typically log normally distributed. In areas around human activities the frequency of contamination is surprisingly consistent. That is, moderate to strong contamination is widespread, typically following human activities. Extreme contamination is more rare.

6.4.1 Range and frequency of contamination in sediments

Prologue. Total concentrations of metals in sediments can range widely within and among water bodies; partly driven by the confounding influences described above and partly driven by differences in anthropogenic contamination.

Where intense human activities have a long history, serious metal contamination is possible.

Table 6.2 shows concentrations of trace metals in British estuaries, some of which have been contaminated by hundreds of years of mining activity or by industrial development (Bryan *et al.*, 1985). Through the 1980s sediment contamination of this level and frequency was probably common in the industrialised world.

Figure 6.6 compares the range of total concentrations of Cu, Pb and Zn from bed sediments from several water bodies and from crustal materials. The higher end of the data range reflects

Table 6.2 *Metal concentrations in sediments (μg/g, Fe%, dw) in British estuaries, some affected by historical mining in their catchments or industrial development along the estuary. Concentrations were typically 'near totals' measured in a nitric acid digest of the <100 μm fraction*

Estuary	Ag	Cd	Co	Cr	Cu	Fe	Mn	Ni	Pb	Zn	As	Hg	Sn
Cornwall and Devon													
Torridge (upper)	0.3	0.8	14	31	27	2.88	725	29	47	145	11	0.50	*37*
Camel (upper)	0.4	0.5	12	33	80	2.76	552	28	64	215	55	0.16	741
Gannel (upper)	2.9	**3.0**	40	29	217	3.32	**1160**	**49**	**2175**	1215	233	0.09	305
Hayle (upper)	1.3	1.0	**28**	36	782	5.15	742	32	218	942	550	*0.06*	**1750**
Helford (upper)	0.6	1.4	8	37	252	2.85	296	32	63	616	23	0.68	1070
Restronguet Creek (upper)	**4.1**	1.2	22	37	**2540**	**5.76**	559	32	290	**3515**	**2520**	0.22	1730
Fal (mid)	0.4	0.4	*3*	*15*	129	*1.21*	*116*	*9*	48	252	56	0.20	125
Fowey (mid)	0.4	0.7	10	29	122	2.79	393	27	74	189	47	0.5	424
West Looe (upper)	1.3	*0.3*	12	25	57	2.12	452	36	256	145	12	0.30	54
East Looe (upper)	1.9	0.4	12	25	36	2.23	464	36	88	151	16	0.13	67
Tamar (upper)	0.9	1.5	23	44	305	2.81	758	**49**	156	392	85	0.90	101
Tavy (upper)	0.8	0.6	19	48	290	3.75	660	42	176	339	131	0.84	151
Avon (mid)	*0.1*	0.3	10	37	*19*	1.94	417	28	*39*	98	13	0.12	28
Dart (upper)	0.3	0.8	10	42	38	3.19	538	27	73	145	24	1.0	58
Teign (upper)	1.0	1.8	18	35	68	2.16	777	30	382	375	74	0.36	176
Southern Bristol Channel													
Avon (upper)	0.4	0.9	16	45	37	3.15	620	37	101	291	9.9	0.57	61
Parrett (mid)	0.3	0.5	16	88	37	2.46	791	35	101	281	—	—	—
South Wales													
Wye (mid)	0.4	1.0	15	43	66	3.20	594	35	96	283	*9.1*	0.48	53
Neath (upper)	0.7	2.1	18	39	89	3.11	524	40	98	327	12	**1.2**	54
Loughor (mid)	0.2	1.1	13	**799**	47	3.81	631	31	77	220	22	0.13	320
Taf (upper)	0.2	0.8	10	32	19	1.73	432	21	43	*118*	14	0.20	50
West Cleddau (upper)	0.4	0.6	13	34	27	2.35	378	28	56	165	12	0.15	40

Note: Highest concentrations are in **bold**, lowest in *italics*.

Source: After Bryan *et al.* (1985).

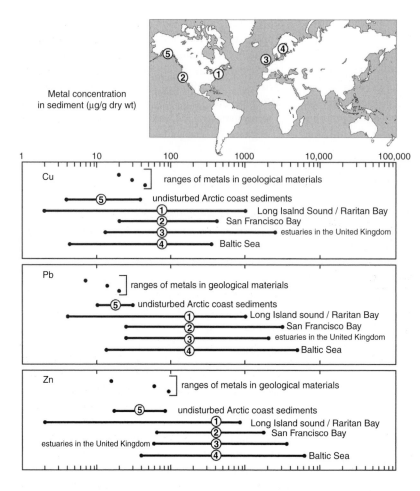

Fig. 6.6 The range of Cu, Pb and Zn concentrations in sediments from an undisturbed Arctic coastal environment and four coastal or estuarine environments surrounded by human activities. Data from San Francisco Bay and UK estuaries are from <63 μm sediments. As a result concentrations biased downward by sands (the lowest concentrations) are absent at those locations. The ranges of concentrations in geological materials are also shown. (After Luoma, 1990, where sources are cited; Baltic data from Szefer, 2002.)

anthropogenic contamination in Long Island Sound/Raritan Bay (New York, USA), San Francisco Bay (California, USA), the Baltic Sea, and 23 smaller estuaries and coastal sites from Great Britain. The data range from individual water bodies can vary by two to four orders of magnitude. A 100- to 10 000-fold difference exists between lowest and highest value for any metal in a water body (Fig. 6.6). This is much greater variability than is typically observed for dissolved metal concentrations, with some exceptions (e.g. Ag). It is instructive that the data range in San Francisco Bay and the UK estuaries (Fig. 6.6) was smaller than in other systems, because analyses were from sediments sieved to <63 μm. Sieving made anomalous low values less influential.

Continent-scale monitoring programmes show that enriched concentrations of metals in sediments occur most frequently near the centres of human activities (Daskalakis and O'Connor, 1995) (Box 6.4). Contamination in sediments shows a relatively consistent frequency distribution in regions affected by widespread human activities. For example, Table 6.3 shows the frequency of different levels of Cu contamination among 22 estuarine sites from the UK and 50 sites from the Baltic. Only 10% to 16% of sites were uncontaminated (<25 μg/g Cu). Human activities clearly have a widespread influence in both regions. From 45% to 62% of sites had less than 50 μg/g Cu. Easily detectable contamination was found at 35% to 36% of sites (50 to 200 μg/g Cu). This is the range of

Schiff and Weisberg (1999) found distributions of contamination in the Southern California Bight (USA) that were similar to that in other coastal water bodies surrounded by human activities. They used Fe normalisation to improve comparability among data. Baseline relationships between Fe and eight trace metals were established from non-impacted sites distant from known point and non-point sources of pollution. Over half of 248 Southern California sites were enriched in at least one trace metal (i.e. deviated from the lower quartile of metal concentrations). All trace metals investigated showed some level of enrichment: Ag, Cd and Cr showed the greatest sediment enrichment, while As and Ni showed the least enrichment. Degrees of enrichment were unevenly distributed. Trace metal contamination was most extensive in Santa Monica Bay, nearest the largest number of people and the receiving waters for the wastes of all Los Angeles. There, 80% of the sites were contaminated by more than three metals.

Table 6.3 *The frequency of Cu contamination (µg/g dw) detected at different levels in extensive surveys of estuaries in the United Kingdom (Bryan et al., 1980) and locations in the Baltic Sea (Szefer, 2002). Human activities have a region-wide influence on contamination in both locations, but extreme contamination is rare (>200 µg/g Cu). The greatest challenge for risk assessors is understanding the risks of the widespread (38% to 40% of sites) contamination in the moderate to highly contaminated range (50 to 200 µg/g Cu)*

Cu concentration range in sediment (µg/g dw)	Percentage of sites in UK estuaries (%)	Percentage of sites in the Baltic Sea (%)
<25	10	16
25–50	35	46
50–100	25	30
100–200	10	6
200–800	5	2
800–2500	5	0

concentrations for which risks are the most difficult to evaluate, and it encompasses a relatively large number of locations. Only 10% of sites were contaminated above 200 µg/g Cu in the UK waters and only one site in the Baltic. Extreme metal contamination, where the probability of ecological risk is relatively unambiguous, was thus rare in these surveys.

6.4.2   Characteristics of distributions

Prologue. Because of the tendency of trace metals to accumulate in sediments, contamination from each source tends to be localised in a hotspot near the input, then dispersed regionally in lower concentrations.

Severe metal contamination typically occurs in hotspots as a result of the tendency of metals to concentrate in sediments near a source of input. Away from the input, metals have been removed from solution, particulate concentrations are diluted by uncontaminated particles and/or the contaminated sediment is redistributed (Fig. 6.7). The area of the hotspot depends upon the load from the source, local hydrology or hydrodynamics and/or local sediment transport. The frequency of hotspots reflects the nature and extent of human activities across a region.

Regional-scale contamination takes the form of a mosaic of hotspots if there are a large number of source inputs over a region (e.g. Table 6.2). In San Francisco Bay, hundreds of hotspots of metal contamination probably existed in the 1980s (Luoma and Cloern, 1982). In southwest England, numerous small mines left behind patches of contamination across the entire region (Chapters 8 and 12). Many such regions occur where watersheds are mineralised, urbanised or industrialised.

Regional-scale metal contamination results if the source of metals is diffuse. For example, a geological source mobilised by human activities, such as Se from irrigation drainage (Chapter 13) can contaminate an entire watershed. Atmospheric inputs of fine particulate material also cause regional-scale contamination. The Sudbury smelters (Chapter 12) contaminated thousands of

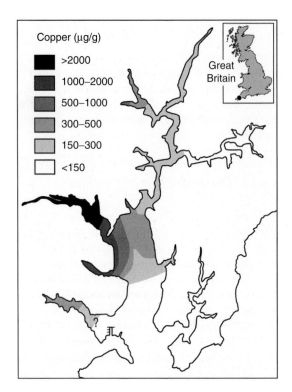

Fig. 6.7 Cu concentrations in fine-grained sediments in the Fal estuary system in southwest England decline progressively from the river, through the estuary and toward the sea. Contamination declines as uncontaminated marine sediments are mixed with contaminated sediment originating from sources (mines) on the river above the estuary. Sources internal to the estuary are minor compared to the mine wastes. (After Bryan et al., 1980.)

lakes in eastern Canada. The widespread dispersal of Pb from motor vehicles burning petrol, or Ag from industrial activities in Asia, detectably contaminate regions of pristine ocean waters (Chapter 5). A long-term atmospheric reservoir of a metal can even enhance contamination on a global scale (as in the case of Hg: Chapter 14).

6.4.2.1 Processes that determine metal distributions

Prologue. The pattern of contamination in a water body is determined not only by anthropogenic inputs and grain size, but dilution processes driven by the physical dynamics of the water body and the transformation of metals between dissolved and particulate forms.

Concentrations in sediments decline with distance away from sources of input more distinctly than do concentrations in water. Important processes differ among rivers, estuaries, bays and oceans. They include:

- dilution or burial;
- resuspension and/or remobilisation away from the original hotspot;
- fluxes of dissolved metal out of the contaminated hotspot.

For example, contaminated sediments in rivers progressively mix downstream with uncontaminated sediment from tributaries (Box 6.2). Mixing is usually complete within 12 channel widths downstream of each tributary. Below that point, the overall concentrations in the main stem will be proportional to the quantity and metal concentration of the incoming tributary sediment, as estimated by the area of the upstream surface area of the watershed (Marcus, 1987) (Box 6.2), as long as the climate and geology are roughly similar (Helgen and Moore, 1996). Forecasts from the basin area model described in Box 6.2 can be useful in projecting the distance over which sediment contamination might be expected from a contamination source, or the rate of dilution downstream.

The contamination gradient in a river is simple when viewed on a broad scale; hundreds of kilometres in the mine-contaminated Clark Fork (USA) and OK Tedi (Papua New Guinea) for example (Chapter 12). But at a closer scale, the detailed features of the gradient can be variable and complex. In the Clark Fork River, a single exponential describes dilution of contamination over 400 km (see Fig. 12.6). But differences in contamination between sites separated by 40 km are not separable. Local influences include bank slumping, variability in deposition and major events like floods, ice jams or remediation activities. These add spatial and temporal variability on small scales (Moore and Landrigran, 1999) and complexity along the gradient.

A declining seaward trend is a typical dilution pattern in estuaries if the source is upstream. Examples include the Rhine and Elbe (Germany)

Box 6.5 Floodplain contamination in rivers

In a river system, floods are the primary mechanism whereby particulate wastes are transported, then deposited. Transport of contaminated sediments is minimal during low flow regimes. During seasonal periods of high river discharge, contaminant transport is enhanced. But the greatest transport occurs during large floods. This is because the mass of sediment transported by a river increases exponentially as discharge increases; so very high river discharges pick up massive quantities of contaminated sediment and leave them behind on floodplains. A few floods can disperse wastes across vast areas. From there they will not be removed by natural processes for centuries.

Metal deposited on the floodplain during floods can be mobilised by overland transport of the particulate wastes into the river or by slumping of banks into the river as the river cuts new meanders into its floodplain. More subtle metal input via the hyporheic zone may also be common. The hyporheic zone is the region of heterogeneous sediments beneath and adjacent to the stream channel that is saturated with a mixture of surface water and groundwater. It connects the terrestrial floodplain, via groundwater to river or stream habitats. In floodplains affected by mine wastes (Chapter 12), for example, zones of metal enrichment exist where biogeochemical activities lower pH (Nimick and Moore, 1991). The contaminations are flushed into the river during rainstorms or carried from the groundwater to the surface water by seasonal subsurface inflows.

Rivers, or river reaches, with extensive floodplains will store and continually remobilise metal contamination. Wherever floodplains broaden, uncontained contaminants will be distributed over them during the large floods or containment failures. Redistribution of these contaminated sediments will extend the scale of contamination and complicate remediation. Rivers with narrow or few floodplains are probably less susceptible to such processes and may be less contaminated along the river (although the wastes eventually are deposited in some receiving waters). Prevention of floodplain contamination is much more important, and more cost effective, than attempting to remediate once floods have spread metal contaminants over vast areas.

and the small, mine-contaminated river estuaries of southwest England (Fig. 6.7). This contrasts with environments where urbanisation surrounds the estuary itself, and patches of hotspots create a regional mosaic as in San Francisco Bay, the Hudson–Raritan system and developed estuaries in Brazil (Luoma, 1990). The gradients around each hotspot are much more difficult to characterise. If inputs are controlled, the hotspots of contamination will gradually be dispersed or buried by new deposition of uncontaminated materials. The rate is governed by local sediment dynamics. In San Francisco Bay, hotspots of contamination began to disappear in the later 1980s as waste discharges were cleaned up in response to the Clean Water Act of 1972. The contamination continued to be dispersed regionally in this strongly mixed ecosystem, however, well after inputs were reduced. The result was that region-wide contamination was sustained by inputs from the hotspots while detectable declines were occurring within the hotspots themselves.

Physical processes can change accessibility and the geographical distribution of contamination (Box 6.5). Remobilisation of sediment-associated contaminants can occur during natural events, or be caused by human activities. Whatever the cause, remobilisation changes the spatial distribution of contamination and the risks it presents. For example, repetitive sediment resuspension and redeposition are common in lakes (Bloesch, 1995). In larger lakes, wind-driven, nearshore sediment resuspension results in focusing and redeposition of sediments in the deep water zone. Daily tidal currents, wind energies and storms in coastal and estuarine systems can cause periodic, frequent remobilisation of surface sediments (Calmano et al., 1994). More turbulent flow conditions, associated with seasonal flooding or storms, can move massive amounts of sediments in rivers. Human activities, including maintenance and dredging, and the disposal of historically contaminated estuarine sediments can result in major sediment disturbances (Eggleton and Thomas, 2004).

Resuspension energies can vary diurnally (tides), seasonally (with hydrology) or interannually (storms or floods).

Some deposits of sedimentary metals can be permanently buried, or sequestered in locations that are not accessible to processes that will redistribute them. Burial is a form of natural remediation if the water body is continuously accumulating sediment. Vertically in the sediment column, metal concentrations integrate temporal variability in input and can provide a record of historic influences.

Changes in sediment contamination often lag behind changes in inputs. The lag may be small if sediment transport processes are dynamic. Surface sediments in shallow waters of San Francisco Bay, for example, are constantly stirred and mixed by winds and tides in summer and autumn; new sediments are deposited by storms in the spring. This dynamic pattern resulted in a rapid decline in Ag concentrations in the sediments near a South Bay waste treatment plant once Ag was reduced in the effluent (Hornberger *et al.*, 2000) (see Box 3.10). On the other hand if sediments are not dynamic, burial of the contamination can be slow (decades).

6.4.2.2 Diffusional flux of dissolved contamination

Prologue. Contaminated sediments can release metals to the water column via diffusion caused by higher concentrations in pore waters than in overlying waters.

Pore waters are the waters within the interstices of the sediment. Metal concentrations in pore waters, like overlying water, are orders of magnitude lower than concentrations in sediment. Therefore, bulk analyses of metal in sediment are not affected by pore water metal contents. But concentrations in pore waters have a strong influence on the exchange of metals between sediments and the water column.

Benthic flux (sometimes referred to as internal recycling) is the term describing the movement of dissolved chemical species between the water column and the underlying sediment. If metal concentrations differ between pore waters and overlying water, a diffusion gradient can occur between the two. Geochemists calculate the direction of metal movement (benthic flux) into the sediments or out of them, depending upon that gradient. Other approaches to determine benthic fluxes include experimental flux measurements made by incubating tubes of sediments collected from the water body; or *in situ* flux chambers that entrain sediment and water at the interface and determine changes in dissolved concentrations.

Where sediments retain historical contamination, but anthropogenic inputs have declined, flux from the sediment pore waters to the water column can sustain a level of contamination in the water column. For example, Ag contamination was common in South San Francisco Bay through the 1980s, but inputs were greatly reduced in the 1990s. In the late 1990s, pore water concentrations of Ag were as high as >0.07 µg/L Ag in these sediments and overlying water concentrations were typically <0.01 µg/L Ag. The gradient caused Ag fluxes from the sediments into the water column of 3.5 to 32.0 kg per year. At its greatest, the outward flux from sediments exceeded the influx to the estuary from rivers (about 12 kg/yr: Rivera-Duarte and Flegal, 1997).

Where gradients are less dramatic, fluxes can be complex. For example, the direction of the flux in some water bodies (into or out of the sediment) can vary with season (Westerlund *et al.*, 1986). Different methods can also yield different results. This variability reflects the complex processes that characterise the interactions between sediments and the water column.

6.5   SEDIMENT CHEMISTRY

Prologue. The reduction/oxidation status of a sediment greatly influences how metals concentrate onto particulate material. Precipitation of Fe oxides and Mn oxides in the presence of oxygen is a key reaction. The oxides are rich in sites that bind metals with strong intensity (high affinity). Organic materials also bind metals strongly, but they are not redox sensitive. In anoxic environments, the primary control is precipitation of metal sulphides.

Explaining why metals partitioned so strongly to particles was an early dilemma for geochemists. Metal concentrations in solution in the oceans were dilute (low). Thus precipitation could not explain their tendency to be in solid form in environments where oxygen was present (Krauskopf, 1956). Ultimately it became clear that the chemistry of major components of the sediment or water column resulted in surfaces that could strongly attract metals. Association with particulate materials varies with the characteristics of the specific metal. For each metal, reduction/ oxidation conditions are also important, as defined by the redox couples formed under different conditions. Redox conditions are themselves influenced by the characteristics of the sediment column.

6.5.1 Redox reactions affect metal partitioning

Prologue. Redox reactions provide a first-order control on metal chemistry. In oxidised sediments, metal ions associate with the surface of particles that have high surface area and a high capacity to exchange with cations: Fe oxides, Mn oxides and organic material. In reducing environments, sulphides dominate the chemistry of metals.

Once it was discovered that the surfaces of particles in oxic environments had characteristics that attracted metals, those surface reactions were studied in detail. Under all conditions, metal ions associate with the surface of particles that have high surface area and a high capacity to exchange with cations (positively charged ions). But the associations that can be formed differ between reducing and oxidising conditions.

Several geochemical phases typical of oxidised particulate material have characteristics that cause them to bind metals strongly. Fe oxides and Mn oxides precipitate onto particle surfaces in the presence of oxygen. They create solid surfaces with many ligands and a strong affinity for metals. Thus any precipitation of Fe oxides in the water column of an estuary or in sediment pore water will bring other trace metals out of solution (Fig. 6.8). Particulate organic materials also have a high cation exchange capacity and strongly sequester metal ions. Dissolved organic material (e.g. humic substances) tend to coat particle surfaces, forming complex, interlayered mixtures with Fe oxides and Mn oxides (Luoma and Davis, 1983). Living microorganisms and organic detritus also are part of the organic material in sediments. Clay surfaces can bind metals, but they tend to be quickly coated with the more reactive oxides and organic materials in the aquatic environment. Clays, therefore, mostly act as high-surface-area carriers for other substrates (Jenne, 1968).

The vast bulk of sediments in nature are not oxygen-rich, however. The chemical differences between oxidised and reduced sediments are well known (Goldberg, 1954; Krauskopf, 1956; Jenne, 1968; Aller, 1978). In anoxic sediments, reduction rather than oxidation dominates chemical reactions (Box 6.6). Soluble Fe and soluble forms of Mn are favoured. The metal-reactive ligands associated with oxidic Fe or Mn are absent. Sulphates are reduced to sulphides, and metals can form insoluble sulphide precipitates or co-occur with iron sulphide (Morse, 2002). The end result is that many metals are concentrated in a highly stable form in anoxic sediment as long as the sediments remain anoxic. These reactions are reversible if oxic conditions begin to prevail; and this reversibility can occur on the time scales relevant in natural waters. Reversals between oxic- and anoxic-driven reactions create a complex cycle within the sediment column that greatly influences exchange between sediment particles and pore waters.

The only major metal-binding substrate in sediments that is not redox sensitive is organic material.

6.5.2 The sediment column

Prologue. The geochemically dynamic and biologically driven heterogeneity of the sediment column is critical to the form, fate and bioavailability of sediment-bound metals. The depth of the zero

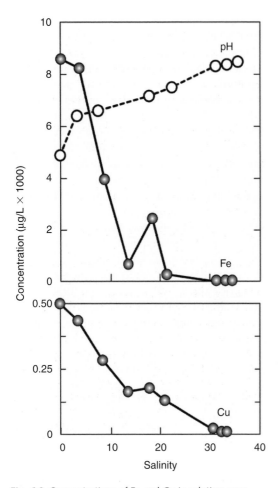

Fig. 6.8 Concentrations of Fe and Cu in solution over distance as the River Carnon mixes with the marine waters of Restronguet Creek (Bryan and Gibbs, 1983). pH increases with salinity because the high concentration of carbonate in seawater consume the H$^+$ ions. The transformation of metals from dissolved to particulate forms is implied by this distribution. Increasing pH causes complete precipitation of dissolved Fe as hydrous oxides. Some other trace metals, including Ag, As, Cu and Pb, become associated with the precipitating Fe hydrous oxides (Cu is shown here), and significant percentages of these metals are removed from solution to be deposited as sediment. Bryan and Gibbs (1983) determined a concentration of 593 µg/L Cu and 5024 µg/L Fe in the incoming Carnon River water at pH 3.8. Using the mean river flow of 0.938 m^2/sec, they calculated that the Carnon would transport a mass of about 21 tonnes of Cu and 280 tonnes of Fe in a year. If all the metals were precipitated, they predicted that mass of metal would turn into a concentration of 56000 µg Fe/g sediment (5.6%), very close to the observed value. If all the Cu were transformed to particulate it would yield 4073 µg/g,

oxygen (redox) interface is determined by a balance between:

- deposition of reactive organic material onto the sediment or its production within the sediment;
- microbial oxygen consumption as microbes decompose this organic material;
- oxygen penetration into the sediment as controlled by physical processes and biological mixing (bioturbation: Aller, 1978);
- microzones that are generated by the activity of organisms.

Together these processes create an ever-changing, open, dynamic, structured biogeochemical system (Sundby, 1990).

Oxygenated sediments cover the surface of bed sediments wherever the water column contains oxygen. But deeper in the sediments, a boundary occurs where the rate of incoming oxygen via diffusion equals the rate at which microbes consume oxygen. This is the redox interface (Rhoads and Boyer, 1983). A predictable zonation of reactions occurs at this zero oxygen boundary. Many of the reactions reflect microbial utilisation of electron acceptors other than free oxygen: the oxygen bonded to compounds like Fe oxides, Mn oxides and other reactions (Howarth *et al.*, 1983; Lovely and Phillips, 1986; Myers and Nealson, 1988). The depth of this zone is also influenced by biological processes like bioturbation (bio-mixing) and physical processes like resuspension (Fig. 6.9).

The vertical depth of the oxic zone can be highly unstable in time and varies widely from location to location, because the controlling factors can differ greatly with the circumstance. The oxic layer of a sediment column is millimetres thick in a highly productive, fine-grained sediment where penetration of overlying water into the sediment is slow, and few macrofauna stir oxygen into the sediment

Caption for Fig. 6.8 (*cont.*)
about twice what was observed in sediment. Thus they concluded that about half of the Cu carried by the Carnon was transformed to particulate mass upon neutralisation of the river by the marine waters, and half remained in solution for transport into the estuary.

Redox reactions play a critical role in the chemistry
of metals in sediments. Most surface waters are
oxidised, meaning that oxygen is present to drive
oxidation reactions. If oxygen is absent (e.g. in
subsurface sediments), pore waters are dominated
by reduction reactions.

Redox potential in the sediment can be reported as
net oxidation/reduction potential or Eh, which is
operationally defined as the potential generated
between a platinum electrode and a standard hydrogen
electrode. In general geochemists find Eh usually of
little value in quantifying the redox chemistry, because
it can seldom be correlated to a single specific redox
couple. The preferred approach is to evaluate the
concentrations of the products of the predominant
redox reactions and infer redox conditions from these.

Redox couples that are important in understanding
the fate of metals in sediments include:

- $O_2 - H_2O$
- $SO_4^{2-} - H_2S$. Sulphide oxidation and sulphate
  reduction convert sulphide (HS^-) to sulphate (SO_4^{2-})
  in the presence of oxygen. In the absence of oxygen
  the important product is sulphide which reacts
  quickly and strongly with metals such as Fe and Mn.

$$2O_2 + HS^- = SO_4^{2-} + H^+.$$

- $Fe^{2+} - Fe^{3+}$. Fe reduction and oxidation. In the
  presence of oxygen, Fe^{3+} is formed which quickly
  converts to the solid reactant Fe_xO_y or Fe oxide.
  Fe oxides have abundant surface sites that strongly
  adsorb metals and some metalloids.

$$O_2 + 4Fe^{2+} + 4H^+ = 4Fe^{3+} + 2H_2O.$$

In the absence of oxygen, soluble Fe^{2+} is present, but
it can be consumed by sulphide to form Fe sulphides
(of several forms). In fact, the amount of free sulphide
in anoxic sediments is usually determined by the
amount of Fe^{2+}. If $[Fe^{2+}] > [HS^-]$, then all sulphide
will be present as Fe sulphide (with some other
inorganic sulphides); if the opposite is true then
hydrogen sulphide will be present and the sediment
will smell like rotten eggs.

$$15O_2 + 4FeS_2 + 14H_2O$$
$$= 4Fe(OH)_3 + 8SO_4^{2-} + 16H^+.$$

- $Mn^{2+} - MnO_2$. Mn oxidation, and reduction. In the
  presence of oxygen, Mn forms the reactive solid
  Mn oxide, MnO_2 as Mn(IV). In the absence of oxygen
  the reactions goes the other way and Mn (IV) is
  reduced to the soluble product Mn^{2+}. Thus Mn is
  found in solution in pore waters if sediments are
  anoxic but is largely insoluble in oxidised waters,
  where it is nevertheless highly reactive in its
  particulate form. Metals also bind strongly to Mn
  oxides in oxygenated sediments. Because it is much
  less abundant than Fe, it is somewhat uncertain how
  metals partition among Mn oxides and Fe oxides in
  sediments. But Mn is certainly much more important
  as a reactant, transforming metals to surface-
  adsorbed forms, than it is as a toxicant in most
  natural waters.

$$O_2 + 2Mn^{2+} + 2H_2O = 2MnO_2 + 4H^+.$$

The only major metal-binding substrate in sediments
that is not redox sensitive is organic material.

column. The oxic zone is metres thick in oligo-
trophic, energetic or sandy environments (Rhoads
and Boyer, 1983). In a sediment rich in macrofauna, the oxic layer can be expanded in depth
(and made heterogeneous) by the stirring created
as organisms move sediment up, down and laterally (bioturbation). Seasonal fluctuations in temperature change the rate of microbial reactions
and can drive a seasonal, vertical migration of the
depth of the oxic zone. Surface sediments can also
be mobilised physically by currents or major climatic events, creating a dynamic exchange of particles between the bed sediment and the water
column (Fig. 6.9).

The location of the redox interface is extraordinarily important to organisms that live in the
sediment (Rhoads and Boyer, 1983), and the organisms are very influential in the structure of the
oxic layer (Fig. 6.10). The vast majority of macrofauna and meiofauna require oxygen for their
metabolism. Meiofauna and microflora form a
microscopic community of animals and plants that
thrive in the interstices between sediment grains.
Epibenthic organisms also live on the surface of

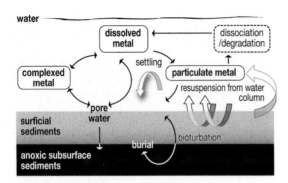

Fig. 6.9 A conceptualisation expanding on Fig. 6.1 and emphasising internal processes that affect the form of particulate metals in the sediment.

Fig. 6.10 The interaction of organism with sediments differs widely among species. The mode of interaction influences how organisms are exposed to contamination.

the sediment and/or within the upper centimetres, and obtain their food near this interface with the water column. Some species move in and out of the sediments every day. Some macrofauna (infauna) live deep in the sediment and are surrounded by reduced sediments. But almost all these species use connections to the surface (tubes, burrows or siphons) to obtain the oxygen they require. Some deposit-feeding animals feed on the subsurface anoxic materials, but most feed somewhere at the surface or from the surface where organic material is most nutritious. Suspension-feeding infauna will draw down water bearing their food supply from the overlying water into their burrows.

The structures in which organisms live create miniature oxic zones within the reduced layers (Fig. 6.10) of the sediment column, on the scale of millimetres to centimetres (Charbonneau and Hare, 1998). Thus sediments at the redox interface and within the upper layers of the reduced zone resemble a mosaic of biogeochemical microenvironments as organisms move oxygenated water in or mix oxygenated and reduced sediments (Aller, 1978, 1980).

The heterogeneity of the sediment reflects processes fundamental to the survival of many species; but it makes the distribution and form of metals in the sediment complex. Collection and handling sediments for experimentation or testing toxicity can change that heterogeneity completely. The result is a poor simulation of natural processes. For example, merely constraining the sediment can create a bottle effect that disrupts oxygen input, affects fluxes and quickly generates a different zonation of redox reactions than occur in the open, undisturbed system (Sundby *et al.*, 1986; Westerlund *et al.*, 1986). Homogenising oxic, interface and reduced sediments completely disrupts all important processes in this dynamic system; yet that is common operating procedure in many experiments. The mosaic-like distribution of different environments, with fluxes among all, must be built into any analysis of metal biodynamics in sediments; the mosaic is just as important as the differences themselves.

The oxidised zone is also the zone where Fe and Mn can precipitate as oxides; the anoxic zone is where they redissolve as their reduced forms. If Fe^{2+} and Mn^{2+} precipitate as they encounter oxidised conditions at the surface of the sediment, this can create a diffusion barrier for other metal solutes migrating upward in the pore waters and reduce exchange between sediment and water. Fe and Mn also cycle in sediments. They are reduced as they are buried by deposition of new sediments, solubilised in the deeper layers, pumped toward the surface by diffusion, reprecipitated where conditions become oxidising near the surface, then re-reduced as those sediments are buried. That cycle constantly refreshes particle surfaces with newly formed oxides in the surface sediments (Fig. 6.9).

6.5.3   Metal form in oxidised sediments

Prologue. It is much more difficult to determine the distribution of metals among forms or phases in an oxidised natural sediment than it is to compute speciation in solution. This is because the metal intensity of binding sites is complex and capacity (number of sites) is variable within and among phases. The complex nature of the particles also adds complexity. Available methodologies are not yet able to fully cope with these complications.

The distribution of metals among ligands (or substrates) in an oxidised sediment is determined by the same processes as in solution:

- the number of metal binding sites;
- the strength with which each site associates with the specific metal involved (the binding intensity);
- the amount of metal available to bind to the site.

Basic characteristics of each trace metal (described in Chapter 3) govern their reaction with particle surfaces. Adsorption of metals from solution (Box 6.7) will follow a predictable order in a medium with few dissolved ligands. For example, the Irving–Williams series was defined from experiments in simple media, to describe the rank order for complexation with organic materials:

$$Zn^{2+} > Cu^{2+} > Ni^{2+} > Co^{2+} > Fe^{2+} > Mn^{2+}$$
$$> Ca^{2+} > Mg^{2+}$$

but the rank order of affinity will differ with different types of surfaces. For example, Cu has an exceptionally strong affinity for organic ligands (binding sites); but Co has a stronger affinity than Cu for Mn oxides, and Cr has a strong affinity for Fe oxides.

The combination of a large number of binding sites with high intensity for binding is the reason that metals are sequestered strongly by oxides and organic materials (Box 6.8). In most sediments, the number of binding sites is dominated by Fe oxides and organic materials, but the intensity with which Mn binds some metals (e.g. Pb and Co) makes it an equally important phase.

Box 6.7 Surface chemistry reactions control metal concentrations on particulate material

Adsorption and complexation are empirically expressed by a simple complexation reaction:

$$M_i^z + S_j \overset{K_{M_iS_j}}{\longleftrightarrow} M_iS_j$$

where M_i^z represents metal ion i, S_j represents a surface site of ligand or component j, and M_iS_j is the surface complex. $K_{M_iS_j}$ is the measure of binding intensity (the equilibrium or stability constant). The equation may be resolved and rearranged to illustrate the factors that affect metal concentration as:

$$[M_iS_j] = K_{M_iS_j} * [M_i^z] * [S_j] * [A_j]$$

where [ ] depicts concentration and $[A_j]$ is the concentration of component per unit mass of sediment. This complex equation simply states that the concentration of metal on the surface of a component (Fe oxide, Mn oxide or organic material) is determined by the intensity of binding, the concentration of metal available for binding and the number of binding sites as estimated by the concentration of the solid phase component. When the equation is simultaneously solved for all components, the distribution of metal among components can be computed. The total concentration of bound metals is predicted in some models by the sum of the metal associated with each component. The value of the equation is that it systematically describes how changes in these different characteristics of the environment can influence concentrations and form of particle (sediment) bound metal and K_d.

Operational methods are available for evaluating metal distribution among solid phases in oxidised sediments. The implicit meaning of 'operational' is that the method is good enough to differentiate two distinctly different sets of conditions, but it does not necessarily give a quantitatively precise measure of the distribution among forms. Operational methods include 'selective' chemical extractions (Tessier et al., 1979), operational models (Luoma and Davis, 1983; Tessier et al., 1984) and statistical techniques (Luoma and Bryan, 1981), or some combination of these.

Box 6.8 Limitations to quantifying the solid phase distribution of metals in sediments

Different approaches are useful in pointing towards tendencies in the partitioning of metals among solid phases in oxidised sediments, but no method achieves the equivalent rigour of speciation models/methods for dissolved metals. There are several reasons that quantification of metal form in oxidised sediments is so difficult:

- The number and intensity of binding sites are variable among types of organic matter. Some types of organic material have few sites per unit surface (e.g. lignins) and others have many (fulvic materials). The intensity of complexation also appears to vary progressively with metal concentration, with high-intensity sites being filled first followed by sites of progressively lower intensity.
- Freshly precipitated Fe oxides have many surface sites, but as Fe oxides age, they become increasingly crystalline and progressively lose binding sites. Recycling of Fe in surface sediments, therefore, is instrumental in continuously creating new amorphous Fe at the surface, with a strong affinity for metals. But standardised methods do not exist for reliably quantifying how much of the different types of Fe are available in a natural sediment to bind trace metals.
- Because of the two considerations above, a single equilibrium constant will not accurately describe the metal association with organic materials or oxides. For the same reason, determining binding sites per phase in any specific circumstance is also problematic.
- The longer metals are associated with sediments, the more metal partitions to sediment as compared to the dissolved phase. It appears that metals migrate to less accessible sites over time in a sediment and more metal is progressively removed from solution (this process occurs over months). This is an especially important factor in experimental studies in which outcomes are determined by the distribution (e.g. Lee *et al.*, 2004).
- Ligands occur mixed together on particle surfaces making their accessibility for metals difficult to characterise precisely.

Selective chemical extractions were designed by soil scientists to remove metals from specific substrates in the sediment, and have a long history of application in that field (Jenne, 1968). But they are less useful in aquatic environments. Tessier *et al.* (1979) developed a set of sequential extractions to progressively remove: (a) ion-exchangeable metal, (b) a carbonate fraction, (c) a reducible fraction (presumably metal associated with hydroxides), (d) an oxidisable fraction (presumably metal associated with sulphides and organic carbon) and (e) a residual fraction. This is one of the most cited papers in the study of metals, illustrating the desirability of the approach. The extractions can differentiate metal form under different conditions (for example, Dollar *et al.* (2001) compared wet and dried wetland sediments). But their selectivity is clearly operational (tendencies but not precise answers). Particular examples of problems include removal of metals from many types of sites with lower pH extractants (Luoma, 1990), and reprecipitation and/or readsorption as metals are released from heterogeneous phases then precipitate or interact with unsolubilised surfaces (Belzile *et al.*, 1989a).

The principles for modelling the relative abundance of metal forms in an oxidised natural sediment are the same as those used for modelling speciation in solution. But the complexities in characterising ligands greatly limit reliability (e.g. Luoma and Davis, 1983). In one of the most successful examples of modelling, Tessier *et al.* (1984) narrowed the computations to three parameters determined in lake sediments from Quebec, Canada: the concentration of metal in the sediment (determined by partial extraction), the concentration of an assumed dominant sediment phase (amorphous Fe oxides, as determined by partial extractions) and a laboratory determined equilibrium constant for amorphous Fe oxides (Tessier *et al.*, 1984). They then predicted the reactive dissolved metal ion concentration in solution that would equilibrate with that sediment, and correlated that to metal uptake by a bivalve. Few authors have attempted to apply this technique more widely.

Luoma and Bryan (1981) reconstructed metal interactions with the major phases in oxidised

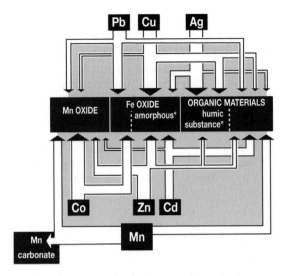

Fig. 6.11 Metal associations with Fe oxides, Mn oxides and organic materials as interpreted from statistical correlations and selective chemical extractions in oxidised estuarine sediments from southwest England (after Luoma and Bryan, 1981). The widths of the arrows are indicative of the fraction of metal partitioned to each phase in the oxidised sediment.

sediments from a combination of chemical extraction and statistical correlations. This statistical approach is most valid in a broad data set from several water bodies (their data were from 17 estuaries in southwest England), with wide ranges in concentrations of metals and phases. The data had few confounding correlations among phases. Their depiction illustrates the complex network by which each metal appeared to interact among the major sediment phases (Fig. 6.11).

6.5.4   Differentiating reduced metal forms

Prologue. Operational extraction techniques and a simple model (AVS–SEM) capture the tendency of metals to be in reduced forms. The exchange of metal between sediment and pore water can be estimated by this approach in homogenised sediments. The spatial and vertical heterogeneity of nature make application of the approach in field sediments more problematic. Metal sulphides also can be converted to the more exchangeable oxidised forms

if conditions change. The AVS–SEM theory captures the processes that affect metal form at the transition between oxic and anoxic sediments; but it is difficult to sample sediments in a way that realistically captures the AVS–SEM status that is relevant to sediment-dwelling organisms.

The biogeochemical differences with the greatest implications for metal chemistry occur between the oxic and anoxic zones. Differentiating reduced from oxidised metal forms is therefore important for understanding processes that influence ecological risks from metals. An extraction with hydrochloric acid will operationally quantify at least some fractions of the amorphous Fe sulphides in sediments (termed acid volatile sulphide or AVS) (Rickard and Morse, 2005). Metal is also released from the sediment sample during the AVS extraction. This is termed simultaneously extracted metal (SEM) (Di Toro et al., 1992). The extractions are operational and not highly specific to these phases (Morse and Rickard, 2004). It is not possible to selectively separate metals associated with AVS with exactness due to the natural instability of the mineral phase (Huerta-Diaz et al., 1998). But the estimates can be useful if interpreted carefully.

The moles of total SEM (the sum of all metals) can be compared to the moles of AVS released from the sediment to estimate metal association with sulphides. In anoxic environments, most cationic metals associate with sulphides, as long as there is enough sulphide present, because of the overwhelming strength of the metal–sulphide equilibrium constant (e.g. Ag, Cd, Cr, Cu, Ni, Pb, Zn). If SEM > AVS, it is assumed that forms of metal other than sulphide are present. If AVS > SEM, it is assumed all metal is associated with sulphides. Molar units are essential so the two can be quantitatively compared on an atom-to-atom basis.

The majority of reduced sulphur in anoxic sediment is present in the form of Fe sulphides. These include amorphous Fe monosulphide, and more crystalline minerals like mackinawite, gregite and pyrite (Cooper and Morse, 1998). Organic sulphides are also common in some environments (e.g. wetlands). Adding cationic metals to

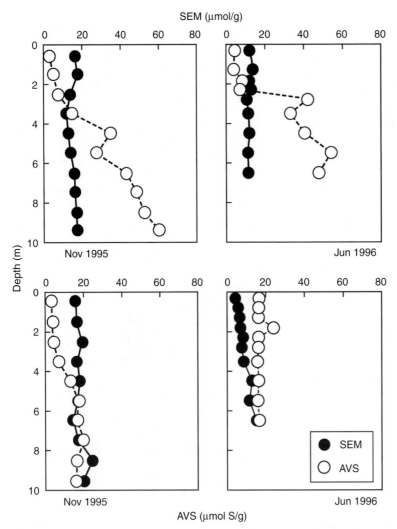

Fig. 6.12 Vertical profiles of acid volatile sulphide (AVS) and simultaneously extracted metal (SEM), in the same units (μm/g dw) in two sediments the River Meuse in the Netherlands, collected in November and June (after van den Berg *et al.*, 1998). A sample of the top 3 cm would collect a different proportion of high AVS and low AVS sediment from every sediment.

a sediment containing amorphous Fe sulphide results in the formation of a metal–sulphide precipitate or a coprecipitate with the Fe sulphide. Chemical thermodynamics predict that the transformation from an oxidised to a reduced form of most metals will occur quickly and will reduce metal exchange to the water column or pore waters.

Vertical profiles of AVS and SEM, determined at a fine scale, follow the trends expected from typical vertical redox conditions (van den Berg *et al.*, 1998; J.S. Lee *et al.*, 2000) (Fig. 6.12). Sediments on the surface of the mud are oxidised and low in AVS.

Below the redox interface, the reduced subsurface sediments have elevated AVS. Thus, SEM is typically greater than AVS in surface oxidised sediments and SEM < AVS in subsurface sediments.

The methodology by which the sediment sample is collected is particularly important to the operational determination of AVS concentrations. If a single sample is taken from a fixed depth (in fact that is the recommended approach in standardised methodology: ASTM, 2003), the proportion of oxidised and anoxic sediment in the sample will vary from place to place. As noted above, it is common for the depth of the redox interface to differ with space and time (Fig. 6.12). Bulk sampling does not consider such heterogeneity. The result is an averaged view of AVS concentrations that is useful to the geochemist or toxicologist, but of questionable relevance to what any organism experiences directly, because organisms are restricted to the oxidised zone whatever its depth. Other aspects of sediment heterogeneity are also missed by bulk sampling (e.g. oxidised microenvironments).

The redox status of metals in sediments is not necessarily a permanent characteristic; in fact AVS determinations alone can be quite variable vertically and spatially (Rickard and Morse, 2005). Resuspension by natural or human activities will cause relatively rapid changes in the redox state of the metals as the mobilised sediments are exposed to a different chemical environment. A change in temperature or biological productivity can also be influential. Sediments cycle from the water column to the sediment surface and back; or from the surface to the subsurface, as a result of physical processes. In such active environments sediments are continually moving in and out of different redox environments. The redox status of metals in a vertical section of a deposit is therefore not predictive of the chemical form of metals in suspension above the sediment, at the sediment surface and/or in environments to which suspended metals are transported.

6.5.5   Effect of redox on metals that behave as anions

Prologue. Behaviour varies among the elements that behave as anions in aquatic environments, e.g. Se, Cr and As. Chemical behaviour is influenced by the characteristics of the particular oxidation state, biological processes and kinetics.

Trace metals that behave as anions in aquatic environments change oxidation state as the redox environment changes. Se, as an example, is discussed in Chapter 13. Cr is another example of an element that strongly reacts with surfaces in its reduced form Cr(III), but the oxidised form

---

Box 6.9 Oxidation/reduction behaviour of chromium

The hexavalent state of Cr, Cr(VI), in oxidised environments forms chromate (CrO_4^{2-}) or bichromate ($HCrO_4^-$). Chromate is very soluble and chromate adsorption on particle surfaces is limited because the ion has a negative charge. In contrast, the reduced state, Cr(III), forms insoluble precipitates under slightly acidic and neutral conditions, and/or is scavenged by formation of Fe oxides. This difference is very important because chromate is toxic and carcinogenic, whereas Cr(III) is typically less bioavailable (with some exceptions) and not as toxic (Blowes, 2002).

Although Cr(VI) is favoured in oxidised waters, it is rarely abundant in surface waters. Its residence time is reduced by incidental contact with microdeposits of reactive Fe or sulphur. Apparently, wherever Fe oxides, for example, are forming on a sediment surface or in the water column, Cr(VI) will be reduced to Cr(III) and remain in that state because the back reaction from Cr(III) to Cr(VI) is slow (days to months). The result is progressive scavenging of Cr(VI) out of solution, onto particulate material, even in predominantly oxidised waters. For example, Abu-Saba and Flegal (1997) showed that Cr(VI) inputs occurred in the watershed of San Francisco Bay, but there was a progressive decline of Cr(VI) concentrations as waters moved from the river to the estuary.

In groundwaters, where Cr(VI) toxicity or health risks are greatest, reductants may be in limited supply. If Cr(VI) concentrations exceed the capacity to reduce the Cr, concentrations may increase as Cr(VI) is added and/or Cr(VI) may be retained in solution for long periods.

Cr(VI) is soluble and the most toxic. Reactions in microenvironments and slow retransformation make Cr(VI) rare in surface waters, however (Box 6.9).

As reacts in an opposite manner to Cr as redox changes. The reduced form, As(III), is more soluble, more mobile (binds less strongly to particle surfaces) and more toxic; As(V) strongly associates with Fe oxides in oxidised environments. Simultaneous solubilisation of Fe and As occurs under

Box 6.10 Microbial processes determine the oxidation state of arsenic

The major oxidation states of As are the oxidised form As(V) and the reduced form As(III). As(III) is soluble and mobile in aquatic environments. It is also more toxic than As(V). Therefore, the processes that determine the oxidation state are critically influential in determining the ecological significance of this contaminant. Like Se (Chapter 13) microbial dissimilatory reduction of As(V) to As(III) represents an important means by which As can be mobilised and transported from the adsorbed solid phase into the liquid phase. This process occurs in anoxic sediments, soils and subsurface aquifers, and takes several forms. One process is driven by microorganisms that are protecting themselves from the deleterious effects of As(III) in As-enriched anaerobic environments. The internal pools of cytoplasmic arsenate [As(V)] are rapidly reduced to arsenite [As(III)] for extrusion from the cell, or external As(III) is oxidised at the cell surface and denied entry into the cells. Other anaerobic bacteria can actually conserve the energy gained via the oxidation of organic compounds (or H_2) with the reduction of As(V) to As(III). This process is a means of anaerobic respiration that supports the growth of a number of microbes, where As is elevated in concentration (the organisms obtain oxygen from H_2AsO_4 in place of free oxygen). When oxygen is present some microorganisms can also conserve energy for growth by oxidising As(III) to As(V). This process is likely to be of importance in retarding the mobility of As. (After Oremland and Stolz, 2003.)

reducing conditions, as microbes breakdown organic material, consume oxygen and/or begin to use the oxygen in the hydrolysed anionic metals (Box 6.10). These are the reactions that have created As-enriched groundwater and, in extreme cases, raised the spectre of human poisoning (as in Bangladesh or the arid western USA: Chapter 1; van Geen *et al.*, 2004). Such reactions also make As difficult to contain in landfills (Delemos *et al.*, 2006), and emphasise the importance of avoiding organic contamination of constrained water bodies with contaminated sediments (e.g. aquifers).

6.6   METAL PARTITIONING: THE DISTRIBUTION COEFFICIENT K_d

Prologue. The distribution of metals between solid and liquid phases is described by a coefficient (K_d) that characterises each metal's tendency to associate with particulate material. It can vary more than two orders of magnitude for a given metal in different circumstances, however. The degree of partitioning between solid phase and dissolved phase is dependent not only upon the characteristics of the specific metal, but also water (and pore water) chemistry. The characteristics of the sediment also play a role.

Total metal concentrations in the particulate phase do not, alone, determine ecological risks. The exchange of metals between the particulate phase (in sediments or suspension) and solution is also important.

Partitioning behaviour can be expressed as a distribution coefficient, K_d. This is the relative concentration of each contaminant in each phase. Assuming an infinite quantity of water, it is mathematically expressed as:

$$C_{sed}/C_{water} = K_d$$

where C_{sed} is the metal concentration in sediment and C_{water} is the concentration in water. The concentrations should be expressed in comparable units. For example, if C_{water} is expressed as µg metal/L water, C_{sed} should be expressed as µg metal/kg sediment.

Each metal has a different characteristic binding affinity for particulate material. Risk assessments would be easier if that was the sole source of variability and a single K_d could be defined for each metal, or at least for most circumstances. But K_d not only differs widely among metals, but absolute values and the rank order can be different in different circumstances.

The rank order and ranges of K_d for estuarine water columns affected by human activities are typical of those reported by Feng *et al.* (1999):

$$Fe(10^{5.6}-10^{8.3}) > Pb(10^4-10^7) > Ag(10^5)$$
$$> Cu(10^{4.2}-10^{5.6}) \geq Zn(10^{3.4}-10^{6.2})$$
$$> Cd(10^3-10^4).$$

Box 6.11 Variability of K_d among estuaries

Shafer *et al.* (2004) investigated metal partitioning in three different estuaries in North America: Cape Fear and the Elizabeth River in Norfolk, Virginia; and San Diego Bay, California. They found that the K_d for Cu in Cape Fear and Norfolk were 5000 to 45000, or 5×10^3 to 4.5×10^4. High river discharge was accompanied by generally higher K_d values. Low K_d values in both systems are associated with sites with exceptionally high quantities of colloidal Cu, which held more Cu in solution. Values for Cu K_d in San Diego were higher and more variable than in the other water bodies. Most K_d values in San Diego Bay were 5.0 to 6.0×10^4; but some locations had Cu K_d values of 1.0 to 5.0×10^5. Across all locations, Zn showed similar variability: the full range of K_d was 7.7×10^3 to 4.7×10^5.

Shafer *et al.* (2004) concluded that variability in the dissolved concentration and dissolved speciation of Cu and Zn was more responsible for the variability of K_d than was variability in metal concentrations on particulate material.

The range within each metal can be up to 1000-fold or three orders of magnitude. (The lower ranges for metals like Ag are the result of limited data.) The ranges overlap among metals, but there is a general rank order. However even the rank order can change with environmental circumstances (Box 6.11).

The wide ranges in values reflect the complexity of processes controlling the transformations: the nature of the particulate material, biogeochemical influences, reducing and oxidised conditions and geochemical conditions that affect speciation.

6.6.1 Effect of water chemistry on K_d

Prologue. Water chemistry can affect the partitioning process in oxidised waters. Speciation determines the degree to which a metal will bind to particulate material. It thereby influences the metal concentration remaining in solution and occasionally affects the metal concentration on particulate material.

Speciation affects partitioning between dissolved and particulate metals because the form of metal in the water affects the efficiency of transformation to particulate material. If the metals in solution are held strongly enough by dissolved ligands they are less effectively transformed to particulate material. For examples Cu will be partitioned less efficiently to particulates if it is held tightly in solution by organic complexes and/or colloids that do not flocculate.

Water chemistry affects K_d in other ways, as well. For many trace metals, thermodynamic principles would suggest a general reduction in K_d (less metal on particle surfaces) as conditions change from a river to the sea. Ocean water has very high concentrations of Ca^{2+}, Mg^{2+} and Na^+ (Chapter 4) which should compete with metal ions for adsorption onto binding sites and reduce their concentration on the particulate phase. Cd and Ag form very strong complexes with chloride in seawater (see Chapter 5). The K_d values of Cd and Ag in brackish (saline) waters are therefore lower than those in freshwater.

On the other hand, freshwaters contain dissolved organic materials and Fe oxides in colloidal form. Some of these aggregate onto particle surfaces or flocculate when freshwater mixes with higher ionic strength seawater (Sholkovitz, 1976). Dissolved metals are sequestered on these particulates reducing dissolved metal concentrations (Fig. 5.4) and increasing K_d. The net effect of these opposing processes was different for different metals in the Mersey estuary, UK (Turner *et al.*, 2002). The K_d of Cd declined threefold at the freshwater/seawater interface, as Cd was desorbed into solution, depleting particulate concentrations and raising dissolved concentrations. The K_d of Ni and Zn increased five- to tenfold from the head to the mouth of the estuary. The changes in K_d were determined by the decline in water column concentrations of the metal as they adsorbed or flocculated onto particles. There was little change in concentrations on particulate material; presumably, not enough of these elements were added to the existing load to detectably affect concentrations. The K_d and concentrations of particulate Cu and Cr both approximately doubled in the same

Rember and Trefrey (2004) studied trace element dynamics in the Sagavanirktok, Kuparuk and Colville Rivers, undisturbed water bodies in the Alaskan Arctic. Concentrations of dissolved Cu, Pb, Zn and Fe were three- to 25-fold higher during snowmelt and peak discharge in the rivers, compared to off-peak discharge. Trends in dissolved metal concentrations followed trends in dissolved organic carbon (DOC) concentrations, suggesting that organic complexation of dissolved metals was responsible for mobilising metals from the watershed and holding them in solution during runoff. The sources of the dissolved metals were thawing ponds and dissolved carbon-rich upper soil layers in the floodplains.

Particulate metal concentrations were more uniform than dissolved concentrations. Particulate metal concentrations correlated well with the Al content of the suspended particles; so variability in particulate concentration was related to changes in the granulometric character of the metal mass being transported. However, concentrations of particulate Al were poorly correlated with particulate organic carbon, presumably because snowmelt mobilised particulate detritus and dissolved organic carbon, independently of the particulate load. Normalisation of metal concentrations to organic carbon would not have corrected for granulometric influences in this system. Seasonality was also important. Results showed that >80% of the mass of suspended sediment and more than one-third of the annual inputs of dissolved Cu, Fe, Pb, Zn and DOC were carried to the coastal Beaufort Sea in 3 and 12 days, respectively, by the Kuparuk and Sagavanirktok rivers.

circumstance. These metals were associated with sedimentary organic matter, which increased from the head to the mouth of the Mersey. Thus additional particulate sites for binding metals were sufficient to increase concentrations detectably. Similar processes affect K_d in undisturbed as in disturbed estuaries (Box 6.12).

6.6.2   Pore waters

Prologue. The redox status of sediments is also important in determining particle–water phase distribution. High sulphide concentrations in sediments reduce pore water metal concentrations to very low levels in experiments and in some sediments in nature. But processes in sediments also affect pore water chemistry in ways that hold metals in solution, or facilitate their diffusion out of the sediments. The net outcome seems to vary with the circumstance. Pore water metal concentrations are not necessarily predictable from chemical theory or laboratory experiments.

Pore water chemistry reflects biogeochemical fluxes of properties within, into and out of the sediments. Where highly insoluble sulphides are the dominant metal form in sediments, thermodynamic or equilibrium predictions dictate that metal concentrations in pore waters should be very low. Such predictions are verified by short-term laboratory experiments. Pore water metal concentrations are consistently very low where AVS > SEM and higher in sediments where SEM $\geq$ AVS. Such experiments are usually conducted using homogenised sediments (e.g. Di Toro et al., 1992; Hansen et al., 1996; van den Berg et al., 1998).

In nature the complexity of the sediment column and the influences of living processes can disrupt these predictable processes. For example, as a result of redox processes, metal concentrations in pore waters should follow AVS concentrations gradients (van den Berg et al., 1998). Metal concentrations should be lower at depth and increase toward the surface of the sediment column. Deviations from that simplistic pattern are common, however. Pore water metal concentrations will increase if an oxidised zone develops in the surface sediments (Fig. 6.13). But the precipitation of Fe and Mn oxides in the oxidised pore waters adds complexity to the profiles. It is not uncommon to see elevated pore water concentrations at the redox interface with reduced concentrations toward the sediment surface (Hare et al., 1994). Such observations are detectable only

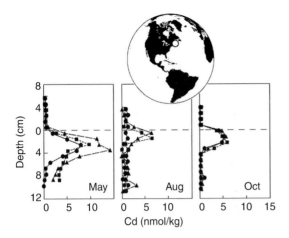

Fig. 6.13 Concentrations of Cd after 60 days, in pore water from replicate trays (different symbols) of sediments placed at the bottom of a lake in Quebec (Canada). The initial Cd : AVS ratio was 0.5 and sediments were homogenised. As an oxidised zone developed at the surface over the 60 day period, Cd increased in pore waters near the redox interface, but decreased again toward the surface, due either to scavenging by Fe and Mn oxide precipitation or diffusion out of the sediment. (After Hare *et al.*, 1994.)

in the most careful studies, using micro-scale sampling. But they reflect the realities of nature.

In nature, pore waters also accumulate colloids, organic breakdown products and complex soluble ligands like polysulphides over long periods of time, as a result of their close association with the sediment. These are all effective ligands for binding metals, and help pore waters retain metals that otherwise might be removed from solution. Careful micro-gradient studies of such sediments (Box 6.13) show that such influences on dissolved speciation have effects analogous to what happens in surface waters. Pore water metal concentrations are higher, at least in patches, than predicted from the thermodynamic characterisation of the surface sediments.

6.7   CONCLUSIONS

Metal contamination in sediments represents a concentrated reservoir that can retain historical inputs of metal contamination and be a source from which that contamination is redistributed to other locations. Wherever contaminated

Box 6.13 Processes affecting pore water chemistry

Relatively new techniques for studying pore waters demonstrate the influence of the intimate contact between pore waters and sediments (Zhang *et al.*, 2002). Thin films are deployed in natural sediments to yield high-resolution profiles of total dissolved and labile trace metals in pore waters (labile metals are those that diffuse into the gel). These are termed diffusional equilibrium (DET) and diffusional gradients in thin films (DGT). DGT exclusively measures labile metal species in the interstitial water (*in situ*) by immobilising them on a resin gel after they diffuse through a diffusive gel that excludes less reactive metal forms. For DET, an equilibrium is established between the DET gel and the pore water.

In general, many constituents (e.g. Fe and Mn) of the pore waters followed the general profiles predicted by redox conditions. But some metal concentrations showed millimetre-scale variability in concentration that would not be predictable from what is known of the thermodynamic controls on metal–sulphide interactions. The vertical profiles indicate that high concentrations of labile metals occur in pore waters at the same places in sediments where localised breakdown of organic material is also occurring. Release of organically complexed dissolved metal would be expected in such circumstances (Motelica-Heino *et al.*, 2003; Fones *et al.*, 2004). The breakdown of organic material and planktonic skeletal material in sediments can also complex Cd, Cu, Ni and Zn and solubilise them into pore water. Releases of particulate metal driven by locally produced dissolved organic material are not necessarily predicted from AVS-type models. Mixing of sediments by living organisms also adds complexity to the profiles. Thus, metal speciation in pore waters acts like metal speciation in the water column. Formation of dissolved metal complexes are as important as sediment redox status in influencing the particulate/water distribution.

sediments are deposited, the contamination will exchange among water, living organisms, detritus and inorganic particulate material. Some deposits are more accessible to food webs than others, in particular surface and suspended sediments. Some deposits are also more accessible to remobilisation than others (e.g. floodplain deposits).

Metal concentrations in sediments cover a very wide range in nature. The lower end of the range for each metal represents the natural background in geological materials. Background can be determined from pre-disturbance deposits. These are located using sediment cores, or from undisturbed watersheds of similar geology. The lowest quantile (10% to 20%) of a statistical regression of metal concentration versus a conservative property of sediment can be used to determine regional baseline concentrations, if it is not possible to determine geological background.

The frequency of low concentrations is accentuated, if samples include sediments dominated by sand-sized materials. Sands dilute trace metal contamination because of their high mass per unit surface area. Correction for grain size biases is essential for unbiased interpretation of metal concentrations in sediment. Useful correction protocols include normalisation, regression, or separation of fine-grained sediments for analysis.

Human activities can be responsible for concentrated, local hotspots of extreme sediment contamination, especially if waste discharges are not constrained. Regional-scale examples of extreme sediment contamination are rare, but moderate to severe regional contamination is common around areas of intense human activities. About 50% of sediments from large water bodies, surrounded by landscapes with intense human activities, are contaminated with metal concentrations in a range for which evaluating risks is a challenge.

Contamination of sediments will usually recede in response to reduction of inputs, but the rate of the return to less contaminated conditions depends greatly upon local sediment dynamics. Large numbers of hotspots of contamination are likely in any water body where human activities are intense and waste inputs are not constrained; these will feed regional-scale contamination.

The form of metal in sediments (distribution among substrates) is also important in determining ecotoxicological risk to biota. Determination of metal distributions among particulate Fe oxides, Mn oxides and organic materials in oxidised sediments is difficult because binding intensities and numbers of available binding sites for the different substrates are difficult to determine. Operational techniques (mathematical and chemical) indicate tendencies toward different types of associations but are not quantitative.

General differences in metal form between oxidised and reduced sediments are more readily discernible than differentiating oxidised forms. Reduced metal forms can be operationally differentiated from oxidised metal forms in homogenised samples by determination of acid volatile sulphide (AVS) and simultaneously extractable metal (SEM). But spatial distributions of sulphide (vertical and horizontal) are complex in natural sediments. The depth of the redox interface and average concentrations of sulphide vary seasonally with temperature and biological activity. Also, sulphides are not stable if chemical conditions of the sediment change or if anoxic sediment is mobilised into an oxic water column. These inherent characteristics of the sediment column add uncertainty to the precise determination of the AVS status of any particular sediment.

Sediment and water contamination are integrated by exchange between the two. Contamination in one is usually accompanied by contamination in the other. Nevertheless, the precise ratio of sediment/water concentrations (K_d) can be highly variable (50-fold to 1000-fold) across a range of environmental conditions. Climate, hydrology, metal concentrations, metal mass discharge, total suspended solids concentrations and organic carbon all affect K_d in both contaminated and uncontaminated situations. Metal speciation can also affect metal distribution between the phases. To address uncertainties K_d can be determined empirically, with little ambiguity, from dissolved and particulate metal determinations.

The spatial distribution of metal contamination is not static; it is dynamic on several scales. It is responsive to both natural processes and human activities. Considering those dynamics is as important in risk assessment, or in managing contamination, as is understanding toxic threats from concentration and form.

Suggested reading

Aller, R.C. (1978). Experimental studies of changes produced by deposit feeders on pore water, sediment, and overlying water chemistry. *American Journal of Science*, **278**, 1185–1234.

Axtmann, E.V. and Luoma, S.N. (1991). Large-scale distribution of metal contamination in the fine-grained sediments of the Clark Fork River, Montana. *Applied Geochemistry*, **6**, 75–88.

Daskalakis, K.D. and O'Connor, T.P. (1995). Distribution of chemical concentrations in US coastal and estuarine sediment. *Marine Environmental Research*, **40**, 381–398.

Elbaz-Poulichet, F., Garnier, J.-M., Guan, D.M., Martin, J.-M. and Thomas, A.J. (1996). The conservative behaviour of trace metals (Cd, Cu, Ni and Pb) and As in the surface plume of stratified estuaries: example of the Rhône River (France). *Estuarine, Coastal and Shelf Science*, **42**, 289–310.

Helgen, S.O. and Moore, J.N. (1996). Natural background determination and impact quantification in trace metal-contaminated river sediments. *Environmental Science and Technology*, **30**, 129–135.

Rickard, D.G. and Morse, J.W. (2005). Acid volatile sulfide. *Marine Chemistry*, **97**, 141–197.

7 • Trace metal bioaccumulation

7.1 BIOACCUMULATION OF TRACE METALS

Prologue. Bioaccumulation describes the net accumulation of a metal into the tissues of an organism as a result of uptake from all sources, and is typically a measure of an organism's metal exposure summed across all routes (diet or water). It is influenced by the chemistry of the specific metal, how that metal reacts with its environment and species-specific physiological and ecological traits. Important misunderstandings have limited the use of bioaccumulation data for metal management in the past. It will become an increasingly important tool in the future.

Bioaccumulation is defined as the *net accumulation of a chemical into the tissues of an organism as a result of uptake from all environmental sources, i.e. both food and ambient water* (ASTM, 2001, cited in Burkhardt *et al.*, 2003). Bioaccumulation is often a good integrative measure of the trace metal exposures of organisms in polluted ecosystems. It is of particular value for metals because they are not broken down into simpler forms. But trace metal bioaccumulation is complex, because of the interplay of:

- metal-specific influences;
- species-specific influences;
- environmental influences;
- exposure route.

The sections that follow will consider how these factors interact to determine metal uptake from different sources, by different organisms and under different conditions. We will consider specific geochemical effects on bioavailability. We will show how physiological processes can result in variable patterns of accumulation among species, so that vastly different concentrations of different metals are accumulated by different species (Eisler, 1981; Hare *et al.*, 1991; Hare, 1992; Rainbow, 2002; Cain *et al.*, 2004). And we will discuss conceptual and quantitative models that can explain why such patterns exist where they do; and make it feasible to begin relatively simple predictions.

Different perspectives on three questions (again the traditions of geochemistry, ecotoxicology and

---

Box 7.1 Definitions

absorbed metal: metal that has been taken up into an organism by crossing the cell membrane (e.g. of an epithelial cell on the surface or in an alimentary tract) with the potential to interact with physiological processes within an organism.

adsorbed metal: metal on the external surface of an aquatic organism that has not crossed a cell membrane (cf. *absorbed metal*) and may be desorbed by physical exchange with the surrounding medium. Adsorbed metal does not interact with the physiological processes within an organism.

assimilation efficiency: the efficiency with which an organism extracts metal from its food and absorbs it into its tissues (expressed as a percentage of the total metal ingested).

bioaccumulation: net accumulation of a metal into the tissues of an organism as a result of uptake from all environmental sources.

bioavailability: a relative measure of that fraction of the total ambient metal that an organism actually takes up when encountering or processing environmental media, summated across all possible sources of metal, including water and food as appropriate.

biodynamic modelling: derives net bioaccumulation for a species from a balance among uptake rate from diet, uptake rate from food, rate of loss from tissue and rate of growth. Derives net bioaccumulation at a location by combining the quantified physiological processes above with site-specific concentrations and geochemical conditions.

biokinetic modelling: see *biodynamic modelling*.

carrier protein: also termed *transporter protein*. Carrier proteins traverse the lipid bilayer of the apical membrane of epithelial cells. They bind metal ions then pass them across the lipid bilayer and release them on the intracellular side of the membrane.

concentration: in tissue expressed here as micrograms metal per gram tissue dry weight or μg/g dw.

content: the mass of metal in an organism expressed as micrograms metal (μg); a weight-free measure.

desorption: The physical release into solution of metal adsorbed on the outside of an organism.

detoxification: the process by which a metal released intracellularly is bound by a ligand of such high affinity that it in effect prevents that metal from binding to other molecules, in either an essential role or to cause toxicity. Trace metal detoxification processes may involve soluble ligands (e.g. metallothioneins, glutathione), insoluble granules or intracellular deposits.

efflux rate constant: a constant defining the proportion of the tissue concentration of metal lost per day, expressed in the units d^{-1}, derived from efflux experiments using a unique tracer. Rate constants of loss are metal- and species-specific, but are constant for each metal and species combination.

facilitated diffusion: diffusion of metal into a tissue as facilitated by a carrier to aid passage across the membrane. Facilitated diffusion does not require energy.

gut passage time (GPT): the time between ingestion of a bolus of food and its first appearance in the faeces.

gut residence time (GRT): the time between ingestion of a bolus of food and the complete elimination of all the undigested material (including metal) from the digestive tract.

major ion channel: formed from a protein which traverses the lipid bilayer of the membrane of epithelial cells. The proteins form temporary aqueous pores of a specific diameter; specifically suited, for example for a major ion like calcium, potassium or sodium. Ion channels open and close according to the cell's requirements.

metabolically available metal: intracellular metal that has not been detoxified and is therefore available to bind with molecules in the cell, in either an essential metabolic role or to cause toxicity.

metallothioneins: low molecular weight cytosolic proteins with high cysteine contents which bind and detoxify metals. Synthesis can be induced by the presence certain trace metals (e.g. Zn, Cu, Cd, Ag, Hg).

metallothionein-like proteins: proteins that appear to be metallothioneins but which await final analytical identification of all characterising features of metallothioneins.

transporter protein: see *carrier protein*.

trophic transfer: transfer of metal from one level of a food web (trophic level) to another.

Trophically Available Metal (TAM): fraction of the accumulated metal concentration in a food (prey) organism that is available for trophic transfer to the feeding (predator) animal.

uptake rate constant: a constant (species- and metal-specific) that describes the uptake rate of metal per unit metal concentration in the medium. For dissolved metal, for example, the units are microgram metal per gram tissue per day, per microgram metal per litre of water (μg/g/d per μg/L). Because formal mathematics requires cancelling units where possible, these rate constants are usually expressed as L/g/d.

natural system studies sometimes collide), have led to some confusion about whether bioaccumulation is a useful measure for managing metal contamination.

- How does metal bioaccumulation respond to changes of metal concentrations in the environment?
- Are bioaccumulated metals representative of toxic exposures?
- What are the proper uses of bioaccumulation in animals from nature in monitoring, assessing or evaluating site-specific metal contamination issues?

In this chapter we will address aspects of all three questions, although later chapters link bioaccumulation, biomonitoring and toxicity in detail. We expect that future managers of metal contamination will find many more uses for bioaccumulation data than was traditional in the past.

7.2   PROCESSES GOVERNING BIOACCUMULATION

Prologue. Net bioaccumulation is the result of a balance between uptake and loss. Growth rates can also affect the concentrations of metal bioaccumulated by an organism.

The bioaccumulation of a trace metal by any organism in any habitat depends on the balance between two processes – the uptake of the trace metal by the organism summated across all sources, and the loss of that metal from the organism summated across all routes of loss. The content or load (often expressed in micrograms – μg) of a trace metal in an organism is therefore the net difference between uptake and loss integrated over time. The concentration of a trace metal is expressed as content per unit weight, for example micrograms per gram (μg/g) or milligrams per kilogram (mg/kg). It is also usually expressed in terms of dry weight (μg/g dw).

Concentration can be affected by changes in the weight of the organism. If weight remains constant, uptake and loss may ultimately balance each other and a steady state concentration may be reached. But weight is variable with time in most

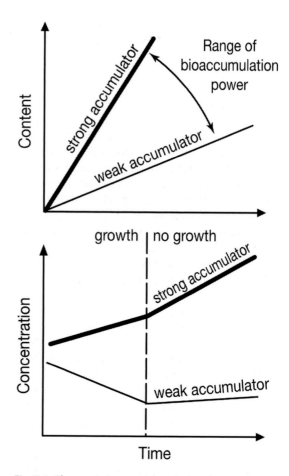

Fig. 7.1 Changes in trace metal content and concentration (content per unit weight) over time in organisms with different strengths of accumulation of that trace metal, during and after a period of growth. As a result of the relative rates of growth and metal accumulation, content may increase while concentration decreases.

organisms. Weight changes may be positive or negative, being controlled by growth, the build-up or loss of gametes in the reproductive season, the build-up or utilisation of energy reserves, etc. If weight changes, the concentration of a trace metal will change if trace metal content stays constant. The degree of change is dependent upon the rate of weight change compared to the rates of metal uptake and loss.

Figure 7.1 illustrates typical relationships between trace metal content and concentration and time in organisms with different strengths of trace metal accumulation, during and after a period of growth.

7.2.1   Uptake processes

Prologue. Uptake is determined by how much metal crosses the cell membrane. In animals this means the surface epithelial cells or cells of the gut epithelium. Adsorption onto the body surface can add to the body content of trace metal; adsorbed metal has not crossed the cell membrane and has not therefore been taken up by the organism. Adsorbed metal plays no role in the ecotoxicity of metals. Bioavailability describes a relative measure of that fraction of the total ambient metal that an organism actually takes up.

More than one uptake route for trace metals is potentially available for the entry of trace metals into multicellular organisms. An aquatic invertebrate can gain metals from solution across the permeable body surface (whether or not such permeable areas are localised into gills), and also from the diet via the alimentary canal (gut). In all cases trace metals need to cross the membrane of a body cell.

An epithelium forms the covering of most internal and external surfaces of an animal's body. Trace metals must cross the apical cell membrane of an epithelium cell, whether that cell lies on the outside of an organism (perhaps on a gill) or within an alimentary tract. Metals that have crossed the apical membrane of a body cell are referred to as metals absorbed into an organism.

Distinct from absorbed metal is metal that is adsorbed onto the surface of an organism. Adsorbed metal has not crossed a membrane and can undergo desorption according to the physicochemical conditions of the habitat. Adsorbed metal may be a measurable component of total body load in the case of surface-active trace metals like manganese on the exoskeleton of an aquatic crustacean, but has no potential to play any ecotoxicological role within the organism concerned.

Processes and mechanisms of uptake of metals into cells are similar whether the cells are on external surfaces or in the alimentary tract. But very different physicochemical factors may control how those metals are presented to the cell surface, for example, in the gut as opposed to at the gill surface. Bioavailability describes a relative measure of that fraction of the total ambient metal that an organism actually takes up when encountering or processing environmental media, summated across all possible sources of metal, including water and food as appropriate (Rand *et al.*, 1995). Thus the total metal bioavailability to an aquatic invertebrate animal includes the fraction of total dissolved metal extracted by permeable epithelial surfaces from water passed across the gills and/or bathing the body (including where appropriate interstitial water in sediment). It also includes the fraction of metal taken up into the alimentary tract epithelium from the total metal content of ingested food particles.

7.2.1.1   Processes at the membrane

Prologue. Protein carriers and major ion channels provide the major (but not always only) means by which hydrophilic metal species cross the hydrophobic cell membrane. Transport of most metals from outside to inside a cell is usually not energy dependent. An inward diffusion gradient is sustained by the rapid binding of intracellular metal in non-diffusible form. Definable membrane traits determine the metal-specific and species-specific characteristics of uptake. Environmental factors are also important. Only specific forms of metal (e.g. the free metal ion) are transported by particular routes. Many forms of metal common in the environment are unavailable for transport.

Figure 7.2 is a simplified schematic of the ways that trace metals may cross the membrane of a cell. Details of trace metal transfer mechanisms into cells continue to emerge in this era of molecular biology research, but will not be covered here. Discussion here is restricted to multicellular eukaryotic organisms, with most examples drawn from invertebrate animals. We exclude the prokaryotic archaea and bacteria, and the unicellular protistans like amoebae.

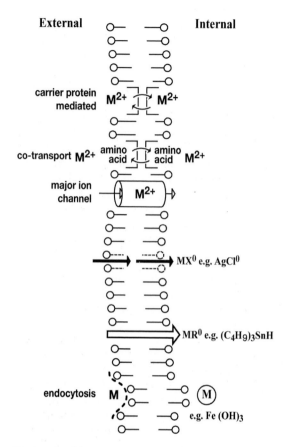

Fig. 7.2 Possible routes by which trace metals may cross the apical membrane of an epithelial cell, for example of an aquatic invertebrate. Such an illustration must necessarily be a simplification in a book aimed at the non-specialist. New details of potential specific trace metal transfer mechanisms into cells (particularly metal carrier proteins) are being described increasingly often in this era of molecular biology research.

The eukaryotic cell membrane consists fundamentally of a lipid bilayer traversed by many proteins (Quinnell et al., 2004). The lipid bilayer acts as a barrier to the passage of water and water-soluble chemicals into the cell. It is the embedded proteins that typically allow (and control) soluble chemicals to traverse the membrane. The cell membrane is impenetrable to dissolved trace metals that are in ionic form (e.g. Cd^{2+}) without the intercession of membrane proteins.

Proteins function either as carriers (transporters) or channels, passing metal ions down concentration gradients (Quinnell et al., 2004). Carrier proteins bind the metal ions, pass them across the lipid bilayer and release them on the other side of the membrane. Channel proteins form temporary aqueous pores of a specific diameter. For example ion channels exist specifically to carry major ions like calcium, potassium or sodium. Ion channels open and close according to the cell's requirements. Proteins of either type often show high degrees of specificity to particular metals.

As discussed in Chapter 4, trace metals have a high affinity for the elements nitrogen and sulphur. These two elements occur in the side chains of several of the amino acids that make up the structure of proteins, like the carriers. Nitrogen, for example, is found in the amino acid histidine and sulphur is common in cysteine (Lippard and Berg, 1994). Trace metals dissolved in the medium bathing an epithelial cell will bind to certain carriers in the apical cell membrane via the amino acids. The actual specificity of such carriers is a topic of continuing research but essential metals like zinc may have specific carriers and a chemically similar non-essential trace metal like cadmium may ride on these zinc carriers.

Perhaps surprisingly at first sight, given the typically low dissolved concentrations of most trace metals in most natural waters, transport via a carrier from the outside to the inside of a cell does not involve the need for energy. The concentration gradient required for the process of passive facilitated diffusion (Box 7.2) is maintained by rapid binding of the metal once it reaches the cell interior. There is not a concentration gradient of total metal; total metal concentrations

---

Box 7.2 Passive facilitated diffusion

Passive facilitated diffusion is 'passive' because the metal ion follows a concentration gradient. It is facilitated because the metal ion crosses the lipid membrane on a protein 'carrier'. It crosses by 'diffusion' down a concentration gradient and so no energy is required.

Box 7.3 Free Ion Activity Model (FIAM)

The Free Ion Activity Model proposes that the activity of the free metal ion is a good predictor of the bioavailability of a dissolved trace metal to aquatic organisms (Campbell, 1995). To be taken up by (traverse the apical membrane of an epithelial cell of) an organism, a metal must interact with a cellular ligand at the cell membrane, forming a metal–ligand complex. By the FIAM, the free metal ion is a good predictor of the form of metal that binds to this cellular ligand. Uptake is assumed to be proportional to the concentration of metal in this cell surface metal–ligand complex. Provided the number of free cellular ligands at the cell surface is approximately constant, then uptake rate will vary as a function of the activity of the free metal ion. Amongst the various assumptions underlying the FIAM (Campbell, 1995) is that metal transport in solution towards the cell membrane occurs rapidly in comparison to the metal uptake process.

Only a proportion of the total dissolved metal concentration may be in the form of the free metal ion, much of the remaining dissolved metal being complexed by inorganic or organic complexing agents – see Chapter 5. (The chemical activity of a dissolved substance is not actually the same as the concentration, being a fraction of the concentration as determined by an activity coefficient. In practice, most experimental data in this field consist of free metal ion concentrations. The difference is negligible in comparison with inherent biological variability (Campbell, 1995).)

If facilitated diffusion via a carrier (transporter) protein is the uptake process, then the FIAM predicts the rate of uptake well. The transport of metals via major ion channels would also be predicted by the FIAM. There are, however, other potential routes by which a trace metal can cross the apical cell membrane of an epithelial cell (see Box 7.4). For example, the FIAM is not a good predictor of the rate of uptake if a complexed metal accompanies a nutrient (like an amino acid) when it is transported (as occurs in the digestive tract), or if the metal crosses the membrane via diffusion of a neutral or otherwise lipophilic species (as proposed for the chloro-complex of silver and some organometal compounds). The success of the FIAM as a first-order predictor of the rate of uptake of a trace metal from solution in most circumstance, indicates the predominance of the carrier protein (and/or major ion channel) route of uptake.

The free metal ion model was first defined for marine waters (Sunda and Guillard, 1976; Anderson and Morel, 1977), which are of relatively stable chemical composition. As more studies developed in freshwaters, exceptions to exact predictions of the FIAM were noted. The exceptions (Campbell, 1995) reflected alternative (otherwise less significant) trace metal uptake routes (Box 7.4), effects on facilitated diffusion by competition with other ions for uptake sites or protonation of the carrier (i.e. at low pH).

are usually higher within cells than they are externally. It is a concentration gradient of the specific chemical form of the trace metal that binds with the carrier protein externally.

Not surprisingly the specific chemical form of the dissolved trace metal that is available for binding with the carrier protein on the membrane has generated much interest. For those metals that behave as cations, experimental results are best explained if this dissolved form is considered to be (or rather, is best modelled by) the free metal ion, strictly the hydrated free metal ion (see Chapter 5). This concept is commonly referred to as the Free Ion Activity Model (FIAM) (Campbell, 1995), considered further in Box 7.3.

On the inside of the cell any free metal ion released from the carrier is not going to remain in free ionic form for long. The protein-rich environment of the cell offers numerous immediate opportunities for a free metal ion to bind strongly. Thus the transported ion is immediately bound within the cell, maintaining an intracellular concentration of free metal ion that is effectively zero. This maintains the concentration gradient of free metal ion from outside to in, and promotes continuing passive uptake of the trace metal by facilitated diffusion via the carrier protein. It is important for the successful functioning of the cell that the entering trace metal does not bind to the nearest protein indiscriminately or toxic

Box 7.4 Trace metal uptake routes

After passive facilitated diffusion via a metal-specific carrier (transporter) protein, the second route that may be used by trace metals involves trespass on the routes used for the uptake of the physiologically 'major ions'. The transport of the major ions sodium, potassium and calcium across cell membranes typically involves the expenditure of energy (ATP) in a membrane-associated ATPase (pump) to maintain an appropriate concentration gradient between the outside and inside of the cell, and specific major ion channels across the membrane.

Cadmium and calcium represent one example of such an interaction between a major ion and a trace metal. The concentration gradient is set up by the active pumping of calcium out of the cell from the base of the gill epithelial cells. This promotes the entry of calcium into the cell from the medium through the apical calcium channels. As it happens, the free Cd ion (0.92 Å) is of a very similar ionic radius to the calcium ion (0.94 Å), and it is chemically inevitable that some Cd from the external medium will enter epithelial cells through any calcium channel (Flik et al., 1995; Rainbow and Black, 2005b). Whether this route of entry for Cd is as significant as that of Cd entry via facilitated diffusion on trace metal carrier proteins will depend on the specific animal in question. The activity of any process that is varying the uptake of calcium is also important. For example, calcium uptake activity is high in some marine gastropod molluscs that lay down a heavy shell; or in a post-moult crab calcifying its soft new exoskeleton.

Similarly in teleost fish, Zn uptake in the gills appears to occur at least partly via a calcium channel in gill chloride cells, supplementing Zn uptake via a Zn transporter (Bury et al., 2003). A proportion of the uptake of dissolved Cu and Ag by teleost fish gills will be via a sodium channel (Bury and Wood, 1999; Bury et al., 2003). Transport of Cu via the sodium channel indicates it is the oxidation state Cu^+ rather than Cu^{2+} that is transported.

Some trace metals, including As, Cr, Mo, Se and V, do not follow the FIAM because their commonly dissolved ionic forms in seawater are anionic (arsenate $HAsO_4^{2-}$, chromate CrO_4^{2-}, molybdate MoO_4^{2-}, selenate SeO_4^{2-} and variations on vanadate e.g. HVO_4^{2-}) (Bruland, 1983) (Chapter 5). Their chemical behaviour offers potential for their entry into cells via routes for sulphate or phosphate (Lippard and Berg, 1994). Silver in a silver-thiosulphate complex (Fortin and Campbell, 2001) or Cd as a cadmium-citrate complex (Errécalde and Campbell, 2000) may also enter cells via anion transport systems.

Non-ionic, non-polar chemical species are more soluble in lipid than their ionic counterparts, and so some trace metals may, enter the cell in this form (Florence and Stauber, 1986; Phinney and Bruland, 1994) (Fig. 7.2). Organometallic compounds are also typically lipophilic and therefore able to cross lipid bilayers relatively easily (see Quinnell et al., 2004). Table 7.1 illustrates how the presence of different alkyl groups, which increase the lipophilicity of mercuric chloride, increases its toxicity.

It is also possible for a metal to enter a cell in particulate form by pinocytosis (Fig. 7.2). Possible examples of this route are the uptake of particulate Fe oxides or hydroxides by the gill cells of the lamellibranch bivalve mollusc, the mussel Mytilus edulis (George et al., 1976), or particulate V in the pharynx of the ascidian tunicates (sea squirts) (Kalk, 1963). The uptake of metals in particulate form is also very likely in the gut epithelial cells of those invertebrates that rely to a significant extent on digestion within cells (intracellular) as opposed to digestion only in the lumen of the gut (extracellular digestion).

Co-transport with amino acids is less well studied, but it is a major mechanism for methyl mercury uptake (Chapter 14) and could be of great importance in metal transport in the gut (Glover and Hogstrand, 2002).

Table 7.1 *The relative toxicities of different organic forms of mercury to nauplius larvae of the barnacle* Elminius modestus *(after Corner and Sparrow, 1957). Note that increased lipophilic character increases toxicity*

Mercury compound	Relative toxicity
Cl – Hg – Cl	1
CH_3 – Hg – Cl	4.7
C_2H_5 – Hg – Cl	6.8
C_4H_9 – Hg – Cl	16
C_5H_{11} – Hg – Cl	20

effects may ensue. The presence of intracellular molecules of extra high affinity for the incoming metal (detoxificatory molecules) will limit the toxicological potential of the new metal and begin the physiological process of metal detoxification.

Although facilitated diffusion via a protein carrier is usually considered to be a major, often dominant, route of uptake of trace metals by aquatic invertebrates, it is not the only possible uptake route, as depicted in Fig. 7.2. Other routes may include transport via the pathways (channels) used by major ions, direct diffusion of non-polar forms through the lipid bilayer (a simplified description of the lipophilic transport processes involved: Quinnell *et al.*, 2004), transport as an organic complex with a molecule for which a specific carrier system also exists (e.g. 'piggybacking' with amino acids) or transport in particle form (Box 7.4).

Trace metals may, therefore, potentially enter an epithelial cell and thence the body of an organism by a variety of routes and indeed by several routes simultaneously. The various membrane transport routes are a consequence of the variety of environmental and physiological conditions organisms encounter. All routes are not equally important at all times, nor is more than one route always quantitatively significant in the total uptake of trace metals. Indeed the relative importance of different routes of entry will change under different conditions, whether these conditions are the physicochemical conditions of the medium bathing the cell surface or the physiological conditions within an organism.

7.2.1.2 Uptake of dissolved forms
Prologue. Concentration, environmental factors and species-specific biological processes control uptake of dissolved metal.

The uptake of trace metals from solution by an aquatic organism is primarily concentration dependent. The higher the dissolved concentration of the trace metal, the higher will be the uptake rate of the metal from solution into the organism, until (if ever under real environmental circumstances) the uptake mechanism becomes saturated.

The concentration dependence is modified by external factors characteristic of the environment, and internal factors characteristic of the organism.

7.2.1.2.1 Uptake rates and K_u
Prologue. The characteristics of membrane transport dictate that metal uptake from solution will be concentration dependent. An uptake rate constant (K_u) characterises metal-specific uptake in a species. K_u is determined from the linear correlation of uptake rate with concentration, under a given set of geochemical conditions. Metals have a consistent rank order of uptake rates, typical of most, if not all species.

Mechanistically, the uptake rate of metal across the membrane reflects the interaction between bioavailable metal concentration (external) and the (internal) characteristics of the biological transport system. This interaction is manifested as an influx rate, defined for each species and metal by an influx rate constant (K_u). For each metal, a specific affinity can be defined for its attachment to carriers (or to channel proteins as appropriate), and a specific capacity (number of carriers) can be defined for transporting the metal. These species-specific membrane transport characteristics are important in driving uptake rates for each metal. The pattern of uptake rate as a function of concentration in the medium is used to quantify the affinity and the maximum uptake rate characteristic of the transport system. These experiments define what is called Michaelis–Menten kinetics.

Experimentally, the rate of passive facilitated diffusion of a trace metal across the membrane via a carrier protein is proportional to the delivery of the bioavailable form of the metal to the carrier. The same is true for other trans-membrane mechanisms depicted in Fig. 7.2. For example, if a trace metal is trespassing on a major ion channel, as in the case of Cd and calcium channels, increases in the free trace metal ion activity will still increase the rate of uptake of the trace metal. The free metal ion activity in solution is proportional to the total dissolved concentration in these experiments because all other physicochemical factors are held equal.

Where uptake rates are measured over environmentally realistic dissolved concentration ranges, they show a linear relationship with concentration. The slope of the relationship defines the uptake rate constant (K_u), expressed as μg/g/d per μg/L, or L/g/d. K_u is now widely accepted in bioaccumulation modelling as a mechanistic term that can be quantitatively compared among species, metals and environmental conditions (Wang, 2002; Luoma and Rainbow, 2005).

Carrier-facilitated membrane transport is potentially saturable. If bioavailable concentrations increase to the point of saturating all carrier sites, then the rate of uptake will reach a plateau (uptake rate will not increase further with concentration). But does the rate of trace metal uptake from solution reach saturation in real environmental situations as opposed to laboratory experimental exposures? Most studies indicate that uptake rates change linearly with concentration over the range of dissolved metal concentrations that might realistically be expected in nature (including heavily contaminated locations: Chapter 5). But rates appear to become non-linear (saturate) at extremely high metal concentrations (Hogstrand *et al.*, 1994; Buchwalter and Luoma, 2005) such as those used to expand the range of toxicity tests (Fig. 7.3).

The major ions, which are the prime users of the channels potentially hijacked by trace metals, are orders of magnitude more abundant than the trace metal ions themselves in all aquatic environments, so saturation of major ion channels by the trace metal ions appears environmentally unrealistic. Saturation might be more likely if a trace metal directly parasitises an enzymatic membrane pump such as an ATPase transporting a major ion, given that all enzymes show saturation kinetics. In oysters (*Crassostrea virginica*) Cd uses only a subset of the available Ca channels, which might also explain why there is a ceiling on its uptake rates (Roesijadi and Unger, 1993).

K_u follows a consistent metal-specific rank order among most if not all species (see Table 7.2 for the rank order of constants among three bivalves). For the most fully studied metals, the order of the uptake rate constant is:

$$Ag > Zn > Cd > Cu > Co > Cr > Se$$

(Wang, 2001; Croteau and Luoma, 2005). Absolute values for uptake rate constants vary among species (internal factors). Environmental conditions (external factors) affect both the absolute K_u and the rank order.

7.2.1.2.2 External factors affecting dissolved uptake (physicochemistry)

Prologue. External factors affect uptake rates by influencing the abundance of the bioavailable form (e.g. the free metal ion activity as dictated by the FIAM) and creating or reducing competition for uptake sites. Dissolved speciation models can predict free ion activities or oxidation state, both of which correlate with bioavailability and uptake rate.

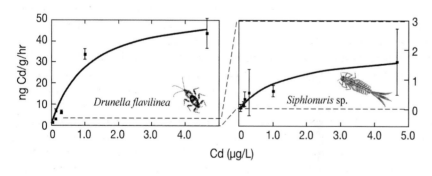

Fig. 7.3 Cadmium uptake rates as a function of dissolved Cd in larvae of two mayflies *Drunella flavilinea* and *Siphlonuris* sp. (Buchwalter and Luoma, 2005). Uptake rates of trace metals from solution can differ by orders of magnitude. Uptake rates appear to become non-linear (saturate) at extremely high metal concentrations such as those used to expand the dissolved exposure range of laboratory toxicity tests.

Table 7.2 *Comparative mean uptake rate constants (L/g per day ± 1 standard deviation) of four metals from solution of three bivalve molluscs*

Metal	*Perna viridis*	*Septifer virgatus*	*Ruditapes philipinnarum*
Zn	0.483 ± 0.051	0.350 ± 0.076	0.191 ± 0.029
Cd	0.182 ± 0.035	0.180 ± 0.053	0.054 ± 0.013
Cr(VI)	0.037 ± 0.012	0.041 ± 0.008	0.020 ± 0.006
Se(IV)	0.019 ± 0.004	0.031 ± 0.007	0.009 ± 0.002

Source: After Wang (2001).

Where competition with major ions is important, the speciation model can be adjusted to predict uptake at the site of membrane transport (the Biotic Ligand Model).

For each metal, external factors other than simple changes in total dissolved concentrations can also affect the rate of uptake of trace metals from solution. The effects of changes in such physico-chemical factors are explicable in terms of how they affect:

- the free metal ions available for interaction with protein membrane carriers (or ion channels, etc.);
- competition for the uptake routes themselves; or
- changes in the relative access to the different uptake routes on the membrane.

7.2.1.2.2.1 Chelation or complexation

The role of the FIAM in understanding the bioavailability of dissolved metal was first explained using organic ligands and effects on phytoplankton in seawater (Sunda and Guillard, 1976). Early experiments showed that addition of well-known water-soluble chelating agents affected the activity of the free copper ion. Increasing dissolved organic carbon (DOC) reduced the activity of the free copper ion through equilibrium partitioning to the dissolved ligands. Many similar experiments have since been conducted with animals and plants. For example, the addition of the chelating agent EDTA (ethylene diamine tetra-acetic acid) reduces the absolute equilibrium concentration

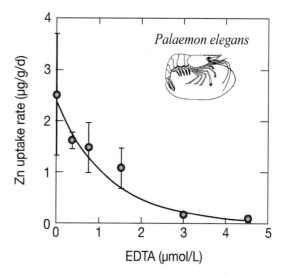

Fig. 7.4 The addition of the chelating agent EDTA (ethylene diamine tetra-acetic acid) reduces the absolute equilibrium concentration of the free Zn^{2+} ion, and reduces the mean rate of zinc uptake (± 1 standard deviation) by the decapod crustacean *Palaemon elegans* exposed to 100 µg/L (1.53 µmol/L) Zn at 10 °C. (After Nugegoda and Rainbow, 1988a.)

of the free Zn^{2+} ion. That reduces the rate of zinc uptake by the amphipod *Orchestia gammarellus* (Rainbow *et al.*, 1993b) and the decapod *Palaemon elegans* (Fig. 7.4) (Nugegoda and Rainbow, 1988a), shown in the latter case to be in direct correlation with the predicted free zinc ion concentration (O'Brien *et al.*, 1990).

Some environmental conditions trigger alternative transport mechanisms to those in which the bioavailable form is predicted by the FIAM. The presence of the lipid-soluble ligands can enhance uptake. For example, k ethyl xanthogenate and oxine increase the toxicity of copper to the amphipod *Allorchestes compressa*, in contrast to the decreased copper toxicity caused by water-soluble chelating agents tannic acid and NTA (nitrilotriacetic acid) (Ahsanullah and Florence, 1984). This is of practical importance in the case of the association of trace metals with some lipid-soluble pesticides. Carbamate pesticides (e.g. 'Ziram') are lipid soluble because of their external structure, but can bind metals within the

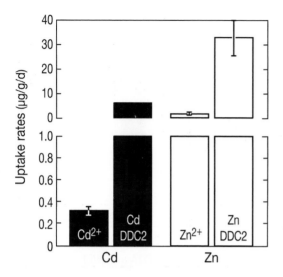

Fig. 7.5 Carbamate pesticides like 'Ziram' (DDC2) are lipid soluble because of their external structure, but can bind metals within the molecule (Zn or Cd). Uptake of dissolved Cd by the bivalve *Corbula amurensis* is accelerated 10-fold, compared to the free ion, when Cd is associated with 'Ziram' (Wellise, 1999).

molecule (Zn or Cd). Uptake of dissolved Cd by the bivalve *Corbula* (*Potamocorbula*) *amurensis* is accelerated 10-fold, compared to the free ion, when Cd and Ziram occur together (Wellise, 1999) (Fig. 7.5). Cd appears to dissociate from the pesticide after it enters the organism, however.

Ag is transported across membranes as the free ion in freshwaters. But the predominant form of Ag in brackish and marine waters is a strong zero-charge, non-polar, chloro-complex. This complex is highly bioavailable (Luoma *et al.*, 1995), and the uptake rate constant is one of the highest for any metal in seawater. $AgCl^0$ is bioavailable presumably because it can diffuse through the membrane (Fig. 7.2).

The rank order of K_u for metals that behave as anions does not follow the FIAM. Their uptake rate is affected by oxidation state. For the best known of these, the bioavailable oxidation states are: Cr(VI) > Cr(III); Se(IV) > Se(VI); As(III) > As(IV). Among the available oxidation states, the K_u for both Cr(VI) and Se(IV) are considerably lower than for other trace metals.

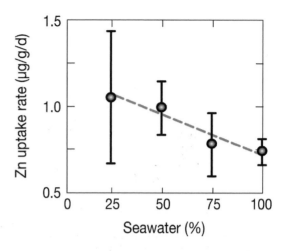

Fig. 7.6 The mean rate of uptake of Zn (± 1 standard deviation) by the decapod crustacean *Palaemon elegans* exposed to 56.2 µg/L Zn at 10°C increased as salinity was reduced from 35 to 7. (After Nugegoda and Rainbow, 1989.)

7.2.1.2.2.2   Major ion concentrations and salinity

Changes in salinity will often affect the rate of uptake of particular trace metals by marine or estuarine organisms in the absence of changes in the total concentration of dissolved metal. Uptake rates of Zn and Cd, for example, increase with decreased salinity. Thus the rate of uptake of Zn by the decapod crustaceans *Palaemon elegans* (Fig. 7.6) and *Palaemonetes varians* increased 50% as salinity was reduced from 35 to 7 (Nugegoda and Rainbow, 1989).

Enhanced bioavailability at low salinity may be explained simply by physicochemical changes in solution causing changes in the free metal ion activity in solution. In particular, the K_u for Cd will decline as salinity increases because the strong, unavailable chloro-complex increase in dominance as chloride concentrations increase.

The changes in the rate of Zn and Cd uptake by *O. gammarellus* at salinities down to 25 (Fig. 7.7) correlated with the modelled free metal ion concentrations in solution (Fig. 7.8) (Rainbow *et al.*, 1993b). The proportion of free ion increases more for Cd than for Zn at the lowest salinities (Fig 5.5). Cd uptake rates therefore change more than Zn uptake rate at very low salinities, for those animals that can survive wide salinity ranges (Box 7.5).

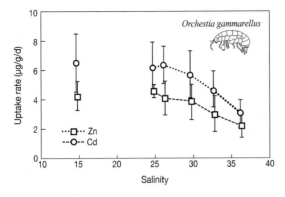

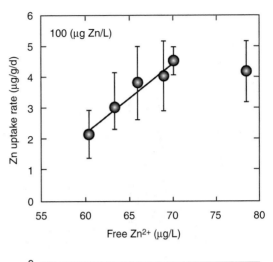

Fig. 7.7 Decreased salinity down to 25 increased the mean rates of uptake of Zn and Cd (±1 standard deviation) by the amphipod crustacean *Orchestia gammarellus* exposed to 100 μg/L Zn or 50 μg/L Cd at 10°C. (After Rainbow *et al.*, 1993b.)

The predictive capabilities of the FIAM are affected in freshwaters by differences in the concentrations of competing cations (Meyer *et al.*, 1999). Competition can decrease the amount of the trace metal bound to receptor sites (e.g. as increasing calcium concentration may do). To create a model that can predict these effects, Pagenkopf (1983) included a biotic ligand as the receptor in a speciation model, allowing consideration of both speciation effects and competition effects (often represented as 'hardness' in freshwater). The incorporation of a biotic ligand into a geochemical speciation model has produced the Biotic Ligand Model (Playle *et al.*, 1993; Chapter 15).

7.2.1.2.2.3 Differences in pH

In freshwater habitats, changes in pH will have profound effects on the speciation of dissolved trace metals (Campbell and Stokes, 1985; Gerhardt, 1993). Decreasing pH reflects increasing hydrogen ion concentrations. Positively charged hydrogen ions protonate (occupy) ligands in solution, replacing metals and causing increased free metal ion activity. The extra H^+ ions at low pH may also compete with metal ions at the membrane binding site and decrease the rate of metal uptake. The

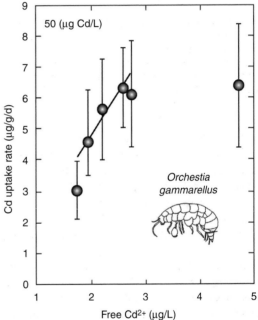

Fig. 7.8 Changes in mean uptake rates of Zn and Cd (±1 standard deviation) by the amphipod crustacean *Orchestia gammarellus* exposed to 100 μg/L Zn or 50 μg/L Cd at 10°C correlate with free metal ion concentrations at salinities down to 25 (at the upper limit of the straight line in each case). Below salinity 25, the correlations break down as the amphipod makes a physiological response to the low salinity (Rainbow *et al.*, 1993b; Rainbow and Kwan, 1995).

Box 7.5 Effects of salinity on uptake rate constants for Zn and Cd

In the absence of dissolved organic material, both Zn and Cd in seawater are predominantly complexed with inorganic ligands such as chloride. Indeed in full salinity seawater, only 47% of the Zn and 2.5% of the Cd exists as the free metal ion (Mantoura *et al.*, 1978; Bruland, 1983; Rainbow *et al.*, 1993b). As salinity decreases, the free ion becomes a larger proportion of the inorganic species in solution (Fig. 5.5) and bioavailability increases. But the change is proportionately greater for Cd than for Zn. From a salinity of 30 to a salinity of 5, free ion concentrations of Zn double, while free ion concentrations of Cd increase 20-fold. The cadmium uptake rates of the bivalves *Corbula (Potamocorbula) amurensis* and *Macoma petalum* (as *M. balthica*) increased about fivefold for Cd with salinity reduction from 30 to 5 (Lee *et al.*, 1998). The greatest increase occurred at the lowest salinities (salinity 5 to 10), as predicted from the speciation model. There was no detectable change in Zn uptake in either bivalve between the highest and lowest salinities, the result of the difficulty of detecting the impact of only a twofold change in free ion concentration on uptake rate (Lee *et al.*, 1998).

balance between these two contradictory effects determines uptake rate. Hare and Tessier (1998) compared the bioavailability of dissolved Cd to an invertebrate predator (*Chaoborus* sp.) across 28 Canadian lakes with different dissolved Cd concentrations, different geochemistry and different pH. They found that low pH enhanced Cd uptake as predicted by the presence of greater free ion activity; but overall uptake was best predicted if they also took into account the counter-effect of competition between free Cd ion and H^+ ion at the biological membrane.

7.2.1.2.2.4   *Other trace metals*
The presence of other trace metals in solution has the potential to affect the uptake of a trace metal by an aquatic organism (Amiard-Triquet and Amiard, 1998). Interaction between trace metals may occur via the disturbance of physicochemical equilibria in solution controlling trace metal

speciation, or at the point of interaction with a carrier protein or a membrane channel with subsequent effects on uptake rates. If accumulated concentrations are used to infer uptake rates, it must be remembered that interaction between trace metals may occur within the cell (after uptake) with consequent effects on subsequent accumulation rates. The interactive effects of competing trace metals are quite clear at very high metal concentrations. Results are somewhat ambiguous as to how important those interactions are at the concentrations that occur in nature (contaminated or uncontaminated).

7.2.1.2.3   Internal (biological) factors affecting dissolved uptake (physiology)
Prologue. Different organisms take up dissolved metal at different rates under the same conditions. Factors that determine differences among species include the number of uptake sites, affinity for metal at the site of uptake, rates at which water is passed over the respiratory surface (clearance rates) and physiological differences in water permeability. Within species, body size and previous exposure to metals also can be influential.

Corrections for geochemical conditions in a water body are increasingly common in risk assessments and site-specific water quality guidelines for each metal. But such corrections cannot account for the differences in uptake and bioaccumulation among species. For that, biology must be considered. Organisms differ widely in their physiological make-up and possess sophisticated means of exploiting and/or physiologically adapting or acclimating to their environment. Some of these adaptations affect how they respond to trace elements. Influences of metal and speciation occur within that context.

7.2.1.2.3.1   *Species specificity of K_u*
Different organisms take up trace metals at different rates under identical physicochemical conditions. These biological differences often are large. For example, uptake rates can differ by orders of magnitude among aquatic larvae of insect species (mayflies, caddisflies and stoneflies); differing

Table 7.3 *Zn and Cd uptake rates from solution by a decapod crustacean (*Palaemon elegans*), an amphipod crustacean (*Echinogammarus pirloti*) and a barnacle (*Elminius modestus*) under identical physicochemical conditions (artificial seawater, 33 salinity, 10°C)*

	Zinc	Cadmium
Dissolved concentration (µg/L)	100	31.6
Uptake rate (µg/g/d)		
Palaemon elegans	1.52	0.074
Echinogammarus pirloti	2.68	0.72
Elminius modestus	26.9	2.75

Source: After Rainbow and White (1989); Rainbow (1998).

Table 7.4 *Comparative Zn uptake rates (±1 standard deviation where shown) from solution of three decapod crustaceans under identical physicochemical conditions*

Conditions	Crustacean	Zn uptake rate (µg/g/d)
20 µg/L Zn 33 salinity, 10°C	*Pandalus montagui* (sublittoral)	0.931
	Palaemon elegans (littoral pools)	0.582 ± 0.155
100 µg/L Zn 16.5 salinity, 10°C	*Palaemon elegans* (littoral pools)	5.27 ± 3.67
	Palaemonetes varians (brackish water)	1.80 ± 0.61

Source: After Rainbow (1998).

strongly even among genera in the same order (Fig. 7.3) (Buchwalter and Luoma, 2005).

Physiological traits differ among species and many of these differences explain the different uptake rates. For example, some organisms acclimate to variable salinities by changing their permeability to water (e.g. estuarine decapod crustaceans: Mantel and Farmer, 1983). Some (which may be the same ones) adapt to changing osmotic conditions by changing major ion transport rates (e.g. estuarine crabs or freshwater insect larvae). Water passage across the gills is also extremely different among species. The combined influence of this array of biological adaptations on trace metal uptake is one reason why K_u of a metal might be so different among organisms.

The number of uptake sites is one of the factors causing differences in uptake rates among species. Among the crustaceans, barnacles have Zn and Cd uptake rates about an order of magnitude greater than the uptake rates of decapods and amphipods (Table 7.3). Barnacles have large areas of permeable surface exposed to solution compared to decapods and amphipods, with correspondingly increased potential for trace metal uptake (Rainbow, 1998).

Differences in Cd uptake rates between the larvae of two mayfly species were attributed to interspecific differences in the numbers of Cd transporters present (Buchwalter and Luoma, 2005). The number of transporters appeared to

be determined by how the animal managed osmoregulation. Some mayfly species exclude water entry by impermeability. Impermeable animals with few ion transporters have slow Cd uptake rates. Others allow more water in, but take up major ions rapidly to manage the composition of their body fluids. Where these animals have abundant specialised cells, called chloride cells, for regulating major ion fluxes, they take up Cd more quickly.

Differences between marine and estuarine species in integument permeability to water also may be related to differences in trace metal uptake rates. For example, the Zn uptake rates of caridean decapod crustaceans (commonly called shrimps or prawns) decline in animals that live at lower salinity. As Table 7.4 shows, the Zn uptake rate of the fully marine *Pandalus montagui* exceeds that of the tidepool-dwelling *Palaemon elegans*, which is greater than that of the brackish water inhabitant *Palaemonetes varians* (Nugegoda and Rainbow, 1989; Rainbow, 1998). This interspecific decrease in Zn uptake rate follows differences in permeability expected as an adaptation to life where low salinity can occur periodically (Mantel and Farmer, 1983).

Large differences in the rate at which an organism moves water across its gills also appear to influence uptake rate. In some organisms this is manifested by how fast the gill is propelled through the water (e.g. swimming speed in fish). In many invertebrates it is manifested by physically propelling water across the gill (filtration or clearance rates). The influence of water movement across the gill is especially noticeable in comparisons of K_u in bivalves that have very different filtration rates. *Corbula (Potamocorbula) amurensis*, for example, is a filter-feeding clam that pumps hundreds of litres of water per gram tissue per day across its gills (a very high filtration rate) (Lee *et al.*, 1998). Its uptake rates of Cd, Cr and Zn from solution are four to five times greater than those of *Macoma petalum* (as *M. balthica*) *Macoma petalum* is primarily a deposit feeder, morphologically constructed to filter very slowly as it takes in sediments. It only pumps a few litres of water per gram per day across its gills.

Wang (2001) compared uptake rate constants for Cd, Cr(VI), Se(IV) and Zn across two species of mussels (*Perna viridis* and *Septifer virgatus*) and a clam (*Ruditapes philippinarum*) (Table 7.2). K_u values were comparable between the two mussels but were two to three times lower in the clams (Table 7.2). Interspecific differences in K_u were strongly related to the weight-specific clearance (filtration) rates of the different bivalves, a relationship maintained when other bivalves were introduced into consideration (Fig. 7.9) (Wang, 2001). Gill surface area (which also affects the clearance rate) and membrane permeability could also have some effect on differences in the metal uptake rates among these bivalve species, but these are not well known (Lee *et al.*, 1998; Wang, 2001).

7.2.1.2.3.2 Intraspecific variation

Intraspecific variation in trace metal uptake rate constants is influenced by body size within species (although it is not a dominant factor among species: Buchwalter and Luoma, 2005). Uptake rates of Cd, Co, Se and Zn decreased with increasing body size in the mussel *Mytilus edulis* (Wang and Fisher, 1997). Similarly uptake rates of Cd, Cr and Zn were negatively correlated with tissue dry weight in two other bivalves, *Corbula (Potamocorbula) amurensis* and *Macoma petalum* (as *M. balthica*) (Lee *et al.*, 1998). Metabolic rates, gill surface area,

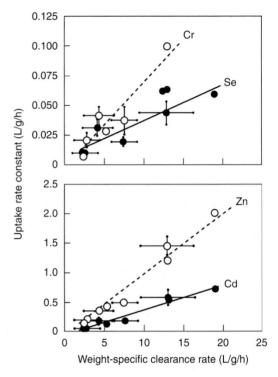

Fig. 7.9 Interspecific differences in uptake rate constant K_u of bivalve molluscs (*Perna viridis, Septifer virgatus, Ruditapes philippinarum, Mytilus edulis, Macoma petalum, Corbula (Potamocorbula) amurensis, Crassostrea rivularis, Saccostrea glomerata*) for four trace metals are strongly related to the weight-specific clearance rates (rates of water flow over the gills) of the different bivalves (Wang, 2001).

body surface area and filtration rates are examples of physiological factors that increase per unit mass in smaller animals. These factors can influence uptake rates, if all other physiological factors are the same.

If water permeability is important to differences in uptake rates between species, it might also be expected that changes in integument permeability might cause changes in the uptake rates of trace metals within a species. There is now evidence that this might be the case as some estuarine organisms acclimate to changing salinity, although it must be stressed that the organism is not making a physiological change in response to changes in trace metal availabilities but to changes in salinity. But those adaptations have consequences for water and major ion balance (Mantel and Farmer, 1983).

Thus the Zn and Cd uptake rates of the common shore crab *Carcinus maenas* actually decrease with decreased salinity (Chan *et al.*, 1992; Rainbow and Black, 2002, 2005a), in contradiction to the expected increase associated with physicochemically driven increases in free metal ion availabilities. The physiological response to decreased salinities probably reflects a decrease in apparent water permeability (Rainbow and Black, 2001) of sufficient magnitude to offset the physicochemical enhancement of trace metal uptake (Rainbow and Black, 2002, 2005a, b). Similarly, below a salinity of 25, the Zn and Cd uptake rates of the euryhaline amphipod *O. gammarellus* no longer correlate with free metal ion activities (Fig. 7.8), as the amphipod makes a physiological response to the low salinity that counteracts the physicochemical speciation effect (Rainbow *et al.*, 1993b; Rainbow and Kwan, 1995).

Physiologically driven changes in the flux of a major ion can also occur. These affect the rate of uptake of a trace metal independently of a change in a physicochemical external factor. Intraspecific changes in calcium uptake rate, for example occur at different stages of the moult cycle of a decapod crustacean. These would be accompanied by changes in the rate of Cd uptake via calcium channels. Large quantities of calcium may be lost when the animal sheds its exoskeleton during a moult, but this calcium is quickly replaced post-moult. Inferentially, this would suggest that species with high Ca uptake rates, like some snails, might also have elevated uptake rates of metals transported via the calcium channel (e.g. Cd).

Previous exposure to raised bioavailabilities of trace metals can occasionally cause intraspecific differences in metal uptake rates from solution, but it is difficult to draw general conclusions (Amiard-Triquet and Amiard, 1998; Wang and Rainbow, 2005). Selection for metal-tolerant populations of aquatic organisms is well known, where a population is exposed to ecotoxicologically significant bioavailable metal concentrations (Chapter 9) (Luoma, 1977; Klerks and Weis, 1987;

Rainbow *et al.*, 1999). Selection for a phenotype with reduced metal uptake rates is, theoretically, one way to reduce metal exposure and thereby acquire greater tolerance to a metal. Wang and Rainbow (2005) reviewed the effect of metal exposure history on trace metal uptake by marine invertebrates. They found that exposure of a marine invertebrate to raised dissolved metal produced inconsistent effects varying with the invertebrate under consideration. Exposure concentration, duration of exposure and the metal are important. Cain *et al.* (2006) suggested that changes in fundamental physiological processes like influx rate might be more conservative than changes in more adaptable processes like the production of metallothionein (see Box 7.11). In freshwater fish (especially the rainbow trout *Oncorhynchus mykiss*), chronic pre-exposure to Cd seemed to result in lower membrane affinity for Cd, but increased capacity to transport the metal. However, there was often no net effect on bioaccumulation (Niyogi and Wood, 2003).

In summary, there is a potential in certain organisms for physicochemical influences on trace metal uptake from solution to be counteracted (or perhaps enhanced) by physiological responses of the organism in response to external or internal drivers (e.g. salinity change). Species for which uptake rates change as they adapt to their environment are probably most common in highly variable environments (estuaries). For the vast majority of organisms it probably remains true that the influence of external physicochemical factors on uptake rate is more important. The effects of physiology within species are also usually small compared to the differences among species.

7.2.1.3 Dietary uptake processes

Prologue. Dietary uptake is an important route of exposure in nearly all aquatic organisms under nearly all conditions in nature. Dietary exposure can be quantified by determining assimilation efficiency (AE) of metal from diet. Uptake rate from diet is determined by ingestion rate, AE and metal concentration in food. Environmental (e.g. available food), ecological (e.g. choice of food) and biological (e.g. digestive strategy; ingestion rate) processes influence uptake rates from diet.

It is increasingly apparent that the diet is an important route for the uptake of trace metals by animals (Wang, 2002). Uptake from food is a function of how much food an animal ingests (ingestion or feeding rate), the metal concentration in the food and how much of that metal is extracted and assimilated by the feeding organism into its tissues (Reinfelder and Fisher, 1991; Wang and Fisher, 1999; Wang, 2002; Wallace and Luoma, 2003).

7.2.1.3.1   Feeding rate
Prologue. Feeding rate is the third factor (with AE and concentration) determining metal uptake rates from food. Feeding rate varies among species; generalisations are possible to characterise feeding rate within each species.

Autecological factors like feeding rate and choice of food are highly influential in determining metal uptake rates from diet (autecology is the study of interactions of an individual organism or a single species with the living and non-living factors of its environment). Feeding rate is an inherent property that differs among species. Feeding or ingestion rate is typically expressed as a weight-specific process: g food taken in / g body weight / day or, if multiplied by 100, as percent body weight per day. If food abundance or temperature, for example, change, then the feeding rate of an organism might change. Nevertheless, feeding rates are defined for many species within a relatively narrow range, but among species there is much variability. Animals with a high metabolic rate require a high ingestion rate and/or efficient digestion. Animals can get the same nutrition from ingesting food rapidly and processing it inefficiently, as from ingesting food more slowly but processing it more efficiently (Jumars, 2000). Feeding rate alone can sometimes explain differences in metal bioaccumulation from diet. The bivalve *Corbula (Potamocorbula) amurensis* accumulates more Se than the bivalve *Corbicula fluminea* where they co-occur (Lee *et al.*, 2006). Only feeding rate differs enough between the two to explain the greater uptake rates from food.

7.2.1.3.2   Assimilation efficiency
Prologue. Assimilation efficiency (AE) is another of the three factors determining metal uptake rate from food. AE can be narrowly defined for a species in any set of circumstances, as driven by the specific traits of that species. But AE can also vary within and among species. The factors that influence AE are known, but their precise contributions in any given situation are not as well known.

Bioavailability of metals from food is also important, but it is affected by a number of biological and environmental factors. These include:

- the nature of the food;
- the factors that control the digestive release of dietary trace metals;
- the form of metal that is released; and
- whether that form can be taken up by gut epithelial cells, the first step on the route to their translocation elsewhere in the body.

Assimilation efficiency (AE) is the most effective quantitative measure of bioavailability from diet, characterised by the efficiency of release of trace metals from ingested food in the gut and uptake into the animal tissues, initially the epithelial lining of the digestive tract. The use of radioactive tracers allowed the development of standard techniques for the measurement of trace metal assimilation efficiencies (Reinfelder and Fisher, 1991; Wang and Fisher, 1999). More recently, stable isotopes techniques have been used for this purpose, eliminating the handling problems that limit the widespread use of radioactive isotopes (Croteau *et al.*, 2004). Assimilation efficiency studies have now been expanded to a variety of invertebrates including marine and freshwater zooplankton (Croteau *et al.*, 1998; Xu and Wang, 2004), bivalves (Luoma *et al.*, 1992; Wang *et al.*, 1996), gastropods (Cheung and Wang, 2005), barnacles (Wang and Rainbow, 2000), decapod crustaceans (Rouleau *et al.*, 2000; Wallace and Luoma, 2003; Rainbow *et al.*, 2006), sipunculids (Wang *et al.*, 2002), echinoderms (Temara *et al.*, 1996) and fish (Reinfelder and Fisher, 1994; Rouleau *et al.*, 2000).

Factors affecting the AE of a metal in animals may be either internal, associated with the animal itself, or external, associated with the diet.

7.2.1.3.3 Taxon specificity of assimilation efficiency

Prologue. Median AE can vary two- to threefold among species for each metal. Trophic level, choice of diet and digestive processing combine to determine the degree of variability. Simple classifications like taxon or feeding guild are not a good basis for generalising about differences in AE.

Assimilation efficiency varies among species for each metal, reflecting the complex mix of processes that influences uptake from the diet (Box 7.6). Figure 7.10 shows the range of AE for Cd and Zn among six species of bivalves. Some species consistently assimilate more of the element than others. For example, the clams *Macoma petalum* and *Ruditapes philippinarum* both assimilate Zn more efficiently than the mussel *Mytilus edulis*. *Macoma petalum* does not assimilate Cd as efficiently as *R. philippinarum*, however (Fig. 7.10).

Box 7.6 Examples of taxonomic differentiation of assimilation efficiency

Because animals are adapted to eat different foods, AE among all species cannot always be compared unambiguously (the influence of the choice of food cannot be separated from the influence of the physiological differences between the species). Where such comparisons are possible, some differences among species are unequivocal. For example, the clam *Ruditapes philippinarum* had higher assimilation efficiencies of Cd and Zn than the mussel *Perna viridis*, when they were fed the same algal food. Cr assimilation efficiencies of the two bivalves were comparable (Table 7.5) (Chong and Wang, 2000). Ng *et al.* (2005) verified these results across a range of food types (Fig. 7.11) and showed the same difference in the case of Ag. The barnacle *Balanus amphitrite* usually had Zn and Cd AE intermediate to those of the two bivalves when fed the same range of algal species (Ng *et al.*, 2005) (Fig. 7.11). Some predators appear to have extremely high assimilation efficiencies (e.g. the whelk *Thais clavigera*: Cheung and Wang, 2005), while others are not particularly efficient at extracting metals: the fish *Menidia menidia* (Reinfelder and Fisher, 1994) or striped bass *Morone saxatilis* (Baines *et al.*, 2002). Some deposit feeders seem to be inefficient at assimilating metals from their food (e.g. sipunculids: Wang *et al.*, 2002), but some are quite efficient (the clam *Macoma petalum*). Generalisations about the species-specific differences are difficult to draw, but will develop as we better understand the linkages among digestive processes, gut physiology and the influence of metal partitioning within foods.

Table 7.5 *Comparative assimilation efficiencies (mean ± 1 standard deviation) in two marine bivalves (the green mussel* Perna viridis *and the clam* Ruditapes philippinarum*) feeding on different phytoplankton and natural seston*

Food type	Assimilation efficiency (%)	
	Perna viridis	*Ruditapes philippinarum*
Cadmium		
Thalassiosira pseudonana	24.7 ± 5.8	55.1 ± 10.4
Phaeodactylum tricornutum	18.2 ± 2.9	43.5 ± 2.5
Chlorella autotrophica	13.9 ± 5.1	25.7 ± 9.9
Tetraselmis levis	11.1 ± 2.0	37.6 ± 8.2
Prorocentrum minimum	11.3 ± 2.9	52.9 ± 8.8
Natural seston	10.8 ± 1.9	22.3 ± 7.8
Chromium		
Thalassiosira pseudonana	12.0 ± 2.9	12.5 ± 3.2
Phaeodactylum tricornutum	16.3 ± 5.6	18.6 ± 2.9
Chlorella autotrophica	13.8 ± 3.1	13.3 ± 4.0
Tetraselmis levis	10.5 ± 4.2	14.2 ± 5.2
Prorocentrum minimum	14.0 ± 3.8	24.4 ± 3.3
Natural seston	9.5 ± 2.0	11.3 ± 4.0
Zinc		
Thalassiosira pseudonana	29.6 ± 7.7	33.7 ± 7.5
Phaeodactylum tricornutum	36.2 ± 7.2	54.3 ± 3.5
Chlorella autotrophica	24.3 ± 4.9	29.2 ± 7.9
Tetraselmis levis	21.3 ± 6.7	40.6 ± 11.7
Prorocentrum minimum	32.2 ± 6.8	59.2 ± 4.7
Natural seston	27.4 ± 6.1	31.3 ± 9.8

Source: After Chong and Wang (2000).

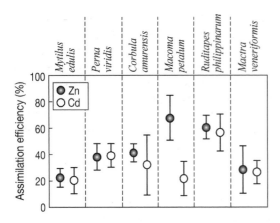

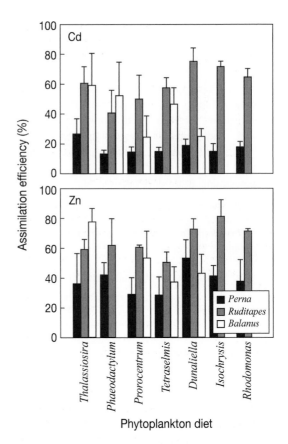

Fig. 7.10 The range of assimilation efficiencies observed for Cd and Zn among six species of bivalves in different studies. These metals are primarily accumulated from food by these species in most circumstances. Some bivalves have persistently low assimilation of Cd and Zn (*M. edulis*); some have persistently high assimilation of the two metals (*R. philippinarum*). AE is more plastic in some species than others (e.g. AE varies most widely in *Corbula amurensis* when foods range from sediments of different type to algae). Many of the bivalves assimilate Zn and Cd with similar efficiencies, but *M. petalum* is an exception. The range of variability in AE of Cd and Zn among the bivalves encompasses the full range seen among all species studies to date (about 20% to 70%).

Fig. 7.11 Mean assimilation efficiencies (AE%, ±1 standard deviation) of Cd and Zn from up to seven phytoplankton species in the green mussel *Perna viridis*, the clam *Ruditapes philippinarum* and the barnacle *Balanus amphitrite*. The clam had higher AEs than the mussel, while the barnacle usually had Zn and Cd AE intermediate to those of the two bivalves. (After Ng et al., 2005.)

The variability in assimilation of Cd and Zn (Fig. 7.10) among the different bivalves species captures the range observed among other animals as well (e.g. Wang and Fisher, 1999). Some organisms are consistently at the low end of the range of AE for both metals (sipunculid worms: Wang *et al.*, 2002); some organisms are consistently at the high end, such as barnacles (Rainbow and Wang, 2001) and whelks (predatory gastropod molluscs) (Cheung and Wang, 2005). However, conclusions about taxonomic differentiation of AE must be drawn carefully at the present level of understanding. Phylogenetic generalisations are probably accurate only to the extent that they follow similarities in choice of diet and digestive processing. Variability within a species is mostly due to differences in food type in different studies (Fig. 7.11). AE from a single food type typically varies less than twofold.

7.2.1.3.4 Range of variability: metal and concentration dependence

The rank order of AE is not as consistent as found for uptake rates from solution, but tendencies can be identified. For example, Fig. 7.12 expresses the frequency distributions of AE among several species ingesting sediments across a number of studies and species. Assimilation efficiencies of metals like Cr and Am are typically very low (1% to 10%), although exceptions can occur (Decho and Luoma, 1991; Luoma *et al.*, 1992). Assimilation of Ag is distributed toward the lower range of

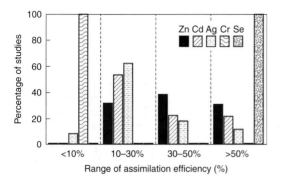

Fig. 7.12 shows the frequency among four to nine studies of AE for Zn, Cd, Ag, Cr and Se for animals fed oxidised sediments (the number of studies was different for each metal). The animals included bivalves, a copepod, a sipunculid worm and a polychaete. The data illustrate that Cr was assimilated with <10% efficiency in all studies and thus was the least effectively bioaccumulated in this data set, although exceptions occur (Decho and Luoma, 1991; Lee *et al.*, 1998). In contrast Se AE were >50% in all studies; it is most efficiently bioaccumulated. Examples occur of AE between 10% and >50% for Ag, Cd and Zn. Zn had the widest range of AEs among the species (11% to 86%).

AE compared to Cd and Zn. Many instances of Zn assimilation efficiencies >50% occur. Assimilation of Se is almost always very efficient (80% to 100% is not uncommon). In the few studies available, Cu assimilation appears to be somewhat similar to Zn. Thus generalisations are possible defining at least a range of AEs expected for each metal. Variability occurs for each metal but the range is smaller than for many other processes.

7.2.1.3.5   Digestive processes
Prologue. Membrane transport in the gut involves carriers typical of other epithelial surfaces. In particular, metals can be transported with small organic molecules like some amino acids. As a result speciation-based models are unlikely to fully predict metal bioavailability in the gut. Residence time in the gut as well as other physical and chemical aspects of digestion are also influential. All contribute to differences in AE among species.

An important cause of the differences in AE among species is differences in digestion. Digestive

processes influence how organisms obtain nutrition and take up pollutants. Metals are typically ingested in an innocuous solid form but are released for transport across the membrane of the digestive tract by powerful solubilising forces in the milieu of the digestive tract (Campbell *et al.*, 2003). Digestion is also a living process: adaptable and variable within the context of each species' characteristics.

Characteristics of digestion that appear to affect AE include:

- extracellular versus intracellular digestion;
- the characteristics of gut fluids;
- residence time in the gut.

Intracellular digestion is the process by which an animal cell engulfs a food particle and digests it within a food vacuole or some other type of inclusion. The particles engulfed by endocytosis occur intracellularly as lysosomes from which nutrients and the accompanying metals can be released. Intracellular digestion is a predominant process in only a few groups of organisms. It is common, for example, in bivalves. Intracellular digestion is thought to be an especially effective process for releasing nutrients and metals compared to extracellular digestion, but it also requires more energy and slower throughput. In animals reliant on intracellular digestion, particulate uptake of metals will also assume a degree of prominence.

More commonly metals are released from food particles by extracellular digestive processes in the gut of animals. They then have the potential to be taken up by gut epithelial cells and so enter the body. The metals may cross the apical membrane of the epithelial cells by any of the mechanisms described for the uptake of dissolved metals (Boxes 7.3 and 7.4), but it is likely that the proportional contribution of the different routes of uptake is different in the gut.

Metals are concentrated into the fluids of the gut medium by complexation with high concentrations of small organic molecules (Chen *et al.*, 2000). In addition, gut fluids can recycle several times to enhance concentrations of nutrients (and metals). Gut epithelial cells are adapted for the uptake of many of the organic molecules (for example amino acids) released by digestion, and the representation of carrier systems for such molecules may be much greater in gut cells than in surface epithelial cells. Experimentally, it is well established that the presence of amino acids like histidine and cysteine enhances the bioavailability of Zn and perhaps of Cu in the gut (Bury et al., 2003; Conrad and Ahearn, 2005). Apparently the metal and the amino acid are co-transported across the intestine by a carrier that complexes both (Conrad and Ahearn, 2005). If the uptake of trace metals bound to amino acids is an important route of trace metal uptake in the gut, then the FIAM and/or the Biotic Ligand Model are not fully applicable to trace metal uptake in the gut.

Organisms differ widely in their gut fluids and these fluids differ in their ability to extract metals from food (Chen and Mayer, 1999b). Digestive enzymes are a common complement in most guts, designed to break down complex proteins. But enzyme activities vary over five orders of magnitude among invertebrate species alone, depending upon major food sources (Mayer et al., 1997). Some species also release surfactants into the mixture in order, apparently, to create nano-sized hydrophobic micelles that help remove nutrients from detritus. Some organisms have a low pH gut (<2: fish). Digestive fluids of invertebrates typically are more neutral in pH. Some polychaetes have a basic gut (pH>8). Gut fluids also vary in redox status; some animals have reducing environments and some have moderately oxidising environments.

For example, mussels (Mytilus edulis) and clams (Macoma petalum) have moderately low gut pHs and a partially oxidised gut environment ('moderate reducing conditions'), which enhances solubilisation of sulphide-bound metals, for example (Griscom et al., 2002); whereas sulphides remain insoluble in other types of gut fluids (e.g. in the polychaete Arenicola marina: Chen and Mayer, 1999b). The clam M. petalum (as M. balthica) assimilates some metals typically resistant to assimilation in other species (e.g. americium, Cr), probably because of their long gut transit time and the high proportion of food subjected to intracellular digestion. Amphipod crustaceans assimilate only 40% to 50% of Se from algal food, apparently because they do not break all algal cells in their digestive tract (Schlekat et al., 2000a). Organisms that efficiently break open algal cells can assimilate >90% of the Se in their food (e.g. copepods or bivalves: Reinfelder and Fisher, 1991).

The length of time food spends in the gut of an animal also varies widely among species, with environmental conditions or with food type in the same species. Gut residence time can be very influential in causing differences in metal assimilation (Box 7.7). In general, more efficient digestion and absorption of the metals by occur when metals were retained longer in the gut (Xu and Wang, 2001).

Intraspecific variation in AE may also occur with body size. Wang and Fisher (1997) fed the diatom Thalassiosira pseudonana to three different size classes of the mussel Mytilus edulis, and found that Cd AE decreased with body size whereas Co AE increased with body size; Se and Zn AE did not vary significantly with body size of the mussels.

7.2.1.3.6 Chemical nature of metal in the diet

Prologue. Metals are assimilated with different efficiencies from different types of food, because of differences in how they are chemically partitioned within particulate food material or prey organisms. In general it appears that there is a progression of bioavailability from soluble forms > insoluble intracellular materials > insoluble metal-rich granules in the food species. It is likely that absolute AE among these fractions varies with the consumer.

Animals tend to choose their foods more or less selectively, but choice of food varies widely among species. Metal bioavailability differs among food choices (Box 7.8). The rate of metal uptake by a species will be proportional to the concentration of metal in the diet, for one type of food. However, if animals are fed different foods, or fed under widely different conditions, uptake rates (driven by AE) might differ. Thus uptake from diet will not necessarily follow concentration linearly in comparisons across a range of conditions.

The median range of AE from different (feasible) foods can be as high as threefold: median AE for

Box 7.7 Gut transit time and assimilation efficiency

The gut transit time is defined by gut passage time (GPT) and gut residence time (GRT). GPT is the time between ingestion of a bolus of food and its first appearance in the faeces. GPT varies from minutes to several hours among species, but is usually relatively constant within a species (Decho and Luoma, 1991). RT is the time between ingestion of a bolus of food and the complete elimination of all the undigested material (including metal) from the digestive tract. GRT typically measures hours to days in invertebrate animals. For example, it can take up to 96 hours for *Macoma petalum* to completely eliminate the undigested material from a gut-full of food (Decho and Luoma, 1991). *Mytilus edulis* can take 3 days, although most is defaecated in the first 24 hours (Wang *et al.*, 1996). Copepods take only a few hours (Reinfelder and Fisher, 1991).

Gut transit time may be an important factor, in at least some species, in causing variability of metal assimilation efficiency. Longer retention times of metals in the gut can be associated with more efficient metal assimilation (Decho and Luoma, 1994; Wang and Fisher, 1996; Roditi and Fisher, 1999). Some bivalve species appear to have particularly variable gut transit times, processing some particle types more than others through intensive intracellular digestion. For example, in a study of the assimilation of Cd, Cr and Zn by the green mussel *Perna viridis* and the clam *Ruditapes philippinarum*, Chong and Wang (2000) found a significant relationship between the Cr AE and its gut transit time in the clams, but not the mussel. No relationship for Cd or Zn was found for either species. Decho and Luoma (1996) also found that Cr and Cd assimilation was enhanced from particle types held longer in the digestive tract of the clam *Corbula amurensis*, as a result of partitioning to intracellular digestion (Decho and Luoma, 1994). Free-living bacteria, for example, were extracted from the water column by *C. amurensis* and processed by intracellular digestion. Assimilation efficiency of Cr was nearly 100%.

In the case of the copepod *Calanus sinicus* feeding on diatoms and dinoflagellates, Xu and Wang (2001) found that GPTs for Cd, Se and Zn are longer than the gut transit times of food particles, implying a decoupling of metal solubilisation from food digestion, as suggested by Chen and Mayer (1999b). Gut transit time and metal assimilation were still inversely related to food concentration and ingestion rate, however.

Cd are 20% to 60%; median AE for Zn range from 22% to 70%; the full range is larger. One of the factors contributing to the differences in AE probably is the chemical form of the metal stored in the food particle, cell or prey organism.

The chemical form of metal stored in the food or prey organism appears to be a major factor affecting the variability in metal assimilation from different foods. An early proposal was that only metal bound to the soluble fraction in prey is available to higher trophic levels (Reinfelder and Fisher, 1991; Fisher and Reinfelder, 1995). In the case of herbivores, Reinfelder and Fisher (1991) observed a linear 1:1 relationship between the metal assimilated by marine copepods from various phytoplankton diets and the metal partitioned in the cytoplasm of the ingested phytoplankton. Further studies show that the relationship is more complex when more species are considered (Hutchins *et al.*, 1995; Wang and Fisher, 1996;

Xu and Wang, 2001, 2002b, 2004). Animals with digestion strategies more complicated than copepods assimilate metal from more fractions than just the metal in cell solution (Decho and Luoma, 1994; Fisher and Reinfelder, 1995). For example, Ng *et al.* (2005) investigated whether the nature of the binding of the trace metals Cd, Ag and Zn accumulated by phytoplankton can affect their subsequent assimilation efficiencies in three filter-feeding benthic invertebrates, the green mussel *Perna viridis*, the clam *Ruditapes philippinarum* and the barnacle *Balanus amphitrite*. Seven phytoplankton species were chosen from a wide systematic range to ensure large differences in the partitioning of their accumulated trace metals into three fractions: (i) exchangeable metal adsorbed on the outside of the cells as defined by extraction with the chelating agent 8-hydroxyquinoline-5-sulphonate, (ii) incorporated metal that is in soluble form and (iii) insoluble incorporated metal.

Box 7.8 Effect of different natural food types on AE in the same species

Mussels and other filter-feeders filter suspended material from the water column to obtain their food. That material includes includes plant cells, detritus and inorganic particles (sometimes with an organic coating). Animals differ in the degree to which they separate these materials; but most filter-feeders ingest some proportion of all types of particles. Ke and Wang (2002) compared Cd, Se and Zn assimilation from three foods typically encountered by filter-feeders: algal cells (diatoms), suspended sediment and mixtures of the two. When the mussels, *Perna viridis*, were fed a mixture of suspended sediment and diatoms, they were able to selectively remove the nutritious algal food and discard most, but not all, of the sediment as pseudofaeces, production of which is a particular trait of some bivalves. If sediment loads were extremely high the bivalve was less effective at separation. In those conditions assimilation from algae was reduced to a small degree. Table 7.6 shows that assimilation efficiencies from the algal food were high, when either mixed with sediment or presented alone. When sediments were fed alone to the bivalves they were ingested but the AE was greatly reduced compared to the algae. Griscom *et al.* (2002) reported similar results for the clam *Macoma petalum*

(as *M. balthica*) for Ag, with an AE of 39% to 49% from algae and 11% to 21% from sediment; and Cd with an AE of 33% to 88% from algae and 8% to 23% from sediment. Thus the addition of a nutritious living organic material to suspended material can double (or more) the efficiency with which an organism can obtain metal from its food. The variability within a food type is typical of the range observed in repeated tests of AE. In many laboratory experiments, sediments are fed to such animals without supplementation by algae, even though the latter is most common in nature.

Table 7.6 *Effect of different natural food types on Cd, Se and Zn AE in the green mussel* Perna viridis *fed on diatoms, suspended sediment and mixtures of the two. Assimilation efficiencies from the phytoplankton food were high, when either mixed with sediment or presented alone. When sediments were fed alone to the bivalves they were ingested but the AE was greatly reduced compared to the AE of the diatom diet (Ke and Wang, 2002)*

Food type	AE (%)		
	Cd	Se	Zn
Sediment	11–18	16–27	20–33
Diatoms	36–48	48–59	31–48

There was no support for a generalised conclusion that any of the three fractions represented the sole form of phytoplankton metal that was bioavailable for trophic transfer to a herbivore. Even trace metals bound to the insoluble fraction in phytoplankton were bioavailable to some herbivores.

At the other end of the spectrum it has also been suggested that if trace metals are detoxified in trace metal-rich granules (MRG) they are not trophically available. Nott and Nicolaidou (1990) showed that trace metals in metal-rich phosphate granules of prey animals remain in the same insoluble form during passage through the gut of molluscan carnivores, but that magnesium, calcium carbonate granules (temporary metabolic stores which do not contain metals other than these major ions), were leached and demineralised by the process of digestion. In the case of the trophic

transfer of metals from the digestive gland of three gastropods as prey tissue to the hermit crab *Clibanarius erythropus*, Nott and Nicolaidou (1994) similarly showed that the bioavailability of metals in the diet was affected by the nature of chemical binding during intracellular compartmentalisation in the prey tissue. Metals in membrane-bound phosphate granules (Mn, Ni, Zn), lysosomes (also Mn, Ni, Zn) and membrane-bound sulphur-rich granules (Cu) were not bioavailable in the diet, in contrast to cadmium in the cytosol (Nott and Nicolaidou, 1994). Cheung and Wang (2005) on the other hand have since shown that the predatory gastropod *Thais clavigera* can assimilate Ag, Cd and Zn from MRG isolated from gastropod and oyster prey.

The differences in availability of different fractions and differences in fractionation among

Box 7.9 What fraction of metal in a prey species is bioavailable?

In the transfer of metal from prey to predator, it appears that the distribution among physicochemical forms of accumulated metal in the prey affects assimilation, but no single metal form dictates bioavailability. This is not surprising given the biological variability in both digestive strategies among predators and the diversity of different metal-binding forms that can occur among prey. The physicochemical form of Cd in the oligochaete worm *Limnodrilus hoffmeisteri* was found to be a major factor controlling the assimilation of Cd by a predator, the decapod crustacean *Palaemonetes pugio* (Wallace and Lopez, 1996, 1997; Wallace *et al.*, 1998). Cd in a population of Cd-tolerant worms from Foundry Cove, Hudson River, New York was predominantly stored in what was considered to be trophically unavailable form in Cd granules. Wallace and Lopez (1997) developed the concept of trophic availability, concluding that while Cd associated with cytosolic proteins in *L. hoffmeisteri* was 100% trophically available to *P. pugio* and Cd bound to metal-rich granules was unavailable, Cd bound to cell organelles was 70% trophically available. Wallace and Luoma (2003) progressed further, examining how the subcellular partitioning of Cd and Zn in the bivalves *Macoma petalum* (as *M. balthica*) and *Corbula amurensis* affected the trophic transfer of these metals to the predatory decapod crustacean *Palaemon macrodactylus*. Wallace and Luoma (2003) could best explain their comparative assimilation results if trace metals bound to cell organelles were added to protein components to form what they termed the Trophically Available Metal (TAM) fraction of metals accumulated in the bivalve prey (Fig. 7.13).

Rainbow *et al.* (2006) studied the trophic transfer of the trace metals Ag, Cd and Zn accumulated by two populations of the polychaete worm *Nereis diversicolor* collected from estuaries with different degrees of metal contamination. *Nereis diversicolor* were fed to the decapod crustacean predator *Palaemonetes varians*, in an attempt to test the general applicability of the TAM fraction. The two populations of worms had different subcellular distributions of radiolabelled metals in the subcellular fractions separable by the technique of Wallace *et al.* (2003). Correlations were sought between AEs of the predator and selected fractions or combinations of fractions of metals in the prey – metal-rich granules, TAM and total protein. The predator (*P. varians*) in fact assimilated dietary metal from a range of the fractions binding metals in the prey (*N. diversicolor*), with different assimilation efficiencies summed across these fractions. The TAM fraction defined by Wallace and Luoma (2003) did not account for all trophically available metal in the diet of all predators. If the TAM fraction in food was high, then AE tended to be high. But, in some cases, high AE occurred when the TAM fraction was low (i.e. some other forms of metal were assimilated).

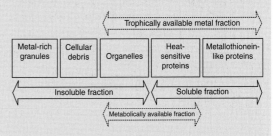

Fig. 7.13 Accumulated metals in a prey organism can be separated into five operational fractions (three insoluble and two soluble), representing the subcellular compartmentalisation of the metal (Wallace and Luoma, 2003; Cheung and Wang, 2005). The trophically available metal (TAM) fraction of metals accumulated in the bivalve prey of a decapod crustacean was best explained if trace metals bound to cell organelles and the two soluble protein components were considered bioavailable (Wallace and Luoma, 2003). For other predators, some assimilation also occurs from metal-rich granules (Cheung and Wang, 2005.)

food types, together with the assimilative powers of the predator, combine to cause different AE from different foods. For example, when predatory gastropod molluscs (whelks) were fed a variety of their natural prey (Cheung and Wang, 2005), AE were lowest for whelks fed on barnacles, because the MRG fraction was much more prominent in intracellular binding than in other prey. The herbivorous snails *Monodonta labio* had the highest percentage of metal distributed in the soluble and organelle fractions (defined as trophically available metal (TAM): Wallace and Luoma, 2003) (Box 7.9) among the five prey species, and correspondingly, the AE for whelks fed on this snail were generally high. But there was not a simple correlation between assimilation efficiency and proportion of any single fraction or any combination of fractions in the prey. Other studies show similar results (Box 7.9).

7.2.1.3.7 Effects of excess metal

Prologue. High concentrations of metal in food can inhibit digestive enzymes, slow feeding rates or affect the movement of food through the digestive tract. Animals may also avoid food contaminated with some metals. Such effects are manifestations of toxicity, and but they can also lead to misinterpretation of both dietary bioaccumulation and toxicity tests.

Very high concentrations of metal can affect gut processes, affecting bioaccumulation and causing adverse effects for the organism. High Cu concentrations in a contaminated sediment ingested by the lugworm *Arenicola marina* and other deposit feeders inhibited digestive proteases in isolated gut fluid (Chen and Mayer, 1998b; Chen *et al.*, 2002). Such effects are not known in whole organisms, however (Campbell *et al.*, 2003).

Organisms fed metal-contaminated food may also reduce feeding rates; and/or the metals can reduce the predator's ability to move food through the digestive system. The littoral amphipod crustacean *Orchestia gammarellus* had decreased feeding rates on a diet of decaying kelp *Laminaria digitata* when these diets were enriched with Cu or Zn (Weeks, 1993). The ingestion rate of the copepod crustacean *Acartia spinicauda* feeding on the planktonic dinoflagellate *Prorocentrum minimum* was reduced by increased concentrations of Cd and Se (but not Zn) in the food particles (Xu *et al.*, 2001). Feeding rates in bivalves fed Cr-contaminated sediments declined when Cr concentrations were increased (Decho and Luoma, 1996).

Reduced feeding may result from behavioural avoidance of the contamination or from poisoning of the digestive tract. Taylor *et al.* (1998) found substantial reduction in feeding rate of freshwater crustaceans *Daphnia* sp. fed algae exposed to 2.5 µg/L Cd. Subsequent investigation indicated that the reduced feeding rate was related to poisoning of digestive processes, preventing gut

transit of the food. Woodward *et al.* (1995) saw a similar effect in trout fed frozen insect larvae collected from a metal-contaminated river. As digestive enzymes or gut transit are inhibited, whole organism responses might include signs of starvation, constipation or reduced growth. These responses can confuse experimental intepretations if they occur in the absence of metal uptake and bioaccumulation (dietary uptake is prevented).

7.2.1.3.8 Exposure history

Prologue. It is not clear if or how a history of metal exposure affects dietary metal uptake.

As in the case of uptake from solution, there is potential for previous exposure to raised trace metal bioavailability to affect subsequent metal uptake from the diet. But the significance of any such effect is not well known, and the literature is somewhat inconsistent (Box 7.10).

7.2.2 Physiological loss or excretion

Prologue. Physiological loss, expressed as the rate constant of loss, strongly affects metal bioaccumulation. The rate constant of loss appears to be a definable trait for each metal and species. It is often the most important contributor of differences among species in metal bioaccumulation.

The accumulation of a trace metal by an organism is the net result of two processes – the uptake of the trace metal integrated across all uptake routes, and the excretion of that metal summated across all excretion routes. Physiological loss is the excretion of metals that have been taken up into a cell and thus the body of an organism. The desorption of metals that have been adsorbed onto the surface of an aquatic organism will also be expressed as part of the loss rate. Metal may also be lost from the digestive tract when a spike of contaminated food is followed by uncontaminated food. Neither desorption of adsorbed metal nor defecation of unassimilated metal in food are considered to be physiological loss. Physiological loss varies most widely among species and for different metals.

Box 7.10 Effects of previous exposure to metals on metal assimilation from food

Examples of effects of pre-exposures come from studies with clams (Decho and Luoma, 1996) and mussels (Wang and Rainbow, 2005). When clams *Corbula amurensis* were pre-exposed to a food source of Cd-contaminated bacteria, they ingested fewer bacteria, processed a smaller proportion by intracellular digestion (60% versus 100%) and thereby reduced their AE from 90% to 55%. This appeared to be a mechanism for avoiding the contamination.

In contrast, pre-exposure of the green mussel *Perna viridis* to high dissolved concentrations of Cd resulted in raised AE of Cd from food. AE correlated with an elevation in the proportion of Cd associated with metallothionein-like proteins (MTLP) (see Box 7.11) apparently induced by the Cd pre-exposure (Blackmore and Wang, 2002). No such effect was observed for Zn in the mussel (Blackmore and Wang, 2002), but it was for Ag, although in this case without an increase in MTLP-bound Ag (Shi *et al.*, 2003). In this case the induction of metal-binding ligands might have had consequences on the rate-limiting step of any uptake process (Fig. 7.14) (Wang and Rainbow, 2005).

Another bivalve, the clam *Ruditapes philipinnarum*, did not show increased Cd or Ag AE after pre-exposure (Ng and Wang, 2004), however. Similarly pre-exposure of barnacles to raised metal bioavailabilities, whether in the field or in the laboratory to diets of phytoplankton enriched in Ag, Cd or Zn, had no significant consistent effect on the AEs of any of these metals (Rainbow *et al.*, 2003, 2004c). As in the case of uptake from solution, there is great variability apparent in the conclusions from such experiments, and in the magnitude of the effects. Exposure details and the animal under investigation

both appear to be influential (Amiard-Triquet and Amiard, 1998; Wang and Rainbow, 2005). Thus it may be that a strong reliance in mussels on the induction of metallothioneins as a detoxificatory mechanism (see later), as opposed to a reliance on metal-rich granules as in barnacles, causes the presence of pre-exposure effects in mussels in particular (Wang and Rainbow, 2005).

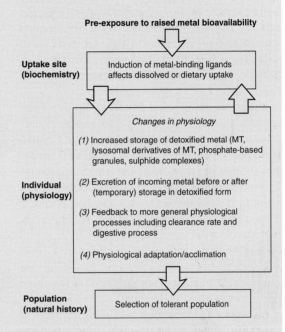

Fig. 7.14 A schematic illustration of the possible effects of pre-exposure to raised trace metal bioavailability on the subsequent uptake and accumulation of trace metals in marine invertebrates. The induction of metal-binding ligands might have consequences on the rate-limiting step of any uptake process. MT, metallothioneins. (After Wang and Rainbow, 2005.)

It does not appear to be affected by geochemical conditions. Different sources of metal (dissolved versus diet) may result in different physiological loss rates in some cases, but these are probably exceptions.

7.2.2.1 Efflux rate constant (K_e)
Prologue. Loss rate is highly influential in determining bioaccumulation. Loss rates are determined by the

concentration in the animal multiplied by an efflux rate constant (K_e). Differences in K_e are a common explanation for differences in bioaccumulation among species or trace metals.

The efflux rate constant (K_e), also called the excretion rate constant or loss rate constant, is a proportional rate constant of the total loss of metal from the body of an organism, usually expressed as 'per day'. (Luoma and Rainbow, 2005). Efflux is an instantaneous function of the concentration in tissues times the rate constant(s) of loss. The constant describes unidirectional efflux of a metal (i.e. in the absence of any influx).

Loss experiments are conducted by feeding an organism metal-labelled food, then removing the metal-labelled spike and following loss from the tissues. It is essential that such experiments use a unique radionuclide or stable isotope label that can be followed independently of the natural background fluxes of the metal. An experiment with an unlabelled metal will underestimate efflux because there is always some influx occurring from the natural metal background.

Efflux can usually be described by first-order exponential decay (Fig. 7.15). The rate constants can vary with route of exposure, but such variability appears to be small at least for mussels (Fisher et al., 1996). If more than one exponential slope appears in the loss curve, the proportion of metal in each of these compartments can be calculated. This is affected by exposure time (Cutshall, 1974), and thus can be experiment-specific. It is useful to assume that exposures in nature are chronic; thus rate constants from the most slowly exchanging compartment(s) can be employed as the efflux term (Luoma et al., 1992).

In general, metal efflux rate constants fall in the range of 0.01 to 0.05 per day for many of the species that have been studied. However, when exceptions occur they are of great importance to bioaccumulation, metal fate and metal effects.

For example, efflux rate constants of Cd, Zn and Cr differed between *Corbula amurensis* and *Macoma petalum* (as *M. balthica*) over the range of 0.01 to 0.05 per day although it was not always the same species with the higher K_e (Lee et al., 1998) (Table 7.7). Efflux rate constants of Cd, Cr and Zn were, on the other hand, comparable between the green mussel *Perna viridis* and the clam *Ruditapes philippinarum* (Chong and Wang, 2001) (Table 7.7). It is not phylogenetic relatedness that produces similar trace metal physiologies and affects efflux. Rate constants for Cd, Se and Zn differed very significantly between two oysters, *Crassostrea rivularis* and *Saccostrea glomerata*, being lower in the latter (Table 7.7), with consequent effects on accumulated body concentrations in the two oysters (Ke and Wang, 2001). Many strong bioaccumulators of metals (see Section 7.3.3 'Accumulation patterns' below) have rate constants of loss of 0.01 to 0.001 per day or slower. *Saccostrea glomerata* is a strong metal bioaccumulator and the rate constants of loss were 0.003 to 0.004 in the experiment above. The same is true of barnacles and Zn (Rainbow and Wang, 2001) and marine gastropod predators, in general (Wang and Ke, 2002).

Many animals that accumulate metals weakly do so because their rate constants of loss are ≥ 0.08 per day. Se accumulates in bivalves because their rate constant of loss are in the typical range 0.01 to 0.03 per day but Se accumulates to a lesser extent in zooplanktonic crustaceans because they lose Se rapidly (K_e 0.1 to 0.2 per day). These differences are propagated up the food chains involving

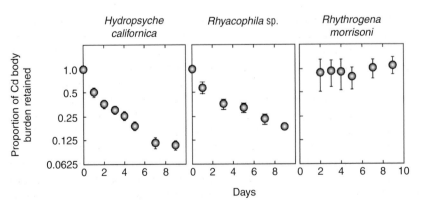

Fig. 7.15 Loss of Cd when all sources of Cd influx were eliminated, from three species of aquatic insect larvae. The rate of loss differs widely among species, although the pattern for *Hydropsyche californica* and *Rhyacophila* sp. is more common than the pattern shown by *Rhythrogena morrisoni* (Data from Buchwalter and Luoma, 2005.)

Table 7.7 *Trace metal efflux rate constants (K_e) (per day, means $\pm$ SD) compared between pairs of bivalve molluscs*

	Cd	Cr	Se	Zn	Reference
Corbula amurensis	0.011 ± 0.002	0.048 ± 0.002		0.027 ± 0.002	Lee *et al.*, 1998
Macoma petalum	0.018 ± 0.001	0.024 ± 0.001		0.012 ± 0.001	Lee *et al.*, 1998
Perna viridis	0.020 ± 0.003	0.012 ± 0.006		0.029 ± 0.006	Chong and Wang, 2001
Ruditapes philippinarum	0.010 ± 0.004	0.010 ± 0.004		0.023 ± 0.007	Chong and Wang, 2001
Crassostrea rivularis	0.014 ± 0.004		0.034 ± 0.007	0.014 ± 0.012	Ke and Wang, 2001
Saccostrea glomerata	0.004 ± 0.003		0.013 ± 0.002	0.003 ± 0.001	Ke and Wang, 2001

these species, resulting in much greater exposure to Se of bivalve predators than zooplankton predators (Stewart *et al.*, 2004). Some marine copepods lose Cd and Zn extremely rapidly (K_e 0.1 to >1.0 per day) and as a result do not accumulate more of these metals than found in the algae they eat (Xu *et al.*, 2001).

Other factors can also affect K_e, but the effects are relatively small. For example the K_e for Cd doubled with decreased size in the mussel *Mytilus edulis*, while there was no significant change with size for K_e of Co, Se and Zn (Wang and Fisher, 1997). In general, size effects on K_e are rare or small (Lee *et al.*, 1998).

7.3   ACCUMULATION

Prologue. All aquatic invertebrates accumulate trace metals in their tissues, and different organisms accumulate trace metals to different concentrations in their tissues, organs and thence their bodies (Eisler, 1981). The significance of a bioaccumulated concentration cannot be judged without considering the accumulation characteristics typical of that organism.

Different degrees and patterns of accumulation are the outcome of the inherent variability in the dynamic parameters of uptake and loss discussed earlier. As a result of interspecific variability in K_u, AE and/or K_e, aquatic invertebrates living in the same habitat may well have very different body concentrations of trace metals (Hare *et al.*, 1991;

Hare, 1992; Phillips and Rainbow, 1994), even within closely related taxa (Moore and Rainbow, 1987; Rainbow *et al.*, 1993a; Rainbow, 1998). The assessment of whether an accumulated trace metal concentration is high or low, therefore, cannot be made on an absolute scale. An accumulated trace metal body concentration that is very high for one species may be very low for another. A low body Zn concentration in an oyster would be high in a mussel (Phillips and Rainbow, 1994), and a high body Zn concentration in a caridean decapod crustacean would be below any ever recorded for a barnacle (Rainbow, 1998, 2002). The crustaceans can be used as an illustrative example (Table 7.8).

When metal first enters the body of an animal it will initially be metabolically available, having the potential to bind to molecules in the receiving cell or elsewhere in the body after internal transport via body fluids (Fig. 7.16). In the case of an essential metal, it is available to bind to sites where it plays an essential role (e.g. Zn in the enzyme carbonic anhydrase or Cu in haemocyanin in malacostracan crustaceans), or, if present in excess (caused by entry at too high a rate), to sites where it may cause toxic effects. Such an excess of essential metal (and all non-essential metal) must be detoxified, i.e. bound tightly to a 'sacrificial' site from which escape is limited, probably in a storage organ beyond the site of uptake. The metal has now entered the second component of accumulated metal – the detoxified store (Fig. 7.16) which may be permanent or temporary. Trace metals taken up into the body may or may not be

Table 7.8 *Variability in Cd, Cu and Zn concentrations among crustaceans: a selection of body concentrations (μg/g dry weight) of three trace metals (Zn, Cu, Cd) in a systematic range of crustaceans from clean and metal-contaminated sites (from Rainbow, 2007). Accumulated concentrations of trace metals in crustaceans vary widely between metals and between taxa (Rainbow, 1998, 2002), as illustrated for Zn, Cu and Cd. Three taxa are shown: barnacles (Cirripedia), and two malacostracan taxa – amphipods and caridean decapods. Barnacles have Zn concentrations an order of magnitude above those of amphipods and carideans, even when the barnacles are from uncontaminated sites. On the other hand, barnacles from uncontaminated sites usually have body Cu concentrations below those of amphipods and carideans, which, as malacostracans, have the copper-bearing respiratory protein haemocyanin which is absent from barnacles. Barnacles can, however, increase their body Cu concentrations well above those of amphipods and carideans when at Cu-contaminated sites, for example Chai Wan Kok, Hong Kong (Phillips and Rainbow, 1988) or Dulas Bay, Wales (Walker, 1977). Differences between body Cd concentrations in the three crustacean taxa are not so marked*

Species	Location	Zinc	Copper	Cadmium	Reference
Cirripedia					
Tetraclita squamosa	Hung Hom, Hong Kong (contaminated)	6963	94.9	2.8	Phillips and Rainbow, 1988
	Tung Chung, Hong Kong	2245	14.9	4.2	Phillips and Rainbow, 1988
Balanus amphitrite	Chai Wan Kok, Hong Kong (contaminated)	9353	3472	7.3	Phillips and Rainbow, 1988
	Lai Chi Chong, Hong Kong	2726	59.3	5.5	Phillips and Rainbow, 1988
Semibalanus balanoides	Dulas Bay, Wales (contaminated)	50280	3750	–	Walker, 1977
	Menai Strait, Wales	19230	170	–	Rainbow, 1987
	Southend, England	27837	232	28	Rainbow *et al.*, 1980
Malacostraca					
Amphipoda					
Orchestia gammarellus	Restronguet Creek, England (contaminated)	392	139	9.8	Rainbow *et al.*, 1989, 1999
	Dulas Bay, Wales (contaminated)	126	105	9.1	Rainbow *et al.*, 1999
	Millport, Scotland	188	77.5	1.6	Rainbow *et al.*, 1999
Talorchestia quoyana	St Kilda, Dunedin, New Zealand	481	31.9	17.2	Rainbow *et al.*, 1993c
	Sandfly Bay, Dunedin,	133	15.6	8.9	Rainbow *et al.*, 1993c
Eucarida					
Decapoda Pleocyemata, Caridea					
Palaemon elegans	Millport, Scotland	80.6	110	0.9	White & Rainbow, 1986
Pandalus montagui	Firth of Clyde, Scotland	57.5	57.4	–	Nugegoda and Rainbow, 1988b

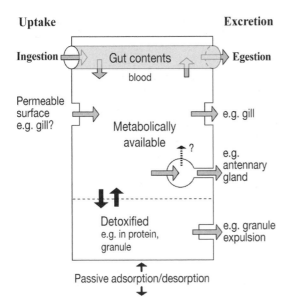

Uptake   **Excretion**

Fig. 7.16 A schematic representation of the body metal content of an aquatic invertebrate such as a decapod crustacean (after Rainbow, 2007). When metal first enters the body, it will initially be metabolically available, before potentially being stored in detoxified form, probably elsewhere in the body after internal transport via body fluids. Detoxified storage may be permanent or temporary. Trace metals taken up into the body may or may not be excreted, either from the metabolically available component or from the detoxified store.

excreted, either from the metabolically available component or from a detoxified store (Fig. 7.16), depending on the accumulation pattern for that particular metal (Rainbow, 1998, 2002, 2007).

7.3.1   Metabolic requirements for trace metals

Prologue. Essential metals bind to internal sites where they play an essential biochemical role. Every organism has a requirement for a minimum concentration of essential metal. At concentrations lower than that requirement adverse effects can occur.

Theoretical estimates can be made of the metabolic requirements of essential trace metals in an attempt to understand the size of the metabolically available component of accumulated metal (Fig. 7.16) in an aquatic invertebrate. White and

Rainbow (1985) made theoretical calculations of enzyme requirements for Cu and Zn, based on the number of Cu- and Zn-bearing enzymes and the total concentration of enzymes in tissues. Making many assumptions, they estimated that metabolising tissue needs approximately 26.3 µg/g Cu and 34.5 µg/g Zn (Table 7.9) to fulfil enzyme requirements (White and Rainbow, 1985). White and Rainbow (1985) also made estimates of the amount of Cu needed by the respiratory protein haemocyanin in decapod crustaceans. Rainbow (1993) refined these theoretical calculations by allowing for the different contributions of soft tissue and blood to total body weight in decapod crustaceans, as exemplified by the caridean *Pandalus montagui* in Table 7.10. In *P. montagui*, the concentration of Cu needed by the haemocyanin in the body was estimated to be 22.9 µg/g Cu, giving a total theoretical body concentration of Cu of 38.1 µg/g Cu (Table 7.10). The measured mean concentration of Cu in *P. montagui* from the Firth of Clyde, Scotland is 57.4 µg/g Cu (SD 18.9) (Nugegoda and Rainbow, 1988b) (Table 7.10), suggesting that the theoretical estimates may be approximately correct.

7.3.2   Detoxified metal

Prologue. If an excess of essential metal (or non-essential metal) accumulates, it must be detoxified to prevent adverse effects. This is usually accomplished by exceptionally strong binding (conjugation) to a site in either insoluble or soluble form, from which escape is limited.

The second component of accumulated metal is that of detoxified metal (Fig. 7.16). Trace metals can be detoxified as inclusions within intracellular organelles (e.g. lysosomes). They can also be sequestered into one of a variety of insoluble granules or deposits (Hopkin, 1989; Mason & Jenkins, 1995; Marigómez *et al.*, 2002). Hopkin (1989) described three types of intracellular granules: type A – consisting of concentric layers of calcium and magnesium phosphates which may contain trace metals such as manganese and zinc; type B – more heterogeneous in shape and always

Table 7.9 *Essential metal requirements in enzymes of metabolising soft tissue*

	Number of metal-associated enzymes	Percentage of total number of enzymes	Average number of metal atoms per enzyme molecule	Estimated enzyme metal requirement in tissue (μg/g dry wt)
Copper	30	1.40%	2.95	26.3
Zinc	80	3.74%	1.41	34.5

Source: After White and Rainbow (1985); Rainbow (1993, 2007).

Table 7.10 *Estimates of the essential requirements for copper in the caridean decapod* Pandalus montagui

Percentage distribution dry weight	
Exoskeleton	40.0%
Blood	2.1%
Soft tissues	57.9%
Haemocyanin	
Blood Cu concentration	44 μg/mL
Blood volume in body	0.52 ml/g dry wt
Blood Cu concentration in body	22.9 μg/g
Enzyme requirement	
Metabolising soft tissue	26.3 μg/g
Whole body	15.2 μg/g
Total body Cu metabolic requirement	
Haemocyanin	22.9 μg/g
Enzymes	15.2 μg/g
Total	38.1 μg/g
Measured body Cu concentration	
Mean $\pm$ SD	57.4 $\pm$ 18.9 μg/g

Source: After Rainbow (1993, 2007) with data from Nugegoda and Rainbow (1988b); Depledge (1989a).

containing sulphur in association with metals that include copper and zinc; type C – often polyhedral with a crystalline form, mainly containing iron, probably derived from ferritin. In crustaceans, the most commonly reported metal-rich granules are type A and B granules (Al-Mohanna and Nott, 1987, 1989; Nassiri *et al.*, 2000), while large ferritin crystals are characteristic of the ventral caecum cells of stegocephalid amphipods (Moore and Rainbow, 1984).

Detoxification also occurs in the soluble phase within cells, for example by the binding of trace metals with glutathione or metallothioneins (Box 7.11) (Mason and Jenkins, 1995).

7.3.3   Accumulation patterns

Prologue. The accumulation pattern of a particular species for a particular trace metal determines the accumulated body concentration, and these accumulation patterns correspondingly vary within and between taxonomic groups (Rainbow, 1998, 2002).

The first accumulation pattern to be considered is that of the essential trace metal Zn in caridean decapods. The caridean *Palaemon elegans* regulates its body concentration of Zn (to about 90 μg/g Zn) when exposed to a wide range of dissolved Zn bioavailabilities (White and Rainbow, 1982; Rainbow and White, 1989), an accumulation pattern for Zn also shown by other carideans, *Palaemonetes varians* (c. 96 μg/g Zn) and *Pandalus montagui* (c. 70 μg/g Zn) (Nugegoda and Rainbow, 1988b, 1989). Zn is taken up by *P. elegans* in significant quantities (14% of total body Zn content per day at 100 μg/L Zn under defined physicochemical conditions at 20°C), but the uptake rate is balanced by the excretion rate. As a result the body concentration remains unchanged over most environmentally realistic Zn exposures (White and Rainbow, 1982, 1984a, b). At a high enough dissolved Zn availability, the excretion rate fails to match the uptake rate (regulation breakdown), and there is a net increase in body Zn concentration to only about double the regulated body concentration, but with lethal toxic effect (White and Rainbow, 1982). The implication here is that much of the Zn

Box 7.11 Metallothioneins

Metallothioneins (MT) are low molecular weight cytosolic proteins which bind, are induced by and are involved in the cellular regulation and detoxification of certain trace metals (e.g. Zn, Cu, Cd, Ag, Hg) (Roesijadi, 1992; Amiard *et al.*, 2006). The most important and original MT characteristic is their high cysteine content. Cysteines account for 33% of the 61 constitutive amino acids of mammalian MTs while crustacean MTs contain 18 cysteines in 58 to 60 amino acids (Binz and Kägi, 1999). The presence of sulphur in the high proportion of cysteine residues in these proteins provides the high metal affinity of the molecule, sequestering metals in the cytoplasm and reducing their metabolic availability. The alignment of Cys–Cys, Cys–X–Cys and Cys–X–Y–Cys sequences where X and Y are amino acids other than cysteine is the criterion that allows the distinction between different structural MT classes and that leads to many isoforms of MT. Different isoforms probably play different roles in the physiology of trace metals in an organism (Amiard *et al.*, 2006), for example in the detoxification of the non-essential metal Cd as opposed to the physiological handling of the essential metal Cu (Dallinger *et al.*, 1997).

The *in vitro* affinity of MT decreases in the hierarchical sequence $Hg^{2+} > Cu^+, Ag^+, >> Cd^{2+} > Zn^{2+}$ (Vasak, 1991) showing that Zn is likely to be displaced from MT by the other metals (Amiard *et al.*, 2006). Thus Roesijadi (1996) has proposed a model for the coupled MT induction and rescue of target ligands compromised by inappropriate metal binding. A more expanded model of MT induction (Haq *et al.*, 2003) is shown in Fig. 7.17.

Metallothioneins, like other proteins, have limited lives, being integrated into the classical cytophysiological 'circuit' in order to be expelled (or stored) as cellular wastes from the cytoplasm via lysosomes where protein degradation occurs (Isani *et al.*, 2000). The concomitant presence of S and trace metals (particularly Cu) in lysosomes probably results from the incorporation of MTs into lysosomes to give the sulphur-rich type B granules described above (Brown, 1982; Martoja *et al.*, 1988; Nassiri *et al.*, 2000; Marigómez *et al.*, 2002). MTs associated with Cu are particularly resistant to lysosomal degradation because of the stability of the Cu sulphide links in their molecular conformation (Bremner, 1991; Langston *et al.*, 1998).

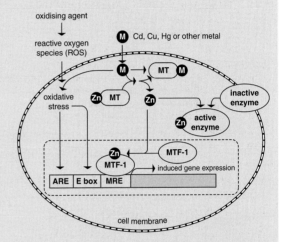

Fig. 7.17 A model for the induction of metallothionein (MT) gene expression. The figure depicts a cell delimited by a plasma membrane, with the nucleus containing a metal-responsive gene. Zn, displaced from MT by other metals of stronger affinity, binds with a transcription factor (MTF) which subsequently interacts with a metal responsive element (MRE) to induce MT gene expression. ARE and E box are DNA sequences in the promoter region that can bind transcription factors. (After Haq *et al.*, 2003.)

in the body is in metabolically available form without detoxification (Fig. 7.18a). Thus, at high Zn uptake rates, the concentration of metabolically available metal builds up sufficiently in the body to cause toxicity. It can also be concluded that in caridean decapods at least, the body concentrations of Zn are regulated to approximately those required to meet metabolic demand with relatively little stored in detoxified form (Rainbow, 1998, 2002).

The Zn accumulation pattern of barnacles is at the other extreme of the possible range of Zn accumulation patterns (Rainbow, 1998, 2002). All Zn taken up from solution by barnacles is accumulated without significant excretion (Rainbow and White, 1989) and any excretion of Zn taken up from the diet is also extremely limited (half-life of 1346 days in *Elminius modestus*: Rainbow and Wang, 2001). Correspondingly accumulated body

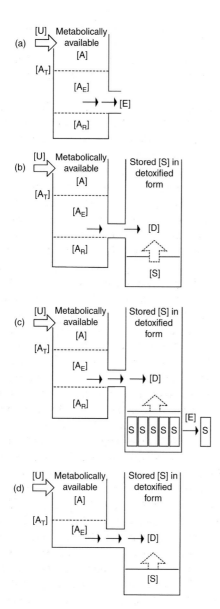

concentrations can reach very high values (e.g. 50 000 µg/g Zn or more: Table 7.8) (Rainbow, 1987, 1998) and the vast majority of this accumulated Zn is inevitably in detoxified form (Fig. 7.18b). Zn in barnacle bodies is bound in Zn pyrophosphate granules (Walker *et al.*, 1975a, b; Pullen and Rainbow, 1991), a Type A granule in the categorisation of Hopkin (1989). Thus accumulation patterns for Zn in these two crustacean taxa could not be more different – (a) regulation to a constant body concentration with apparently very little stored in detoxified form (caridean decapods), and (b) stored in detoxified form without significant excretion (barnacles) resulting in some of the highest accumulated concentrations of any trace metal in any animal tissue (Eisler, 1981; Rainbow, 1987).

The accumulation patterns of amphipod crustaceans for many trace metals including Zn are also those of net accumulation with detoxified storage in type B granules in the ventral caeca with the potential to be excreted with the faeces (Figs. 7.18c and 7.19, Box 7.12) (Rainbow and White, 1989; Rainbow, 1998, 2002). The ferritin crystals that characterise Fe accumulation in stegocephalid amphipods are similarly lost into and then from the alimentary tract (Moore and Rainbow, 1984).

No crustacean regulates the body concentrations of non-essential trace metals like Cd or Pb (Amiard

Caption for Fig. 7.18 (*cont.*)
the decapod crustacean *Palaemon elegans*. (b) The trace metal accumulation pattern of an aquatic invertebrate that is a net accumulator of an essential metal without significant excretion of metal taken up. Metabolically available metal in excess of requirements is detoxified [D] to be stored [S] as the detoxified component of accumulated metal with no upper concentration limit. This accumulation pattern is exemplified by Zn and Cu in barnacles. (c) The trace metal accumulation pattern of an aquatic invertebrate that shows net accumulation of an essential metal in detoxified form, but excretes some of that accumulated metal in the detoxified component. Examples of this accumulation pattern are those of Cu and Zn accumulated from the diet by amphipod crustaceans. (d) The trace metal accumulation pattern of an aquatic invertebrate that shows net accumulation of a non-essential metal in detoxified form with no significant excretion. An example of this accumulation pattern is that of Cd in barnacles.

Fig. 7.18 Trace metal accumulation patterns of aquatic invertebrates (after Rainbow, 2002). (a) The trace metal accumulation pattern of an aquatic invertebrate which regulates the total body metal concentration of an essential metal by balancing uptake [U] with excretion [E]. All metal is accumulated in the metabolically available component [A], itself subdivided into the essential metal required for metabolic purposes [A_R], and excess metal [A_E] over and above this metabolic requirement. There is a threshold concentration [A_T] of metabolically available metal, above which the accumulated metal is toxic. An example of such an accumulation pattern is that of Zn in

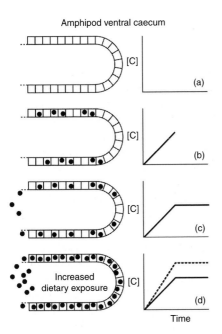

Amphipod ventral caecum

Fig. 7.19 (a, b, c) The accumulation of detoxified metal (e.g. Zn or Cu) in the ventral caecal cells of an amphipod crustacean exposed to a trace metal in the diet, and the corresponding changes in total body metal concentration [C] over time. (d) The effect of increased dietary exposure to the metal. (After Galay Burgos and Rainbow, 1998; Rainbow, 2002.)

et al., 1987; Rainbow, 1998, 2002). For example, in caridean decapods (*Palaemon elegans*), amphipods (*Echinogammarus pirloti*) and barnacles (*Elminius modestus*), all Cd taken up from solution is accumulated with no detectable excretion over at least a 28 day period (Rainbow and White, 1989) (Fig. 7.18d). The accumulated Cd is necessarily detoxified, typically as metallothionein which will be broken down in lysosomes (Langston *et al.*, 1998). Cd, however, is rarely visualised in lysosome residual bodies (type B granules), unlike Cu or Zn.

7.4 UNIFYING THE PROCESSES THAT DETERMINE BIOACCUMULATION (BIODYNAMIC MODELLING)

Prologue. Biodynamic modelling offers a tool to unify the complex influences on bioaccumulation of species, metal, environment and uptake route. It is the most effective approach for quantifying dietary and dissolved contributions to bioaccumulation for any species in any environmental circumstance.

Earlier sections have illustrated the complex factors that influence the bioaccumulation of trace metals. Unifying these detailed complexities into a manageable model requires considering the most important four factors: metal, species, environment and uptake route. If, for a species, the metal-specific uptake rates from both food and water can be quantified, along with loss rates, it becomes feasible to model metal bioaccumulation for that species under defined environmental conditions (Luoma and Rainbow, 2005). This approach is termed biodynamic modelling (Box 7.13, Fig. 7.20).

A great advantage of biodynamic modelling is that the outcomes can be unambiguously validated by comparison to bioaccumulation in independent field observations of the species of interest (Fig. 7.21). Across widely ranging environmental conditions, metals and species, biodynamic forecasts seem to agree well with observations in nature.

Outcomes can also explain differences among species in responses to metal exposure (Luoma and Rainbow, 2005). For example, biodynamic modelling explains the consistent differences in accumulated concentrations of Zn between barnacles and mussels, irrespective of the contamination status of their site of origin (Fig. 7.22, Table 7.11).

Biodynamics also explain differences in bioaccumulation among metals. For example, bioaccumulated concentrations of Cd and Ag are nearly always lower than bioaccumulated Zn concentrations in all environments and among all species. In part, this is because Zn is more abundant in the environment than Cd and Ag. But Cr is as abundant as Zn in most sediments, and Cr concentrations are typically low in animal tissues. This is because uptake rates of Cr from all dissolved forms, even Cr(VI), are very slow compared to Zn uptake rates, and bioavailability from diet is relatively low as well (Decho and Luoma, 1996; Wang *et al.*, 1997; Rainbow and Wang, 2001).

One of the most important issues in unifying our understanding of bioaccumulation is the relative

Box 7.12 Crustacean bioaccumulation patterns: copper

Accumulation patterns for Cu also vary among crustaceans (Rainbow, 1998). The caridean *Palaemon elegans* appears to regulate the body concentration of Cu (c. 130 μg/g Cu) over a wide range of dissolved Cu exposures as for Zn (White and Rainbow, 1982; Rainbow and White, 1989). The lack of a suitable radiotracer for Cu has as yet prevented confirmation that the rate of Cu excretion matches that of Cu uptake. The measured body Cu concentration of *Pandalus montagui* from the Firth of Clyde, Scotland (Nugegoda and Rainbow, 1988b) approximates to estimated theoretical requirements (Rainbow, 1993) (Table 7.10). During regulation most accumulated Cu will be in metabolically available form and excretion will be from this component (Rainbow, 2002). Once the rate of Cu uptake exceeds the rate of Cu excretion at high dissolved Cu bioavailabilities and regulation has broken down, the accumulated concentration in *P. elegans* can reach higher levels (c. 600 μg/g Cu) than in the case of Zn (White and Rainbow, 1982). Under these circumstances, the hepatopancreas of *P. elegans* contains type B Cu-rich granules, probably residual bodies from the lysosomal breakdown of metallothionein binding Cu (Rainbow, 1998). These granules have the potential to be excreted via the gut as hepatopancreatic epithelial cells complete their cell cycle. The accumulation of Cu by carideans may therefore show two excretory routes – excretion from the metabolically available component during Cu regulation, and excretion from an insoluble detoxified store during the net accumulation that follows regulation breakdown at high Cu bioavailabilities (Rainbow, 2002).

As shown in Table 7.8, Cu concentrations in barnacles from uncontaminated sites are generally low, approximating to theoretical estimates of Cu enzyme requirements (Table 7.9), and barnacles lack haemocyanin. Barnacles do, however, have the potential to accumulate high concentrations of Cu (Table 7.8) and show strong net accumulation of Cu from solution in the laboratory with no suggestion of regulation (Rainbow and White, 1989). It is likely that, as for Zn, all incoming Cu is accumulated with the necessity for detoxification (Rainbow, 1998, 2002) (Fig. 7.18b). Barnacles from Cu-contaminated sites such as Dulas Bay have many type B Cu-rich granules (Walker, 1977), probably again resulting from lysosomal breakdown of metallothionein binding Cu (Rainbow, 1987).

The Cu accumulation pattern of amphipods appears to be intermediate between the apparent Cu regulation of carideans and the strong net accumulation of barnacles (Rainbow and White, 1989; Weeks and Rainbow, 1991; Rainbow 1998, 2002). There is no evidence for regulation of Cu body concentrations and amphipods exposed to a range of dissolved Cu exposures show net accumulation at all exposures (Rainbow and White, 1989; Weeks and Rainbow, 1991). Cu is accumulated in the cells of the ventral caeca (equivalent to the hepatopancreas of caridean decapods), again in the form of type B Cu-rich granules (Nassiri *et al.*, 2000), presumably derived from metallothionein. Cu detoxified in these granules will be excreted on completion of the cell cycle of the ventral caeca epithelial cells (Figs. 7.18c, 7.19) (Galay Burgos and Rainbow, 1998). This is not a process of regulation for the body concentration of Cu in the amphipods reaches a new steady state level as Cu bioavailabilities change, the availability of Cu being reflected in the number of granules in (and hence the Cu concentration of) the ventral caeca (Fig. 7.19). As the Cu concentration increases the proportion of accumulated Cu in the detoxified component increases, from a starting point at which most Cu in an amphipod from an uncontaminated site will probably be in metabolically available form (Table 7.9). Amphipods have haemocyanin and their metabolic requirements for Cu might therefore be of the order of 38 μg/g Cu as calculated for the caridean *Pandalus montagui* (Table 7.10). Indeed *Talorchestia quoyana* from St Kilda, New Zealand had 31.9 μg/g Cu (Table 7.8); the lower Cu concentration in the same species from Sandfly Bay may reflect a temporary absence of haemocyanin in these amphipods, the haemocyanin content of talitrid amphipods being known to vary seasonally (Rainbow and Moore, 1990).

Box 7.13 Biodynamic models

In a biodynamic (kinetic) bioaccumulation model, a set of rate expressions is used to relate chemical fluxes among biological 'compartment(s)' and the environment (Fig. 7.20). Kinetic models are flexible and there is as much focus on the biological processes that influence contaminant exposures as on geochemical complexities. The drawback is a requirement for both physiological and geochemical data (Landrum *et al.*, 1992; McKim and Nichols, 1994).

Transport physiologists originated the concept that accumulation of required chemical constituents (elements, amino acids, etc.) occurs as a balance of gross influxes and effluxes (biodynamics). Radioecologists were the first to apply these principles to aquatic ecosystems, quantifying the bioaccumulation of radionuclides using simple exponential equations. Pentreath (1973), for example, noted that ^{65}Zn bioaccumulation by mussels (*Mytilus edulis*) could be expressed using a linear differential equation with constant coefficients. If the activity in water was maintained constant, accumulation of a dissolved radioelement could be expressed most simply as

$$C_t = C_{ss}(1 - e^{-kt})$$

where C_t = Concentration of nuclide at time t in the organism,

C_{ss} = the steady state concentration in the organism,

and k = the rate constant of loss (he called it excretion).

The model for accumulation from water alone, then, was

$$dC_t/dt = (k_u * C_w) - (K_e * C_t)$$

which states that the change in concentration in an organism (C_t) over time (t) is a function of uptake minus loss. The uptake rate is defined by an uptake constant (K_u) in units of $\mu g/g_{tissue}/d$ per $\mu g/L_{water}$ (or L/g/d), multiplied by the concentration in water (C_w). The loss rate is defined by a proportional rate constant of loss (K_e per day) multiplied by the concentration in the organism.

Biodynamic models account for both dietary and dissolved pathways of accumulation. In a first-order rate coefficient model, accumulation from the two sources is determined by summing unidirectional influx rates from water plus unidirectional influx rates from food minus the rate constant of loss and the growth rate:

$$dC_t/dt = (I_w + I_f) - (k_e + g)(C_t).$$

I_w is the unidirectional influx rate from water as derived above, and I_f is the unidirectional influx rate from food where

$$I_f = AE * IR * C_f.$$

That is, uptake rates from food are determined by how much food is ingested (IR in $g_{food}/g_{tissue}/d$), the concentration of metal in that food (C_f = metal concentration in food such as phytoplankton, suspended particulate matter, sediment, prey, etc.) and what proportion of that mass of metal is assimilated (AE = assimilation efficiency in %).

The differential equations describing these processes can be solved to determine metal concentrations at steady state (C_{ss}):

$$C_{ss} = [(k_u \times C_w) + (AE \times IR \times C_f)]/(k_e + g)$$

where g is growth rate per day. The equation can also be solved for any exposure time.

For their data, biodynamic models use experiments that quantify each of the physiological processes fundamental to bioaccumulation: rate constants of uptake from solution, assimilation efficiencies from food, rate constants of loss (Table 7.11). The physiological constants are determined for each species of interest under a range of conditions typical of the site of interest. The physiological model is combined with concentrations and environmental conditions that characterise a specific site. Bioaccumulation is forecast for scenarios that cover the possible environmental conditions at the site. Once physiological constants are determined for a species, they can be used over and over again. Generalisations about geochemical influences (e.g. change in free ion activity) or ecological influences (e.g. choice of food) can be developed from experiments and included in scenarios.

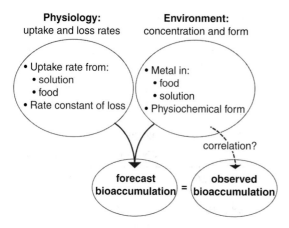

Physics:
uptake and loss rates

Environment:
concentration and form

- Uptake rate from:
  - solution
  - food
- Rate constant of loss

- Metal in:
  - food
  - solution
- Physiochemical form

correlation?

forecast bioaccumulation = observed bioaccumulation

Fig. 7.20 Biodynamic modelling. Trace metal bioaccumulation can be forecast for any species, in any specific field situation, from a combination of physiological and geochemical data. The forecasts can be compared to independent determinations of bioaccumulation in that species from that environment, for validation.

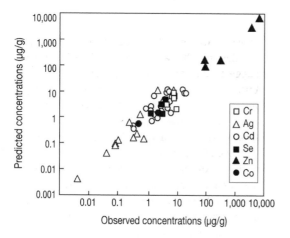

Fig. 7.21 Comparison of trace metal bioaccumulation predicted by biodynamic modelling in 15 different studies to independent observations of bioaccumulation in the species of interest from the habitat for which the physiological parameters were developed (Luoma and Rainbow, 2005). The data covered six metals, thirteen species of animals, three phyla, and eleven marine, estuarine and freshwater ecosystems. The agreement between predictions and independent observations from nature was strong.

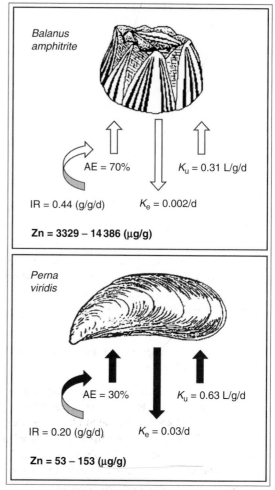

$Zn = 3329 - 14386$ (µg/g)

$Zn = 53 - 153$ (µg/g)

Fig. 7.22 Zn concentrations and physiological coefficients for uptake and loss for a barnacle (*Balanus amphitrite*) and a mussel (*Perna viridis*) collected simultaneously from the same locations in Hong Kong coastal waters. AE, assimilation efficiency; IR, ingestion rate; K_u, dissolved uptake rate constant; K_e, rate constant of loss. More assimilation from food and much slower excretion explain the consistently higher Zn concentrations in barnacles than in mussels, observed everywhere barnacles and mussels co-occur (metal-contaminated or not). (After Luoma and Rainbow, 2005.)

contribution of food and water. An assumption key to the biodynamic model is that pathways of uptake are additive. The role that dietary exposure plays in bioaccumulation differs. For some trace metals (e.g. Se), bioaccumulation is overwhelmingly from food. Forecasts that eliminate dietary exposure greatly underestimate internal exposure to Se (Luoma *et al.*, 1992) (Fig. 7.23). For most metals the role of diet varies with the circumstance

Table 7.11 *Comparisons of accumulated Zn concentrations (μg/g) in barnacles and mussels collected simultaneously from the same location*

Hong Kong site[a]	Hang Hau	Chai Wan Kok	Kwun Tong	Tai Po Kau	Lai Chi Chong
Balanus amphitrite	11990	9353	7276	4381	2726
(barnacle)	10220–14070	7411–11800	5269–10050	4195–5201	967–7688
Perna viridis	111	153	115	61	53
(mussel)	75–147	59–247	79–151	42–80	39–67
Gulf of Gdansk site	Puck	Mechelinki	Gdynia	GN Buoy	Vistula plume
Balanus improvisus	3293–14106	4466–14386	6088–10048	4197–7448	5610–12217
(barnacle)					
Mytilus trossulus	83.8–130	103–192	98.1–153	61.1–136	96.1–187
(mussel)					

Note: [a]Data (with 95% confidence limits) for *Balanus amphitrite* (concentration in barnacle body of 4 mg dry wt) and *Perna viridis* (mean soft tissue concentration) in Hong Kong waters are from Phillips and Rainbow (1988). [b]Data (ranges of weight-adjusted mean concentrations) for *Balanus improvisus* and *Mytilus trossulus* in the Gulf of Gdansk, Baltic are from Rainbow *et al.* (2004a).
Source: From Luoma and Rainbow (2005).

(e.g. Cd: Fig. 7.23). Some of the factors that influence that variability include:

- **Metal.** K_u varies somewhat predictably among metals (Table 7.12). Metals with a very low K_u (e.g. Cr and Se) are accumulated almost entirely from diet in almost all conditions. Metals with the highest K_u (like Ag, Cd, Zn or Cu) can be accumulated from either food or water, depending upon assimilation efficiency.
- **Ratio of concentrations.** The greater the concentration of metal in particulate material or a prey organism relative to concentration in water, the greater the contribution of uptake from food. In conditions where metals partition particularly strongly to particulate material (including phytoplankton), diet will dominate uptake. In conditions where a prey bioaccumulates high concentrations of metal, food is more likely to predominate. Prey that have high bioaccumulation factors will pass more metal to their predators (and the likelihood that food is the source of the metal in predators increases). The ratio of concentrations is very important in comparing laboratory and field studies. Many laboratory studies spike food by exposure to elevated

concentrations of dissolved metal, and compare uptake from the two in those media. Observations suggest that solute/solid concentration distributions in such situations can be an order of magnitude greater or more than occurs in nature.
- **Species.** Some species have slow rates of uptake from water (e.g. marine polychaete or sipunculid worms). Diet is more likely to be the dominant pathway of bioaccumulation for these. Some species have low assimilation efficiencies from their food (e.g. some amphipods). These species are more likely to accumulate proportionately more metal from water.
- **Geochemical conditions.** Metal bioaccumulation from both pathways is affected by geochemical form. In particular, Cd uptake from water is substantially faster in freshwater than in brackish or marine waters because formation of chloride complexes slows K_u. As salinity declines, Cd uptake from solution by aquatic invertebrates will switch from the food vector toward the water vector. As dissolved organic material increases, uptake from water declines and diet is likely to be increasingly important (Croteau and Luoma, 2005).

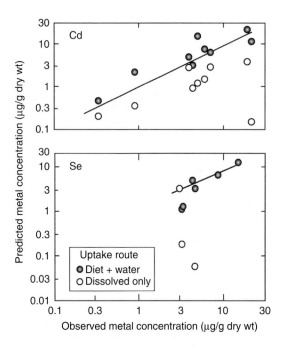

Fig. 7.23 Accumulation of Cd and Se by a variety of species as forecast by a biodynamic model for a specific environment, with and without the dietary vector of bioaccumulation. Modelled Cd bioaccumulation is compared to Cd concentrations observed in the animals from that environment. For Cd some forecasts accurately depict Cd bioaccumulation using dissolved data alone, and some forecasts are more than an order of magnitude too low. In the absence of considering diet, forecasts of Se bioaccumulation are more than an order of magnitude lower than observed in nature. Consideration of diet is critical to an understanding of metal exposure.

- **Ratios of physiological dynamics**. Tendencies for metals to bioaccumulate from food can be generally forecast from comparisons of biodynamic constants like K_u and AE. These are affected by choice of metal, species and environment (Box 7.14). But some generalisations are possible.

7.4.1   Trophic transfer of metals

Prologue. Although metal biomagnification may be quite specific to the circumstance, and difficult to generalise about, it is remains an important consideration in assessing metal exposures. When comparisons are carefully constructed, it is clear that

instances occur where metal concentrations in predators or consumers can be greater than those in their food. Identifying the specifics of the question being addressed is always critical before comparisons are begun.

Early studies of chemical contamination emphasised the role of food chains in carrying contaminants to the larger, longer-lived, higher trophic level organisms in the ecosystem (Woodwell *et al.*, 1971). This conceptual model is unequivocal for a few contaminants, especially methyl mercury and organochlorine pesticides. For most other chemicals the situation is much more complex.

Obviously, metals are assimilated from the diet by consumer organisms. Nevertheless, it is difficult to establish a linkage between metal bioaccumulation and position in the food web (trophic level: Wang, 2001) in simple tests or a simple set of observations. Conventional wisdom held that food web transfer of metals was of limited interest (Timmermans *et al.*, 1989), except in the case of mercury, and/or that metal concentrations in animals are not related to the trophic level in the food chain. Recent work suggests that, if feeding relationships are carefully identified, a number of instances occur where trace metal bioaccumulation (and toxicity) is greatly influenced by choice of food (e.g. Box 7.15; Larison *et al.*, 2000) and/or concentrations increase in organisms with each trophic transfer (Croteau *et al.*, 2005).

There are three possible outcomes of metal transfer from prey to predator (Newman and Unger, 2003): biomagnification (predator concentrations > prey concentrations), no difference in concentration between prey and predator, or biominification (prey concentrations > predator concentrations). Historically metal concentrations in animals from different trophic levels were compared to identify which outcome characterised trophic transfer. Recent work has shown, however, that some important qualifications are necessary to obtain unbiased results in such studies:

- species-specific metal accumulation physiology must be considered;
- food webs must be considered separately;

Table 7.12 *Tendencies that determine the relative importance of food and water as sources of bioaccumulation depend upon species-specific characteristics of uptake from water (K$_u$) and uptake from food (AE)*

	K_u	AE	
Ag	High	Highly variable	Contribution of food or water will vary widely, depending upon K_u /AE.
Cd	High in FW Low in SW	High	Dissolved Cd is important in FW, but diet still dominates in some surprising circumstances. Food is usually dominant in marine systems and salinities above 5.
Cr	Very slow	Low	Diet dominates in all situations, despite the low AE.
Cu	Moderate	Moderate	Diet dominates where it has been studied.
Ni	Unknown	Unknown	Unknown.
Pb	Unknown	Low	Likely that trophic transfer is low; but has not been studied.
Se	Very slow	Very high	Diet is entirely responsible for bioaccumulation in all known circumstances.
Zn	Moderate	High	Food dominates despite moderately high K_u.

Note: FW, freshwater; SW, seawater.

Box 7.14 Examples of how metal, species and environment combine to determine the relative importance of food and water as sources

Uptake from food is likely to be dominant where the metal uptake rate constant from solution is relatively low, because of the physiological traits of the organism, the nature of the metal, or the characteristics of the environment. This is the case for both Cr and Se; uptake rate constants from solution are the lowest among all the metals (Wang, 2002).

Dietary and dissolved sources vary in importance for Cd. Dissolved Cd is dominated by the free ion in freshwater and uptake rate constants are very high. Uptake from water is therefore very important (60% to 70% of total uptake) for lower trophic level organisms like the common zooplankter, *Daphnia magna* (Barata *et al.*, 2002). Dissolved Cd dominates uptake (50% to 98%) by the freshwater crustacean *Asellus aquaticus*, but this is partly facilitated by the low assimilation of Cd from an important food source, the aquatic plant *Elodea* sp. (Hare *et al.*, 2003). However, dietary uptake exceeds dissolved uptake in many of the invertebrate predators that have been studied, because of a combination of very efficient assimilation and high concentrations in prey. Examples are the predaceous zooplankter *Chaoborus* sp. (Munger and Hare, 1997) and the alderfly *Sialis* sp. which lives in the sediment (Roy and Hare, 1999). Cd bioaccumulation from water in *Chaoborus* is negligible; and *Sialis* receives 80% of its Cd from food.

- animals must be analysed in ways that are comparable;
- feeding relationships must be identified from site-specific data.

The manner in which organisms handle metals (biodynamics), as well as the food they select, can greatly influence trophic transfer (Reinfelder *et al.*, 1998; Wang 2001). In theory, the potential for trophic transfer (defined by a trophic transfer factor or TTF) can be defined by influx rates from food divided by efflux rate from the organism. That equation resolves to:

$$TTF = AE * IR * C_{prey}/K_e,$$

where C_{prey} represents metal concentration in prey species. Species-specific characteristics affect AE and IR as well as K_e. Biological considerations like complicated feeding behaviour add complexities to generalisations about TTF (Amiard-Triquet *et al.*, 1993). Nevertheless, theoretical TTF suggest that

Box 7.15 Examples of efficient trophic transfer of metals

Examples of efficient trophic transfer of metals are growing. Biomagnification of Hg and Zn (but not Pb and As) were observed in lake ecosystems (C.Y. Chen *et al.*, 2000). Concentrations of both metals increased in top predators with the length of the food web. Cd, Cu and Zn concentrations increased when the predatory gastropod (snail) *Thais clavigera* was specifically compared to its prey (Blackmore, 2000).

Croteau and Luoma (2005) used the stable isotope approach to study Cd biomagnification in a tidal lake with large beds of macroflora. There was no evidence of biomagnification when data from all species in the lake were all aggregated. However, when invertebrates and fish from the macrofloral food web, specifically, were separated, it was clear that higher trophic level invertebrates bioaccumulated more Cd than lower trophic level invertebrates. Higher trophic level fish also accumulated more Cd than lower trophic level fish. Fish accumulated less Cd overall than invertebrates, however; possibly because the assimilation efficiency of Cd is typically lower in fish than in predatory invertebrates (Wang, 2002). Once the food webs were sorted out, Cd biomagnification appeared (Fig. 7.24).

The widespread adverse effects of Cd on a bird (ptarmigan) living in riparian zones impacted by mine tailings is an example of how choice of food influences toxicity (Larison *et al.*, 2000; Box 3.8). Among the riparian plants Cd accumulated to the highest concentrations in willows. Adverse effects of Cd were seen in ptarmigan specifically feeding on willows. Deterioration of bone strength was found in ptarmigan from the metal-enriched ore belt, coincided with high Cd bioaccumulation and was explained by Cd substitution for calcium in the bones. Ptarmigan were least abundant where such effects were seen.

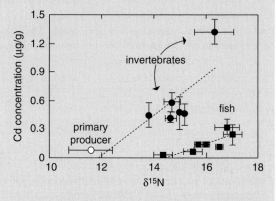

Fig. 7.24 Bioaccumulation of Cd in the invertebrate and fish food webs of the littoral zone in a freshwater tidal lake as a function of δ^{15}N. The latter is an indicator of trophic position (higher δ^{15}N indicates higher trophic level). Cd biomagnified in both the invertebrate and fish food chains within the littoral food web, but it did not biomagnify between invertebrates and fish. (After Croteau *et al.*, 2005.)

there are instances when Zn, Cd, perhaps Cu and Ag, and certainly Se and Hg could biomagnify or at least be transferred with a TTF near 1. On the other hand, biominification along a classic marine phytoplankton food chain might be the most common result (e.g. high efflux rate constants characterise metals like Cd). Food web transfer for some metals might also be diminished by low assimilation efficiencies in fish.

Identification of specific feeding relationships is critical. For example, Stewart *et al.* (2004) found no evidence of Se biomagnification between lower trophic level feeding guilds and higher trophic level guilds in San Francisco Bay (USA), when broad comparisons were made across data aggregated from many organisms. When the two most important food webs (benthic and water column) were separated, however, it was clear that both biomagnified Se. But bivalves from the benthic food web bioaccumulated more Se than zooplankton from the water column food web. This was because the rate constant of loss for Se was about fivefold slower in bivalves than zooplankton. Predators of bivalves also had higher concentrations of Se than predators of zooplankton. Thus the effect of greater bioaccumulation at the base of each food web was propagated up that food web. Biomagnification was specific to each food web, but was not detectable if the specific food of each predator was not considered.

How animals are analysed can also add great ambiguity to a comparison, especially when small prey are compared to large predators. For invertebrates, whole bodies are usually analysed. That is more difficult for fish, however. Muscle contributes the most mass to the whole body of a fish, proportionately much more than in most invertebrates. But fish physiologically exclude most trace metals from their muscle tissues (Hg and Se are exceptions). Trace metals accumulate as a function of exposure in liver, but trace metals do not accumulate in fish muscle (Wiener and Giesy, 1979). A comparison of fish muscle metal concentration to whole invertebrate metal concentration is irrelevant. A comparison of metal concentration averaged over whole bodies might be relevant for some questions (what is the exposure of a predator? how does the mass of metal compare between trophic levels?), but not for others (are adverse effects more likely at higher trophic levels than at lower trophic levels?).

7.5   CONCLUSIONS

The mechanisms of trans-membrane metal transport underlie our understanding of metal bioaccumulation from both dissolved and dietary sources. Suffice to say trace metals will enter cells via protein carriers; the characteristics of carrier facilitated diffusion explain most metal uptake behaviours.

Understanding of species-specific bioaccumulation attributes (geochemistry, physiology and ecology) is the pathway to a more sophisticated characterisation of where and when we might expect metal effects. Bioaccumulation and trophic transfer are key elements along that pathway.

Bioaccumulation of metals can differ widely from element to element within a single species. Bioaccumulation of the same metal can differ widely among species. For the same metal and the same organism, bioaccumulation can be strongly affected by environmental conditions like water chemistry, sediment chemistry or type of food accessible to the organism. Bioaccumulation is also the net outcome of uptake from multiple routes of exposure, especially diet and water. The relative importance of route of exposure varies with the circumstance, but generalisations are available to explain likely outcomes. Very few situations exist where dietary exposure is not important.

The effects of geochemistry on dissolved metal bioaccumulation are the subject of much study. The Free Ion Activity Model (FIAM) in all waters and competition with major ions in freshwater explain much of the uptake of dissolved metals.

Ingestion rate and assimilation efficiency are the drivers for bioaccumulation of trace metals from food. Assimilation efficiency is an aggregate measure of all the processes that are involved in digestion. It is not surprising that AE is variable among species, within species and among types of food, given the degree of the variability in food type, digestive processes and feeding rates. Nevertheless, tendencies are sufficiently common that generalisations are possible. In summary, it appears that there is a progression of metal bioavailability in food from soluble forms > other intracellular materials > insoluble materials. It is likely that absolute AE among these fractions varies with the consumer or predator. Insoluble forms contribute to the metal uptake from the diet by at least some predators, but probably to a lower extent than metals bound in soluble form and with variation according to the nature of the insoluble binding.

Although bioaccumulation varies widely among taxa, these differences often reflect basic biological differences that can be captured by variations in the combination of a few physiological rates, including ingestion rate, growth rate, unidirectional accumulation rates from food and water and the rate constant governing unidirectional loss. While these factors seemed to add myriad complexity when considered individually, there is great simplifying advantage to a quantitative approach to estimating their net effect (biodynamic modelling).

Using scenarios for different environmental conditions, biodynamic models appear to be quite effective in forecasting the bioaccumulation expected under different exposure regimes. Thus far, comparisons of biodynamic modelling forecasts against observations of bioaccumulation in rivers, lakes, estuaries and coastal waters agree remarkably well.

Suggested reading

Campbell, P.G.C. (1995). Interaction between trace metals and aquatic organisms: a critique of the free-ion activity model. In *Metal Speciation and Aquatic Systems*, ed. A. Tessier and D.R. Turner. New York: John Wiley, pp. 45–102.

Luoma, S.N. and Rainbow, P.S. (2005). Why is metal bioaccumulation so variable? Biodynamics as a unifying concept. *Environmental Science and Technology*, **39**, 1921–1931.

Rainbow, P.S. (2002). Trace metal concentrations in aquatic invertebrates: why and so what? *Environmental Pollution*, **120**, 497–507.

Reinfelder, J.R. and Fisher, N.S. (1991). The assimilation of elements ingested by marine copepods. *Science*, **251**, 794–796.

Sunda, W.G. and Guillard, R.R.L. (1976). The relationship between cupric ion activity and the toxicity of copper to phytoplankton. *Journal of Marine Research*, **34**, 511–529.

Wallace, W.G., Lopez, G.R. and Levinton, J.S. (1998). Cadmium resistance in an oligochaete and its effect on cadmium trophic transfer to an omnivorous shrimp. *Marine Ecology Progress Series*, **172**, 225–237.

Wang, W.-X., Fisher, N.S. and Luoma, S.N. (1996). Kinetic determinations of trace element bioaccumulation in the mussel, *Mytilus edulis*. *Marine Ecology Progress Series*, **140**, 91–113.

8 • Biomonitors

8.1 MONITORING AND BIOMONITORS

Any analysis of the potential detrimental effects of toxic metals in aquatic environments needs comparative measures of how much potentially toxic metal is present at a particular time and place. Every measure of exposure has its strengths and weaknesses. Just as perspective on metal concentrations in the water column (Chapter 4) and in sediments (Chapter 5) is necessary to interpret contamination, so is perspective on concentrations accumulated by living organisms.

'Monitoring' involves intensive, systematic data collection in time and/or space with the purpose of characterising occurrence, distribution and trends of (in our case) metal contamination. Explaining the contamination, or at least raising testable questions about its causes, can also be an objective of monitoring. Monitoring programmes typically follow predesigned protocols so that observations are comparable.

Monitoring has a mixed reputation in environmental science. Discussions about assessing risk, managing contamination or even about use of tools like adaptive management usually speak of the need to monitor chemical exposures. In practice, however, early monitoring programmes (especially) involved routine data collection, sometimes with minimal cost being an important goal, and often with little interpretation. Data sets resulted that were either underutilised or too superficial to be of much use. While the need for data on natural waters is recognised in the science community, monitoring lost much of its credibility because of these early efforts. Many assessment programmes now exist, however, that are carefully designed, periodically reviewed and regularly interpreted. These are observational programmes and have all the limitations inherent in that approach. But, increasingly, valuable data on metal contamination in water, sediments and biota are now available from many of the major aquatic environments of the world (e.g. Szefer, 2002).

Box 8.1 Definitions

bioavailability: describes a relative measure of that fraction of the total ambient metal that an organism actually takes up when encountering or processing environmental media, summated across all possible sources of metal, including water and food as appropriate.

biomonitor: an organism that accumulates trace metals in its tissues, the accumulated metal concentration of which provides a relative measure of the total amount of metal taken up by all routes by that organism, integrated over a preceding time period.

cosmopolitan biomonitor: a biomonitor that is geographically widespread, so enabling the construction of an extensive database to identify high and low bioaccumulated trace metal concentrations indicating the presence or absence of local raised trace metal bioavailabilities to that organism.

suite of biomonitors: a selected group of biomonitors that take up and accumulate metals from a variety of sources (e.g. solution, seston, deposited sediment, prey, etc.), so providing comparative information on the relative importance of different bioavailable sources of metals in a specific habitat.

Designing a cost-effective monitoring programme is much more difficult than typically recognised. Recognition of budgetary constraints, clear goals and a broad framework are necessary to guide the many choices that must be made. In this chapter we talk about one of those choices: which media to use to develop characterisations essential to risk assessments or policy decisions. In particular we focus on an underutilised approach: biomonitoring. We dedicate a chapter to this subject because of its great potential for narrowing uncertainties in risk assessments and risk management in the next generation of management approaches. We will present specific ideas in this regard in Chapter 17.

8.2   DISSOLVED METAL CONCENTRATIONS

Prologue. Accurate dissolved metal monitoring requires an extensive and expensive programme. The data obtained provide traditional but relatively insensitive baseline observations of differences in contamination. The data are measures of total, not bioavailable, metal concentrations. Either instrumental determinations of relative speciation or manipulation of the data using physicochemical speciation modelling are necessary to obtain more accurate estimates of relative dissolved bioavailabilities and risk.

Regulation of toxic metals in aquatic environments was based historically on the assumption that the ecotoxicological effects of trace metals in water bodies are ultimately caused by high dissolved concentrations. The implicit assumption was that metals taken up from solution are the significant source of toxicity to the exclusion of metals taken up by animals from the diet. When ecotoxicity testing relies exclusively upon laboratory exposures of organisms only to dissolved metals, then it follows that an appropriate comparison in the environment is the dissolved concentration. If dietary exposures are important, then additional measurements must be added to environmental monitoring.

The technology and instrumentation now exist to measure extremely low dissolved concentrations of trace metals. Nevertheless, dissolved concentrations of metals in natural waters are still close to the limits of detectability. More importantly, nearly all measurements are at risk from contamination during collection and analysis procedures. Extreme precautions need to be taken with appropriate sampling and clean laboratory techniques to produce bona fide results. Furthermore, dissolved concentrations usually vary over time (see Chapter 5) in all types of water bodies. For example in estuaries variations occur with differential inputs of river and seawater at different states of the tide, and differential river flow according to recent rainfall in the catchment which may vary seasonally. Similarly variable flow in a river receiving metal-rich effluents will affect the dilution capacity of the receiving water body. It follows also that each measurement represents a single point in time that may be different from the dissolved concentration present at that exact location the day before or the day after. Unless sampling programmes are intensive over time, the power to detect meaningful differences in contamination can be low. Effective dissolved metal sampling programmes are achievable but are expensive. As the only source of data, they can be of limited sensitivity (alone) unless differences among locations are large.

Crucially, it is necessary to ask what such data mean in terms of ecological risk, however accurately they have been measured. As discussed in Chapter 7, the physicochemistry of the medium will affect the rate of uptake of a metal from solution by biota. In short, the bioavailabilities of metals in different water bodies may vary even when their total dissolved concentrations are identical. For example the uptake rates of Zn and Cd by biota from solution typically increase with decreased salinity over the salinity range found in an estuary even if the dissolved concentrations are unchanged. Thus comparisons of dissolved concentrations between habitats can be misleading with regard to exposures of organisms. If the physicochemistries of the two media compared are established, it is then possible to use a model such as the Free Ion Activity Model (FIAM: see Chapter 7) to narrow uncertainties (Campbell,

1995). This comes at more expense and brings some uncertainties of its own.

It does remain true, of course, that gross differences in total dissolved metal concentrations will coincide with differences in bioavailability. But uncertainties will remain about ecological risks. One reason is that the diet of animals cannot be ignored as a potential route of ecotoxicologically significant metal input in the field (Wang, 2002; Luoma and Rainbow, 2005) and partitioning between water and food varies with environmental circumstances. Dissolved concentrations thus provide valuable perspective on metal contamination in an environment, but do not supply the comparative information on ecological risk that their cost would imply.

8.3   SEDIMENT CONCENTRATIONS

Prologue. Sediments concentrate metals, respond sensitively to changes in contamination and integrate the variability of contaminant inputs. However, concentrations are susceptible to bias from grain size and physicochemical differences. Chemical methods are not available to account for all the important processes that affect metal bioavailability from sediments; so, used alone, metal concentrations in sediments are unlikely to be suitable, sensitive measures of bioavailable exposure and risk.

A second possible measure of exposure to metals is their concentration in the local sediments. Metal concentrations in aquatic sediments are typically high enough to be easily measured, and there are few dangers of significant contamination during collection and analysis. Sediments also show some integration of accumulated metal over time and changes in response to contamination are large. Because concentrations are accumulated over a period of time, they integrate the variability typical of the water column. Power to detect differences is much greater than for water concentrations.

Problems with the use of sediment metal concentrations as comparative measures do, however, remain. Physicochemical characteristics of sediments, especially grain size, affect their ability

to accumulate trace metals so that different sediments will reach different concentrations from identical dissolved sources of metal (Chapter 6; Luoma, 1989, 1990; Bryan and Langston, 1992). Coarse sediments are poor indicators of contamination; and susceptible to falsely underrepresenting contamination. Separation of fine material or normalization to conservative constituents is necessary to interpret differences in contamination in time or space (Chapter 6).

Sediments in different aquatic habitats can be originally derived from the erosion of rocks with different mineral composition, including constitutive trace metal contents. Accurate determination of the baseline of metal concentrations (under undisturbed conditions) can be a source of controversy. Such constitutive metal contents will contribute to total metal concentrations measured (according to the nature of extraction or digestion used in the preparation of the sediment for analysis), and detract from their accuracy as comparative measures of the more recent supply of trace metals in different aquatic habitats.

Sediments can be a direct source of metals to sediment-dwelling or sediment-ingesting animals. The trophic availability of metals in ingested sediments will also vary with the physicochemical characteristics of the sediment. Again, therefore, total metal concentrations may not be good measures of the relative bioavailabilities of trace metals in different sediments. Different approaches are available either to directly determine (e.g. extractions) or model metal bioavailability from sediments; but none is especially sensitive to all of the conditions that can influence risks (see Chapter 16).

Much experience with assessing metal contamination and its effects in nature suggests that chemical measures (metals in water or sediment), alone, cannot provide a sufficiently robust definition of contamination to confidently forecast ecological risks. Consideration of the metal accumulated into the biological component of the aquatic environment can greatly reduce uncertainties about bioavailability and can translate directly into evaluations of risk.

8.4   BIOMONITORS

Prologue. A biomonitor is an organism which accumulates trace metals in its tissues, the accumulated metal concentration of which provides a relative measure of the total amount of metal taken up by all routes by that organism, integrated over a preceding time period

In general, trace metals equilibrate between particulate material (including sediments), dissolved forms and the biota of a water body (Fig. 8.1). This is evidenced by correlations between biota and either sedimentary or dissolved metal concentrations, if a very broad data range is considered (Fig. 8.1). The variability in such relationships is driven by differences in metal bioavailability among locations and species. Metal concentrations in neither sediments nor water unambiguously define metal bioavailability. Therefore a complementary line of evidence is necessary to refine the definition of exposure: determination of accumulated trace metal concentrations in selected organisms, termed biomonitors. The term *biomonitor* refers to an organism in which the accumulated concentration of

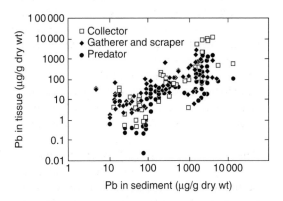

Fig. 8.1 Concentrations of Pb in sediment and aquatic macroinvertebrates from (mostly) mine-contaminated streams. Three feeding guilds are represented and concentrations of metals are broadly correlated with sediment in all three. Across such a broad data range the figure illustrates that lead in the organisms increases as lead in sediment increases, but with much variability at any one concentration. The variability reflects effects of speciation in water, form in sediment and species-specific biological process. (After Goodyear and McNeill, 1999.)

metal represents metal taken up by all routes, integrated over a preceding time period.

Metal concentrations in biomonitors are typically high enough to be easily measured without significant risk of laboratory contamination. The total accumulated metal concentration of an organism will include any metal adsorbed onto the surface of the organism that has not therefore been taken up into the organism (see Chapter 7); such adsorbed metal is usually considered to be a negligible proportion of total accumulated metal for most trace metals in most organisms, particularly in strong accumulators (see Chapter 7; Rainbow, 1988, 2007). (Discussion later in this chapter (Section 2.3.4) addresses this aspect in insect larvae.) Thus, pragmatically, the accumulated metal content of a biomonitor represents bioavailable metal, the fraction of ambient metal in a habitat that can be taken up, and therefore with the potential to cause any ecotoxicological effects. The total metal bioavailability to an aquatic invertebrate animal consists of the fraction of total dissolved metal extracted by permeable (usually respiratory) epithelial surfaces from water passed across the gills and/or bathing the body (including where appropriate interstitial water in sediment) and ingested into the alimentary tract, plus the fraction of metal taken up into the alimentary tract epithelium from the total metal content of ingested food particles (including ingested sediment where appropriate) (see Chapter 7).

Because the accumulated metal concentration of a biomonitor is an integrated proxy measure of the total bioavailability of metals to that organism in that habitat, the use of biomonitors overcomes the major disadvantage of the use of either dissolved or sediment metal concentrations in the monitoring of toxic metal availabilities in different aquatic habitats. At the very least, biomonitoring data are an important complement to dissolved and/or sediment metal data.

8.4.1   Requirements and uses of biomonitors

Prologue. The use of carefully selected taxa as biomonitors facilitates comparisons of local trace metal bioavailabilities over space and time. The accumulated concentration of a trace metal in a particular biomonitor will be a measure of bioavailability of the metal, summed across all sources, to that specific species, over some time period. Using more than one biomonitor increases the strength of generalised conclusions about different degrees of metal contamination over different time periods from different sources of metals (e.g. solution, suspended material, deposited material, etc.).

Despite their important advantages, biomonitors have been underutilised in many regulatory applications. Misunderstandings about biomonitors (e.g. the frequency with which organisms regulate internal metal concentrations) have led to implication that biomonitoring data are not useful for helping understand ecological risks (e.g. Chapman, 1997). In fact biomonitoring addresses questions other tools cannot, and as such is a crucial component to understanding metal exposure in natural waters, if the question and the details of the approach are appropriate.

- **The question must be clear**
  In recognition of the shortcomings of laboratory experiments alone, biomonitoring is best suited to address the question of bioavailable metal exposure in natural waters. Typically, bioaccumulation in indigenous aquatic organisms from metal-contaminated waters is compared to bioaccumulation in specimens from reference (undisturbed) environments. Comparisons in space and/or time are typical of the most effective studies. The principles of biomonitoring also can be used to explore distribution of contamination among the organisms of a food web or compare internal accumulation of metals in different species (as part of an assessment of potential vulnerability to metal effects).
- **The choice of species is crucial**
  In broad terms, nearly all species will generally have enriched metal concentrations in a contaminated environment. Metal concentrations in any organism are an indication that metals are bioavailable (i.e. metals are penetrating the food web). But the degree of enrichment will vary among species. If the goal is to compare bioavailabilities

among sites or otherwise evaluate the degree of bioavailability resulting from a specific contamination sources, then the best biomonitor is an organism that is widely present along the profile of metal contamination. Crucially, of course, the biomonitor must be a *net accumulator* of the trace metal in question (see Chapter 7).

Interspecific variation in metal uptake rates from solution and from the diet (in the case of an animal), as well as metal loss rate constants, cause the different bioaccumulation patterns among species (see Chapter 7). So long as some net accumulation occurs, accumulation is correlated with total uptake rate and the accumulated concentration will be a relative measure of exposure. Strong net accumulation will enhance the power to discriminate between metal bioavailabilities at different sites (Rainbow et al., 2004a). The use of weak net accumulators as biomonitors need not be ruled out as long as their biology is recognised. In those cases where there is effectively no significant excretion of accumulated metal (see the case of barnacles in Chapter 7), accumulated concentrations are arguably absolute measures of all metal taken up by the organism. Indeed the more information that can be compiled on the accumulation pattern of a biomonitor for a trace metal, the better the interpretation of data from a biomonitoring programme.

- **The chosen species should fit the well-established criteria for a suitable biomonitor**
  A suitable biomonitor (Phillips and Rainbow, 1994) needs to be sedentary, or at least representative of the area under study. It should be abundant, easily identified and sampled, long-lived and large enough for (tissue) analysis. It should be hardy and be robust enough to withstand experimental handling, both for laboratory measurement of bioaccumulation kinetic parameters and possibly for use in local transplant field experiments.
- **If spatial or temporal comparison of bioavailable metal is the goal, then cosmopolitan species are most desirable**
  Comparisons are necessarily intra-taxon. Bioaccumulated concentrations in different taxa

are comparable only if a quantitative relationship between bioaccumulation in the two species is established. Thus the use of selected cosmopolitan (widespread) species is a requirement for determination of relative metal bioavailability. The ideal biomonitor can tolerate raised metal bioavailabilities and persist through broad areas in the presence of physicochemical variation (e.g. salinity changes: Phillips and Rainbow, 1994; Rainbow, 1995b). Data from globally cosmopolitan taxa are increasingly available from coastal, estuarine, lake and freshwater stream environments. The US Mussel Watch Program, which uses biomonitoring data from mussels and oysters from the coastal waters of the subcontinental USA typifies this approach (Farrington et al., 1983; Goldberg et al., 1983; Lauenstein et al., 1990; Cantillo, 1998).

- **Time integration can and should be quantified to better understand exposures**
  An understanding of the bioaccumulation kinetics, particularly the efflux rate constant, of a particular trace metal in a particular biomonitoring species will provide information on the time period over which the metal content is accumulated or integrated. The metal efflux rate can be quantified by biodynamic calculations. Time integration is assumed not to be highly variable within a species, especially between populations of a single species over a restricted geographical region (the concern of most biomonitoring programmes). The time period will, however, necessarily vary between species with different metal efflux rates.

- **The use of a suite of biomonitors provides information on the relative importance (or behaviours) of bioavailable metal from different sources (e.g. water, sediment, suspended material, other organisms)**
  It is highly desirable to use more than one biomonitor in a programme in order to increase the strength of generalised conclusions to be made about different degrees of metal contamination over different time periods (Phillips and Rainbow, 1994; Rainbow, 1995b). The use of a suite of different biomonitors allows the introduction of different time periods over which

bioavailability is integrated. The well-chosen suite of biomonitors also provides information on the relative importance of different sources of toxic metals to the biota, typically solution and, in the case of an animal, the diet. Dissolved sources may be the overlying medium or interstitial water in a sediment, or indeed a combination of the two according to the burrowing behaviour and/or irrigation powers of an infaunal animal. The diet may, for example, consist of macrophytes fed on by a grazing herbivore, the animal prey of a carnivore, suspended particles (detritus, phytoplankton and/or zooplankton) in the case of a suspension feeder, and sediment (deep or surface deposited) for a sediment-ingesting deposit feeder. Implicit in the choice of a suite of biomonitors is a knowledge of the biology of the chosen taxa in order to understand the sources of metals available to each (Rainbow, 1995b). Biodynamic modelling can also help identify the partitioning of the contribution of different routes of metal uptake to the total accumulated concentration of a biomonitor (Wang, 2002; Luoma and Rainbow, 2005), again important in the interpretation of biomonitoring data. Use of multiple species thus exploits the species-specific nature of 'bioavailability'. Metal bioavailability to each species is determined by the species' biology (determining the specific sources of metal to which that organism is exposed) as much as the geochemistry of the environment which affects each source.

- **Biomonitors are used to quantify bioavailable exposure in relative terms, the first step in analysis of risk**
  Bioaccumulation, alone, is not a simple, direct measure of ecological risk from metals or risk of toxicity to the species. Under some circumstances relationships exist between bioaccumulation and risk; but those should be interpreted with care. On the other hand, the link between bioaccumulation and ecological risk has, arguably, fewer uncertainties than the links with other measures of exposure (water and sediment).

  Thus, biomonitoring as discussed here does not provide direct information on the ecotoxicological effect of the different degrees of metal exposure. It simply provides information on where and when exposure to high bioavailabilities of toxic metals is occurring. This still represents a vital and simple first step in any investigation of the possible ecotoxicological effects of metals in an aquatic habitat. Biomonitoring using appropriate fauna could, for example, replace the analysis of sediments in the first stage of the 'Sediment Quality Triad' (Chapman and Long, 1983; Long and Chapman, 1985). The traditional triad characterises the contamination of aquatic sediments and subsequent ecotoxicological effects in three stages: (i) analysis of trace contaminants in sediments; (ii) bioassays of sediment toxicity in lethal and sublethal laboratory tests; and (iii) investigations of infaunal community assemblages (Phillips and Rainbow, 1994). Similarly biomonitoring will identify the habitats (and thence the constituent organisms) with the highest toxic metal exposures, making more efficient any subsequent investigation of ecotoxicological effects to be assessed (see Chapter 9).

8.4.2   Cosmopolitan biomonitors of marine waters

Prologue. Suitable cosmopolitan biomonitoring species are available from nearly every kind of environment. For many species substantial data are available to allow comparative interpretations of the degree of contamination. Notable species include mussels (*Mytilus* spp.; *Perna* spp.) from coastal waters, tellinid clams from estuaries, hydropsychid caddisfly larvae from rivers, and both perch and the phantom midge larva *Chaoborus* from lake environments. Broad correlations exist across large data ranges between biomonitors with similar exposure histories, and between biomonitor metal concentrations and concentrations in sediment and water. The relationships themselves reflect the system-wide response of aquatic environments to contamination. The variability in the relationships reflects variability due to bioavailability, biology and hydro-physicochemistry.

Biomonitoring data are most effective when the same taxon is compared across time and/or space. Examples of potential biomonitor species to include in a suite suitable for the investigation of different bioavailable sources of trace metals in coastal waters might be as follows (Rainbow, 1995b; Rainbow *et al.*, 2002).

8.4.2.1   Solution
Dissolved bioavailabilities are best measured with organisms that take up metals only from solution, without any confounding uptake from the diet. Thus the most appropriate biomonitors to serve this purpose are plants, specifically macrophytic algae lacking the roots of flowering plants which provide the potential to take up metals from sediments. A suitable macrophyte in marine and estuarine waters, for which data are available across a wide range of conditions, is the brown alga *Fucus vesiculosus* (Bryan and Hummerstone, 1973a; Bryan and Gibbs, 1983; Barnett and Ashcroft, 1985; Bryan *et al.*, 1985; Rainbow and Phillips, 1993; Rainbow, 1995b), making sure that the algal frond is not in contact with sediment that might provide a metal source by surface adsorption (Luoma *et al.*, 1982), and that a specified region of the plant (representing older tissue) is selected for analysis (Bryan and Hummerstone, 1973a; Bryan *et al.*, 1985). Table 8.1 shows the range of concentrations that might be expected in this biomonitor from undisturbed to highly contaminated waters. Other macrophytic algae can be used as trace metal biomonitors, although less extensive data exist. These include other fucoids (e.g. *Ascophyllum nodosum*), other brown algae (species of *Sargassum*) and green algae (species of *Enteromorpha* or *Ulva*, e.g. *E. intestinalis* or *U. lactuca*) (Bryan *et al.*, 1985; Rainbow and Phillips, 1993; Rainbow, 1995b).

8.4.2.2   Solution and suspended particles
Any assessment of the bioavailability of trace metals from the water column requires the use of a suspension feeding animal as a biomonitor. Uptake by such species occurs from solution and suspended material, both of which contribute to the integrated accumulated metal concentration in the biomonitor.

Table 8.1 Accumulated concentrations of trace metals (µg/g dry weight) in the macrophytic alga Fucus vesiculosus, categorised here as 'typical' or 'high'. The low end of the typical range is indicative of uncontaminated conditions; the high end is indicative of concentrations representative of moderate contamination on a regional scale (see Chapters 5 and 6 for comparable ranges for water and sediments). The 'high' concentrations are indicative of atypically raised bioavailability of that metal in the local habitat

	Ag	As	Cd	Co	Cr	Cu	Ni	Pb	Zn	Reference
Typical	0.3–1.0		0.2–1.5	0.6–5.3	0.6–4.8	8–32	1.8–10	3–20	99–560	Bryan and Gibbs, 1983
			2.3			29–71		5–11	405	Barnett and Ashcroft, 1985
		11–55								Langston, 1980
			1.0–1.9			1–19			23–175	Rainbow et al., 1999
			1.9–3.3			13–21		10–15	230–575	Rainbow et al., 2002
High				16–20		200–1450		68	1000–4200	Bryan and Gibbs, 1983
		138–160	7.7							Langston, 1980
										Rainbow et al., 1999

176

In temperate coastal waters extensive biomonitoring data are available for mussels (bivalve molluscs of the family Mytilidae, e.g. *Mytilus edulis* and *M. galloprovincialis*). These are the classic biomonitors in this category, integrating metals taken up from solution and suspended material filtered from the water column (Rainbow and Phillips, 1993; Phillips and Rainbow, 1994; Rainbow, 1995b). Suspended material will potentially include detritus, phytoplankton and zooplankton, and, under particular hydrodynamic circumstances, may also contain surficial sediments resuspended in the medium. Mussels filter suspended matter at the bottom end of the range spectrum, smaller than most zooplankton.

Mussels have long been used in mussel watch programmes, not only in the USA (Farrington *et al.*, 1983; Goldberg *et al.*, 1983; Lauenstein *et al.*, 1990), but also elsewhere in the world (Cantillo, 1998) including Scandinavia (Phillips, 1977, 1978) and France (Amiard-Triquet *et al.*, 1999; Chiffoleau *et al.*, 2005). Thus there is an established literature identifying what is a high or low accumulated concentration of a defined trace metal in *Mytilus edulis* (Table 8.2), and in several of its close taxonomic relatives within the genus *Mytilus* or in related genera (e.g. *Perna*). There is also a growing literature on the biodynamic modelling of the relative strengths of different metal sources to mussels (Chapter 7; Wang *et al.*, 1996; Chong and Wang, 2001), aiding the interpretation of the results of biomonitoring programmes involving mussels.

Other bivalves are also appropriate to play a biomonitoring role, perhaps differing from mussels in their range of suspended particle size filtered, or in the characteristics (salinity range, degree of wave action, etc.) of the coastal habitat occupied. Substantial data exist for oysters (family Ostreidae) (Table 8.3). They are used in the USA (*Crassostrea virginica*) and other (*C. gigas*) mussel watch programmes (Cantillo, 1998; Chiffoleau *et al.*, 2005), with the advantage of living across broader ranges of salinity and being stronger accumulators of metals that mussels do not accumulate strongly (e.g. Zn and Cu: Amiard-Triquet *et al.*, 1999; Geffard *et al.*, 2004). Table 8.4 compares the median and 85th percentile of concentration data from the World Mussel Watch database. The differences between mussels and oysters (compare Tables 8.2 and 8.3) reflect the influence on bioaccumulation of important differences in physiology and autecology. Data from either oysters or mussels will result in a similar interpretation about bioavailable contamination if judged against what is typical for that species.

Fewer data are available, but barnacles are also used as biomonitors of the bioavailable metals taken up from solution and suspended material (Table 8.5). The suspended material filtered by all barnacles will typically include zooplankton as well as larger phytoplankton such as diatoms. Species of *Balanus* are also capable of filtering very small phytoplankton (flagellates) and fine detritus (Anderson, 1994). In northern Europe, *Balanus improvisus* has been used in biomonitoring programmes in the Thames estuary, England (Rainbow *et al.*, 2002) and the Gulf of Gdansk, Poland (Rainbow *et al.*, 2004a). The extensive data set produced in the latter study provides a benchmark against which other barnacle monitoring data can be assessed, particularly if the significant supply of trace metals to the Gulf from the River Vistula is reduced by reduction of local industry and/or by local environmental protection action.

8.4.2.3 Solution and sediment

While mussels and oysters filter suspended material from the water column, other bivalves can act as biomonitors of metals associated with deposited sediments. Substantial data exist from a few systems for tellinid bivalves (Table 8.6). The tellinids are mainly deposit feeders. They primarily collect newly deposited organically rich particles from the top of the sediment, while being isolated by the shell and mantle from interstitial or pore water in the deeper sediment in which they are burrowed (Rainbow, 1995b). The tellinids *Scrobicularia plana* and *Macoma balthica* are strongly represented in British biomonitoring studies of estuaries (e.g. Fig. 8.2a, Table 8.6), *S. plana* occurring higher up an estuary in lower salinity than *M. balthica* (Bryan *et al.*, 1980, 1985).

The accumulated metal content of *S. plana* is derived then from the overlying medium and from

Table 8.2 Accumulated concentrations of trace metals (μg/g dry weight) in the soft tissues of mussels, categorised here as 'typical' or 'high'. The low end of the typical range is indicative of uncontaminated conditions; the high end is indicative of concentrations representative of moderate contamination on a regional scale (see Chapters 5 and 6 for comparable ranges for water and sediments). The 'high' concentrations are indicative of atypically raised bioavailability of that metal in the local habitat

	Ag	Cd	Co	Cr	Cu	Ni	Pb	Zn	Reference
Mytilus edulis									
Typical	0.1–1.1	0.4–2.7	0.3–6.3	0.2–2.5	6–11	0.5–3.6	2–19	45–119	Bryan et al., 1985
	0.1–2.4	0.8–3.4			4–19	0.4–3.7	0.2–3.3	51–150	Goldberg et al., 1983
		0.7–4.6	0.9–6.5		5–13	1.4–3.0	7–25	78–139	Boalch et al., 1981
		0.7–4.7			3–17	1.3–7.1	2–22	32–150	Gault et al., 1983
High	17	21–65			262	12	105	198–579	Bryan et al., 1985
		36		7.2			58	235–440	Gault et al., 1983
Perna viridis									
Typical		0.2–1.5		1–10	10–39		1–15	53–115	Phillips and Rainbow, 1988
High				17–38	60–219		38–48	141–153	Phillips and Rainbow, 1988

Table 8.3 *Accumulated concentrations of trace metals (µg/g dry weight) in the soft tissues of oysters, categorised here as 'typical' or 'high'. The low end of the typical range is indicative of uncontaminated conditions; the high end is indicative of concentrations representative of moderate contamination on a regional scale (see Chapters 5 and 6 for comparable ranges for water and sediments). The 'high' concentrations are indicative of atypically raised bioavailability of that metal in the local habitat*

	Ag	As	Cd	Cu	Hg	Pb	Se	Zn	Reference
Crassostrea virginica									
Typical	0.2–3.0	3–10	1–7	30–150	0.01–0.20	0.2–1.0	1–4	400–3000	Presley *et al.*, 1990
			1–4	6–55				230–1200	Lytle and Lytle, 1990
High	4.0–7.5	25–95	10–15	300–450	0.35–0.43	3–10	6–10	7000–9000	Presley *et al.*, 1990
				300–1000+				10000–50000+	Wright *et al.*, 1985
Crassostrea gigas									
Typical	1.6			216		5.1			Szefer *et al.*, 1997
			4	200		1		1700	Ayling, 1974
				80					Han and Hung, 1990
High	46			5110		14.5			Szefer *et al.*, 1997
			134	1700		135		14000	Ayling, 1974
				1000–2700					Han and Hung, 1990
Crassostrea rhizophorae									
Typical			0.7–2.5	8–39				233–11500	Silva *et al.*, 2003, 2006
			0.9						Reboucas do Amaral *et al.*, 2005
High				281				3949	Silva *et al.*, 2001, 2006
				1567				3006	Wallner-Kersanach *et al.*, 2000
								9770	Reboucas do Amaral *et al.*, 2005

Table 8.4 *Median and 85th percentile accumulated concentrations of trace metals (µg/g dry weight) in the soft tissues of combined species of mussels and combined species of oysters from the World Mussel Watch database*

	Mussels		Oysters	
	Median	85th percentile	Median	85th percentile
Ag	0.25	1.0	1.3	2.6
As	7.1	16	5.7	14
Cd	2.0	7.5	4.1	21
Cr	1.6	6.5	2.5	10
Cu	7.9	21	160	680
Hg	0.32	0.99	0.27	0.70
Ni	2.2	5.0	2.2	4.7
Pb	5.0	20	2.5	8.6
Se	2.2	3.9	–	–
Zn	130	260	1600	4500

Source: Cantillo (1998).

surface deposited material, which is presumably newly deposited or possibly recently bioturbated. The oxidised surface sediment appears to be the principal source of several metals to *S. plana* (Bryan *et al.*, 1985), and significant soft tissue to sediment concentration relationships have been observed for Ag, Co, Cd (Luoma and Bryan, 1982), As and Hg (Langston, 1982), Pb (Luoma and Bryan, 1978), Zn (Luoma and Bryan, 1979) and Cr (Bryan *et al.*, 1985). In fact the best fit correlations often needed to take into account the effects of physicochemical characteristics of the sediments. Normalisation to iron oxides for Pb (Fig. 8.3) (Luoma and Bryan, 1978) and As (Langston 1980, 1985) greatly improves correlations between sedimentary and bioaccumulated metal. Normalisation for organic matter contents improved the relationship for Hg (Langston, 1982) in the sediments. This reinforces the uncertainties in relying on total metal concentrations in sediments, alone, as measure of bioavailable metal concentrations.

Although *S. plana* is an excellent biomonitor of the bioavailabilities of many trace metals in the sediments, it must be remembered that dissolved metal in the overlying water column still represents a potential metal source to the bivalve. Bryan *et al.* (1985) showed a good correlation (Fig. 8.4) between accumulated concentrations of Cd in *S. plana* and the algal macrophyte *Fucus vesiculosus*. Since the latter obtains all metal from solution, this could be interpreted to mean that Cd taken up from solution contributes significantly to the total Cd concentration accumulated by *S. plana*. It is more likely, however, that Fig. 8.4 reflects the broad relationship between Cd in sediments, water, the deposit feeding bivalves and epibenthic seaweed that occurs over the 100-fold range in contamination among the estuaries studied by Bryan *et al.* (1985). Modelling could provide important information about the quantitative contribution of sources in such instances.

The related tellinid bivalve *Macoma balthica*, and closely related species of the genus *Macoma*, occur further down estuaries than *Scrobicularia plana*, and are widespread in middle latitudes, including the estuaries of England and France, the (low salinity) Baltic Sea (Hummel *et al.*, 1996; Sokolowski *et al.*, 1999), and the Pacific coast of North America as far south as San Francisco (where it was recently re-identified as *Macoma petalum*: Väinölä, 2003). *Macoma balthica* similarly takes up metals from the overlying water and from surface-deposited sediment, although it will also show facultative feeding on suspended material (seston) including resuspended surficial sediments (Lin and Hines, 1994). *Macoma balthica* has been used as a biomonitor of sediment-associated metals in northwest Europe (Bryan and Hummerstone, 1977; Bryan *et al.*, 1980, 1985) (Table 8.6). Data are comparable to those from *Macoma petalum* in San Francisco Bay (Luoma *et al.*, 1985). In the case of this latter species, biodynamic modelling has identified the relative importance of dissolved and dietary sources of Ag, Cd and Co (Griscom *et al.*, 2002) and the uptake and loss kinetics of Cd, Cr and Zn (Lee *et al.*, 1998). These studies emphasise the importance of ingested sediment as a source of metals. For example 60% to 90% of Ag was derived from food and 40% to 80% of Cd. The ranges represent different scenarios for responses of physiological parameters to different habitat conditions and

Table 8.5 *Accumulated concentrations of trace metals ($\mu g/g$ dry weight) in the bodies of barnacles, categorised here as 'typical' or 'high'. The low end of the typical range is indicative of uncontaminated conditions; the high end is indicative of concentrations representative of moderate contamination on a regional scale (see Chapters 5 and 6 for comparable ranges for water and sediments). The 'high' concentrations are indicative of atypically raised bioavailability of that metal in the local habitat*

	Ag	As	Cd	Co	Cr	Cu	Ni	Pb	Zn	Reference
Balanus amphitrite										
Typical	0.7–9.0	10–71	0.8–10.2	0.1–2.1	1–5	52–415	1.3–8.2	0.4–8.3	2860–8530	Rainbow and Blackmore, 2001
			2.1–10.1		1–8	60–486		1.7–9.2	2730–9350	Phillips and Rainbow, 1988
High	22.5			9.0–11.0		494–1810	99		14000–23300	Rainbow and Blackmore, 2001
		457			17–38	1010–3472		37–39	12000	Phillips and Rainbow, 1988
Balanus improvisus										
Typical			3–11			24–71	7–35	13.5	3300–7500	Rainbow et al., 2004a
	8–11		7–9			140–240		27		Rainbow et al., 2002
High			17					76	14400	Rainbow et al., 2004a
									19000–28000	Rainbow et al., 2002

Table 8.6 Accumulated concentrations of trace metals (µg/g dry weight) in the soft tissues of tellinid bivalves, categorised here as 'typical' or 'high'. The low end of the typical range is indicative of uncontaminated conditions; the high end is indicative of concentrations representative of moderate contamination on a regional scale (see Chapters 5 and 6 for comparable ranges for water and sediments). The 'high' concentrations are indicative of atypically raised bioavailability of that metal in the local habitat

	Ag	As	Cd	Co	Cr	Cu	Ni	Pb	Zn	Reference
Macoma balthica										
Typical	0.3–12.3 0.2–1.1	11–33 5–28	0.1–2.2	0.7–6.8	0.8–6.3	30–88 17–32	0.3–7.7	2.3–10.8	396–747 377–692	Bryan et al., 1980, 1985 Bordin et al., 1992
High	100–122	46	9.4		16.3	224	12.7	36.5 33–124	1510–1790	Bryan et al., 1980, 1985 Bourguin et al., 1991
Scrobicularia plana										
Typical	0.1–4.6	12–31	0.2–9.1	2.0–13.3	0.5–7.1	9–61	1.0–5.8	5–109 14–60	256–1514	Bryan et al., 1980, 1985 Luoma and Bryan, 1978 Langston, 1980
High	259	98–191 97–190	31.5–42.7	33.0–106	16.3–23.8	136–752	14.4–22.4	225–3000 120–1016	2060–4920	Bryan et al., 1980, 1985 Luoma and Bryan, 1978 Langston, 1980

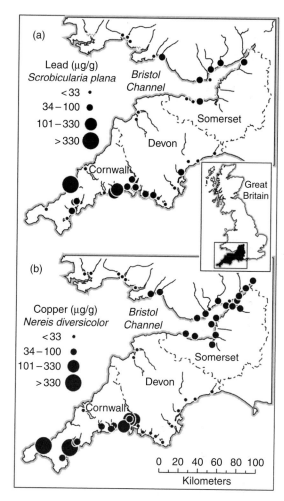

Fig. 8.2 (a) Biomonitoring of lead bioavailability to the tellinid bivalve *Scrobicularia plana* in estuaries of southwest England and South Wales (after Bryan *et al.*, 1980). (b) Biomonitoring of copper bioavailability to the polychaete worm *Nereis diversicolor* in estuaries of southwest England and South Wales (after Bryan *et al.*, 1980).

food choices. They also showed that the tellinids had a relatively slow rate constant of loss for Zn, explaining why this group, like oysters, is especially responsive to changes in bioavailable Zn contamination in the environment, in contrast to mussels.

Another member of the northwest European estuarine infauna often used in biomonitoring

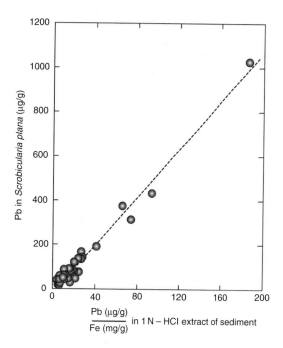

Fig. 8.3 Correlation between concentrations of Pb in the soft tissues of the tellinid bivalve *Scrobicularia plana* and the ratio Pb/Fe extracted with 1 N HCl from sediments from 17 estuaries of south and west England. (After Luoma and Bryan, 1978.)

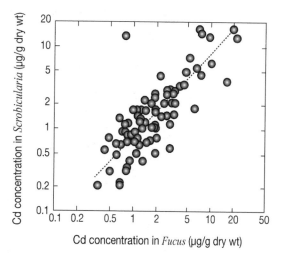

Fig. 8.4 Correlation between accumulated concentrations of Cd in the tellinid bivalve *Scrobicularia plana* and the macrophytic brown alga *Fucus vesiculosus* from a variety of estuarine and coastal sites in England and Wales. (After Bryan *et al.*, 1985.)

bioavailable sediment contamination is the poly-chaete worm *Nereis diversicolor* (Fig. 8.2b). Accumu-lated concentrations of many trace metals in *N. diversicolor* correlate significantly with those of surface sediments (Fig. 8.5) making it a good bio-monitor of bioavailability from the sediment (Bryan *et al.*, 1980, 1985). These trace metals include Ag, As, Co, Cu, Hg and Pb (Langston, 1980; Luoma and Bryan, 1982; Bryan *et al.*, 1985), but not Fe, Mn and Zn (Bryan and Hummerstone, 1971, 1973b; Bryan *et al.*, 1985).

Body concentrations of Zn, at least, are regulated by *N. diversicolor* when exposed in the field (Fig. 8.5) or laboratory to raised bioavailabilities of this element (Bryan and Hummerstone, 1973b; Amiard *et al.*, 1987), negating its use as a biomonitor for this metal. The accumulated concentrations of Pb in *N. diversicolor* and *S. plana* from the same south-west England estuaries correlate well, indicating that both species are responding to contamination in a similar manner (Luoma and Bryan, 1982). In fact, *N. diversicolor* shows a remarkable range of feeding strategies including deposit feeding (sediment ingestion), sediment browsing, carnivory, scavenging and even suspension feeding via a secreted mucus net (Harley, 1950). But it appears that sediment ingestion dominates as a source of dietary metals to this worm in (metal-contaminated) estuaries.

Other polychaete worms also have potential as biomonitors of the availabilities of trace metals in marine sediments, although fewer data are available. These include members of the family Nephtyidae which are easy to clean prior to analy-sis, widespread and accumulate most trace metals (Bryan and Gibbs, 1987). Knowledge of the feeding habits of different polychaetes will make them valuable additions to a suite of biomonitors to be used to investigate the bioavailabilities of trace metals in different components of a sediment. Ampharetid and terebellid polychaetes collect newly deposited organically rich particles with tentacles spread over the sediment surface (Rainbow, 1995b). They therefore exploit the same dietary source as tellinid bivalves, but, unlike tell-inids, infaunal polychaetes have the potential to take up metals from interstitial water across the body wall. Knowledge of the irrigation behaviour of such burrowing polychaetes will provide information on the relative contributions of inter-stitial water and water drawn down from the over-lying water column as potential metal sources to the polychaete concerned. Arenicolid polychaetes ingest sediment below the sediment surface and therefore sample a sedimentary source of metal potentially older and more sulphide-rich than that sampled by tellinid bivalves or terebellid polychaetes.

The relatively recent appreciation that infaunal lucinid bivalves obtain nutriment from symbiotic chemoautotrophic bacteria in the gills and do not carry out suspension feeding suggests that they might be suitable biomonitors of trace metals in the interstitial water of sediments without the confounding of a dietary source of metal. Lucinid bivalves access oxic water from the water column but also hypoxic hydrogen sulphide-rich water from the organically rich sediment (Distel, 1998). Such sulphide-rich interstitial water may also be rich in dissolved trace metals in reduced chemical form. Bioavailabilities of trace metals dissolved in the water column could be monitored in macrophytic algae for comparison against the

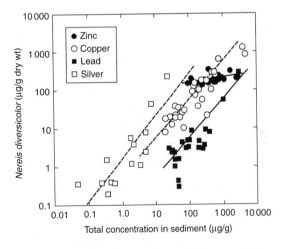

Fig. 8.5 Relationships between accumulated concentrations of Ag, Cu, Pb and Zn in the polychaete worm *Nereis diversicolor* and local sediments of estuaries in southwest England. (After Bryan *et al.*, 1985.)

results obtained from lucinids with their dual water sources. Relatively common littoral lucinid bivalves in British coastal waters include *Loripes lucinalis* and *Lucinoma borealis*.

8.4.2.4 Herbivores and detritivores

Herbivorous grazing animals or detritivores that occupy unique habitats have been included in biomonitoring programmes when there is a limited choice of species. For example, herbivorous littoral gastropod molluscs like limpets or littorinid winkles may be abundant in a habitat such as a pier installation lacking accessible sediments. On the other hand, detritus-feeding talitrid amphipod crustaceans are the only common accessible littoral invertebrate on the sandy shores of the Gulf of Gdansk at the south end of the Baltic Sea (Fialkowski *et al.*, 2000), where access to mussels, barnacles and infaunal bivalves requires subtidal dredging (Rainbow *et al.*, 2004b).

The herbivorous gastropod *Littorina littorea* (common winkle) is abundant on northwest European shores and is suitable as a trace metal biomonitor (Bryan *et al.*, 1983; 1985; De Wolf *et al.*, 2005). *Littorina littorea* grazes on algae such as the bladder wrack *Fucus vesiculosus* (see above). Very significant relationships between accumulated concentrations of Ag, As, Cd and Pb in winkle and wrack indicate that, directly or indirectly, concentrations in *L. littorea* reflect the dissolved bioavailabilities of these metals in the overlying water (Bryan *et al.*, 1983, 1985). However, Ag concentrations in the winkle were relatively low when Cu concentrations in the seaweed were high (Bryan *et al.*, 1983, 1985), suggesting competition of the two metals for uptake routes from the diet (see also Chapter 7). Significant relationships between winkle and wrack accumulated concentrations were also found for Cu, Fe, Hg and Zn but the regression slopes were relatively shallow, particularly in the case of Zn for which the winkle body concentration appears to be regulated (see Chapter 7) (Bryan *et al.*, 1983, 1985).

Another common Atlantic littoral herbivore is the limpet *Patella vulgata*. Limpets can be used as trace metal biomonitors (Bryan and Hummerstone,

1977; Bryan *et al.*, 1985), but particular care is needed to allow for the contrary effects of size on their accumulated metal concentrations (see below) (Boyden, 1974, 1977). Nevertheless, they respond to environmental contamination like other biomonitors across broad gradients (e.g. Fig 8.6).

Talitrid amphipod crustaceans occupy the strandlines of shores feeding on decaying macroalgae deposited by the tide, and are easily accessible for collection. Talitrids potentially accumulate trace metals from solution and from the diet. Their macrophyte diet reflects dissolved metal availabilities. *Orchestia gammarellus* has been used as a biomonitor for trace metals in British coastal waters (Rainbow *et al.*, 1989, 2002; Moore *et al.*, 1991). *Orchestia gammarellus* is a net accumulator of Cu and Zn from water and diet, with the latter contributing more to the accumulated concentrations of both metals (Weeks and Rainbow, 1991, 1993). The moult cycle has no effect on body concentrations of either metal (Weeks and Moore, 1991; Weeks *et al.*, 1992). Another talitrid, *Talitrus saltator*, was successfully used as an accessible littoral trace metal biomonitor in the southern Baltic (Fialkowski *et al.*, 2000, 2003c).

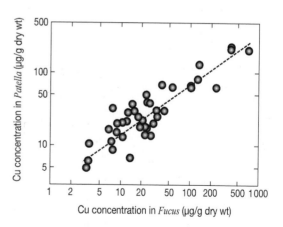

Fig. 8.6 Correlation between Cu concentrations in the limpet *Patella vulgata* and the macrophytic brown alga *Fucus vesiculosus* from a variety of estuarine and coastal sites in south and west England and Wales. (After Bryan *et al.*, 1985.)

8.4.3 Tropical coastal biomonitors

The examples above have their ecological counterparts suitable for biomonitoring programmes elsewhere in the world, including tropical waters.

Macrophytic brown algae are more typical of temperate shores than tropical ones, although species of the genus *Sargassum* have potential as biomonitors of dissolved metal bioavailabilities (Rainbow and Phillips, 1993). Macrophytic green algae can be common on tropical shores, particularly in the presence of eutrophication. Species of the genera *Enteromorpha* and *Ulva* are likely contenders for use in biomonitoring programmes, although a major caveat remains concerning the taxonomy and correct identification of species in these genera (see below, Section 8.2.3.5.6) (Rainbow and Phillips, 1993).

It is easier to recommend tropical equivalents of the mussels used so extensively for biomonitoring in the northern hemisphere (Rainbow and Phillips, 1993; Rainbow, 1995b). The outstanding candidate in the IndoPacific is the green lipped mussel *Perna viridis* (Table 8.2), now well established in biomonitoring programmes in the Far East and Southeast Asia (Phillips, 1985; Phillips and Muttarasin, 1985; Phillips and Rainbow, 1988). From the same mussel genus, *Perna perna* is a trace metal biomonitor available for use on both sides of the South Atlantic, round Africa (Rainbow, 1995b; Banaoui *et al.*, 2004) and in the Gulf of Aden (Szefer *et al.*, 2006) (Fig. 8.7).

Barnacles have much to offer tropical biomonitoring programmes. *Balanus amphitrite* flourishes in the IndoPacific, and is widely spread as a fouling agent in other tropical and subtropical waters. *Balanus amphitrite* has been extensively used as a trace metal biomonitor in the Far East (Table 8.5) to follow changes in trace metal bioavailabilities over space and time (Phillips and Rainbow, 1988; Rainbow and Blackmore, 2001). While *B. amphitrite* is abundant in sheltered waters, another Indo-Pacific barnacle, *Tetraclita squamosa*, prefers more wave action, including that caused by the constant passage of ships in the otherwise relatively sheltered waters of Hong Kong Harbour (Phillips and Rainbow, 1988; Rainbow and Blackmore, 2001). In waters of reduced salinity in the Far East and Southeast Asia where it is widespread, the barnacle *Balanus uliginosus* (synonymous with *Balanus kondakovi*) has relatively unexploited potential as a trace metal biomonitor (Rainbow *et al.*, 1993a; Rainbow and Phillips, 1993). The barnacle *Fistulobalanus citerosum* has proved a successful trace metal biomonitor in the mangrove-lined estuaries of tropical Brazil (Silva *et al.*, 2006).

The coastal waters of Hong Kong are a fruitful natural laboratory for the extension of biomonitoring techniques to tropical coastal waters (e.g. Phillips and Rainbow, 1988; Rainbow and Blackmore, 2001). The strong gradients of toxic metal bioavailabilities (in 1986) provided an opportunity to compare results from different biomonitors,

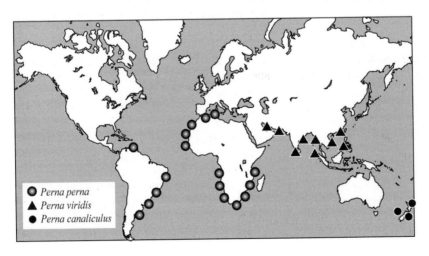

Fig. 8.7 The global geographical distributions of three species of the mussel *Perna*: *P. canaliculus*, *P. perna* and *P. viridis*. (From Rainbow, 1995b.)

○ *Perna perna*
▲ *Perna viridis*
● *Perna canaliculus*

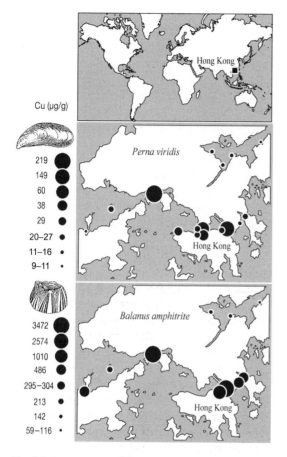

Cu (μg/g)

Perna viridis

Hong Kong

219
149
60
38
29
20–27
11–16
9–11

Balanus amphitrite

3472
2574
1010
486
295–304
213
142
59–116

Hong Kong

Fig. 8.8 A comparison of the geographical distributions of accumulated Cu concentrations (μg/g dry weight) in the green lipped mussel *Perna viridis* (mean soft tissue) and the barnacle *Balanus amphitrite* (weight-adjusted mean body) in Hong Kong coastal waters in April 1986. (After Phillips and Rainbow, 1988.)

in this case mussels (*P. viridis*) and barnacles (e.g. *B. amphitrite*) (Fig. 8.8) (Phillips and Rainbow, 1988). This 1986 survey aimed to plot geographical variation in trace metal bioavailabilities, but it also provided a baseline against which bioavailability changes over time could be followed. Thus Rainbow and Blackmore (2001) were able to investigate changes in the bioavailabilities of trace metals to barnacles in Hong Kong waters from 1986 to 1989 to 1998, as well as analysing their geographical distributions in 1998. Bioavailable exposures (to barnacles) of most trace metals in Hong Kong coastal waters fell from 1986 to 1998, possibly as a result

of local environmental protection action but also possibly as a result of the movement of local industry from Hong Kong into the adjacent Guangdong province of China (Rainbow and Blackmore, 2001).

In the case of oysters (Table 8.3), the Pacific oyster *Crassostrea gigas* has been spread widely through the world for mariculture purposes and is a good biomonitor showing strong net accumulation of many trace metals (Rainbow and Phillips, 1993; Rainbow, 1995b). In South America, the mangrove oyster *Crassostrea rhizophorae* is also a good trace metal biomonitor (Silva *et al.*, 2001, 2003, 2006). The intertidal rock oyster *Saccostrea cucullata* can also be used as a biomonitor in IndoPacific coastal waters, although specific identifications in this genus can be extremely difficult (Rainbow and Phillips, 1993; Rainbow, 1995b).

The talitrid amphipod crustacean *Platorchestia platensis* has exciting potential for it is remarkably widely distributed (Fig. 8.9) and has been used as a biomonitor of Cu and Zn (Weeks, 1992; Rainbow and Phillips, 1993).

8.4.4   Freshwater biomonitors

Global, cosmopolitan biomonitors exist for lakes and streams, and extensive databases are available for a few species.

The use of aquatic plants as biomonitors of trace metals in freshwater systems is essentially comparable to that of macroalgae in coastal environments. As before, algae lack roots and will obtain all metal from solution while flowering plants have the potential to access sediment-associated metal via the roots. The green alga *Cladophora* and the red alga *Lemanea* are geographically and ecologically widespread and have good biomonitoring potential, although care needs to be taken with correct species identifications in each case (Phillips and Rainbow, 1994). Few data exist for these species.

Forage fish can be effective biomonitors in lake environments. The mobility of fish is less of a confounding problem to data interpretation in a small lake than in a stream or more extensive coastal habitat. The yellow perch (*Perca flavescens*) is an example of a commonly used biomonitor in

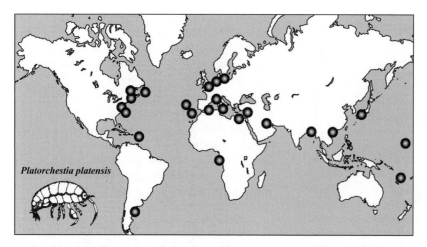

Fig. 8.9 The global distribution of the talitrid amphipod crustacean *Platorchestia platensis*. (After Rainbow, 1995b.)

metal-contaminated lakes in northeastern Canada (e.g. Campbell *et al.*, 2003; Pyle *et al.*, 2005). Perch tolerate a wide range of environmental conditions and are among the most widely distributed fish across the northern hemisphere. Campbell *et al.* (2003) described the reasons for choosing the yellow perch as a biomonitor in their study of metal contamination in lakes from Quebec and Ontario:

> ubiquity; abundance; relative immobility; ease of sampling; metal tolerance; metal bioaccumulation capacity; dynamics of metal accumulation; capacity to synthesize metallothionein; available physiological and behavioral data; and ecological role . . . this species does not travel over long distances and therefore its metal body burden tends to represent local sources.

Yellow perch are numerous in smelter-contaminated lakes in eastern Canada because they are among the most acid-tolerant fish species.

It is notable that in spite of the scarcity of competing species and an abundant supply of zoo-plankton, the yellow perch rarely exceed 15 cm in length in the contaminated lakes (Campbell *et al.*, 2003; Chapter 11). In undisturbed environments the perch feed upon benthic invertebrates in the nearshore (littoral) environment, mainly amphipods and burrowing mayfly larvae. Larger, less stressed perch live in these environments. But in the contaminated lakes, they feed mainly on zooplankton and

tiny chironomid midge larvae, which are resistant to metal contamination. The surviving perch are not only good biomonitors of metal exposure, but their growth and physiology provide co-varying indications of metal stress in the contaminated environments (Campbell *et al.*, 2003; Pyle *et al.*, 2005; Chapter 11).

Some species of zooplankton also proliferate under moderate contamination and have great potential for use in biomonitoring lakes. Larvae of species of the phantom midge *Chaoborus* occur over a wide range of chemical conditions (e.g. pH, ionic strength), and are abundant and widely distributed in lakes throughout North America. They are easy to collect and identify. They are also are able to accumulate and tolerate high concentrations of trace metals (Croteau *et al.*, 1998). This predator is of sufficient size for analysis and abundant in the presence of metal contamination. Croteau *et al.* (1998) recommended separating species of *Chaoborus* in biomonitoring. But when species were pooled there was only a moderate loss of predictive power. Munger and Hare (1997) showed that Cd concentrations in the phantom midge larva can be correlated with those in lake water. Croteau *et al.* (1998) further showed a strong relationship between Cd bioaccumulation in *Chaoborus puncti-pennis* and Cd free ion or total concentrations in water, after correction for H^+ ion concentrations (Fig. 8.10). In fact this correlation is not derived from Cd taken up directly from the water by

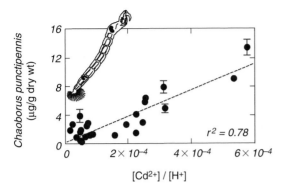

Fig. 8.10 Correlation of Cd bioaccumulation in larvae of *Chaoborus punctipennis* with free Cd ion concentrations normalised to hydrogen ion concentrations (pH) across a series of lakes in Quebec. (After Croteau *et al.*, 1998.)

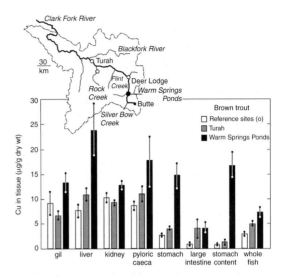

Fig. 8.11 Concentrations of Cu in different tissues of brown trout (*Salmo trutta*) from sites with different degrees of contamination in the basin of the Clark Fork River, Montana. (After Farag *et al.*, 1994.)

C. punctipennis, but is a consequence of Cd uptake from water by organisms lower in the food chain followed by dietary uptake by the phantom midge (Munger and Hare, 1997); again an example of a system-wide response to contamination that can cause concentrations to broadly co-vary over large data ranges.

Croteau *et al.* (1998) showed that the competitive effects of the other metal ions had few implications for bioaccumulation in *C. punctipennis*. But *Chaoborus* larvae are not useful to monitor bioavailable Cu or Zn because bioaccumulation of these essential metals is regulated.

In streams and rivers, physical processes assure access of many different kinds of benthic organisms to contaminated sediments. Fine-grained, contaminated sediments are spread throughout all riverine habitats. Contaminated sediments will settle in pools and backwaters. In fast-flowing, rocky bottomed streams, pockets of fine-grained contamination will accumulate in small pools within biologically productive riffle areas or within the matrix of filamentous algae and periphyton that coat rocks (Cain *et al.*, 1992). These microhabitats of bottom-dwelling animals provide a source of contaminated food or exchange with pore waters. Geochemical and biological factors add variability to simple relationships between metal concentrations in environmental media (sediment, water) and bioaccumulated metal. But

some correspondence to other monitors of contamination (e.g. sediments or water) is common in the contamination profiles typical of metal-contaminated stream environments.

Ideally, a charismatic pelagic fish species like trout, suckers, whitefish and carp would be the ideal biomonitor in stream environments. In Chapters 11 and 12 we show that bioaccumulation of metals in the most metal-tolerant species of trout (brown trout, *Salmo trutta*) do vary broadly between contaminated and uncontaminated environments. But many stream fish migrate over large distances into and out of contaminated streams. Some avoid contaminated waters unless they have no choice. The result is that bioaccumulated metal concentrations in fish can be extremely variable. One result is loss of power to differentiate differences among locations or times.

Another consideration is that the choice of tissues to analyse is crucial for fish. Concentrations vary widely among internal organs (Fig. 8.11). Cationic metals (Ag, Cd, Cu, Pb, Zn) are effectively trapped in the liver of trout; thus this is the organ that most sensitively reflects changes in environmental metal exposure (e.g. Farag *et al.*, 1994)

Table 8.7 *Concentrations of Cu, Cd and Zn in species of the free-living caddisfly biomonitor* Hydropsyche *from four mine-contaminated streams, compared to local reference areas where available. Concentrations in sediment are also presented where available. Differences in the ratio of sediment to tissue concentrations reflect (probably geochemically driven) differences in bioavailability among the environments. All values in* $\mu g/g$ *dry weight*

	Cadmium		Copper		Zinc	
	Sediment	Biota	Sediment	Biota	Sediment	Biota
Guadiamar River (Spain)						
Reference		0.13	150	17		125
Mine-impacted		6.0		98	1900	2300
Coeur d'Alene River (USA)						
Reference		0.3				172
Mine-impacted	24	6.7			503	642
River Lynher (UK)						
Mine-impacted				539		
Clark Fork River (USA)						
Reference		0.2		18		112
Mine-impacted	6.6	2.8	779	204	1073	270
Sacramento River (USA)						
Reference		0.06		14		113
Mine-impacted	3.1	2.2	240	38	330	169

Source: Data from Gower *et al.* (1994); Maret *et al.* (2003); Cain *et al.* (2006); Solà and Prat (2006).

(Fig. 8.11). These metals are not translocated efficiently to muscle tissue. Great loss of power to distinguish differences occurs if whole body fish are analysed because concentrations reflect the integrated concentration among body organs essentially diluted by muscle mass. Pelagic stream fish might be added to a biomonitoring programme to add relevance or address questions about human or predator exposure. But additional data from more suitable biomonitors are more useful as comparative measures of ambient metal bioavailabilities.

Some insects, for example corixid bugs and dytiscid beetles, spend their whole life cycles in freshwater. But some of the most abundant invertebrates in rocky bottom rivers and streams are the larval forms of flying (terrestrial) insects. These include caddisflies, mayflies, midges, phantom midges and alderflies. The larval insect community provides a cosmopolitan, globally available set of biomonitoring taxa for which biomonitoring protocols and comparable data sets are available (Table 8.7) (Hare *et al.*, 1991; Cain *et al.*, 1992; Hare,

1992; Hare and Campbell, 1992; Kiffney and Clements, 1994).

Although they increase in size over time, insect larvae do not develop gametes nor have gender differences that might otherwise confound the interpretation of body concentration data (see below). Some live as long as a year or two in larval form. Larval insects are an important source of food for upper trophic levels (e.g. fish). Downstream drift of invertebrates is common in rivers and streams, but does not seem to disturb the response of populations to metal gradients. Some species are metal tolerant and some are very metal sensitive (Gower *et al.*, 1994; Cain *et al.*, 2004).

Two potential problems may arise in the interpretation of measured body concentrations in insect larvae: the contribution of gut contents and that of metal adsorbed onto the exoskeleton. Both are especially important for these species because they tend to be small, with large surface area/mass and gut volume/mass ratios. Metals adsorbed onto the exoskeleton have not been taken

(a)

(b)

Fig. 8.12 Fe oxide particles adsorbed to (a) the whole body and (b) the gills of larvae of the caddisfly *Rhyacophila acropedes* The specimen was collected at a site in the headwaters of the Clark Fork River, Montana, USA, a mine-impacted river, heavily contaminated with metals. (Photo by M. Hornberger, US Geological Survey, Menlo Park, CA, USA.)

up across epithelial cells of the insect larva, and therefore are of no ecotoxicological significance to that larva. They may well be trophically available in the diet of a predator feeding on that larva, but nevertheless are not a true measure of bioavailable metal supply to the original larva. Fig. 8.12 shows Fe oxides precipitated on the gill of a caddisfly larva *Rhyacophila acropedes* at an up-stream station in the Clark Fork River, Montana, USA. The prevalence of such conditions is probably restricted to stations closest to sources in mine-contaminated rivers. It is uncertain how such adsorbed coatings affect whole body concentrations even there. In practice, there is probably no need to allow for adsorbed metal concentrations *per se* in any interpretation of insect larval accumulated metal concentrations. Adsorbed components will serve to accentuate accumulated metal concentration differences between larvae from habitats of different metal contamination levels, but are unlikely to disturb any ranking of sites in order of contamination.

Free-living caddisfly larvae (Order Trichoptera) are examples of common survivors in mine-impacted streams (Cain *et al.*, 1992; Gower *et al.*, 1994) that show a strong response to metal contamination in sediment and water (Fig. 8.13). They are common inhabitants of streams and rivers

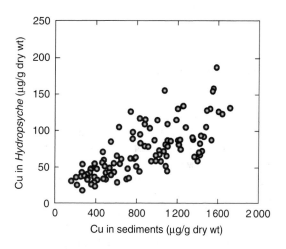

Fig. 8.13 Correlation of Cu concentrations in the detritus feeding (net spinning) caddisfly *Hydropsyche* sp. with metal concentrations in fine-grained sediments from the Clark Fork River, Montana (USA) (M. Hornberger, unpublished data). Metal bioaccumulation is also correlated with dissolved concentrations, which are more difficult to determine. Metal concentrations in the fine-grained sediments are measures of the distribution of the contamination; bioaccumulated metal is evidence that the metal is bioavailable.

through much of the world. *Hydropsyche* species, for example, are free-living, relatively sedentary, widespread in occurrence and of sufficient mass for analysis. They live for a year or more as aquatic larvae, and inhabit riffle zones where they are an important food for local fish. Caddisfly larvae are net accumulators of many trace metals. They are weak net accumulators of Zn but accumulate Cu in response to contamination. Thus in most ways they are excellent biomonitors. Comparisons of *Hydropsyche* species show a consistent response to mine wastes across different regions of the Earth (Table 8.7).

The contribution of metal-rich gut contents to total body burdens is a consideration in detritus-feeding caddisflies, particularly in environments with substantial contamination of fine-grained materials. Cain *et al.* (1995) compared caddisfly bioaccumulation between animals with the gut removed and untreated animals in a mine-contaminated river. The main effect of undigested gut contents was to increase slightly the variability in metal concentrations among samples, and slightly bias the values upward. Concentrations of Cu in whole caddisflies were 63, 43, 22 and 26 µg/g Cu dw at four sites. Concentrations in animals after gut removal were 48, 38, 25 and 25 µg/g Cu dw, respectively. The differences were statistically significant, but did not affect conclusions about bioavailability, trends in contamination, comparisons among sites or comparisons among species (Cain *et al.*, 1995). Hydropsychid caddisflies lose metals very rapidly (rate constant of loss for Cu and Cd>0.1/day); and require incoming food to defecate their gut content. Thus correction for gut content, or quantifying and recognising this relatively small bias, is probably the best strategy.

The ambiguities caused by adsorbed and undigested gut content can be avoided by analysing metal concentrations in some internal organs, or preferably, by determining concentrations within the cell solution (the cytosolic material inside cells as isolated by homogenisation and ultracentrifugation). Cytosolic bioaccumulation is direct evidence that the contaminant metal has been assimilated into tissues, and the metal in cell solution is also probably readily available for trophic transfer to predators (Wallace *et al.*, 2003) (see Chapter 7). Cytosolic concentrations of Cu in *Hydropsyche* sp. in the Clark Fork River were about 20% to 60% of whole body concentrations and correlated strongly with whole body concentrations (Cain and Luoma, 1999). The proportion of Cd in the cell solution is even higher. In the Clark Fork, Cu and Cd in the cytosol of *Hydropsyche* sp. followed the same general gradient as did whole body and sediment concentrations of Cu and Cd. Thus cytosolic analysis unambiguously demonstrated that bioavailable metals accumulate internally. But separation of the cytosolic metal is not always necessary to characterise trends in contamination or the contamination gradient.

Bioaccumulation in mayflies (Ephemeroptera) has also been widely studied. Goodyear and McNeill (1999) concluded that concentrations of Cd, Cu, Pb and Zn in mayfly larvae are directly proportional to those in sediments, demonstrating their ability to reflect ambient metal availabilities in accumulated body concentrations. In a study of larvae of *Baetis rhodani*, one of the most common mayflies in running waters of central European highlands, Rehfeldt and Söchtig (1991) similarly found that the concentrations of Cd, Cu. Pb and Zn in the larvae correlated with those in river sediments in Lower Saxony, Germany. Fialkowski *et al.* (2003b) used accumulated metal concentrations in larvae of two mayfly species of the genus *Baetis* (*B. rhodani* and *B. vernus*) to investigate spatial and temporal variability in the bioavailabilities of the trace metals Cd, Cu, Pb, Zn and Fe in streams in the catchment of the river Biala Przemsza draining an area of Pb and Zn mining in Upper Silesia, Poland. Both species identified significant local geographical variability in the bioavailabilities of Zn, Fe, Pb and Cd but not Cu in this river system. The genus *Baetis* is an example of a mayfly for which some species can tolerate metal contamination. But many mayflies are especially sensitive to metals and will not persist through a metal gradient (Clements, *et al.*, 2000; Cain *et al.*, 2004). Bioaccumulation in mayflies is informative if used to compare internal exposures of different species under the same conditions; but only a few mayflies are suitable biomonitors because limited tolerances are so common in this group (see also Chapter 11).

Chironomid midge larvae are typical inhabitants of organically rich sediments in freshwater streams ingesting the sediment as a diet. The small size and an especially difficult taxonomy limit the usefulness of some species as biomonitors. Larger species of *Chironomus* are widely used in experiments, although systematic data sets on their use as bioaccumulation monitors do not seem to be widely available.

The bioaccumulation of trace metals by freshwater amphipods has also been comparatively well studied, and they have been widely employed in biomonitoring studies (Plénet, 1995). There are data available among others on species of the amphipod genera *Gammarus* (Amyot *et al.*, 1994, 1996), *Hyalella* (Borgmann, 1998; Borgmann and Norwood, 1995), *Diporeia* (Song and Breslin, 1998) and *Niphargus* (Plénet, 1999). Fialkowski *et al.* (2003a) investigated spatial and temporal variability in the bioavailabilities of the trace metals Cd, Cu, Pb, Zn and Fe by measuring accumulated metal concentrations in the amphipod crustacean *Gammarus fossarum* in the catchment of the river Biala Przemsza in Upper Silesia, Poland. The data presented (Fialkowski *et al.*, 2003a) have the potential to act as a reference for surveys involving this widespread and abundant freshwater inhabitant in Central Europe.

Similarly, widespread freshwater mussels like the zebra mussel, *Dreissena polymorpha*, offer opportunities for biomonitoring (e.g. Mersch *et al.*, 1996) using protocols developed for marine and estuarine bivalves.

8.4.5 Interpretation of biomonitoring data

Prologue. Design and interpretation of biomonitoring studies require consideration of sample size, gut content, animal size, temporal variability and systematics. A strategy should be devised for each that is specific to type of environment and the species chosen as biomonitor.

The basic tenet of the use of biomonitors is that the accumulated metal concentration in a biomonitor represents an integrated relative measure of bioavailable metal taken up by that biomonitor across all routes. If interpretations of bioavailable contamination are to be valid, it is important that the metal accumulated by the biomonitor is measured accurately and that experimental design, statistical techniques and interpretations are appropriate. Several points need to be considered.

8.4.5.1 Gut contents
Prologue. Protocols exist for gut clearance for many biomonitor species, if it is suspected or shown that gut content will bias results. Some species will not clear their gut without incoming food; and a few biomonitors lose metals sufficiently rapidly that there is no net benefit to gut clearance (e.g. *Hydropsyche* sp.). Correction for gut content without clearance is appropriate after quantifying the bias.

Metal concentrations determined in the invertebrates can be affected by unassimilated food in the digestive tract which by definition has not (yet) been assimilated for accumulation, as noted above. Gut contents are especially problematic in the case of sediment deposit feeders, particularly in mine-affected water bodies because metal concentrations in ingested particulate material can be much higher than concentrations in animal tissues. As noted above, Cu and Cd concentrations in the caddisfly *Hydropsyche* sp. were only about 30% of the concentrations in sediments in the Clark Fork River, Montana, USA. Theoretically, even a small amount of undigested sediment could affect interpretations of bioavailability in such circumstances.

To correct the bias, species like bivalves can be starved for 24 to 48 hours before analysis, allowing them to defaecate their gut contents. NAS (1980) recommended that bivalve molluscs used in biomonitoring programmes that may contain significant amounts of inorganic particulates should be depurated for 36 to 48 hours before analysis. Bryan *et al.* (1980, 1985) recommended that depuration should be undertaken for all trace metals (particularly for Al and Fe) in biota, and used different protocols for different species. Thus, in the cases of three bivalves, they maintained *Scrobicularia plana* for 7 days in regularly changed 50% seawater, *Mytilus edulis* were exposed to clean seawater for 1 day, and the cockle *Cerastoderma edule* were depurated in seawater for several days (Bryan *et al.*, 1980, 1985). The polychaete worms

(*Nereis diversicolor*) were allowed to depurate in acid-washed sand covered with 50% seawater for 2 to 6 days, before transfer to 50% seawater for 1 day (Bryan *et al.*, 1980, 1985). Similarly, Robinson *et al.* (1993) recommended depuration of *M. edulis* in clean seawater for 36 hours before analysis.

In the case of freshwater examples, Brooke *et al.* (1996) concluded that a period of 12 to 24 hours in clean water was sufficient for depuration of sediment from the guts of the mayfly larva *Hexagenia limbata*, the midge larva *Chironomus tentans* and the oligochaete *Lumbriculus variegatus*. Hare *et al.* (1989) also investigated the effect of gut contents on trace metal concentrations in the burrowing mayfly larva *H. limbata*. Trace metals in the gut contents of *H. limbata* represented up to 22% of the whole body metal burden, and (in contrast to Brooke *et al.*, 1996) Hare *et al.* (1989) concluded that even 48 hour depuration in water is not sufficient to ensure a complete emptying of the gut. Hare *et al.* (1989) compared three approaches to compensate for the metal held in the gut contents: (i) direct removal of gut contents by dissection, (ii) induced removal of gut contents (depuration) and (iii) analysis of whole animal (including gut contents) followed by a correction for the contribution of the gut contents. The third approach offered simplicity, savings in time and reasonably accurate metal concentration estimates, yet ultimately the choice of method will depend on the goals and means available to an investigator (Hare *et al.*, 1989).

Certain species either do not eliminate their gut contents without incoming food and/or have such fast rates of metal efflux that starvation will affect their body burden significantly (both are true for free-living caddisflies like *Hydropsyche* spp.). Feeding caddisfly larvae (*Plectrocnemia conspersa*) had 60% higher Cu concentrations than starved animals in a mine-affected stream in the southwest UK. The influence of gut content was probably only a part of this, however, because physiological efflux may have affected concentrations when the animals were starved (Gower and Darlington, 1990).

The metals in unassimilated gut contents appear to contribute less significantly to total body loads of animals that are not sediment ingestors (deposited or resuspended) (Phillips and Rainbow,

1994), as in the case of detritivorous talitrid amphipod crustaceans (Weeks and Moore, 1991). Furthermore gut content metal burdens become of decreasing significance as the strength of net metal accumulation (see Chapter 7) by the biomonitor increases. To generalise then, it is necessary to confirm experimentally that there is no significant difference between the metal concentrations of animals before and after gut depuration. Thereafter it is acceptable to analyse samples without recourse to depuration. If there is a significant difference between metal concentrations of depurated and non-depurated specimens, the investigator can choose to submit all specimens to a depuration protocol or apply a correction factor (Phillips and Rainbow, 1994).

8.4.5.2   Size, weight and age
Prologue. Metal concentrations may vary as a function of the size of the individuals that comprise a sample of a biomonitoring species. The form of the relationship depends upon the metal, the degree of contamination, the species and the circumstances. The most effective strategy for avoiding size bias in expression of the representative concentration at a site is to collect individuals from a range of sizes, and determine the relationship between size and metal concentration for each species at each site. This can be done by analysing individual animals separately or by compositing individuals of a similar size (to reduce the number of analyses).

The size, weight or age of biomonitors may affect their accumulated concentrations of trace metals (see Chapter 7). It is not practically possible to know accurately the age of samples collected in the field, and therefore it is size effects that must be investigated (Boyden, 1974, 1977) and allowed for in the interpretation of the results of any biomonitoring programme (Phillips and Rainbow, 1994). Dry weight (soft tissue dry weight in the case of molluscs) is the usual parameter measured, although a parameter such as shell length in a bivalve or length of a defined part of the exoskeleton in an arthropod may also be suitable.

Boyden (1974, 1977) explored potential metal content (µg) and dry weight (g) in shellfish. From the slopes of these relationships he derived several possible relationships between metal concentration (µg/g) and dry weight (g), with regression coefficients varying from positive to zero to negative (Boyden, 1974, 1977). Strong and Luoma (1981) described four conditions that can affect the relationship of size and concentration:

(1) The metal uptake rate of by smaller individuals may be faster than uptake rates by larger individuals. Increased surface area to volume ratios in smaller individuals may result in increased weight-specific uptake rates from solution, for example. This would cause a negative relationship between size and concentration.

(2) Growth may dilute tissue concentrations if tissue is added faster than net metal uptake occurs; and smaller animals tend to grow more rapidly than larger animals. Again, a negative relationship between size and concentration is the outcome.

(3) If metal loss rates are extremely slow, or metals are progressively added to a slowly exchanging detoxified pool (like granules: Chapter 7), net bioaccumulation might occur throughout an organism's life. The positive relationships that this situation would inspire are more common in metal-contaminated habitats with metals that are strongly bioaccumulated.

(4) Insignificant relationships might occur when fluxes balance or bioaccumulated metals reach steady state. For no size relationship it is also necessary that the animal adds metal content as fast as it adds tissue mass. The balance between accumulation of metal content and the potential effect of growth determines whether accumulated concentration increases or decreases with growth even in a net accumulator (Fig. 8.14).

All three relationships between size and metal concentration are seen across species, metals and environments, depending upon the circumstance. All three may even be observed within the same species at different seasons or under different contamination regimes (Fig. 8.14).

Physiological factors may also contribute to relationships between bioaccumulated metal concentration and size. The permeability of the exoskeleton of a crustacean, for example, may decrease

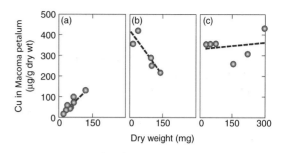

Fig. 8.14 Relationships between Cu concentrations in the soft tissues of the bivalve *Macoma petalum* (as *M. balthica*) and the weight of the animal. Positive, negative and insignificant relationships were seen depending on the growth phase, degree of contamination, (a): a contaminated situation, (b): period of rapid growth in weight. Each point represents a pool of two to five similarly-sized individuals. (After Strong and Luoma, 1981.)

with size with increased tanning and calcification of the cuticle, particularly if there is a change in habit from planktonic to benthic as for most marine decapods. This would lead to a decreased weight-specific uptake rate with size. There may be a change in diet with age, for example from microphagic to macrophagic, with consequences for metal uptake rates from the diet.

A change of accumulated metal concentration with size, thus, is expected and must be allowed for. One method is to choose specimens for analysis from a limited size range, on the assumption that they are of a similar age. This is not always feasible across a range of field conditions, however. Alternatively a range of sizes might be collected from every habitat and a metal concentration versus dry weight relationship determined for each site (Luoma and Bryan, 1978). These data can be fitted to a regression model and the concentration at a chosen weight interpolated for each site.

For example, one particularly useful model that fits many concentration (C in µg/g) to dry weight (W in g) relationships is the power model (Boyden, 1974, 1977; Rainbow and Moore, 1986; Phillips and Rainbow, 1988; Ridout *et al.*, 1989),

$$C = aW^b$$

in which a and b are constants. The constant b is positive when metal concentration increases with dry weight, and negative when metal concentration decreases with dry weight. The power model also has the advantage of producing a straight line equation on logarithmic transformation,

$$\log_{10}C = \log_{10}a + b\log_{10}W.$$

Such double log straight line regressions of different intraspecific sample sets can be compared by analysis of covariance (ANCOVA) (e.g. Phillips and Rainbow, 1988; Fialkowski et al., 2000; Rainbow and Blackmore, 2001; Rainbow et al., 2004a). The first stage of ANCOVA compares the parallelism of the regression lines, comparing the regression coefficients b. If there is no significant difference between regression coefficients (usually the case in such comparisons), the elevations of the lines (representing accumulated metal concentrations) are then compared in the second stage of ANCOVA. These comparisons can be made a posteriori using such tests as Tukey's post hoc. Significant differences between regression coefficients indicate that the metal accumulation patterns of the samples under comparison are different anyway, a conclusion that is of biomonitoring significance.

In the case of macrophytic algae, growth causes differences in metal concentrations in different parts of the thallus, with metal concentrations being low in the growing tips of Fucus vesiculosus and increasing towards the older part of the thallus (Bryan and Hummerstone, 1973a). While the analysis of younger tissue at the thallus tip provides information on recent metal exposure, analysis of older portions of the thallus provides an integrated picture of metal exposure over several months in some cases (Bryan et al., 1985). In a biomonitoring programme, it is therefore important to use a standard region of the thallus for all samples (Barnett and Ashcroft, 1985; Rainbow et al., 2002).

8.4.5.3   Season
Prologue. It is necessary to check for seasonal effects (e.g. associated with the reproductive cycle) if monitoring metal concentrations over time in a biomonitor. If there is a seasonal cycle in metal concentrations, comparisons among years must be from the same season or otherwise take the cycle into account.

Changes in weight of an organism are not brought about only by growth. Weight changes are common during the reproductive cycle with development and subsequent expulsion of gametes as well as the build-up and subsequent utilisation of energy reserves (e.g. glycogen and lipid). Both reproduction and feeding are often seasonal in aquatic invertebrates, and indeed may be interrelated. Associated weight changes can be rapid enough to affect accumulated trace metal concentrations. It is important, therefore, to be aware of, and if necessary allow for, seasonal changes in accumulated concentrations. For example comparisons between the accumulated metal concentrations of samples collected in a biomonitoring programme should not be made between samples collected at different times of the year unless seasonal variation in accumulated metal concentrations has already been ruled out.

Seasonal effects on accumulated metal concentrations typically associated with spawning have been demonstrated for the mussel Mytilus edulis, although there is inconsistency between studies at different locations across its geographical range (Cossa et al., 1979; Boalch et al., 1981; Popham and D'Auria, 1982; Gault et al., 1983; Amiard et al., 1986). Similar effects of spawning and therefore season have also been identified in oysters (Boyden and Phillips, 1981; Talbot, 1986; Páez-Osuna et al., 1995). The San Francisco Bay Macoma petalum (as M. balthica) showed strong seasonal variation in metal concentrations associated with seasonal variation in soft tissue weight (Cain and Luoma, 1990), although seasonal changes in local metal input and hydrological factors affecting metal bioavailabilities also affected the accumulated metal concentrations in this clam (Chapter 11; Luoma et al., 1985). In freshwater, certain insect larvae will show temporal patterns in accumulated metal concentrations, varying for example with seasonal changes in prey availability or progressive change in instar number through the seasons. These complicate interpretation of direct and indirect changes in metal bioavailabilities (Hare and Campbell, 1992).

If sampling is either restricted to a specific season, or equally represents all seasons confounding effects on interpretation can be minimised or interpreted. Nevertheless in long-term comparisons of changes in bioavailabilities over years at specific sites, it is important to make sure that seasonality is not a confounding factor. Rainbow and Blackmore (2001) sampled barnacles in Hong Kong coastal waters in April 1998, for direct comparison with samples collected in April 1986 and April 1989. Cain and Luoma (1991) sampled with equal frequency every year making sure that each season was similarly represented.

8.4.5.4   Moult cycle
Prologue. It is important to use only intermoult animals when using arthropods as biomonitors.

In the case of aquatic arthropods, the moult cycle can still affect metal concentrations even after the cessation of growth. Moulting in decapod crustaceans can be associated with changes in the internal distribution of trace metals, particularly Cu (Depledge and Bjerregaard, 1989), and metals may be lost with the cast moult. Furthermore, the cuticle of newly moulted arthropods is relatively permeable and a period of enhanced metal uptake from solution may occur before the more typical permeability or impermeability is restored by tanning and/or calcification of the new cuticle (Phillips and Rainbow, 1994). When using arthropods, particularly adult malacostracan crustaceans, in biomonitoring programmes, it is important to use only intermoult animals unless the effect of moulting on accumulated concentrations has already been explored (e.g. Weeks and Moore, 1991; Weeks et al., 1992).

8.4.5.5   Intrasample variability
Prologue. The number of individuals in a sample should exceed 10 and preferably be >20 if statistical power is to be great enough to usefully interpret differences among samples. To reduce the number of analyses composites of similar-sized animals can be used.

Even after all known sources of variation have been accounted for, there remains residual variation ('inherent variability': Boyden and Phillips, 1981) in samples of biomonitors (Phillips and Rainbow,

1994). This inherent variability can sometimes be striking even when comparing accumulated concentrations within intraspecific samples from the same location (Lobel and Wright, 1983; Wright et al., 1985; Lobel, 1987). In effect, the inherent variability creates a background noise that reduces the power to detect any signal caused by altered metal bioavailability (Phillips and Rainbow, 1994). The simplest way to increase power is to increase sample size. For shellfish (and it is a good general rule for all biomonitors) the percentage difference between two samples that can be detected is very low until 10 to 20 individuals are included in each sample (Gordon et al., 1980) (Fig. 8.15); above that number power increases more slowly. In general, this means samples of fewer than 10 individuals may be difficult to interpret. Brown and Luoma (1995) used 8 to 10 composites of 4 to 10 individuals each to study different sources of variability in the biomonitor *Corbula amurensis* from San Francisco Bay. Changes in time (seasonal and interannual) and space were readily separable; small-scale spatial variability was small.

The use of parametric statistical techniques such as analysis of variance (ANOVA) of mean accumulated concentrations of individual biomonitors in the absence of a size effect, or ANCOVA in the presence of a size effect, takes into account

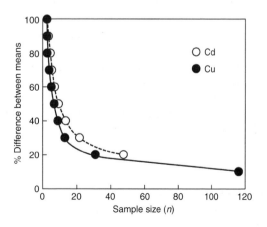

Fig. 8.15 Relationship between the number of animals analysed and the percentage difference that can be detected between samples for Cd, Cu and Cr. Data from *Mytilus edulis*. (After Gordon et al., 1980.)

individual variation in any comparison. Logarithmic transformation (e.g. Rainbow and Blackmore, 2001) is usually sufficient to transform the distributions of concentration and weight data sets to normality, which should be checked before the application of parametric statistics. Wright *et al.* (1985) addressed the problem of individual variability in trace metal concentrations in the oyster *Crassostrea virginica* in Chesapeake Bay, USA, particularly the presence of outliers ('superaccumulators') in concentration distributions, resorting to reciprocal square root transformation to convert the skewed metal distributions to normality before the application of parametric statistics (see also Lobel and Wright, 1983). The use of non-parametric statistics for comparisons avoids the need for transformation of data to fit a normal distribution.

8.4.5.6 Systematics
Prologue. It is possible that closely related species have significantly different accumulated trace metal concentrations when collected from the same site. Thus correctly identifying biomonitors to the species level and analysing different species separately are important for reducing variability and improving interpretability of data.

It is often important to be absolutely sure of the specific identifications of biomonitors used in biomonitoring programmes, particularly if two closely related species are sympatric and small differences in bioaccumulated metal are important to interpretations. Even closely related species belonging to the same genus can accumulate different trace metal concentrations from the same habitat (Rainbow and Phillips, 1993), as a result of specific differences in the kinetic parameters (e.g. rate of uptake from solution, assimilation efficiency, rate of efflux) contributing to their accumulation patterns (see Chapter 7).

Cain *et al.* (1992) first analysed separately several species of aquatic insects from the Clark Fork River, Montana (USA); then mathematically pooled different combinations of data to show the distortions in interpretation that can result if different species were combined into a measure of 'invertebrate' bioaccumulation. Even closely related species can

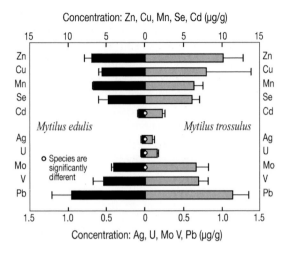

Fig. 8.16 Concentrations (means and standard deviations, μg/g dry weight) of 10 trace metals in the whole soft parts of the mussels *Mytilus trossulus* and *Mytilus edulis* collected from the same site at Bellevue, Newfoundland in summer 1988. (From Phillips and Rainbow, 1994, after Lobel *et al.*, 1990.)

sometimes have different bioaccumulation patterns. For example, the two mussel species, *Mytilus edulis* and *Mytilus trossulus*, collected from the same site in Newfoundland in 1988 had significantly different accumulated concentrations of Ag, Cd, Mo and U (Lobel *et al.*, 1990) (Fig. 8.16). The two talitrid amphipod crustaceans, *Orchestia gammarellus* and *Orchestia mediterranea*, can occur on the same shore, the former above the latter, yet it is important to distinguish between them for *O. gammarellus* has a higher accumulated Cu concentration than its relative (Moore and Rainbow, 1987). It is also important to distinguish between two common coastal Indo-Pacific barnacles, *Balanus amphitrite* and *Balanus uliginosus*, when their distributions overlap, although the latter is usually found in lower salinity water than the former (Rainbow *et al.*, 1993a). As Table 8.8 illustrates, the two species collected from the same site may have different accumulated Cu concentrations. Similarly two species of the cirratulid deposit-feeding polychaete *Tharyx* collected at the same site had accumulated arsenic concentrations that differed by an order of magnitude (Bryan and Gibbs, 1987).

Table 8.8 *Weight-adjusted mean Cu concentrations (μg/g dry weight, with 95% confidence limits) in the bodies of barnacles* Balanus amphitrite *and* B. uliginosus *as estimated from best-fit double log regressions, collected in 1991 in the region of Xiamen, Fujian Province, China*

Site	Balanus amphitrite		Balanus uliginosus	
	Mean	95% CL	Mean	95% CL
Song Yu	43.1	36.8, 50.6	70.7	49.0, 102
Outer Houzhu	37.1	26.8, 51.4	61.3	39.6, 94.9
Hai Cang	34.4	22.3, 53.1	61.5	47.6, 79.6

Source: After Rainbow *et al.* (1993a).

In contrast, Fialkowski *et al.* (2003b) used accumulated metal concentrations in larvae of two mayfly species, *Baetis rhodani* and *B. vernus*, to investigate spatial and temporal variability in the bioavailabilities of Cd, Cu, Pb, Zn and Fe in streams in the catchment of the river Biala Przemsza, Upper Silesia, Poland. Failure to identify correctly and consequently separate the larvae of these two species of *Baetis* did not affect the conclusions drawn.

Given that it may be important to be able to distinguish between closely related sympatric species in the design of a biomonitoring programme, it is important to check with appropriate local systematic experts either that only one species is present, or how to distinguish between the relevant species. Fortunately it is often the case that only one species is present or that closely related species occur in recognisably different habitats. Thus in the case of the two species of the talitrid amphipod *Orchestia* above, which are found at different tidal heights on the shore, appropriate sampling can avoid any area of distribution overlap.

Of particular concern are actually the two groups of bivalves, mussels and oysters, that are commonly used in biomonitoring. Not only are these sometimes difficult to identify to species with confidence, they have been spread anthropogenically around the world, often for mariculture purposes, and previously reproductively isolated species may have come into contact and produced hybrids (Rainbow and Phillips, 1993; Rainbow, 1995b). The smooth-shelled mussels, *Mytilus edulis, M. galloprovincialis* and *M. trossulus*, are actually difficult to distinguish from each other without the use of multivariate techniques (Seed, 1992), so special care is needed when more than one mussel species is known to be present. The rock oysters of the genus *Saccostrea* are also challenging to identify correctly (Rainbow and Phillips, 1993), and expert taxonomic help may be needed to verify any identification.

8.4.5.7 Intraspecific comparisons of absolute trace metal concentrations

Prologue. For optimal sensitivity, comparisons of metal bioavailability in time and space should be restricted to data from the same species. Combining even similar species has the potential to sacrifice statistical power in interpretations.

From the discussion above, it is clear that comparisons of accumulated metal concentrations should only be made intraspecifically in any biomonitoring programme. Where there are significant differences between accumulated concentrations of a single biomonitor within a biomonitoring programme after allowance for size effects, it can usually be concluded that such differences have been caused by different local bioavailabilities of the trace metal concerned. Nevertheless it is probably wise to check whether there are significant differences in size or condition index (see Chapter 9; Phillips and Rainbow, 1994) between samples from different sites in case concentration differences have been caused by weight differences.

It is also useful where possible to put the concentrations measured (and by extension the local metal bioavailabilities) into a wider geographical context. There are now sufficiently extensive databases to do this for several biomonitors discussed earlier. Tables 8.1 to 8.7 provide perspective for such comparisons. Comparison of these tables accentuates the point that what is a high accumulated metal concentration in one species may be a very low concentration in another.

Databases for Mussel Watch programmes typically have relevant accumulated concentration data for separate species of mussels (and often oysters),

Table 8.9 *Concentrations of trace metals (μg/g dry weight) in the soft tissues of mussels and oysters proposed by Cantillo (1998) to be indicative of local metal contamination*

Trace metal	Concentration
'Mussels'	
Ag	0.75
Cu	10
Zn	200
'Oysters'	
Ag	5
Cu	300
Zn	4000
'Mussels and oysters'	
As	16
Cd	3.7
Cr	2.5
Hg	0.23
Ni	3.4
Pb	3.2

but published syntheses of collected data do still have a tendency to lump mussel species together and oyster species together. Thus Cantillo (1998) synthesised a mass of data from Mussel Watch programmes around the world, presenting median and 85th percentile values of metal concentration frequency distributions collated from several mussel species combined and several oyster species combined (Table 8.4). Cantillo (1998) went as far as to propose accumulated concentrations in 'mussels' and 'oysters' that are indicative of raised local metal contamination (Table 8.9). These represent a valid starting point, but they would benefit from analyses of single species concentrations to avoid the inevitable confounding of the combined data by interspecific differences (compare Tables 8.2 and 8.3 with Tables 8.4 and 8.9).

8.4.5.8   Interspecific comparisons
Prologue. If biomonitoring data from multiple species are available, useful interpretations can be expanded. If the same species does not occur at all sites of interest, protocols are available to compare data among species.

Multivariate techniques can be used to harmonise interpretations from several biomonitors, adding power to interpretations.

It is recommended above that suites of biomonitors should be used in biomonitoring programmes and yet comparisons of accumulated concentrations should only be made intraspecifically. How can the accumulated concentration data for separate species be compared or synthesised?

If sites within a biomonitoring programme are ranked, for example in order of decreasing accumulated concentration of a trace metal, rank orders for species can be compared by Spearman's rank correlation, a non-parametric test. For example, accumulated concentrations of Cr, Cu, Pb and Zn, but not Cd, were correlated using that approach in data from the green lipped mussel *Perna viridis* and the barnacles *Balanus amphitrite, Tetraclita squamosa* and *Capitulum mitella* in Hong Kong coastal waters (Phillips and Rainbow, 1988).

Alternatively accumulated metal concentration data (appropriately transformed to fit normal distributions) from two species can be correlated using parametric statistics. The fact that barnacles will grow on mussels, ensuring sampling of exactly the same location, was put to good effect in the Gulf of Gdansk at the south end of the Baltic Sea. The mussel *Mytilus trossulus* and the barnacle *Balanus improvisus* were sampled from five sites between February 2000 and September 2001 to follow seasonal and geographical variation (Fig. 8.17) in the bioavailabilities of Cd, Cu, Fe, Mn, Ni, Pb and Zn (Rainbow *et al.*, 2004a). Accumulated concentrations of Cu, Fe, Ni, Pb and Zn, but not Cd or Mn, were correlated in the mussel and the barnacle. It is likely, therefore, that the bioavailable sources of the former metals to *B. improvisus* and *M. trossulus* are very similar, both from solution and from suspended material.

The lack of correlation of Cd concentrations in the sympatric barnacles and mussels in both Hong Kong and the Baltic Sea probably reflects a difference in the relative importance of solution and diet as sources of Cd for barnacles and mussels. Barnacles (exemplified by *Elminius modestus*) obtain the vast majority (>97%) of their Cd from the diet

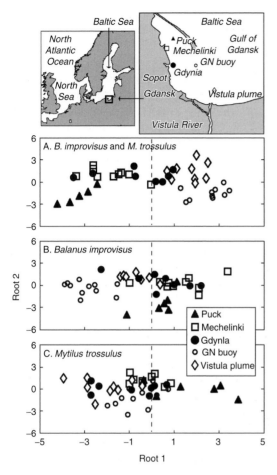

Fig. 8.17 Discrimination analysis can be used to distinguish local differences in trace metal bioavailabilities to the mussel *Mytilus trossulus* and the barnacle *Balanus improvisus* from five sites in the Gulf of Gdansk, Poland (Feb 2000 to Sept 2001). The combined data sets of weight-adjusted accumulated trace metal concentrations gave the best discrimination, and the barnacle data alone led to greater discrimination than the mussel data alone. (After Rainbow *et al.*, 2004a).

one biomonitor can tease out the differential significance of dissolved and dietary sources of bioavailable metal.

Discrimination analysis can be used to distinguish between sites on the basis of accumulated metal concentrations in one or more biomonitors. Thus discrimination analysis of a combination of the data from *Mytilus trossulus* and *Balanus improvisus* in the Gulf of Gdansk clearly distinguished between four of the five sites investigated (Rainbow *et al.*, 2004a) (Fig. 8.17). The barnacle data alone were better able to distinguish local differences in trace metal bioavailabilities than the mussel data alone (Fig. 8.17), in probable reflection of the stronger powers of net accumulation of trace metals in barnacles than mussels (Rainbow *et al.*, 2004a). Similarly discrimination analysis showed that mayfly larvae of the genus *Baetis* had better discriminatory power than the amphipod crustacean *Gammarus fossarum* to distinguish local bioavailabilities of Cd, Cu, Fe, Pb and Zn at sites in streams draining a Pb and Zn mining area of Upper Silesia, Poland (Fialkowski and Rainbow, 2006).

Rainbow and Blackmore (2001) on the other hand used multidimensional scaling to compare similarity between up to 18 sites in Hong Kong coastal waters with respect to accumulated metal concentrations in the two barnacles *Balanus amphitrite* and *Tetraclita squamosa*. Discrimination between sites was clearer when using the *B. amphitrite* data set, possibly because the smaller suspended particles ingested by the former had the more significant geographical differences in metal loadings available for uptake and accumulation (Rainbow and Blackmore, 2001).

8.4.6   Translocation

Prologue. Translocation of biomonitors can be a very effective approach for toxicity testing (*in situ* toxicity testing). But it is not a good substitute for sampling indigenous fauna when assessing metal exposure.

The most successful biomonitoring programmes make use of indigenous species: collections of biomonitors living in the habitats to be compared,

(Rainbow and Wang, 2001), while mussels (e.g. *Mytilus edulis*) get more Cd from solution than the diet (Wang *et al.*, 1996). In the case of Zn, for which the barnacle and mussel concentrations were correlated (Rainbow *et al.*, 2004a), both barnacles and mussels gain the majority of the metal from the diet (Wang *et al.*, 1996; Rainbow and Wang, 2001). Thus the comparative use of more than

taking advantage of their ability to integrate trace metal bioavailabilities over time. Biomonitors are of course even more informative if the biology and age of the specimens and the accumulation kinetics of the particular trace metals are known. Nevertheless much information on differences in local trace metal bioavailabilities can still be gained from a single collecting programme.

A variant of biomonitoring is to translocate specimens of an appropriate biomonitor into sites, leave them for a defined period and then collect the same specimens again for analysis. This procedure has been referred to as Active Bio-Monitoring (ABM) and its rationale has been dealt with in detail by de Kock and Kramer (1994). Translocation has the great advantage that start and finish concentrations in the translocated biomonitors can be measured over a defined period, and that time period is the same for all sites under consideration; i.e. it more closely resembles an experimental than an observational programme. The specimens to be translocated are often collected initially from a site known to be uncontaminated, or preferably may have been specifically reared under uncontaminated conditions. It is vital that only species known to be already present in the general area are used. The most important reason is to avoid the unfortunate ecological consequences of the introduction of alien species into a habitat. But there is also precedent for translocating species into habitats unsuitable to their unstressed survival. One of the most important issues with translocation is assuring that the translocated species feed and move water similarly to indigenous natives. If translocated species do not feed at normal rates their exposure to metals predominantly bioaccumulated via diet will be an underestimate of what happens naturally (e.g. Linville et al., 2002).

Translocation studies have been carried out using a variety of species (see de Kock and Kramer, 1994). The mussels Mytilus edulis, M. californianus and M. galloprovincialis have figured strongly in such studies, not only in the USA (Smith et al., 1986; Koepp et al., 1987; Salazar and Salazar, 1995) but elsewhere such as in France (Geffard et al., 2004). Oysters are also used, for example Crassostrea gigas in France (Geffard et al., 2004), the choice of bivalve

being affected for example by the time period under consideration. Deposit-feeding bivalves such as Scrobicularia plana (Bryan et al., 1985) and the San Francisco Bay Macoma petalum (as M. balthica) (Cain and Luoma, 1985) have also been used with labelling and/or caging as appropriate.

Translocation is best used to address a very specific question (e.g. will specimens from an uncontaminated site react to contamination similarly to specimens resident at the contaminated site?). It can be an extremely effective way to conduct in situ toxicity tests if the protocols are carefully designed (e.g. Salazar and Salazar, 1995). As a means of addressing adverse effects of metals at a site, translocation has many advantages. However, as a routine tool for biomonitoring metal exposure it is less advantageous. There is no comparable database to which data might be compared. Most important, behavioural anomalies (e.g. failure to feed) are not uncommon, confounding results compared to local specimens (e.g. Linville et al., 2002).

8.5   CONCLUSIONS

Metal concentrations in water, sediment and all resident biota respond to contamination of a water body. Thus any component of the environment can be monitored and detect the contamination if gradients are large. Correlations occur among all components, but those correlations are typically variable over narrow subsets of the data range. Because the goal of monitoring often involves interpreting change across relatively narrow differences, careful choice of the suite of monitoring tools is important. Furthermore some approaches cannot define important considerations like bioavailability. Use of a biomonitor or a suite of biomonitors has a long history of success in identifying and differentiating instances of and gradients in bioavailable metal contamination, when the tool is used and interpreted appropriately.

Although a number of factors can affect biomonitor data, protocols exist to facilitate careful design and meaningful interpretation. When analysing comparative data on accumulated metal

concentrations in biomonitors, it is most important to make only intraspecific comparisons of
measured concentrations, for even closely related
species may show different metal bioaccumulation
kinetics resulting in different accumulated concentrations in identical situations. There is, therefore,
advantage in identifying suitable species that
exhibit widespread geographical distributions and
therefore can be used as cosmopolitan biomonitors
(Rainbow and Phillips, 1993).

In coastal waters there are several candidate
cosmopolitan biomonitors widespread across the
globe, often coincidentally as a result of human
activities deliberately spreading species with mariculture potential or inadvertently spreading
fouling organisms. The mussels *Mytilus edulis* and
Mytilus galloprovincialis have been spread anthropogenically from the north Atlantic to the Pacific and
Indian oceans (Rainbow and Phillips, 1993). The
related mussels *Perna viridis* and *Perna perna* are also
very widely distributed in their own right (Fig. 8.7).
Amongst the oysters, *Crassostrea gigas* in particular
has been spread anthropogenically across many
parts of the world, and *Saccostrea cucullata* is widespread from Hong Kong to Australia (Phillips
and Yim, 1981; Talbot, 1986). The barnacle *Balanus
amphitrite* is a major fouling organism on ships and
man-made shore installations in the tropics and
subtropics, with an extensive literature available
on high and low accumulated metal concentrations (e.g. Rainbow and Blackmore, 2001). Amongst
the macrophytes, species of the green algal genera
Enteromorpha and *Ulva* do have cosmopolitan
biomonitoring potential, if they can be correctly
identified (Rainbow, 1995b).

In freshwater rivers, hydropsychid caddisfly
larvae are excellent candidates for global, cosmopolitan biomonitors. The same is true for perch in
some lakes and, especially, larvae of species of the
phantom midge *Chaoborus* in many lakes.

As data sets develop for these biomonitors,
comparative analyses of the degree of bioavailable
contamination will become possible for any water
body. Furthermore, correlation of degree of contamination in such biomonitors against field-
derived observations of metal effects in these or
other biota (e.g. Chapters 11 and 17) might provide
a new tool for assessing risks and managing
contamination.

Suggested reading

Bryan, G.W. and Langston, W.J. (1992). Bioavailability,
    accumulation and effects of heavy metals in
    sediments with special reference to United Kingdom
    estuaries: a review. *Environmental Pollution*, **76**,
    89–131.
Hare, L. (1992). Aquatic insects and trace metals:
    bioavailability, bioaccumulation and toxicity. *Critical
    Reviews in Toxicology*, **22**, 327–369.
Croteau, M.N., Hare, L. and Tessier, A. (1998). Refining and
    testing a trace metal biomonitor (*Chaoborus*) in highly
    acidic lakes. *Environmental Science and Technology*, **32**,
    1348–1353.
Phillips, D.J. H. and Rainbow, P.S. (1988). Barnacles and
    mussels as biomonitors of trace elements:
    a comparative study. *Marine Ecology Progress Series*,
    49, 83–93.
Rainbow, P.S. (1995). Biomonitoring of heavy metal
    availability in the marine environment. *Marine
    Pollution Bulletin*, **31**, 183–192.

9 • Manifestation of the toxic effects of trace metals: the biological perspective

9.1 INTRODUCTION

Prologue. Adverse effects of trace metals are defined by expressions of stress at different levels of organisation. These are operationally determined by:

- understanding and detecting what the signs of stress are;
- using toxicity tests to determine experimentally concentrations at which the more severe signs of stress, including lethality, might occur;
- incorporating field observations at higher levels of organisation to provide the final piece of information necessary to understand ecotoxicological risks from metals.

The next three chapters consider adverse effects of metals in aquatic environments, and how they are determined. Chapter 9 describes the processes by which adverse effects of metals are expressed at different levels of organisation. It takes a biological approach to understanding how metals adversely affect aquatic life. Chapter 10 defines trace metal toxicity and considers toxicity testing and the concentration dependence of toxicity; the traditional

approaches to determining adverse effects. The terminology, approaches and methods used in toxicity testing have a strong influence on our perceptions (and knowledge) of trace metal toxicity; perhaps more so than with other stressors. Chapter 11 describes observations of toxicity in nature and what they mean to managing metal contamination (and assessing risks). It uses examples that tie field observations together with methodologies from Chapters 9 and 10.

9.2 TOXICITY AND DOSE

Prologue. The general principles of essentiality and non-essentiality are known for trace metals, but there is some uncertainty about the shapes of the curves describing the stress response. Essential metals are not less toxic than non-essential metals.

All trace metals are potentially toxic to living organisms at high availabilities, with the apparent paradox that many of them are essential to life in smaller doses. As the availability of either an

---

Box 9.1 Definitions

biomarker: a biological response (e.g. a biochemical, cellular, physiological or behavioural variation) that can be measured at the lower levels of biological organisation, in tissue or body fluids or at the level of the whole organism.

cellular energy allocation: the amount of energy at a cellular level available to an individual organism for growth and reproduction, calculated from

the difference between measured energy reserves (glycogen, protein and lipid contents) and energy consumption quantified from the activities of a selected set of intermediary metabolism mitochondrial enzymes.

genome: the whole hereditary information of an organism, encoded in the DNA (or, for some viruses, RNA) and including both the genes and the non-coding sequences of the DNA.

genomics: the study of an organism's entire genome.

lipofuscin: finely granular yellow brown pigment granules composed of lipid-containing residues; a product of lysosomal digestion.

lysosome: a membrane-delimited intracellular organelle containing hydrolytic enzymes which break down substances within a cell.

metabolites: the intermediates and products of metabolism, usually restricted to small molecules.

metabolome: the collection of all metabolites in an organism, which are the end products of its gene expression.

metabolomics: the systematic study of the metabolome of an organism.

metallothionein: a non-enzymatic protein with low molecular weight, high cysteine content, no aromatic amino acids and heat stability, induced by and binding to particular trace metals such as Ag, Cd, Cu, Hg and Zn. One function is metal detoxification.

micronucleus: the smaller of two nuclei in a cell.

oxyradicals: toxic intermediates such as peroxyl radicals, hydroxyl radicals and peroxynitrite, produced in small quantities during normal metabolic activities

(aerobic respiration) or inflammatory activities. Some metals catalyse oxyradical formation.

proteome: originally the entire complement of proteins expressed by a genome, cell, tissue or organism; subsequently this term has been specified to contain all the expressed proteins at a given time point under defined conditions.

proteomics: the study of the proteome.

scope for growth (SFG): energy available to an organism for growth and reproduction, calculated from the difference between the energy absorbed from food and that used in respiration.

stress proteins or heat shock proteins: proteins, present in all cells, that are induced when a cell undergoes various types of environmental stresses like heat, cold and oxygen deprivation.

total oxyradical scavenging capacity (TOSC): an integrated measure of the several chemicals produced by the body to scavenge oxyradicals. A measure of antioxidant defence activity. More oxyradicals produce greater defence activity.

transcriptome: the set of all messenger RNA molecules (transcripts), produced in one or a population of cells, which can vary with external environmental conditions.

essential or a non-essential metal to an aquatic organism increases, first sublethal and ultimately lethal effects become observable (Fig. 9.1). A similar increasing deterioration of health occurs with increasing deficiency of an essential metal (Fig. 9.1).

For every essential metal there is presumably an optimum range of trace metal availabilities over which sufficient metal is entering the body of the organism to meet metabolic requirements; but not so much as to cause toxic effects (Fig. 9.2). For both essential and non-essential metals, there will certainly be a range of availabilities over which toxic effects become apparent (Fig. 9.2). It is arguable whether or not there is a lower range of availabilities of a non-essential metal over which there is no detrimental effect on performance at all (Fig. 9.2).

All incoming non-essential metal will need to be detoxified (see Chapter 7), but presumably the energetic costs of any detoxification will only become significant to the energy budget of the organism at higher metal availabilities.

The positions of the curves in Fig. 9.2 (and the relative proportions of the different divisions of the curves) along the availability axis will vary both between metals and between organisms for the same metal. Differences are also caused by the physico-chemical characteristics of different habitats and are taken into account by considering bioavailabilities as opposed to concentrations (see Chapter 7). It is by no means the case that all non-essential trace metals are more toxic than all essential metals, and the rank orders of toxicity of metals will vary between

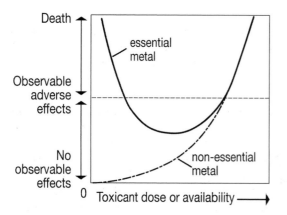

Fig. 9.1 Biological response of an organism to increasing dose or availability of an essential or non-essential trace metal. (After Connell *et al.*, 1999.)

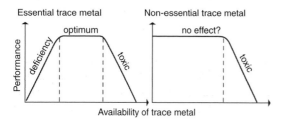

Fig. 9.2 Relationships between performance (e.g. growth, fecundity, survival, etc.) and availability of an essential or non-essential trace metal to an organism. (After Hopkin, 1989, and Walker *et al.*, 1996.)

organisms. Nevertheless a typical order of toxicities of trace metals to aquatic organisms might be:

Hg > Ag > Cu > Cd > Zn > Ni > Pb > Cr > Sn.

9.3   UPTAKE, BIOACCUMULATION AND TOXICITY

Prologue. The internal metabolic availability of trace metals depends on the relative binding affinities of 'detoxifying' and 'target' ligands competing for the favours of the trace metal concerned. The factors that affect the rate of uptake of trace metals affect the toxicity of the metal. Toxicity ensues once the threshold of metal availability has been passed at which the rate of uptake exceeds the rates of excretion and

detoxification combined. Metal toxicity results when metal accumulates at an undesirable site(s) in the organism and disrupts an important molecular function.

Trace metals can only act as toxicants after uptake into the organism, be it a single cell or, in the case of multicellular organisms, via an epithelial cell on a permeable external surface or in the absorptive region of the alimentary tract (Chapter 7). The chemical characteristics that make trace metals an invaluable resource in the evolution of metabolic pathways is their affinity for sulphur and nitrogen and consequential strength of binding to proteins. Those same characteristics make them dangerous potential toxins – binding in the wrong place at the wrong time, disrupting the metabolic function of proteins and other molecules crucial to the organism concerned.

All organisms therefore have evolved biochemical and physiological mechanisms to control the internal binding of essential trace metals, detoxification mechanisms that are often also used to limit the internal metabolic availability of non-essential trace metals (see also Chapter 7). Indeed it is possible to conceptualise the partitioning of trace metals accumulated by organisms into detoxified and metabolically available forms. But the conceptualisation is a simplification. It represents what is really a complicated dynamic series of reactions that bring about the detoxificatory binding of trace metals to different degrees rather than an absolutely black and white division into two extreme forms.

Toxic effects ensue when an excess of metal is present internally in a form that is not detoxified and therefore in a metabolically available form. Of concern here, therefore, is the interaction of three rate processes – the rate of uptake, the rate of detoxification and the rate of excretion (if any) of the trace metal. Kinetics are key. If the rate of uptake of a trace metal into an organism is greater than the combined rate at which it can be excreted or detoxified, then that trace metal will accumulate internally in metabolically available form – that is a form that is available to bind inappropriately to the 'wrong' internal molecule and cause toxic effects. If the build-up of trace metal in metabolically available form continues, then sublethal

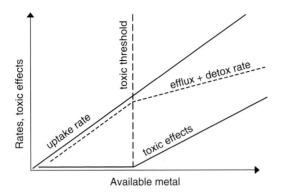

Fig. 9.3 Schematic representation of how the uptake rate (combined across all routes of uptake) of a trace metal and hence (after a threshold) the manifestation of toxic effects will increase with the availability of the trace metal to an aquatic organism. Toxic effects occur when the uptake rate exceeds the combined rates of efflux and detoxification.

toxic effects of increasing strength and severity will ensue, followed by lethality. It is likely that both trace metal excretion rate and detoxification rate of many organisms will increase with increased uptake rate (see Chapter 7), but there will still be a threshold when uptake rate exceeds the combined rates of excretion plus detoxification (Fig. 9.3).

In Chapter 7, we discussed the factors that affect the uptake of trace metals by aquatic animals from solution and diet. In the field either or both routes of uptake will be important and it is the combined rate of trace metal uptake that is of concern. Trace metals from either uptake route have the potential to bind at sites of toxic action. The sites of toxic action may be key proteins in metabolic pathways in all parts of the body or binding sites specific to a particular 'target organ'. For fish, two examples of metal-specific reactions in target organs are:

- the ATPase transport system in the gill, which is responsible for major ion regulation;
- the vitellogenin produced in the liver to aid egg maturation and successful reproduction.

In typical laboratory toxicity tests, high dissolved concentrations are used. It can usually be safely assumed that under these circumstances the dissolved uptake route is delivering the vast majority or all of the incoming metal that is causing the toxic effect. Furthermore if the site of toxic action is actually integral to the site of uptake from solution then the rate of dissolved uptake alone will correlate well with manifestation of toxic effects. The Biotic Ligand Model (BLM; see Chapter 15 for detail) is an example of a predictor of toxicity that focuses on the site of dissolved metal uptake (Paquin *et al.*, 2002). Its predictions are verified in dissolved metal toxicity tests. In field conditions the BLM needs adapting to account for any uptake via the diet and for possible multiple sites of toxic action.

If the rate of uptake controls toxicity of a trace metal, then the onset of toxicity is not controlled by the accumulated concentration of trace metal in an organism unless that accumulated trace metal is all (or for the vast majority) in metabolically available form (Box 9.2). Laboratory toxicity experiments on aquatic invertebrates, or extreme exposures to high dissolved availabilities in natural waters, may cause such a high rate of uptake that little accumulated metal is detoxified (i.e. detoxification processes are rapidly overwhelmed). In such circumstances the organism is killed (concentration reaches a critical body concentration) with little bioaccumulated metal in the body. In typical field situations, however, with chronic contaminant input, the accumulated concentrations of trace metals in aquatic organisms usually contain a large proportion of stored detoxified metal that builds up over time. The rate of accumulation will differ in the same organisms between sites with different metal availabilities, as will the point (reflected in total body concentration) at which the concentration of the metabolically available component becomes critical.

A corollary to the above is that there will be no constant critical body total concentration of a toxic metal in an aquatic organism that is applicable across different contaminated habitats, if the animal employs metal detoxification processes. In Baudrimont *et al.* (1999), the accumulated metals in the clam *Corbicula fluminea* at the two most contaminated sites were approximately the same, with contrastingly different effects on mortality as a result of the different proportions of accumulated metal held in detoxified or metabolically available forms (Baudrimont *et al.*, 1999).

Box 9.2 Toxicity depends upon the rate of uptake of trace metal, combined across all uptake routes

Baudrimont *et al.* (1999) provide convincing field evidence of the importance of the rate of trace metal uptake in determining toxicity. Specimens of the Asiatic clam *Corbicula fluminea* were translocated from a clean lacustrine site to four stations along a polymetallic contamination gradient in the River Lot, France, downstream of an old Zn ore treatment facility. They measured Cd and Zn accumulated concentrations and concentrations of the detoxificatory trace metal-binding protein metallothionein (MT: see Chapter 7) (strictly MT-like proteins) in the bivalves after 0, 21, 49 and 150 days. At the most contaminated station, MT concentrations did not increase despite very rapid metal accumulation; all bivalves died between days 49 and 85 indicating that metal detoxification mechanisms had been overwhelmed at this station (Baudrimont *et al.*, 1999). At the next station downstream, the final accumulated concentrations of the trace metals were as high as those reached earlier in clams at the first station, but in this case the MT concentrations had increased progressively with the more slowly accumulating metal concentrations. No mortality was observed. In the extreme conditions where mortality occurred, metal bioaccumulation had been too rapid. The rate of increase in soft tissue concentrations of Cd and Zn in the clams exceeded the rate at which metals were detoxified by MT (Baudrimont *et al.*, 1999) and death ensued.

The important causative role in toxicity of the binding of trace metals in the wrong place at the wrong time is often referred to as the 'spillover model' (Mason and Jenkins, 1995). There is not consensus about the shape of the curve implied by the term 'spillover' (Fig. 9.3). But there is consensus that metal toxicity at the cellular level arises from the non-specific binding of metals to physiologically important ligands rather than to a detoxificatory ligand such as the detoxificatory protein metallothionein (Winge *et al.*, 1974; Brown and Parsons, 1978; Mason and Jenkins, 1995). Usually, subcellular metal partitioning and metallothionein

synthesis have been related to toxic effects in timecourse experiments involving abrupt changes in trace metal exposure (Perceval *et al.*, 2004). The non-specific binding of trace metals to cytosolic ligands other than MT has been associated, for example, with lipid peroxidation (Couillard *et al.*, 1995b), growth reduction (Sanders and Jenkins, 1984), decreased reproduction (Jenkins and Mason, 1988), modification of behaviour (Wallace *et al.*, 2000) and mortality (Baudrimont *et al.*, 1999).

9.4   HIERARCHY OF RESPONSES THROUGH LEVELS OF BIOLOGICAL ORGANISATION

Prologue. The effects of metal accumulation at undesirable sites are expressed by malfunctions at a variety of levels of biological organisation. The degree of disruption and malfunction (toxicity) is ultimately a function of the bioavailable metal, but it is also influenced by biological vulnerabilities and compensation mechanisms.

The first toxic effects of trace metals occur at the level of molecules within cells at the point when a trace metal binds inappropriately and disrupts a metabolic function. With increasing availability of the toxin, effects are manifested at higher levels of biological organisation (McCarthy and Shugart, 1990). These include the level of the cell (biochemistry, cytology), organ-specific targets, response of the whole organism (morphology, physiology, behaviour) and effects on recruitment to the population or loss from the population due to mortality. Ultimately the loss of species changes the nature of the community and provision of ecosystem services (via degradation of ecosystem function). It is at the higher levels of biological organisation that we see effects usually considered of significance to risk assessment.

Simplistically, the greater the metal exposure, the further along the cascade the responses will proceed. In reality, however, the responses, or effects, are not necessarily detected (or even expressed) in a linear fashion; nor are all responses necessarily adverse. This is especially the case for metals compared to other toxins. For metals, compensatory

Box 9.3 Example of compensation

Populations of the zooplankton *Daphnia galeata mendotae* from Lake Michigan compensated for Cd mortality with their high reproductive capacity. Animals were exposed for 22 weeks (>20 generations) to concentrations ranging from 1.0 μg/L Cd to 8.0 μg/L Cd in water (Marshall and Mellinger, 1980). Live births were reduced fourfold at 1.0 μg/L Cd; it was projected that they would be reduced twofold at 0.15 μg/L Cd. At the lowest doses, the number of egg-bearing females and eggs per female all increased. The greater effort at reproducing compensated for the reduced proportion of eggs that resulted in live births. Thus population size was reduced only at exposures greater than 2.0 μg/L Cd. Extinction occurred only in populations exposed to 10 μg/L Cd. Concentrations of 0.15 μg/L Cd were determined in Lake Michigan. These would probably reduce live births by the daphnia, but were not projected to have a detectable effect on the numbers of *D. galeata*. However, Marshall and Mellinger (1980) noted that compensation could be less effective in the lake than in the laboratory. Factors that compete for compensatory capabilities occur in nature, such as predation or reduced food. Natural disturbances might also more severely affect a population already employing compensatory capabilities than an undisturbed population. Natural disturbances are inevitable in the long term and would be a strong test of the resilience offered by this population level compensatory response.

reactions can occur at every level of organisation, analogous to detoxification at the molecular level (Box 9.3). Interactive malfunctions can also occur that are more complex than a simple adverse effect. Thus, even at levels of organisation higher than the molecular, adverse effects occur when the compensatory mechanism is overwhelmed and/or when compensation imposes secondary costs.

Complexity increases from lower to higher levels of organisation in the hierarchy. Population sizes, community structures and ecological function are the high-level responses that environmental regulations are designed to preserve. These are the responses that risk evaluations address. But, at higher levels

of organisation, it is a great challenge to detect malfunctions with sensitivity, much less to link their cause to metal exposure (we will discuss this in detail in Chapter 11). The lower-level responses are more readily detectable and easier to experiment with, but their linkage to the most relevant high-organisation risks is more difficult to determine. Study at each level of organisation tends to have its advocates. But, every choice of focus in the hierarchy has trade-offs (e.g. inherent ambiguities in the results against relevance to protecting nature); it is the body of work that ultimately establishes the important findings.

9.5   BIOMARKERS

Prologue. Biomarkers are signs at lower levels of biological organisation that compensation is occurring or that metal binding at undesirable sites has overwhelmed compensatory capabilities.

A biomarker is a biological response (for example a biochemical, cellular, physiological or behavioural variation) that can be measured at the lower levels of biological organisation, in tissue or body fluid samples or at the level of the whole organism. Biomarkers provide an early, sensitive response, indicative of exposure to and/or effects of a contaminant (Depledge, 1989b; Peakall, 1992; Depledge *et al.*, 1993, 1995; Depledge and Fossi, 1994). They can act as early warning signals of biologically significant trace metal pollution in the environment.

A whole range of biomarkers has been proposed from investigations of pollution by both organic contaminants and toxic metals. Many of the biomarkers investigated lack specificity, being indicative of the general status of health of an organism. Generic stress responses are relevant, but a biomarker that is specific in its response to a single toxin or limited range of toxins has particular diagnostic value in ecotoxicology. That biomarker would have even greater ecotoxicological importance if links (probably necessarily correlational) could be established between its detection in exposed organisms in the field and consequent ecotoxicological effects at higher levels of biological organisation.

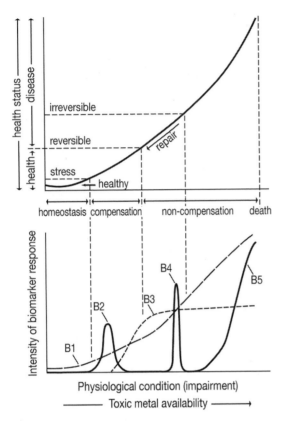

Fig. 9.4 Relationships between exposure to toxic metal, health status and biomarker responses. The upper graph shows the progression of the health status of an organism with increased exposure to toxic metal availability. The lower graph shows the response of five hypothetical biomarkers (B1 to B5) associated with changes in physiological condition and used to assess the health of the organism. (After Depledge, 1989b, and Depledge *et al.*, 1993.)

A biomarker approach based on physiological alterations in an organism is illustrated in Fig. 9.4 (Depledge, 1989b; Depledge *et al.*, 1993). Starting from the position of good health, there is progressive deterioration in health status (eventually fatal) as the organism is exposed to greater availability of the toxic metal. Early departures from health are not overt and are associated with compensatory responses as shown on the physiological condition scale (Fig. 9.4) (Depledge *et al.*, 1993). One set of biomarkers may signal this onset of compensation. These are meaningful stress signals, because

survival potential may have already begun to decline as the organism expends energy. However, the organism may still be able to recover (even if its physiological condition enters the non-compensation phase) if conditions improve sufficiently and quickly enough, but eventually the poor physiological condition becomes irreversible and death ensues (Depledge *et al.*, 1993). A second set of biomarkers might detect deteriorating physiological condition.

Figure 9.4 illustrates the response of five hypothetical biomarkers (B1 to B5) associated with changes in physiological condition and used to assess the health of the organism. B1, B2 and B3 represent different biomarker responses in the compensatory phase. These responses signal a change as compensatory mechanisms are overwhelmed. Biomarker B4 is a signal of the reversible non-compensation phase where the organism is still curable with adequate excretion or detoxification of the toxicant. Biomarker B5 is a signal of the incurable phase of health impairment. The greatest challenges in this approach arise from the need to relate responses at lower levels of biological organisation to those at the higher levels.

Below we identify a number specific and non-specific responses that fit within this scheme.

9.5.1   Molecular biology

Prologue. If metals bind to DNA and cause enough damage, effects are manifested as mutations and alterations in gene function. Experiments show that metals feasibly can cause DNA damage. Observations of damaged DNA in nature indicate such effects might occur in a contaminated environment.

DNA plays a fundamental role in the metabolism of all cells and ultimately in reproduction. Damage caused by a toxin to DNA can interrupt normal metabolic functioning, and even lead to carcinogenesis (the development of tumours), heritable mutations, and teratogenesis (the malformation of embryos).

The first stage in the action of a genotoxin is the formation of an adduct, the covalent binding to DNA

Box 9.4 Methods for determining DNA damage

The comet assay is a common technique to measure an early stage of damage to DNA in order to detect the possible genotoxic effects of contaminants (Pellacani *et al.*, 2006). Comet assays are sensitive and rapid. They provide micro-electrophoretic assays of DNA strand breaks in individual cells (Singh *et al.*, 1988; Fairbairn *et al.*, 1995). For example blood cells, which are easily sampled, are suspended in a gel on a microscope slide, lysed, electrophoresed and stained with a fluorescent DNA binding dye; the electrophoresis current pulls charged DNA from the cell nucleus and the broken DNA fragments migrate further, giving the appearance of a comet (Fairbairn *et al.*, 1995). The tail length of the comet is measured as a measure of DNA damage.

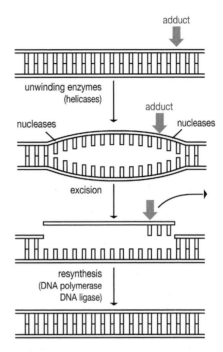

Fig. 9.5 The repair of DNA after the formation of an adduct. (After Walker *et al.*, 1996.)

of the toxin, or a metabolite, such as a highly reactive free radical generated in the cell by the presence of the toxin. A second stage might be DNA strand breakage (Box 9.4). Adducts (Fig. 9.5) and strand breakages can be repaired up to a point. If the structural perturbations to the DNA become fixed, however, affected cells often show altered function (the third stage: Walker *et al.*, 1996). Ultimately damage to cell division can lead to mutations and consequent alterations in gene function (Walker *et al.*, 1996).

At very high concentrations, experimental exposure to trace metals like Cu can cause DNA damage (Ozawa *et al.*, 1993). The red blood cells of fathead minnows (*Pimephales promelas*) and large-mouth bass (*Micropterus salmoides*) exposed to the relatively high concentration of 2 mg/L hexavalent Cr showed the formation of DNA–protein cross-links (Kuykendall *et al.*, 2006).

The high concentration experiments show that DNA damage by metals is feasible, but signs of DNA damage in metal-contaminated waters provide the evidence that this might be a viable biomarker. The gills of mussels (*Mytilus galloprovincialis*) from the metal-impacted Cecina estuary in Tuscany, Italy showed DNA damage measured by the comet assay in comparison to control mussels (Nigro *et al.*, 2006). Laboratory reared daphnids (freshwater crustacean zooplankton) caged in a river receiving trace

metal-containing discharges also showed clear DNA damage (De Coen *et al.*, 2006).

9.5.2   Biochemistry

Prologue. Biomics provide biomarkers of the induction of genetic or other biochemical responses to metal contamination. Some of these will be metal-specific responses. The potential for monitoring ecotoxicological effects of trace metal contaminants using an 'omics' approach is enormous but investigations are still at an early stage (Watanabe and Iguchi, 2006). The best known metal-specific response to contamination is induction of the metal-specific binding proteins, the metallothioneins (MT). Induction of these is a compensatory response. Increasing concentrations of metal associated with MT and non-MT proteins have been related to signs of physiological stress of significance to populations in a few studies. General stress responses have also been tied to metal exposure. The most common are inhibition of enzymes, generation of oxyradicals and expression of stress proteins. Attributing these biomarker responses to metal contamination is the greatest challenge in complex natural waters.

9.5.2.1 Biomics
Prologue. There is great hope that the ability to detect gene expression patterns, or indicators of gene expression, will lead to metal-specific measures of stress from contamination.

In traditional molecular biology, specific genes or their associated proteins are analysed. Modern approaches are more concerned with the whole genome. It is the age of 'omics' – studies on the genome, the transcriptome, the proteome and the metabolome. While proteomics is the study of the full set of proteins (the proteome) encoded by a genome, metabolomics measures changes in the quantities of a metabolite present, for example in a biological fluid such as urine or blood, either by nuclear magnetic resonance (NMR) (Viant, 2003) or mass spectrometry (MS) (Soga *et al.*, 2003). In essence, these studies investigate which genes are in action under different circumstances, for example at different stages of the life cycle or reproductive cycle. Thus it is becoming possible to understand which genes are induced or show reduced transcription and which metabolites are present when concentrations of trace metals are raised (Lopez *et al.*, 2002; Watanabe and Iguchi, 2006). Because of the specificity of such responses, there is much optimism that these techniques will yield metal-specific diagnostic tools.

Microarrays are already in use to detect ecotoxicological effects in the aquatic environment. For example, Dondero *et al.* (2006) used a low-density oligonucleotide microarray, employing 24 different genes involved in both cellular homeostasis and stress-related responses, in an investigation of caged mussels (*Mytilus edulis*) along a Cu pollution gradient in the Visnes fjord, Norway. Microarray analyses showed that gene expression patterns were severely altered in mussels at the two innermost, most polluted sites, with MTs and catalase (see below) exhibiting a linear activation response along the Cu gradient (Dondero *et al.*, 2006).

9.5.2.2 Metallothionein
Prologue. Elevated metallothionein induction is a well-studied and sensitive indicator that an organism is responding to metal contamination with compensation, the first step toward adverse effects on the organism.

The most commonly discussed candidate for a specific biomarker of trace metal exposure at the biochemical level is expression of the cytosolic detoxificatory protein metallothionein (MT), which is induced by and binds to particular trace metals such as Ag, Cd, Cu, Hg and Zn (see Chapter 7). Metallothioneins are non-enzymatic proteins with a low molecular weight, high cysteine content, no aromatic amino acids and heat stability. They are found in a wide range of invertebrates and vertebrates, playing a role in the homeostasis of essential trace metals and the detoxification of both essential and non-essential trace metals (Amiard *et al.*, 2006).

Metallothionein induction has long been known from laboratory studies (George *et al.*, 1979; Viarengo *et al.*, 1984; Roesijadi, 1986; Langston *et al.*, 1989; Amiard *et al.*, 2006). More important, a number of examples exists of induction in natural waters (Amiard *et al.*, 2006) (Box 9.5).

Induction of MT is a reliable response to metal contamination, but it can be variable. Species, tissue, MT isoform, metal and concentration, must all be considered (Amiard *et al.*, 2006). For example, trace metal detoxification in mussels depends more on MT than does detoxification in oysters which depend heavily on trace metal detoxification in insoluble form (Wang and Rainbow, 2005; Amiard *et al.*, 2006). This is one of many examples of differences among species. MT can also be induced by other (stress) factors unrelated to metal contamination such as handling, starvation, anoxia, freezing and the presence of antibiotics, vitamins or herbicides, albeit to a lower level of induction than that caused by trace metal exposure (Amiard *et al.*, 2006). It is also a feature of MTs, as of other cellular proteins, that they are turned over in a cell cycle of synthesis and breakdown (in lysosomes). It is possible, therefore, for increased MT synthesis to be reflected in increased rate of MT turnover (Couillard *et al.*, 1995a; Mouneyrac *et al.*, 2002) without any increase in MT concentration. Thus determination

Box 9.5 Examples of MT induction in natural waters

The amount of Cd bound to MT in the gastropod *Littorina littorea* was markedly increased in samples from Cd-contaminated as opposed to uncontaminated sites (Langston *et al.*, 1989). The concentration of metallothionein-like proteins (MTLP) binding Cu in the digestive gland of the mussel *Mytilus galloprovincialis* was three times higher in samples from a metal-polluted habitat than a clean one (Viarengo *et al.*, 1982). Similarly MT concentrations in the digestive gland (but not the gills) of the very closely related mussel *M. edulis* differed significantly between mussels translocated to a metal-rich site and those left at the original uncontaminated site (Fig. 9.6) (Geffard *et al.*, 2005). In the same translocation experiment (Geffard *et al.*, 2001, 2002, 2005), MT concentrations in the gills of another bivalve, the oyster *Crassostrea gigas*, showed better correlations with accumulated tissue metal concentrations than in the digestive gland, but in both tissues natural (seasonal) variations in MT concentrations were high enough to mask intersite differences.

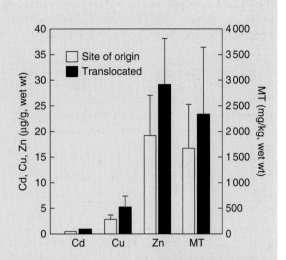

Fig. 9.6 Grand means of metallothionein and trace metal concentrations in the digestive glands of the mussel *Mytilus edulis* collected over a period of up to 7 months (March to October) from the uncontaminated site of origin (Bay of Bourgneuf, France) or translocated to the metal-contaminated Gironde estuary, France (Geffard *et al.*, 2005).

of MT concentration alone, the usual method of measurement, can sometimes be misleading. Nevertheless, with careful experimental design, particularly choice of species and tissue analysed, MT concentration can be a reliable biomarker that the cell is responding to raised exposure to the trace metals Ag, Cd, Cu, Hg and Zn (Amiard *et al.*, 2006). There is some discussion about whether the onset of compensation (elevated MT induction) is an indicator of stress or an indicator of exposure. If it is assumed that compensation requires energy, then elevated MT induction is certainly a sensitive indicator of the first step on the path toward adverse effects on the organism.

9.5.2.3 General biochemical stress responses
Prologue. Metals are among the stressors that can cause build-up of oxyradicals in cells and subsequent signs that enhanced activities to defend against these toxins have been induced in the cell.

Trace metals can exert their toxic effects by inhibiting the action of enzymes and thereby catalyse the

generation of oxyradicals or slow defences against oxyradicals. Accumulation of oxyradicals can cause damage in cells (e.g. by damaging DNA), or induce expression of stress proteins. Useful indicators of this kind of stress include inhibition of enzymes by exposure to raised toxic metal availabilities, measurement of antioxidant defences or determination of stress protein expression (Box 9.6). Unfortunately these are all general indicators of stress; many different factors can cause similar responses. Conclusions that trace metals have caused such effects in nature can be confounded by the possibility of other explanations. Nevertheless, studies exist to show such responses are feasible. If those responses are observed, trace metals should be considered as one explanation.

9.5.3 Cytology

Prologue. Micronucleus formation and changes in intracellular lysosomes are both responses to metal stress. Both have been demonstrated in the

Box 9.6 General biochemical stress responses are induced by metals

Enzyme inhibition

Exposure of the European shore crab *Carcinus maenas* to raised dissolved availabilities of Cu, for example, will reduce the activities of gill Na,K-ATPase and disrupt osmotic and ionic regulation (Bjerregaard and Vislie, 1986; Hansen *et al.*, 1992a) and inhibit activities of enzymes involved in intermediary metabolism (Hansen *et al.*, 1992b). In both cases, however, the dissolved Cu concentrations needed to elicit any effect are in the range of 1000 to 10000 µg/L Cu, well above concentration ranges in most metal-contaminated coastal waters (see Chapter 5) for which a biomarker survey would be relevant.

Antioxidant defences

Trace metals can catalyse the generation of oxyradicals that can cause damage to DNA, and so the measurement of antioxidant defences in organisms is of particular interest (Frenzilli *et al.*, 2001; Regoli *et al.*, 2002). Primary antioxidant enzymes such as superoxide dismutase and catalase, for example, have been measured as biomarkers for some time (e.g. Funes *et al.*, 2006), as have the concentrations of malondialdehyde, a breakdown product of lipid peroxidation (Livingstone, 2001) and glutathione, an oxyradical scavenger (Nicholson and Lam, 2005). The antioxidant

defences of mussels such as *Mytilus galloprovincialis*, *M. edulis* and *Perna viridis* have come in for particular attention (Cavaletto *et al.*, 2002; Geret *et al.*, 2002; Mourgaud *et al.*, 2002; Camus *et al.*, 2004; Nicholson and Lam, 2005), as general assays have been developed for their assessment (Gorinstein *et al.*, 2006). Total oxyradical scavenging capacity (TOSC), for example, is proving a useful integrated measure of antioxidant defence (Regoli, 2000; Moore *et al.*, 2006).

Stress proteins

Stress proteins constitute a set of protein families (with molecular weights from 16 to 90 kDa) long called 'heat shock proteins' because they are inducible by heat shock stress. These stress proteins are highly conservative, and are known to exist in all animal and plant groups studied with the probable normal function of binding to proteins for modulation of protein folding, protein transport and protein repair (Connell *et al.*, 1999). It is now appreciated that their induction is not specific to heat shock but is also responsive to many stressors including exposure to trace metals and organic compounds, salinity changes and ultraviolet radiation. Trace metal exposure can therefore cause raised concentrations of stress proteins (e.g. Stress-70 and Stress-60 proteins), and they do have potential as trace metal biomarkers so long as allowance is made for the potential induction effects of other more general stressors (Kammenga *et al.*, 2000).

---

laboratory and associated with metal exposure in natural waters. Lysosomal responses are best known, and include destabilisation of the lysosomal membrane, changes in size and number of lysosomes in a cell and the accumulation of insoluble lipofuscin granules.

9.5.3.1 Micronucleus formation

Prologue. Proliferation of micronuclei can be seen in cells from organisms found in or translocated to metal-contaminated environments.

One manifestation of the genotoxic damage caused by contaminants (or their metabolite derivatives), in this case at the chromosomal level, is the formation of micronuclei, displaced from the main nucleus. Blood cells again provide a suitable choice

for investigation. Galloway *et al.* (2004) adhered blood cells from the bivalve *Cerastoderma edule*, the cockle, and the crustacean *Carcinus maenas*, the common shore crab, onto glass slides, fixed and then stained them with Giemsa stain to show up nuclei and any micronuclei, the frequency of which could be estimated. The blood cells of mussels (for example *Perna viridis*: Nicholson and Lam, 2005) are also amenable to this technique. The gill cells of mussels (*Mytilus galloprovincialis*) from the metal-impacted Cecina estuary in Italy showed a fourfold increase in frequency of micronucleus formation in comparison to control mussels, in correlation with DNA damage measured by the comet assay (Nigro *et al.*, 2006). The frequency of micronuclei in the gill cells of mussels transplanted into the Cecina estuary doubled after 30 days deployment (Nigro *et al.*, 2006).

9.5.3.2 Lysosomes

Prologue. Metal exposure can cause lysosomes to become more abundant and become larger. Lysosomal membranes may lose their integrity and the lysosome may accumulate a lipid breakdown product, lipofuscin, in response to such exposures.

The most useful cytological biomarkers of metal stress are centred on lysosomes (Lowe *et al.*, 1981; Regoli, 1992; Au, 2004; Moore *et al.*, 2004; Guerlet *et al.*, 2006). Lysosomes are membrane-delimited cellular organelles containing hydrolytic enzymes which break down substances within a cell (autophagy) or substances that have been taken in from outside the cell by endocytosis (heterophagy). Lysosomes are present in almost all cells of eukaryotic organisms, and their function includes the degradation of redundant or damaged organelles (e.g. mitochondria and endoplasmic reticulum) and longer-lived proteins as part of autophagic cell turnover (Moore *et al.*, 2006). The lysosomal system also participates in detoxification, involving the sequestration and intracellular accumulation, of a wide range of trace metals and organic xenobiotics (Moore *et al.*, 2004).

The functioning of the lysosomal system is an indicator of general stress, as a result of its crucial roles in cellular turnover and detoxification. Several responses to contaminant exposure and other stress factors are known. Most is known for bivalve molluscs, in which the digestive cells of the digestive gland diverticula are especially rich in lysosomes. These lysosomes are practical to study because they are considerably larger than typically found in mammal cells (Lowe *et al.*, 1981; Au, 2004).

9.5.3.2.1   Change in size and number

Enlargement and increase in number of lysosomes in digestive cells of bivalve molluscs (and in the livers of fish) have been observed in response to environmental contaminants including trace metals (Etxeberria *et al.*, 1994; Marigoméz *et al.*, 1996; Au, 2004; Giambérini and Carajaville, 2005). Etxeberria *et al.* (1994) showed that increasing bioavailabilities of Zn in the field were associated with enlarged digestive cell lysosomes in the mussel *Mytilus galloprovincialis*, the lysosomal size decreasing when Zn-exposed mussels were depurated in clean sea-

water. Laboratory exposure of the field Zn-exposed mussels to Cu and Cd in the laboratory caused further enlargement of the lysosomes. Etxeberria *et al.* (1994) concluded that lysosomal enlargement should be considered as a non-specific biomarker of environmental stress induced by sublethal bioavailabilities of metals. Digestive gland cells of mussels (*Mytilus galloprovincialis*) from the metal-impacted Cecina estuary in Italy had significantly larger lysosomes than control mussels (Nigro *et al.*, 2006). Giambérini and Cajaraville (2005) extended the use of this biomarker to freshwater, demonstrating an increase in size and number of digestive gland cell lysosomes upon laboratory exposure of the freshwater mussel *Dreissena polymorpha* to raised dissolved Cd availabilities.

9.5.3.2.2   Destabilisation of the lysosomal membrane

The functional stability of the lysosomal membrane is a good indicator of lysosomal integrity (Moore *et al.*, 2006). Lysosome membrane permeability generally changes quantitatively in relation to the degree of contaminant-induced stress, and is responsive to a range of contaminants including trace metals (Moore, 1985). The assessment of membrane permeability is, therefore, considered a useful non-specific biomarker of contamination. An important advantage of the technique is that determination of the response is relatively simple and quantification straightforward (Box 9.7).

9.5.3.2.3   Generation of lipofuscin and/or lipid peroxidation

Lipofuscin is a lipopigment end product of lipid peroxidation (for example of cell membrane components) brought about by free radicals or other reactive oxygen species (Moore, 1988; Regoli, 1992; Au, 2004). Increased production of lipofuscin granules indicates increased turnover of organelles and proteins (which may include MTs). It is a biomarker of cellular well-being, affected by exposure to trace metals as well as to organic contaminants. Lysosomal accumulations of lipofuscin are reported in molluscan digestive gland cells after exposure to organic and trace metal (e.g. Cd, Cu) contaminants or to oxidative stressors (Viarengo *et al.*, 1985, 1987;

Box 9.7 Determination of lysosomal membrane instability

A technique based on the dye neutral red was developed for determining lysosomal membrane instability in the blood cells of the mussel *Mytilus edulis* (Moore, 1985). It has been extended to other mussels (e.g. *M. galloprovincialis*: Regoli, 1992; *Perna viridis*: Nicholson and Lam, 2005), other bivalves (*Cerastoderma edule*: Galloway *et al.*, 2004) and other invertebrates (the crab *Carcinus maenas*: Galloway *et al.*, 2004). The technique typically involves incubating blood cells (in suspension or adhered to a glass slide) with neutral red which accumulates in the lysosomes. Neutral red only stays for any length of time in healthy cell lysosomes, leaking from the lysosomes into the cytosol at an increasing rate as lysosomal stability decreases (lysosomal membrane permeability increases) in unhealthy cells. Cells are viewed down a microscope. The measure of stress is the time taken for 50% (for example) of the cells to show staining of the cytosol as evidence of leakage of neutral red from the lysosomes (Nicholson and Lam, 2005; Galloway *et al.*, 2006).

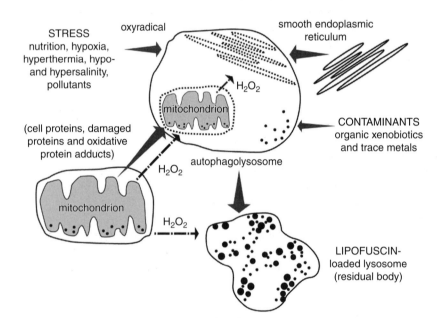

Fig. 9.7 The formation and accumulation of lipofuscin in lysosomes and its association with oxyradical-mediated cell injury. (After Moore *et al.*, 2006.)

Moore, 1988; Regoli, 1992; Moore *et al.*, 2006; Nigro *et al.*, 2006). Moreover lipofuscin granules may contribute to the detoxification of trace metals which may be trapped both chemically and mechanically (George, 1983; Regoli, 1992) during lysosomal autophagy of metallothioneins (Chapter 7). Lipofuscins are recognised under the microscope as hydrolysis-resistant residual bodies remaining in lysosomes (Moore *et al.*, 2006); (Fig. 9.7). Nigro *et al.* (2006) included digestive gland lipofuscin accumulation in a battery of biomarkers used in their study of mussels (*Mytilus galloprovincialis*) in the metal-impacted Cecina estuary, Italy. Lipid peroxidation was also used as one of a suite of signs that trace metals had effects on fish populations in the Clark Fork River, Montana, USA (Chapters 11 and 12; Farag *et al.*, 2003).

9.5.3.3 Phagocytosis

Prologue. Suppression of the immune system is another response indicative of metal stress.

The immune system of mussels is comprised of an integrated response whereby phagocytic blood cells remove foreign particles that are subsequently degraded in lysosomes (Nicholson and Lam, 2005). This response is generally suppressed by exposure of mussels to contaminants including trace metals (Nicholson and Lam, 2005), providing a potential biomarker of immunotoxicity (Galloway *et al.*, 2004).

9.5.4   Histopathology

Prologue. Histopathological alterations in bivalve tissues are receiving attention in marine pollution monitoring, but not as extensively as fish histopathology. Histopathological changes in bivalves in responses to metals are better known than in fish, where the presence of organic chemicals often confounds interpretation. Dose–response relationships to bioavailable metal contamination exposure as well as the confounding effects of biotic and non-biotic factors are still relatively poorly known.

Genotoxins can produce tumours or neoplasias (carcinogenesis), and there is an extensive literature on their occurrence (particularly in fish) (Depledge, 1996; Au, 2004). Correlations between the occurrence of neoplasias and organic xenobiotics concentrations are more common (Au, 2004). Reports that incriminate trace metals often cannot separate their effects from those of coincident organic chemicals. For example, the elevated prevalence of epidermal hyperplasias or papillomas in a fish, the dab *Limanda limanda*, was related to high concentrations of trace metals (Cr, Ti) in dab tissues, sediments and the water column in the North Sea (Dethlefsen, 1984), but also to concentrations of organic contaminants in the dab liver (Vethaak *et al.*, 1992; Au, 2004).

Histopathological changes in the livers and gills are known from fish exposed to contaminants, but most reports again accentuate the role of organic compounds (see review by Au, 2004). Fish gill histopathology in particular has potential as a biomarker for general environmental contamination (Au, 2004). Histopathological changes have also been reported in invertebrates, particularly bivalves, exposed to trace metals (Wedderburn *et al.*, 2000; Au, 2004) (Box 9.8).

9.5.5   Morphology

Prologue. Morphological responses to metals are known in natural waters in small species like hydroids in marine environments and chironomid larvae in freshwater lakes, but are not widely used as biomarkers. The best-known responses are metal-specific and highly diagnostic: imposex as a response to tributyl tin exposure and teratogenesis in response to Se exposure

Hydroids (the sedentary polyp stages of Cnidaria) show altered morphologies in response to exposure to contaminants including trace metals (Stebbing,

---

Box 9.8 Examples of histopathological changes in invertebrates related to metal exposure

Long-term exposure of the mussel *Mytilus edulis* to dissolved Cu caused histopathological changes to the epithelium of the digestive gland which consisted of non-ciliated cuboidal cells instead of ciliated columnar epithelial cells (Calabrese *et al.*, 1984). There were histopathological changes in the digestive gland, gills and ventral foot epithelium of the bivalve *Macoma carlottensis* sampled close to a previous discharge site of Cu mine tailings in British Columbia, Canada (Bright and Ellis, 1989). Bivalves collected from nearer the discharge site had more extensive lesions in the tissues including vacuolation and increased fragmentation of digestive gland cells (Bright and Ellis, 1989). In a study of the digestive gland of mussels (*Mytilus edulis*) transplanted into a Cu pollution gradient resulting from mining in the Visnes fjord, Norway, Zorita *et al.* (2006) found atrophy of the digestive gland epithelium and changes in the relative frequency of different cell types in mussels from the site associated with the highest environmental Cu concentrations. Blood cells can infiltrate connective tissue in contaminant-exposed bivalves (Nasci *et al.*, 1999), for example in oysters *Crassostrea virginica* exposed to Cd (Gold-Bouchot *et al.*, 1995).

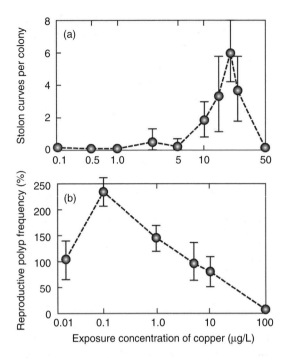

Fig. 9.8 (a) Mean stolon curving frequencies (with SE) of colonies of the hydroid *Campanularia flexuosa* exposed to different dissolved Cu concentrations for 14 days. (b) Mean frequencies (with SE) of reproductive polyps in colonies of *C. flexuosa* after exposure to different dissolved Cu concentrations for 11 days, expressed as a percentage of the ratio of reproductive polyps to total number of colony polyps in control samples. (After Bayne *et al.*, 1985.)

1976; Bayne *et al.*, 1985). The colonial marine hydroid *Campanularia flexuosa* shows increased curving of stolons during growth if exposed to unfavourable conditions, deviating from the more typical colony form of linear radiation, and curvature increases with toxin concentration (Stebbing, 1979) (Fig. 9.8a). A more sensitive morphological response of *C. flexuosa* to increasing trace metal availabilities is the increased formation of reproductive polyps (Bayne *et al.*, 1985) (Fig. 9.8b).

Deformities of mouthparts have also been observed in freshwater insect larvae from the taxon Chironomidae, the midges (Wiederholm, 1984). The frequency of deformed animals is often relatively low. For example, Wiederholm (1984) found that the percentage frequency of deformed

mouthparts in recent and subfossil material of mostly *Chironomus, Micropsectra* and *Tanytarsus* increased from less than 1% of the larvae at unpolluted sites or time periods to figures in the range of about 5% to 25% at strongly polluted sites. More recent work indicates that these are not necessarily metal-specific responses and that the frequency of occurrence can be even lower than observed in the early work.

Trace metals at very high availabilities can cause skeletal deformities in fish (Bengtsson, 1979; Bengtsson and Larsson, 1986). However, skeletal deformity in fish is particularly sensitive to exposure to chlorinated hydrocarbons with a good dose–response relationship, and induction of skeletal deformity can be confounded by biotic factors (e.g. hereditary defects, parasite infection) and abiotic factors (e.g. vitamin deficiency, gear damage) not related to contamination (Au, 2004).

The most notorious morphological responses to a trace metal contamination include imposex (see chapter 14) in gastropod molluscs in response to tributyl tin (TBT) exposure (Chapter 14) and teratogenesis (deformities) observed in the young of birds and fish exposed to Se (Chapter 13). Both will be discussed in detail in these later chapters.

9.5.6   Physiology

Prologue. Physiological responses to trace metal exposures integrate responses from lower levels of organisation. Determination of energy reserves or growth rates represent integrated measures characterising status in a given phase of the life cycle. Determination of instantaneous rates can be combined into an assessment of energetics termed 'scope for growth'. All these measures can be affected by metals, but factors like stage of life cycle, nutritional status or other stressors can also be important.

Physiological biomarkers are measures at the level of the individual organism such as rates of feeding, digestion, respiration and excretion. All can be affected by exposure of aquatic animals to high availabilities of toxic metals with consequences for their energy budgets, and thus growth and reproduction. It is not surprising, therefore, that physiological biomarkers are advocated as a means of integrating responses at lower levels of biological organisation.

> one cannot establish the connections between the subtle biochemical changes within an organism and the ecological changes unless studies at the level of the individual are included. Studies at the molecular level provide no insight into consequences for higher levels of organization, and studies of populations and communities do not shed any light on how the effects are caused.
>
> (Weis *et al.*, 2001).

Perhaps the longest history exists for studies of metal exposure on rates of growth. The hydroid *Campanularia flexuosa* shows reduced growth rate of the colony with increasing dissolved metal concentrations (Stebbing, 1976; Bayne *et al.*, 1985) (Fig. 9.9). Mussels (*Mytilus edulis*) in Trondheimsfjorden, Norway showed a reduction in growth rate when growing in the water contaminated with Cu and Zn near a local mining industry, in comparison with mussels from 'unpolluted' parts of the fjord (Lande, 1977).

Condition index is a simple but crude measure of the accumulated effects of the nutritional, reproductive and energetic status of an organism. The assumption is that the detoxification of pollutants is an energy-consuming process with the consequence that organisms inhabiting contaminated waters have retarded growth and limited energy reserves (Bayne *et al.*, 1985; Nicholson and Lam, 2005). In bivalves, for example, condition index is measured as the ratio of soft tissue weight to shell weight, shell volume or shell length. Preferably the weights are dry weight. It is also possible to determine the relative contribution of the weight of an organ to total soft tissue weight as a body component index, for example for the digestive gland of a mussel which stores energy reserves and responds to changes in the animal's energy demands (Phillips and Rainbow, 1994). Hummel *et al.* (1997) found a significant negative correlation between accumulated Cu concentrations in the clam *Macoma balthica* from Dutch and French

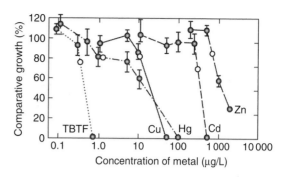

Fig. 9.9 The growth rates (means with SE) of the marine colonial hydroid *Campanularia flexuosa* decrease with increasing concentrations of tributyl tin fluoride (TBTF) and trace metals. Growth is measured by percentage comparison to controls; open circles indicate thresholds of a significant reduction in growth rate. (After Bayne *et al.*, 1985.)

estuaries and a condition index defined as the ratio of soft tissue dry weight to shell length. As illustrated by Hummel *et al.* (1997), and in Chapter 11, understanding the seasonal cycle of condition index is critical to separating effects of pollutants, nutrition and the reproductive cycle in natural waters.

Direct measurement can be made of body contents of energy reserves such as glycogen, lipid and protein, which can also be expressed in terms of their integrated energy contents (Smolders *et al.*, 2004). These data are included in an overall measure of energy allocation at a cellular level, an approach referred to as cellular energy allocation (CEA) (De Coen and Janssen, 1997, 2003). CEA reflects the amount of energy available to an individual organism for growth and reproduction. In addition to the indicators of energy reserves, the activities of a selected set of intermediary metabolism enzymes are determined, from which metabolic flux is characterised through the major catabolic pathways (glycogen breakdown, glycolysis, lactic acid metabolism, hexose monophosphate shunt and Krebs cycle). The difference between the measured energy reserves and the quantified energy consumption is the CEA. CEA has the advantage of incorporating accumulated or integrated measures of energetic status (the energy reserves). It also must be interpreted within the context of reproductive and nutritional cycles. The utility of the CEA approach has been shown in experiments with daphnid crustaceans (De Coen and Janssen, 1997, 2003) held in metal-rich effluents (De Coen et al., 2006), and mysid crustaceans exposed to TBT (Verslycke et al., 2003).

Instantaneous measures of physiological processes are also used to determine physiological well-being. These depend upon the assumption that the measured rate reflects the condition of the organism in nature (i.e. making the measurement does not affect the rate).

Heart rates of many invertebrates including bivalves and crabs can be measured non-invasively (Depledge and Andersen, 1990), and are affected by exposure to contaminants including trace metals. Both bradycardia, the slowing of heart rate (Scott and Major, 1972; Curtis et al., 2001), and tachycardia, the elevation of heart rate (Akberali and Black, 1980; Nicholson, 2003a, b), have been reported as an effect of trace metal exposure of bivalves. Given that mussels close their valves as a behavioural response to high trace metal concentrations (Davenport and Manley, 1978; Nicholson, 2003b), Nicholson and Lam (2005) concluded that tachycardia in mussels is indicative of enhanced metabolism during the detoxification of contaminants in contaminated coastal waters.

Respiration rate (the rate of oxygen consumption) can be measured in animals collected from different habitats or in animals exposed to trace metals in the laboratory. A common approach is to seek correlations between respiration rates and accumulated concentration of the contaminant as a measure of previous exposure (see Chapter 8), as depicted in Fig. 9.10a for mussels (Mytilus edulis) exposed to TBT.

Feeding rates (clearance rates in bivalve molluscs) are also affected by exposure to trace metals (Fig. 9.10b) and have been measured widely (Bayne et al., 1985; Nicholson and Lam, 2005). The clearance rate of M. edulis was maintained at a rate not significantly different from controls in TBT-exposed mussels until TBT accumulated concentrations reached more than 4 µg/g TBT, above which there was a rapid decline (Widdows and Page, 1993) (Fig. 9.10b). Exposure to raised availabilities of dissolved Cu inhibits clearance rate in the green lipped

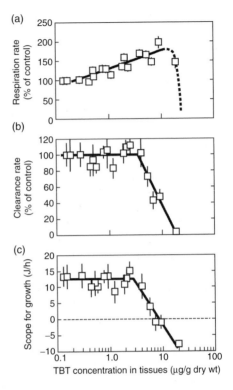

Fig. 9.10 Relationships between (a) respiration rate, (b) clearance (feeding) rate (both expressed as percent of control), and (c) scope for growth (all with 95% confidence intervals) and tributyl tin (TBT) concentrations (µg/g dw) in tissues of Mytilus edulis experimentally exposed to TBT. Respiration rate (a) rises steadily as energy is used in TBT detoxification until it declines, ultimately as a result of the cessation of pumping rate as reflected in clearance rate (b). The relationship between scope for growth (c) and TBT concentration reflects that of clearance rate and TBT concentration. (After Widdows and Page, 1993.)

mussel Perna viridis (Nicholson, 2003b). The freshwater amphipod crustacean Gammarus pulex had a reduced feeding rate when exposed to 101 µg/L Cu for 3 hours or 8.3 µg/L Cu for 96 hours (Taylor et al., 1993).

Scope for growth (SFG) integrates several instantaneous physiological measures to calculate the energy balance of an organism (Widdows et al., 1984; Bayne et al., 1985; Widdows and Page, 1993). SFG is an estimate of the energy available to an organism for growth and reproduction, calculated from the difference between the energy absorbed from food and that used in respiration (presumed to increase in contaminant-exposed organisms to meet contaminant detoxification and/or excretion requirements). The physiological measures needed in the calculation of SFG in a mussel (*M. edulis*) for example are respiration rate, clearance rate and assimilation efficiency, measured after the transfer of mussels into the laboratory (Phillips and Rainbow, 1994). SFG is considered a robust indicator of the health status (best developed for mussels) in both experiments and test for stress in natural waters (Allen and Moore, 2004). In the example of the TBT-exposed mussels depicted in Fig. 9.10, SFG decreased rapidly when the accumulated TBT concentration exceeds 4 µg/g TBT, as for clearance rate (Widdows and Page, 1993). Similar results were determined in *M. edulis* exposed to Cu in a mesocosm study (Widdows and Johnson, 1988), *M. edulis* along a pollution gradient in Narragansett Bay, Rhode Island, USA (Widdows, 1985), *M. edulis* along North Sea and Irish Sea coasts (Widdows *et al.*, 1995, 2002), *M. galloprovincialis* from the Venice Lagoon, Italy (Widdows *et al.*, 1997) and *Perna viridis* in Southeast Asia (Nicholson and Lam, 2005).

Other organisms can also show reduced SFG in response to metal exposure. For example the freshwater amphipod crustacean *Gammarus pulex* (Maltby *et al.*, 1990a, b), had reduced SFG when exposed to 0.3 mg/L Zn. The reduced SFG was associated with a decrease in the size of offspring released in the subsequent brood (Maltby and Naylor, 1990).

Both accumulated and instantaneous measures of physiology are quite variable (Chapter 11) and can be affected in complex ways by a myriad of factors. Attributing changes in physiology to metal exposure in natural waters requires consideration of alternative explanations and understanding of the life cycle of the organism (Chapter 11).

9.5.7 Linking responses in suites of biomarkers to physiological change

Prologue. Batteries of relatively easily measured biomarkers (particularly those based on lysosomes such as lysosomal membrane stability) are shown to correlate with higher-level physiological changes in response to pollutant exposure in aquatic ecosystems.

Many biomarkers exhibit a response in only part of the health status space (Allen and Moore, 2004) (Fig. 9.4). The use of a suite of biomarkers in the field is, therefore, required to cover the full range of physiological condition of resident biota, and is a first step to seeking correlations between the expression of biomarkers at different levels of biological organisation (Bayne *et al.*, 1988b; Galloway *et al.*, 2004; Nicholson and Lam, 2005; Nigro *et al.*, 2006; Dondero *et al.*, 2006). Nigro *et al.* (2006) used a suite of cellular biomarkers in native and transplanted mussels (*M. galloprovincialis*) in an investigation of the metal-impacted Cecina estuary in Tuscany, Italy. Genotoxic biomarkers (comet assay, micronucleus formation) and cytological biomarkers (lysosomal membrane stability, lysosome size, lipofuscin accumulation) all indicated the correlated presence of ecotoxicological effects in the estuary. Dondero *et al.* (2006) also used a suite of biomarkers in their study of transplanted mussels (*M. edulis*) along the Cu pollution gradient in the Visnes fjord, Norway, including a microarray of 24 genes, cytological biomarkers (e.g. lysosomal membrane stability, lysosome relative size, lipofuscin accumulation) and biochemical biomarkers (e.g. MT, catalase), with consistent results for several of these biomarkers at different hierarchical levels of biological organisation.

Lysosomal stability reflects toxicant induced cell pathologies caused by a variety of chemicals (Moore, 2002), while scope for growth, for example of the mussel *M. edulis*, is a robust biomarker at the physiological level of organisation (Allen and Moore, 2004). Data from a number of laboratory and field studies of *M. edulis* have been combined (Allen and Moore, 2004) to show that there is indeed a strong correlation between these two biomarkers (Fig. 9.11). Using various published data, Moore *et al.* (2006) extended these correlations, showing a significant positive correlation between lysosomal stability and total oxyradical scavenging capacity (TOSC) in mussels (Fig. 9.12), and a significant negative correlation between lipofuscin accumulation and lysosomal stability in digestive gland cells of mussels (Fig. 9.13). Strong correlations between lysosomal parameters support the hypothesis that stress or contaminant induced lysosomal responses are

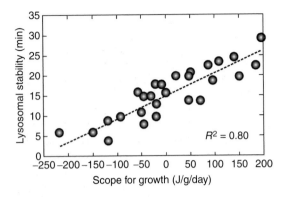

Fig. 9.11 Significant positive correlation between scope for growth and lysosomal stability in the mussel *Mytilus edulis* using data from laboratory and field studies. (After Allen and Moore, 2004.)

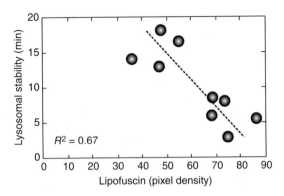

Fig. 9.13 Significant negative correlation between lysosomal stability and lipofuscin accumulation in digestive gland cells of mussels *Mytilus edulis*. (After Moore *et al.*, 2006.)

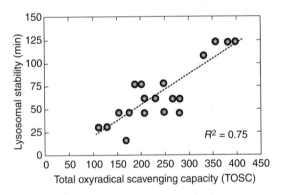

Fig. 9.12 Significant positive correlation between lysosomal stability and total oxyradical scavenging capacity (TOSC) in mussels. (After Regoli, 2000, Moore *et al.*, 2006.)

mechanistically linked to increased cellular autophagy and consequent cell degradation (Moore *et al.*, 2006). The correlations of lysosomal parameters with TOSC and scope for growth support the efficacy of the use of lysosomal biomarkers as indicative of toxic effects at higher biological levels.

Beyond simple correlations, useful methods of integrating biomarker data into a health status index are the ordination techniques of principal components analysis (PCA) and multidimensional scaling (MDS) (Allen and Moore, 2004; Galloway *et al.*, 2004) (Box 9.9). Galloway *et al.* (2004) investigated a suite of biomarkers in the crab

Carcinus maenas and the cockle *Cerastoderma edule* in Southampton Water, southern England. PCA revealed a gradient of detrimental impact to biota from head to mouth of the estuary coincident with high sediment concentrations of trace metals, poly-aromatic hydrocarbons and biocides. MDS was then used to identify which of the organisms investigated showed the greatest discrimination between sites. MDS identified the crab as the more sensitive to contaminant exposure, and carboxy-lesterase activity, metallothionein and total haemolymph protein in the crabs were the most discriminating biomarkers (Galloway *et al.*, 2004). Galloway *et al.* (2006) have taken this multi-biomarker approach further to develop an evidence-based approach in which a suite of biomarkers is analysed by MDS to assess the health of coastal systems through the general condition of individuals, in this case using a field study in the Humber estuary, England based on the mussel *M. edulis* and again the crab *C. maenas*.

Moore *et al.* (2006) also used MDS in their exploration of the use of lysosomal stress responses as predictive markers of animal health status, confirming good cluster separation of 'impacted' and 'healthy' sites by the lysosomal parameters investigated. Moore *et al.* (2006) propose a conceptual mechanistic model linking lysosomal damage and autophagic dysfunction with injury to cells, tissues and whole animals, as a putative operational tool

Box 9.9 Multivariate techniques for analysing complex data

Ordination techniques, such as principal components analysis (PCA) and multidimensional scaling (MDS), aim to replace a complex matrix of original variables (e.g. sites, species, abundances) into a smaller set of derived variables which still maintain most of the relevant information (Scott and Clarke, 2000; Rice, 2003). The new variables are often referred to as ordination axes and results can be presented by plotting graphs of these new variables against each other to visualise the relationships between sites, samples, species and other variables (e.g. sediment metal concentrations) more easily (Scott and Clarke, 2000).

Thus PCA aims to replace complex data sets with a smaller set of derived variables that still retain most of the relevant information; the derived variables in PCA are presented as axes that represent a best-fit line through the data (Galloway et al., 2004). The process is continued through successive principal components until all of the variation has been explained; successive principal components are contributed to by different variables (e.g. the sediment concentration of a particular metal) to different measurable extents (Eigenvalues), identifying which measured variables contribute most to the explanation of the variation

in the data set. Principal components are plotted against each other to show, for example, how data for different sites fall out along the different principal axes (e.g. Galloway et al., 2004). PCA, however, will only perform effectively as an ordination technique if the first few principal components (ideally two) account for a substantial proportion of the total variance in the original data. If the original variables are uncorrelated then the components derived from the PCA turn out to be just the original variables arranged in order of variance (Scott and Clarke, 2000). Furthermore PCA assumes that abundance data are normally distributed, zero abundances are rare and gradient(s) are well sampled across their full range (Rice, 2003).

Non-metric MDS provides an alternative ordination technique without the same assumptions about the distributions of abundance data. MDS estimates similarities (or dissimilarities) between sites, relying on the weaker assumption that the rank order of values is informative, but the actual quantitative data might not be (Rice, 2003). MDS increases robustness in the face of irregular distributions of abundance and high sampling variance, and as a result has become a preferred technique for ecological ordinations (Clarke and Ainsworth, 1993; Cao et al., 1996; Rice, 2003).

to link lysosomal stability in mussel digestive gland cells to the 'health status' of the mussel as a sentinel organism of marine environmental health (Fig. 9.14). The major challenge remains to develop computational models to bridge the gap between the 'health status' of individual organisms and ecosystem-level functional properties (Moore et al., 2006).

9.6 RESPONSES THAT ARE INTEGRATIVE OF LOWER-ORDER EFFECTS

9.6.1 Behaviour

Prologue. Sublethal effects on behaviour can occur in environmentally impacted populations. Avoidance mechanisms, reduced feeding rates, or reduced prey

capture efficiency have implications for the energy status of affected animals and ultimately populations.

Behaviour is a response at the level of the individual organism that has clear links to the biochemical level (e.g. neurotoxic effects) and clear connections to effects on the population and community level (Weis et al., 2001). Metal exposure causes changes in the behaviour of animals at potentially achievable environmental metal availabilities and in contaminated environments (Weis et al., 2001). But behavioural parameters may not always be the most sensitive response to trace metals.

The effects of trace metal exposure on feeding rates are well known (see above and Chapter 11). Many bivalves respond to high dissolved trace metal concentrations by shutting their valves, temporarily in the case of M. edulis when exposed

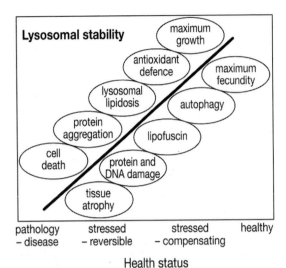

Fig. 9.14 Idealised diagram of a putative tool to interpret the significance of a biomarker response (in this case lysosomal stability in digestive gland cells of mussels) and pathological reactions in relation to the animal's health status (see Fig. 9.4). (After Moore et al., 2006.)

to 21 µg/L Cu in seawater, then permanently at 200 µg/L Cu or higher (Davenport and Manley, 1978). *Macoma petalum* (as *M. balthica*) also closes its valves in response to elevated dissolved Cu for as much as several days, eventually reopening them, presumably when it is not longer capable of anaerobic metabolism (D. J. Cain and S. N. Luoma, unpublished observation). *Macoma petalum* does not respond behaviourly to Ag in a similar way (although it is highly toxic). Another marine bivalve, *Macomona liliana*, changed its burrowing behaviour in sand dosed with the Cu or Zn (Roper et al., 1995). In Cu-dosed sand, there was a significant decrease in the number of *M. liliana* burrowing after 10 minutes at sediment concentrations of 25 µg/g Cu; Zn-dosed sand slowed burial at 80 µg/g Zn (Roper et al., 1995). The very low effect concentrations may reflect the relatively bioavailable form of metal in sand and the dilution of concentrations by the low surface area/mass ratio of sands compared to muds (Chapter 6).

Similar avoidance behaviour is observed in fish. Trout are notoriously sensitive to Cu, for example. Some species (for example bull trout *Salvelinus*

confluentus) will avoid waters with concentrations not much above that typical of undisturbed waters (Woodward et al., 1997; Chapter 12).

In the freshwater amphipod *Gammarus pulex*, the male will mate-guard a female until she is ready to mate. These precopula pairs will separate at 23.5 µg/L Cu under conditions when the 96 hour LC_{50} is 37 µg/L Cu (Pascoe et al., 1994; see Chapter 10).

The Norway lobster *Nephrops norvegicus* lives in muddy marine sediments rich in Mn (a potential neurotoxin) which in low oxygen conditions is released as bioavailable Mn^{2+}. Environmentally realistic concentrations of Mn (0.1 and 0.2 mmoles/L) doubled the reaction time of the lobsters to food odour stimuli and reduced their success in reaching the source of the stimulus (Kräng and Rosenqvist, 2006). Thus the ability of *N. norvegicus* to detect and find food can be reduced in sediment areas of high Mn concentration, with potential consequences at individual and population levels (Kräng and Rosenqvist, 2006).

The grass shrimp *Palaemonetes pugio*, an inhabitant of low-salinity creeks, showed a reduced rate of capture of prey when fed on laboratory exposed Cd-contaminated prey (*Artemia salina*) or field-exposed Cd-contaminated oligochaete worms (Wallace et al., 2000). Similarly *P. pugio* from a creek impacted by historic smelting activities had reduced prey capture success in comparison with control grass shrimp (Perez and Wallace, 2004).

9.6.2   Tolerance

Prologue. The presence of a metal-tolerant population of an organism in a particular habitat is evidence that toxic metal availabilities there are of ecotoxicological significance, a conclusion that is of practical significance in developing a body of evidence that toxic metal availabilities in a habitat are sufficiently high to be of concern to environmental managers.

Like any toxin, metal contamination can act as a form of selection pressure, selecting for the subset of physiological (and lower order) traits that are the most metal tolerant in a gene pool (Box 9.10).

Box 9.10 Trace metal exposure can select for specific genotypes

Within a population there are inter-individual differences in susceptibility to contaminant exposure (Depledge, 1990; Lopes et al., 2006). Trace metal contaminants may therefore exert selection pressures which are reflected in changes in the genetic make-up of a surviving population (Depledge, 1994a, 1996). Different studies show increases or decreases in heterozygosity in response to the exposure of a population to trace metals, but it remains the case that trace metal exposure can alter the genetic structure and diversity of populations (Depledge, 1996).

A positive relationship was observed between the growth rates of individual mussels (*Mytilus edulis*) and heterozygosity at polymorphic enzyme loci (Hawkins et al., 1989; Depledge, 1996). Those individuals expressing greater heterozygosity survived longer (Hawkins et al., 1989). Nevo et al. (1986) also reported that genetically rich (highly heterozygous) species of marine gastropods display significantly higher survivorship after exposure to multiple inorganic and organic contaminants than genetically poor species. On the other hand, laboratory exposure of five marine gastropod species to lethal Cd concentrations selected genotypes homozygous for the enzyme phosphoglucose isomerase, with significantly more heterozygotes among the dead animals (Lavie and Nevo, 1986). Similarly Patarnello et al. (1991) reported evidence of the selection due to metal exposure of three polymorphic allozymes in a population of the barnacle *Balanus amphitrite* in the lagoon of Venice, Italy, expressed as a significant reduction of genetic polymorphism. Field studies of the mosquitofish *Gambusia holbrooki* populations at clean and metal-contaminated sites also showed reduced heterozygosity (of three allozyme loci) in the populations under trace metal exposure (Guttman, 1994).

This selection pressure can ultimately lead to the establishment of metal-tolerant populations (Luoma, 1977; Klerks and Weis, 1987). Indeed the presence of a metal-tolerant population of an organism in a particular habitat is evidence that toxic metal availabilities there are of ecotoxicological significance, not only to that species but also to others (Luoma, 1977).

A metal-tolerant population may have reduced uptake rates, increased loss rates, greater detoxification capabilities or changed ecological attributes that facilitate survival of trace metal exposure (e.g. Cain et al., 2006). Sometimes the attributes of a tolerant population may seem counter-intuitive until considered carefully. For example, an elevated Cu uptake rate (and consequently metal accumulation rate) was found in the Cu- and Zn-tolerant population of the polychaete *Nereis diversicolor* from metal-contaminated Restronguet Creek, Cornwall, UK (Bryan and Hummerstone, 1971, 1973b). But these surviving worms had especially well-developed physiological systems to detoxify and store excess accumulated trace metals, with atypical proportions of accumulated metals in different detoxified subcellular components, particularly in insoluble forms (Berthet et al., 2003; Mouneyrac et al., 2003; Rainbow et al., 2004b). Both Cu and Zn tolerance in these Restronguet Creek *N. diversicolor* are inheritable (Grant et al., 1989; Hateley et al., 1989).

Restronguet Creek also hosts a population of the brown seaweed, *Fucus vesiculosus,* that is tolerant to Cu, but not Zn (Bryan and Gibbs, 1983) (Fig. 9.15); crabs, *Carcinus maenas*, that are tolerant to both Cu and Zn, (in both cases via reduced metal uptake rates from solution: Bryan and Gibbs, 1983); and populations of the polychaete *Nephtys hombergi*, the bivalve *Scrobicularia plana* and the amphipod crustacean *Corophium volutator* that are tolerant to Cu (Bryan and Gibbs, 1983).

Cd-tolerant populations of the oligochaete *Limnodrilus hoffmeisteri* were reported from Foundry Cove, Hudson River, New York, in association with sediments with extraordinarily high Cd loadings (Klerks and Levinton, 1989). These Cd-tolerant worms showed enhanced detoxification, in this case of the excess Cd taken up from the heavily contaminated sediment (Klerks and Bartholomew, 1991). The incoming Cd in the oligochaetes is first bound to soluble metallothionein and then in Cd-rich granules probably resulting from lysosomal breakdown of the Cd-MT (Wallace et al., 1998). The Cd-tolerant worms had much higher concentrations of Cd-MT than their control conspecifics,

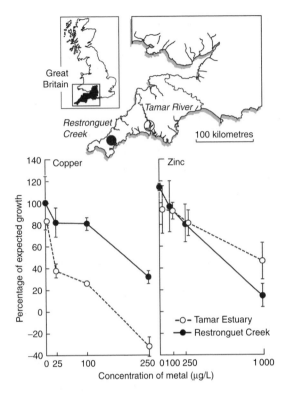

Fig. 9.15 The seaweed *Fucus vesiculosus* from Restronguet Creek is tolerant of Cu, but not Zn. This seaweed grows significantly better in high Cu concentrations than control seaweed from the Tamar estuary, but there is no difference in growth between the two seaweed samples in high Zn concentrations. The effects of the metal exposures were assessed by comparing the measured growth during metal exposure to that predicted from a pre-exposure period. Vertical lines indicate range of results. (After Bryan and Gibbs, 1983.)

the animal concerned. It is logical therefore that the physiological processes underlying tolerance also have an energetic cost. In support of this view is the typical observation that tolerant strains are outcompeted by non-tolerant strains in the absence of exposure to the contaminant. This is indeed the case for *N. diversicolor* in Restronguet Creek where Cu and Zn tolerance is restricted to the top of the creek, and non-tolerant populations of the polychaete take up residence throughout the rest of Restronguet Creek (Grant *et al.*, 1989; Hateley *et al.*, 1989).

9.6.3   Population

Prologue. Use of biomarkers within an ecological risk assessment framework involves extrapolating from lower levels of organisation to populations of the same species. A better understanding is needed of the modifying factors that influence the biological responses to stress at the different levels; especially in the more realistic lifetime exposures typical of organisms exposed to elevated trace metal availabilities in their natural habitats.

Effects of contaminants, in addition to increased mortality, that are observable at the population level include the numbers of individuals present, the age structure of a population, the reproductive output of a population or the recruitment rates of the organisms involved. In addition contaminant exposure may alter gene or genotype frequencies in a population, in the extreme leading to the selection of tolerant strains (see above). Changes in any of these processes can be just as significant for a population (and ultimately a community) as increases in mortality. The links between these processes and lower level biomarkers can be difficult to establish, however (Perceval *et al.*, 2004); (Box 9.11).

Many examples of metal effects on complex population processes exist. Exposure of the freshwater amphipod *Gammarus pulex* to increasing dissolved concentrations of Cu over 100 days reduced the rate of population increase, and eventually, the population density itself to below the initial

a feature that was also observed in the second-generation offspring of Foundry Cove worms that had no previous history of Cd exposure (Klerks and Levinton, 1989; Klerks and Bartholomew, 1991). Experiments suggest that tolerance mechanisms became a dominant characteristic of the population within four generations. When the contamination was removed from the cover, tolerance was lost within a few generations (Levinton *et al.*, 2003).

The conceptual basis of physiological biomarkers like scope for growth is that the detoxification and/or excretion of a toxic metal has an energetic cost reflected in the resultant energy balance of

Box 9.11 Linking MT induction to higher-order effects

While many studies of MT induction exist, only a few investigations have related expression of the metal detoxificatory protein metallothionein (MT) and population-level responses. Results from the existing studies are often conflicting or not clear (Couillard *et al.*, 1995b; Perceval *et al.*, 2004). Metallothionein expression has been linked to decreased density and biomass in fish (Farag *et al.*, 2003) and shifts in age structure in bivalves (Blaise *et al.*, 2003), but Schlenk *et al.* (1996) could not correlate increased MT levels with the occurrence of deleterious effects at the population level. Giguère *et al.* (2003) measured steady state Cd and MT concentrations and determined the subcellular partitioning of Cd in gills of the bivalve *Pyganodon grandis* living in lakes that differed markedly in the degree of environmental Cd exposure. Cd was present bound to cytosolic ligands in addition to MT and the increased accumulation of Cd in the gill cytosol was associated with symptoms of cellular toxicity. In a follow-up study Perceval *et al.* (2004) investigated the hypothesis that the toxic effects observed at the cellular level in this system could affect higher levels of biological organisation. In practice it was difficult to assign to subcellular metal partitioning measurements any predictive role for toxic effects at higher levels of biological organisation because of the presence and influence of confounding variables such as temperature differences (Perceval *et al.*, 2004). Perceval *et al.* (2004) sound a warning on the validity of environmental studies that do not explicitly consider habitat characteristics, together with those components related to anthropogenic activities.

level, with a lowest observed effect concentration (LOEC) of 14.6 µg/L Cu (Maund *et al.*, 1992). Cu exposure also affected the age structure of the final populations. In the control and lowest treatment concentration, the final population was mostly composed of juveniles; with increasing Cu exposure, there was a decrease in the number of juveniles (LOEC 14.6 µg/L Cu), then adults (LOEC 18.2 µg/L Cu) (Maund *et al.*, 1992).

Salice and Miller (2003) conducted a life-table response experiment to ascertain the demographic effects of low-level Cd exposure on two strains of the freshwater gastropod mollusc *Biomphalaria glabrata* from the embryonic stage through to adulthood. Cd exposure significantly affected several individual-based parameters (percentage hatch, juvenile survival, adult survival) in both strains, and others (fecundity, time to maturity) in one strain. As a result, population growth rate was significantly affected by Cd exposure with reduced juvenile survival being the greatest contributor to this effect (Salice and Miller, 2003). A detailed discussion of the effect of contaminant exposure on population dynamics is given by Walker *et al.* (1996).

Effects of contaminant exposure on the reproductive rates of populations can involve compensation. In an example cited above, the marine hydroid *Campanularia flexuosa* produced more reproductive polyps in the presence of low dissolved availabilities of trace metals (Stebbing, 1979). It is usually reduced fecundity, however, that results from contaminant exposure. In a marine member of amphipod genus *Gammarus*, *G. locusta*, exposed to contaminated sediments, chronic toxicity was associated with a bias towards females in the survivors but with severely impaired offspring production (Costa *et al.*, 2005). Reish (1978) showed significant reduction in the reproductive rates (numbers of eggs or offspring produced) of polychaetes exposed to sublethal concentrations of various trace metals, with sensitivities varying between species (see also Langston, 1990). Marine free-living flatworms *Stylochus pygmaeus* exposed to 25 µg/L Cu for 10 days laid fewer egg batches with reduced hatching success (up to 80% reduction in reproductive success) (Lee and Johnston, 2007). Trace metals also affect the success of fertilisation and embryological development, for example in echinoderms which are commonly used in water quality bioassays (Kobayashi, 1984; Langston, 1990). In experiments with dietary toxicity of Ag, Cd and Zn, impaired reproduction is the most sensitive response of zooplankton (Hook and Fisher, 2002; Bielmyer *et al.*, 2006).

Reduced reproductive output may not always affect the abundance or density of a population (Underwood and Peterson, 1988). Marine invertebrate planktonic larvae have a naturally high mortality rate (Mileikovsky, 1971). In some cases, small additions to that rate may have little effect on recruitment of adults, if other factors control the ultimate density of the population (density-dependent regulation of recruitment: Connell, 1985). On the other hand, additional toxicity from a metal may be significant in years when recruitment fails due to other natural factors. Effects of metals on reproduction will be especially dangerous to populations with low fecundity that depend upon high survival of young life stages. The fact that many benthic invertebrates have dispersive larval phases and recruit from stock in areas beyond those which are contaminated add to the likelihood of the persistence of a local population, even when reproduction is unsuccessful (Phillips and Rainbow, 1994). The toxicity of trace metal contamination at the recruitment stage of a population (e.g. toxicity to settling larvae) may be of greater ecological significance than reproductive success in such situations (Box 9.12). These are but a few of the complexities that confer benefit for some species and enhance the vulnerability of others in a metal contaminated habitat.

9.6.4   Community

Prologue. In situations where exposure to high trace metal availabilities has caused changes at the population level, for example changes in numbers of individuals or even the complete loss of a species, there will be subsequent changes at the level of the community. Altered community structure clearly coincides with elevated bioavailable metal, especially where concentrations are extreme. Cause and response are complex in communities in nature, however.

Changes in the number of species and/or the abundance of individuals within each species are effects on community structure. There are many instances in the literature where elevated or extreme metal contamination is associated with an apparent alteration in the nature of the biological community. We cite examples of these below to illustrate the feasibility of effects on community structure. We also show that effects can be complex. In Chapter 11

> **Box 9.12** Effects on recruitment
>
> Recruitment (addition of young animals to the adult population) appears to be an especially susceptible stage when considering the ecotoxicological significance of trace metal exposure to populations of aquatic invertebrates. For example, Restronguet Creek, Cornwall, England has extraordinarily high sediment concentrations of trace metals including As, Cu, Fe, Pb, Mn and Zn (Chapter 6) (Bryan and Gibbs, 1983). The observed impoverishment of bivalve molluscs in Restronguet Creek has been attributed to the toxicity of local Cu and Zn dissolved and sediment bioavailabilities to bivalve larvae and juveniles with consequent effects on bivalve recruitment (Bryan and Gibbs, 1983). Similarly, in a field experiment on the colonisation of contaminated sediments in Oslofjord, Norway, sediments with high concentrations of Cu (400 to 1500 µg/g) clearly had a negative effect on the colonisation of several taxa, including the polychaetes *Glycera alba* and particularly spionid polychaetes such as *Polydora caulleryi* (Trannum *et al.*, 2004). In a study of the effect of high Cu availabilities on the development of marine fouling communities on settlement plates at Williamstown, Victoria, Australia, Johnston and Keough (2000) showed that reductions of populations of sponges and polyp stages of scyphozoan cnidarians were greatest when pulse-pollution events occurred at times of high settlement.

we will address in more detail separation of cause and effect in such instances and mechanistic attributes of the changes.

Lande (1977) investigated the benthic fauna and flora of sites differentially contaminated by trace metals in Trondheimsfjorden, Norway, showing decreases in abundance and species diversity at sites adjacent to mining areas. Other early studies were essentially correlational. Rygg (1985) showed correlations between the species diversity of benthic faunal communities in Norwegian fjords and local sediment concentrations of Cu (strong), Pb (moderate) and Zn (weak) (Fig. 9.16). Of the 50 most frequently occurring species, 29 were absent or only rarely present where the sediment

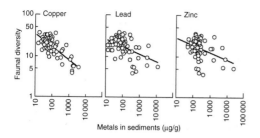

Fig. 9.16 Negative correlations between faunal diversity in benthic communities of Norwegian fjords and the concentrations of copper (strong, $R = -0.76$, $p < 0.001$), lead (moderate, $R = -0.49$, $p < 0.001$) and zinc (weak, $R = -0.37$, $p < 0.01$) in local sediments. (After Rygg, 1985.)

Table 9.1 *Pollution-induced community tolerance (PICT). The median survival times in 75% artificial seawater with added 50 µg/L Cu of nematode assemblages of sediments in Restronguet Creek and Percuil Estuary, Cornwall, England, are significantly different ($p < 0.001$)*

Site	Median survival time (h)
Percuil Estuary	66.3
Restronguet Creek	135.8

Source: After Millward and Grant (1995).

concentration of Cu exceeded 200 µg/g (Rygg, 1985). But in a number of cases, occurrence of tolerant populations adds complexity to the response.

Restronguet Creek with its extraordinarily high sediment concentrations of trace metals including As, Cu, Fe, Pb, Mn and Zn (Chapter 6) (Bryan and Gibbs, 1983) might be expected to show evidence of changes in the faunal composition of the sediment benthos in comparison with adjacent estuaries lacking such metal contamination. Sites in Restronguet Creek do indeed have a nematode meiofauna that has a lower abundance, lower generic richness and lower species diversity than local uncontaminated sites (Somerfield *et al.*, 1994). On the other hand, Restronguet Creek has a surprisingly strong representation of local macrobenthos, except perhaps for a relative impoverishment of bivalve molluscs attributed to the toxicity of local Cu and Zn bioavailabilities to bivalve larvae and juveniles (Bryan and Gibbs, 1983). The major factor maintaining the diversity of the local fauna and flora in Restronguet Creek appears to have been the selection of Cu- and/or Zn-tolerant strains over the last 150 years or so (see above) (Bryan and Gibbs, 1983). The presence of tolerant populations, in this case of the Cu- and Zn-tolerant population of the polychaete *N. diversicolor* has in fact been used to define threshold metal concentrations in the sediments of Restronguet Creek that are of ecotoxicological significance: 1000 µg/g Cu and 3500 µg/g Zn (Grant *et al.*, 1989). Other ecotoxicological consequences of the high sediment trace metal

concentrations in Restronguet Creek have not been ruled out. The local tolerant strains often contain atypically high trace metal concentrations (for example Cu in *N. diversicolor*) with the potential to be passed on to predators like decapod crustaceans, fish and birds (Bryan and Gibbs, 1983; Rainbow *et al.*, 2004b, 2006).

Blanck *et al.* (1988) extended to communities the argument that occurrence of a metal-tolerant population in a particular habitat is evidence that toxic metal availabilities there are of ecotoxicological significance (Luoma, 1977). Blanck *et al.* (1988) suggested that the selection pressure of an ecotoxicologically significant availability of a toxicant will lead to an increase in the average tolerance of individuals among all species within that community to that toxicant. Such pollution-induced community tolerance (PICT) might be used as an ecotoxicological tool, linking contaminant availability directly to functional change within communities (Blanck *et al.*, 1988). Millward and Grant (1995) demonstrated that PICT could be used in the field to evaluate the biological impact of chronic pollution on coastal benthos, using toxicity tests on the whole nematode communities present to show that nematodes in Restronguet Creek sediment were more resistant to Cu than nematodes from an adjacent, less metal-contaminated estuary (Table 9.1).

Over the last 20 years or so there has been a decided shift from the use of univariate measures (e.g. abundance, diversity index, species richness and evenness) to multivariate methods (e.g. classification, ordination and discrimination tests) to discriminate between communities at sites with (suspected) differences in the availabilities of contaminants including trace metals (Warwick and Clarke, 1991; Phillips and Rainbow, 1994; Cao et al., 1996; Scott and Clarke, 2000; Rice, 2003). Clarke and Ainsworth (1993), for example, illustrated the use of non-metric MDS on three coastal data sets – a macrobenthic community transect across the sewage sludge disposal ground off Garroch Head in the Firth of Clyde, Scotland, a meiobenthic nematode community in the Exe estuary, Devon, England, and a planktonic diatom community off western Greece. McRae et al. (1998) provide another example of the use of non-metric MDS to link patterns in benthic infaunal communities to environmental data in estuaries in Florida as part of the estuaries component of the Environmental Monitoring and Assessment Program (EMAP-E) initiated by the US Environmental Protection Agency in co-operation with the National Oceanic and Atmospheric Administration.

9.7   CONCLUSIONS

Sublethal adverse effects of metals are well established in experiments. The greatest issue lies in establishing the significance of such responses: what are implications for populations? While there is certainly a hierarchy of effects across the different levels of organisation, empirically demonstrating integration across the hierarchy of biological organisation is often a challenge, especially in natural waters. Management of environmental risks from metals requires sensitive measures of stress, so as to avoid misdiagnosis of no effect when one is present. But it also requires addressing the question of whether the contamination is of sufficiently high bioavailability to cause an ecologically important ecotoxicological effect. A similarly vexing issue is establishing cause and effect in the natural setting, when several potential stressors could cause similar biomarker responses.

The solution to both dilemmas seems to lie in understanding processes in a suite of repeated observations built around understanding the system(s) of interest. Interpretations of ecological

risk are most convincing when supported by a systematic body of evidence consistently linking effects with the same cause. In fish, functional response groups recommended for a suite of observations include: detoxification enzymes, organ dysfunction, histopathology, feeding and nutrition, and condition indices (Adams et al., 1999). In invertebrate survivors, sublethal signs of stress specifically from metals, with the potential for population damage, include metallothionein induction (if it is an energy cost: Farag et al., 1995), lysosomal destabilisation, lipid peroxidation, histopathological lesions, reduced growth or energy balance, gut malfunctions and disruption of reproductive processes. Avoidance of contamination or other behavioural changes can also be important in both groups. Interpretive techniques, including multivariate analyses, can be used to separate responses in a complex data set.

Characterisations of biomarkers (and attempts to link them to higher trophic level processes), however, often undervalue an understanding and estimation of bioavailable metal exposure (Bayne et al., 1988b). A fundamental tenet for demonstrating linkages with signs of disease is understanding and demonstrating exposure to the agent causing the disease (Chapter 11). Biomonitoring, the measurement of accumulated trace metal concentrations in selected organisms as integrated measures of the total bioavailabilities of metal to those organisms over a preceding time period (Chapter 8), is an underused tool that can provide vital information on the geographical and temporal variations in distributions of the bioavailabilities of toxic metals. Biomonitoring is a valuable complement to traditional geochemical data, when it is necessary to focus investigation on bioavailability. It is a similarly valuable complement to biomarker data to address increased probability of metals being the cause of an ecotoxicological effect.

Of course, it is subsequent effects monitoring that has the potential to answer the question whether any toxic effects have actually been caused by the high metal availabilities. But the third great issue is that effects monitoring in nature relies upon determining responses in the survivors of the contamination. Disappearance of a species is the ultimate effect along a gradient in increasing sublethal stress, but that effect precludes determining bioavailable metal from the bioaccumulated concentration in that species. On the other hand, there seems little reason why correlation between level of exposure in a widespread biomonitor and level of stress in other (survivor) species is not a suitable dose–response measure. Further linking to population and community responses could have a similar base. Any relationship between metal bioaccumulation in a biomonitor and biomarker-to-community responses must be calibrated in contaminated systems where it is by definition too late to prevent problems. But extrapolation of such relationships to other circumstances at least holds the potential to be predictive.

In such a context, long-standing arguments about whether biomarkers are measures of exposure or meaningful effects become irrelevant. Environmental chemistry, biomonitoring data, lower-level biomarkers, physiology, behaviour, signs of population response and signs of community response are the tools available to assess and manage risks from metals. The problem is sufficiently complex that we cannot afford to discard the evidence from any of them.

Suggested reading

Amiard, J.-C., Amiard-Triquet, C., Barka, S., Pellerin, J. and Rainbow, P.S. (2006). Metallothioneins in aquatic invertebrates: their role in metal detoxification and their use as biomarkers. *Aquatic Toxicology*, **76**, 160–202.

Galloway, T.S., Brown, R.J., Browne, M.A., Dissanayake, A., Lowe, D., Jones, M.B. and Depledge, M.H. (2004). A multibiomarker approach to environmental assessment. *Environmental Science and Technology*, **38**, 1723–1731.

Hook, S.E. and Fisher, N.S. (2002). Relating the reproductive toxicity to five ingested metals in calanoid copepods with sulfur affinity. *Marine Environmental Research*, **53**, 161–174.

Klerks, P.L. and Weis, J.S. (1987). Genetic adaptation to heavy metals in aquatic organisms: a review. *Environmental Pollution*, **45**, 173–205.

Levinton, J.S., Suatoni, E., Wallace, W., Junkins, R., Kelaher, B. and Allen, B.J. (2003). Rapid loss of genetically based resistance to metals after the cleanup of a Superfund site. *Proceedings of the National Academy of Sciences of the USA*, **100**, 9889–9891.

Viarengo, A., Pertica, M., Mancinelli, G., Palmero, S., Zanicchi, G. and Orunesu, M. (1982). Evaluation of general and specific stress indices in mussels collected from populations subjected to different levels of heavy metal pollution. *Marine Environmental Research*, **6**, 235–243.

10 • Toxicity testing

10.1 INTRODUCTION

Prologue. Managing trace metal contamination requires understanding the concentration dependence of toxicity. Traditional methods for defining that concentration dependence are dominated by toxicity testing. The answers are greatly influenced by choices made in the testing approach. Understanding the assumptions of those tests is critical to using such data to manage metal contamination and evaluate risks.

In a simple world, we should be able to describe the potential toxicity of metals, relative to one another and relative to other forms of chemical contamination. And we should be able to identify the threshold concentration beyond which each trace metal begins to become harmful in natural waters. Scattered work on these questions began in the mid nineteenth century. Studies in the 1930s and 1940s provided the original insights on biological testing and laid the groundwork for direct tests of metal toxicity (Adams and Rowland, 2003). The body of work on metals grew dramatically after the 1950s. Yet today, the question of relative hazard and harmful concentrations remains controversial. These controversies at least partly result from the complexity of the problem, different perspectives on toxicity and the choices of data.

The most widely used platform for developing data to address metal toxicity is the toxicity test. Toxicity testing provides the underpinning for

---

Box 10.1 Definitions

acute toxicity: involves harmful effects in an organism through a single or short-term exposure.

chronic toxicity: the ability of a substance or mixture of substances to cause harmful effects over an extended period, usually upon repeated or continuous exposure, sometimes lasting for the entire life of the exposed organism.

community: a group of species residing in the same location. Community structure is defined by the number of species and the relative abundance of each. Community function is defined by processes that the entire community contributes to (e.g. community metabolism, organic matter breakdown, secondary production, or the biomass production rate above the first trophic level).

dose: the amount of a substance that will reach a specific biological system, and is a function of the amount to which the individual is exposed, namely the exposure, taking account of the fact that some proportion is not bioavailable.

EC_{50}: the 50% effects concentration is actually the concentration of trace metal required to provoke a response halfway between the baseline response (control response) and the maximum response; any endpoint qualifies as an 'effect'.

endpoint: any quantifiable response that can be related to chemical dose or exposure. A *measurement endpoint* is the quantitative responses determined in an experiment study. Usually the measured response is extrapolated to predict some effect at a higher level of organisation that is socially and biologically relevant, the *assessment endpoint*.

exposure: the concentration of the substance in the relevant medium (food, water).

hazard: the potential to cause harm.

LC_{50}: lethality is the endpoint (lethal concentration at 50%); the definition is otherwise the same as the EC_{50}.

LOEC: the lowest observed effect concentration. The lowest concentration for which the effect is significantly different from that of controls. It is the first concentration tested after the NOEC.

MATC: maximum acceptable toxic concentration; the mean of the LOEC and the NOEC.

NOEC: the no observed effect concentration; the lowest concentration in the experiment at which there is no effect or the highest concentration tested for which the effect is not significantly different from controls.

populations: groups of individual organisms that interbreed, or with a substantial amount of genetic exchange (i.e. individuals of the same species in the same location). Such populations are also called demes, and the study of their vital statistics is termed demography.

risk: quantification of the likelihood of harm occurring. Risk is assessed from consideration of the likelihood of exposure, the dose and the inherent toxicity of the substance to which the organisms may be exposed.

subchronic toxicity: the ability of a toxic substance to cause effects for more than one year but less than the lifetime of the exposed organism.

sublethal toxicity: harmful effects that are less than immediately lethal.

toxicity: the degree to which a substance can harm humans or animals.

---

traditional regulatory approaches for all chemicals (not exclusive to metals) and is an important part of many risk assessments. A large base of standardised toxicity test data now exists. Those data established the basic principle that metals can be as toxic in nature as other chemicals (Chapter 2). Toxicity testing data were influential in the management decisions that resulted in improvements in water quality after passage of the Clean Water Act in the USA and equivalent legislation in other developed countries. But modern risk assessments and risk management now appear to need more sophisticated information than the traditional approaches to toxicity testing can supply. How to address those needs is one of the most important questions in managing metal contamination in aquatic environments. In this chapter we explain trace metal toxicity as defined in toxicity tests. We consider the factors that influence the outcomes of traditional toxicity tests, some of which are particularly important for trace metals compared to other chemicals. We also discuss alternative assessments of the concentration at which metals are harmful.

10.2   HISTORY AND GOALS OF TOXICITY TESTS

Prologue. Historically, toxicity tests were expected to be reliable, repeatable and inexpensive so as to screen toxicity a vast array of unknown situations (e.g. effluents, new chemicals). Simple, repeatable controlled tests are best suited for simple goals that demand comparability among treatments, but they compromise relevance to nature and predictability to achieve those pragmatic goals.

Evaluation and management of ecological risks from metals are often built around the question: what is the relationship between the amount of trace metal that enters a water body, and the influence of the resulting contaminant concentration on life in that environment? There are many ways to address this question. Regulatory science relies primarily on data from controlled experiments that directly test toxicity at the level of the individual organism.

Toxicity tests take many forms and can have many objectives. Calow (1992) cited three Rs for toxicity testing: Relevant, Reliable and Repeatable. The original underlying belief was that testing organisms under controlled laboratory conditions was the most direct way to address whether contaminants were affecting natural ecosystems (Adams and Rowland, 2003). It was also felt that toxicity tests were necessary to predict future effects from human inputs of pollutants, rather than just observing effects after they have occurred. Toxicity bioassays are designed to determine the concentration at which a chemical elicits a response in an organism. The exact approach can vary widely, however, depending upon the trade-offs chosen during design. Thus, the bioassay tool can be flexible. It is consistent with the long scientific tradition of controlled laboratory experimentation.

The complexity of the test influences the degree of extrapolation possible and which of the above goals it is best suited to address. The most simple, repeatable bioassays (single species dissolved tests of short duration) are well suited to screen a large number of chemicals for toxicity on a comparable basis. They provide a sound basis to screen natural or effluent waters on a comparable basis. Simple,

repeatable tests are ideally suited to test specific questions about test conditions (e.g. how does the geochemistry of the water affect the concentration that is toxic?). More complex tests are more relevant to conditions under which metals will be toxic to a species or will cause ecological change in a complex natural water body. More complexity is required to predict which species will be affected by a metal and/or how an environment will change in response to a given metal input.

Standardisation allows precision, practicality and repeatability, the most commonly cited advantages of toxicity testing (Cairns and Mount, 1990). Simple bioassay approaches are the least expensive and most amenable to standardisation. Standardised procedures for such tests were first published in the 1950s through to the 1970s. Regulatory guidelines now exist for a variety of such tests with a variety of species (see listings in Adams and Rowland (2003) for example). These are the data with the greatest influence on the regulatory process. The dominant paradigm in such tests was the single species test: testing for dissolved toxicity with adult individuals of a single species for a short duration (Box 10.2).

Box 10.2 History of development of toxicity tests (Cairns, 1989)

In a 1989 foreword to a special issue of the journal *Hydrobiologia* dedicated to bioassay techniques, Professor John Cairns wrote:

Most environmental assessment in the 1940s and 1950s was carried out by sanitary engineers and chemists on purely chemical/physical measurements. Bioassays were considered a luxury or novelty. (At one time) technological solutions for environmental protection, such as best practicable technology (BPT) were thought adequate. In short, if the best treatment system were installed the environment would inevitably be protected . . . Ultimately there was realization that no instrument devised by man can measure toxicity – only living material can be used for this purpose. A period followed of developing more sensitive end points than lethality, extending the duration of the tests,

utilizing more life history stages in the tests, and expanding the number of species routinely used . . . More recently attention has been given to predictive quality of bioassays including extrapolation from laboratory to field conditions, extrapolation from one level of biological organization to another . . ., extrapolation from short exposure times to generational exposure and extrapolation from one set of conditions to another.

Cairns added that 'the correspondence of tests in simple laboratory systems to responses in complex natural systems . . . is a problem that has yet to be resolved'. Cairns' laboratory produced a body of work that characterised the limits to predictions from toxicity testing, while at the same time recognising that all other alternatives also have limits. He wrote in 1989: 'I have never wavered in my conviction that decisions made on the crudest bioassays are more defensible and produce far better results than those made without any bioassay information.'

The databases that have developed from toxicity testing reflect decades of effort; thousands of tests with thousands of chemicals. The Aquatic Toxicity Information Retrieval (AQUIRE) database or the ECOTOX database (http://mountain.epa.gov/cgi-bin/ecotox) of the US Environmental Protection Agency are examples of databases that store these data. Thus extensive descriptions of toxicity exist for virtually every trace metal and tens of species amenable to laboratory testing.

Tests more complex than the traditional standardised approaches have developed more slowly. Because it is not always clear when to insert new scientific knowledge into the regulatory process, the data from some of the more complex tests sometimes seem slow to be applied. The choices made in determining what bioassays were acceptable for regulatory use embedded trade-offs in defining toxicity that are often left implicit. Moving to the next testing paradigms will require overcoming the inertia built into the regulatory system by the massive legacy of investment in single species testing.

10.3   PRINCIPLES OF TOXICITY TESTING

Prologue. Choices in test design are determined by the question that the toxicity test addresses. These choices always involve trade-offs. The trade-offs affect interpretations and extrapolations.

The fundamental principle of the toxicity test is that the response of living organisms to the presence of a toxic agent is dependent upon the exposure level of the agent (Adams and Rowland, 2003). The result is a concentration–response curve, also termed *dose–response* (Fig. 10.1). Response can be expressed in different ways. Cumulative mortality is one common expression (Fig. 10.1); or its inverse, declining survival. As dose increases, mortality (or some other measure) at a chosen time is expressed in an increasing proportion of the population, until all individuals are affected. The typical curvilinear response can be converted to a linear relationship if the y-axis is converted to a log scale, as shown in Fig. 10.1.

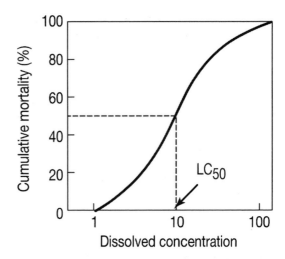

Fig. 10.1 Percent mortality as a function of exposure concentration defines a typical dose–response curve, in this case on arithmetic scales, showing the lethal concentration for 50% of individuals.

Toxicity testing is more complex than it seems at first glance. Every study of toxicity must begin with an explicit definition of the question to be addressed (Box 10.3). Then choices are made to determine the circumstances under which to address the question. Trade-offs are also associated with every choice of methodology or design.

For example, Fig. 10.1 addresses the question: at what concentration (dose) does a metal cause mortality in a population? The response variable (endpoint) for this question is lethality. *Acute toxicity* is lethality at the whole organism level after a short exposure. If the exposure time is lengthened, the results are defined as *chronic toxicity*. If the endpoint is a response less fatal than lethality, the effects are termed *sublethal*. Chronic tests often evaluate sublethal effects on measures such as growth, reproduction, or behaviour. Tests grow in complexity and cost as the more subtle responses are studied over longer periods of time. Thus more data exist for acute toxicity than for chronic or sublethal testing.

Lethal concentration is usually defined from mortality in a chosen proportion of the individuals, in a test over a given time period. For statistical reasons, the centre of the response curve is

Box 10.3 Choices for toxicity tests

Choices must be made in deciding how to test for toxicity. Trade-offs and outcomes associated with many of these choices are more complex with trace metals than with other contaminants. Every choice has an influence on what concentration causes toxicity. Some of the choices include:

• Media
• One species or many
• Endpoint
• Test species
• Life stage
• Time of exposure
• Environmental conditions
• Exposure route.

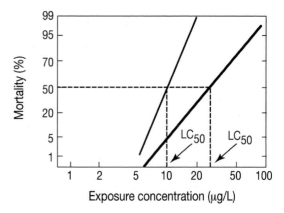

Fig. 10.2 Two different dose–response curves for species with different sensitivies. In this case a typical probit plot is presented (see scale differences with Fig. 10.1). The LC$_{50}$ for one species occurs at a lower concentration than that for the second species, and therefore the chemical is more toxic to species A than species B.

the most reliable single number describing the toxicity of the chemical under the specific conditions of the test. This is the concentration at which 50% of the individuals are affected, over a given time period, or the LC_{50} (the EC_{50}, or effective concentration, is an analogous term if an endpoint other than mortality is used). Figure 10.2 compares LC_{50}s in two organisms with different sensitivities to the metal. If there is a baseline level of response in experimental control animals, the EC_{50} is the halfway point between the baseline and the maximum response.

Often risk assessors or managers want to know the lowest concentration of a chemical that elicits a toxic effect. Thus the lower left-hand end of the curve is interesting from the point of view of environmental management. An LC_{10}, for example, is the lethal concentration to 10% of individuals. Alternatively, some studies define the threshold for toxicity by the lowest concentration at which a detectable effect is observed or the lowest observed effect concentration (*LOEC*). Also of interest is the highest concentration with no effect, or the no observed effect concentration (*NOEC*). Stated in statistical terms this is the highest concentration at which the response is not significantly different from the control. Still other studies use the mean of the LOEC and the NOEC to define the

maximum acceptable toxic concentration or *MATC*. Additional terminology and statistical interpretive techniques for toxicity tests stem from these basic measures (Newman and Unger, 2003).

The advantage of the LC_{50} is that it is statistically robust; this is the portion of the curve with the least uncertainty or variability (smallest confidence intervals). The LC_{10}, the NOEC, the LOEC and the MATC are statistically less reliable than the LC_{50}. Small differences in the lowest-level responses are therefore more difficult to differentiate than small differences in the less protective LC_{50}.

The LC_{10} is also defined as the concentration at which detection of a difference between treatment and control becomes unlikely (Adams and Rowland, 2003). The calculated values are influenced by the dosing regime (Fig. 10.2); i.e. how the test is designed, the range of concentrations and the number of data points within different concentrations ranges (Box 10.4). Thus the threshold that would protect a species can become a point of great dispute in assessments of metal risk (see examples in Chapters 13 and 15), often centred around differentiating small differences at the low end of the curve.

Some regulatory bodies attempt to mandate 'good toxicity test design'; but the lack of flexibility

Box 10.4 Calculation of NOECs and LC$_{50}$s

Data distributions and methods of analysis are important considerations in interpreting toxicity tests. For example US Environmental Protection Agency guidelines for toxicity tests call for six test solutions, ranging from zero (control) to a concentration that will elicit the maximum response. EC$_{50}$ results compare reasonably well among laboratories when such guidelines are followed (results differing only by a factor of two or so: Chapman *et al.*, 1996). Chapman *et al.* (1996) found that the agreement among NOECs in tests of this design were 'decidedly poorer' – differences of five- to ninefold. In a dilution series, the concentration that affects 50% of the test organisms is interpolated (calculated) from the entire data set. In contrast, the NOEC is based upon one value: the highest concentration tested that gives results not significantly different from the control. Chapman *et al.* (1996) argued that the choice of concentrations used in the test will have a larger effect on NOECs than on EC$_{50}$s. Thus, EC$_{50}$s or other point estimates are more consistent, more reliable and less variable estimates than NOECs and can be compared between tests. But, of course, they are not as environmentally protective.

   Figure 10.3 provides a hypothetical example. The LC$_{50}$s and the LOECs for experiment 1 and experiment 2 are the same, but the NOECs differ by 10-fold because several more data points were collected in experiment 2 at the low end of the experimental range. The two experiments would also result in very different LC$_{10}$s because of the choice of exposure concentrations affects the shape of the curve.

   The ideal study design might spread treatments over 10 or more concentrations, and/or emphasise a greater number of replicates around the 'target' effects level. Chapman *et al.* (1996) suggest a design, for example, as follows: if range-finding or other information suggests an EC$_{50}$ at 100 mg/L, and one wanted to determine an EC$_{20}$, then one might have six replicates at 50 mg/L, five replicates each at 25 and 75 mg/L and four replicates at 15 and 100 mg/L. Higher concentrations may have only two replicates. The high concentrations could still drive the statistical interpretation, but the greater data density in the region of greatest interest will give greater interpretive power where it is most needed.

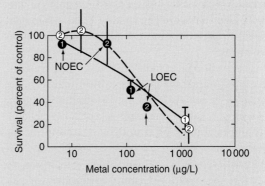

Fig. 10.3 Two typical types of (hypothetical) dose–response curves illustrate the NOEC (no observed effect concentration) and LOEC (lowest observed effect concentration) in two experiments with different distributions of data. Although toxicity, overall, is similar in the two experiments, a lower NOEC is found in experiment 1 and than in experiment 2. (Concept from Kaputska *et al.*, 2003.)

in such mandates can do as much harm as good (Chapter 15). Multiple tests are usually needed to narrow the uncertainty in NOECs, LOECs and LC$_{\geq 10}$s. It is sometimes necessary to recognise that some uncertainties in this area of the curve may not be resolvable within narrow limits.

10.3.1 Media

Prologue. Trace metal toxicity tests are dominated by the use of water as the exposure medium.

Standardised sediment toxicity tests also exist. There is less consensus on standardised protocols to evaluate dietary toxicity. The dissolved metal medium is more amenable to simple testing procedures than other media, but less relevant to nature than are approaches that include other reaction and exposure pathways.

Media are the water, sediment, food, organisms and mixes of these that are included in a toxicity test. Toxicants can be spiked in known concentrations into water only, mixtures of water and sediment or mixtures of food and water. Alternatively, the toxicity of (sometimes poorly known) ambient water/sediment mixtures from an ecosystem may be tested directly or organisms may be translocated to the field to test their responses in ambient conditions (*in situ* toxicity tests). Similarly, prey from a contaminated environment might be fed to a predator to test ambient dietary toxicity (Wallace *et al.*, 2000). The most complex choices of media attempt to simulate an environment (artificial streams), transplant communities of organisms to a natural system (mesocosms) or experiment with an intact ecosystem (experimental lakes). Toxicity tests can also be used in tandem with ecological studies and chemical analyses (e.g. the Sediment Quality Triad: Chapman, 1986) to provide a more complete interpretation of effects.

The traditional standardised test chose water as the most important medium within which to test toxicity. In general, tests for dissolved toxicity are the most feasible to achieve cost-effective reliability, repeatability and simplicity of operation. For metals, this was an important choice because it allowed definition of consistent conditions to control the chemically complex processes of partitioning (between particulates and water) and speciation (within the water column). These are the tests that were used to screen chemicals and establish the toxicity of metals compared to other chemicals. The controversies in the use of dissolved-only tests come when the data are used to predict what concentrations would be toxic in a water body, where the medium of exposure is affected by factors such as speciation or transformation to particulate forms; and where routes of exposure may involve more than one medium (e.g. dietary as well as dissolved exposure).

Other choices of media, like spiked food or spiked sediments, add complexity to a test. At least some are less feasible for uses that require many tests to be done quickly and inexpensively. Some control of experimental conditions is sacrificed as sediments or food are added. On the other hand, the results of test with media more complex than just water can contain more information relevant to nature than do simple dissolved tests.

10.3.2   One species or many?

Prologue. Single species toxicity tests are relatively easy to manage, control and standardise. Data exist for a number (tens) of species. Repeatability, reliability and flexibility are the strength of this approach. Unfortunately, no amount of single species data can capture environmental sensitivity if key species are untested. Single species data do not represent the interactions among species that can be a critical component of how communities respond to metals.

Single species bioassays are the most widely used tests in ecotoxicology. They test lethality or sublethal stress in adults or in individuals from a single life stage. It has long been argued that single species tests are the best way to meet the ideal characteristics of a bioassay designed to meet management requirements: 'a response predictive of organisms in the . . . system to be protected', 'a clear end point for the bioassay', the test organism must be easy to cultivate and assay, and it must have 'a satisfactory response to all relevant toxicants' (Blum and Speece, 1990).

Thus, the greatest advantage of single species tests is that they are pragmatic. This is the largest single group of tests in the toxicity testing database. These dependable tests are responsible for many successes in screening large numbers of chemicals over a reasonable time span and for reasonable costs (Calow, 1992).

Calow (1992) argued 'repeatability must have primacy of relevance and realism', in defending standardised single species tests. That statement defines the most controversial aspect of the single species toxicity testing approach. Antagonists argue that single species tests are not protective of all species, are not designed to measure natural population and community responses and do not address issues of contaminant bioaccumulation (La Point and Perry, 1989). Protagonists suggest they are the most flexible, powerful and pragmatic choice for an experimental approach.

Batteries of single species tests can be conducted to evaluate simultaneously responses of several different types of test species, but maintain the advantages of testing one species at a time.

Even using the single species approach, data are now available for a variety of species under a variety of conditions, although the richness of the suite of available data varies from metal to metal and between freshwater and seawater. The ability to accumulate data sets for many species counters some the limits of the individual experiments.

Multiple species bioassays allow study of responses to metal exposure at higher levels of biological organisation, but still in a controlled environment. Multi-species tests range from a microcosm of meiofauna in a beaker to a manipulated lake (Box 10.5). A few species can be introduced to the system, or a natural community can be enclosed, manipulated or transplanted. The goal of the experiment is to hold the complex system constant as different concentrations of metal are added.

For the closest simulations of nature, multi-species bioassays should include multiple pathways of metal transfer and the influences of ecological processes like immigration, emigration, predator–prey interactions, competition or changes in nutrition (Cairns, 1986). The duration of the exposure should be sufficient for lags in feedback processes to become operative. Multi-species bioassays are also especially well suited for studying how effects of metals on some species in a community can cause secondary effects on other species in the community. Well-designed multi-species bioassays have many similarities to nature. They are most practical to accomplish when the target organisms are sufficiently short lived that it is feasible to study several generations.

Multi-species tests often trade off economy, simplicity and control in order to offer understanding of complex ecological processes. One cost of realism is that many important variables or interactions (especially biological) are difficult to control (Box 10.5). Multi-species tests can yield unexpected results, and sometimes even widely different outcomes between replicates of a treatment. It can be difficult to reproduce the same effect in separate experiments. Another basic problem is that elaborate multi-species test systems cannot, pragmatically, incorporate very many replications of treatments. It can therefore be difficult

to separate an effect from variability. Decisions must be made about the mix of species to employ, and this can influence results.

These disadvantages can be overcome by reducing scale. Control is thereby improved, replication can increase, and treatment effects become more feasible to detect. Increasingly, important results are coming from the investments in this approach to testing. But, as with every choice, there are trade-offs involved. Outcomes with small species (e.g. microbial or meiofaunal communities) are difficult to extrapolate to the larger charismatic species of greatest regulatory concern. The replication of nature is also less than complete in the absence of predation and migration which greatly affect such assemblages in nature.

Multi-species tests generally provide unique knowledge of how metals affect natural water bodies; knowledge that could not have come from single species tests. Two important generalisations that could only come from such tests are:

- Toxicity and ecological change come from ecological reactions more complex than mortality to individuals; and
- When interactions of multiple species are considered, unexpected changes occur and toxicity is often observed at concentrations lower or higher than predicted from single species tests (Box 10.4).

The appropriate uses of multi-species tests depend upon the question the investigator wants to answer. They are less simple, less repeatable and typically have an abbreviated dose–response curve compared to single species tests. Thus multi-species test outcomes are often not considered in deriving regulatory standards (Calow, 1992), even though the insights they generate could be quite valuable at least as context. Standardised multi-species testing protocols are not likely to be appropriate to screen large numbers of chemicals on a routine basis. But only multi-species tests can help us understand the important secondary interactions among species initiated by the toxicity of metals (Box 10.5). Assessments of risks (and risk management) cannot afford to not understand these interactions.

Box 10.5 Multi-species testing in defining the toxic concentration of Cd

Two different mesocosm approaches (in a whole lake and in a beaker) demonstrate Cd toxicity at a low concentration (<0.5 µg/L) and illustrate that complex ecological reactions, not predictable from single species exposures, can accompany influential metal exposure.

Cairns et al. (1986) demonstrated differences in sensitivity between simple measures of mortality and integrated community processes. They tested the effects of Cd exposure on the colonisation of an uninhabited surface by meiofauna, microscopic biota that occupy surfaces and the interstices of sediments. The potential sources of toxicity that could affect colonisation included mortality in the source community, impairment of reproduction, disruption of migration, poor survival of migrants, changes in competition among migrants and disruption of succession on the uninhabited surface. A source community of protozoans was collected by allowing them to colonise plates (colonising surfaces) in a pond. The plates were then transferred to a series of Cd solutions, and movement from the colonised surfaces onto new surfaces was studied. The tests were run for 3 to 4 weeks. The meiofauna reproduced at rates approximating once per day. Thus the processes were studied for up to 20 generations. The number of species that migrated onto the new surfaces was extremely sensitive to Cd. The highest concentration with no effect on colonisation was 0.4 µg/L Cd, and colonisation was reduced by 20% at 0.9 µg/L Cd. In contrast, a 10 µg/L Cd exposure was necessary to reduce the number of species in the source community. Thus the processes involved in colonisation were about 10-fold more sensitive to Cd than was direct lethality. Conducting the study with very small organisms allowed treatments and controls to be replicated. Multiple generations could be studied, illustrating the importance of subtle, but important, processes like immigration. On the other hand, working with meiofauna probably reduced the influence of the work, because it did not directly address the welfare of the larger species usually considered most important to conserve. These studies are rarely cited in regulatory documents, but they teach us a great deal about both the sensitivity of nature to Cd and the processes by which effects might be manifested.

It is interesting to compare the Cairns et al. (1986) mesocosm study to a whole lake experiment, also conducted with Cd (Lawrence and Holoka, 1991). Experimental Lake 382 in the Canadian experimental lakes district was injected with Cd to test if the Canadian water quality standard for this metal would protect an ecosystem. Sufficient Cd was added to the lake to reach dissolved concentrations of 0.05 to 0.20 µg/L Cd. The Canadian water quality standard for Cd is 0.2 µg/L. Mesocosms were also studied: natural communities of zooplankton impounded in flow through systems injected with 0.4 to 3.0 µg/L Cd.

The most sensitive response was a change in the assemblage of species in both the lake and the mesocosms. Some species disappeared (suffered some form of toxicity) while others did not. The biomass of one type of crustacean zooplankton, the cladocerans, was reduced more than that of calanoid copepod crustaceans (Lawrence and Holoka, 1991). The latter were more sensitive than cyclopoid copepods. The abundances of sensitive species of zooplankton, the cladocerans *Daphnia galeata mendotae* and *Holopedium gibberum*, were reduced at 0.2 µg/L Cd. One species, *H. gibberum*, was reduced or eliminated at the lowest Cd exposures (0.1 µg/L) in all three studies. These effects occurred at lower concentrations than would be predicted in single species bioassays, especially under the geochemical regime of the lake. Nor could single species bioassays predict the change in species assemblages, since most species had never been tested individually.

Replication of the whole lake experiment was not feasible, however; although the mesocosm results were replicated. One of the disadvantages of this inflexibility was that elevated levels of dissolved organic carbon (DOC) that were typical of the lake could not be varied or controlled. The authors argued that high DOC in Lake 382 could have ameliorated additional toxic effects of Cd; in other words effects at lower concentrations might be expected in lakes with less DOC. The authors suggested that the Canadian Water Quality Guideline for Cd would not ensure the protection of the most sensitive aquatic organisms in lakes where DOC is low (like similar lakes on the Precambrian Shield in eastern Canada: Lawrence and Holoka, 1991). But they were not able to demonstrate that experimentally because of the limited flexibility of the approach.

10.3.3   Endpoint and level of organisation

Prologue. 'The simplest toxicity tests require . . . the most complex methods for translation into prediction of environmental outcomes. The most complex test methods measure endpoints that are most closely related to management (assessment) goals and are more simply translated into site-specific effects . . . (but) toxicity tests with few components (*sic*, simple endpoints) provide simpler and less expensive tools . . .' (Cairns, 2003).

10.3.3.1 Measurement, assessment and performance
Prologue. Measurement endpoints from a quantitative study of toxicity must be relevant to assessment endpoints used to derive regulatory criteria.

In a toxicity test, an *endpoint* is any quantifiable response that can be related to chemical dose or exposure. Measures of effect, or *measurement endpoints*, are defined as the quantitative responses determined in the experiment or the study; and these can take many forms (Table 10.1). But usually the measured response is extrapolated to predict some effect that is socially and biologically relevant (Suter, 1990). This is termed the *assessment endpoint*. Assessment endpoints justify environmental policies and public values. Assessment endpoints must be clear to decision makers and the public, and they require real units of actual environmental properties. Cairns (2003) described the desirable characteristics of an ideal endpoint (Box 10.6). Norton *et al.* (2003) presented some examples of assessment endpoints from the regulatory perspective:

- extrapolations between taxa (e.g. using experimentally derived bluegill mortality to infer rainbow trout mortality);
- extrapolation between temporal scales (using 96 hour exposures to infer effects from lifetime exposures)
- extrapolations between responses (using lethality tests to infer effects on growth or reproduction);
- extrapolation from laboratory to field (using bluegill mortality in the laboratory to infer bluegill mortality in the field);
- extrapolation between geographical areas (where, for example, water chemistry may be very different).

Both lists suggest that application of toxicity test data to management requires the assumption that the data can be extrapolated broadly. Extrapolation is fundamental to connecting measurement and assessment, and to using toxicity testing data for risk assessment or risk management.

A long-standing debate exists about whether data from natural waters provide appropriate endpoints for risk assessments. Suter (1990) suggested

Table 10.1 *Some examples of endpoints that might be used to address the implications of metals in a hypothetical metal-impacted river (for real world examples, see Chapter 12)*

Assessment endpoint	Performance measure	Measurement endpoint	
		Metal (Cu) toxicity	Field
Sustained production of native trout	Bull trout density; cutthroat trout density	LC_{50} of Cu to bull trout and cutthroat trout in water and diet	Bioavailable Cu in river water and food web; biomarkers in trout
Eliminate toxic sediments	Determination of bias-free Cu concentration in sediment	*In situ* toxicity of sediment to benthic indicator species	Cu exposure of benthic fauna; recovery of benthic community
Sustain biodiversity	Diversity of benthic and fish communities	LC_{50} to most sensitive species of benthos and fish (water and food)	Presence/absence of Cu-sensitive species in nature

Box 10.6 Some desirable characteristics for endpoints in a toxicity test (after Cairns, 2003)

Ideally a test should also be cost-effective, diagnostic (specific to the chemical of interest) and integrative of other unmeasured endpoints; but these come with larger trade-offs that must be considered with each test.

- *Relevant*: to the well-being of the organism(s) being tested;
- *Sensitive*: to the hazard being tested;
- *Feasible*: to measure with the facilities available;
- *High signal-to-noise ratio*: response is detectable and not hidden by large variability;
- *Interpretable*: can distinguish acceptable from unacceptable condition;
- *Anticipatory*: provides indication of harm that can be extrapolated to other situations or warns of serious harm to come;
- *Appropriate scale*: to the management goal;
- *Timely*: relative to the need for feedback information;

- *Potential for continuity in measurement over time*: is important if measurement endpoint is to be tied to performance.

Cairns (2003) also listed some other characteristics that are ideal. However, these come with trade-offs that may or may not be acceptable to the goal of the test.

- *Socially relevant*: sometimes measurement endpoints do not need to be directly relevant to social decisions, but they must be justifiable on the basis of the information that they provide about assessment endpoints or performance measures.
- *Broadly applicable to many hazards and sites*: the former is an especially difficult requirement.
- *Historical data are available*: ideal if tied to a performance measure, but not always feasible.
- *Cost-effective*: saving in short-term costs can lead to more expenditure in the long term. If the problem is important enough and the stakes are high enough, choosing the endpoint most likely to yield interpretable results is a better choice than choosing an inexpensive endpoint with too many trade-offs in interpretation.

that field observations are 'peripherally concerned about causal relationships, while risk assessment is devoted to elucidating causal relationships'. But Suter (1990) goes on: 'As a result, risk assessments may use the results of monitoring studies, but only after disaggregating the indicators to their components and choosing those that are appropriate.' Equivocation about if and how field-derived endpoints should be used in a regulatory context continues to have important implications for protective criteria, risk assessments and risk management, as we will see when discuss those criteria in Chapter 15.

More recent analyses conclude that measurements from nature and the laboratory are inseparable in a risk assessment (Burton *et al.*, 2002). Assessment endpoints must be accompanied by a measure of performance. Performance measures must come from observations in the field. A toxicity test is one way to address how a specific stressor (e.g. a metal) affects that performance measure. But the relevance and validity of the toxicity test depend upon related field observations. Together

these legitimise policy much more strongly than any one approach alone.

Schiller *et al.* (2001) evaluated endpoints most positively received by the public. They found that these were descriptions of the kinds of information that various *combinations* of indicators provide about broad ecological conditions. Descriptions that respondents found most appealing contained general reference to both the set of indicators from which the information was drawn and aspects of the environment valued by society to which the information could be applied.

In our discussion of toxicity we will focus on measurement endpoints. A measurement endpoint can be drawn from any level of organisation in the hierarchy of effects. The proper endpoint is determined partly by what the test is trying to accomplish (the question). For example the majority of toxicity tests are conducted at the level of the individual because it is relevant to population-level effects, and cause and effect can be easily demonstrated (Maltby, 1999) in addition to the advantages of control and repeatability.

10.3.3.2 Traditional lethality tests

Prologue. Acute toxicity is a scientifically valid approach to evaluating toxicity in that it directly tests the *question* of toxicity, but it is difficult to justify scientifically as an assessment endpoint for long-term metal inputs.

From a precautionary point of view, it would seem that acute mortality goes directly to the question of the most obvious adverse effect of a pollutant. Lethality or mortality is also intuitively simple and the response is visible and unambiguous. Directly testing for toxicity eliminates the necessity of considering many poorly understood complexities about the biology of the organism and the behaviour of the metal. Experience proves that mortality is repeatable, precise and reliable in well-run tests. The shorter the test, the more repeatable are the results (less time for the biota to misbehave). Thus *acute toxicity* is the most common lethality endpoint; defined by extremely short exposures (48 to 96 hours). The exact choice of exposure time was derived for the convenience of the investigator (4/5 of a working week), rather than on any biologically justifiable basis (Cairns and Mount, 1990).

Despite its advantages, it is widely accepted that mortality as an endpoint is a somewhat crude measure of toxicity; and acute toxicity is especially problematic. In nature, acute lethality to adults is relevant when pulse inputs of metals occur, or after dramatic spills of heavily contaminated waste. In updated regulations and risk assessments, acute testing data are used to regulate acute die-offs, by limiting short-term exposures to acute levels. But acute episodes of toxicity are probably rare events. Unless they occur repeatedly, they will not have long-lasting effects. It is probably much more common that metals permanently eliminate species by much less visible means, even in environments where acute events might periodically occur. Reduced growth rates, inhibition of an aspect of reproduction or behaviour changes can result in a reduced ability to replenish the population, inability to compete with other species or inability to escape predators (Chapter 9). These signs of reduced fitness do not leave dead bodies behind as evidence, but the reduced fitness is just as effective as mortality in ultimately exterminating the population.

Despite its limitations, the acute toxicity database is the largest data set available on metal toxicity. Therefore these data continue to be used widely, usually with 'corrections' applied to make up for the insensitivities.

10.3.3.3 Sublethal responses

Prologue. Sublethal responses occur at lower concentrations than lethality. Such data are critical to assessing risks from metals. Endpoints like growth, feeding rate, reproduction and behaviour depict responses that eventually could eliminate species in nature. Non-traditional tests (e.g. dietary toxicity) may be necessary to obtain realistic dose–response data on at least some of these responses.

Ultimately, adverse effects from metals are expressed in a cascade of responses: biochemically within individuals, in whole organisms, in populations and communities (Chapter 9). In theory, sublethal toxicity tests attempt to identify the metal exposure at which the first sign appears of metal-imposed stress. The most common standardised sublethal toxicity tests focus on influential responses to metals at the level of the individual: growth, reproduction and sometimes behaviour. It is well known that sublethal responses occur at lower bioavailable metal exposures than lethality (e.g. Table 10.2, after Milani *et al.*, 2003). The sensitivity of sublethal endpoints is what suggests that organisms probably succumb to metals via stresses that are more subtle than acute lethality.

Recent attempts to understand dietary toxicity has expanded uses of the traditional sublethal endpoints. For example, trace metal effects on reproduction are an important endpoint in dietary toxicity tests. Reproductive effects can be tested with dissolved metals in small animals (daphnid zooplankton), but the most realistic tests include contaminated diets (e.g. Hook and Fisher, 2001a; Chapter 11). For example, after dietary exposure to Se (Chapter 13), inhibition of hatching success

Table 10.2 *A comparison of EC$_{25}$ in lethality and growth tests across a range of Cd exposures, for three species of freshwater invertebrates commonly used in toxicity tests: the amphipod* Hyalella azteca, *the midge* Chironomus riparius *and the mayfly* Hexagenia sp. *Tests were run for 28 days. Growth was more sensitive than lethality in all cases, but reproduction was less sensitive than either response (data not shown). Effective concentrations are shown based upon bulk sediment concentration, overlying water concentration and pore water concentrations, of cadmium*

| | Hyalella | | Chironomus | | Hexagenia | |
Media	Lethality	Growth	Lethality	Growth	Lethality	Growth
Sediment (µg/g Cd)	21	10	28	16	560	14
Overlying water (µg/L Cd)	23	1.0	3.3	1.4	480	3.2
Pore water (µg/L Cd)	6.9	1.0	18	5.6	790	9.6

Source: Data after Milani *et al.* (2003).

in bird or fish eggs and deformities in young animals are readily demonstrated. In nature, where these sublethal signs of Se stress are evident, populations suffering such effects eventually disappear. Reduced feeding rate is another endpoint that is specific to many metals (Decho and Luoma, 1996; Woodward *et al.*, 1995; Chapter 11 (Box 10.7), but can only be observed in dietary toxicity tests. Feeding rates have implications for growth of individuals and eventually for populations (Maltby, 1999).

Sublethal endpoints also have trade-offs. Biomarker responses are sensitive and relatively easy to detect, but extrapolation to a process that threatens the population is sometimes difficult. A whole field of study exists to understand metal responses at the lower levels of organisation (Chapter 9). But regulatory documents rarely cite dose–response relationships from such tests. For example, determination of metallothionein response is usually of great interest in hypothesis testing and field assessments of metal effects, but is rarely influential in derivation of regulations or risk assessment. Sublethal tests are also more complex to conduct than acute toxicity tests because they require longer exposures, more sophisticated physiological measurements or consideration of multiple exposure routes. The data are critical to interpreting risks from metals, but fewer acceptable sublethal toxicity test data exist than acute data (e.g. see USEPA, 2002c).

10.3.4   Test species

Because finances limit the number of bioassays that can be run, effective surrogate testing organisms must be found whose responses can be related to target organisms or ecosystems ... ecosystem toxicity (testing) is not a reasonable option, surrogate organisms are required ...
Blum and Speece (1990)

A number of assumptions are inherent in the most sensitive species approach, although these are not always explicitly stated. [It is assumed that]:

- ... the response of the most sensitive species at least partly corresponds to that of a much larger array of exposed organisms in natural systems.
- There are no significant responses ... that are more sensitive than the endpoints chosen for the most sensitive species toxicity tests.
- Savings that result from using the 'most sensitive species' approach are not offset by the cost of making bad management decisions.
- A species shown to be most sensitive to a limited array of toxic substances will invariably be so for a much larger array of toxic substances.
Cairns (1986)

The single species toxicity test was developed around the assumption that toxicity can be tested against a limited number of standard test organisms, and the results will be representative of a

Box 10.7 Effects of cadmium on feeding and growth in mayflies

Baetis tricaudatus (Ephemeroptera: Baetidae) is an ubiquitous mayfly in North American aquatic systems. Irving *et al.* (2003) investigated the response of this grazing nymph (insect larva) when feeding upon periphyton spiked with either 4 or 10 μg/g Cd. Dietary toxicity was investigated in the short term, following the addition of Cd-contaminated food, and then during a substantial portion of the *B. tricaudatus* life cycle. The mechanistic response that the authors studied was feeding inhibition. The response endpoint that they measured was reduced growth. Short-term feeding experiments indicated that *B. tricaudatus* nymphs did not initially avoid grazing on Cd-contaminated diatom mats. But after 13 days, the feeding rate of the mayflies and their growth were reduced in the 10 μg/g Cd treatments (Fig. 10.4). Feeding inhibition was the likely mechanism that inhibited growth (i.e. through reduced energy intake). Metal inhibition of growth affected the size of the mayfly when it emerged from its pupal stage (Irving *et al.*, 2003). The reproductive capabilities (fecundity) of adult mayflies (*B. tricaudatus*) are determined by size of the pupae. Ultimately, the population size of the mayfly is eventually affected by the reduced input of young animals to the population. Models also suggested that the fitness of predator species would be affected by the loss of mayflies from the prey community.

The sublethal response to Cd was more sensitive than lethality. When exposed to waterborne cadmium using lethal toxicity test procedures, *B. tricaudatus* nymphs were relatively tolerant (96 hour median lethal concentration 1611 μg/L). If a K_d of 1000 is assumed for Cd, the concentration in water that would result in a periphyton concentration of 10 μg/g would be 10 μg/L.

This concentration is 100-fold lower than the dissolved concentration that causes lethality. Thus the sublethal effect on feeding, and eventually effects on the population, occurred at a much lower exposure concentration than lethality from dissolved Cd.

Irving *et al.* (2003) used an exposure approach more relevant to nature than the typical testing procedures. It demonstrated a much greater sensitivity to Cd via the sublethal response than suggested from the dissolved concentration of Cd that was lethal. The study also demonstrated the mechanistic cause of the sublethal response. Relating effects on grazing rate and growth to population responses provided a convincing tie between the organism-specific response and an assessment endpoint/performance measure of relevance to risk assessment.

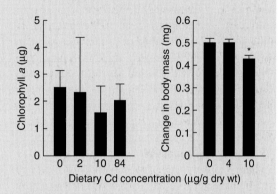

Fig. 10.4 (a) Weight of algal biomass eaten by the mayfly larva *Baetis tricaudatus*, as a function of different Cd concentrations in the algae. (b) Change in body weight of *B. tricaudatus* after 13 days feeding on algal mats with different concentrations of Cd. (After Irving *et al.*, 2003.)

larger group of other organisms (surrogate species). Responses in the most sensitive of these surrogates would then be used to establish a lower limit of metal exposure that would protect other species. Cairns (1986) described these early requirements as the 'myth of the most sensitive species' (see quote above). Much has been learned about metal toxicity, in particular, since Cairns' proclamation of skepticism about the concept:

10.3.4.1 Species differ widely in their susceptibility to metals

It is clear from the toxicity testing database that there is great variability in toxicity testing outcomes when different species are employed. Acute toxicity data for Cd, produced in acceptable standardised tests, are available for 59 freshwater species and 54 saltwater species (USEPA, 2000). Among the freshwater species, at a hardness of

50 mg/L, the acute toxicity of Cd to a salmonid fish (*Salmo* sp.) was 47451 times lower than the acute LC_{50} to an aquatic insect from the midge genus *Chironomus*. The mean acute value for salt-water invertebrates ranges from 41 µg/L for a mysid (zooplankton) to 135000 µg/L for an oligochaete worm. A similar database is available for Cu. In freshwater, LC_{50} values for acute lethality of Cu range from 5 to 20000 µg/L among species (Landner and Reuther, 2004). Chronic NOEC values range from 2 to 20000 µg/L, although the ranking of species is different between chronic and acute tests, and the chronic testing database is smaller.

It is unclear how much of the variation in toxicity among species is due to inherent differences among species and how much is due to differences in test conditions (probably some of both). For example, daphnids (freshwater cladoceran crustacean zooplankton) tend to be sensitive to at least some metals and are quite amenable to laboratory experimentation. They are one of the organisms most widely used in single species bio-assays. But different clones of the same species (*Daphnia magna*), tested under similar conditions, had 48 hour acute LC_{50}s for Cd that ranged about 30-fold, from 0.8 to 25.8 µg/L (Baird *et al.*, 1989). Thus, biological variability is clearly an issue.

Some authors cite geochemical differences in test conditions as another cause of variability among species (Landner and Reuther, 2004). For example, one species of fish (rainbow trout, *Oncorynchus mykiss*) had 30 day LC_{10}s for Zn that ranged about 30-fold, from 38 to 900 µg/L, in tests under which geochemical conditions varied in unspecified ways.

In addition, it is possible that time of exposure causes differences in response among species. Toxicity is partly determined by the duration of exposure relative to the generation time (lifetime of a generation). For example, 96 hour toxicity tests with phytoplankton species cover several generations of individuals because phytoplankton replicate rapidly. Tests for 96 hours with a small organism with a short lifespan, like *Daphnia*, also cover a percentage of the generation time. An exposure of 96 hours to a bivalve or a fish that lives for years is trivial compared to the time that species would be exposed to a metal in nature.

10.3.4.2 Hardiness of test species might affect their representativeness

Many of the species well suited for reliable, repeatable, standardised laboratory testing are inherently hardy organisms. Hardiness may enhance their ability to tolerate contaminants. Thus sensitivities among the most common test species may not be representative of at least some of the most sensitive species in natural waters.

10.3.4.3 At least some test organisms can only be tested in narrow range of conditions

Most organisms are limited to life in a narrow range of conditions. If natural waters or sediments are to be tested, species with narrow requirements are unlikely to yield representative results. Species from one range of conditions will not reflect toxic responses in species that live in a different set of conditions, especially if the test conditions are stressful to one or the other (Box 10.8).

10.3.4.4 If sufficient data are available, statistical tests can help select the most sensitive species

Where a large database from multiple species is available, sensitivities (e.g. EC_{50}s) can be displayed as a cumulative frequency distribution. From the distribution it is possible to either determine the number of (tested) taxa affected by any concentration of the metal, or, more commonly, determine the concentration at which a given proportion of species is affected. The US Environmental Protection Agency's choice for regulation is to protect 95% of species. Then, the concentration is derived graphically at which the most sensitive 5% of the species are affected (Fig. 10.5).

The number of species for which data are available is typically a very small sample of nature, even in the largest databases. The availability of data from different species also differs among metals and media. The choice of test species is thus crucial to the presumption that species representative of the lower range of sensitivities in nature were actually tested (Box 10.9). The outcomes of species

Box 10.8 Most sensitive species in sediment bioassays

In sediment bioassays, the test species is exposed to a mixture of contaminated bed sediment and water. In developing the methodologies, it was presumed that the test organism should be from one of the important groups of animals that live in uncontaminated sediments and should be pollutant sensitive. Phoxocephalid amphipods were found to be extremely pollutant sensitive, and also amenable to laboratory culture and manipulation. The best-developed amphipod sediment bioassay used a species common on the west coast of North America, *Rhepoxynius abronius* (Swartz et al., 1988). Experimentation, however, showed that *R. abronius* was indeed sensitive to pollutants, but also was sensitive to salinities outside a narrow range and would often die if the particle size of the sediments was not within a narrow range. Many sediments fall outside both ranges. Therefore, the pollutant sensitivity of *R. abronius* was not useful for applying to organisms that lived outside

that range of conditions (because those conditions might also affect contaminant bioavailability). In fact, studies with *R. abronius* yielded falsely positive predictions of contaminant effects in some of the early studies.

Bioassays with amphipods that could tolerate a wider range of environmental conditions were developed to address conditions where *R. abronius* could not be used. For example, *Eohaustorius estuarius* tolerates a broader salinity range and a broader range of particle types than *R. abronius*. However, *E. estuarius* is more tolerant to Cd than *R. abronius*, i.e. it is not the 'most sensitive species' to all toxicants.

Diverse biological sensitivities to natural conditions (and sometimes poor knowledge of these sensitivities) make it difficult to extrapolate pollutant sensitivity under one set of conditions to other conditions and other species. Species that are thought to be universally 'most sensitive' may be protective surrogates only for other species that live under similar conditions; extrapolation across conditions can be a source of uncertainty.

sensitivity analyses thus depend upon what species are available (or chosen) for toxicity testing (see also Forbes and Calow, 2002); and relevance to nature depends upon what species are not represented.

10.3.4.5 Some groups of species in different types of habitat appear to be highly sensitive to metals

The differences in sensitivities among species appear to have some systematic basis (see Chapter 11). Observations from nature and multi-species tests have identified some mayfly larvae as especially sensitive to Cu, Cd and Zn in streams (Winner et al., 1980). Observations from metal-contaminated lakes, and mesocosm studies in those lakes (Marshall and Mellinger 1980), first pointed out the extreme Cd sensitivity of some zooplankton, including those from the cladoceran genera *Daphnia* and *Ceriodaphnia*. Where trophic transfer is important (e.g. Se and methyl mercury) the most vulnerable species are upper trophic level species. For Se, bioaccumulation by the predators' choice of food determines the ultimate degree of exposure to Se. Predators that prey on bivalves, for example,

are exposed to more Se than predators that eat crustaceans (Stewart et al., 2004; Chapter 13). Better knowledge of species-specific vulnerabilities in nature provides a systematic basis for focusing efforts in toxicity testing.

10.3.5 Life stage

Considerable improvements in the sensitivity (of bioassays) may be achieved by employing the most sensitive stages of the life cycle (e.g. embryos and larvae).

Wright and Welbourne (2003)

Prologue. Development of eggs, embryos and larvae, and recruitment of those into adult populations, is often the most sensitive of the life processes to metals. Defining the sensitivities of these stages often requires understanding detailed aspects of the life cycle. Early life stage tests exist for a few species, although not all are especially sensitive. For all but a few species such tests are difficult to implement effectively at the present state of knowledge.

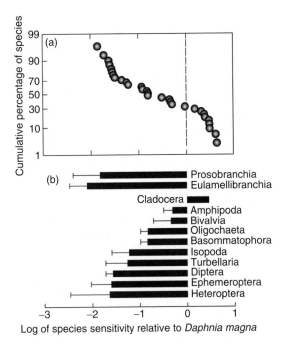

Fig. 10.5 The cumulative ranking of species that have been tested for Cd toxicity plotted against sensitivity compared to *Daphnia magna*. (a) The species that are the least sensitive to Cd compared to *D. magna* are at the upper end of the curve (left). The species that more sensitive than *D. magna* are at the lower end (right). Positive numbers signify greater sensitivity. About 30% of species are more sensitive than the daphnia. (b) Sensitivity ranking of specific kinds of organisms. (After von der Ohe and Liess, 2004.)

Toxicity at any point in the reproduction, development and maturation of a species will determine the fate of the population (Fig. 10.6). Vulnerabilities to metals within life cycles are well enough known for it to be possible to state that some life stages are more sensitive than others, although it is not always clear which life stages are most sensitive for which species or for which metal. For example, an analysis of the toxicological literature on Cd concluded that no conclusion could be drawn about the life stages that were most sensitive to that metal (USEPA, 2002c). But some generalisations are available to lead choices about life stages in toxicity tests.

In general, the adult is often described as the least sensitive of the life stages, a conclusion usually consistent with acute toxicity data. But

Box 10.9 Ranking of species susceptibility to all metals in freshwaters

Because *Daphnia magna* is widely used as a model organism in toxicity tests, von der Ohe and Liess (2004) tested whether this species was, indeed, the most sensitive species to all metals. They compared the ranked sensitivity of other species with those of *D. magna* for 37 different metals and metal formulations (Fig. 10.5). The endpoint that they considered was the lethal concentration (LC_{50}) value. Their comparison was restricted to freshwater laboratory tests with an exposure duration of 24 and 48 hours. Data from 672 tests were available from 106 publications. They found that 30% of the investigated taxa were more sensitive than *D. magna*. The most sensitive 10% of taxa were at least fourfold more sensitive than *D. magna* to metals. The least sensitive taxon was a midge, *Chironomus thummi*, and the most sensitive was a species of *Ceriodaphnia*, another cladoceran. Thirteen taxa belonging to the Crustacea were the most sensitive, whereas the five least sensitive taxa were members of the Insecta. They noted that these rankings did not necessarily apply to chronic tests.

sublethal effects on adults can occur at low metal concentrations and have important implications for future generations (i.e. affect the ability to propagate). Effects on reproduction, defined by exposure of the reproductively active adult and evaluated as the number of eggs per female, are quite sensitive to metal effects in zooplankton, for example (Hook and Fisher, 2001a; De Schamphelaere and Janssen, 2004).

Conventional knowledge holds that gametes (eggs and sperm), embryos and larvae are the sensitive life stages in aquatic organisms. The most sensitive animals in the Cu toxicity testing database for marine waters are the early life stages of bivalves, although bivalve adults are among the least sensitive species to metals in standardised tests.

There are theoretical reasons to expect reproduction and development of early life stages to be particularly vulnerable. But the point in the life cycle when the test is conducted seems very important (Box 10.10). In early development, damage of

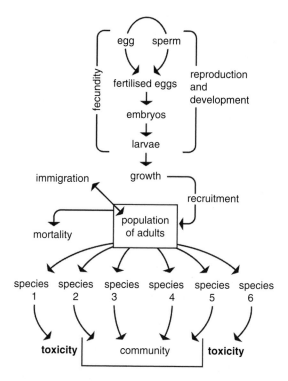

Fig. 10.6 Steps in reproduction and development that will ultimately affect the abundance of individuals in a population and the number of species in a community. Reducing the success of any stage in reproduction or development will ultimately affect fecundity (the potential to produce young) and recruitment (the addition of individuals to the population). Reducing recruitment to a population can cause extirpation as readily as increasing mortality. If some species are eliminated by a toxin, the richness of the community suffers.

a small number of cells can be extremely important, because each cell will eventually become many cells.

The rank order of toxicity among young life stages also differs between species, where tests are available:

- directly exposed sea urchin sperm cells (*Paracentrotus lividus*, the most common sea urchin in the Mediterranean Sea): Hg > Cu > Zn > As > Cr > Cd > Pb > Ni;
- embryo development for *Paracentrotus lividus*: Hg > Pb > Cu > Zn > Cd > Ni > As > Cr;
- development of abalone (*Haliotis* sp.) from fertilised egg to larvae: Cu > Hg > Zn >> Cd > Pb (Novelli *et al.*, 2003; Gorski and Nugegoda, 2006).

Box 10.10 Toxicity depends upon when in the developmental cycle the test is conducted

Development proceeds in progressive stages and this is likely to influence when and how metal effects are manifested. Vulnerability can be linked to what is developing at the time the metal is presented to the organism. For example, exposure to Cu has the greatest effects in abalone as development proceeds from the fertilised egg through the embryo to the larva. Once the larva is developed it appears to be less sensitive. When the abalone *Haliotis rubra* was exposed for the entire cycle of egg–embryo–larva, deformities in the larvae appeared at a Cu concentration of 4 µg/L (LC$_{50}$ 7 µg/L: Gorski and Nugegoda, 2006). In the related species *Haliotis rufescens*, under a similar exposure regime, the EC$_{50}$ was 9 µg/L during development. But when *H. rufescens* was exposed after the larval stage was developed, the EC$_{50}$ was greater than 80 µg/L.

Studies with mussels, *Mytilus edulis* (Luoma and Carter, 1990) also showed that larval survival (LC$_{50}$ = 360 µg/L) was relatively insensitive to Cu but embryonic survival was very sensitive (LC$_{50}$ = 5.6 µg/L). Similarly, the number of eggs, survival of eggs, and released larvae in the polychaete *Ophryotrocha diadema* are affected at concentrations of Cd 5 to10 times lower than the 96 hour LC$_{50}$ concentration for adults (Akkeson, 1983).

The approximately 10-fold differences between sensitivities of different life stages dramatically illustrate the sensitive timing issues involved in identifying when an organism is most vulnerable to a metal and/or what the presence of the metal at different times might mean.

Many developmental processes occur over a relative short time period. Thus, it should be feasible to target a specific phase of development to evaluate vulnerability to metals. Tests of this sort are more relevant to nature than acute toxicity to adults. Unfortunately, targeted tests of development require robust knowledge of the life cycle of the species and an ability to culture the organism. Despite decades of interest in metal effects, few such basic studies are available. Culturing organisms through their life cycle is a delicate process;

Table 10.3 *Toxicity of five metals to larval development in several species of marine organisms. Sea urchin data are from 10 species of sea urchins and 22 publications on development in this species, as summarised by Novelli* et al. *(2003). Data are also presented for the abalone* (Haliotis rubra), *mussels* (Mytilus edulis) *and an oyster* Crassostrea gigas *as summarised by Gorski and Nugegoda (2006). In general the range of dissolved concentrations that affect development in different sea urchin species is wide, as is common among species (similar species do not necessarily have similar vulnerabilities). The rank order of dissolved Hg and Cu toxicity differs among the species, as well. Dissolved Cd and Pb would have little effect on development in any of these species (such concentration would never occur in marine waters, even near sources of contamination). But there could be numerous instances of contamination where Cu might affect development of abalone, mussel or oyster larvae (these concentrations are within those found in contaminated marine ecosystems; although speciation would influence the results). Extreme contamination would be necessary for effects from dissolved inorganic Hg or Zn*

Species	Copper (µg/L)	Mercury (µg/L)	Zinc (µg/L)	Cadmium (µg/L)	Lead (µg/L)
Sea urchin NOEC	32	10	100	1600	
Sea urchin EC_{50}	20–110	0.8–4.6	20–580	230–11400	68
Abalone NOEC	1.0	8	8	320	320
Abalone EC_{50}	7–8	14–28	18–22	4316–4821	3891–4398
Mussel EC_{50}	6	6	175	1200	758
Oyster EC_{50}	5	11	209	611	476

in fact it is an art in itself. Adding toxicants to the process makes it more challenging. Nevertheless, there are standardised early life stage test data available for several freshwater and marine fish (Wright and Welbourne, 2003). Fecundity and life cycle tests are well developed for zooplankton from both freshwater and marine waters. Standardised larval bioassays exist for several marine mollusc species (oysters, mussels, abalone), polychaetes, brittle-stars and sea urchins (Table 10.3).

10.3.6   Time of exposure

Chronic assays, which were originally designed to cover the complete life cycle of the organism, or a substantial portion of it, have been largely replaced by shorter tests, sometimes no more than a few weeks in duration . . . because of ease and economy. Acute toxicity data . . . is much more commonly and widely available than chronic data.

Wright and Welbourne (2003)

Comparing the tolerance of a (complete) key developmental stage . . . may be more ecologically relevant than simply comparing exposure over a fixed period of time, because for an embryo to survive to adulthood it must successfully complete all development stages.

King and Riddle (2001)

The time compromises of standardised (and non-standardised) toxicity tests are usually made for pragmatic reasons. It is argued that it is simply unrealistic to test all organisms over their full life cycle. But the choice of exposure time is crucial in determining the concentration at which metals are toxic and the relative sensitivity of species.

Acute toxicity tests are defined as acute because of the short exposure period. The 48 hour and 96 hour LC_{50} toxicity data are the most abundant of any data on metal toxicity. Nevertheless, it is well known that the expression of toxicity reflects both exposure concentration and exposure time. In a toxicity test, the proportion of organisms affected increases with time if exposure is held constant. Shorter exposure times always require higher concentrations to elicit a response (Fig. 10.7).

Chronic toxicity tests either use mortality as an endpoint, at exposures longer than 96 hours, or address sublethal endpoints over extended exposures. Chronic does not seem to have a precise definition

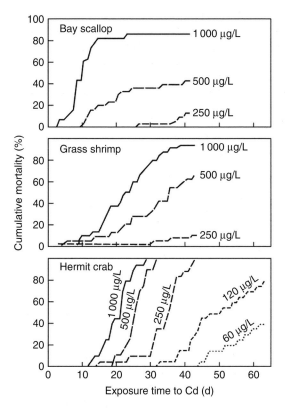

Fig. 10.7 A classic study by Pesch and Stewart (1980) demonstrated how increasing exposure time (x-axis) resulted in progressive response of three marine invertebrates to ever lower Cd concentrations. The longer the exposure time, the lower the LC_{50}. For example, all three species showed essentially no response to 25 µg/L Cd until 30 days of exposure. An exposure time of 20 days could lead to the conclusion that 25 µg/L Cd had no effect. Many extreme LC_{50}s for metal exist in the acute toxicity database. It is uncertain to what extent these are the result of metal resistance or extremely short exposure times. Bay scallop, *Argopecten irradians;* grass shrimp, *Palaemonetes pugio;* hermit crab, *Pagurus longicarpus.*

in terms of time, however (Wright and Welbourne, 2003). The original intention was that exposures should include at least a substantial fraction of one generation (the generation time) of the population (Adams and Rowland, 2003). The exposure time necessary to fully simulate chronic natural exposures is influenced by how effects progressively accumulate over time. For metals, the time dependence of toxicity would be influenced by uptake rate,

redistribution of the metal among organs and sustained rates of detoxification and excretion.

Lifetime bioaccumulation studies using radio-isotopes of Zn showed that a complete equilibration of all sites within a small fish or a mussel takes at least several months (Willis and Jones, 1977); uniform labelling of the small forage fish *Gambusia affinis* with ^{65}Zn occurred only after lifelong exposure. Lifelong exposures are impractical in most experimental settings, however. Almost no such data exist. Modern versions of chronic tests tend to standardise the time (e.g. to 28 days for sediment toxicity tests), with less regard to the life cycle of the organism than convenience to the investigator/decision maker.

If time is standardised, toxicity tests of different species are probably not comparable, because a fixed time represents a different proportion of the total life cycle of each organism. Generation or life cycle times for phytoplankton, zooplankton and meiofauna are hours to months. A 96 hour test with phytoplankton toxicity includes several generations; no extrapolation to longer timescales is necessary. However, extrapolation includes some difficult assumptions if the 96 hour test is used to determine effects on an organism that lives for years. Nor are the two approaches robustly comparable. For example, exposure of the Antarctic sea urchin, *Serechinus neumayeri*, during development from embryo through larval stages yielded less sensitive results when exposures were 6 to 8 days than when the duration of exposure was extended to the entire development period (embryo to two-armed pluteus stage, 20 to 23 days: King and Riddle, 2001). The EC_{50} for Cu in the longer-term exposure was 1.4 µg/L; the EC_{50} for the 6 to 8 day test was 11.4 µg/L. Most sea urchin embryo tests use the latter methodology, despite the difference.

To make matters even more complex, one of the most sensitive endpoints for metals is success of the population. Effects on populations may only become apparent after multiple generations are exposed. Life cycle tests are a means to take such long-term perspective. Influences on age structure, reduced longevity, changes in development period or delay of reproductive maturation can be used to model the long-term fate of the population.

Box 10.11 Toxicity of Cd in long-term life cycle tests

Two population bioassays with amphipods have demonstrated the sensitivity of population processes when they are addressed in long-term tests. Sundelin (1983) held the marine amphipod *Pontoporeia affinis* in sediments for 460 days (several generations) and held the dissolved concentrations constant in the system at 6.5 µg/L Cd. Sediments equilibrated with the dissolved Cd to reach a concentration of 5 µg/g dw.

Juveniles survived when they were introduced to the bioassay. They matured to adults, grew normally and went through reproduction. However, only a small proportion of their young survived, apparently because the embryos could not develop to juveniles. Eventually the population died out. Population models predicted that the failure to recruit young animals into the

populations was the cause. But extinction took several generations.

Similar results were observed in the estuarine amphipod *Ampelisca abdita* (Scott and Redmond, 1989). A population was exposed for 56 days to dilutions of a contaminated sediment. Survival of young (birth rate) was reduced at a lower level of contamination than that causing acute or chronic mortality in adults. Growth rate and time to first reproduction were also affected at the lower concentration. The effects of the suppressed birth rate were compounded with every generation. Extinction was predicted after several generations, because the rate of births had fallen below the natural death rate.

Neither result would have been evident if the tests had not been conducted long enough to analyse effects on multiple generations.

Box 10.11 cites two examples of studies that illustrate such effects. It is interesting that the sensitivity of amphipods (*Pontoporeia affinis*) to Cd in a life cycle 460 day test suggested extirpation of the population at 6.5 µg/L Cd in water and 5 µg/g Cd dw in sediment). Adult amphipods exposed to only dissolved Cd in traditional short-term lethality tests succumb at concentrations orders of magnitude higher (230 to 4300 µg/L Cd: McGee *et al.*, 1998; USEPA, 2002c). Endpoint and species differences could contribute to these differences, but extrapolation from short to long timescales is also undoubtedly a factor.

10.3.7 Environmental conditions (physical and chemical choices)

Models based on sound physical-chemical and biological principles, such as the free-ion activity model, have . . . been effective in explaining the central role of the free-ion concentration (or activity) as a regulator of interactions (uptake, toxicity) between metals and aquatic organisms.

Hare and Tessier (1996)

Adopting a biotic-ligand modeling approach could help establish a more defensible, mechanistic basis for regulating aqueous discharges of metals.

Meyer *et al.* (1999)

An important source of variability in the toxicity of metals is the influence of environmental conditions (Chapter 7). Metal speciation in solution is particularly influential in the outcomes of toxicity tests and such effects are well studied. Traditional regulatory bioassays typically used formulated water or water sources of low ionic strength to standardise geochemical effects. This approach minimises complexation with ligands and maximises dissolved bioavailability. This set of conditions was chosen intentionally to provide the most environmentally protective values for metal toxicity.

It is now relatively well known how speciation affects bioavailability (Chapter 7; Table 10.4) and is linked to toxicity. Some dissolved forms or species of trace elements are inaccessible or of negligible bioavailability or toxicity to aquatic species. Other forms are highly bioavailable and toxic. Bioavailability of dissolved metals that behave as cations is typically proportional to the free ion activities

Table 10.4 *The links between speciation and toxicity of metals have been studied for decades and are relatively well known. Speciation models or operational methods of defining speciation are necessary to quantify speciation in any specific circumstance. Conventional knowledge has developed that can provide some simple, consistent insights about the linkages*

| | Controls on speciation and links to toxicity of dissolved metal | |
Metal	Seawater	Freshwater
Arsenic	Oxidation state. As(IV) toxic.	Oxidation state. As(IV) toxic.
Cadmium	Chloro-complexes dominate and greatly reduce bioavailability and toxicity.	Free ion.
Chromium	Oxidation state. Cr(VI) toxic. Slow uptake rate reduces dissolved toxicity.	Oxidation state. Cr(VI) toxic. Slow uptake rate reduces dissolved toxicity.
Copper	Free ion. Strong complexes with organic matter reduce toxicity.	Free ion. Strong complexes with organic matter reduce toxicity.
Mercury	Free ion and methylation. Methyl Hg most toxic. Chloro-complexation dominates form of inorganic mercury but does not reduce bioavailability to the extent of Cd.	Free ion and methylation. Methyl Hg most toxic.
Nickel	Free ion.	Free ion.
Lead	Free ion. May form strong complexes that reduce toxicity.	Free ion. May form strong complexes that reduce toxicity.
Selenium	Oxidation state. Se(IV) toxic. Very slow uptake rate and very low dissolved toxicity.	Oxidation state. Se(IV) toxic. Very slow uptake rate and very low dissolved toxicity.
Silver	As chloride concentrations increase to levels found in estuaries, Ag chemistry changes and soluble Ag-chloro-complexes are formed. In contrast to Cd, the dominant $AgCl^0$ appears to penetrate membranes (is bioavailable) and is very toxic.	At low chloride concentrations Ag precipitates atom for atom as insoluble AgCl. Ag can also form strong organo- or sulfide-complexes that are of low bioavailability. Ag^+ controls bioavailability.
Zinc	Free ion.	Free ion.

in most marine waters. In freshwaters, free ion activity is a primary consideration, but major ion concentrations and pH can also be important influences. Quantification of how speciation affects metal bioavailability in solution is relatively advanced. The Free Ion Activity Model (FIAM) and the Biotic Ligand Model (BLM) represent quantitative approaches for incorporating consideration of speciation into toxicity testing. Use of the free ion concept to define toxic concentrations of dissolved metals was first demonstrated in the late 1970s (Sunda and Guillard, 1976), but the early approaches were not adapted by regulatory science and were rarely used in risk assessments. However, with new abilities to predict toxicity from geochemical modelling (e.g. the Biotic Ligand Model), site-specific geochemical corrections are increasingly incorporated into extrapolations of toxicity data to natural waters (Box 10.12). The details of how these are applied are described in Chapter 16.

10.3.8 Exposure route

it is perhaps worth pointing out that practically all the toxicity data relate to concentrations in solution when we know that in many animals food and particulates are the major sources for accumulation.

Bryan (1984)

Box 10.12 Demonstration of the need for geochemical correction of toxicity testing data

Geochemists have long used toxicity tests, rather than bioaccumulation, as a measure of the response of organisms to different geochemical conditions. The earliest studies of the effects of metal speciation on bioavailability were, in fact, conducted with bioassays using reduced phytoplankton growth rates or zooplankton life cycle tests as the endpoint (Sunda and Guillard, 1976; Anderson and Morel, 1977). For example, Sunda *et al.* (1987) compared the toxicity of Cu and Zn to the estuarine copepod *Acartia tonsa* and to the two diatom food species *Thalassiosira pseudonana* and *T. weissflogii* while controlling speciation with a nitrilotriacetate-trace metal ion buffer system. They expressed the toxicity on the basis of free ion activities. They showed that *A. tonsa* was more sensitive to cupric and zinc ion activity than either of the diatoms. They also verified that sensitivity varied among the different life stages of the copepod. Egg-laying rate was most sensitive to Zn, and survival of the naupliar early life stages was extremely sensitive to Cu. Perhaps most importantly, they demonstrated that the metal ionic activities in some polluted estuaries were at similar levels to those that caused the more sensitive toxic responses in the bioassays when geochemical differences were accounted for (Chapter 11).

the field of aquatic toxicology entered the 1990s with the general belief that although dietborne exposure of aquatic organisms to common metals could increase body burdens, the toxicological consequences of this exposure were relatively minor.

Mount (2005)

A matter of current, intense debate with regard to the effects of metals on biological systems is the potential toxicity of metals associated with food. Regulatory science and ecosystem science have historically taken very different views of the influence of dietary exposure on toxicity. As we noted in Chapter 7, experiments demonstrated the feasibility of metal accumulation from diet in the 1970s (Luoma and Jenne, 1976, 1977). At the same time

models and observations established that the existing metal concentrations in marine animals from nature could not be explained by uptake from solution; diet had to be important (Benayoun *et al.*, 1974). So it is not surprising that field biologists like G.W. Bryan (quote above) assumed that if diet was a source of internal exposure it also could be a source of metal toxicity. Mount (2005) notes, however, that the focus on waterborne metal exposure in regulatory toxicology was not arbitrary. There was experimental evidence that contradicted the feasibility of metal uptake from diet (i.e. diet did not dominate uptake in all circumstances). It was also thought that biomagnification (higher concentrations with each trophic level) was necessary for diet to have an impact; and it was thought that this was a rare situation with metals.

Quantification of dietary exposure of metals is now possible, and protocols are well developed (e.g. Luoma *et al.*, 1992; Wang *et al.*, 1996). It is clear that diet is the predominant source of most (although not all) metal exposures for many aquatic animals (Wang, 2002). It is unequivocal that diet controls the most important aspects of the toxicology of Se and Hg (see Chapters 13 and 14). But translation of exposure via diet into toxicity with other trace metals has been challenging. It is interesting that regulatory guidelines remain driven by traditional toxicity tests, even for Hg and Se.

The greatest impediments to understanding dietary toxicity are probably pragmatic. Short, acute toxicity tests are not feasible using dietary exposure; rarely are effects found in 3 days from food that is palatable to test organisms. Most effects are sublethal malfunctions resulting from chronic exposures (Handy *et al.*, 2005). The paradigm that simple, short tests are a requirement for testing is not compatible with dietary exposure at our present state of knowledge. Dietary toxicity experiments are inherently complex in several ways:

- Food or, preferably, prey must be first allowed to build a dose of metal, then the transfer to the consumer organism can be tested.
- The nature of metal distribution in the food can influence results.

Table 10.5 *Effects of dietary metals on various endpoints in various organisms after exposure via diet from a thorough compilation of available data in 2003 (Handy et al., 2005; references in that volume). The purpose of the table is to demonstrate that responses to metal occur at dietary concentrations that are reasonable for contaminated environments; thus some of the most sensitive responses are listed here. Less sensitive responses are also found for some species under some conditions, and no effects were observed in some studies. Data for mixtures of metals, Hg and Se are not shown (the latter are covered in their own case studies, Chapters 13 and 14)*

Metal	Organisms	Response	Concentration range (μg/g)
Cadmium	Zooplankton, fish	Biochemical change	10–125
	Zooplankton	Growth	10
	Zooplankton, crab	Reproduction	14–20
Copper	Fish	Biochemical signs of impairment	16–2400
Lead	Fish	Biochemical signs of impairment	100
Silver	Fish	Feeding rate	3–3000
	Zooplankton	Reproduction	4–21
Zinc	Fish	Decrease food assimilation	2500–10 000
	Zooplankton	Reproduction	235

- Conditions are more difficult to control when food is introduced to the experiment.
- Experiments for a month or longer are necessary to elicit effects in longer lived species (Forrow and Maltby, 2000).
- Exchange of metal back into solution often results in both dissolved and dietary presentation of the metal, confounding conclusions.
- Feeding rates should be documented (Clearwater et al., 2002).
- The ratio of dissolved concentrations to dietary concentrations must be similar to nature if the routes are to be compared realistically (Lee et al., 2004).
- A sufficient range of exposures to achieve a justifiable dose response is essential, but represents a challenge when the investigator is spiking living organisms as prey.
- The metal presented in the diet must be nutritionally sufficient and desirable to the test species.
- As is the case with dissolved speciation, experiments must simulate bioavailability that might occur in nature. At a minimum the metal in the diet must be bioavailable before toxicity can be expected.

As a result of such complexities, our understanding of dietary toxicity is at an early stage (Mount, 2005). Nevertheless some data are available, and even the early data show that dietary toxicity occurs within the range of concentrations that can be found in prey or food in metal-contaminated waters (Table 10.5).

Some of the studies that illustrate adverse effects from dietary metals were controversial, partly because the stakes were high for the issues the studies addressed (Box 10.13).

Results to date suggest that important endpoints for dietary toxicity are impairment of behaviour, energetics, gut function and impairment of reproduction. A workshop of experts in 2005 concluded that it was unequivocal that metals were bioavailable via the diet of aquatic animals and they probably caused toxicity. But no one yet fully understands the implications of this form of metal uptake (Meyer et al., 2005). However, it is important to recognise that this is a potentially important deficiency in the existing regulatory approaches, and, arguably, it may be the most important deficiency. Dietary toxicity is a subject toward which research emphasis can be expected to grow if risk assessments are to reach the level of sophistication that will be needed in the future.

Box 10.13 Effects of ingestion of metal contaminated insect communities on trout

Benthic macroinvertebrates are an important source of food for fish. Woodward *et al*. (1994) fed invertebrates collected from the contaminated Clark Fork River to early life stage trout. After eating the contaminated diet, the fish showed elevated biochemical dysfunctions typical of metal-induced stress: concentrations of products of lipid peroxidation and histological abnormalities (effects on hepatocytes, pancreatic tissue, and the mucosal epithelium of the intestine), as well as reduced growth and survival (Farag *et al*., 1994; Woodward *et al*., 1994, 1995). They reproduced the approach in a series of experiments using different trout species, and with benthos from different contaminated rivers. Similar toxicological responses usually were observed: reduced survival (Woodward *et al*., 1994; Farag *et al*., 1999), decreased growth (Woodward *et al*., 1994, 1995; Farag *et al*., 1999), reduced feeding activity (Woodward *et al*., 1995; Farag *et al*., 1999), and histopathological abnormalities (Woodward *et al*. 1994, 1995; Farag *et al*., 1999). Farag *et al*. (1994, 1995) determined that metals from these diets were accumulated in liver, pyloric caeca and large intestine. Lipid peroxidation (Farag *et al*., 1994; Woodward *et al*., 1994, 1995) was also observed in the liver, pyloric caeca and large intestine of brown trout *Salmo trutta* resident in the Clark Fork River (Farag *et al*., 1999). Farag *et al*. (1995) defined a liver concentration between 238 and 480 μg/g Cu (dry wt) as indicative of detriment to growth and reproduction.

Attempts were made to repeat these experiments with prey spiked in the laboratory. Mount *et al*. (1994) tried to gain greater control over exposure variables by using brine shrimp (*Artemia*) as a food source. They showed no effects on growth or survival after 60 days. Mount *et al*. (1994) attributed some adverse effects from the water to metals leaching from the diets into the surrounding water. Later experiments showed that important differences existed in the manner in which metals are incorporated into the prey species used in the different experiments (Meyer *et al*., 2006). Woodward *et al*. (1995) and Mount *et al*. (1994) also both documented effects on the intestinal tract of fish fed diets contaminated with metals. Gut impaction was observed in 3% to 9% of brown trout fed diets from the Clark Fork River, and constipation was observed in nearly 50% of the trout fed diets from the most contaminated reaches of the Clark Fork River. Mount *et al*. (1994) found that 7 of 18 mortalities observed were associated with actual rupture of the body cavity. As observed in zooplankton, the metal contamination appeared to disrupt intestinal function. Later experiments (Hansen *et al*., 2004) exposed juvenile rainbow trout (*Oncorhynchus mykiss*) to diets of the earthworm *Lumbriculus variegatus*, cultured in metal-contaminated sediments from the Clark Fork River basin. They observed significant growth inhibition in trout fed the contaminated diets; growth inhibition was associated with reductions in conversion of food energy to biomass rather than with reduced food intake. Growth inhibition was negatively correlated with As in trout tissue.

Although some disparity can occur in results of such experiments, it was clear in all cases that ingestion of contaminated prey can be a significant source of stress for predatory fish.

10.4   VALIDATION OF EXTRAPOLATIONS FROM TOXICITY TESTING

Prologue. Traditional single species bioassay approaches are effective predictors of toxicity where contamination is extreme. False negative and false positive predictions of toxicity are expected in less than extreme conditions, however. Balancing (or being aware of the need to balance) uncertainties arising from the different trade-offs in toxicity testing will continue to be a consideration in use of toxicity data in risk assessment and management.

Many regulatory successes resulted from risk management approaches built upon toxicity testing. Many of the most severe cases of metal contamination in the developed world were controlled by the regulations that stemmed, at least partly, from reliable, repeatable data derived from simple toxicity tests. But complex testing approaches demonstrate that simplified toxicity tests cannot consider many of important responses to metals of populations in nature, and can lead to predictions that miss important compensatory capabilities or underestimate important vulnerabilities of natural

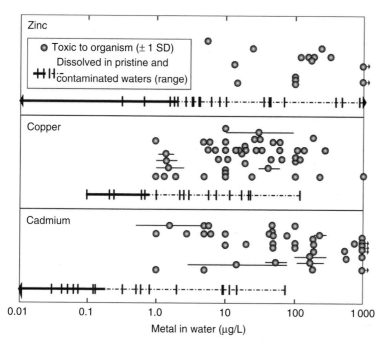

Fig. 10.8 G. W. Bryan (1984) conducted one of the first comparisons of metal concentrations in natural waters with concentrations found toxic in dissolved toxicity tests for three metals (Zn, Cu and Cd). Bryan's review of the toxicity literature, although early, was particularly thorough for tests that included extended exposures and/or early developmental stages. The dissolved metal concentrations at the higher end of the range may also be suspect (Chapter 5). Nevertheless, his conclusions were instructive. In particular, he noted that toxic concentrations of Cu were well within the range found in contaminated waters; but he noted the importance of dissolved organic matter in reducing Cu toxicity. Many toxic concentrations of Cd were far above concentrations ever reached in even the most contaminated waters.

populations. Geochemical simplifications overstate metal bioavailability from solution unless speciation is considered. But better empirical extrapolation to nature will require trade-offs against cost, simplicity and economy.

In the early stages of chemical testing, the toxicity of most substances was unknown. So the problem of screening the relative toxicity of chemicals took precedence over understanding toxicity in ways that promoted extrapolation. Knowledge of metal toxicity has grown substantially; so screening unknown chemicals is no longer the major bottleneck. The demands on toxicity testing are changing. It is now essential to consider whether the trade-offs in the traditional test data are beginning to impede accomplishing the assessment goals expected by the regulatory community (Norton *et al.*, 2003).

Empirical agreement between ecological studies and toxicity test predictions is typically best at the severe end of the spectrum of ecological effects. Dickson *et al.* (1992) compared eight freshwater studies where both ecological and toxicity test data were available and concluded that: 'a high level of ambient toxicity assisted in making it possible to

observe a relationship between ambient toxicity and community response. Where ambient toxicity is low or marginal . . . it will be more difficult to elucidate a relationship.' The most inconclusive results appear where contamination is less than extreme. Bryan (1984) showed overlap between toxicity test results and metal concentrations in natural waters using the literature and available data of the time (Fig. 10.8). Bryan (1984) did recognise that at least some of the analytical data available at the time probably overestimated dissolved metal concentrations. Nevertheless, his comparisons show that chronic toxicity could be demonstrated at concentrations found in contaminated environments.

Early regulators recognised the various limitations stemming from their choice of toxicity testing methods. But it seems, at least, that a pragmatic decision was made to balance the sources of scientific uncertainty. Insensitivities stem from compromises in exposure time, routes of exposure, choice of life stage and choice of species. Oversensitivities were built in to compensate for these, however. Geochemical conditions were standardised to maximise metal bioavailability. Some agencies used unfiltered water samples to define environmental exposures (overestimating bioavailability, sometimes greatly). The regulatory criteria promulgated in the 1980s also include application (safety) factors; these were precautionary multipliers that arbitrarily adjusted the regulatory value to be more protective. We are now in the process of correcting for geochemistry, thus reducing the arbitrary but protective cushion provided by maximising bioavailability in the tests. Filtered samples are used to define exposures in nature (the only scientifically defensible approach). Cairns and Mount (1990) predicted that application factors would become less important in risk management, and that prediction seems to be coming true (Chapter 16). The question for the future is whether efforts to reduce clear oversensitivities in one aspect of procedures (e.g. bioavailability, geochemistry and precautionary reductions) must be matched by equal efforts to reduce insensitivities in the tests (e.g. in biology) if the delicate balance is to be maintained.

Traditional toxicity testing was a successful management tool when contamination was extreme and/or conditions/decisions were relatively simple. We are moving into an era where decisions are more difficult, stakes are higher (solutions are expensive; the environment is less resilient to mistakes) and site-specific risk assessment and management are more important. Field observations are also more sophisticated and more common. Multi-species, *in situ* and mesocosm tests are increasingly sophisticated and credible. It seems increasingly difficult to justify direct forecasts from standardised toxicity testing in these circumstances. Additional testing with more complex goals and integration of the more complex testing approaches, geochemistry and the ecosystem-process knowledge seem increasingly important for tomorrow's difficult decisions.

10.5   CONCLUSIONS

Many advances have been made in understanding metal toxicity and the effects of geochemical conditions on toxicity. This has led some to conclude that protective criteria might be directly derived from geochemically corrected toxicity test results. But what is less often considered is that few of the biological limitations of toxicity tests are resolved. Despite decades of study, limited data exist on the most sensitive biological responses to metals: long-term, sublethal, sensitive life stage studies from metal-sensitive species or multi-species assemblages. Limited data are also available on toxicity via exposure routes other than dissolved metal. These biological limitations still apply to the dominant data available for ecological risk assessments and the data used to set standards.

The existing toxicity testing data are used to derive the general concentration ranges around which metal will begin to have adverse effects. These data were historically considered adequate for derivation of regulatory standards, largely because liberal, but somewhat arbitrary, protective correction factors were incorporated to assure that the environmental standards were protective. But traditional toxicity testing has not taught us which species are most likely to disappear under what kinds of circumstances and at what metal concentrations; much less why such species might disappear. More complex tests and field observations are beginning to answer those questions, however.

Most metals and reasonably worrisome metal formulations have now been screened for toxicity by traditional methods. But the new demands on risk management will require managing metal contamination across the boundaries that divide toxicity testing, process studies and ecosystem-level understanding. The body of available science shows that metal concentrations in contaminated water bodies approach levels that cause adverse effects, at least for some metals in some circumstances. The quantitative frequency distribution of such effects is unclear, however; and challenges remain in judging effects at any specific location. New, improved risk management will require better understanding of the distribution of contamination, linked to better understanding of the details of how organisms are affected, linked to better understanding of which species are most sensitive in different environments. Interactive effects of environmental conditions and different exposure routes must also be better taken into account. This lateral approach to management is necessary to address the increasing demand for resolution of site-specific issues and resolve the demand for defensible science-based risk assessments.

Correction factors are probably still necessary in extrapolating results from single species toxicity tests. Geochemical corrections may limit tendencies to overprotect natural waters. But site-specific assessments must also address how the important biological trade-offs have been considered in order to prevent underprotection. Better understanding and better methods to deliver results in a site-specific and species-specific way must be among the goals of future trace metal toxicity testing. Those goals will probably require moving beyond the priority that simplistic, inexpensive testing is essential and that only standardised data are suitable to set regulations.

Suggested reading

Adams, W.J. and Rowland, C.W. D. (2003). Aquatic toxicology test methods. In Handbook of Ecotoxicology, ed. D.J. Hoffman, B.A. Raffner, G.A. Burton Jr and J. Cairns Jr. Boca Raton, FL: CRC Press, pp. 19–45.

Cairns, J. (1986). What is meant by validation of predictions based on laboratory toxicity tests? *Hydrobiologia*, **137**, 271–278.

Forbes, V.E. and Calow, P. (2002). Species sensitivity distributions revisited: a critical appraisal. *Human and Ecological Risk Assessment*, **8**, 473–492.

Hook, S.E. and Fisher, N.S. (2001). Sublethal effects of silver in zooplankton: importance of exposure pathways and implications for toxicity testing. *Environmental Toxicology and Chemistry*, **20**, 568–574.

Irving, E.C., Baird, D.J. and Culp, J.M. (2003). Ecotoxicological responses of the mayfly *Baetis tricaudatus* to dietary and waterborne cadmium: implications for toxicity testing. *Environmental Toxicology and Chemistry*, **22**, 1058–1064.

Suter, G.W. (1990). Endpoints for regional ecological risk assessments. *Environmental Management*, **14**, 9–23.

11 • Manifestation of metal effects in nature

11.1 INTRODUCTION

Prologue. Studies of populations and communities in natural waters can characterise the effects of metal stress, but information on how effects are caused is more difficult to obtain (Maltby, 1999). Co-occurring stressors are common in nature and natural variability can disguise effects or, itself, cause change. Inter-actions among contaminants and how effects at one level of biological organisation affect processes at the next higher level are also considerations. These challenges are not insurmountable, however, as our knowledge of the relevant processes grows. A body of evidence is accumulating that increasingly clarifies influences of metals in streams, lakes, estuaries and coastal ecosystems.

Toxicity tests (Chapter 10) and indicators of stress from the cellular to the organismal level (biomarkers: Chapter 9) make it clear that metals *can* have substantial or important ecotoxicological effects in aquatic environments. But there remains some controversy about whether important ecological effects occur in natural waters, how common they are and how they are manifested.

Are trace metals only ecological stressors of significance in a few local instances of extreme contamination? Or are they pervasive contributors to ecological change? We know that moderate to strong trace metal contamination is ubiquitous. Metal contamination is typical of any location where human activities surround or release wastes to a water body (O'Connor, 1996). Severe contamination is now rare in jurisdictions where environmental controls are in place. But new technologies and the spread of modern economies across the globe will introduce unprecedented challenges

for managing metal pollutants. Linking our knowledge of these exposures to conclusive evidence of where effects occur is the next difficult step.

In this chapter we consider the challenges of uncovering metal effects in complex circumstances; and review the evidence that supports our understanding of how and where metals play a role as an ecological stressor. It is nature that we must protect, so the ultimate sources of evidence about the stresses imposed by metal contamination are studies from nature, or studies that minimally simplify the complexities of nature. Observations from most types of ecosystems are now available, but they are most persuasive when accompanied by experiments, sophisticated methodologies, multiple lines of evidence and persistent study to reduce uncertainties (Burton *et al.*, 2002; Adams, 2003; Forbes and Calow, 2004). Insufficient validation of experimentally derived regulatory criteria remains a problem, but a growing body of work shows that it is possible to improve the ecological perspective of these influential regulatory decisions (Clements and Newman, 2002).

11.2 EXPERIMENTAL EVIDENCE DEFINES THE FEASIBILITY OF METAL EFFECTS IN NATURAL WATERS

Experimental testing is necessary to define if it is feasible that metal effects might occur in modern contaminated water bodies. Dietary toxicity studies, life cycle studies and complex tests such as those conducted in mesocosms or whole ecosystems reduce some of the most important uncertainties associated with extrapolating experimental testing results to natural waters (Chapter 10). Yet such lines

Box 11.1 Definitions

ecoepidemiology: study of ecological influences on the health of the environment.

functional ecology: branch of ecology that focuses on the roles or functions that species play in the community or ecosystem in which they occur, emphasising physiological, anatomical and life history characteristics of the species.

of evidence are often excluded from derivation of water quality criteria. The complexity of these underutilised approaches may preclude obtaining rigorous dose–response relationships. But here we submit that the body of evidence from all approaches provides a clearer view of how metals manifest their effects in natural waters than if evidence is restricted to a single predefined paradigm.

11.2.1   Dietary exposure

Prologue. Metal effects are feasible in contaminated waters when exposure of animals is via their diet. Dietary exposure to metals affects feeding, growth and/or egg production in freshwater and marine animals including fish and invertebrates. These effects occur at concentrations that are not unprecedented in contaminated conditions found in aquatic environments.

Traditional toxicity tests do not account for the possibility that toxicity can result from dietary intake (Chapter 10), but evidence that dietary exposure is common in nature is unequivocal (Meyer *et al.*, 2005). Demonstrating toxicity from these exposures requires complex experiments (Box 11.2), but evidence is growing that such effects occur in contaminated waters.

Dietary toxicity is typically expressed as effects on behaviour, reproduction, or feeding ecology.

- **Effects on behaviour.** Wallace *et al.* (2000) observed reductions in prey capture in grass shrimp fed Cd-contaminated oligochaetes from a heavily contaminated site in the Hudson River. Use of prey from a contaminated water body in nature and choice of an endpoint unaffected by the nutritional value of the prey unequivocally linked exposure and effects in this study.

- **Depressed feeding rate.** A common effect of ingestion of contaminated food is depression of feeding rate. Taylor *et al.* (1998) found that the LC_{50} of dissolved Cd to *Daphnia magna* in the absence of food was 120 µg/L. But when algal cells were spiked with dissolved Cd at 2.5 µg/L, feeding rates in their zooplankton consumers were strongly inhibited. Dissolved Cd alone did not inhibit feeding rates. The mechanism by which algal-bound Cd reduces feeding rates was associated with poisoning of the digestive system or feeding mechanisms. The mechanisms of toxicity might involve interference with peristalsis, nutrient uptake, functions of endocrine cells or mechanisms that control ion transport in the gut epithelium. Irving *et al.* (2003) observed similar results with larvae of the mayfly *Baetis tricaudatus* at Cd concentrations typical of substrates in contaminated ecosystems (Box 10.7). Feeding rates were reduced when the diet contained 10 µg/g Cd dw. Periphyton collected from metal-contaminated streams ranged in Cd concentration from 0.2 to 33.3 µg/g; presumably disturbance of feeding was occurring in the more contaminated of these environments. Dissolved toxicity would require Cd concentrations that would never be observed in such waters. Feeding or digestive tract dysfunction has also been shown for larvae of the mayfly *Leptophlebia marginata* exposed to dietary Cd (Maltby, 1992).

- **Effects on reproduction.** Hook and Fisher (2001a, b) studied effects of dietary exposure to Ag on reproduction in zooplankton from both freshwater and marine environments. The acute toxicity of dissolved Ag to marine zooplankton occurred at 40 µg/L and to freshwater zooplankton at 10 µg/L. These are concentrations far above what would ever occur in any contaminated aquatic environment. Algal cells were exposed to a range of Ag concentrations and then fed to the zooplankton. When the zooplankton consumed algae exposed to 0.1 µg Ag/L in

Box 11.2 Reproductive endpoints: an understudied response to metal contamination

Metals affect growth (which can affect reproductive success) and directly affect reproductive processes. Direct reproductive toxicity may stem from both the way exposures occur (at least in some species) and the way metals associate with active sites in organisms. In fish, for example, metal accumulated via diet accumulates in the liver. The portal vein in fish carries nutrients (and toxins) from food, first to the liver. There many metals are efficiently removed before potential transport to other organs or tissues. The liver is also the site of synthesis of many reproductive products used in the developing egg, for example the egg yolk protein, vitellogenin. Trace metals may substitute for essential elements (e.g. Se for sulphur) in the vitellogenin and be passed to the developing embryo; or the metals may inhibit vitellogenin synthesis itself.

There are several ways to study effects on reproduction:

(1) Direct exposure of reproductive products. For example, eggs are injected with the metal, which is probably all in metabolically available form (Linville, 2006). The route of exposure is unrealistic in these tests but they have great advantages in obtaining data rapidly over broad concentration ranges. Results can be validated by comparison with more complex maternal exposure studies over smaller ranges.

(2) Naturally contaminated adults. Adult organisms can be taken from a contaminated habitat and spawned artificially; then hatch is monitored. Development is a highly sensitive process, so the details of the protocol can be very influential in the outcome of the experiment (e.g. Hamilton and Palace, 2001).

(3) Following a natural experiment. For example, effects of Se on birds were studied by randomly identifying hundreds of nests during the breeding season with suspected different Se exposures. Then hatching success was followed over time (Skorupa, 1999). Large sample sizes are necessary to obtain results useful in interpreting dose–response relationships; and a wide-ranging study may be necessary to obtain a range of exposures. Results also may be influenced by factors other than the trace metal of interest.

Endpoints are an important consideration in reproductive tests. To define exposure, Heinz (1996) concluded that Se concentrations in eggs were better predictors of adverse effects on reproduction than were concentrations in the livers of birds. For metals that are passed from the mother to the young, powerful measures of developmental effects include teratogenesis (deformities), hatchability and survival of young (Fairbrother *et al.*, 1999). Either chick mortality (in birds) or hatchability is sensitive to Se, for example, at about one-half to one-quarter of the concentration that induces deformities (two- to fourfold more sensitive) or growth effects.

marine waters and 0.05 μg Ag/L in freshwaters reproductive success was reduced by 50% (respectively). Very similar effects were observed with Cd, Hg and Zn (Hook and Fisher, 2001b, 2002). The dissolved metal concentrations with which algae were spiked are not unprecedented in contaminated water bodies, although the dissolved concentrations that directly cause toxicity are unprecedented (100-fold higher than those causing toxicity from diet). For example, concentrations of Ag in estuarine waters like San Francisco Bay range from 0.005 μg/L to contaminated concentrations of 0.1 μg/L Ag. The concentration associated with the dietary toxicity described above was similar to the dissolved Ag concentration in San Francisco Bay when Ag toxicity was observed on a local mudflat (Hornberger *et al.*, 2000). The assimilated Ag depressed egg production by reducing yolk protein deposition and ovarian development. The dissolved Ag did not have that effect because most Ag was adsorbed to the exterior of the animal. Ingested Ag was assimilated into the tissues of the zooplankton (Hook and Fisher, 2001a).

- **Reduced growth rates.** Ball *et al.* (2006) fed the freshwater amphipod *Hyalella azteca* with algae raised in Cd-contaminated water. They found reduced growth at an EC_{25} for the amphipod when the algae had been exposed to 0.5 µg/L Cd. They cited dissolved Cd concentrations in Ontario lakes (near Sudbury Canada) as ranging from 0.05 to 0.55 µg/L, with concentrations as high as 4.3 µg/L reported in the past. Again, the effects concentrations occur within the range typical of contaminated natural waters. Like other studies, extremely high dissolved Cd concentrations, unprecedented even for smelter-contaminated lakes in Canada, were necessary to cause lethality in *H. azteca*.

Sublethal responses like feeding inhibition and reduced growth influence the success of populations (Maltby, 1999). For example, adult mayfly recruitment into the population is determined by the size of the larva when it emerges to become an adult. Feeding inhibition causes larvae to be small, and increases the probability of extirpation of the population by reducing recruitment of new individuals into the population. The subtle nature of these effects is one reason for the difficulty in detecting such effects in natural waters.

Effects from dietary exposure have also been observed in upper trophic level organisms, as cited in Chapter 10 for trout (Box 10.13: Woodward *et al.*, 1994). When fish experienced metal exposure typical of contaminated river systems they showed reduced survival (Woodward *et al.*, 1994; Farag *et al.*, 1999), decreased growth (Woodward *et al.*, 1994, 1995; Farag *et al.*, 1999), reduced feeding activity (Woodward *et al.*, 1995; Farag *et al.*, 1999), and histopathological abnormalities associated with lipid peroxidation (Woodward *et al.*, 1994, 1995; Farag *et al.*, 1999). All such responses are typical of metal toxic effects (Chapter 9). These experiments also raised the possibility that metals may exert effects on fish by changing the types of prey species that are present (see later example), and thus changing the dietary sufficiency available to fish in the system. A number of metal-sensitive species are typically missing from the benthic community of metal impacted streams (see Section 11.4.2 below). Some of the dietary toxicity results may reflect deficient nutritional conditions in such streams.

11.2.2 Algal tests and speciation

Prologue. Bioassays that do not deviate greatly from natural exposure conditions are more feasible with phytoplankton than with longer-lived animals. Extensive comparisions of metal sensitivities were made in the 1980s among marine phytoplankton species. Results show that toxicity of Cu is feasible at the high end of dissolved concentrations in undisturbed environments, and toxicity to Cd is feasible in contaminated environments. These data were not used in deriving today's regulatory standards, nor have regulatory institutions supported further work with these powerful protocols.

Some of the most important caveats stemming from standardised toxicity testing with higher-order animals can be accounted for in well-run phytoplankton bioassays. Phytoplankton are typically autotrophic and most are photosynthetic protistans that do not feed (dietary exposure is not a consideration). Their generation time is very short, so uncertainties about the influences of time of exposure are eliminated by exposing several generations. The most reliable experiments are conducted under ultra-clean conditions, and in continuous culture, so the media are not depleted of any substances. If metal speciation is to be considered, concentrations can be chemically buffered so that free ion concentrations (Chapters 5, 7 and 10) are held constant throughout the experiments. Such protocols greatly reduce the uncertainty about concentrations at which adverse effects from metal contamination might be feasible.

Brand *et al.* (1986) used free ion protocols to compare effects of Cu and Cd on reproductive rates (divisions per day) of 38 different marine phytoplankton, including a range of sensitive and insensitive species. They found that cyanobacteria (blue–green algae) were the most sensitive to Cu; diatoms were the least sensitive, with coccolithophores and dinoflagellates of intermediate sensitivity (Fig. 11.1). Toxicity was expressed on the basis of free ion concentrations but reasonable assumptions allow extrapolations to total Cu concentrations in natural waters. For example, the most sensitive 5% of species in Brand *et al.* (1986)

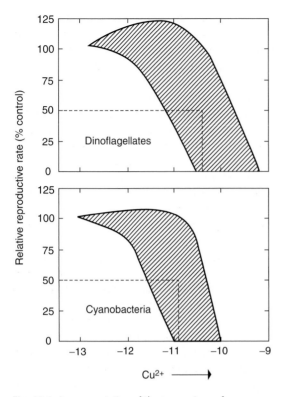

Fig. 11.1 A representation of the comparison of reproductive rates in several species of dinoflagellates and cyanobacteria (blue–green algae) exposed to a range of Cu levels based upon free ion concentrations. Dark lines represent the most and least sensitive species within a group; an EC$_{50}$ was estimated for each group to provide perspective. Copper concentration is presented as the negative molar concentration of free ion Cu (also known as pCu), so concentrations increase on the x-axis toward the right as negative numbers decline. Blue–green algae are about 10 times more sensitive to Cu than dinoflagellates. The most sensitive species show no growth at 0.65 µg/L total dissolved Cu in marine waters, whereas the most sensitive dinoflagellate species show no growth at about 3 µg/L total dissolved Cu under conditions of minimal organic complexation. (Data after Brand *et al.*, 1986.)

(the cyanobacteria) showed reduced reproduction at a free ion concentration of about 10^{-11} M or 0.00065 µg/L free ion Cu. Under conditions where dissolved organic carbon concentrations are low (as in the open ocean) it is reasonable to assume that 0.1% of Cu is the free ion (Brand *et al.*, 1986).

Then Cu toxicity to the most sensitive phytoplankton might be expected at 0.65 µg/L total dissolved Cu. Indeed, Brand *et al.* (1986) suggested that Cu was a natural selective factor and eliminated some species from the open oceans when dissolved organic matter concentrations were low. One might use a similar calculation with these data to suggest that dinoflagellates might be eliminated at 1 to 5 µg/L and diatoms at around 5 to 10 µg/L total dissolved Cu. These are all concentrations typical of coastal waters affected by human activities, although the direct extrapolation to any specific circumstance should be made on a free ion basis (because organic matter concentrations vary widely in natural waters and are very influential in Cu speciation).

The speciation of Cd in coastal waters is less uncertain than Cu in coastal waters, so extrapolations from free ion experiments to nature are more reliable. As we show in Chapter 16, Cd toxicity to the most sensitive phytoplankton species could be expected at about 0.20 µg/L, a concentration that can be observed in contaminated marine or estuarine waters. Cd is not a natural selective force, but it can be an ecologically significant contaminant for phytoplankton in anthropogenically disturbed waters.

The important point is that the gap between toxicity and ecosystem concentrations of bioavailable metal is extremely narrow in the oceans for larval fish, ciliates and copepods (Sunda *et al.*, 1990). Small changes in bioavailable trace element concentrations as a result of anthropogenic contamination have the potential to change the type of phytoplankton that compose natural communities.

Sunda *et al.* (1990) also conducted bioassays, as described above, to define the free Cu ion concentrations that affect zooplankton (copepods) in estuaries. They determined estimated free Cu ion concentrations in the Elizabeth River, Virginia, USA, and found concentrations that were near the levels that caused toxicity in the bioassays. They further showed that survival of copepods was reduced when bioassays were conducted with aliquots of Elizabeth River water (ambient bioassays). Then they added complexation agents to the water to reduce metal availability and found increased copepod survival. The lines of evidence converged to support the assertion of metal effects in the river water itself. These were not life cycle tests, but the exposure times of the short-lived zooplankton were sufficient to provide convincing results. Cu contamination was sufficient to affect zooplankton in this water body.

These bioassay protocols have since been used in other natural waters to demonstrate Cu toxicity in moderately contaminated natural waters (Box 11.3).

11.2.3 Sublethal toxicity and population effects

Prologue. Determining metal effects in natural waters requires recognition of the links between levels of biological organisation. One of the most important (and complex) of those is the linkage of sublethal signs of stress to population dynamics. Extirpation of populations is an important risk of metal exposure; certain sublethal signs of stress can signal the threat of extirpation.

Population dynamics involve the balance between individuals leaving the population (via mortality and emigration) and individuals entering the population (recruitment) via reproduction and immigration (Chapters 9 and 10). Population models assume that a fixed proportion of energy is spent on maintenance and growth, with the remainder being allocated to reproduction (Calow and Sibly, 1990). Any change in energy intake and/or greater expenditure on maintenance can be reflected in less recruitment and less success as a population. Metal exposure can cause increased energy consumption via the production of detoxification products (like metallothioneins or intracellular granules), decreased energy intake or increased maintenance costs. Each of these means that fewer resources are available for growth and reproduction (Maltby, 1999). Changes in age structure, reduced longevity, changes in development period or delay of reproductive maturation are also population-level processes sensitive to metals. Out of context these may seem insignificant

lower-order effects. Over several generations, however, reducing the rate of input of new individuals to the population (recruitment) can lead to extinction as readily as increasing the rate of mortality.

The autecology of the species is also important. If the species has a long life, low fecundity (produces few reproductive products) and thereby low recruitment, additional disruption of recruitment with metal stress is likely to be important. In San Francisco Bay, for example, the two fish species exposed to the most Se are sturgeon (*Acipenser transmontanus*) and Sacramento splittail (*Pogonichthys macrolepidotus*). Splittail are extremely fecund and vary widely in population size (habitat for breeding is reduced during dry years). Except when populations are at a low ebb, we would not expect splittail populations to be greatly affected by anything but extreme reductions in hatching success; recruitment does not appear to be a bottleneck. Sturgeon on the other hand are extremely long-lived and do not become reproductively mature until they are 10 years old. Populations are much more sensitive to effects on recruitment than are splittail. If sturgeon lose a proportion of their recruitment class to a contaminant every year, the long-term result could be devastating for the species, although the effects might not be visible for a long time.

Populations also have mechanisms to compensate for stress. Populations of white sucker (*Catostomus commersoni*), a freshwater fish, in metal-contaminated Canadian lakes show reductions in mean age, reduced spawning success, smaller egg size, poor survival of eggs or larvae and an absence of older age classes (Bendell-Young et al., 1986). However, compensatory increases in the growth of survivors and increased abilities of surviving females to produce eggs also occurred. The populations were not eradicated by the contamination because of the compensatory changes. They ultimately may have been more vulnerable to natural perturbations. Testing for compensatory responses has also proven a sensitive way to determine, *post-factum*, if a population has been affected by metals (Munkittrik and Dixon, 1989).

Box 11.3 Cu effects on phytoplankton in marine harbours

Cu speciation was studied in 1994 in four harbours on the south coast of Cape Cod, Massachusetts, USA (Moffet *et al.*, 1997). The four were exposed to varying degrees of Cu contamination. Cu in waters outside the harbours was complexed by very strong chelators, presumably organic matter, that occurred in concentrations twofold higher than ambient Cu concentrations. But in two of the harbours, total dissolved Cu concentrations were sufficiently elevated to saturate the strong chelators and the free Cu increased by 1000-fold. Cell densities of cyanobacteria (blue–green algae) were very low in the harbours with elevated free Cu ion activities, compared to harbours where less bioavailable Cu was present (Fig. 11.2).

The disappearance of blue–green algae in the contaminated harbours was consistent with culture studies that showed Cu toxicity to these organisms at the free ion concentration determined in the harbours (Fig. 11.2). The dominant phytoplankton species present in the affected harbours was the diatom *Skeletonema costatum*. Culture studies showed that this species was tolerant of Cu and could survive the free Cu ion concentrations observed; thus it was a survivor of the contamination. In harbours that showed no significant Cu contamination or saturation of strong ligands, cyanobacteria were a dominant component of the phytoplankton.

The authors concluded that significant anthropogenic inputs of Cu may overwhelm processes occurring in seawater that often reduce Cu bioavailability. Relatively small changes in total Cu concentration (7 to 10 times) led to much larger changes (>1000-fold) in concentrations of free Cu ions once organic complexation was saturated. The authors also considered other factors that might have affected blue–green algae abundances, but they could find no evidence unique to the two affected harbours that would support potential effects from salinity, production of other algae, grazing or viruses. Moffat *et al.* (1997) concluded that growth rates of other algal species and small zooplankton should also be affected by the Cu ion concentrations in the contaminated harbours. Their study and others (e.g. Brand *et al.*, 1986) suggest that the effect of Cu on the species composition and physiology of phytoplankton assemblages becomes apparent above a threshold free Cu^{2+} ion concentration of 10^{-11} M.

This study fits all the criteria for conclusively establishing an effect in natural waters (see later discussion). The decline of the blue–green algae coincided with the concentration of the free Cu ions (*co-occurrence*). The effect was confirmed experimentally (*feasible*), and was consistent with stressor-specific effects (Cu eliminates blue–green algae before diatoms: *specificity*). Two *mechanisms* explained the results (group-specific sensitivity and sensitivity for free ion Cu). The results were repeated in two harbours but did not occur in two uncontaminated systems (*consistency of replication*). The authors also *eliminated alternative co-occurring explanations* with their study design.

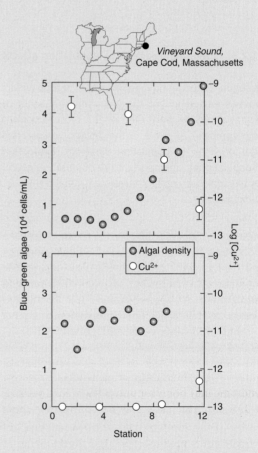

Fig. 11.2 Cyanobacteria (blue–green algae) abundance and free ion copper concentrations at two sites in Cape Cod, Massachusetts, USA. Abundance of blue–green algae is reduced wherever free Cu ion concentrations exceed (indices lower than) 10^{-11} M or 0.000 067 µg/L (top graph). (After Moffet *et al.*, 1997.) Because free ion Cu is less than 10^{-11} M everywhere in the harbour in the bottom graph, there is no change in the algal abundances. Note that free ion Cu is only a small fraction of the total Cu in both places; nevertheless it is the correlate with toxicity.

11.2.4 Multiple sources of stress

Prologue. Multiple sources of stress can confound interpretation of responses to metals or accentuate their effects.

Multiple sources of stress also add to the complexity of responses to trace metals in aquatic environments. Metal contamination in natural waters often involves more than one trace metal. As we will see later, there are methods for considering these influences, although attributing specific effects to a specific metal can be very difficult.

Effects of other stressors on responses to metals are less recognised, but could be important. For example, Lemly (1993b) hypothesised that stresses associated with the winter season might increase the vulnerability of fish to Se. The feeding of juvenile centrarchid fish is inhibited at cold temperatures, as body fat is depleted by metabolic demand. It was hypothesised that juvenile bluegill (*Lepomis macrochirus*) would be more stressed under winter-like conditions while they were fed Se contaminated food (5 µg/g dry wt). Indeed, in low water temperatures, elevated Se lowered the concentration that caused non-reproductive mortality in juvenile bluegill (reduced feeding, depletion of 50% to 80% of body lipid, and significant mortality: Lemly, 1993b). Responses that would be sublethal during much of the year became lethal during the time of such metabolic stress.

Even though an environmental stressor may impinge on a population only part of the year, this may become a bottleneck inhibiting recruitment. Lemly (1993b) suggested that Winter Stress Syndrome may occur for only a few weeks, yet 20% to 30% of a year class may be lost in that short period. The result could be depleted populations (often seen in polluted habitats) or slowed recovery after Se contamination is eliminated (Lemly, 1993b).

11.3   ESTABLISHING FIELD-BASED EFFECTS

Prologue. Establishing the feasibility of metal effects with experimental studies must be followed by detecting such effects in a water body and separating the influences of metal from the influences of other factors. Formal criteria frame the types of evidence necessary to establish that metals are influential in ecological risk in a water body.

Demonstration of metal effects in nature ultimately will be necessary to justify high-stakes environmental policies. However, simplifications incorporated into toxicity testing (Chapter 10) can conspire with the complexities of nature to make evidence of effects difficult to interpret. Formal criteria can be useful in systematically characterising the body of evidence necessary to narrow uncertainties about effects. Koch's postulates (Suter, 1993) are used in the medical sciences for these purposes and are termed 'ecoepidemiology' when adapted for environmental studies (Sinderman, 1996; Adams, 2003; Forbes and Calow, 2004).

Such criteria include:

(1) **Identification of response.** Identifying that there is an effect involves separating stressor-induced change from background fluctuations in the process being affected.
(2) **Co-occurrence.** The adverse effect must be associated with exposure to the metal in the field; and/or elevated concentrations of the metal must be found in the tissues of the affected receptor (e.g. organism, population, community).
(3) **Plausibility.** Eliminating confounding variables in the co-occurrence relationship makes it plausible that one explanation supersedes another. Correlations do not conclusively demonstrate cause and effect, although a stronger correlation between a metal and its effect is less ambiguous than a weaker correlation. Confounding variables, such as natural gradients in a property, can either mask effects of the metal or co-vary with them. Careful sampling design and quantifying the co-varying properties can help clarify potential influential factors. Time−order association, where one variable changes but others are dynamically stable over time (e.g. during recovery of an ecosystem from stress) is an example of a design that can aid understanding influences of confounding variables.

(4) **Feasibility or experimental confirmation (field or laboratory).** The adverse effect must be inducible under controlled experimental conditions. If so, it is feasible that the effect observed in the field could be caused by this contaminant. This can be made more convincing if the concentration known to cause adverse effects is exceeded; but how that level is determined is important (e.g. remember the uncertainties in traditional toxicity testing).

(5) **Specificity (stressor causes unique effect).** If effects are specific to the metal of interest, that can help narrow the list of potential causal agents. Effects that are specifically diagnostic to a metal can occur as biomarker responses (Cu-enriched lipofuscin in lysosomes: Chapter 9), at the whole organism level (teratogenic effects of Se: Chapter 13), or at the community level (the loss of species known to be particularly sensitive to a metal or metals). For example, occurrence of imposex in certain gastropod species is almost exclusively specific to tributyl tin exposure (Chapter 14). Once this was understood, wherever the disappearances of molluscan species coincided with occurrence of imposex in surviving species, it reinforced the conclusion that the disappearances could be attributed to tributyl tin.

(6) **Mechanistic coherence.** Understanding the mechanisms behind the effect adds credibility to conclusions about cause and effect. Many examples exist to show that evidence becomes more powerful, and ambiguities in cause–effect relationships can be reduced, as mechanisms are better understood (although this continues to come as a surprise to some practitioners).

(7) **Replication.** Demonstrating an effect that meets the above criteria repeatedly in different times and places increases the credibility of conclusion about causes.

Rarely does a single study meet all of the criteria for defining causation. But assignment of the cause of an effect to metals becomes less ambiguous as the body of evidence develops. Persistent study can delimit pollution-induced change from natural change and continuously narrow uncertainties.

Directly describing how the body of work addresses (or does not address) the inferential criteria above will improve the credibility and transparency of the interpretation (Sindermann, 1996). Ultimately, policy decisions require some judgements about the weight of scientific evidence (Forbes and Calow, 2004). But the uncertainties in a body of evidence from nature are rarely greater than uncertainties in extrapolating the results from controlled laboratory experiments to complex natural environments. A combination of the different approaches narrows the uncertainties inherent in both.

11.3.1 Understanding exposure and natural variability

Prologue. The two challenges in understanding metal effects in a natural setting are: (a) detecting the response to the trace metals, and (b) establishing causal relationships between the metal contamination and the effect. Detecting an effect is made complex by the variability in many biological processes. Detecting an effect and understanding its cause both must be linked to metal exposure.

Meeting the criteria for demonstrating effects in complex settings requires a systematic approach. Variability in natural processes can make an effect difficult to detect. If a response is detected, its cause may be difficult to separate among environmental or ecological factors that can modulate the response (Adams *et al.*, 1999). Some underappreciated requirements for an effective course of study include:

- a clear conceptual model defining expectations for metal-specific responses;
- explicit consideration (or at least conceptualisation) of:
  - spatial variability of exposure and response,
  - temporal variability of exposure and response,
  - variables that might provide alternative explanations of a detected response;
- adequate statistical power to detect expected responses and/or separate causes;

- progressive evolution of multiple lines of evidence that address the critical links in the metal conceptual model;
- explicit explanation of how the body of evidence satisfied ecoepidemiological criteria.

11.3.1.1 Distributions in time and space
Prologue. Exposures and biological responses to those exposures are variable in space and time. The importance of carefully documenting exposure in field studies is often underestimated. Careful documentation of variables like habitat, at multiple sites, can help separate causes of a complex response.

Experimentally, metal effects are deduced from responses to different exposure concentrations. In nature, differences in exposure occur at different scales through time or space. Such differences offer opportunities for 'experiments in nature' or systematic comparisons. But if a complex exposure regime is not well documented, it can confound detection of response and/or determination of cause and effect.

Spatially, typical exposure scenarios include (see also Burton *et al.*, 2002) distinct gradients and spatial mosaics. Gradients in contamination (e.g. downstream in a river or stream) are prone to covariance of several potential stressors or influential variables. If the distribution of contamination is complex (a mosaic), careful characterisation of exposures on the appropriate scale becomes especially important. Metal contamination of San Francisco Bay, for example, resembled a mosaic of hotspots superimposed on more moderate regional contamination in the 1980s. As metals were removed from effluents, the historic contaminants were dispersed across a broader (but less contaminated) region (Chapter 6). Failure to carefully characterise a complex distribution of exposure can lead to uncertainties about the causes of responses (e.g. Bayne *et al.*, 1988a).

Comparison of reference and test points (sites) represents a typical approach to understanding implications of contamination, whatever the distribution. Nature is complex, however. Replicating test points, replicating reference points and careful study of environmental characteristics other than contamination is essential (Box 11.4).

Biological and geochemical factors, on large and small scales, have an underappreciated influence on metal exposure. For example, each species' exposure to pollutants is determined by how that species 'samples' the complicated water, suspension, sediment milieu. A subsurface benthic deposit feeder experiences metal exposure from a microhabitat defined by the water and sediment in its burrow deep in the sediments. An epibenthic suspension feeder experiences, in the same habitat, metal exposure from suspended material, as influenced by the chemistry at the bottom of the water column. A predatory benthic fish integrates its exposure from food and water over a large area. Thus the linkage between environment, metal distributions and metal bioavailability is species-specific, in addition to environment-specific (Chapter 8). Conclusions about metal effects need to appreciate those exposure differences driven by biology as much as those driven by chemical distributions.

11.3.1.2 Temporal variability of exposure
Prologue. Metal inputs and ecosystem attributes can all vary with time at multiple scales, affecting metal exposures. The variations can take the form of dynamic stabilities (predictable fluctuations) on one scale, non-directional variability on longer scales, unidirectional long-term change, or episodic events. An understanding of the variability of exposure can provide opportunities to design observations to better understand metal exposures and effects.

Exposure and biological responses to contamination also vary with time. The simplest cases involve changes before and after an event. Long-term recovery is another type of change, if chronic contamination is gradually eliminated. Both of the above are unidirectional changes. The more common type of temporal variability is not unidirectional (or not obviously so). This variability adds complexity to exposures and adds uncertainties to characterisations of risks from contamination.

Human inputs of metals are more variable in time than is typically appreciated (Box 11.5).

Box 11.4 Using reference sites to narrow uncertainties about cause and effect

Convincingly demonstrating an effect of metal contamination on fish abundance in streams is difficult using reference sites if large differences in geology, geomorphology, channel conditions and habitat exist among the streams. Streams are also affected by a variety of stressors, especially if there are human activities in their watersheds.

So identifying the existence of a stressed population and separating metal influences from those of other stressors is a challenge. Hillman *et al.* (1995) addressed this issue when considering metal effects in the metal-contaminated Clark Fork River, Montana, USA and its headwaters (Silver Bow Creek) (see also Chapter 12). An approximately 200 km segment of the contaminated system was first classified into 19 discrete segments in terms of ecoregion, geology, geomorphology, land type association, valley bottom type, stream state type, elevation, valley grade, stream sinuosity, stream grade, dominant substrate, riparian vegetation type, and channel type. Segments in three creeks and three rivers were found that were uncontaminated but matched the classification system for a Clark Fork segment. Habitat surveys were conducted at each paired reference and treatment site. Habitat measurements included channel width, wetted perimeter width, riffle, run and pool widths, pool rating, bank angle, average and thalweg depths, substrate, bank cover, vegetation overhang, canopy cover, bank alteration, woody debris, sun arc, and bank undercuts. In addition, water quality characteristics were determined (dissolved oxygen, dissolved organic carbon, temperature, nitrate, conductivity, hardness, and alkalinity). A model was used to calculate the weighted useable area (WUA) of trout habitat for a given segment. Numbers and biomass of trout were divided by WUA to account for differences in habitat.

A comparison of adjusted trout populations showed that adult and juvenile trout densities in reference segments averaged 5.8 times greater than densities in the similar Silver Bow/Clark Fork segments. Three years later the studies were repeated and similar results were observed. Although there were fluctuations in trout densities over the course of the summer in each of the Clark Fork segments, similar patterns of change were observed in the reference segments, and the difference between Clark Fork and reference segments remained significant at all times. Total trout densities in the Clark Fork segments ranged from 7 to 188 trout/ha. Trout populations in reference segments ranged from 39 to 528 trout/ha.

These careful comparisons come as close as possible to documenting that population differences are not accounted for by physical differences in the environment. Repetition of the study further narrowed uncertainties about cause and effect. Overall, the population studies substantiated the uniquely low abundance and diversity of trout in the metal-contaminated Clark Fork River, compared to streams affected by stressors typical of Montana streams but unaffected by mine wastes.

The variability of human inputs is superimposed on the variability of nature, driven by climate, hydrology, water movement, water/sediment interactions and/or other physical, geochemical or biogeochemical processes. Documenting such processes requires consideration of several scales in time and space.

Many properties of environmental variability are difficult to predict. But *dynamic stabilities* also exist, when a factor varies in a repeated, somewhat predictable fashion (i.e. variability is non-directional but consistent on some time scale; see Ni and V fluctuations in Fig. 11.3). To some degree dynamically stable variability can be characterised by observation. Prediction of dynamically stable variability, from empirical or theoretical models (e.g. Thomson *et al.*, 1984), is possible if the processes driving the variability are understood. Variation of natural processes (including extreme episodes of natural disturbance) can be stressful themselves, if ecosystem properties range outside the limits to which the organism, population or community is adapted.

Fluctuations of environmental factors may have dynamically stable similar patterns on one scale (e.g. within each year) but may vary in amplitude on a longer scale (e.g. from year to year). For many attributes of ecosystems (or organisms) there is

Box 11.5 Fluctuation of dissolved metal concentrations

Fluctuations of dissolved metal concentrations are well known for streams impacted by mine wastes in their floodplains and/or their bed sediments. Traditionally, data collection occurs on timescales much longer than some fluctuations. For example, dissolved metal concentrations are typically determined quarterly in large-scale US monitoring programmes like the National Water Quality Assessment (US Geological Survey), or San Francisco Bay's Regional Monitoring Program. Most programmes monitor less consistently or less frequently. But even seasonal sampling is insufficient to characterise fluctuations. For example, targeted studies in the Clark Fork River, Montana, USA show the variability in dissolved metal concentrations on diurnal, weekly, monthly or seasonal scales. The hyporheic zones under the bed of the contaminated stream, or in the subsurface of contaminated floodplain deposits, are microbially and geochemically heterogeneous environments (Wielinga *et al.*, 1999) that contain acidic, metal-rich pore water (Benner *et al.*, 1995). Snowmelt, early spring flushing events or post-rain surges can transport these metals to the stream (Nagorski *et al.*, 2003) in pulses. Generally higher inputs of precipitation and snowmelt during spring and winter cause differences in flushing on a seasonal scale. Concentrations in a contaminated stream also fluctuate regularly from day to night, presumably as the activities of plants fluctuate between photosynthesis and respiration (and thus change pH: Nagorski *et al.*, 2003). Concentrations can be two- to threefold higher at night than during the day (Brick and Moore, 1996). The short-term variations may be as great as the longer-term variability; and both may have important implications for toxicity. Interestingly the US Environmental Protection Agency's water quality criteria include values that protect against pulses of a few hours to a few days duration (acute toxicity) as well as against more chronic exposures. But no field programmes exist that evaluate the former.

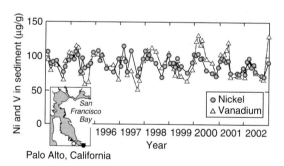

Palo Alto, California

Fig. 11.3 Dynamically stable seasonal variability in concentrations of V and Ni in sediments at a South San Francisco Bay site, as depicted by near-monthly sampling. The time series followed similar patterns for the two metals, suggesting that their variability was linked to the same driving force. In this case, the driving force was seasonal variation in fine sediment inputs. January to March runoff brings in particles with higher metal concentrations and a greater dominance of finer-grained particles. In summer and autumn, fine-grained sediments are progressively winnowed from the bed by strong daily winds (Thomson-Becker and Luoma, 1985). Sampling monthly, at a scale smaller than the unit of interest (season), was necessary to characterise this dynamic stability.

no directional change in this variability, however. In a situation where trace metal contamination increases or is remediated and declines, for example, long-term directional change may appear in the time series. If so, the long time series becomes the equivalent of a controlled experiment. The metal contamination shows unidirectional change, but everything else stays dynamically stable (Fig. 11.4). Metal contamination becomes a treatment under fluctuating (in the short term) but constant (in the long term) conditions. Changes caused by the metal contamination become easier to separate out in such circumstances.

Episodic events are another form of variability and may be one of the most important features of exposure to metals. For example, Hg concentrations in sediments at some sites in South San Francisco Bay (USA) often increased from 0.3 to 3.0 μg/g Hg dry wt within a month during the rainy seasons because of erosion of the tailings of an historic Hg mine in the watershed. That pulse of contamination was dissipated into the greater Bay

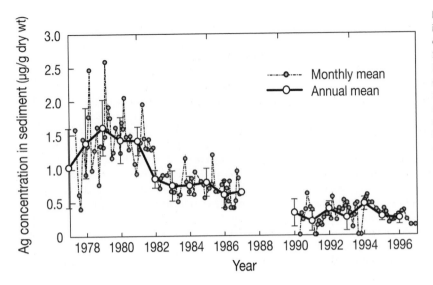

Fig. 11.4 Long-term trends in annual mean concentrations of Ag in sediments at a site in South San Francisco Bay. Each bar represents the standard deviation from the running mean of 9 to 11 samplings in each year. The unidirectional downward trend in Ag contrasted to organic carbon concentrations in the sediments which fluctuated seasonally and differed in peak concentration from year to year, but showed no unidirectional trend. Organic carbon is a simple indicator of food availability for benthic fauna. Improved reproductive capabilities of a resident bivalves in this case was attributed to the decline in Ag contamination because availability of food was constant in the long term, eliminating it as a causative factor in the biological trend. (Data after Hornberger *et al.*, 2000.)

over the succeeding months (Luoma *et al.*, 1997). Such events can be difficult to detect and/or characterise if sampling frequency is inadequate.

11.3.2 Fluctuations in biological properties

Prologue. Appreciation of temporal variability is important in detecting sublethal responses of organisms as well. Understanding the natural variability of the response is essential to separating changes in the response from fluctuations that occur in the absence of stress. Related measurements at the appropriate time or space scales can improve the probability of identifying relationships between metal exposure and biological stress.

Biological properties are also variable; but a repeatable 'dynamic stability' can be an element of the variability. Characterising both stochastic variability and dynamically stable variability is critical to detecting a response to metals (e.g. Box 11.6).

In order to determine the nature of temporal variability it is necessary to sample more frequently than the expected cycling. In other words, a seasonal cycle like that of condition index cannot be reliably determined by sampling seasonally (two to four times per year). In the above example, 10 samplings per year allowed a convincing demonstration of the seasonal cycle. Similarly, long-term trends are difficult to characterise if samplings of seasonal cycles are not equivalent from year to year. For example, annual mean condition index cannot be compared between a single determination of condition index in the spring of one year and a single determination in the autumn of the next year.

Experimental evidence is growing that many metals have effects on reproduction, either directly or indirectly. But detecting reproductive effects in nature is more challenging than often recognised. Seasonal variability in condition index is tied to the seasonality of reproduction. Thus the same principles that apply to understanding metal effects

Box 11.6 Fluctuation in a biological variable: condition index

Condition index is defined in estuarine bivalves as the weight of tissue within a standard shell-length animal. Condition in bivalves increases as glycogen and lipid are added to the tissue mass during the early stages of the seasonal maturation of the reproductive tissues. Then these energy reserves are lost as gametes are released and the animal reproduces. Condition index is also responsive to any environmental change that affects the energetics of the animal (Chapter 9). Reduced availability of food or exposure to pollutants, for example, results in reduced condition index. But detecting the response of condition index to stressors requires separating the stress response from the natural cycle (Cain and Luoma, 1990). For example, condition index fluctuated with a similar seasonal pattern in the bivalve *Macoma petalum* (as *M. balthica*) at two sites in South San Francisco Bay between 1988 and 1998 Luoma *et al.*, 1997 (Fig. 11.5). Near-monthly data collection was important to characterise the seasonal trend. With such data it is possible to average the proportionate difference from the annual for each month (called detrending the interannual data) then find the mean deviation for each month over several years. This calculation illustrates the 'typical', dynamically stable, seasonality of condition at a site (shown in Fig. 11.6a for Palo Alto).

Condition index is predictable in bivalves because it is linked to the reproductive cycle, which is

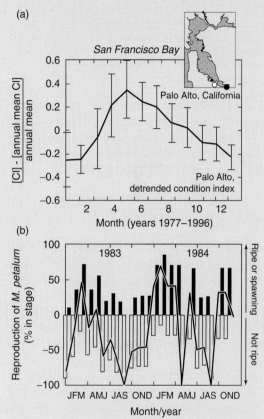

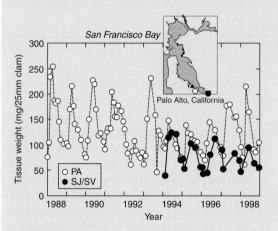

Fig. 11.5 Condition (soft tissue weight in a standard shell-length clam) in *Macoma petalum* (as *M. balthica*) at two sites in South San Francisco Bay from near-monthly determinations over 10 years shows the dynamically stable seasonal cycle. The cycle is similar at both sites but the amplitude of the maxima usually is not as great at the second site (animals are not as fat). Sites: PA, Palo Alto; SJ/SV, San Jose/Sunnyvale. (S. N. Luoma *et al.*, unpublished data.)

Fig. 11.6 (a) The central tendency for the annual trend in the seasonal cycle of condition index (CI) in *Macoma petalum* over the period 1977 to 1996. The central tendency for each month is calculated from the deviations from the annual mean of each month's measurement; then the mean and standard deviations (vertical bars) of all 12 months are combined on one plot showing the trend in proportional deviation from each annual mean. (b) The link between condition index and reproductive maturity is shown at an uncontaminated site 30 km from Palo Alto. The portion of a 10 animal sample that was ripe (ready) for reproduction or spawning is shown as a positive bar; the portion that was not ripe is shown as a negative bar, for each month over 2 years. Line, condition index. (S. N. Luoma *et al.*, unpublished data.)

usually tied to seasonal processes in the habitat (e.g. availability of food). An annual addition of glycogen and lipid occurs from March to July as *M. petalum* adds energy reserves before reproduction, and as reproductive tissues mature. Then energy reserves are consumed, partly in association with the release of gametes and recovery from reproduction. The pattern in both condition index and in maturity of the animals' gonadal tissue is shown in Fig. 11.6b for an uncontaminated site about 40 km north of Palo Alto.

Condition index in bivalves also affects the internal concentration of metal in the organism. Rapid addition of lipid energy reserves and gonadal tissue dilutes metal concentrations (weight increases faster than metal concentrations), partly because metals are not bound by the fat in these tissues. So the seasonal increase in weight produces a seasonal, dynamically stable, fluctuation of metal concentration. Changes in condition index, reproductive status and even metal bioaccumulation in bivalves are most interpretable if this seasonality is accounted for.

on condition must be applied to understanding effects on reproduction in natural waters. Most importantly, sampling through the reproductive cycle is essential, and interpretations must be made within the context of the season (Box 11.7).

11.3.3   Statistical power

Prologue. Statistical power is determined by the number of observations and the variability in the data set. More samples and/or less variability yield a better ability to detect a statistically significant response (i.e. more statistical power) and vice versa. Expertise in making observations is a less appreciated, but important, consideration in detecting effects.

Statistical power is a critical consideration in detecting metals effects in a natural setting. For example, laboratory studies show that primary effects of Se on fish and waterbirds are failure of eggs to hatch and deformities in young animals. Such effects are amenable to study in the field, but interpretations are very dependent upon the statistical sensitivity of the study design. If nests are not sampled randomly or are studied more intensively in one area than another, this can compromise comparisons and affect the probability of understanding effects (Skorupa, 1999).

In order to detect effects of Se on deformities and/or hatchability, the number of observations must be large. For example Skorupa (1998) developed a dose(Se concentrations in eggs)–response curve for hatching success for a species of wading

birds (stilt, *Himantopus mexicanus*) from field data. Data were collected from 410 nests across an area of variable Se contamination. A regression was developed between concentrations in eggs and reduced hatchability. The dose–response curve was used to forecast what concentration of Se would cause a 5% depression in the proportion of nests at a poorly known site. Reliably detecting such an effect would require study of 225 nests in the unstudied area. Deformities or reduced reproductive success were not found in the 23 nests that were sampled at the unstudied site. But because the sampling did not have sufficient statistical power, no conclusions could be drawn from the field effort.

11.3.4   Multiple lines of evidence

Prologue. If alternative explanations for effects are not eliminated, interpretations are subject to Type I error (concluding that effects are occurring when none exists). If the field studies are simplified to reduce costs or make them timely, they are subject to Type II error (concluding that there are no effects when effects exist). Common short-cuts that cause Type II uncertainties include inadequate characterisation of exposure regimes (in time or space), inadequate understanding of natural cycles in biological processes, oversimplified ecological surveys (e.g. one-time surveys of limited scope), inadequate reference sites, inadequate replication and simplistic analyses of community structure or function (Luoma, 1996). Multiple lines of evidence can narrow the likelihood of either Type I or Type II error.

Box 11.7 Detecting and interpreting sublethal signs of stress in a variable environment

At a contaminated site in South San Francisco Bay (USA) metal effects on condition were not strongly evident, but reproduction was affected by metal contamination. At the contaminated site the clam *Macoma petalum* was able to prepare its energy reserves for reproduction, but gonads never reached maturity. A near-monthly time series shows their reproductive status for three periods between 1974 and 1989 (Fig. 11.7). In this case reproductive status was defined by the proportion of 10 clams with gonads that were either ripe or spawning. When contamination was most severe (1974 to 1981), it was rare that reproductively mature individuals were present. As the contamination

receded, the seasonal occurrence of reproductively mature individuals became more similar to patterns observed at unaffected sites (Fig. 11.7b versus Fig. 11.6 and Fig. 11.7c; Box 11.6). There was no unidirectional trend in organic carbon, Fe, Mn, dissolved oxygen, phytoplankton blooms or salinity over the 15 year period at this mudflat (Hornberger et al., 2000). Nor were contaminants other than Ag and Cu present at concentrations exceeding regional levels. The reproductive status of *M. petalum* was impaired when metal contamination was at its worst, and improved as contamination declined. Multiple variables and frequent sampling over a period of years were key to uncoupling confounding sources of effects, and thus, demonstrating which biological variables were most vulnerable to effects from metal contamination. (Partly after Hornberger et al., 2000; partly unpublished data from S. N. Luoma et al., 1997.)

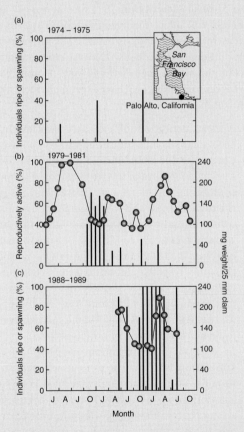

Fig. 11.7 Reproductive status in near-monthly samplings of the bivalve *Macoma petalum* (as *M. balthica*) on the contaminated Palo Alto mudflat in South San Francisco

Caption for Fig. 11.7 (*cont.*)
Bay at three intervals between 1974 and 1989. Contrast cycles to Fig 11.6 for a spatial comparison. (a) In 18 of 24 months sampled between January 1974 and January 1975, ripe individuals were found only three times. Mean Cu concentrations were 225 µg/g dw and mean Ag concentrations were 80 µg/g dw in the surviving animals. (b) Gonadal samples were available in 20 of 28 months between June 1979 and October 1981; condition data (tissue weight in a standard shell-length clam) were also available (line). A small proportion of reproductively ripe individuals were found in the typical reproductive season in 1979–80, but few were found in 1980 or 1981. Mean Cu concentrations were 266 µg/g dw; mean Ag concentrations in the bivalves were 90 µg/g dw over this period. (c) Reproductive cycles were similar to those expected from uncontaminated sites in 1988 (compare to Fig. 11.6). Nearly all individuals were reproductively mature during the expected season and some were mature during most of the year. In this case, 11 of 14 months were sampled between June 1988 and Sep 1989. Mean Cu concentrations had fallen to 35 µg/g dw and mean Ag concentrations were 11 µg/g dw in the bivalves in 1989. (After Hornberger et al., 2000.) Because no other characteristics of the mudflat changed during this period, the improvement in reproduction in 1988 to 1989 was attributed to the decline of contamination (i.e. the earlier loss of reproductive capability was caused by the contamination). The improved reproductive status continues until the present, in coincidence with the greatly reduced bioavailable contamination.

Evaluating the weight of evidence from multiple sources usually narrows uncertainties about cause and effect. Each type of evidence, alone, might be subject to a considerable amount of uncertainty. But if the aggregated evidence points toward the same conclusion (e.g. one factor as a cause), the total uncertainty is reduced (assuming that the evidence is analysed within a systematic, transparent and logical framework: Burton *et al.*, 2002; Forbes and Calow, 2002).

Some studies have suggested combining multiple lines of evidence into a single number that describes the degree of impairment. Because the individual lines of evidence convey different types of information, however, simplistic indices can lose much of the richness necessary for a convincing interpretation (Burton *et al.*, 2002).

Even with multiple lines of evidence, sensitivity can be limited if the approach is simplistic. For example, one of the early contaminant risk assessments in San Francisco Bay employed a triad of methods to evaluate effects across several locations (Chapman *et al.*, 1987). One-time collections of contaminant concentrations in sediments and indices of benthic community structure were accompanied by toxicity tests with local sediments. Bioassays, chemistry and community structure all indicated stress in areas of extreme degradation (with multiple stressors). But results were ambiguous in moderately contaminated or uncontaminated areas. Greater understanding of the Bay system later showed that all three approaches yield quite variable results in this water body. The statistical power of the 'one shot' approach was sufficient to detect extreme adverse effects but not sufficient to resolve impacts where stress was less extreme.

11.3.5 Conceptual model of ecological change caused by metals

The presence of a species in a polluted area may be more a question of life history strategies than the tolerance of adverse environmental conditions. If so, considerable doubt must be placed on the ecological relevance of toxicity tests.

(Gray, 1979)

Prologue. As the stress increases, a community increasingly 'simplifies' as species become rare or disappear. Species vary widely in the likelihood that they will be affected by trace metal contamination. Physiological and ecological traits determine each species' phylogenetic window of vulnerability. In addition, different trace metals affect different species. The concentration at which effects occur differs among environments. The crucial questions are which are the most vulnerable species in a particular setting, and how much trace metal causes which species to disappear.

An explicit description of expectations of how metals will affect biodiversity and ecosystem services in a water body is essential to interpreting the influences and risks of metal contamination. Examples where metal effects were demonstrated with limited ambiguity have several characteristics in common:

- A clear definition of exposure was essential; ecological responses followed those exposures.
- Changes in the abundance of all organisms is not a reliable measure of the stress imposed by metal contamination. Along a stress gradient, it seems the total biomass and total number of organisms would initially rise then fall; and this may happen in extreme conditions. But this sequence of events is rare in natural waters. One reason is that the response in an affected community is that some species actually increase in abundance while others become rare or disappear (Gray, 1979). Elimination of some species appears to favour the survival of other species.
- In general, the elimination of species by contamination results in a well-known pattern of response in the community, termed ecosystem 'simplification' (Woodwell, 1970). With very strong contamination, species numbers fall and/ or species diversity declines as stress increases. But species numbers and species diversity are not especially sensitive measures of the changes that

occur in the types of organisms that survive and the types that are extirpated by contamination. Elimination of sensitive species occurs at concentrations lower than those that cause measurable loss of species diversity. Species can disappear with no detectable change in the more general community measures.

- The differing sensitivity among species is partly driven by an inherent physiological (and/or phylogenetic) window of tolerance to metal enrichment typical of each species. Life evolved with potentially toxic metals in the environment. Some physiological traits undoubtedly evolved specifically to mitigate effects of natural toxins like metals (e.g. Chapter 9; Rainbow, 2002; Vijver *et al.*, 2004). The specific mechanisms deployed, and the energetic cost of deployment, determine each species' window of vulnerability to metals and/or to different types of metal exposure. That window is probably fixed (at least to some degree) for the species; i.e. it helps define the phenotype of the species. In a polluted environment, depending upon the type of pollution, ecological effects of metals are manifested as these windows are exceeded. Those species for which costs become too great will eventually disappear; those with wide windows of tolerance to that metal in that setting will survive. These differences in tolerance combine with ecological conditions to create typical responses along a contamination gradient in the abundance of sensitive, moderately sensitive and tolerant species (Fig. 11.8). These distributions can be quite detectable.

- Functional ecology (or life history characteristics) as well as inherent sensitivity to the contaminant together determine which species are extirpated and which thrive (Gray, 1979; Forbes and Depledge, 1992) (Box 11.8). For example, flexible life history strategies may be more advantageous than specialised strategies, in general (Gray, 1979). Where sediments are locally contaminated, organisms with less intimate contact with sediments may have advantages: direct development, brooding (avoiding contact of young life stages with sediments), suspension versus deposit feeding. Pelagic larvae may facilitate survival

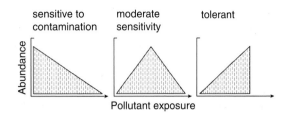

Fig. 11.8 Typical species abundance patterns can develop along a gradient of contamination, as depicted in this theoretical conceptualisation. Sensitive species will never be most abundant where contamination is present above a threshold. Species of moderate sensitivity will be absent where contamination is most severe but may be very abundant in areas of moderate contamination, if they react favourably to the absence of sensitive species. Tolerant species may be abundant where contamination is severe, but decline in abundance if ecological interactions with other species affect them unfavourably in areas where contamination does not limit the community.

of an adult population if young animals cannot develop locally. Examples exist of all of these (e.g. Calow *et al.*, 1997; Box 11.8).

- Sublethal signs of stress may be evident in moderately tolerant and/or tolerant species (the survivors). These species may also be good sources of biomarkers.

- Reductions of ecosystem services and a cascade of ecological effects can have implications for species not directly affected by metals. Changes in a community can occur within one generation if toxicity occurs to young life stages of a vulnerable species. But it might take several generations of effects on recruitment to eliminate (or reduce) the population of another species. Changes can also occur over decades if there are effects from the loss of critical species that cascade through the food web. Over long periods of time, upper trophic level species ultimately can be affected by the latter (Box 11.9).

11.4   CASE STUDIES FOR METAL EFFECTS IN NATURAL WATERS

Prologue. Co-occurrence is necessary but not sufficient to prove metal effects.

Box 11.8 Inherent physiological sensitivity and life history strategies determine which species will be eliminated by metal contamination and which will survive

The 30 year study of the metal contaminated mudflat (hotspot) in San Francisco Bay provided examples of how life history influences what species survive metal contamination and which return when contamination is eliminated (Hornberger *et al.*, 2000). As seen earlier, contamination was identified by tissue analysis of resident species and elevated concentrations of Cu and Ag in sediments. The contamination declined after the mid-1980s and recovery provided evidence of what had been affected (Box 3.10). Biochemical signs of stress were observed in surviving species during the period of contamination. These included induction of metallothioneins in a surviving clam (*Macoma petalum* (as *M. balthica*)) (Johansson *et al.*, 1986), along with the reproductive anomalies in that species (Hornberger *et al.*, 2000). Surviving *M. petalum* were also tolerant to metals (Cu and Ag) compared to individuals from an uncontaminated mudflat (Cain and Luoma, 1985). This organism apparently survived because it successfully reproduced on other mudflats and its life history included a pelagic larval stage that was able to immigrate regularly to the contaminated mudflat.

Recovery of the community also occurred after the contamination receded. The ecological sign of recovery was an increase in the number of species that are present in abundance (Fig. 11.9). Contaminated environments are typically characterised by a few species that are very abundant and a large number that are more rare. Less stressed communities have more species that are abundant. The type of species that recovered in San Francisco Bay was a function of life history. Those organisms whose life history involved direct exposure to the contaminated sediments were most affected by the contamination (Shouse, 2002; Fig. 11.9):

- Suspension feeders survived. The contaminated community was predominantly suspension feeding animals that could feed on the less contaminated material brought in by physical processes from other mudflats.
- Surface-dwelling species survived. Species that dwelt in the subsurface sediments were often absent

during the period of contamination. Surface-dwelling species presumably had access to less contaminated suspended sediments during the period of contamination.

- Species that laid their eggs directly in the sediments were affected. Species that brooded their young (away from the sediments) or had pelagic larvae were predominant when the area was most contaminated.
- Predators disappeared (became more abundant when the contamination receded). This is an increasingly common observation in metal-contaminated environments, but the mechanism is not known.

Thus life history at least partly determined which species survived contamination in the South Bay hotspot, as predicted by Gray (1979); natural history traits also determined the nature of the ecological recovery.

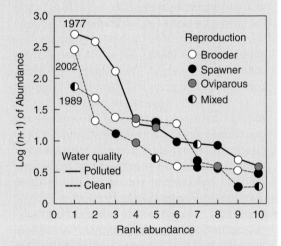

Fig. 11.9 Rank analysis of most common species (categorised by life history) at the South San Francisco Bay mudflat described in Fig. 11.7, before and after contamination receded. A number of species increased in abundance as the contamination began to recede; the most abundant species during contamination declined in relative abundance. Thus the community became more evenly distributed among species with different life histories; one attribute of greater diversity. The rank of species also changed, depending upon their functional ecology. (After Shouse, 2002.)

Box 11.9 Ecological cascades can affect upper trophic level species

One of the most enlightening studies of how pollution is manifested over time was the open system studies of the acidification of Lake 223 in the Experimental Lakes Area of Canada (Schindler *et al.*, 1985). The lake was acidified gradually and studied over an 8 year period. Even this was not sufficient to see the complete manifestation of ecological change. The most rapid response of the lake system was that some sensitive species disappeared (some phytoplankton and small fish). In most cases these were replaced by other species, with no detectable net loss of diversity, production and biomass. The most vulnerable were short-lived species with poor powers of dispersal. Within a few years the assemblage of phytoplankton and small fish was drastically different from the original. As time went on, algal mats eventually formed, probably because of the food web disruption. Crayfish began to disappear because of chemical effects and related parasitism eroded recruitment. As the ecosystem simplified, food organisms for larger fish disappeared. Eventually the top predators, lake trout and white sucker, declined in condition, then began to decline in numbers. Direct toxicity, failure of young to survive, disruption of

food chain relations and disappearance of spawning sites (because of the algal mat formation) all contributed to simplification and ultimately the demise of the top predators. The top predator in these lakes, the lake trout (*Salvelinus namaycush*), was not directly sensitive to acidification, but its major prey items were progressively eliminated as the lake pH declined. The lake trout exhibited severe physiological and reproductive effects that could be linked to malnutrition. Such food web mediated effects are not accounted for in the aquatic toxicity testing approach (Campbell *et al.*, 2003), but need to be accounted for in an accurate ecological risk assessment.

The process of predator extinction was projected to take 10 years or more. Simple measures like phytoplankton production, nutrient cycling and organic matter decomposition were not affected. It was the more subtle malfunctions that vastly disrupted the original system. Sufficient stress from any source (including metals) ultimately can be the cause of such a cascade of effects; although the specific signs of effects will differ with the stress. The complexity of such events is the reason that it remains important to be alert to Type II error before concluding 'no effects' from what otherwise appears to be a significant metal exposure.

A first step in testing for adverse effects from metals in nature is to determine if contaminant concentrations in the environment (particularly in biota or other environmental media such as sediments) can be correlated with endpoints indicative of biological response. Finding a meaningful or convincing correspondence between simple toxicity tests or simple measures of exposure and ecological change is relatively straightforward in extremely contaminated environments (Luoma, 1996). Uncertainties are relatively easy to eliminate where exposure and effects are very large. Chronic or episodic acid mine drainage are examples (Chapter 12). In Chapter 9, we also cited instances where simple, strong correlations were observed between community structure (e.g. number of species) and metal contamination in sediments (Lande, 1977; Rygg, 1985). But as situations become more complex, uncertainties multiply and confound

conclusions. Co-occurrence is a necessary first step linking metals to ecological change. But further evidence must follow because correlations do not prove cause and effect. Oversimplification of approach (Chapter 2) can add to the difficulty of unambiguously determining causes of change in metal contaminated natural waters.

11.4.1 Co-occurrence of contamination and metal effects in marine communities

Prologue. Multiple lines of evidence show that metal contamination can simplify marine shore communities at both the small (meiofauna) and large (rocky intertidal) scales. Physiological traits and functional ecology determine which species are most vulnerable. Physical factors influence how effects are manifested. Elimination of keystone species is possible.

Box 11.10 Effects of mine wastes on intertidal marine habitats: Chañaral, Chile

Meiofauna

The El Salvador mine on the slopes of the Andes Mountains in northern Chile discharged more than 150 million tones of untreated metal-contaminated tailings into the marine intertidal zone between 1938 and 1990 (Lee and Correia, 2005). The tailings from the mine visibly affect approximately 100 km of sandy shore and rocky intertidal shoreline. Local authorities claimed to achieve a reduction in the input of tailings in 1990 by building a ponding system to settle out particulate material. Historic particulate waste materials remained in the system, and appear to be a source of continued metal inputs. Effluents from the ponds continue to be discharged into the sea. Despite the removal of particulate material from the discharge, no decline in Cu concentrations in the ocean water was evident in 2002, compared to 1990.

Metal contamination and meiofaunal communities were compared at 12 sites of varying contamination on the high-energy sandy shores (Lee and Correia, 2005). Meiofauna are predominant type of community in such environments: the community of small organisms living within the sand grains of the high-energy beaches. Larger fauna are not suitable for study on sandy shores because few live there exclusively.

The most important changes from clean-up of the tailing discharges were persistently high Cu concentrations in coastal waters and beach gradation. At their maximum, Cu concentrations in water were 1449 µg/L Cu, nearest the site of mine tailings input. Labile Cu was determined by Diffusive Gradient Thin Films (DGT; Chapter 5) to be 41 µg/L Cu. Sites north of the visible tailings had labile Cu concentrations of 6.43 µg/L and sites >100 km south had 1.93 µg/L labile Cu.

The meiofaunal assemblages at the contaminated sites had significantly lower densities (Fig. 11.10) and taxa diversities when compared to the least contaminated reference sites. Multivariate analysis (Chapter 9) showed that the assemblage of species that made up the community also changed with the degree of contamination (Fig. 11.11). Unique assemblages occurred in the contaminated zone, the moderately contaminated area and the least contaminated area. The authors considered factors other than Cu that might explain the results. Owing to the desert nature of this area other forms of contamination, such as industrial and agricultural effluents, were absent. Although several metals were enriched in

concentration, only the Cu concentrations appeared to be associated with the distribution of the tailings. Another possible confounding variable in the area was sediment grain size, which greatly affects the diversity and density of meiofauna. The authors established that the variation in natural sediment grain size from beach to beach was not a significant factor in the observed differences in the meiofaunal assemblages. One group of animals, known to be physiologically sensitive to chemical contaminants, was consistently absent in the contaminated areas: harpacticoid copepods. They found that another group, turbellarian flatworms, was more abundant when the harpacticoids were absent. Some foraminiferans were also reduced in the contaminated zone. The sensitivity of the harpacticoids and the abundance of turbellarian flatworms are observed in other contaminated situations (e.g. Sunda *et al.*, 1990).

Toxicity tests with the local waters verified direct toxicity to harpacticoids, a similar effect as observed in the field. Thus a metal-specific impact correlated with

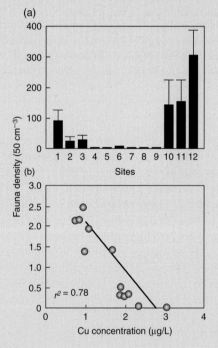

Fig. 11.10 Abundance of the sensitive meiofaunal harpacticoid copepods as a function of distance along the shore in Chile (a) showing that abundance declined in the area where mine wastes were deposited. (b) Abundance of harpacticoids also correlated inversely with Cu concentrations in water: where Cu was higher fewer, harpacticoids were found. (After Lee and Correia, 2005.)

exposure. Expectations of the conceptual model were met with regard to species distributions. Alternative explanations for the co-occurrence of exposure and effect were eliminated. The field results were consistent with an experimental explanation. The suggestion that metal contamination was responsible for simplification of the meiofaunal community was convincing.

Macrofauna

Study of macrofaunal diversity in the rocky intertidal zone is typically more difficult than the study of more compact communities like the meiofauna; but of course it is more relevant to the interests of most risk assessors and managers. Many factors affect the diversity of rocky intertidal communities, making demonstration of causation particularly challenging.

Total dissolved Cu concentrations were determined in water, which are not directly comparable to the labile Cu determined in the meiofaunal studies. Total Cu ranged across the intertidal sites from 8.7 μg/L Cu to 34.1 μg/L Cu (Medina *et al.*, 2005). An intact community of survivors occurred throughout the study area 13 years after tailings discharge ceased. However, less than half of the species recorded at the control sites were found at the metal-impacted sites; a typical response to stress. The site with the highest Cu concentration had the least number of species. In particular, the diversity of sessile organisms (organisms that do not move over substantial areas) correlated negatively with dissolved Cu concentration (Fig. 11.12). The diversity of mobile organisms did not. Presumably mobile organisms could move in and out of the zone of contamination and/or

could avoid contamination. Thus effects were not especially evident unless functional ecology was considered. A multivariate statistical analysis was the most effective approach to illustrate the statistically significant differences in community composition between reference and contaminated sites.

A major factor in the simplified nature of the rocky shore community was the absence of ecologically key species. Starfish typical of the Chilean rocky shore (*Heliaster helianthus* and *Meyenaster gelatinosus*) were absent from sites along about 100 km of shoreline surrounding the waste discharge. These starfish were found at two sites 100 km apart, to the north and south of the discharge; but were not present in the most contaminated zone. Starfish are voracious predators that play an important role in determining the relative abundances of herbivorous species in the rocky intertidal zone, thereby promoting diversity of undisturbed marine rocky shore communities (species with such effects are called keystone species). The lack of predators may also have indirectly regulated algal abundance, suggested by the absence of the kelp *Lessonia nigrescens* in the same zone. *Lessonia nigrescens* is also a keystone species with respect to biological diversity owing to the rich diversity of organisms living within its holdfast and its influence in predator–prey interactions. The authors emphasised that their observations were mainly qualitative and represented only the starting point for future assessments of links between contamination and biodiversity at these impacted areas. Elimination of other explanations for variability would be an important next step.

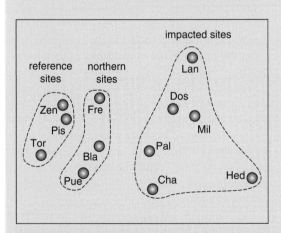

Fig. 11.11 Similarity analysis among assemblages of meiofauna. Assemblages differed in areas of different contamination on the sandy shore affected by mine wastes in Chile. (After Lee and Correa, 2005.)

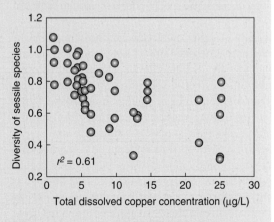

Fig 11.12 Diversity of rocky intertidal species in communities associated with copper mine tailing discharges in northern Chile as a function of copper concentration in seawater, shown as a simple correlation. (After Medina *et al.*, 2005.)

Box 11.10 provides two detailed examples of eco-logical change in a contaminated coastal location: a site of marine disposal of mining wastes on the northern coast of Chile. Additional coastal and estu-arine examples occur in the chapters that follow. The Chilean studies take advantage of unique envir-onmental features and accumulate lines of evidence beyond experimental toxicology. They show how studies from natural waters can raise mechanistic questions that would never result from experiments.

11.4.2   Stream benthos: sensitive species, patterns of distribution and effects on ecosystem services

Prologue. At lower levels of contamination, metal effects on streams are manifested as absence of a pre-dictable suite of species. In streams from North America, the UK and New Zealand, some mayfly (Ephemerop-tera) species, particularly of the families Heptageniidae and Ephemerellidae are sensitive to metals. Those species (especially mayfly species) that tended to accu-mulate high concentrations of internal Cu and Cd, in a form available for binding to sites active in causing toxicity, tended to be absent from contaminated stream sites. Generic indices, like invertebrate abundance, diversity or species richness, are indicative of extreme effects, but subject to Type II error (underestimation of effects) at moderate levels of contamination. Elimin-ation of vulnerable species affects ecosystem services.

A body of study on stream benthos confirms the conceptual expectations of how ecological impair-ment by metals is manifested. The benthic commu-nity in rocky bottom streams around the world is primarily composed of aquatic insect larvae. Multi-species toxicity studies, observations from nature, and mechanistic studies all point to a similar pat-tern of ecological response to metal contamination in this group of organisms. Similar ecological responses are observed along single gradients as within spatial mosaics of contamination and are repeated in different stream environments from widely different areas. The consistency of findings leaves little uncertainty about whether and how metals affect such communities.

11.4.2.1 Toxicity tests
Prologue. In situ toxicity tests with stream insects defined the vulnerable groups of organisms and thus the pattern of effect expected in metal contaminated communities from rocky bottom streams.

A crucial step in understanding metal effects in stream environments was the development of multi-species experimental microcosms to test tox-icity (Clements et al, 1988; Kiffney and Clements, 1994). Specimens were obtained for the toxicity tests by placing boxes of stones in Rocky Mountains (USA) streams for 40 days. The boxes were colonised by local fauna, including the insect larvae, and were then taken to the laboratory intact. The entire box was submerged in microcosm streams for con-trolled exposures to dissolved metals for 10 days.

Results from a variety of experiments established a hierarchy of vulnerability among the different taxa that typically inhabit streams. Mayflies and stoneflies were reduced in abundance or richness at the lowest metal concentrations. In particular heptageniid mayflies were highly sensitive to trace metals; especially some of the taxa that are common in pristine streams, including species of the genera Epeorus, Rhythrogena and Drunella. Numbers of taxa (overall), numbers of mayfly taxa, and summed numbers of Ephemeroptera (mayflies), Plecoptera (stoneflies) and Trichoptera (caddis-flies) taxa (EPT) showed statistically significant concentration–response relationships with metal concentrations. But EC_{10} values were generally two to three times greater for such taxon-lumped measures as compared to measures of metal-sensi-tive taxa (like abundance of heptageniid mayflies).

When the same approach was used with organ-isms from New Zealand streams (Hickey and Clem-ents, 1998) the results were similar. Abundance and species richness of mayflies were a sensitive meas-ure of metal effects. Such studies suggested that responses to metal contamination in streams are predictable and should be amenable to validation in natural waters.

The bioassays also addressed the question of potential interactions among metals. Community-level responses were compared among Zn alone, Zn + Cd, and Zn + Cu + Cd (Clements, 2004).

Responses of macroinvertebrates to a mixture of three metals were generally greater than responses to individual metals. Thus an index that assumed additive toxicity appeared to be justified to define exposures in water bodies where mixtures of metals were observed. To this end, a cumulative criterion unit (CCU) was defined as the ratio of each measured metal concentration to its hardness-adjusted US Environmental Protection Agency criterion. The ratios were then summed. For each metal:

$$CCU = \Sigma\; m_i/c_i,$$

or

$$CCU = Zn_w/Zn_{criterion} + Cd_w/Cd_{criterion} + Cu_w/Cu_{criterion}$$

where m_i is the measured concentration of the ith metal and c_i is the hardness-adjusted criterion value for the ith metal. Criterion values are adjusted for water hardness via a formula provided by the US Environmental Protection Agency. This parameter can influence the toxicity of metals to aquatic organisms (Chapter 16).

This body of work established the pattern expected for metal-specific effects in nature, although short exposures and lack of a formal dietary exposure may somewhat limit quantitative extrapolation of toxic concentrations (Clements, 2004).

11.4.2.2 Mechanisms
Prologue. Species that tended to disappear in metal contaminated streams were those that showed enhanced uptake with limited capabilities to detoxify metals.

The differences in sensitivities among stream species are explained mechanistically by differences in how insect species bioaccumulate and detoxify metals. Buchwalter and Luoma (2005) showed that metal uptake rates and loss rates varied widely among aquatic insect larvae; Martin et al. (2007) showed that wide differences in detoxification capabilities also existed. Some of the aquatic insects that were most likely to disappear in metal contaminated environments (e.g. species of the mayfly genera *Rhythrogena* and *Ephemerella*) had rapid uptake rates and extremely low rate

constants governing Cd and Zn loss. Insect larvae that tended to be abundant in metal-contaminated streams (some of the trichopteran caddisflies like species of *Hydropsyche*) had slow uptake rates, rapid loss rates or strong detoxification capabilities (Cain et al., 2004, Martin et al., 2007). These predictions were consistent with the nature of the insect larvae communities found in field studies in both the Clark Fork River, Montana (USA) and the streams of Colorado (USA) (Cain et al., 2004; Martin et al., 2007). Bioaccumulation of Cu and Cd was reduced in species that were common in the metal contaminated Clark Fork River, such as *Hydropsyche* spp. and *Arctopsyche grandis*. Bioaccumulation was considerably greater in species typically absent from the Clark Fork River. The latter include ephemerellid and heptageniid mayflies like *Serratella tibialis* and *Timpanoga* sp. (Fig. 11.13). Sensitive strong bioaccumulator species, like *S. tibialis*, did not sequester metals as efficiently into detoxified fractions within their cells (metal specific binding proteins or intracellular granules: see Chapter 7),

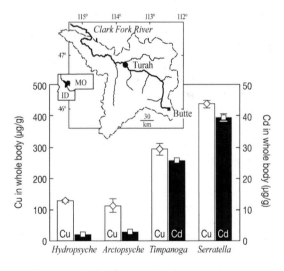

Fig. 11.13 Cu and Cd concentrations in four taxa of insect larvae at a moderately contaminated site in the Clark Fork River, Montana, USA. Mayfly taxa (*Timpanoga* sp. and *Serratella* sp.) accumulated the highest Cu and Cd concentrations and disappeared upstream as metal contamination increased. Caddisfly taxa (*Hydropsyche* sp. and *Arctopsyche* sp.) accumulated less Cu and Cd than the mayflies and persisted at the more contaminated sites. (Data from Cain et al., 2004.)

Box 11.11 Changes in benthic community structure along the contamination gradient in the Arkansas River, Colorado, USA

Elevated metal concentrations in the Arkansas River were correlated with alterations in benthic community structure (Clements *et al.*, 2002). Where metal concentrations in water, sediment, and periphyton were greatly elevated, statistically significant reductions occurred in species richness of mayflies and abundance of heptageniid mayflies. Exposure of the community to bioavailable metal in the Arkansas River was verified by occurrence of high bioaccumulated metal concentrations in several metal-tolerant biomonitor species, including the caddisfly *Arctopsyche grandis*. Clements *et al.* (2002) explained their results and conclusions:

> Critics argue that although inferential statistics can be used to show differences among locations, these differences cannot be attributed to a specific cause (e.g. metal effects). This is a serious issue because the typical motivation for conducting biomonitoring studies is to attribute differences to suspected stressors. We used inferential statistics (e.g. ANOVA) in this study to show that observed differences in ecological indicators between (uncontaminated) upstream and (contaminated) downstream sites were greater than those expected by chance alone. Correspondence between the level of metal pollution and impairment of benthic communities is a necessary but not sufficient condition to demonstrate causation because factors other than metals cannot be eliminated. Our hypothesis that heavy metals caused reduced species richness and abundance of heptageniid mayflies was bolstered by an exposure assessment, microcosm experiments, and field experiments. However, the strongest support for this hypothesis was provided by temporal correlation between improvements in water quality and benthic community structure after remediation. Based on these analyses, we conclude that heavy metals in the Arkansas River caused alterations in benthic community structure.

as did *Hydropsyche* spp. (Cain *et al.*, 2004). These findings specifically linked species absences with physiological vulnerability to metal effects.

11.4.2.3 Validation of toxicity testing with field observations

Prologue. Gradients of contamination, regional mosaics of contamination and recovery from metal contamination all offer opportunities to determine co-occurrence of trace metal exposure and effects. In all three types of exposure the patterns of change are consistent with predictions from mesocosm toxicity tests and mechanistic understanding.

Studies of contaminated stream gradients repeatedly show the same ecological response to metals in invertebrate communities whatever the location; loss of sensitive mayfly species is the first response to metal contamination. Complementary findings have come from single large watersheds such as the Coeur d'Alene in Idaho, USA (Maret *et al.*, 2003), the Clark Fork River, Montana, USA (Cain *et al.*, 2004) and the Arkansas River, Colorado,

USA (Clements, 2004) (Box 11.11). It is also common along metal contamination gradients that some species are not affected. Maret *et al.* (2003) suggested that these may make the best biomonitors; for example, free-living caddisflies (*Hydropsyche* spp., *Plectrocnemia* spp). The metal concentrations in biomonitors were determined where the number of mayflies was halved in the contaminated Coeur d'Alene River, compared to reference sites (EC$_{50\text{-field}}$). The 50% reduction in mayfly abundance was coincident with tissue contamination in the tolerant insect *Hydropsyche* sp. at mean concentrations 707 µg/g Zn dw, 142 µg/g Pb dw and 6.7 µg/g Cd dw. D. J. Cain (unpublished data) made a similar comparison in the Clark Fork River (Montana), where Cu was the most important contaminant. A 50% reduction in heptageniid mayfly abundance occurred at sites where *Hydropsyche* spp. accumulated concentrations of 60 to 80 µg/g Cu dw.

Correlations among confounding variables are a disadvantage of stream gradient studies (several factors often vary simultaneously from upstream to downstream). Confounding variables are not as problematic in large regional mosaics of metal contamination. Clements *et al.* (2000) used stepwise multiple regression analyses to investigate the relationship among metal contamination, physicochemical characteristics of several streams and benthic community measures and physicochemical characteristics at 78 randomly selected sites in the southern Colorado ore belt (Box 11.12). They found that trace metal concentration was the most important predictor of benthic community structure at these sites, and that effect of trace metals was an important factor in large-scale spatial patterns in benthic macroinvertebrate communities in Colorado's mountain streams. Furthermore, the types of changes in benthic communities were similar to those found in gradients and experiments, and similar to the effects predicted from physiological studies of sensitive and tolerant species.

The stream studies were examples of the differences in measures of community structure. Overall abundance of all animals (aka number of individuals) is not a reliable measure of effects in any community. Reduced abundance is a typical response in controlled experiments where migration is not possible; and in some instances of extreme contamination. But in nature, tolerant (usually opportunistic) species immigrate into a contaminated site and/or become highly productive, replacing species that have suffered toxicity (Clements, 2004).

The number of species present (species richness) is a reliable measure of severe impacts, but is not especially sensitive. Consistently, the species richness of the benthic community is reduced in the most contaminated areas. As contamination declines (in time or space), species richness may recover, but important changes in the nature of the community may be present still. Sensitive, moderately sensitive and tolerant species show distributions expected from the conceptual model for how metal effects are manifested (compare Figs. 11.8 and 11.15).

The source of exposure differs among species. Some of the sensitive species appear to respond to metal contamination in the water column (Courtney and Clements, 2002), while others appeared to be affected by metal contamination in their diet (Prusha and Clements, 2004). Functional ecology is also important in determining which species

survive in streams. For example, similar to indications in marine environments, predator species might also be reduced in number by metal contamination in streams (Fig. 11.14).

11.4.2.4 Loss of ecosystem services
Prologue. Metal-driven changes in what species are present can have effects on processes driven by the community as a whole (community function).

The disappearance of sensitive species also appears to have important effects on the functions performed by the invertebrate communities in streams (the services provided by the ecosystem to support a complex food web). Leaf litter breakdown declines (Carlisle and Clements, 2005) in contaminated mountain streams. This is similar to the contaminant-induced reductions in detritus processing observed in shallow gradient English streams (Maltby *et al.*, 2002). In the latter, the effect appeared to be the result of a reduction in feeding by detritivores like the freshwater amphipod crustacean *Gammarus pulex*. *Gammarus pulex* suffer direct toxicity when in contact with contaminated sediments. Loss of this process has implications for the availability of energy to the rest of the food web.

A significant reduction in community respiration is another effect on ecosystem services (Carlisle and Clements, 2005). The most likely cause is effects on periphyton communities, which are highly sensitive to metals (Clements, 2004). Carlisle and Clements (2003) concluded that total production attributable to algae and animal prey declined in contaminated streams. A specific reduction in herbivory was attributed to effects on the highly sensitive heptageniid mayflies, which are an important group of herbivores in undisturbed cobble bottom environments.

11.4.2.5 Links to upper trophic levels
Prologue. Although not widely appreciated, the disappearance of predators often accompanies metal contamination. The most likely mechanism is a cascade of losses of ecosystem functions critical to the predator, although much remains to be learned about this phenomenon.

Box 11.12 Effects of trace metal contamination on stream benthic macroinvertebrate community structure

Trace metals and other materials released from 10 000 abandoned mines are evident over 2600 km of streams in southern Colorado, USA (Clements et al., 2000). Data were available from 95 sites in 78 randomly selected streams in this region. Thirty-one of the streams were relatively uncontaminated and 64 were contaminated. Data included physicochemical characteristics, trace metal concentrations, and benthic macroinvertebrate community structure. Each site was placed into one of four metal categories: background (CCU less than 1), low (CCU less than 2), medium (CCU 2–10; 19 sites) and high metals (CCU >10; 13 sites).

According to correlations and multivariate statistics, both the number of heptageniid species and the abundance of heptageniid individuals were significantly reduced when summed metal concentrations exceeded 10 times the CCU. Comparable concentrations in the Arkansas River were characterised by an absence of heptageniid mayflies. The abundance of invertebrates that were predators also declined with the contamination (Fig. 11.14). Interestingly, these same species disappeared in laboratory microcosm tests, but at concentrations about 10 times higher (Martin et al., 2007). The limited time of exposure (10 days) and the lack of explicit dietary exposure probably reduced the sensitivity of the microcosm study compared to nature.

Not all mayflies were metal-sensitive; in fact abundances seemed to follow the patterns for different sensitivities shown in Fig. 11.8. Some species showed the expected sensitivity (Fig. 11.15a, b, c); some species

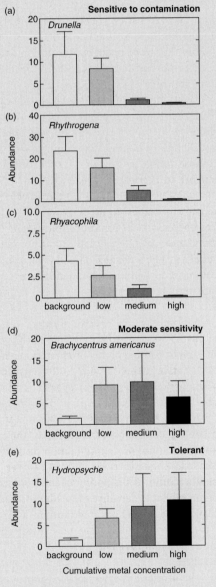

Fig. 11.15 Abundance of five insect larvae in relation to the sum of metal contaminants in streamwaters of the southern Rocky Mountains, USA. Contamination is represented by the sum of the ratio of each metal relative to its regulatory criterion (CCU). Some species were consistently sensitive to the contamination (a–c) while others were not (d–f). (Data from Clements, 2000.)

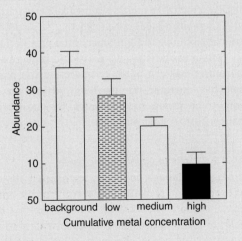

Fig. 11.14 Abundance of predator species in streams of the southern Rocky Mountains, USA, as a function of the degree of metal contamination. (After Clements, 2000.)

were moderately sensitive (Fig. 11.15d); and some species were tolerant (Fig. 11.15e). The distributions were consistent with mechanistic studies that show that at least some of these species have physiological traits that facilitate such survival (low net uptake and/or high detoxification capabilities); while others have traits that appear to facilitate sensitivity (Martin *et al.*, 2007). Clements *et al.* (2000) concluded that metal contamination was a primary force in structuring the benthic communities of the southern Rocky Mountains region. Martin *et al.* (2007) suggested that the structuring that occurred might be predictable from physiological traits.

Gower *et al.* (1994) found similar results in a spatial mosaic of contamination among streams in the historic mining districts of southwest England. They surveyed 46 sites on 12 streams in Cornwall, UK (Chapters 4 and 5). Where Cu concentrations were highest, selected species of mayflies were absent. Some tolerant species, on the other hand, were abundant at even the most contaminated sites. Examples of tolerant animals included turbellarian flatworms, selected chironomid larvae and, as in Colorado and Montana (USA), a net-spinning caddisfly (in this case *Plectrocnemia conspersa*).

Invertebrate predators decline in abundance in the presence of metal contamination. Although direct toxicity is a possibility, loss of benthic biodiversity and ecosystem services suggests that predators will also be at a disadvantage in a metal-contaminated stream because of reduced services and a disturbed supply of food. A contaminated stream is less productive, less diverse, and missing attributes of likely (but unquantified) importance to predator diversity and productivity. Thus stream environments seem vulnerable to cascades of ecological conditions that could indirectly affect upper trophic levels, in addition to any direct effects on predators.

A loss of major fish predators is also common in metal-contaminated streams and seems a feasible outcome of metal contamination. In Chapter 12 we discuss in detail the influence of mine wastes on upper trophic fish species that appear to follow the concepts typical of cascading ecological effects from metal contamination. In the next section we show a similar set of circumstances for lakes. Cascading effects, as compared to direct toxicity, are particularly difficult to demonstrate conclusively and/or to link to metal contamination; but they deserve further study.

11.4.3 Effects in lakes: elimination of species and cascades to upper trophic levels

Metal contamination affects lake communities in generally the same way as stream and marine communities; but different groups of species are involved. Most of the studies of metal influences on lake communities consider water column organisms. It is thought that zooplankton may be one of the most sensitive groups of organisms. In the mid twentieth century, many lakes near Sudbury, Canada, were severely contaminated by acid (sulphur dioxide) and metal emissions from local smelters (see also Box 12.10). Smelter activities similarly affect the area in Quebec around Rouyn-Noranda. Both of these areas are in a geological regime known as the Canadian Shield where topsoils are thin and lakes are numerous. Most such lakes have low major ion concentrations and low dissolved organic concentrations, making them especially vulnerable to trace metal contamination.

As in streams and rivers, metals simplify lake communities too; and the effects are propagated up food webs both directly and indirectly. The studies in Canada (Box 11.13) demonstrated:

- Metal inputs from smelters increased metal bioavailabilities to both invertebrates and fish coincidentally. The higher exposures correlated with measures of stress (Yan *et al.*, 2004; Pyle *et al.*, 2005).
- Effects are characterised by the disappearance of metal-sensitive taxa of invertebrates (zooplankton: Box 11.13). The most sensitive zooplankton are cladocerans, which disappear in contaminated environments and are the last to recover (if at all) as contamination recedes (Yan *et al.*, 2004).

Box 11.13 Case studies on metal-contaminated Canadian lakes

Several thousand lakes have been affected by the combination of acidification and metal contamination around the Sudbury smelter district (Yan and Miller, 1984). Metal contamination was documented in the early 1970s, and continues to the present time. Middle Lake in Ontario had a pH of 4.2, 500 µg/L Cu in solution and 1000 µg/L Ni in the 1970s (Yan et al., 2004); Swan Lake was a mesotrophic acidified lake (pH 3.97) with Cu concentrations of 64 µg/L and Ni concentrations of 300 µg/L (Yan et al., 1991). More recently, 20 affected and unaffected lakes near Rouyn-Noranda were sampled (Campbell et al., 2003). Dissolved metal concentrations ranged from 0.011 µg/L to 0.77 µg/L Cd; 2.1 to 17 µg/L Cu; 0.14 to 170 µg/L Zn. When these metals are considered together, the CCU was <1 in only five of the 20 Rouyn-Noranda lakes; obviously it had been much higher in the Sudbury lakes.

Early studies compared affected and unaffected lakes in the regions. After the 1970s, smelters in Sudbury began to remove both sulphur dioxide and metals from their emissions. The trajectory of lake communities after emission reductions also provided evidence about metal effects.

Field studies demonstrated that metal contaminants were bioavailable to biota in the lakes, and thus metal exposure was a feasible source of stress. For example, in a survey of 33 unaffected or moderately affected lakes Cd concentrations in water ranged from 0.002 to 0.112 µg/L. Concentrations of Cd in zooplankton (range 0.16 to 29.8 µg/g dw) correlated with concentrations of Cd in solution (Yan et al., 1991). Influences from dissolved organic material, major ions and pH affected Cd bioavailability consistent with modern knowledge of geochemical influences on dissolved Cd bioavailability.

The feasibility of metal effects was further established by chronic (22 week or several generations) toxicity tests with zooplankton conducted under relatively realistic conditions (Marshall, 1978). The cladoceran zooplankter Daphnia galeata was fed phytoplankton that were held for 24 hours in Cd-spiked media and exposed to dissolved Cd. The lowest effect concentration for reduction of live births (prenatal mortality) was 0.15 µg/L; live births were reduced 50% at 1 µg/L. Zooplankton populations were not affected at 0.15 µg/L because of their ability to compensate for the small increase in rates of mortality (Box 9.3). But, in nature, effects would likely become significant between 0.15 and 1.0 µg/L because of interactions with other sources of mortality or reduced recruitment. Such concentrations were well within the range of concentrations observed in contaminated lakes. Marshall (1978) also showed that zooplankton from the order Copepoda accumulated lower Cd concentrations than zooplankton from the order Cladocera. Cladocerans would thus need better detoxification capabilities than copepods to survive the same level of contamination.

Several types of stressors affected the Sudbury and Rouyn-Noranda lakes (Yan et al., 2004):

(1) Acidification (pH < 6). Many Sudbury lakes were severely acidified when first studied in the 1970s, some with pH levels near 4.
(2) Metal contamination with Cu, Ni and Cd. This stress reached back at least to the 1950s, if not much earlier.
(3) Physicochemical changes in the lakes and watershed related to acidification.
(4) Elimination of fish, which altered predator–prey interactions.

Field studies showed that unusual water column communities resulted when metal contamination and acid conditions were present. The phytoplankton community of both Middle and Swan Lakes, for example, were simplified and dominated by a few species of green algae and dinoflagellates. In Middle Lake, all fish disappeared, and biodiversity was reduced at all levels of the food web (Yan and Miller, 1984). On average, vertical plankton hauls yielded only two to three zooplankton species. In Middle Lake, a single small cladoceran, Bosmina longirostris, comprised 99% of the zooplankton mass (Yan and Strus 1980). In Swan Lake the zooplankton community was strongly dominated by rotifers, which were 10 times more abundant than the typically dominant crustacean zooplankton. Crustacean biomass was dominated by Bosmina longirostris. Normally abundant cladocerans, known to be acid-sensitive (Daphnia spp.) and/or metal-sensitive (Holopedium gibberum: Lawrence and Holoka, 1987), were absent. The absence of fish and the success of small zooplankton appeared to allow explosive growth of an invertebrate predator, the larvae of the phantom midge Chaoborus sp. Chaoborus is now known to be tolerant to metal contamination, common in metal-contaminated lakes and a good biomonitor for metal exposures (Croteau et al., 1998).

Selective *Chaoborus* predation on crustacean zooplankton appeared to be the main reason that rotifers were more dominant than the crustaceans (a secondary effect of the contamination).

Other secondary effects from the contamination also contributed to ecological simplification. For example, destruction of local soils reduced dissolved organic carbon (DOC) inputs to the lakes (Yan *et al.*, 1996). Combined with a weak phytoplankton standing crop, this meant that the lakes became exceptionally clear. This allowed ultraviolet radiation (sunlight) to penetrate deeply into these lakes. The lakes warmed, as a result, and summer thermoclines became unusually deep. Much of the cold-water habitat that was essential for many species disappeared (Yan and Miller, 1984). As a consequence fish disappeared, facilitating the large numbers of macroinvertebrate predators. Predator–prey relationships were destabilized. The result was a great reduction in the number of all biological species.

To determine if crustacean zooplankton could recover from such severe and chronic damage, Middle Lake was neutralised in 1973 (Yan *et al.*, 2004). Metal concentrations declined slowly over the 30 years after the change in pH. Many remain contaminated still, although not as severely as the original problem. For example, the mean dissolved Cd concentration in unaffected lakes in central Ontario (Canada) is 0.009 µg/L. Lakes near Sudbury range from 0.045 to 0.552 µg/L Cd, although 4.3 µg/L was still determined in one lake (Ball *et al.*, 2006).

Phytoplankton community composition recovered fairly quickly judging from the return of major taxa (Yan *et al.*, 2004). Although pH stress was removed, there was little evidence of recovery of crustacean zooplankton in Middle Lake by 1989, with the exception of an acid-sensitive species of *Daphnia*. As metal concentrations declined in more recent years, some important groups of fauna returned. The copepod zooplankton community in Middle Lake was comparable to 22 non-acidic reference lakes in the same zoogeographical region in 2004. Six species of copepod colonists founded large, stable populations in Middle Lake by 2004. In contrast, zooplankton in the cladoceran assemblage improved but did not recover. The cladocerans and copepods colonised the lake with equal frequency, but six separate colonisations by cladoceran species failed.

Yellow perch (*Perca flavescens*) also returned to many recovering lakes, although fish communities remained generally simplified (Pyle *et al.*, 2005). Twenty fish species were identified in 12 lakes from the Sudbury region across a range of metal exposures (Pyle *et al.*, 2005). Only one species, yellow perch, was common to all 12 lakes. Yellow perch are known for their metal tolerance (Pyle *et al.*, 2005). Rainbow trout (*Oncorhynchus mykiss*) occurred in none of the lakes, fathead minnows (*Pimephales promelas*) in only three. Fish diversity was correlated with sediment metal concentrations but no other characteristics across the lakes.

Continuing metal toxicity appeared to be one factor in the failure of cladoceran zooplankton to recover; and probably affected the fish as well. This was most evident in young fish, which showed morphological, physiological, biochemical and metabolic signs of stress (Campbell *et al.*, 2003; Couture and Kumar, 2003). But the simplified invertebrate community also affected the yellow perch. Because benthic species in the shallow zones of lakes and other fish populations were reduced in the contaminated lakes (Pyle *et al.*, 2005), yellow perch were forced to prey upon zooplankton their entire lives (Campbell *et al.*, 2003). In contrast yellow perch in uncontaminated lakes gradually shift their prey from zooplankton, to benthos, to other fish as they grew older. Yellow perch in the metal-contaminated lakes were stunted (small) for their age compared to those able to go through the typical dietary shift (Fig. 11.16). Thus metals affected the ability of this fish to reach its maximum size via effects on invertebrates and other fish; and fish affected the nature of the zooplankton community because they had no other sources of food. Yellow perch also live shorter lives in these lakes (Campbell *et al.*, 2003; Pyle *et al.*, 2005), apparently because of sublethal stress imposed by metals early in life.

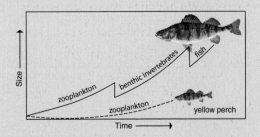

Fig. 11.16 Conceptualisation of yellow perch growth as a function of food availability. (After Campbell *et al.*, 2003.)

Particular species, usually common in these lakes, appeared to be highly metal sensitive (e.g. the cladoceran *Holopedium gibberum*: Lawrence and Holoka, 1991) or perhaps *Daphnia* species if acidification is not a confounding variable. The absence of these species is potentially a metal-specific indicator of stress.

- Tolerant species were abundant where contamination was less than extreme. The phantom midge *Chaoborus* sp., the fish the yellow perch *Perca flavescens*, and the cladoceran zooplankter *Bosmina* sp. were examples of metal-tolerant taxa.
- Effects were propagated to upper trophic levels. This occurred in several ways in the Canadian lakes: (a) a reduction in overall fish diversity (sensitive species like rainbow trout were not present); (b) the lifespan and size of tolerant species (yellow perch) was shortened by physiological and metabolic stress (Campbell *et al.*, 2003; Couture and Kumar, 2003); (c) reduced invertebrate diversity affected the food supply of the surviving fish and forced them to feed upon a less than optimal food source. The simplified invertebrate assemblage appeared to be the cause of the shortened lifespan and stunted growth of yellow perch in the metal-contaminated lakes (Box 11.13, Fig. 11.16).
- Multiple stresses characterised the lakes, many of which were related to the contamination. The relative effects of the individual stressors tended to become more clear when aspects of the contamination receded at different rates, followed by differential recovery of the community (Box 11.13).

11.5   INDIRECT EFFECTS

Prologue. Indirect ecological effects of metals occur as top-down or bottom-up changes in communities in response to loss of key species to metal contamination. Indirect effects amplify the ecological risks of metal exposure and are probably common. But understanding of the specifics of many of those responses is just beginning. Single species, laboratory based toxicity tests cannot detect or predict indirect contaminant effects (Fleeger *et al.*, 2003).

Observations of metal effects in nature must be alert to such possibilities; studies must extend beyond the simplest ecotoxicological variables to the mechanisms that drive community ecology.

The examples in our case studies illustrate several complex ways that metals affect aquatic environments. In many of these instances, direct effects on sensitive species initiate indirect effects on tolerant species via changes in competition, predation or function (Fleeger *et al.*, 2003). Or effects on predators occur as a result of simplification of their prey community. It is well documented that changes in species composition of a community, or even loss of a single species, can cause such a cascade of indirect changes through a food web. This may partly explain why disappearance of predator species seems to be a common response to metal contamination (in stream invertebrates, marine invertebrate communities, and fish communities from rivers and lakes). Shifts in abundance are also common and similarly explained (Box 11.14). In some cases changes in predator–prey relationships cause changes in the invertebrate community, a top-down effect (e.g. in contaminated lakes). In other cases, changes in the invertebrate

---

Box 11.14 Changes in interspecies competition

In the case of marine epifaunal communities of hard substrates, field experiments in Port Philip Bay, Victoria, Australia showed that pulse episodes of high dissolved Cu availability decreased the densities of large solitary ascidian tunicates (e.g. *Pyura stolonifera* and *Ascidiella aspersa*) while increasing the densities of serpulid polychaetes (e.g. *Ficopomatus enigmaticus*, *Hydroides* sp. and *Pileolaria pseudomilitaris*) as an indirect effect mediated through competition for space (Johnston and Keough, 2003). Thus the interaction between a toxicant disturbance and competition for a limited resource modified the community-level response of organisms to that toxicant, an interaction that can be shown in manipulative field experiments but cannot be predicted from conventional laboratory based toxicity testing (Johnston and Keough, 2003).

community may be the cause of some of the effects on predators. These are called bottom-up effects (again lakes provide a good example). Clearly metal contamination can set off a sequence of events that are complex, varied and not yet fully predictable (Box 11.14).

11.6   CONCLUSIONS

Both the structure and function of aquatic communities change in most water bodies subjected to contamination from metals. Three lines of evidence best demonstrate such adverse effects: (a) field observations, (b) toxicity tests, particularly sophisticated approaches such as *in situ* microcosm studies and (c) mechanistic studies that explain why the effects occurred. Most generic indices describing community structure can identify the most extreme effects; but recognising important subtle effects requires appreciation of the concept that some species are more sensitive than others. At lower levels of contamination, effects are manifested as missing species. The important question is what species will disappear and what effects will cascade to the rest of the food web when those species disappear.

Some common generalisations from streams, lakes and marine environments are now reasonably well supported, and may apply to other environments where impacts are less studied.

(1) Measures of community structure vary in their sensitivity in detecting adverse effects. Overall abundance of all taxa (number of individuals) is not a reliable measure of metal ecotoxicological effects in nature. It is a typical response only in controlled experiments where migration into the experiment is not possible. In all but the most extreme circumstances, tolerant species can become highly productive, or immigrating, opportunistic species can rapidly replace species that have suffered toxicity. Then abundance is either unaffected or increases.

(2) The number of species present (species richness) is a reliable measure of severe impacts, but is not especially sensitive. As contamination declines

(in time or space), species richness may recover, but important changes in the nature of the community may be present still.

(3) Every environmental setting is characterised by species that are more sensitive and species that are less sensitive to trace metal contamination. When the environment is contaminated, the presence or absence of the metal-sensitive groups of species is the most effective way to evaluate ecological change caused by metals. These effects can be detectable over distances of tens to hundreds of kilometres away from a large source of contamination. The metal-sensitive groups are becoming well known for streams, lakes and some marine habitats.

(4) Effects at upper trophic levels are the most subtle and most difficult to detect in metal-impacted systems, but may also be the most important from both an ecological and socio-cultural point of view. Effects on predaceous fish (like trout) are well known (see also Chapter 13). Field observations also suggest that predators, in general, may be less common in water bodies affected by the metal contamination.

(5) It is essential to accompany ecological study with measures of contaminant effects and of bioavailabilities. Biomonitor species are useful in the latter respect, and perhaps, occasionally, in the first also. Defining exposure with a biomonitor and response with abundance of a sensitive species may offer great promise in many types of ecosystems.

(6) A mixture of metals also may be more toxic than any one metal alone. Indices that capture the combined concentrations of metals are developed and calibrated to metal effects (Clements, 2004).

(7) Recovery of community structure and function has been observed in all habitats once waste inputs are removed. But as long as elevated metal contamination remains, recovery of community composition will be slow. Recovery rates of upper trophic levels are probably slowest, but are not well known for metal contamination situations.

(8) The rate and nature of recovery depend very much on local conditions. In some cases

centuries may be necessary for natural recovery; in others it may happen in years (see also Chapter 13). The probability that biota will quickly recover from disturbance is greatly reduced when the damage is severe, widespread and/or chronic; especially if effects are accentuated by multiple stressors (Yan et al., 2004). Recovery is slowed if nearby colonist sources are depleted, and habitat conditions that fostered the pre-disturbance community are incompletely regenerated.

The now well-established principle that the sensitivity of some species to metals determines the response of a water body to metal contamination raises important questions about traditional methods of risk assessment. Risk assessments are prone to error if they exclusively use compliance with water quality regulations and/or depend heavily upon traditional bioassay data. Ultimately it will be essential that risk assessments include some characterisation of bioavailable exposure, use biomarker responses in survivors to help link cause and effect, and include a measure(s) of community response that considers estimates of relative sensitivity among expected species. Knowledge of mechanisms whereby risks are manifested for each metal is important.

Ultimately the ecological complexities of metal effects at any location require persistent study, mechanistic understanding, and multifaceted, complementary efforts in the field and laboratory: a lateral approach to risk assessment and risk management. One type of study, such as a bioassay or even a battery of bioassays, alone cannot convincingly predict the array of possible responses and in fact can be misleading. This assignment was once considered to be so large and complex as to be impractical. It becomes more practical as we learn more and more about the processes that determine metal effects, and couple that with programmes that emphasise understanding local environments (e.g. the European Water Framework Directive). In the growing number of circumstances where the stakes are large, implementation of the lateral approach is now thwarted more by a lack of will than by a lack of means.

Suggested reading

Brand, L.E., Sunda, W.G. and Guillard, R.R.L. (1986). Reduction of marine phytoplankton reproduction rates by copper and cadmium. *Journal of Experimental Marine Biology and Ecology*, **96**, 225–250.

Campbell, P.G.C., Hontella, A., Rasmussen, J.B., Giguere, A., Gravel, A., Kraemer, L., Kovasces, J., Lacroix, H. and Sherwood, G. (2003). Differentiating between direct (physiological) and food-chain mediated (bioenergetic) effects on fish in metal-impacted lakes. *Human and Ecological Risk Assessment*, **9**, 847–866.

Clements, W.H. (2004). Small-scale experiments support causal relationships between metal contamination and macroinvertebrate community responses. *Ecological Applications*, **14**, 954–957.

Martin, C.A., Luoma, S.N., Cain, D.J. and Buchwalter, D.B. (2007). Cadmium ecophysiology in seven stonefly (Plecoptera) species: delineating sources and estimating susceptibility. *Environmental Science and Technology*, **41**, 7171–7177.

Woodwell, G.M. (1970). Effects of pollution on the structure and physiology of ecosystems. *Science*, **168**, 429–433.

Yan, N.D., Girard, R., Heneberry, J.H., Keller, W.B., Gunn, J.M. and Dillon, P.J. (2004). Recovery of copepod, but not cladoceran, zooplankton from severe and chronic effects of multiple stressors. *Ecology Letters*, **7**, 452–460.

12 • Mining and metal contamination: science, controversies and policies

'To one approaching the city the general appearance is most desolate. Bare, brown slopes, burnt and forbidding, from which all vegetation was long ago driven ... rise from an almost equally barren valley ... great heaps of gray waste rock from the mines form the most conspicuous feature of the landscape ... Heaps of waste are everywhere prominent, attesting by their great size to the extent of the underground workings ... ore was being roasted outside in the grounds of the reduction works, the fumes rising in clouds of cobalt blue, fading into gray, as it settled over the town like a pall ... The driver reined in his horse as we entered the cloud of stifling sulfur and cautiously guided them up the hill. A policeman, with a sponge over his mouth and nose, to protect him from the fumes, led us to a little hotel, for we could not see across the street.'

(US Geological Survey Geologist J. H. Weed, describing Butte, Montana, USA in 1912; from J. N. Moore, University of Montana)

Planning must account for environmental protection ... and for reclamation ... The cost of environmental protection is minimized by incorporating it into the initial design, rather than performing remedial measures to compensate for ... deficiencies (at a later time), ... negative publicity or poor public relations (from ill-considered environmental policies) may have severe economic consequences.

Mine Engineer.com, 2005 (http://66.113.204.26/mining/coal/opn_pit.htm)

12.1   INTRODUCTION

Prologue. This chapter describes metal contamination issues associated with metal mining. Mining and smelting are unique metal management and risk assessment issues. The landscape around a mining operation is intensely disturbed, but the size of the footprint of impact is dependent upon how the operation is managed. The major problems stem from secondary dispersion of wastes. Large operations can affect entire watersheds if wastes are not contained. Similarly, numerous, poorly managed, small mining operations can have cumulative contamination effects that cause regional-scale ecological change. But successes in environmentally responsible management of modern mines and smelters are growing in number, as the long-term pay-offs are recognised by operators, investors and stakeholders.

Controversies can develop over both risk assessments and risk management of past, present and future mining projects. The technical issues underlying these controversies stem from uncertainties in the general scientific principles we have presented in earlier chapters. Addressing ecological risks from mining provides lessons in modern management of metal contamination. Most of the lessons come from understanding the implications of historic practices. Modern mines, whether well managed or not, are poorly represented in the open literature.

Operations that extract metals from the Earth are required to satisfy the many demands of modern societies for these materials. Mining operations are managed at many scales. Global corporations are groups of large-scale operations that follow unified environmental policies, dictated by the corporation. Small-scale artisan mines are run independently and often are less concerned with environmentally responsible practices. Metal contamination is a common impact from mining, although not the only one. The degree of impact is greatly influenced by how the operation is managed.

Mining is confined to regions in the vicinity of appropriate geological formations. But the associated metal contamination can range from quite localised to regional in scale. In 1989, mining activities affected about 240 000 km^2 of the Earth's surface, an area about the size of Oregon in the USA (Moore and Luoma, 1990).

Both large-scale mining and smaller operations progressively shift in location as deposits are exhausted relative to the technology of the times. Spain was a major source of copper and sulphur in ancient times, exploited by the Romans. The first evidence of mining in Cornwall in the south-west of England also comes from 3 to 4 BC. Cornwall became the dominant source of Cu for Europe in 1800 to 1900; then the Cu mines closed. Mines in the USA proliferated in mineralised areas between 1870 and 1970, but many have closed since the mid-1980s. Today major suppliers of Cu for the world are mines in Canada, Australia, Africa and South America.

Every individual metal mine eventually reaches a point at which economics dictate closure. Nevertheless, absolute depletion of metal ores is not likely to be a resource issue of significance (Goedkoop and Dubreuil, 2004). Recycling, substitution and the adaptations of extraction technologies are likely to sustain the availability of most metal resources (although not necessarily keep prices low). When mining activities moved on, historically they often left behind an undesirable environmental legacy. The constant shift towards better mining opportunities has created a patchwork of abandoned mines across the world, from which we are learning the environmental risks that mining can potentially present.

Mining, smelting and processing of metal products will grow in importance as an issue in the

---

Box 12.1 Definitions

cut bank: the outside bank of the channel wall of a meander in a river. This vertical bank of riparian sediment is continually undergoing erosion as the river cuts into its floodplain to expand the length of the meander.

dispersion train: the distance from an ore body over which metal concentrations in water, soil or sediment decrease toward background.

footprint: the area disturbed by a mining operation.

mining: the process of obtaining useful materials from the Earth's crust.

ore: a mineral or combination of minerals from which a useful substance, such as a metal, can be extracted and marketed at a price that will recover the costs of mining and processing, and yield a profit (Encarta, 2001).

primary contamination: contamination from a mining/smelting operation that occurs at the site of mining

activity, including disturbed landscape (the mine), waste rock (tailings), tailings ponds, smelting (fumes, flue dust, slag), acid mine drainage and flooded open pits.

secondary contamination: contamination in soils, groundwater and rivers as contaminants are moved in streams or through the atmosphere; secondary spread defines the ultimate environmental footprint of the mining operation, for example tailings materials on floodplains.

tailings: fine-grained rock particles, typically derived from ores by milling and flotation at a site close to the mine, constituting potentially the single most important ecological hazard from mining.

tailings pond: a pond or lagoon, typically behind a newly constructed dam, containing (permanently) the water and tailings wastes derived from mining activity.

tertiary contamination: contamination remobilised from a site of deposition (secondary contamination), retransported and dispersed or deposited even further from their origin.

years ahead. New products and new technologies (e.g. nanotechnologies) will expand traditional uses and involve novel uses of metals. Demand for metals is growing rapidly across the world where economies are rapidly expanding (e.g. China and India). The number of active mines is already increasing rapidly. The disturbed surface area per unit metal extracted could increase as smaller quantities of metal are extracted from larger quantities of earth. Or this could be balanced by better extraction technology, more underground mining and better controls from the beginning of the operation. Thus, it is crucial that policies for managing the contamination from mining continue to grow in sophistication as mining itself expands.

Today, the impacts of mining and smelting can be greatly moderated. Few individual human activities visually disturb the earth surface more dramatically than a mine. But the environmental risks are less a function of disturbed area or volume of waste, and more determined by how the mining wastes are dispersed. 'Sustainable' mining means a mining industry that minimises long-term environmental damage, partly by controlling waste dispersal, and is therefore viable in the eyes of a public conscious of past environmental costs.

Historical operations spread wastes over large areas and/or dispersed them, uncontained, into waterways or the atmosphere. Sustainable mining involves minimising and containing wastes. Most modern smelters in the developed world remove air pollutants before they are released. Containment of tailings is standard best practice; although not necessarily achieved everywhere. In the Global Mining Initiative of 2002, several of the largest global mining corporations pledged to no longer release tailings into rivers (although many have not made that pledge).

Historical operations often abandoned mines when the ore body was depleted, leaving the environmental degradation for society to clean up. Environmentally responsible modern operations begin planning for closure when the mine is opened. Advanced planning involves measures to prevent dispersion of wastes, measures to reduce contamination at the primary site of activity, ongoing mitigation that makes up for disturbances

by fostering conservation in other areas, and, ultimately, reclamation of disturbed lands. Mining is a long-term proposition. Responsible companies take a long-term view of the factors that determine their economic success and factor in the cost of adverse environmental impacts. One benefit of environmentally responsible behaviour is assurance of access to ore deposits in the future. Emphases on environmental awareness and human health are recognised by forward-thinking operators (and their shareholders) as necessary to sustain the industry. Mining is, therefore, a case where cooperation and collaboration among government, industry and the public (i.e. shareholders and stakeholders) allow the environmental ethic to extend beyond just compliance with local regulations; and it seems to be beginning to work.

Supply and demand for metal products are likely to tighten as populations grow and societies modernise. This could create great pressure to justify environmental policies. Governments will be asked to justify regulatory policies, and shareholders will require corporations to better justify costs for their environmental policies. Better knowledge of the implications of past mistakes might help justify the investments necessary for the future. One purpose of this chapter is to characterise those historical environmental impacts. Such understanding will also help remediation of historical mining damages that many risk managers have inherited. And it is essential to preventing recurrence of past mistakes.

Ecological risks from mining follow a conceptual model similar to that for other metal risks (Fig. 12.1). Interpreting those risks therefore builds from science we have discussed in earlier chapters. In more specific terms, mining takes many forms and covers many kinds of environments. For the sake of simplicity, we emphasise metal mining from pyrite deposits in this chapter, one of the most widespread of mining activities. In other chapters and case studies we consider metal contaminants from other major mining activities. Coal and phosphate mining will be considered in the chapter on Se contamination (Chapter 13). Gold mining will be considered in the chapter on Hg (Chapter 14).

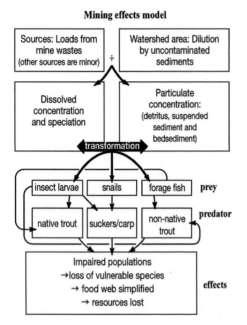

Fig. 12.1 The generic conceptual model framing processes that contribute to ecological risks from mining activities.

12.2   SOURCES OF METALS: NATURE OF MINES AND MINING ACTIVITIES

Mining in its broadest sense is the process of obtaining useful minerals from the earth's crust. A mineral is generally defined as any natural accruing substance of definite chemical composition and consistent physical properties. An ore is a mineral or combination of minerals from which a useful substance, such as gold, can be extracted and marketed at a price that will recover the costs of mining and processing, and yield a profit.

Encarta (2001)

Prologue. Mining involves six sequential stages: exploration, development, extraction, concentration, processing or refining, and decommissioning. Historically, each advance in mining technology expanded the footprint of the operation, and the potential for dispersing contamination. Recent advances in mine management allow shrinking the primary footprint and limiting dispersion of the contamination.

Mining involves a sequential set of processes, each of which has the potential to generate metal-bearing wastes (Lottermoser, 2003). *Exploration* involves finding and delineating ore bodies. The main impacts are building roads for access. *Development* prepares the mine site for production. It includes the initial excavation of rock overlying the deposit (waste rock), building roads, constructing facilities near the mine, and even building ports or rail lines. In modern mining activities, development can include preparing communities for sustained economic success after the mine closes (e.g. by improving educational opportunities or setting aside land for other economically viable activities), and beginning mitigation activities or associated research (Dorward-King, 2007).

Extraction involves removing ore from the earth and crushing it (as necessary). Concentration or benefaction is a step in base metal mining that involves milling ores to a fine particle size at a location near the site, then separating the desired product from the waste materials (tailings), usually using some type of chemical extraction. Smelting, refining and metallurgical *processing* then use high-temperature processes to separate and concentrate the pure metal that is the target of the operation. Extraction, concentration and smelting are the operations that generate the most environmental contamination.

Every ore body is finite. Economic limits are reached before physical limits, typically. Ultimately the mine will be decommissioned (*closure*). In modern operations this can involve working to return the disturbed area to its original, or some other useful, state (Box 12.2). Plans for closure must start before mining begins, or environmental remediation is unlikely to be successful. The exact processes involved in closure are determined by the ore and the ore body, as well as economic considerations.

Ore is the material that contains the minerals targeted for exploitation. Sulphide (more specifically pyrite or porphyry) ore bodies constitute more than half of the major source of several common metals, including Cu, Ni, Pb and Zn. Sulphide deposits are formed in anoxic environments deep within the Earth's crust. When these deposits are

Box 12.2 Closure or decommission of a mine: the World Bank Guidance for mine reclamation

Project sponsors are required to prepare and implement a mine reclamation plan (before the mine opens). The plan should include reclamation of tailings deposits, any open pit areas, sedimentation basins, and abandoned mine, mill, and camp sites. The main objectives of the mine reclamation plan are:

(a) return the land to conditions capable of supporting prior land use or uses that are equal to or better than prior land use, to the extent practical and feasible,
(b) eliminate significant adverse effects on adjacent water resources.

Mine reclamation plans should incorporate the following components:

(a) conserve, stockpile, and use topsoil for reclamation,
(b) slopes of more than 30% should be recontoured to minimise erosion and runoff,
(c) native vegetation should be planted to prevent erosion and encourage self-sustaining development of a productive ecosystem on the reclaimed land.

exposed to air, the potential exists to form acid mine drainage (see later discussion). A variety of trace metals concentrate in sulphide deposits, some of which are exploited and all of which become contaminants in waste products. The metals targeted for extraction are also present in the waste products (e.g. Cu, Pb or Zn). Their concentrations depend upon the efficiencies of the extraction process. Other metals that can become constituents of the waste products may include Sb, As, Bi, Cd, Co, Ge, Au, Mo, Ag, Se and Te. The actual contaminants differ widely among ore bodies.

Ore deposits other than sulphide deposits have slightly different contamination characteristics. Haematite deposits contain uranium (U) ores, fewer contaminants, and are less likely to form acid drainage. Carbonate deposits are mined for Pb, Zn and Hg. Environmental concerns related to mining and processing of silica-carbonate Hg deposits, for example, consist primarily of Hg contamination of soil and water from mine waste rock and tailings (calcine) and Hg vapour released during ore processing. If pyrite is interspersed in such ores, acid mine drainage can be generated, however.

Different types of operations result in different yields of waste and different potentials for environmental impact. Underground mining removes less waste rock and leaves a smaller area of disturbance (footprint) than does surface mining. But subsidence, access, worker safety, groundwater disturbance and oxidation of ores are potential issues. In surface or open pit mining, access is easier and operations involve less risk for workers. But surface mining expands the area disturbed (the footprint) and thereby expands the potential environmental impacts. Large amounts of waste rock, mine drainage and ore oxidation are prevalent in surface mining. A few carbonate deposits can be mined by injecting acidic water into earth, and recapturing the dissolved metal, if the deposit is porous (termed *in situ* leach mining). Undersea mining is also possible, although much less is known about its environmental management or implications.

The procedures used to mill, concentrate and smelt ores are especially important in determining the potential to disperse wastes. The Romans smelted Cu at the mine by finding ways to set ore veins on fire in place, then shattering them with cold water. In the early nineteenth century large new furnaces were developed that roasted the ore, then removed the impurities by blowing air through the molten matte to remove Fe sulphides. In 1873 a British company, working at Rio Tinto in Spain, discovered how to set ore heaps on fire outside the mine, then leach them with water to isolate the Cu more quickly. Some environmental effects of this technology at Butte, Montana, USA were described at the beginning of this chapter (Weed, 1912). In the late nineteenth century it was discovered that Cu, Au, Ag, Zn and Mo could be separated from their host rock by milling and flotation. Usually this was done at a site near the mine so the bulk of the ore did not have to be transported. Such operations create the fine-grained tailings that are the most damaging legacy from such operations. The tailings are usually of a much greater mass than the product that is captured and, unless contained, constitute the single most important ecological hazard from mining.

At least some evidence from historical studies of contamination (usually from dated sediment cores) shows that the greatest mine-related contamination of rivers, streams and estuaries came with the advent of mechanised mining, milling, flotation and smelting. For example, regional-scale production of Cu was at its peak in Cornwall, southwest England between 1800 and 1900, but mine-related metal contamination in the sediments of estuaries in the region did not increase rapidly until 1870 (Box 12.3). Similarly, Hg contamination from the Clear Lake Mercury mine (California, USA) did not become extensive until mechanisation and open pit mining began, late in the history of the operation (Suchanek *et al.*, 2004).

12.3   SOURCES: PRIMARY CONTAMINATION

Prologue. Mine and smelter sites are typically constrained in area but the disturbance is visible and dramatic. Contaminants spread over the site of activity are termed primary contamination. Effects at the mining site constitute one ecological risk from primary contamination, but the most important effects occur if this contamination is dispersed, creating secondary and tertiary waste deposits. Environmentally responsible mining reduces the primary contamination and contains its dispersal.

12.3.1   On-site disturbance

Primary contamination from a mining/smelting operation occurs at the site of activity and includes the most concentrated wastes (Fig. 12.2). The disturbance can be dramatic (Fig. 12.3). For example, the historical mine and smelter complex at Butte

---

Box 12.3 Effects of Cu mining in the Tamar Valley, southwest England

Major contamination did not appear in the estuaries of the region until late in the mining period. Before the twentieth century, high-grade deposits of Cu (e.g. 1% to 15%) were worked in small underground mines. The early mines were 'scratching' at the surface to expose brightly coloured Cu ores; or small open pits called 'coffins' only a few metres deep (Atkinson, 1987). After 1900, new methods allowed deposits of 0.5% Cu or less (porphyry ores) to be mined economically; resulting in much greater quantities of waste material. Another technology that dispersed contamination was the use of adits or tunnels that drain water away from a mine so the ore can be exploited. The early adits started as small ditches. But by 1900, the County Adit in Cornwall was 50 km long and drained 46 mines. Tailings and the acid mine drainage from sources like the County Adit probably contributed to the rapid growth of contamination in the Tamar River estuary in the late nineteenth century.

Table 12.1 *Production of Cu from the Tamar Valley and concentrations of Cu contamination in the sediments of Plymouth Harbour, which receives river inflows from the valley*

Year	Production of Cu from mines in the Tamar Valley (1000 tonnes of ore mined)	Cu in sediments of Plymouth Harbour (µg/g dry wt)
1830	12	15
1850	23	15
1875	28	25
1900	4	150
1925	0	150
1950	0	140

Source: Data are from sediment cores (Clifton and Hamilton, 1979).

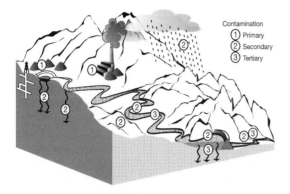

Contamination
① Primary
② Secondary
③ Tertiary

Fig. 12.2 A cartoon depicting the types of waste input from a mining operation and showing primary, secondary and tertiary contamination. (After Moore and Luoma, 1990.)

and Anaconda created severe disturbance of the landscape over 50 km² (Helgen and Moore, 1996). The open pit mine created 300 million m³ of waste rock spread over 10 km² that was enriched in metals compared to unmineralised areas, but less contaminated than other wastes.

Historically tailings were either discarded around the site, or sluiced into streams to get rid of them. In modern operations, containing the massive quantities of tailings is a major aspect of the operation. Dams are constructed to contain the water and tailings wastes permanently in lagoons termed tailings ponds. The lagoons serve as settling basins to hold the particulate waste materials. There are some risks with tailing ponds: destruction of local habitat, leakage to groundwater and overflows

Fig. 12.3 A scene of primary contamination from the Tamar Valley, England, in about 1850. Sites of primary contamination are intensely disturbed in ecological terms (after Dines, 1956). The southwest of England (counties Cornwall and Devon) is a mineral-rich region first exploited by the Romans, perhaps in the fourth and seventh centuries (Thorndycraft et al., 2004). In the nineteenth century, southwest England had more than 1000 active mines and was one of the most important metal-producing regions in the world. The Tamar Valley alone, at its height (1800 to 1900), produced half the world's annual output of Cu. By the end of the nineteenth century, output had declined substantially and only a few tin mines staggered through to the end of the twentieth century. This photograph illustrates the intensity of disturbance caused by one such operation.

or breaches, especially during floods. But these are less severe than the problems caused by long-term uncontained release of tailings into streams, rivers or the sea. The historical mining activity in Butte (Box 12.4) disposed tailings into the local stream from the inception of the mine in the mid nineteenth century until the 1950s. These activities created a visibly and intensely disturbed floodplain for about 80 km of stream, although contamination extended much further. Tailings ponds were finally completed after 1950. These ponds are thought to hold 200 million m^3 of highly contaminated tailings. They cover at least 35 km^2, around Butte and Anaconda.

Box 12.4 Case study: a legacy mining operation

In 1805 Meriweather Lewis and William Clark began exploration of what is now Montana in the USA. In the basin of the Clark Fork River they described 'a unique landscape of primitive beauty' filled with vast resources. Extraction of these resources began in 1864, with an aggressive rush for gold. The emphasis quickly shifted to Ag as the Au played out, then to Cu. By 1896, over 4500 tonnes of Cu ore per day were being mined and smelted near Butte, Montana, in the headwaters of the Clark Fork. One of the world's largest smelters was constructed at the turn of the century, in Anaconda, 40 km west of the mining operations (Fig. 12.4). In 1955, underground mining of high-grade ores in Butte was superceded by large-scale open pit mining. Eventually, the richest ores were depleted. Underground mining ceased in 1976, depressed

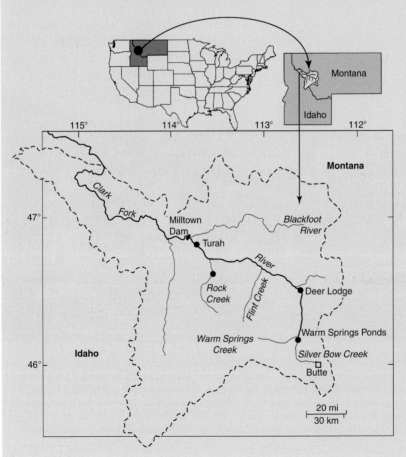

Fig. 12.4 Map of the Clark Fork Basin, home to one of the world's largest Cu mines between 1850 and 1983. In 2005 it was the site of the largest hazardous waste site in the USA.

Cu prices forced closure of the smelter in 1980 and mining in the largest open pit slowed in 1983. Anaconda Company, which owned the mine from its inception, went bankrupt and the operation was purchased by an oil company. Mining has resumed in recent years in adjacent pits along with limited underground operations, as the prices of Cu and Mo have risen.

When the smelter at Anaconda stopped production, over 1 billion tonnes of ore and waste rock had been mined from the district. The Butte district was once touted as the 'richest hill on earth'. The mining and smelting operations that produced this vast wealth left behind deposits of waste covering an area one-fifth the size of Rhode Island, USA (Moore and Luoma, 1990). The waste complex comprised, in 1998, the largest 'Superfund' hazardous waste site in the USA. Litigation

among the mine owner, the state and federal government has been under way for 15 years over who pays for the clean-up.

The Butte/Anaconda complex is an example of many legacy mining issues that exist today. Environmental quality was not a consideration during the first century, or more, of mining. As a result modern societies are left with an intense, widespread hazardous waste problem. Proponents disagree about the exact risks, but clean-up costs have already consumed hundreds of millions of US dollars and litigation settlements are in that range as well. Disagreements over technical issues and responsibility probably delayed remediation by decades. The best that can be made of such situations is to learn from them and use that knowledge to assure that they are not repeated.

12.3.2   Smelting

Smelting is another source of primary contamination. Fumes, flue dust and slag accumulate around a smelter, often leaving land barren of plant growth. Massive piles of slag are a common site. Trace metal contamination from the smelter at Anaconda included As, Sb, Cd, Cu, Pb and Zn. In some smelter flue residues, Cu concentrations were as high as 37100 μg/g dw and As concentrations were as high as 2960 μg/g dw (Moore and Luoma, 1990). As a comparison, Cu concentrations are 20 μg/g dw and As 10 μg/g dw in unmineralised soils in western Montana.

12.3.3   Generation of acid mine drainage

Acid mine drainage is also a common primary contamination issue. It mobilises toxic metals, extends contaminant distribution beyond the site of formation and is very difficult to control. Pyrite (Fe sulphide) ores or porphyry ores (chalcopyrite is the porphyry ore that is the most common source of Cu, for example) are the most likely to generate acid mine drainage.

Unmined ore bodies typically contain little exposed material (most of the ore body is underground).

That on the surface has reached an equilibrium with local conditions over thousands of years, or longer, so dispersion is limited. Mining involves excavating the ore (formed in the absence of oxygen) and its surrounding rock and transporting it to the surface. The surface area of the particulate material is increased through blasting, grinding and crushing. All these activities increase access of the original material to air and water.

When the insoluble sulphides in the ores are exposed to air on the surface they are oxidised. Products of the reaction include reduced iron (Fe^{2+}), sulphate, and hydrogen ions. Simplistically, the reaction can be described as:

$$4FeS_2 + 14O_2 + 4H_2O \rightarrow 4Fe^{2+} + 8SO_4^{2-} + 8H^+.$$

Because the reaction results in an excess of hydrogen ions, sulphuric acid is formed in water accompanied by a sharp decline in pH (increased acidity). Natural waters in contact with the ore, or primary wastes, become acidic and are termed acid mine drainage (AMD).

Acidophilic iron and sulphate oxidizing bacteria, such as *Leptospirillium ferrooxidans* and *Thiobacillus ferroxidans*, are found in all acid mine drainage (Kelly, 1988). They are able to use pyrites as an electron acceptor (energy source), and thereby catalyse the reaction above. They are able to rapidly generate Fe^{3+} from the pyrite and keep the ratio of Fe^{3+} to Fe^{2+} high, so the reaction rapidly moves toward producing acid.

Acid mine drainage has a distinctive look. Nearest the primary waste, pH can be <2.0 and the water is high in dissolved Fe^{2+}. As pH increases from 2 to 4, the bacterial colonies appear as a white 'streamers' or slime, on rocks covered with orange-hued flocs of ferric (Fe^{3+}) precipitate that the bacteria generated. The low pH is initially accompanied by increased concentrations of dissolved metals. This is because trace metals other than Fe are very soluble at the low pH of these situations. As pH increases toward neutral (7.0), Fe precipitates as Fe oxyhydroxide (or Fe oxide: Chapter 6), which adsorbs many metals out of solution. In some cases evaporation also results in precipitation of metal minerals.

Both precipitated minerals and Fe oxide-bound metals are much more available for exchange with the environment than the original insoluble pyrite forms. The combination of low pH and high concentrations of metals is toxic for most macrofaunal and macrofloral forms of life.

Another form of primary contamination occurs when open pit mines are abandoned. Pumping of drainage water is terminated and the pit is allowed to fill with water. The open pit mine in Butte, was abandoned in 1983, and pumping operations to remove water were discontinued. The pit, which was 2 km deep and 1 km in diameter at the surface, began to fill with water. Underground shafts connected to the pit also filled. These waters quickly turned acidic (pH < 2.5), and contained extreme concentrations of dissolved metals. Episodes occurred where flocks of migratory birds mistook the pit lake for a natural water body, landed in the acid waters of the open pit, and quickly died. But the largest environmental hazard is probably the potential for accidental release of the massive volumes of water that accumulate in open pits either through groundwater or malfunction of the storage.

The increase in pit mining since the 1970s, across the world, could lead to the formation of numerous such pit lakes over the next 50 years (Castro and Moore, 2000). In 1985, for example, USEPA (1985) estimated that there were about 50 open pit mines in the USA. Remediation of the contamination in an acid pit lake is a massive problem, once it is created.

Modern mining operations can control acid mine drainage. Sulphide-bearing waste material is isolated from other wastes. Acid production is either contained in ponds or controlled on the isolated wastes by covering the waste heaps or neutralising them (covering with lime, for example). Approaches are also proposed for limiting formation of acidic waters or preventing metal mobilisation in pit lakes (Castro and Moore, 2000). Although these activities require effort and cost, prevention is probably much more economically and ecologically effective than remediation (Tremblay, 2004) (Box 12.5).

Box 12.5 Acid formation: Iron Mountain, California

The Iron Mountain Mine is an abandoned Cu mine near Lake Shasta in northern California, USA. It is characterised by tens of kilometres of underground tunnels running through a sulphide ore body, as well as several streams peripheral to the ore body. Ag, Au, Cu, Fe, Zn and pyrite (for sulphuric acid production) were recovered beginning in the early 1860s. Open pit mining was terminated in 1962. Before remediation, about 300 tons per year of dissolved Cd, Cu and Zn drained from the site into the Sacramento River. Massive fish kills were common in the river when

runoff carried in sudden surges of acid mine waters. More than 20 fish kill events were documented between 1963 and 1986 (when major remediation began), although many more undoubtedly occurred. At least 47 000 trout were killed during a single week in 1967, in the only episode that has been studied in depth (Nordstrom and Alpers, 1999).

Conditions at Iron Mountain are nearly optimal for the production of acid mine waters. The tunnels and shafts allow rapid exposure of the ore to oxygen and water; and high temperatures stimulate movement of air. Primary oxidative dissolution and creation of acid conditions is associated with the main ore body and apparently catalysed by the acidophilic bacterium

Leptospirillium ferrooxidans (Schrenk *et al.*, 1998). The result is accumulation of waters that are some of the most acidic and metal-rich reported anywhere in the world (Nordstrom and Alpers, 1999). For example, the chemical compositions of five of the most acidic waters found underground in one branch of the mine during 1990 to 1991 had the highest concentrations ever recorded in a natural water for As, Cd, Fe, and SO_4 and nearly the highest for Cu. About 8 million tons of sulphide remain in the mine. At current weathering rates it would take about 3200 years for the pyrite in the major ore body to fully oxidize.

One of the options for remediation of one of the largest underground mines was to plug it. Nordstrom and Alpers (1999) estimated that if Iron Mountain were plugged, a pool of about 600 000 m^3 of water would form at or near the top of the groundwater table, all with a pH at or below 1, and with many grams of dissolved metals per litre; in a rock with almost no neutralisation capacity. Hydrological flow in this system could follow fractures, excavations and drill-holes, with a high degree of risk for potentially dangerous results. They cited other attempts to plug abandoned or inactive mines without monitoring, modelling or considering the physical and chemical consequences. Major leaks or failures occurred at most, resulting in widespread and disseminated seeps of enriched acid mine waters, and increases in subsurface head pressures. Remedial actions in recent decades have reduced metal loads into the Sacramento River by capturing the major acidic discharges in reservoirs and routing them to a lime neutralisation plant.

12.4   SECONDARY AND TERTIARY CONTAMINATION

Prologue. Movement of the waste and contamination away from the primary mining/smelting site in streams or through the atmosphere generates secondary contamination in soils, groundwater and rivers. The degree to which such wastes are spread over the landscape determines the ultimate environmental footprint and the ecological risks of the operation. The scale of the contamination can be immense and long term, especially if operations do not prevent acid mine drainage, air pollution, dispersal of tailings materials and/or contamination of floodplains. Tertiary contamination occurs when wastes are remobilised from the site of deposition. The highest concentrations of metals cited in natural waters and sediments are typically the result of secondary and tertiary contamination from mine or smelter wastes.

12.4.1   Contamination of rivers

Prologue. The dispersion of contamination discharged to rivers is usually a result of the geochemical, hydrological and hydrodynamic processes that mobilise contaminants. Ultimately the largest mass of dissolved metal is transformed to particulate form, stored then repeatedly remobilised and progressively transported downstream. The metals exchange between water, particulates and biota. The exchange is influenced by physical, geochemical and biological conditions.

Rivers are perhaps the most common type of receiving water for metal contaminated wastes from mining operations. Geochemical mobilisation and physical mobilisation, as well as hydrological processes that transport, deposit and remobilise metals, are the critical determinants of metal fate and ultimately metal effects.

12.4.1.1 Dispersal of metals in mine-impacted rivers

Prologue. Metal inputs to rivers from mining occur in particulate and dissolved form. Contamination trends downstream depend upon forms of input and hydrological processes. Formation of acid in the primary or secondary deposits can extend dispersion distances and ecological damage. But contamination in particulate form is ultimately the form that is dispersed most widely. Large operations that do not contain their wastes can contaminate hundreds of kilometres of a river bed and floodplain, as well as be the source of contamination for lake bottoms and estuaries. Once metal-contaminated wastes escape the area of primary contamination, remediation can be problematic, controversial and/or inconceivably expensive, where it is possible at all.

Once contamination from mine wastes enters a river system, it can be dispersed far downstream, whether the contamination occurs originally as acid mine drainage or as particulate contamination. If acid is formed in primary and secondary deposits, the resultant dissolution of potentially toxic metals out of their immobile and insoluble forms enhances the potential for metals transport away from the mining site. This is why controlling acid formation is one of the most challenging but essential elements of environmentally acceptable mining practices. As tributaries dilute the contaminated water, pH is neutralised and metal contaminants adsorb onto the particulate Fe oxides (Fig. 12.5). This zone of chemical transition is typically extensive, heavily contaminated and ecologically

impaired. Even though the contaminants are redeposited in the environment bound to Fe and Mn oxides and concentrations in solution are reduced, metals continue to exchange among all components of the ecosystem. Box 12.6 illustrates that metal contamination can be widespread if large masses of acid mine drainage flow unchecked into a major river system.

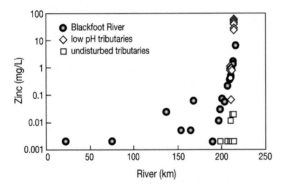

Fig. 12.5 Zn concentrations in solution as a function of km from the mouth of the Blackfoot River, Montana, USA. Moore *et al.* (1989) documented the very high concentrations of metals in low pH drainage waters (tributaries and the uppermost Blackfoot) and the downstream decline in dissolved concentrations as pH was neutralised by dilution. Multiple mines on small streams add metals and low pH water to the headwaters of the Blackfoot. But pH increases rapidly, and dissolved metal concentrations rapidly decline when the river is neutralised upon entry into extensive wetland (old beaver ponds). In this transition zone, metal concentrations in sediments rose as the metals were removed from solution in the wetland. In this case, the wetland greatly inhibited the downstream transport of the contaminants.

Box 12.6 Case study: the Iberian pyrite belt and the Mediterranean Sea

The Iberian pyrite belt is a massive sulphide ore body in Spain and Portugal that covers a region 250 km long and 30 km wide. It has been mined since 4 BC. Cu and sulphur were extracted in the largest quantities (250 million tons) between 1850 and 1970. The best known mining sites in the region are drained by two rivers, the Rio Tinto (Red River) and the Rio Odiel. They carry acid mine drainage to the Atlantic Ocean, just north of the Straits of Gibraltar. Rio Tinto waters are at pH 2.5 when they reach the sea, with dissolved Cd, Zn and Cu concentrations 10 000 to 1 000 000 times higher than concentrations in the uncontaminated surface of the receiving waters in the Gulf of Cadiz. It has long been recognised that the estuary, where the rivers meet the Atlantic Ocean, is also enriched in Cd, Cu and Zn. By comparing the ratios among the three contaminants typical of the river, it was possible to show that the contaminated rivers were the source of that contamination (van Geen *et al.*, 1997). The consistent relationships among concentrations of the three metals were also constant through the Straits of Gibraltar, and for 300 km into the Mediterranean Sea. The intense mining of the pyrite belt, and discharge of these wastes to two small rivers, apparently left a detectable legacy of contamination in a plume 300 km long. Sediment cores in the region indicate that the metal enrichment began in the mid nineteenth century with the intensification of mining in the Iberian pyrite belt. The sediment cores also indicated a reduction of contamination in the two decades before the study was conducted, suggesting that an even larger plume might have existed in earlier times.

Mass loadings to rivers can also occur from particulate metal mobilised from primary or secondary sites. Even if acid formation is controlled, dispersion of particulate contamination is a serious ecological risk, extending the legacy after mine closure. One of the important impacts of uncontained mine wastes is the increased suspended loads of sediment and sedimentation in a river. For example, the Fly River, downstream from the OK Tedi Mine in Papua New Guinea (see Box 12.11 below) carried 60 to 80 mg/L sediment in 1979. After an upstream Cu mine opened in 1984, sediment loads were 330 to 600 mg/L (Hettler *et al.*, 1997). Every year, this river now deposits 5 to 10 years of the natural supply of sediment onto its floodplains, into the river bed and ultimately into the Bay of Papua New Guinea. High sediment loads raise the stream bed (aggrade the stream), make it more susceptible to flooding and accentuate the deposition of fine material in floodplains.

By continuously equilibrating with their surrounding environment, contaminated sediments perpetuate sustained environmental contamination in water and biota which can persist long after inputs to a water body have ceased. Mass balance studies, using total metal loads (dissolved plus particulate metal combined with hydrological discharge as described in Chapters 5 and 14) are one way to detect the degree to which mining activities increase the metal load of a river (Hornberger *et al.*, 1997). A downstream gradient in metal contamination in water, sediment and biota can also characterise rivers where mine wastes are not contained. Characterisation of the gradient is an important first step to understanding the extent, severity and source of any problems.

Gradients in water where waste inputs are in particulate form can be more difficult to detect than gradients in sediments or organism tissues. When generated from secondary or tertiary sources, dissolved metal concentrations are variable on several timescales, especially if acidic, metal-rich pore water accumulates in the hyporheic zones under the bed of a contaminated stream (Fuller and Avies, 1989; Benner *et al.*, 1995; Wielinga *et al.*, 1999; Nagorski *et al.*, 2003). One of the characteristics of the variability in dissolved or particulate metal concentrations in mine-impacted streams is a positive correlation between stream discharge (precipitation-driven) and metal concentrations (Nagorski *et al.*, 2003). When discharge increases, contamination concentrations increase if inputs are from contaminated floodplains, acid mine drainage or contaminated groundwater.

Gradients in sediment contamination are an effective diagnostic in identifying mining as the source of contamination in a river. One approach is to compare the range of contamination and the pattern of downstream dilution to predictions from the cumulative area of the basin upstream from the measurement point (Chapter 6). Another is to use stable isotopes to determine the source of contamination. Ore bodies typically have different stable isotope ratios for elements like Pb than do the rock and soils of the surrounding watershed. Therefore the specific isotope ratios from an ore body can be diagnostic of its contamination. Surprisingly, reservoirs on a river do not necessarily interrupt the downstream dilution gradient. Reservoirs always capture contaminated sediments, but enough fine material apparently passes most reservoirs to allow contamination to spread downstream. The Clark Fork has four impoundments over 500 km. Sediments in the bottom of each impoundment are contaminated similarly to river sites above and below them (Johns and Moore, 1985), and the contamination gradient is not interrupted below any reservoir (Axtmann and Luoma, 1991).

The contamination gradient in a river may be simple when viewed on a broad scale (e.g. hundreds of kilometres) but complex at a smaller scale (e.g. tens of kilometers). In the Clark Fork, a single exponential function describes dilution of contamination over 350–450 km (Fig. 12.6). But adjacent sites separated by 40 km are not statistically separable as significantly different in concentration. Slumping of contaminated cut banks can add local variability, as can inputs from contamination in tributaries (if some have their own mining activities). Events like floods, ice jams or remediation activities can add year-to-year variability (Moore and Landrigan, 1999). The complexity

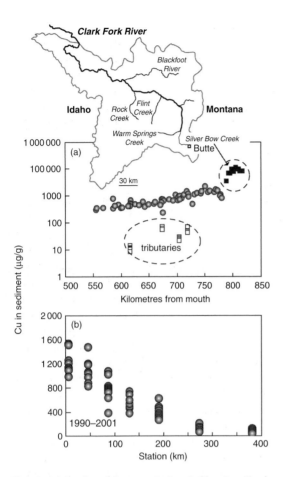

Fig. 12.6 (a) sediment Cu concentrations in Silver Bow Creek, the Clark Fork River and tributaries. Copper concentrations are plotted on a log scale against river kilometre, determined as distance from the downstsream confluence with the Columbia River. (Data from J. N. Moore *et al.*, University of Montana; after Luoma *et al.*, 2007.) (b) Cu concentrations in sediments collected in August of seven different years at seven sites in the mainstem of the Clark Fork river to test variability in time. The *y*-axis is kilometres from the headwaters of the Clark Fork, at the confluence of Silver Bow and Warm Springs Creeks. The downstream gradient is unambiguous overall, but metal concentrations at adjacent stations are not significantly different on a statistical basis. In both cases sediments were sieved to eliminate particles >63 μm grain size and thus reduce grain size biases. (Data from Axtmann and Luoma, 1991; Dodge *et al.*, 2003; assembled by M. Hornberger *et al.*, US Geological Survey, unpublished).

at smaller scales can confound assessments of risks (e.g. USEPA, 1999b), if either is not appreciated. An initial understanding of the broad-scale gradient can help put such variabililty in context,

however. Detection of contamination, its trends and its spatial distribution can also be irresolvable, or misinterpreted, if particle size biases are not removed from sediment data (Horowitz *et al.*, 1993; also see Chapter 6).

12.4.1.2 Pre-mining dispersion of metals: what is background?

Prologue. One of the most contentious questions asked about mining impacts is how much of the contamination was there 'naturally' before the mineral deposit was disturbed, and how much was a result of the mining activity. Responsibility for clean-up often depends upon defining the source of the sustained contamination. Simple models allow estimation of the pre-mine and post-mine dispersal of enriched metal concentrations. The models can provide a basis for estimating the contribution of mining to the metal concentrations at any specific site.

Water, soils and sediments associated with mineral deposits are naturally enriched. The major difference between such deposits and a mined area lies in the distance that the contamination is transported away from the primary ore deposit. As discussed in Chapter 6, the transport of contamination depends upon the metal concentration at the source, the size of the footprint of primary contamination, the area of uncontaminated watershed downstream from the ore body and hydrological characteristics.

Mining greatly expands the footprint of the ore body on the surface of the Earth. Studies of unmined deposits show that metal concentrations in water decrease to background within hundreds of metres of an undisturbed ore body (the dispersion train); perhaps extending a few kilometres if the ore body is large (see examples in Helgen and Moore, 1996). Similarly, contaminated soils and sediments have dispersion trains of 0.4 to 2.4 times the exposed area of the ore body (see examples from Helgen and Moore, 1996). Mining expands the area covered by exposed ore and amplifies the contamination (e.g. by creating fine-grained particles). Historical mining operations increase the size of the primary footprint from about 100 to 500 times the natural anomaly (Helgen and Moore, 1996). Where large active mines allow their

Box 12.7 Distribution of contamination in the Butte/Anaconda complex compared to the original anomaly

In the Butte mines, the major ore body had a surface area of approximately 4 km^2. This would suggest an undisturbed surface anomaly of 1.6 to 9.6 km^2. There were no historical studies to document this, but models that predict that distribution are accurate for modern undisturbed ore bodies. After 150 years of mining activity, primary wastes adjacent to the open pit mine cover an area about 50 km^2. In total, about 1400 km^2 of land is contaminated in the Clark Fork basin from the wastes generated from mining, processing and smelting at Butte and Anaconda (Helgen and Moore, 1996).

Comparing post-mining metal concentrations with the pre-mining estimate of dispersion shows that there was approximately 2200 times more metal released into the upper Clark Fork basin from mining than would be expected from the original mineral deposit. Before mining in the basin, sediment metal concentrations in the streams of the area would have reached background concentrations in about 20 to 30 km, based upon cumulative basin area modelling (Helgen and Moore, 1996). After mining, contaminated sediments extend downstream into Lake Pend Oreille, about 600 km downriver from the mine (Axtmann and Luoma, 1991).

sediment contamination to reach streams and rivers, contamination can be detectable hundreds of kilometres from the activities (Box 12.7).

Thus 'background' concentration at any location in a watershed containing an ore body is determined by the size of the exposed, unexploited ore body, the concentration of metal in the exposed area of the ore body and the area of the uncontaminated watershed between the unexploited ore body and the location. To determine the contribution of mining to contamination at that site, the same formula can be used, substituting values for the exploited area of exposed, uncontained contamination. The important point is that undisturbed ore bodies do not contaminate entire watersheds; but mining can if not managed carefully.

12.4.1.3 Floodplains

Prologue. Floodplains store uncontained contamination from mining, then release particulate and dissolved contamination as hydrological processes operate on or within the riparian zone. The effect is to perpetuate and extend the scale of the contamination. It is usually unclear, and therefore very contentious, how to remediate or even control extensive floodplain contamination once it is incorporated onto large areas. Preventing dispersal of the contamination to floodplains is critical, and more cost effective, than attempting to remediate once floods have spread wastes over vast areas.

Once they leave the mine or smelter site, metal contaminated sediments and soils can be remobilised over and over again, extending their dispersal (secondary and tertiary contamination). A few floods may be responsible for the greatest transport of mine wastes away from an area of primary contamination. For example, a significant flood event in 1908 in the Butte area in Montana appeared to move approximately 2.9 million tons of tailings-laced sediments (Nimick and Moore, 1991). Those tailings were deposited along the floodplain, in banks and bars, for 200 km of the river (Box 12.8).

Metal deposited on the floodplain during floods can be mobilised by overland transport of the particulate wastes into the river or by slumping of banks into the river. Processes that create acidic metal-rich waters also operate within the pore waters and groundwaters of a contaminated floodplain, creating zones of metal enrichment where biogeochemical activities lower pH (Benner *et al.*, 1995). During periods when the streams or rivers gain water from the ground (low flow seasons or flushing events during storms), these metals are carried from the water within the floodplain soils (the hyporheic zone) into the surface water environment.

Floodplains also act to store the contamination. Rivers with broader, more extensive floodplains and frequent floods are, therefore, more likely to store more contamination. Redistribution of these contaminated sediments will occur whenever the river cuts new banks from the contaminated floodplain. Even if the mine site is cleaned up, floodplains can remain a massive

Box 12.8 Silver Bow Creek

Silver Bow Creek is a small stream (annual average discharge 4 to 10 m^3/sec) that extends 40 km between the Butte mine in Montana, USA (Box 12.4) and its downstream-most tailings ponds. Two million m^3 of tailings were dumped directly into the creek between 1876 and 1924. The massive addition of sediment clogged the original stream and increased the potential for flooding, by filling the stream channel (aggradation) (Weed, 1912). Unvegetated, low pH tailings deposits (termed slickens) lined Silver Bow Creek over its entire length until serious remediation began in the late 1990s. The slickens contain contaminated soil up to 2 metres thick (Nimick and Moore, 1991). They are devoid of most signs of life except occasional patches of metal-tolerant grasses and willows. Some of the slickens were over 200 metres wide. Contaminated floodplain sediments contained up to 21 µg/g Cd dw, 1660 µg/g Cu dw and 5690 µg/g Zn dw (Dodge *et al.*, 2003). Pre-mining concentrations, determined from pits dug to pre-mining depths on the floodplain, were <1 µg/g Cd dw, 16 µg/g Cu dw and 49 µg/g Zn dw (Ramsey *et al.*, 2005).

The lack of vegetation in the slickens also increased the erosion potential of the banks and adjacent floodplain, resulting in continual recontamination of the stream as it meandered through the floodplain. Even though tailings ponds were eventually constructed to contain the mine waste, the Clark Fork River remained contaminated. Continuous input from the floodplains of Silver Bow Creek was the primary source, with some additional inputs from cut banks on the downstream floodplains. Remediation of the contaminated river and stream first involved constructing ponds to capture the entire flow of Silver Bow Creek and thus prevent dispersal of the sediments from its floodplain. Ultimately, nearly all the major deposits of contaminated floodplain soils along Silver Bow Creek will be removed, and moved into tailings ponds. The cost for the entire remediation of the area is estimated to be US$200 to US$300 million. Controversy continues over disposal of the tailings; objections come mainly from residents of small towns near the site of soil disposal. These scenarios are (and will be) repeated at sites all over the world where mining has not followed environmentally responsible practices.

source of contamination (Box 12.8). Eventually the river will move all this sediment downstream, but that can take thousands of years.

12.4.1.4 Accidental releases
Prologue. Tailings ponds have an essential role in containing mine wastes, but they can fail. The local damage from ponding tailings, and the ecological risks from failure of a tailings pond are severe, but less than the risks from continuous release of uncontained wastes.

Modern, well-run mining operations reduce the disastrous scale of floodplain contamination by capturing and containing tailings in (sometimes extremely large) pond systems. The largest risk from tailings ponds is their susceptibility to failure because of floods, earthquakes, landslides, poor management or flawed construction. Even a single failure can seriously contaminate a river and its floodplain, and require massive investment in remediation (Box 12.9).

Ponds can be inspected to minimise the likelihood of failure. Pond failures can sometimes be recontained. Measures can be taken to minimise damage to the most valuable natural resources or remediate the one-time damage. Thus, a tailings pond failure is a dramatic event with a sizeable economic cost; nevertheless its danger is not of the magnitude of ongoing contamination. Tailings ponds are an essential component of modern mining.

12.4.2   Estuaries and lakes as receiving waters

Prologue. In-stream lakes and reservoirs, as well as estuaries, capture the entire flow of a river system and can accumulate wastes inputs from the watershed.

As contaminated sediments are transported downstream, the influence of the mining can extend from riverine watersheds to the rivers' receiving waters in lakes or estuaries. Devon, Cornwall and

Box 12.9 The Aznalcollar tailings spill from Los Frailes mine

On 25 April 1998 the retaining wall of a tailings reservoir collapsed at the Los Frailes Zn mine near Aznalcollar, Spain. The reservoir released 5 000 000 m^3 of acid sludge into the Guadiamar River, which runs parallel to the Tinto/Odiel River system (Box 12.6). Zn concentrations in sediment 10 km downstream from the spill were 12 000 µg/g dw one week after the spill (van Geen et al., 1999), compared to a typical uncontaminated river sediment at 80 µg/g. Dissolved Zn concentrations were 429 µg/L compared to 2 µg/L in a nearby uncontaminated stream. The pH fell to 3.8, from a typical near-neutral value for the river. Zn was removed from solution as pH increased when the river neared the sea (40 km downstream from the spill). Van Geen et al. (1999) estimated that the spill added 40 000 to 120 000 tons of Zn to the watershed in one event. This is comparable to 1 year's production of Zn at the mine, and 0.8 to 2.4 year's input to the sea from the nearby, heavily contaminated Tinto/Odiel Rivers.

Dykes were quickly built after the spill to minimise exposure to remobilised contamination of a nearby ecological reserve, Doñana Park. The rest of the Guadiamar River area became the site of one of the largest remediation projects in the world. After the spill, the fine-grained sediments from the tailings ponds covered the floodplain/riparian zone and the entire river bed to depths from a few centimetres to several metres (Solà et al., 2004). Restoration activities began 3 months after the spill. The tailings were removed from the riparian zone in the first year. Two metre high weirs were then constructed within the river, creating multiple ponds in the river to retain the metal-polluted sediments. After removal of the settled sediments, the weirs were removed in 2000, and the typical riverine habitat reappeared. Spring floods washed the remaining sediment into the sea. Ecological recovery is still far from complete, however.

Wales in the western United Kingdom are characterised by numerous small estuaries. Some rivers had no mines in their watershed; others had extensive mining. The nature of the mines also varied from watershed to watershed. The sediments of each estuary continue to carry a metal contamination signature from the historical activities; although most mining ended more than a century ago. The nineteenth-century mining thus left a patchwork of contamination over the entire region (see maps in Chapters 6 and 8).

Milltown reservoir, about 200 km from the mine at Butte, had sufficient As contamination in its sediments to contaminate the water supply of a nearby town that obtained water from an aquifer connected to the river. The source of this contamination was not evident until the contamination gradient in the river was understood. Planning is under way to remove Milltown reservoir to allow upstream access to fish. The mine-contaminated sediments trapped behind it will add complexity to the removal process and could affect outcomes.

The extent of lake (or estuary) contamination depends upon the magnitude of sediment contamination when it reaches the lake, and the sediment load. One of the best-known cases of lake contamination from mining is Lake Coeur d'Alene in Idaho, USA. This large lake is within 80 km of the historic Bunker Hill mining complex. The entire lake bottom is contaminated with mine wastes, and the contamination extends downstream into the river draining the lake (Box 12.10). In contrast, the Clark Fork River does not appear to have heavily contaminated Lake Pend Oreille 500 km downstream. The extensive watershed between the mine and the lake has apparently trapped and/or diluted most of the contamination before it reached the lake. Similarly, the Bay of Papua New Guinea is not contaminated by an extremely large mine in its watershed, the OK Tedi mine (see Box 12.15 below) because of the 1 000 km between the mine and sea; and the extensive dilution of Fly River contamination by the heavy, uncontaminated sediment loads added at about 600 km by the Strickland River (Salomons and Eagle, 1990).

12.4.3 Atmospheric dispersal: smelter wastes

Prologue. Long-range transport of metal contaminants (and sulphur dioxide) is well-documented from smelters. Soils and entire water bodies can be

Box 12.10 Mine wastes and Lake Coeur d'Alene, Idaho

Impacts from the Bunker Hill mining complex in northern Idaho, USA, illustrate how uncontained contamination can disperse from rivers, through in-stream lake waters and beyond. Other human activities (e.g. construction of dams) can accentuate the effects of the contamination.

The Bunker Hill mining/smelting complex was operational between 1880 and the 1970s. Most of the wastes from the operations were drained into the South Fork of the Coeur d'Alene river until 1968, when containment ponds were completed. The sediments of the river are contaminated with Ag, As, Cd, Cu, Pb, Sb and Zn. The floodplain resembles other large complexes with a similar history (Box 12.8). The South Fork enters Lake Coeur d'Alene 80 km from the mining/smelting complex. The lake is a natural submerged river bed, 3.2 km wide by 40 km long; it is one of the scenic wonders of the western USA.

As early as 1923 an extensive plume of sediments was observed, originating from the mining. Sediment cores from the lake showed a layer of metal-enriched sediments 17 to 119 cm thick extending across 85% of the bed of the lake (Horowitz et al., 1995). The fine-grained contaminated sediments infiltrated the lake, apparently, by continual deposition, remobilisation and redeposition caused by currents (e.g. during high river discharges) and winds. Below the surface, sediment cores were banded with different layers, indicative of geochemical recycling. Sediments deposited before the mining (deep in the cores) contained extensive evidence of biological reworking; all signs of life disappeared in the contaminated sediments.

From the cores it appears that 75 million tonnes of contaminated sediment were deposited on the lake bed, containing as much as 468 000 tonnes of Pb and 260 tonnes of Hg. The geographical distributions of the contamination, the dates of deposition, and ratios among metals were consistent with inputs from the mining/smelting complex.

The contamination extended beyond the lake, into its outlet, the Spokane River (Table 12.2). Sediment cores indicated that metal enrichment in the middle of Spokane River began at a similar time to the contamination of Lake Coeur d'Alene (Grosbois et al., 2001). In the most downstream part of the Spokane River, however, sediment cores indicated that the onset of contamination was delayed, beginning in 1930 to 1940. It seems most likely that the later contamination was the result of the construction of the Grand Coulee Dam in about 1930, which slowed the river and allowed the build-up of contamination. (After Horowitz et al., 1993, 1995; Grosbois et al., 2002.)

Table 12.2 *Typical metal concentrations (μg/g dw) in fine-grained sediments in the Coeur d'Alene and Spokane River basins. Values are medians from modern sediments (Horowitz* et al., *1993, 1995; Grosbois* et al., *2001)*

Site	Cu	Pb	Zn	Cd
Upper South Fork, Coeur d'Alene	670	71000	9000	19
Lower South Fork, Coeur d'Alene	240	5600	4500	64
Lake Coeur d'Alene	70	1800	3500	56
Upper Spokane River	49	320	2400	30
Lower Spokane River	51	325	2400	28

contaminated and ecological processes seriously disrupted over hundreds of square kilometres around large smelter operations. Well-documented examples exist in Sudbury, Ontario, Canada; Puget Sound, Washington, USA; and Bristol on the Severn estuary, United Kingdom.

Smelters are a main source of atmospheric emissions of As, Cu, Cd, Sb and Zn in the global budget,

and they contribute substantially to the overall emissions of Cr, Pb, Se and Ni (Nriagu and Pacyna, 1988). The spatial patterns of contamination can be complex, but exponentially declining contaminant concentrations are found with distance, if wind patterns are considered. The degree and scale of contamination depend on factors such as the magnitude of smelter emissions, particle sizes within the smelter plume, wind patterns, and background

Box 12.11 Secondary atmospheric contamination from the Sudbury smelters, Ontario, Canada

Sudbury Basin is a concave geological formation about 60 km by 26 km in size on the Canadian Shield (granitic, poor topsoil geology in Ontario and parts of Quebec, Canada). The area is rich in lakes and poor in topsoil. Copper–nickel mineralisation was discovered in 1880. Surface roasting was used to extract ores until 1920, when the first smelters were built. Gases were vented from the smelters via ever-higher stacks as the smelters developed. High stacks do not eliminate contamination, but they disperse it further, and dilute it more. That reduced the most extreme impacts in the immediate vicinity of the smelter, but it dispersed moderated impacts over a larger area.

Several thousand tons per year of Cu, Ni and Pb were released from these stacks in the late 1970s, along with 1 million tons per year of sulphur dioxide. In 1971 alone, 192 tons of Ni, 145 tons of Cu, 1130 tons of Fe and 4.5 tons of Co were released in 28 days from two of the smelters as airborne pollutants (Hutchinson and Whitby, 1976). Hutchinson and Whitby (1976) reported 'An area in excess of 100 mi^2 is now almost devoid of vegetation, and damage to the forest vegetation is visible over an area of approximately 1800 mi^2.'

The increased stack heights to dilute local pollution problems have spread the problem more widely. The plants that did grow in the area were metal tolerant.

Yan et al. (2004) estimated the number of lakes acidified by historical sulphur emissions to be 'several thousand'. Metal levels in water and sediments from such lakes were highest nearest the smelter and decreased with distance (Yan and Miller, 1984). The increase in contamination of Ni, Cu, Pb, Co and Zn in age-dated sediment cores coincided with the onset of smelting activities; and the metal-to-metal ratios were characteristic of smelted ores (Nriagu et al., 1982).

Metal contaminant concentrations in soils and rain water followed a gradient away from the smelter (Freedman and Hutchinson, 1980). Within 3 km of the smelters, lake pH ranged from 3.3 to 6.1, depending upon chemical composition (capacity to neutralise the acid formed from the sulphur dioxide inputs). Beyond 24 km, lake pH ranged from 4.6 to 7.4. Middle Lake was a well-studied example. In 1974, it had a pH of 4.2, 500 μg/L Cu and 1000 μg/L Ni in solution (Yan et al., 2004). Smelter closures in 1977 to 1979 resulted in dramatic reductions in contamination, clearly demonstrating that this smelter complex was the source of the contamination.

soil concentrations and variability. Removal of air pollutants before their discharge is a feasible technology, and is now widely practised across the developed world. Such practices are not necessarily common in developing countries.

Illness and deaths in domestic animals, or decimation of plant life, can be the first sign of serious metal contamination from a smelter (usually caused by Cd, Pb or As). Outbreaks of As poisoning occurred in cattle, sheep and horses over an area of 260 km^2 in 1902, after the opening of the new smelter at Anaconda, Montana, USA. Even after a flue was built to capture the particulate air pollution at the Anaconda smelter, release of 27 000 kg As per day was documented from that operation (Harkin and Swain, 1907). Surface soil As concentrations over an area of more than 600 km^2 around Anaconda were between 29 and 1856 μg/g (i.e. all above background levels). The pattern matched primary wind direction. The scale of the effects

from the Anaconda smelter was typical of large operations of its time, but were small compared to those from some of the later historical smelter operations (Box 12.11).

12.5 ECOLOGICAL RISK: BIOAVAILABILITY, TOXICITY AND ECOLOGICAL CHANGE

Prologue. Ecological risk assessments for mining activities must consider the nature of the environment; they should evaluate the sum of all exposures; and multiple lines of evidence should be built around an expectation (e.g. conceptual model) of how ecological change might be manifested (see also Chapter 11).

Ecological risks from metals are complex to evaluate at mining sites. The most obvious risks are acid mine drainage and acute toxicity or periodic fish kills in the area of greatest contamination. More subtle (and resistant to remediation) risks occur when particulate wastes are dispersed away from the site of primary contamination. In rivers, for example, pulse inputs of metals from the hyporheic zone, ingestion of a contaminated diet, avoidance of contaminated conditions and secondary effects on habitat all influence the biological community.

12.5.1 Acid mine drainage

Prologue. Abiotic conditions characterise acid mine drainage environments. Domination by a few metalophiles and local extinction of a substantial portion of the biological community is typical of acid mine drainage and the area of transitional pH.

The differences in bioavailability among the different types of mine waste result in great differences in toxicity and effects. In acid mine drainage, the combination of low pH and high metal concentrations is especially toxic. Streams dominated by acidic mine wastes (pH < 4) are immediately toxic to animals, in both tests and the field (note the migratory bird toxicity that happens in pit lakes). Strongly acidic streams and lakes are, typically, devoid of nearly all macrofauna.

Evidence for the high metal bioavailability and toxicity of acid mine drainage also comes from fish kills that accompany acidic runoff from tailings deposits on the floodplains of rivers. Along the Clark Fork River, metal sulphates precipitate on the floodplain surface as by-products of sulphide weathering and are concentrated by evaporation of soil moisture. These salts dissolve during rainstorms, readily releasing high concentrations of Al, As, Cd, Cu, Fe, Mn, Zn and H^+ to overland runoff. When the runoff enters the river, fish kills occur (Nimick and Moore, 1991).

The Upper Clark Fork River contained no trout for nearly a century because of the continuous inputs of wastes from mining, milling and smelting operations (Johnson and Schmidt, 1988). Fish began to reappear in the 1950s, but periodic kills followed summer storms almost every year. In the 1950s, storms would cause the river to become

visibly red (Fe-rich) for as much as 200 km downstream from the mine (Averett, 1961). Trout did not survive 24 hour *in situ* toxicity tests in the river during such episodes. Fish would reappear shortly after the episode subsided, probably because of immigration from tributaries. Controlling tailings inputs to the river reduced the downstream extent of the contamination; and allowed the beginning of recovery. Building berms on upstream floodplains to prevent overland runoff ended the most obvious major fish kills in the 1990s. But pulses of toxic, metal-rich, acidic inputs can be common at mine waste sites, associated with periods of high runoff and especially after prolonged dry periods (Moore and Luoma, 1990; Nordstrom and Alpers, 1999).

Fe and Mn precipitates typically coat all substrates (including organisms) in the transition zone where pH is neutralised. The coatings are very effective in sequestering metals. The transformations also create an unusual abundance of colloidal-sized particles which transport metals downstream. In laboratory tests with such water, metal bioavailability is reduced compared to pristine waters typically used in toxicity tests (e.g. Welsh *et al.*, 2000); but high internal concentrations in the organisms that survive in this zone suggest at least some forms of the metal contamination are bioavailable.

Severe adverse effects on the biological community also provide convincing evidence that metals are bioavailable and toxic in the zone of transitional pH (Box 12.12). For example, the historic description of the waters of Silver Bow Creek (described in Box 12.8) were as 'the consistency of thick soup, made so by the tailings it receives from the mills at Butte (in 1892). No fish could live in such a mixture' (from Johnson and Schmidt, 1988). In 2003, prior to the most intense period of remediation, fish were still completely absent from the 40 km of creek habitat, even though the pH was not reduced. Fish were abundant in surrounding, less contaminated streams.

A few metal-tolerant species can sometimes be found in the transitional environments. Invertebrates were absent over the entire 40 km of Silver Bow Creek before 1975, after which the earliest

Box 12.12 Bioavailability and effects of mine wastes at low pH and the zone of pH transition: the Los Frailes spill

Metal concentrations and benthic communities were studied in 2000 in the Guadiamar River, Spain, 2 years after the Aznalcollar tailings pond rupture in 1998 (see also Box 12.9). The aquatic community of the Guadiamar River was eliminated over the entire 60 km of the river during the spill (Pain *et al.*, 1998; Solà *et al.*, 2004). Some recovery was evident in 2000, after the remediation was complete.

The community was studied in the first 15 km downstream from the spill site in 2000. Three control stations were located upstream of the spill. Ten sites were positioned downstream of the spill. Dissolved metal concentrations were 10- to 3000-fold higher at the six stations nearest the spill than at control stations; pH was 4 to 5.5. Four taxa were present consistently at the most contaminated six stations: two genera of Heteroptera (bugs) and two of Coleoptera (beetles). These were all air-breathing insects. Presumably their limited exposure to the water allowed them to tolerate both the reduced pH and extreme metal concentrations. None of these was common at other stations.

Dissolved metal concentrations declined, pH increased, Cd (and Zn) concentrations in sediments increased (Fig. 12.7) and the caddis fly larva *Hydropsyche* sp. appeared in the stream at the downstream-most four sites. Metal concentrations in *Hydropsyche* sp. were enriched, compared to uncontaminated sites and were comparable or exceeded concentrations in other mine-contaminated streams (see data in Chapter 8). Thus Cu, Cd and especially Zn were highly bioavailable to the food web.

Sixty-two different invertebrate taxa were present among all stations in 2000. Nearly all of these taxa were present at the uncontaminated control stations. Six taxa dominated the contaminated, but neutral reach of the stream; 44 species were found amongst all 10 contaminated sites. Thus 22 of the original 55 species appeared to be unable to survive the contamination, even after remediation.

In summary, at pH<4 to 5, nearly all fauna were eliminated. As metals begin to be adsorbed out of solution, a few tolerant fauna could survive. Even after pH was neutralised, metals remained highly bioavailable. One-third of the expected species had not yet recolonised the river two years after the spill (and after substantial remediation). The progression of these effects can be expected in any stream subjected to inputs of acid mine drainage.

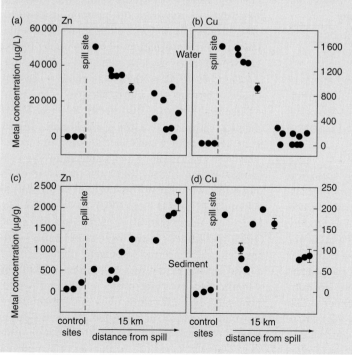

Fig. 12.7 Concentrations of Cu and Zn in water (a, b) and sediments (c, d) in the Guadiamar River (Spain) in 2000, two years after a tailing pond dam burst in 1998 (Box 12.8). Three sites are upstream from the tailings spill, and the remainder are downstream. Metal concentrations in water were high where pH<4 to 5.5. No *Hydropsyche* caddisflies were present in this region. Metals precipitated into the sediment at the four downstream sites where pH was neutral. (After Solà *et al.*, 2004.)

treatments of mine wastes began to take effect and metal-tolerant species began to appear (Chadwick *et al.*, 1986). A similar sequence on a shorter time scale occurred in the Guadiamar River after the Los Frailes tailings spill (Box 12.12). Similarly, only a few species of aquatic macrophytes and algae were found in lake waters nearest the smelters in the Sudbury region when pH was 3.3 to 6.1 (Chapter 11; Box 12.11). The algae that did grow were demonstrated to be populations that had developed metal tolerance (Stokes *et al.*, 1973), one of the first demonstrations that such tolerance could occur in nature. The low pH, high Cu (>6 μg/L) and high Ni (>150 μg/L) concentrations were also toxic to fish.

In lakes with transitional conditions, but high metal concentrations, a few species recover, but most do not until metal concentrations recede (Chapter 11). Yellow perch (*Perca flavescens*) appear to be the only dominant fish species in neutral pH, moderately metal-contaminated lakes near Sudbury in Canada (Campbell *et al.*, 2003; Pyle *et al.*, 2005); cladoceran zooplankton also are not found in such lakes (Yan *et al.*, 2004; Chapter 11).

12.5.2   Ecological risks from secondary particulate contamination: complicating factors

Prologue. Metal form in sediments, pulse inputs of metals, complex exposure routes, behavioural avoidance of contamination and changes in ecological function all add complexity to determining ecological risks from metal mining.

Part of the complexity in evaluating ecological risks lies in the complexity of how the physical and geochemical conditions combine with the ecological properties of the specific system to produce effects. Metal bioavailability from sediments is one property affected by such influences. As sediments are dispersed away from primary sources, the form and bioavailability of metals change. Ores are dominated by insoluble minerals formed under anoxic conditions. Presumably, metals in this form are of low bioavailability. But once these minerals are in the aquatic environment and/or contact oxygenated water, diagenesis (the geochemical processes that transform metals) appears to influence metal form. The original insoluble mineral forms begin to disappear as the wastes are transported. Over the long period of time such wastes are in contact with air and water, Fe is released from the original sulphides (sometimes in microenvironments), then reprecipitated as Fe oxides. Metals are also released, then resorbed onto the surfaces of Fe oxides or complexed with natural organic materials. Once the minerals are transformed, the metals are much more available for ongoing environmental exchange than were the original minerals in the ore body.

For example, surface sediments from Lake Coeur d'Alene (Box 12.10) contained some mineral grains typical of the ore body, but most of the minerals were covered with a surface coating of Fe oxides (Horowitz *et al.*, 1993). The recalcitrant metal sulphides also declined in abundance, overall, with distance from the original ore body. 'Most of the contaminants' in the Coeur d'Alene river were associated with Fe oxides, with a 'markedly smaller percentage' associated with sulphides or recalcitrant phases (Horowitz *et al.*, 1993). Vertical gradients in oxygen concentrations and physical redistribution appeared to promote transformation of minerals in Coeur d'Alene lake sediments, downstream. Bioturbation of sediments by organisms could also accelerate diagenesis; although in Lake Coeur d'Alene there was no evidence that an active benthos was able to survive.

Dissolved inputs are also affected by physical and chemical properties of the environment. Inputs from acidic pore waters in the flooplains occur whenever the stream gains water from the subsurface (USEPA, 1999b); but such inputs occur in pulses (e.g. with rainstorms) and can be difficult to document (Nagorski *et al.*, 2003). To determine the toxicological implications of such pulses, Marr *et al.* (1995) performed laboratory experiments with trout. Metal concentrations were increased over a 1 or 2 hour period and held constant for 6 or 4 hours, then decreased over a 1 or 2 hour period. Alkalinity, pH and conductivity were maintained constant, increased, or decreased during the pulse. Marr *et al.* (1995) found that fry and juvenile trout were equally sensitive to the pulse exposures.

Box 12.13 Toxicity testing and ecological risk in the Clark Fork River

From toxicity testing data, the US Environmental Protection Agency generally established that site-specific, acute lethality to trout would be expected in streams at dissolved Cu concentrations greater than 30 to 40 µg/L (USEPA, 1999b). To determine whether or not metals were responsible for reduced fish populations in the Clark Fork River this standard was compared against environmental Cu concentrations. Dissolved metal data had been collected quarterly (once in 3 months) for more than a decade at monitoring stations. Unfortunately variability in dissolved metal concentration occurs on daily, weekly and monthly scales in areas of heavy contamination (Nagorski *et al.*, 2003), so seasonal sampling was not considered adequate to detect the pulses of exposure that resident organisms probably experienced. Dissolved Cu concentrations in the quarterly data for the Clark Fork

River ranged from 2 to 20 µg/L, with only two of 232 observations exceeding the acutely lethal concentration. However, trout fry and fingerlings placed in cages in the Clark Fork River for several months in the 1990s did not survive, although the same procedures in a reference stream produced no toxicity (Phillips and Lipton, 1995). Thus, the field observations and field toxicity test contradicted the toxicity test predictions of no toxicity expected from the dissolved metal concentrations. The Clark Fork Ecological Risk Assessment (USEPA, 1999b) concluded that trout survival was probably affected by 'exposures to pulses of Cu input, or other high concentration events', even though such events were rarely documented. The necessity for conjecture about pulse inputs reflects the complex behaviour of metals in mine-impacted environments, the difficulty of realistically sampling environmental waters and the uncertainties of toxicity testing. The conclusions of the risk assessment were quite contentious, however, given the weak direct documentation.

Rainbow trout fry (*Oncorhynchus mykiss*) were particularly vulnerable (more so than brown trout, *Salmo trutta*) when elevated metal concentrations were present with depressed hardness and pH, conditions that mimic rainstorm events. The authors concluded that if an event supplied metal-rich, acidic runoff to a river it would limit the survival of trout, especially rainbow trout. These kinds of results led regulatory agencies to conclude that undetected pulse inputs of Cu were the major cause of depressed trout populations in the Clark Fork River (Box 12.13).

Once mineral forms of metals undergo diagenesis in sediments or the floodplain pore waters, multiple pathways of metal exposure are enhanced. Contamination increases in water, plants, detritus, sediment and prey organisms. Exposures certainly occur via contaminated water and ingestion of the various contaminated materials (Farag *et al.*, 1999). But the long history of relying on dissolved metal toxicity as a predictor of effects has impeded appreciation of such processes. For example, studies of dietary exposure to metals were another contentious aspect of the Clark Fork Ecological Risk Assessment (USEPA, 1999b; Chapter 10). In the Clark

Fork River, invertebrates have 2 to 100 times greater Cd, Cu and Pb concentrations than those collected from tributaries, and are an important source of food for the trout that are present. When early life stages of several trout species were fed invertebrates collected from the contaminated river, the fish showed reduced feeding activity, biochemical abnormalities, reduced growth and reduced survival (Farag *et al.*, 1994; Woodward *et al.*, 1994, 1995; Chapter 11, Box 10.11). These experiments were controversial because of questions about nutritional content and species composition of the diet. But Hansen *et al.* (2004) fed rainbow trout (*O. mykiss*) with earthworms (*Lumbriculus variegatus*) grown in contaminated Clark Fork soils and worms grown in clean soils (the same nutritional content). They found histopathological abnormalities in the trout fed the contaminated diet, along with unambiguous inhibition of growth. The growth reduction was associated with reductions in conversion of food energy to biomass (rather than reduced food intake); and correlated with As bioaccumulation.

Another complication in determining ecological risks from mine wastes is that some organisms avoid unfavourable concentrations of metals, like Cu and Zn. Macroinvertebrate drift increases in contaminated streams (Carlisle and Clements, 2005), indicative that invertebrates avoid such habitats to the extent that they can. Avoidance behaviour may also cause reduced fish populations by displacing individuals from preferred habitats. For example, dissolved Cu in the Clark Fork River is often at or above the concentrations that cause avoidance by brown trout and rainbow trout (Woodward *et al.*, 1995). Even after 45 days of acclimation rainbow trout preferred clean water over water with elevated concentrations of metals. This may be particularly important for species that have very specific habitat requirements. For example, bull trout *Salvelinus confluentus* (a native species once abundant in the Clark Fork River) can live in tributaries, but depend upon larger rivers for migration routes during reproduction. If the larger river is contaminated, reproduction is halted by behavioural avoidance of the migratory route.

Finally, the disappearance of sensitive species also appears to have important effects on the ecological functions performed by the invertebrate communities in streams. Carlisle and Clements (2003) concluded that total production attributable to algae and animal prey declined in contaminated streams. This is partly because community and microbial respiration are reduced (Carlisle and Clements, 2005); and leaf litter breakdown declines (Carlisle and Clements, 2005). All could affect food availability for other fauna. In the mining district of southern Colorado, USA, much of the reduction in productivity was a result of reduced numbers of herbivorous insects in the contaminated streams (Chapter 11). Thus, predators must depend upon a disturbed benthic community for their food. The benthos will be less productive, less diverse and missing attributes of likely (although unquantified) importance to predator diversity and productivity.

12.6   SIGNS OF METAL EFFECTS

Prologue. The greater sensitivity to metals of some species than others determines the potential ecological response of biological communities in mine-

contaminated environments. Risk assessments that rely exclusively upon compliance with water quality regulations are unlikely to diagnose such impacts. Risk assessments of existing activities must document bioavailable exposures, consider *in situ* toxicity, evaluate signs of sublethal stress in surviving species and characterise the existing biological communities with knowledge of relative sensitivity among species in mind. These multiple lines of evidence are most likely to result in the least ambiguous evaluations.

Once it is recognised how metal contamination exerts its ecological effects (Chapter 11), risk assessments for mining activities can be built from that knowledge. Important ecological attributes can be disturbed in environments subject to mine waste inputs, in circumstances that are not acutely toxic. Detectable changes in community structure, as described by generic indices like diversity or number of taxa, are common where contamination from mine wastes is severe. But disturbances also occur at lower levels of contamination, manifested as missing species and/or depauperate populations (deviations in community composition). The perceived cost of accumulating a convincing body of evidence for a risk assessment is often an impediment to a thorough study of ecological risks. In most controversies about risks from mining, however, so much is at stake that accumulating a body of evidence is ultimately necessary, and preferable (in time and cost) to litigation and/or contentious dialogue that can delay risk management for years.

12.6.1   Ecological signs of risk in marine environments

Undersea mining (e.g. for Mn nodules) and seabed disposal of mine waters (to replace discharges of tailings into rivers) are much discussed, but only a few examples exist where ecological effects have been studied (Box 12.14; see also the Chilean case study in Chapter 11). On the seabed, geographical distance from the outfall is particularly unreliable as an indicator of exposure in the sea, because currents (driven by tides and winds) often move

Box 12.14 Effects of seabed tailings disposal: Rupert Inlet, British Columbia, Canada

The Island Copper Mine, at Rupert Inlet, northwestern Vancouver Island, British Columbia (Canada), is one of several mines in British Columbia that discharged their wastes onto the seabed between 1970 and the mid-1990s. The submarine (undersea) discharge began in 1971 and the mine closed in 1995. The discharge was at a depth of 50 m of water, into a well-flushed fjord where soft-bottom sediments were common. Fifty thousand tons of tailings slurry were dispersed per day (about 3 million gallons per day). An extensive programme of seabed monitoring accompanied the mining. From 23 to 26 stations were sampled annually (ecological, chemical and physical data) from 1970 to 1998. The sampling procedures were not especially effective at sampling larger organisms, however.

Like many coastal situations, the hydrodynamics of Rupert Inlet are complex. Tidal currents are strong and the water column was well mixed. Tailings were dispersed in a heterogeneous pattern because of the complicated deposition, scouring, resuspension and compaction processes. Eventually, layers of tailings were visible in the bottom sediments (from shallow sediment cores) up to 20 km from the outfall, generally declining in depth with distance from the outfall. At the deposition site nearest the outfall the depth of tailings was 5 to 10 m. Three zones of visible tailings and Cu contamination were defined (see table below). Cu concentrations in undisturbed sediments of the region (and pre-mining) were 20 to 45 µg/g dw.

Zone of influence	Distance from outfall	Cu in sediment (µg/g dw)	Mean depth of visible tailings
Near-field	<5 km	>4000	40–60 cm
Mid-field	5–16 km	up to 600	5–30 cm
Far-field	>20 km	<170	None visible

The ecological studies showed that the tailings disposal initially eliminated all fauna (probably by physical burial) but opportunistic species rapidly recolonised the tailings once disposal ceased. However, statistical analyses of community similarity showed that some species were consistently absent as long as tailings layers and/or Cu contamination were evident.

The number of species was a good indicator of ecological effects when tailings disposal was active (in the near- and mid-field areas) (Fig. 12.8). But neither overall abundance of invertebrates nor the number of taxa was the best measure of ecological effects in later years.

The number of species in the area of disposal increased rapidly for 3 years after discharges ceased; but community composition remained very different from before the mining. Amphipod crustacean species were highly sensitive to changes in the sediments and sedimentation rate caused by tailings deposition. They started to increase in abundance after mine closure. Certain taxa of bivalves and polychaetes were apparently sensitive to Cu contamination. For example, the deposit feeding bivalve *Axinopsida serrata* is one of the most common in marine bottom communities of northwest North America. It disappeared at near-field and some mid-field stations, and never recovered during the 3 years of post-mining sampling. Another sensitive, otherwise dominant species was an ophiuroid echinoderm, *Ophiura sarsi*, which was very abundant in the far-field, but was never present in the near-field. Several polychaetes were tolerant of heavy tailings deposition (*Cossura pygodactylata*, *Lumbrinereis luti* and *Pista cristata*). In general polychaetes rapidly recolonised the tailings sites when the discharge stopped, but recolonisation by bivalves and the ophiuroid had not occurred after 3 years. (After Burd, 2002.)

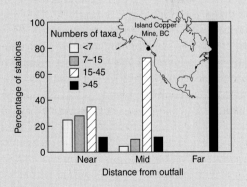

Fig. 12.8 The percent of stations with the designated number of taxa (species) in the near-, mid- and far-field zones around the tailings disposal outfall of the Island Copper Mine. In general, the number of surviving species increased with distance from the outfall. All stations had >45 species at a distance of 20 km from the outfall. (After Burd, 2002.)

tailings in unpredictable ways. In estuaries, land-to-sea gradients can be found where the mine is located on the land; but contamination may also be moved 'upstream' (i.e. landward) of an estuarine mine location by tidal currents.

Generation of acid is unlikely in marine waters. Nevertheless, mobilisation of metals into bioavailable forms occurs, as evidenced by metal bioaccumulation by the species that can inhabit the habitats disturbed by mine waste discharge. Bivalves (in particular tellinid bivalves) and polychaetes (e.g. nereids) were particularly good biomonitors of metal contamination in the estuaries of southwest England (see Chapter 8).

Both the sediment disturbance caused by tailings disposal and the metal toxicity from the tailings contribute to ecological disturbance in marine environments. Undersea disposal of tailings changes the fundamental characteristics of marine sediments. Sedimentation rates are high for tens of kilometres around a large tailings outfall or input. Bed sediments become unstable and organic carbon concentrations will be greatly depleted, to as much as about 10% of the organic carbon content of undisturbed sediments. Very high sulphide concentrations (especially in fine-grained sediments) can also disturb the habitat. After inputs cease, visible layers of tailings may disappear in physically active regions (e.g. Box 12.14) as currents and bioturbation move and mix sediments, respectively. This does not necessarily eliminate the contamination, but it does physically stabilise the sediments. In other types of situations, contamination may remain in place for decades to centuries (e.g. the estuaries of southwest England: Box 12.3).

Changes in composition of the community can occur over surprisingly wide areas. The most sensitive way to detect effects is from statistical analysis for similarity among stations, or zones of similar habitat. But the most effective monitor of effects is the present/absence, or abundance of sensitive species. One sensitive group identified in studies of marine disposal is the harpacticoid copepods, tiny crustaceans (meiofauna) that live in the interstices of sandy sediments (Lee and Correia, 2005). Among rocky intertidal invertebrates, sessile forms might be more affected by marine mine

contamination than mobile forms (Medina et al., 2005). Predators and some other large invertebrate fauna might also be especially susceptible to effects from marine mine waste inputs (Burd, 2002; Medina et al., 2005), although such species are difficult to sample quantitatively. The disappearance of starfish and kelp from the intertidal zone in Chile (with no recovery 12 years after particulate inputs ceased) suggests that such habitats might be particularly sensitive to serious ecological disturbance (Lee and Correia, 2005; Medina et al., 2005; Chapter 11). Recovery of community composition can be slow after the disturbance is removed. Recovery times of 9 to 15 years are documented in British Columbia after a relatively small operation of undersea mine waste disposal was halted (Burd, 2002).

12.6.2   Ecological signs of risk in streams and rivers

Prologue. The signs of ecological damage from mine contamination in streams and rivers are better known than in many environments. Individual case studies may not always include all lines of evidence, and the relationship between the onset of damage and metal concentrations in water or sediments (dose response) is probably variable from site to site. But if a body of evidence is accumulated, uncertainties about the influence of mine wastes and metal contamination can be greatly reduced.

The body of work on mine-impacted streams illustrates how biological communities change when subjected to the metal contamination in mine wastes (see also Chapter 11). Similar results are observed in a number of mine-contaminated watersheds (e.g. the Coeur d'Alene in Idaho, USA: Box 12.10; the Clark Fork River, Montana, USA; and in regional assessments of areas with many small mines). For example, in Colorado, trace metals and other materials released from 10 000 abandoned mines affect over 2600 km of streams (Clements et al., 2000). A survey of 79 randomly selected streams in Colorado showed that 25% were ecologically degraded by metal contamination, largely from the regionally dense mining activities (Clements et al., 2000). This is also the area where ptarmigan populations are affected over large areas of the riparian zone (Box 3.1).

A combination of several signs is diagnostic of the effects of mining on streams and rivers.

(1) Metal contamination is detectable in water, sediments and fauna. Characterisation of exposure is the first step in diagnosing a risk from metal contamination; linking the metals to mine inputs can be done using gradients and stable isotopes.

(2) The greater the bioavailable metal contamination, the fewer species that can survive. In particular, species richness within the order Ephemeroptera, or the number of mayfly individuals (abundance of Ephemeroptera alone) are sensitive indices of metal stress (Clements, 2000). Maret et al. (2003) found half the expected number of mayflies in the contaminated Coeur d'Alene River, coincident with substantial tissue contamination (mean body concentration) in the metal-tolerant biomonitor, the caddisfly larva Hydropsyche sp. Ecological effects were clear when concentrations of Zn reached 707 µg/g dw, Pb reached 142 µg/g dw and Cd reached 6.7 µg/g dw in Hydropsyche sp.

(3) Absence of specific sensitive species (mayflies like Drunella sp., Epeorus sp., Rhythrogena sp. or the caddisfly Rhyacophila sp. in the Clark Fork River, for example) is also an indication of damage from metals (Cain et al., 2004; Clements, 2004) (Fig. 12.9) if accompanied by exposure data. These and similar species are diagnostic of metal effects because they show specific attributes that make them susceptible to metal contamination (Chapter 11).

(4) Invertebrate predator species might be reduced in number as a typical response to metal impacts in streams (Clements, 2004) (see Chapter 11).

(5) Predator fish are of low abundance; and surviving fish might be small or have abbreviated life spans. For example, trout are native to the upper Clark Fork River, as elsewhere in the Rocky Mountains of Montana (USA). Reduced numbers of trout were attributed to the mining activities historically, and fish kills were obviously related to the runoff from the contaminated floodplain. A careful comparison of Clark Fork River sites to reference sites (methods discussed in Chapter 11) showed that trout densities in the latter averaged 5.3 times greater than densities in contaminated segments of the Clark Fork (Hillman et al., 1995) (Fig. 12.10).

(6) Fewer species in the fish community is also a particular characteristic of mine-impacted rivers; the specific species which survive can be a valuable diagnostic. It is not unusual to find only brown trout (Salmo trutta) in a mine-impacted North American streams, although the habitat might be suitable for a variety of trout species. Rivers similar to the Clark Fork River in Montana typically contained brown, rainbow, cutthroat, bull and brook trout in the absence of contamination. Extremely contaminated streams, like Silver Bow Creek nearest the mining/smelting activities in the Clark Fork basin, support no fish. Downstream from Silver Bow Creek, the trout in the upper Clark Fork River were almost entirely brown trout. Rainbow trout (Oncorhynchus mykiss) first appeared in abundance 170 km from the mining/smelting activities. Typical native species, bull trout Salvelinus confluentus and westslope cutthroat trout Oncorhynchus clarki lewisi (Malouf, 1974) were virtually eliminated from the main stem upper river.

Changes in ecological characteristics are difficult to attribute to any one cause without supporting evidence. But when a body of evidence is assembled and viewed as a whole, the nature of the ecological risks becomes much less ambiguous than when each study is viewed in isolation. The body of work demonstrating metal effects on fish in the Clark Fork River is instructive. The contamination of sediments, water and food sources for trout corresponded to fish abundance. The resident fish in the river also contained elevated metals in their tissues (Farag et al., 1995), verifying internal exposure to metals by the predators. In situ bioassays showed that juvenile rainbow trout could not survive more than a month in upstream waters (Phillips and Lipton, 1995). In experiments with a contaminated diet from the river, brown trout

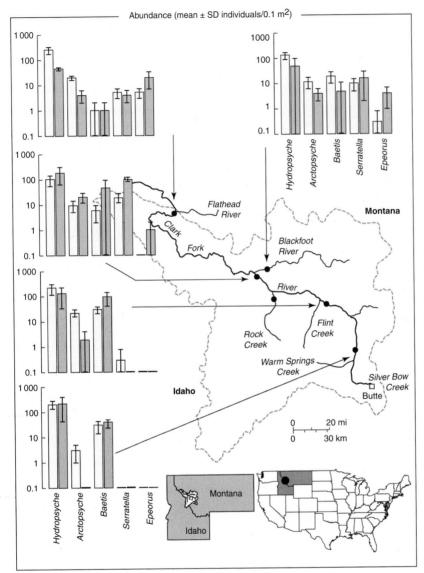

Fig. 12.9 The abundance of five different insect species from two different (replicated) surveys of the Clark Fork River and the less contaminated Blackfoot River at its mouth. The two mayfly species disappear from the community nearest the mining and smelting activities, where contamination is greatest. As contamination declines from the upstream Clark Fork through the Blackfoot, the relative abundances of the different species become more similar, and similar to the distribution in the control stream (the Blackfoot River). (After Cain *et al.*, 2004; Luoma *et al.*, 2007.)

showed elevated concentrations of products of lipid peroxidation and histological abnormalities (effects on hepatocytes, pancreatic tissue and the mucosal epithelium of the intestine), as well as reduced growth (Farag *et al.*, 1994; Woodward *et al.*, 1994, 1995; Chapter 9). Lipid peroxidation was also found in the liver, pyloric caeca and large intestine of brown trout resident in the river (Farag *et al.*, 1999). Brown trout in the Clark Fork River

were also smaller than fish of equivalent age from carefully designated reference sites (Tohtz, 1992). These characteristics were similar to reduced growth and survival seen in experiments and at other contaminated locations (see similar data for fish in lakes: Campbell *et al.*, 2003; Chapter 11).

Finally, toxicity tests showed the complex reasons that brown trout (*S. trutta*) were more tolerant than, for example, rainbow trout (*O. mykiss*) in the river

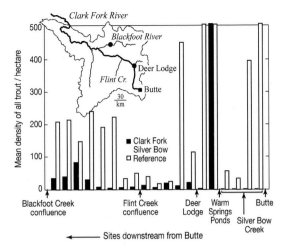

Fig. 12.10 A comparison of trout abundance at 19 sites on the Clark Fork River (solid bars) compared to reference sites (open bars) carefully matched with similar physical and chemical attributes (for methods see Box 11.4). The only site with an equivalent density of trout in the contaminated and uncontaminated streams was immediately below Warm Springs Ponds, where metals were removed by sulphide formation in the ponds and invertebrates were some of the least contaminated in the river for a very short reach (several kilometers). The ponds and a less contaminated stream provided a source of in-migrating trout at this site. As banks recontaminated the Clark Fork downstream, trout densities plummeted again. (After Luoma *et al.*, 2007; data from Hillman *et al.*, 1995.)

(Chapter 11). The simplest tests indicated that brown trout might be slightly more sensitive to Cu than rainbow trout. But brown trout appeared to acclimate to contamination effectively whereas rainbow trout were less effective at detoxifying metals by production of metallothioneins (Marr *et al.*, 1995; Chapter 9). Rainbow trout were also more sensitive than brown trout to pulses of Cu, if combined with reduced pH and hardness (conditions typical of storm-driven inputs). Finally, rainbow trout avoided mixtures of metal contamination at lower concentrations than did brown trout (Marr *et al.*, 1995). When the data from all the studies above are viewed across the boundaries of toxicology, geochemistry, biology and ecology, it is difficult to avoid the conclusion that metals are a crucial factor in determining the depauperate trout populations in the upper Clark Fork River.

12.7   SOCIO-ECONOMIC CONSEQUENCES OF ECOLOGICAL IMPACTS

Prologue. Appropriate diagnosis of ecological risks is especially important in mine-impacted streams. Often substantial socio-economic resources are at stake.

The OK Tedi Mine in Papua New Guinea represents a modern example of the outcome when mines are not managed to contain wastes (Box 12.15). It includes all the elements that we have discussed in earlier cases. It provides a warning about the chemical contamination, ecological disruption and socio-economic issues that a modern mine can generate in the absence of strict consideration of environmental challenges.

The economic and socio-political implications of the ecological damage from the OK Tedi contamination were also significant. But they also reflect the importance of diagnosing risks appropriately. For example, the intake of dietary Cu by humans consuming fish taken from the river was well within guidelines set by the World Health Organisation and did not constitute a health risk (Swales *et al.*, 1998). This will usually be the case in a mine-impacted system, unless Se, Hg or As are important contaminants, or people eat fish organs or shellfish from the environment. Cu, Cd, Pb and Zn do not accumulate to high concentrations in fish muscle.

But the disappearance of fish and the disruption of riparian habitat along hundreds of kilometres of river were very significant to the indigenous people. In 1991, internal mining company reports revealed that fish stocks in the upper OK Tedi had declined between 50% and 80% from pre-mine levels. The first 70 km of the river was 'almost biologically dead, and species diversity over the next 130 kilometres had been dramatically reduced. Fertile river bank subsistence gardens, plantations and approximately eight square kilometres of forest were destroyed' (Ghazi, 2003). Such observations are consistent with other case studies.

Starting in the late 1980s, the indigenous people and their downstream neighbours petitioned mining company and government officials to address the pollution and flooding that were destroying traditional subsistence lifestyles and

Box 12.15 Case study: sediment contamination in the OK Tedi River

Uncontained sediment contamination can be distributed across even a massive watershed, if physical conditions are conducive to secondary and tertiary dispersal and deposition. The OK Tedi mine is one of the world's largest Cu–Au deposits, situated in extremely steep, arduous terrain in the highlands of western Papua New Guinea. The open pit mine began operations in 1984, and was projected to continue operation approximately 2010. Initial attempts at containing wastes failed because of the steep terrain and intense rainfall, so 66 million tons of mining residues per year have been released into the OK Tedi River since the mid-1980s. Over a 20 year life for the mine, this will amount to 1320 million tons of waste (compared to discharges from the Clark Fork River of 500 million tons). Wastes are carried by the OK Tedi River for 200 km, where they enter the 800 km Fly River system, and end up 1000 km away, deposited in the Gulf of Papua (Hettler *et al.*, 1997). In 1997 the heavy sediment loads had raised the channel of the OK Tedi 10 m (30 feet) in the upper reaches. Downstream 200 km, the river bed is projected to aggrade 2.2 m during the life of the mine.

The middle Fly River lies between 200 and 600 km from the mine. It runs through a swampy lowland with connecting tributary channels and oxbow lakes. The floodplain is typically 10 to 30 km wide. Extensive flooding and over bank deposition occur during the wet season of the year. Water returns to the river when the floods recede, but the suspended load is trapped on the floodplain by vegetation and swamps. Lakes cut off by meanders in the old river bed (oxbow lakes) also collect settled mine waste. During drought conditions, the river slows even more and sediments settle to the bed of the river.

The primary metals in the ore deposit are Cu and Au, with very little of the other contaminants typical of some porphyry deposits. Separation of Cu from the ore is about 85% efficient, leaving an average of 1300 µg/g Cu in waste residues. The Cu content of waste rock is typically 1100 µg/g. The chemical composition of these wastes dominates river sediments, floodplain and lake bottoms for 610 km downstream from the mine. Very little dilution of the mine wastes occurs over this distance, presumably because tributary inputs of sediment are small compared to the increased sediment loads. Pre-mining sedimentation rates in the floodplain lakes were <0.1 cm/yr of sediment. They increased to 4 cm/yr after the mine opened. In 29 lakes that were studied by 1997, 20 to 70 cm of sediment had accumulated with Cu concentrations of 800 to 1000 µg/g at their river end. Near surface sediments along the river channel contain up to 1100 µg/g Cu, with a mean of 620 ± 280 µg/g. Extensive studies of the floodplains show an average of 453 ± 332 µg/g Cu.

About 610 km from the mine, the Fly River is joined by a similar-sized tributary, the Strickland River. By the time the river reaches the estuary in the Gulf of Papua, the Cu contamination is diluted to low concentrations (<30 µg/g Cu). In 1990, there was no sign in sediment cores of differences from historical levels in the Gulf sediments (Alongi *et al.*, 1991). Anthropological, toxicological and ecological studies illustrate the controversies detailed throughout this chapter. In 1990, the toxicology of the mine waste was tested using traditional approaches (Smith *et al.*, 1990). Test species were selected on the basis of ecological relevance: two freshwater prawns, *Macrobrachium rosenbergii* and *Macrobrachium* sp., a catfish, *Neosilurus ater* and a cladoceran, *Ceriodaphnia dubia*, were tested for acute toxicity and bioaccumulation of Cu from particulates. There was no evidence of acute toxicity to prawns or fish, nor for bioaccumulation by prawns. The authors concluded that acute toxicity to *C. dubia* and bioaccumulation by *N. ater* were probably due to dissolved Cu in the test environments. They concluded that there should not be toxic effects of particulate-associated Cu in the Fly River unless dissolved Cu levels increased above predicted levels. Later studies of fish populations found contradictions with toxicological predictions as is common in other circumstances. Significant reductions in fish catches were found at most riverine sites in the OK Tedi and the upper and middle Fly River. No significant declines in fish catches were found below the confluence of the Strickland or in the marine environment. Catches in some floodplain habitats also declined but the cause of was more difficult to decipher. Levels of Cu, Zn, Pb and Cd were found to be elevated in liver and kidney tissues from a range of fish species taken from riverine and floodplain sites sampled in the OK Tedi and Fly Rivers. There was a general trend for metal concentrations to decrease with distance downstream from the mine, suggesting a mine-related effect (Swales *et al.*, 1998).

forcing some villagers to relocate. Anecdotal reports suggested that 'the animals living along the river banks—the pigs, cassowaries, pigeons and bandicoots—have all disappeared' (Ghazi, 2003). The case became an international *cause célèbre* when the indigenous peoples living along the rivers launched a series of lawsuits against the international company that operated the mine. The 'David and Goliath' legal case against one of Australia's biggest corporations received widespread media coverage. In 1996 the two sides reached an out-of-court settlement, which included compensation and a company commitment to contain mine tailings. The company agreed to pay affected villagers a total of US$28.6 million. In February 2002, the company transferred its 52% equity to the Papua New Guinea Sustainable Development Program Company in exchange for indemnity from future pollution liability (Ghazi, 2003). 'The mine is not compatible with our environmental values and the company should never have become involved' was the statement from its chief executive officer.

This mine once constituted 15% of the profits of the mining company involved. That was lost with the withdrawal. The Papua New Guinea government lost revenue critical to the country's development. Although the mine is scheduled to be viable until 2010, it is not yet clear if it can be run profitably under present ownership. With the withdrawal of the corporate partner in the mine, revenue was also lost that might have been dedicated to remediation or rehabilitation, or replacing some of the lost resources. The resources of hundreds of kilometres of river and its inhabitants may not be replaceable. There do not appear to be any winners in a situation that deteriorates to this extent.

12.8 RECOVERY, REMEDIATION AND REHABILITATION

Prologue. Lessons from mining suggest constructive goals for international metal policies: (a) recovery from past problems, based upon understanding of the nature of those problems; (b) continued growth in environmentally responsible corporate/government mining and smelting policies, based upon understanding

of the processes most important in minimising environmental damage; (c) understanding the recovery of ecological properties, as (or after) impacted areas undergo remediation (natural or human-driven).

Case studies illustrate how little we know about recovery after metal contamination is widely dispersed by mining activities. It is clear that ecosystems are expensive to restore. Hundreds of millions of US dollars were invested in the Clark Fork River, the Sudbury Lakes and after the Los Frailes spill. To some degree, local contamination was reduced by these investments. Ecological recovery is more uncertain, however. Studies from Sudbury suggest that decades might be the necessary timescale to evaluate remediation in such systems; and even then some species may not return (Chapter 11). In other systems the problem is even more complex. Sediment contamination has not receded in the estuaries of southwest England since 1900. The ecological implications of the contamination are not visually evident, and are only slowly being studied. It is difficult to envision remediation of the bed sediments of Lake Coeur d'Alene. In any of the systems, the rate of ecological recovery probably depends upon the nature of the contamination, the physical processes in the watershed, human dedication, and the extent of contaminated area. Much more published research is necessary to understand what works and what does not in mine remediation.

Prevention of the dispersion of the contamination and rehabilitation at the primary sites of activity are more certain solutions than attempting to remedy past errors (Tremblay, 2004). Modern mines can be designed and operated to limit environmental impacts during operation and following closure (van Zyl, 2004). Land values can be restored through reclamation (Box 12.16). Mining research has improved containment of wastes; and the area affected beyond the primary site has been reduced, or, in a growing number of cases, eliminated (Bourassa, 2004). Notable successes exist for prevention and ever better practices for defining careful environmental stewardship are developing. Initiatives have shifted toward better participation and communication with stakeholders. All parties are beginning to

Box 12.16 Environmentally responsible mining: the Flambeau copper mine in Wisconsin, USA during the mining operation and 2 years after closure

The Flambeau Mine in Wisconsin, USA began operations in July 1991 and mined Cu until 1997. About 1.9 million tons of ore were removed, containing about 9.5% Cu; 3 300 000 ounces of Ag and 0.175 ounces of Au per ton were also mined and shipped from the site. The project was controversial because the 181 acre mining site was within 140 feet of the Flambeau River, 500 feet of a major highway, within 1000 feet of private residences and 2500 feet of a liberal arts college, a hospital and a nursing home. After an ambitious proposal in the 1970s the mine was redesigned to eliminate a mineral separation and concentration facility, tailings pond and a tailings disposal site. Only the very high grade Cu in the upper 225 feet of the deposit was mined through a small open pit mining operation. It was economically feasible to ship the ore to facilities in Canada for further

processing (concentration and smelting), thereby avoiding the need for construction of a tailings disposal facility. Permitting was a 3 year process of baseline data gathering, negotiations with local communities, and project design and review. Reclamation activities were completed by the end of 1999 (Fig. 12.11). For reclamation, the open pit was backfilled by blending stockpiled waste rock with a prescribed amount of limestone, to minimise the potential for the development of acid conditions. Backfilling of the pit took about 1.5 years. The site was graded to approximate the pre-mining condition and provided with surface drainage. Topsoil was reapplied to the site and grasslands, forest and wetlands were restored. The specified, post-mining land use for the site was light recreation and wildlife habitat. In all, over 170 different species of plants were seeded on the mining site during the reclamation phase. The company also installed a trail system through the reclaimed site, which is now available for public use. In 2000, the site was stabilised, native species of vegetation were starting to thrive and survivorship of the woody

Fig. 12.11 Aerial photos of the Flambeau copper mine in Wisconsin, USA (a) during the mining operation and (b) 2 years after closure. (From Wisconsin Department of Natural Resources website: www.dnr.state.wi.us/org/aw/wm/mining/metallic/flambeau/.

Fig. 12.11 (*cont.*)

vegetation was greater than 80%. The Flambeau Mining Company regularly monitored groundwater levels, groundwater quality, air quality, surface water quality, waste water effluent quality and flow, mine inflow, wetlands, aquatic ecology characteristics including fish, macroinvertebrates and sediment, stockpile leachate quality and meteorology. Throughout the life of the project, the company remained in substantial compliance with all permit conditions and applicable standards. The Wisconsin Department of Natural Resources (http://www.dnr.state.wi.us/org/aw/wm/mining/metallic/flambeau/) will continue to monitor the conditions at the reclaimed Flambeau Mine for many years to come.

understand the value of a consistent body of knowledge and information (Udo de Hias, 2004).

But, there are mines that are still operated with little concern for the environment. Site remediation or rehabilitation is at the core of modern mine management; but a serious legacy of mining impacts and a significant number of operations that do not enthusiastically adhere to modern management approaches are the reality (Bourassa, 2004). The impact of mining on the aquatic environment is no longer just a function of total material moved or the kind of material. It is a function of management of that material. Managing metal contamination from mine wastes is possible, but there is need for better definition of the boundaries of environmental models, concepts and indicators that document specific environmental implications (Bourassa, 2004). A constructive scientific dialogue about the ecological implications of past and present practices is an essential part of that solution.

12.9   CONCLUSIONS

Many of the least ambiguous case studies of the ecological changes caused by metals come from mining sites. Sources of mining/smelting contaminant inputs are usually clear and a gradient of contamination can be detectable. Mines are often located in remote regions with only a few other stressors, and their impacts overwhelm potentially confounding sources of stress. Recovery from effects can also add some verification of causation (or can allow separation of different effects from different stressors); and can be critical in guiding risk management. Because there is so much at stake where mine effects are severe, however, risk evaluations can be contentious and uncertainties can also generate advocacy debates that last years. Understanding how mining activities cause ecological change is the most effective way to determine if such change has occurred or whether it will. And that understanding is advancing rapidly.

Large mining operations or even a regional patchwork of smaller mines and smelters have great potential for creating ecological and socioeconomic damage. If wastes are not contained, primary, secondary and tertiary contamination can be spread over immense areas. Exposures to organisms are complex, but methods are available to determine sources of metal contamination in water, sediments and biomonitor species. The specific kinds of ecological changes that occur in streams, rivers, lakes and coastal waters have been documented (see Chapter 11 as well as this chapter). Thus generalisations are possible about ecological signs that are diagnostic of effects from the metal contamination caused by mining. Risk assessments may become less contentious in the future as diagnostic signs become better known.

It seems inevitable that demand for metals in human activities will grow in the years ahead. It is not inevitable that ecological and socioeconomic damage from mining those metals accompany that growth. Sustainable mining and smelting involves investing in management that is environmentally and socio-economically responsible; waste containment is especially important. Examples of successes in sustainable mining and smelting are growing in number. Assuring that this trend is continued into the future could depend upon finding ways to reward those miners whose practices are environmentally responsible as well as penalising those whose practices are not.

Suggested reading

Clements, W.H., Carlisle, D.M., Lazorchak, J.M. and Johnson, P.C. (2000). Heavy metals structure benthic communities in Colorado mountain streams. *Ecological Applications*, **10**, 626–638.

Farag, A.M., Stansbury, M.A., Hogstrand, C., MacConnell, E. and Bergman, H.L. (1995). The physiological impairment of free-ranging brown trout exposed to metals in the Clark Fork River, Montana. *Canadian Journal of Fisheries and Aquatic Sciences*, **52**, 2038–2050.

Moore, J.N. and Luoma, S.N. (1990). Hazardous wastes from large-scale metal extraction. *Environmental Science and Technology*, **24**, 1279–1285.

Pyle, G.G., Rajotte, J.W. and Couture, P. (2005). Effects of industrial metals on wild fish populations along a metal contamination gradient. *Ecotoxicology and Environmental Safety*, **61**, 287–312.

Smith, R.E. W., Ahsanullah, M. and Batley, G.E. (1990). Investigations of the impact of effluent from the OK Tedi copper mine on the fisheries resource in the Fly River, Papua New Guinea. *Environmental Monitoring and Assessment*, **14**, 315–331.

van Geen, A., Adkins, J.F., Boyle, E.A., Nelson, C.H. and Palanques, A. (1997). A 120 year record of widespread contamination from the Iberian mining belt. *Geology*, **25**, 291–294.

13 • Selenium: dietary exposure, trophic transfer and food web effects

13.1 INTRODUCTION

Prologue. For no other trace metal are regulatory approaches as globally incoherent as they are for Se; and for no other metal do at least some regulations diverge more from the state of knowledge. To this extent, understanding hazards and risks from Se provides lessons for all other trace metals.

We dedicate an entire chapter to Se mainly because it has a potential for risk as an environmental toxin second only to Hg (although this is not widely recognised). The high hazard ranking for Se accompanies a unique environmental biogeochemistry. Bioavailability from diet is at the higher end of the continuum for trace metals, as is its trophic transfer and threat to upper trophic level charismatic species. Selenium issues are also one of the few sets of issues that have not shown rapid improvement where they have been addressed by environmental policies. Most important, Se issues illustrate the great importance of thinking laterally, across disciplinary and historical boundaries, when managing risks. Unless approaches to managing risks change, Se problems could grow in the future. But Se issues also offer important lessons for managing trace metal contamination, in general.

In formal chemical terms selenium (Se) is considered a metalloid. We have included it under our wide definition of trace metals (Chapter 1) for the purpose of this book because its environmental behaviours (including biological behaviour), as well as demands for management of contamination, fall within a continuum relevant for all trace metals.

Basic questions continue to arise in the discussion of Se policy. Is it a global threat to aquatic environments or one limited to a few local circumstances? How hazardous is Se compared to other trace metals? What aspects of our knowledge of Se are most relevant in evaluating existing risks and in anticipating future risks? In this chapter we show how the present state of science addresses these questions.

Considered appropriately, Se is one of the most hazardous of the trace metals, following Hg. Universal protective criteria are difficult to derive, however, because traditional toxicology does not address the aspects of the behaviour of Se that make it a serious hazard. Exposures are affected by speciation and dominated by dietary uptake and bioaccumulation. The margin of safety is narrow between concentrations that are essential and those that are toxic; and Se can biomagnify through food webs, accentuating the threat to wildlife. Traditional dissolved toxicity tests literally are irrelevant to natural waters, and surrogate species may not adequately simulate the species which are most threatened by Se in nature. Most confusing, however, traditional approaches to ambient water quality criteria are as likely to overprotect ecosystems as underprotect them. Overprotective regulations can be as detrimental to environmental well-being as underprotection, because Se problems are among the most difficult to remediate; and poorly considered remediation can make problems worse.

Approaches to managing Se seem to differ widely across the world. Se is not listed among chemicals of concern in the European Commission's Dangerous Substance Directive or on the List II Water Quality Standards for Protection of Aquatic Life. Canada has the most stringent ambient water quality guideline in the world (1 μg/L Se), although the guideline does do not have regulatory authority (Outridge *et al.*, 1999). The USA and Australia have

> **Box 13.1** Definitions
>
> **bioaccumulation factor (BAF):** the ratio of a substance's concentration in an organism to its concentration in the environment where the organism lives. BAFs measure a chemical's potential to bioaccumulate in tissue through exposure to both food and water.
>
> **bioconcentration factor (BCF):** the ratio of a substance's concentration in an organism to its concentration in the water where the organism lives. BCFs measure a chemical's potential to bioaccumulate through exposure to only water.
>
> **biomagnification:** a cumulative increase in the concentration of a trace metal in successively higher trophic levels of the food chain. It is the process by which the concentration of a substance increases in different organisms at higher levels in the food web.
>
> **biotransformation:** the process whereby a substance is changed from one chemical form to another (transformed) by a chemical reaction and enzyme catalysis within an organism.
>
> **dissimilatory reduction:** the bacterial reduction of an element from a higher to lower oxidation state, accompanied by deposition of the end product external to the organism; for example the bacterial reduction of selenate (Se(VI)) to elemental Se (Se0) and deposition of the latter in sediments.
>
> **microbial biogeochemistry:** the study of the role of microbes in driving chemical biotransformation of substances in the biosphere.
>
> **peptide:** molecule consisting of two or more amino acids. Peptides are smaller than proteins, which are composed of chains of peptides sometimes termed polypeptides.

an ambient criterion for freshwaters of 5 µg/L Se. The US marine criterion would allow continuous concentrations as high as 58 µg/L Se. Australia's marine criterion is 70 µg/L Se. Both US and Australian marine criteria are under review. In 2004 the USA set a new Acute Se Standard in marine waters of >100 µg/L Se. There is probably no other chemical for which there is such a wide international disparity in protective approaches.

13.2   CONCEPTUAL MODEL

Prologue. Uncertainties about effects of Se are progressively narrowed as the full conceptual model is considered. Knowledge of speciation, transformation, the species-specific attributes of Se accumulation into food webs and links to reproductive toxicity are especially important.

In general, the biological effects of Se are linked to source inputs through the sequence of relationships typical of all trace metals (Fig. 13.1). Like most metals, Se is regulated by predictions of toxicity

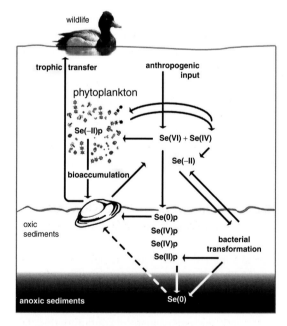

Fig. 13.1 Conceptual model showing processes determining ecological implications (risks) from Se for a benthic, bivalve-based food web. Roman numerals show the different oxidation states of Se; p indicates particulate form.

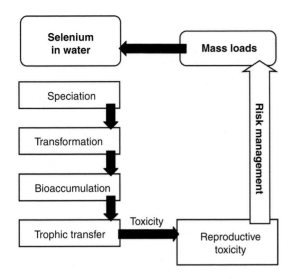

Fig. 13.2 In natural waters, the toxicity of Se is defined by effects on reproduction and development that result from the aggregated influences of concentration in water, speciation, transformation to particulate forms, bioaccumulation and trophic transfer. But traditionally, toxicity is tested by directly linking dissolved exposure to acute mortality, skipping the influences of the other processes.

from dissolved Se concentrations, corrected for speciation. This is the basic reason for incoherence in the regulations. Transformation, bioaccumulation, trophic transfer and reproductive toxicity are not considered in those predictions (Fig. 13.2). The poor correlation between Se concentrations in natural waters and the effects it manifests in nature are now well known; forecasting toxicity requires a full consideration of the conceptual model.

13.3   SOURCES: WHY IS SELENIUM AN ELEMENT OF GLOBAL CONCERN?

Prologue. The biogeochemical cycle of Se does not include a large compartment in the atmosphere, so individual source inputs do not spread globally, as with Hg or polychlorinated biphenyls (PCBs). Nevertheless, Se is a problem of global significance because sources are spread over wide regions of each continent. Perhaps most important, Se contamination is not declining; indeed, it may increase in the years ahead and in many geographical locations.

The global impact of Se stems from its occurrence in relatively high concentrations in organic carbon-rich marine sedimentary rock formations (Presser *et al.*, 2004). These are present on all continents and reflect regions of high biological productivity from the Precambrian through to the Pliocene eras. Any human activity that disturbs or exploits these rocks or their associated soils will mobilise Se.

Major sources of Se include irrigation of lands whose soils are derived from Se-rich rocks; and exploitation of oil, coal, phosphate and, perhaps, uranium from deposits that originate in organically rich shales. Se is also enriched in some of the pyrite ores from which Au, Cu and Ag are mined. It can be a contaminant in smelter and refinery wastes. Contamination from commercially growing animals for food is also possible.

In summary, Se contamination issues might be expected if a region is semi-arid or arid, rich in energy sources, mined, and/or usage of fossil fuels, particularly coal, is intense. Thus these issues are associated with three of the most challenging environmental problems of the future: water shortages, food production and shifts in energy production, particularly to coal. It is a problem that is more likely to grow than disappear (Cumbie and Van Horn, 1978; Presser, 1999; Piper *et al.*, 2000).

Coal exploitation is of particular concern (Fig. 13.3). Se contamination is a relatively recent issue in streams and rivers affected by coal mining in West Virginia (USA) and British Columbia and Alberta (Canada). Extensive coal exploitation and the unique nature of many Australian environments might also make that continent particularly vulnerable to the effects of Se (e.g. Peters *et al.*, 1999).

Storage and washing of coal release some Se. But extracting the energy from this hydrocarbon creates the most intense contamination, whether by burning, gasification or liquefaction. High concentrations of Se occur in ash from burning many (but not all) types of coal and in the wastes of the processes that remove air pollutants. Equipment like precipitators, stack scrubbers or flue gas desulphurisers limit the air pollution from coal, but can produce wastes with very high Se concentrations. If facilities for extracting energy from coal are centralised, the Se issues will be intense but localised.

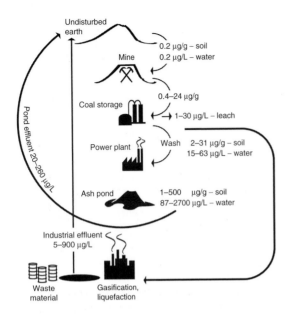

Fig. 13.3 Se concentrations in various waste streams and process streams from coal exploitation. Rivers downstream from coal mining have concentrations of up to 40 µg/L Se in their waters. Lakes receiving waste waters directly from ash waste heaps can suffer adverse ecological effects (Box 13.8). Sequestration of wastes away from natural waters (e.g. tailings ponds, solid waste disposal) are effective means to remediate such problems.

However, if coal use becomes decentralised, as it was traditionally, regionally diffuse contamination is possible, which will be more difficult to control or remediate.

Some deposits of petroleum also have high concentrations of Se. But episodes of contamination are documented mostly from refinery wastes. Two specific examples are discharges of such wastes into San Francisco Bay, USA (Cutter, 1989) and the Scheldt estuary, the Netherlands.

The most widespread source of Se is irrigation of semi-arid farmland. Climatic and geological factors contribute to the potential for Se contamination in association with agriculture. Contamination is possible if (Presser, 1999):

(1) the source rock for the area includes organic rich shales or marine sedimentary rocks,
(2) the climate is sufficiently dry that evaporation greatly exceeds precipitation,

(3) drainage of soils is impeded by a subsurface impermeable layer (e.g. of clay).

When water cannot drain from soils, salts are leached out and accumulate. Where the geology is appropriate, these salts can contain extreme concentrations of Se. All of these processes are exacerbated by irrigation. Across the western USA, $640\,000$ km^2 have the appropriate climate, geology and soils to be Se-contaminated; $8\,100$ km^2 of these are irrigated (Seiler *et al.*, 1999). Seleniferous conditions are documented in Saskatchewan and Alberta in Canada, although the total area affected is smaller than in the USA (Outridge *et al.*, 1999). Drainage of irrigation return flows into wetlands is also less common in Canada than in the western USA. The Middle East, South America, Russia and Asia also have arid regions that would seem vulnerable to Se contamination, but no studies are available.

The Central Valley of California (Box 13.2) and the Colorado River in the western USA (Box 13.3) present two case studies that illustrate the potential importance of Se in conflicts between environment and development (in these cases agriculture). These large-scale issues and their implications for biodiversity could potentially be repeated in many parts of the world.

13.4   TOTAL MASS LOADING

Prologue. Analysis of mass inputs from the natural baseline and from each specific human activity is the first step to determining what sources of Se are the most important. Loads that are important to determine include: (a) the natural baseline, including the baseline influenced naturally by seleniferous geology, and (b) the enrichment above those baselines caused by human disturbance. Defining the historical geological baseline from seleniferous regions is complicated, but may be a valuable contribution to management decision making.

An important challenge in managing Se was, historically (and sometimes continues to be), the poor reliability of analytical results (Box 13.4).

Box 13.2 Kesterson Reservoir and National Wildlife Refuge

The story of Se toxicity in Kesterson National Wildlife Refuge is an infamous and thoroughly documented case of ecological damage that illustrates the potential magnitude, environmental complexity and political challenges of a Se issue (Presser, 1994).

The Coast Range of northern California (USA) lies between the Pacific Ocean and the agriculturally rich San Joaquin Valley. The Range is composed of marine sedimentary rocks that are enriched in Se. Runoff and erosion from the Coast Range have carried salt and Se to the San Joaquin Valley for more than 1 million years. Both build up on soils because the climate of the region is arid. Less than 25 cm of precipitation combine with greater than 225 cm of evaporation per year to result in a negative water balance. Poor drainage also contributes to the build-up. Clay layers in the subsurface soils in the valley impede downward movement of water, causing waterlogging of the root zone. When the soils are wet, Se and salts are leached from them. In the dry summer and fall, salts are precipitated and move toward the surface. The next year the cycle begins again, gradually building an internal reservoir of salt and Se (Presser and Ohlendorf, 1987).

The surface and groundwaters of the San Joaquin Valley are part of a complex, hydrologic system that houses the Central Valley Project, a massive, engineered complex of dams, off-stream storage reservoirs, pumping facilities, and canals. Irrigation of the crops supported by the Central Valley Project accelerates the accumulation and leaching of salt and Se into aquifers and surface waters (Presser, 1994). Agriculture will have to cease within a few decades if salination of the soils continues. An alternative is to sustain agriculture by washing and draining salts and Se from the soil. Subsurface drains can be buried in the fields to catch irrigation waters and carry it to collector drains. In 1960, the Federal government committed to provide disposal of the irrigation drainage along with irrigation water. Planning for a single drain to carry salt-laden irrigation return water (and the accompanying Se) out of the San Joaquin Valley began in 1955.

An 85 mile section of the collector drain was completed in 1975. The ultimate plan was to extend the drain to San Francisco Bay, but Kesterson Reservoir was chosen as interim disposal site, until the remaining two-thirds of the canal could be built.

The drain began discharging concentrated drainage water to Kesterson in 1981. The reservoir was adjacent to Kesterson National Wildlife Refuge, a heavily populated bird sanctuary on the Pacific Flyway. In 1983 it was discovered that >60% of nestling ducks, coots, grebes and stilts were deformed in Kesterson Reservoir and the Refuge (Presser and Ohlendorf, 1987). The symptoms were typical of Se poisoning. A multi-species assemblage of warm water fish resided in the marsh before discharges of irrigation drainage waters. Only mosquitofish (*Gambusia affinis*) persisted after Se contamination was introduced. The drain was ordered closed by the US Department of Interior in 1985 due to the threat of severe damage to the western flyway for migratory waterfowl. Canada considered it a threat to the biodiversity of their migratory waterfowl. The reservoir and low-lying parts of Kesterson National Wildlife Refuge were buried under 46 cm of imported topsoil in 1988. Nevertheless, elevated Se concentrations persist in the remediated terrestrial ecosystem.

The salination and Se problems in the Central Valley affect 700 000 acres of agricultural lands, projected to grow to over 1 000 000 acres (Presser and Luoma, 2006). There is still no resolution to the soil salination issue, after 50 years of debate. Numerous treatment options for the massive volumes of drainage have failed. The drain project remains on hold, because no receiving waters can safely assimilate its waste. The US Bureau of Reclamation is caught between its promise to provide drainage and its inability to resolve the Se issue. The landowners are caught with gradually salinating soils and a groundwater resource accumulating Se. The Pacific Flyway has lost valuable habitat and is threatened by more effects in the centre of the principal major route between Canada and Mexico. California is caught between a serious water supply shortage and expanding contamination in a major aquifer and river system. At the heart of this is the inability to resolve regulatory and management issues for Se.

Typical approaches to trace element analysis (e.g. flameless atomic absorption spectrophotometry) do not give repeatable results for Se. One of the most reliable methods for concentrations typical of natural waters remains hydride generation atomic absorption spectrophotometry, wherein wet digestion methods are used to convert all Se to a form that will volatilise for detection. However, even

Box 13.3 Large-scale ecological effects of Se: the Colorado River basin

Hamilton (1999) assembled a compelling body of evidence suggesting that Se contamination is quite probably an underappreciated factor in the decline of native fish species in the upper and lower Colorado River, USA. The Colorado River basin, in central western USA, had perhaps the most distinctive fish fauna in North America, when it was first explored in the mid to late nineteenth century. The basin of this massive river extends from mid Wyoming in the northern USA into Mexico where it drains into the Gulf of California. Most of its watershed is arid or semi-arid.

The fish fauna of the watershed was uniquely adapted to the challenges of life in a river with a complex topography, temperature regime, climate and hydrology. As an example, one of six species formally listed as endangered in the river is the razorback sucker, *Xyrauchen texanus*, a predator with a lifespan of 40 years or longer in the wild. Hamilton (1999) cites historical records from several sources that indicate that razorback sucker were abundant in the upper Colorado River before 1900. Between 1989 and 1995 only 10 razorback sucker were found in surveys of the river. The last one was captured in 1995. Artificially cultured populations do still exist.

As with most threatened rivers in the world, authors typically relate the decline of the native fishes in the Colorado to stream alteration (e.g. dams, channelisation and dewatering), loss of habitat (e.g. spawning sites), physical changes associated with alteration (temperature, flow regime, blocked migration routes) and competition with exotic species. However, Hamilton (1999) points out that historical evidence indicates that razorback sucker and other native fish in the Colorado River rapidly declined before the dam building and stream alteration reached major proportions in the 1930s. Razorback sucker were already rare in the river in the 1920s, although surveys had found them abundant in the decades before. Activities prior to 1920 best explain at least the initiation of the disappearance. Irrigation projects in the basin began in 1886, shortly after settlement in 1881. Major, nationally sponsored irrigation projects came on-line in 1915. Dam building did not accelerate until the 1930s and later.

As noted above, Se contamination is associated with irrigation projects throughout the arid and semi-arid lands of the American west, including the Colorado River basin. Although early analyses are somewhat suspect, massive Se concentrations were reported in canals and streams within irrigation projects (>1000 µg/L Se in some cases; similar to drains in the San Joaquin Valley) in the 1930s. Concentrations were undetectable above the projects. Concentrations in the Colorado River (near Grand Junction) were reported at 30 µg/L Se. In the 1960s Se concentrations at the end of the irrigation season were 10 to 400 µg/L Se in the Colorado River and some of its tributaries. Since then, Se contamination has been reported in water, sediments, invertebrates and fish tissues in both the upper and lower Colorado basin, at levels higher than associated with effects on reproduction. Se concentrations in fish tissues were consistently the highest in North America in monitoring programmes.

Hamilton (1999) also studied the effects on cultured razorback sucker larvae of eating native zooplankton from three Colorado River (or tributary) sites near Grand Junction, Colorado. He found that a diet of native zooplankton increased the mortality of razorback sucker larvae at every site. Hamilton's hypothesis, therefore, was that inputs of Se-contaminated irrigation drainage to the river initiated the disappearance of native predators from the Colorado River between 1890 and 1930. Adverse effects of Se on the survival of razorback sucker are supported by: (a) documentation of Se contamination in water, (b) Se concentrations in fish tissue high enough to affect development and survival of young, (c) experiments showing that existing Se concentrations in some parts of the river limit survival of larvae, (d) persistent absence of young fish in surveys of the river, and (e) population declines occurring before stresses other than inputs of irrigation drainage took place. Thus most criteria for demonstrating cause/effect are satisfied (Chapter 11).

Arid lands similar to the Colorado River basin occur on every continent. It seems imperative to consider Se contamination as a potential factor among threats to the biodiversity of these watersheds as they are developed.

Box 13.4 Proper Se analysis and problem detection: Kesterson Reservoir

Reliable analytical chemistry was slower to develop for Se than for most other elements. The problem is that some forms of the element are difficult to break down into a form that can be detected. Another problem is that some forms of Se are very labile and may be lost into the atmosphere if sample preparation involves heating (as in flameless atomic absorption spectrophotometry). As a result, traditional flame or flameless atomic absorption spectrophotometry, used for most trace metals, does not yield repeatable results. Poor analyses contributed greatly to a delay in identifying the problem described in Box 13.2 at Kesterson National Wildlife Refuge. When the deformed birds were found in the Refuge, toxicologists recognised the symptoms as potentially from Se. But government analysts could not detect Se in any of the waters. Dr Ivan Barnes and Theresa Presser from the US Geological Survey recognised that the analysts were using approved (at the time) but inadequate wet digestion techniques. When Barnes and Presser did the analyses properly, the Se suddenly appeared, at the highest concentrations ever seen in a natural environment.

this method is subject to difficulties (Box 13.4). Reliable analytical approaches are now well known, but the requirement for non-standard methodologies and special care in obtaining complete recoveries at environmentally realistic concentrations remain challenges for many laboratories.

Dissolved Se typically comprises most of the Se in the water column of a water body. In San Francisco Bay, dissolved Se comprises 80% to 93% of the total (dissolved plus particulate) Se (Cutter, 1989). So analyses of total (unfiltered) concentrations are less affected by suspended sediment load than most trace metals; and the mass of metal in a river, stream, estuary or anthropogenic discharge can be determined reliably from analyses of Se in the water.

Separating human inputs from natural inputs is important in any analysis of Se sources. In San Francisco Bay, for example, the natural baseline of Se is defined by inputs from the Sacramento River, which drains a Se-poor watershed to the east. Refinery wastes also represent an input to the Bay. Cutter and San Diego-McGlone (1990) used mass balance calculations to show that the refinery inputs were more important than the river inputs during low flows (Fig. 13.4). But during high river flows in spring, the refineries represented a smaller fraction of Se input. Then the Sacramento River provided most of the mass input to the Bay; not because the river was contaminated, but because of the very large volume of very low concentration water that comprised the inflow.

The mass balance study in San Francisco Bay was initiated with the assumption that the Se-rich watershed of the San Joaquin River would be the cause of the contamination in Bay animals. But the study showed that river was also not an important source in 1979. This was because the waters from the San Joaquin were entirely consumed or recycled back to agricultural lands (literally) before they reached the Bay, especially during the dry season. Because there was no net inflow, the San Joaquin River could not be an important source.

A difficult aspect of many environmental issues is that the problems are not static; both human activities and nature are constantly changing. For example, refinery inputs to San Francisco Bay were reduced by half in 1995. But new water management plans for California anticipate allowing San Joaquin water to enter the Bay to improve the status of fish populations in that river. If the latter happens, new mass balance studies will be necessary to redetermine the significance of inputs from the Se-contaminated river.

13.5   WATER CONCENTRATIONS

Prologue. Concentration defines contamination. This is because biota respond to Se concentrations, not to the mass of Se that constitutes the load in a water body. Mass loadings identify sources, but risk

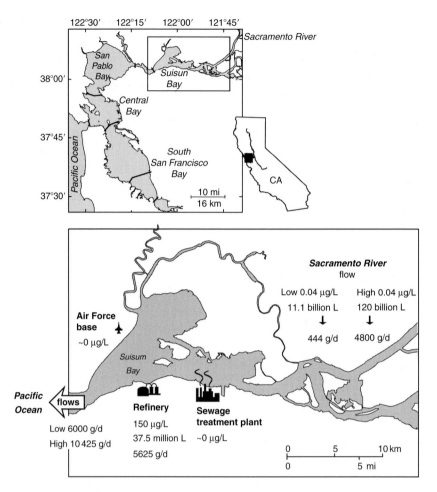

Fig. 13.4 Mass balance. Cutter and San Diego-McGlone (1990) calculated the contribution of refinery wastes to Se contamination in Suisun Bay, a bay in northern San Francisco Bay, California, USA. They used a *mass balance* approach in which input from the rivers plus the input from refineries were compared to output from the estuary under high and low flow conditions. They used field measurements to determine the daily load of Se coming into the estuary from the Sacramento River and used a simple model and measurements to determine the amount leaving the Bay. They then analysed effluents from refineries. The volume of waste released from the refineries was much smaller than the volume of river inflows. In 1994 the refineries discharged 37.5 million L of water per day compared to average river inflows of 11.1 billion L per day during the low flow, dry season. But Se concentrations in the effluents were much higher than in the river (average 150 µg/L versus 0.04 µg/L in the Sacramento River). Using the volumes and concentrations typical of September 1994, total Se inputs from the rivers averaged 444 g/d, whereas Se output from the estuary was about 6000 g/d. About 5600 g/d of output could not be explained by Sacramento River input. Se loads from the refinery effluent were 5625 g/d. This explained the entire 'undefined' internal source term. Since only the refinery inputs were necessary to balance the mass inputs, the results also suggested that sources of Se other than the refineries were trivial. Data for high flow periods are also shown in the figure. The refineries produced about 60% of the Se load carried at high flow because the volume of river input was much larger.

Table 13.1 *Se concentrations typical of the Earth: in waters and sediments from typical water bodies, contaminated ecosystems and effluents compared against ambient criteria*

Baseline	Dissolved concentrations (µg/L)	Particulate (sediment) concentrations (µg/g)
Earth's crust		0.2
Undisturbed surface waters	0.07–0.19	
Some wastes		
Smoke stack scrubber ash		73–440
Fly ash settling ponds	87–2700	
Fly ash pond effluent	2–260	
Irrigation drainage canals,	82–4200	
San Joaquin Valley, CA (USA)	Median = 132	
Contaminated environments		
Kesterson Reservoir: entrance pond	350	
Kesterson Reservoir: terminal evaporation pond	14	
Benton Lake wetland: irrigation drainage input pond	26	0.4–10
Benton Lake wetland: terminal pond	0.7	
Constructed treatment wetland Ecosystems	5–30	2.1–6.7
Typical rivers in Eastern N. America	0.05–0.4	
Sacramento River (uncontaminated)	0.07	
San Joaquin River: 1986–1997 (irrigation inputs)	0.8–4.8	
San Francisco Bay: estuary (refineries and irrigation)	0.1–0.5	0.3–4.5
Great Marsh, Delaware (urbanised estuary)	0.2–0.35	0.3–0.7
Ambient criteria (1996)		
Freshwater	1–5	2–4
Marine	60–150	

management must be based upon concentrations. Concentrations of Se in water are determined by how inputs of Se from natural and anthropogenic sources interact with the hydrological environment.

Some perspective on Se concentrations in water can be useful in evaluating contamination at any specific site. Table 13.1 shows typical Se concentrations in some different types of environmental situations, and the range of their variability. From these data it is clear that Se concentrations in waste streams from human activities can be thousands of times higher than occur in undisturbed waters, soils or sediments draining non-seleniferous geology. For example Se concentrations in oil refinery effluents (Fig. 13.4), coal processing effluents (Fig. 13.3) and irrigation runoff are of a similar order of magnitude. At Kesterson Reservoir, for

example, Se concentrations were 100 to 350 µg/L in the incoming irrigation drainage, although they varied widely elsewhere in the watershed, depending upon proximity to highly enriched geological sources and groundwater conditions. In contrast, concentrations in undisturbed waters are as low as 0.07 µg/L Se; concentrations above 0.2 µg/L Se reflect contamination, whether enhanced by anthropogenic activities or natural. Concentrations in undisturbed waters are usually relatively constant (e.g. Cutter and Cutter, 2004), despite high variability in the natural baseline load.

Se from the Sacramento River mixes with Se from the ocean (e.g. in San Francisco Bay) to form the natural baseline concentration in an estuary. Although mass inputs change with season, the baseline concentrations does not change at high or low river inflows; the different mass loadings do not affect baseline concentrations because water volume changes are the cause of the differences in mass input.

Se concentrations in water can be variable if high-concentration human loadings are discharged daily into receiving waters and riverine inflows change (in this case a diluting factor changes). Se contamination in San Francisco Bay was most evident in summer, for example, but was diluted to low concentrations in winter by the large volume of the river flows (Fig. 13.4).

13.6   SPECIATION

Prologue. The speciation of Se in natural waters does not correspond to predictions based upon chemical thermodynamics alone. Biotransformation reactions and slow kinetics are predominant influences. Se occurs in four oxidation states. If selenate, Se(VI), dominates speciation, the risk of local adverse ecological effects is less than if selenate is accompanied by substantial quantities of selenite, Se(IV), and organic forms of selenide, Se(−II). The likelihood increases for selenite or organo-selenide production if the environment provides contact time between sediment and solution adequate for recycling (lakes, estuaries, wetlands, backwaters). Thus it is especially important that a site-specific risk assessment for a system like a river considers the ultimate fate of Se in all the types of areas downstream from the site of concern.

The biogeochemical cycle of Se is more complex than that of most trace elements, as noted in Chapter 5. The major oxidation states of Se can be determined analytically at the concentrations that exist in natural waters, if modern techniques are carefully employed (Cutter, 1982; Cutter and Bruland, 1984; Cutter, 1989). Therefore, the distribution and cycling of Se species are well known. The most common form of Se in many waste discharges is selenate. However, some combination of selenate (Se(VI)), selenite (Se(IV)), and organo-selenide (Se(−II)) will eventually be found in receiving waters. The exact proportion of the different forms and phases depends upon the opportunities for

biological transformation. For example, in flowing streams with few backwaters, in drainage ditches or in channelised small streams, the primary form of dissolved Se can remain mostly selenate. But, recycling occurs when the residence time of the water body increases. So if a river enters a wetland, both selenite and organo-selenide are formed. Key factors are plant productivity and the contact time between sediment and water. The back reaction from selenite or organo-Se to selenate is extremely slow (hundreds of years: Cutter and Bruland, 1984). Therefore, wherever selenite or organo-Se are formed they accumulate.

Large rivers with complicated backwaters and adjoining wetlands, the oceans and lentic (lake-like) environments all show accumulations of selenite and organo-Se. For example, Se in the uncontaminated Sacramento River includes 47% selenate, 40% organo-selenide and 13% selenite, by the time waters reach San Francisco Bay. Wetlands gradually accumulate organo-selenide and selenite as water moves through them. In the Benton Lake wetland in Montana, USA, Zhang and Moore (1996) found that selenate was 90% of total dissolved Se entering the ponds through Lake Creek (the source was irrigation drainage). The percentage of selenate decreased substantially as water worked its way through the pond system, from pool 1 (55%) to pools 3 and 5 (32%). Organo-selenide rose from less than 10% in pond 1, to 43% to 59% of total dissolved Se in pools 3 and 5. In the oceans, 80% of Se occurs in the organic form Se(−II).

Some sources of contamination from human activities discharge selenite directly. These are probably the wastes most likely to pose the greatest risks, because selenite reacts more readily in all aspects of the biogeochemical cycle. Examples of wastes that generate selenite include ash from coal-fired power plants, air pollution residuals or wastes from refineries (Cutter, 1989). For example, in the heavily polluted Scheldt estuary (the Netherlands), selenite was 85% of the total dissolved Se. Such a high proportion of selenite is probably a result of industrial (refinery) discharges (van der Sloot et al., 1985). Speciation sometimes can be used to trace sources of Se if it is different between anthropogenic and natural Se inputs (Cutter, 1989).

13.7   PHASE TRANSFORMATION

Prologue. Dissolved speciation is important because it determines the nature of phase transformation (from dissolved to particulate phase); and particulate Se is the bioavailable form of toxicological importance. Thus characterising the concentration of Se per unit mass of particulate (plant material, sediment or suspended detritus) is critical to assessing and managing risks from Se contamination. All oxidation states of Se in solution can be transformed to particulate material. But the process of transformation differs with speciation and environmental circumstances.

Several reactions can transform Se from dissolved to particulate form (and back). The likelihood of each reaction depends upon the oxidation state of Se in solution, hydrological conditions and any other environmental conditions that affect biogeochemical cycling (Fig. 5.4).

13.7.1   Plant uptake and transformation

Prologue. Plants play a key role in increasing the bioavailability of Se in aquatic ecosystems, by reducing inorganic forms to particulate organo-selenides, which are eventually lost to the water column or transferred to animals (because organo-selenide is highly bioavailable).

Plants take up selenate, but slowly because of competition with sulphate. Uptake of selenite and some forms of dissolved organo-selenide is as much as 10 times faster than uptake of selenate (Fowler and Benayoun, 1976), probably because there is not competition with sulphate. If Se in wastes originates as selenate, slow plant uptake of selenate and release of the transformed products will build a pool of selenite and organo-selenide in solution. The pool of the forms that facilitate bioavailability grows with time because of the slow rate at which selenite and organo-selenide retransform to selenate. The transformation process in plants was discovered in phytoplankton (Box 13.5) but the larger plants (macroflora) of wetlands also take up and transform Se and play a role in its sequestration.

In Benton Lake (Zhang and Moore, 1997b), wetland plants had Se concentrations 360 to 1200 times higher than found in the water. In a constructed wetland system, Hansen et al. (1998) found that Se accumulated in the roots and shoots of a variety of the wetland plants to concentrations 1000 to 10000 times greater than concentrations in pore waters.

Phytoplankton detritus ultimately settles to the sediments. Large plant (macroflora) tissues are also ultimately incorporated into sediments. Thus, one outcome of transformation is the accumulation of a reservoir of organic Se in the sediment (Zhang and Moore, 1996). Some of the organo-Se in sediment is eventually captured by sulphur minerals and buried. Recycling of some dissolved organo-Se out of the sediments is also common. Mesek and Cutter (2006) showed that selenite and selenate move into sediments in San Francisco Bay, but this is balanced by a movement of dissolved organic selenide out of the sediments. Recycling from sediments accounted for one-third of the dissolved organic Se in the water column of the Bay at any one point in time. A third fate is uptake by animals that ingest the sediment and organic detritus it contains.

13.7.2   Adsorption

Prologue. Most forms of dissolved Se react slowly or weakly with particle surfaces, except in special circumstances. Thus adsorption probably plays a minor role in sequestration of Se into sediments or suspended particulates, unlike many trace metals.

Selenate is not especially geochemically reactive with most particle surfaces, although some authors cite adsorption of selenate to Fe oxides in the laboratory (presumably in the absence of sulphate). Selenite is reactive, but the rates of geochemical reaction are slower than biological transformation processes; as slow as 0.006 per day. The proportional importance of the geochemical reactions in sequestering Se into sediments in natural waters is a subject of some controversy. Adsorption onto Fe oxide is used in industrial treatments to remove

Box 13.5 Biotransformations of Se
by phytoplankton

A number of different studies have characterised Se recycling and biotranformation. Wrench and Measures (1982) first observed removal of selenite and a small decrease in dissolved selenate concentrations during a phytoplankton bloom, in Bedford Basin, Nova Scotia, Canada, implying uptake by the growing plants. Apte *et al.* (1986) later showed similar results during a bloom induced in an experimental enclosure in Loch Ewe (Scotland). Riedel and Sanders (1998) showed that selenite uptake was four to five times faster than selenate uptake in phytoplankton cultures. But when selenate, alone, was injected into the culture, selenite accumulated in solution. So enough selenate uptake occurred to initiate recycling. Selenite uptake varies among phytoplankton species sufficiently that the population dynamics of different species could be a source of variability in the Se concentrations of suspended material (Doblin *et al.*, 2006).

When selenite is taken up by marine phytoplankton in culture, it is rapidly converted within the cells to cellular protein-associated Se (Wrench, 1978). Then selenite and organo-selenide are released to solution as the cells break down (Wrench, 1978; Cutter and Bruland, 1984). The rapid breakdown of phytoplankton cells can immediately release as much as a quarter of the Se originally transformed, thus allowing efficient recycling (Lee and Fisher, 1994). The bioavailability of the dissolved organic selenide is a source of some controversy. If Se is bound to free amino acids (like methionine), it is probably rapidly taken up and recycled into plant bodies. But at least some organo-Se appears to occur in less bioavailable forms, probably in larger organic molecules like polypeptides. The persistent occurrence of organic selenide in oceans, for example, indicates a build-up of a form of low availability. On the other hand, Se released from filtered lysates of the phytoplankton *Thalassiosira pseudonana* were found to be taken up to the same extent as selenite by five other species of phytoplankton (Baines *et al.*, 2001). If organo-Se occurs in a mix of organic forms in most natural waters the first-order assumption is probably that organo-Se is equally bioavailable to plants as selenite.

selenite from wastes, but it is typically very inefficient, in that large masses of Fe are necessary to remove small masses of selenite.

13.7.3 Volatilisation

Prologue. Volatilisation of a fraction of Se from wetland sediments is well known. The effect is to slow the rate of accumulation in sediments below what would be expected from a mass balance calculation. But volatilisation will not completely remove Se from a sediment.

When selenate is taken up by plants or microbes, some volatile forms are created. Biogeochemical volatilisation of Se to the atmosphere has clearly been demonstrated from wetland soils (Cooke and Bruland, 1987; Thompson-Eagle and Frankenburger, 1992) and in evaporation ponds (Fan and Higashi, 1998). Volatilisation rates depend upon physical/chemical conditions, vegetation, water management or other rate limiting factors (Flury *et al.*, 1997; Zhang and Moore, 1997b; Hansen *et al.*, 1998). For example, Se volatilisation in Benton Lake appeared to be greatest at higher temperature and in aerobic situations. Thus they were highest in seasonal wetlands where sediments periodically dry, for example (Zhang and Moore, 1997c). Cooke and Bruland (1987) originally observed that, on balance, approximately 30% of the incoming Se was volatilised at Kesterson Reservoir. Results consistent with such a figure, as a generalisation, have since been reported by Zhang and Moore (1997c) and Hansen *et al.* (1998) for other wetland systems.

The primary effect of volatilisation is to reduce the rates of accumulation in wetland sediments, but the reductions are relatively small compared to what would be necessary for effective remediation of Se contamination. There is no known example where volatilisation removed all the Se that had been accumulated in a wetland.

Table 13.2 *Proportion of different forms of Se in sediments from different environments*

	Elemental Se (Se(0))	Organic Se (Se(II))	Adsorbed Se (Se(IV))
Central Valley, California (Martens and Suarez, 1997)	Nearly all		
Kesterson Reservoir: receiving ponds, (Presser, 1994)	Predominant		
Great Marsh, Delaware (Velinsky and Cutter, 1991)	58%	10–33%	9–32%
Benton Lake, Montana (Zhang and Moore, 1997a)	<30–58%	30–48%	6–14%
San Francisco Bay (Doblin et al., 2006)	54%	41%	17%

13.7.4 Dissimilatory reduction

Prologue. Selenate is sequestered into sediments via a microbial biotransformation reaction, called extracellular dissimilatory reduction. The transformation is enhanced by prolonged contact with sediment, partly explaining why longer residence time stimulates recycling of Se and generation of selenite and organo-Se.

If selenate dominates speciation, the most important reactions removing Se from the water column are microbial. Oremland *et al.* (1989, 1990) were the first to show that nitrate-reducing microbes living in sediment take up selenate in a process that is not affected by sulphate. These bacteria reduce selenate or selenite to elemental Se, Se(0), and some organo-Se, which they deposit as particles, outside their bodies into the sediment. This is termed extracellular dissimilatory reduction. The reaction occurs just below the interface between the water column and the sediments. Transformation appears to be enhanced by prolonged contact between a parcel of water and the sediment. A sharp gradient between oxidised water and reduced sediment is also conducive to the occurrence of the microbes that conduct the reaction. Rates as rapid as 0.015 per day in systems are observed where ponded water remains in contact with sediments (Tokunaga *et al.*, 1998).

Once sequestered into sediments, Se is not necessarily retained in its original form. While wetland sediments retain some Se(0) (Tokunaga *et al.*, 1996), it can also be oxidised to Se(IV) or Se(−II). In surface sediments from natural waters, combinations of Se(0), adsorbed Se and particulate Se(−II) are often

found (Velinsky and Cutter, 1991; Zhang and Moore, 1997b) (Table 13.2).

Sequestration by processes including dissimilatory reduction and plant uptake can result in accumulation of a large inventory (mass) of Se in sediments. For example, 92% of the total Se that was discharged to Kesterson Reservoir was retained in the sediments (Tokunaga *et al.*, 1996). Sequestration of Se in wetland sediments is so effective that some authors (Hansen *et al.*, 1998) suggest that this might be a solution for treating Se-laden wastes. Hansen *et al.* (1998) showed that 89% of the Se entering a constructed estuarine wetland was trapped by the wetland system. Unfortunately, wetlands also are very biologically productive and attract wildlife. Study of 26 wetlands in areas of potential Se contamination across the western USA showed that all retained high concentrations of Se (Seiler *et al.*, 1999). But, in all cases the wetlands increased exposure of wildlife to this toxin.

13.7.5 Partition coefficient K_d

Prologue. Variability in Se speciation, which reflects variable opportunities for recycling, causes variability in the ratio of particulate to dissolved Se (the partition coefficient or K_d). The K_d is one of the critical measures influencing Se bioavailability to food webs, so is an important piece of data for any environment.

The distribution or partitioning of Se between dissolved and particulate states is empirically described by a distribution coefficient (K_d), which is the ratio of Se per unit mass particulate material

---

Box 13.6 Se transformations in a wetland

Zhang and Moore (1996, 1997a, b, c) carefully characterised the Se cycle in Benton Lake, Montana (USA), where transformation was a dominant reaction.

Se in dissolved phase entered the wetland as selenate. Higher concentrations, 26 µg/L Se, occurred near Se inputs, but progressively decreased as waters passed through the Benton Lake marsh system. The decrease in dissolved Se was paralleled by increasing Se in sediment. The range of sediment Se concentrations was 0.4 µg/g Se near the input, building up to 26 µg/g Se. K_ds ranged from 3×10^2 to 2.5×10^3. Sediment cores were used to sample sediments deposited before the inputs from irrigation drainage. They showed that concentrations were 0.2 to 0.3 µg/g Se, historically.

Adsorbed Se made up 6.2% to 14% of total Se in Benton Lake sediments; of this selenite appeared to be negligible. Elemental Se was positively related to total Se concentrations. At sites where Se concentrations were high in sediments and water, elemental Se made up about 58% of total Se. This suggested deposition by microbial dissimilatory reduction. In sediments from sites with low concentrations of dissolved Se, elemental Se was <30%, with an average of 43.6% for the system. The percentage of bioavailable, organic Se was inversely proportional to total Se in sediments. High Se sites contained less than 30% organic Se in sediment; in low Se areas, organic Se was 48% of the total. The shift in partitioning probably resulted from enhanced recycling and gradual conversion to organic Se forms. The average proportion of organic forms of Se for all ponds was 38%. Adsorbed Se in Benton Lake was about 2.5% of the total particulate Se, and thus was of little importance in this wetland environment (Zhang and Moore, 1997a). Volatilisation appeared to be enhanced if sediments were annually dried.

---

to Se per unit volume water, in equivalent units (Chapter 6). Values of K_d vary widely for Se, as for most trace metals. The full range is 10^2 to 10^5, but, for most systems, the range is narrower: ~5×10^2 to ~1×10^4. The diversity of processes that can sequester Se contributes to this variability. The highest K_d values seem to occur where plant biomass dominates the particulate material; the lowest where the contact between water and sediment is of short duration (e.g. flowing waters) or recycling has not built up organo-Se in the sediment (e.g. where irrigation drainage first enters a wetland). Clearly, however, there is not one universal K_d that represents Se sequestration across all environmental situations.

Values of K_d are more variable among environments than within an environment. So, for site-specific estimates of risk, it is possible to directly measure K_ds from the system, and thereby bound the range of variability empirically (Box 13.6).

13.8   BIOACCUMULATION BY CONSUMER SPECIES

Prologue. Dietary bioaccumulation of Se is the predominant route of uptake because assimilation of Se from most food sources is very efficient and uptake rates from dissolved forms are very slow. Differences in loss rates among species can result in large differences in bioaccumulation among species at the lower trophic levels, even when they are exposed to the same concentration of Se. These fundamental aspects of Se bioaccumulation ultimately determine ecological risks and suggest risk management strategies.

There are no known instances where Se bioaccumulation by animals is not dominated by uptake from diet (that is why transformation to particulate Se is such an important reaction). Dominance by dietary bioaccumulation has important implications for management and risk assessment. It means that traditional toxicity tests with dissolved Se do not reflect toxicity in nature (the dissolved toxicity testing literature is not a useful guide for protecting natural waters). Universal bioconcentration factors (BCFs) or bioaccumulation factors (BAFs) are also of questionable value. Any choice of one value for a BAF will be highly uncertain because bioaccumulation is quite variable. At similar total dissolved Se concentrations, BCFs can vary 10 to 100 times, reflecting the variability of speciation and biogeochemical transformation. BCFs underestimate exposure of animals, and toxicity, by more than orders of magnitude, if derived from traditional experimental data on dissolved exposures. Variability in prey bioaccumulation of Se is also important for management, because it means that some predators might be eating very contaminated prey, while others from the same system are eating prey that are not seriously contaminated.

13.8.1   Uptake

Prologue. Uptake rates of Se from even the most bioavailable forms in solution is so slow that it cannot explain Se bioaccumulation in animals. The Se in the animals in natural waters must come from a source other than dissolved forms; i.e. diet. Assimilation from food is typically quite efficient, but is ultimately a function of speciation, because speciation determines the predominant forms and, therefore, the bioavailability of particulate Se.

Dissolved selenite (Se(IV)) is the most bioavailable form of dissolved inorganic Se. The uptake rate constant for dissolved selenate, by mussels *Mytilus galloprovincialis* for example, is one-tenth the constant for selenite uptake (Fowler and Benayoun, 1976). Se is bioaccumulated when it is associated with an amino acid such as Se-methionine. But the forms of Se predominant in natural water are also likely to include Se bound to complex polypeptides, at least some of which are unlikely to be bioavailable. But the mix of compounds of organic selenide released from phytoplankton cells is taken up at a rate similar to that of selenite (Baines *et al.*, 2001).

The uptake rates of even the most bioavailable forms of dissolved Se are slow in all aquatic species that have been studied, including freshwater and marine organisms (Table 13.3). Fowler and Benayoun (1976), for example, found a bioconcentration factor for selenite (from water exclusively) of <500 in mussels. In contrast, mussels from natural waters, like San Francisco Bay, have a ratio between tissue concentration and water concentrations (BCF) of 10^4 to 5×10^4.

The most notable aspect of Se biodynamics is that assimilation efficiencies from food are consistently high. Assimilation of 60% to more than 90%

is typical for aquatic invertebrates, ingesting plant materials. This means that uptake rates from natural food sources are relatively rapid. Calanoid copepods assimilated 97% of ingested Se from the diatom *Thalassiosira pseudonana* (Reinfelder and Fisher, 1991). Assimilation efficiencies in bivalves fed phytoplankton ranged from 70% to 92% in oysters (*Crassostrea virginica*), clams (*Macoma petalum* (as *M. balthica*) and *Mercenaria mercenaria*) and mussels (*Mytilus edulis*) (Reinfelder *et al.*, 1997). Assimilation efficiencies may be slightly less in fish ingesting invertebrate prey (40% to 60%), but are still highly efficient compared to most trace metals (Table 13.3).

As a result of slow uptake from solution and efficient assimilation of Se from food, dissolved Se typically contributes less than 1% of the Se taken up by animals in natural waters. Luoma *et al.* (1992) demonstrated that 99% of the Se taken up by the clam *Macoma petalum* (as *M. balthica*) was from food, under conditions typical of San Francisco Bay. Wang *et al.* (1996) concluded that Se uptake by mussels (*Mytilus edulis*) was dominated by food under all environmental conditions in Long Island Sound or San Francisco Bay. Both studies found that predictions of uptake from food alone matched bioaccumulation observations from these water bodies.

Lemly (1996) showed that the principal source of Se to fish was their food, in an episode of toxicity in Belews Lake, North Carolina, USA. Xu and Wang (2002a) showed that dietary uptake always dominates Se accumulation in the predatory mangrove snapper (the fish *Lutjanus argentimaculatus*). In every study listed in Table 13.3, a similar conclusion was drawn: virtually all the Se found in the tissues of animals originates from dietary bioaccumulation. Dissolved Se determined how much contaminant entered the food web at its base, but contributed little or nothing, directly, to uptake by the animals.

Speciation in solution is important even though uptake is from diet. The more selenite in a natural water, the more Se will cycle through the phytoplankton to consumer animals. Assimilation efficiency varies with the form in food. As noted above, assimilation is very efficient when the source is organo-Se, as in phytoplankton or their detritus. But when selenite is mixed with, and thereby adsorbed to, natural detritus assimilation efficiency

Table 13.3 *Biodynamic constants for Se bioaccumulation in different types of aquatic organisms. A notable consistency among all organisms is the slow uptake rate from solution. For example, uptake rates for dissolved Ag under similar conditions are 10 to 100 times faster (M. edulis 1.8, T. longicornis 10.4, N. succinea, 1.9). Se assimilation efficiencies were always greater than 20% and often exceeded 50%; Se is the only trace metal, other than methyl mercury, where it is common that >50% of the ingested element is assimilated. The variability within species reflects different food sources*

Species	Dissolved uptake rate constant (µg/g/d per µg/L)	Assimilation efficiency (%)	Rate constant of loss (/d)	Reference
Freshwater bivalves				
Corbicula fluminea	0.0025	29–81	0.010	Lee *et al.* (2006)
Dreissena polymorpha	0.14	20–30	0.03	Roditi *et al.* (2000)
Estuarine bivalves				
Corbula amurensis	0.009	35–54	0.023	Lee *et al.* (2006)
Mytilus edulis	0.035	20–70	0.022	Wang and Fisher (1999)
Polychaete worms				
Nereis succinea	0.006	30–50		Wang and Fisher (1999)
Copepods				
Acartia tonsa	0.024	55–97	0.16	Wang and Fisher (1999)
Temora longicornis	0.024	50–60	0.11	Wang and Fisher (1999)
Mixed estuary species	0.024	51	0.16	Schlekat *et al.* (2004)
Mysid crustacean				
Neomysis mercedes	0.027	68	0.25	Schlekat *et al.* (2004)
Fish predators				
Striped bass, *Morone saxatilis*		42	0.08	Baines *et al.* (2002)
Mangrove snapper, *Lutjanus argentimaculatus*	0.008	32–67	0.03	Xu and Wang (2002a)

declines (40% to 50%: Wang *et al.*, 1996). When particulate elemental Se (Se(0)) was produced from the microbial reduction of ^{75}Se-selenate and the sediments were fed to clams or mussels, the assimilation efficiency was only 20% to 40% (Luoma *et al.*, 1992; Schlekat *et al.*, 2000b). But such forms are not abundant in most natural settings.

Food choices may also affect the uptake rate of Se from diet in invertebrates, by affecting the concentration of Se to which the consumer is exposed. For example, Baines *et al.* (2004) found that 42% to 56% of selenite uptake from a natural lake water community was done by bacteria (in the dark). They suggested that bacteria-feeding invertebrates might be more exposed to Se than plant-feeding invertebrates.

13.8.2   Physiological loss rates

Prologue. Differences in rate constants of loss among species are large. In particular, at least some arthropods lose Se relatively rapidly, while molluscs tend to lose Se slowly.

Comparisons of rate constants of loss of Se among animal species are also informative (Table 13.3). Bivalves and fish lose Se at relatively slow rates, from 0.015/d in some clams to 0.03/d in the mangrove snapper *Lutjanus argentimaculatus* (Wang, 2002). Half the body burden of these species is lost in about every 20 to 40 days. Crustacean zooplankton, like copepods and mysids, as well as amphipods (perhaps all crustaceans), lose Se much more rapidly. Rate constants are 0.16 to 0.23/d. These animals lose half their Se burden every 3 days (Table 13.3). The implication is that crustaceans will accumulate lower burdens of Se in their tissues than will bivalves. The internal exposure to Se is less in these crustaceans than in bivalves, and they will pass less Se to their predators. Thus the differences in rate constants of loss among invertebrates translate to different risks for different types of food webs.

13.8.3 Variability of bioaccumulation in natural waters

Prologue. As a result of variable loss rates, variable food choices and differences in assimilation, bioaccumulated concentrations of Se can vary widely among species within the same water body.

Se concentrations vary widely among species in the same habitat. Table 13.4 compares several molluscs from an uncontaminated site in Australia. Concentrations range from 0.1 to 3.3 µg Se/g dw. Feeding habit and, probably, rate constants of loss, are important influences. In one environment in San Francisco Bay (a contaminated site), Se tissue concentrations were 2 µg/g in some zooplankton and amphipod crustaceans, but reached nearly 20 µg/g in a resident bivalve (*Corbula (Potamocorbula) amurensis*) (Purkerson *et al.*, 2003; Stewart *et al.*, 2004). Zooplanktonic crustaceans and amphipods lose Se rapidly, and as a result they do not accumulate high body burdens. *Corbula amurensis* loses Se slowly (Lee *et al.*, 2006) and has a very high ingestion rate, causing efficient bioaccumulation. Se concentrations in the Dungeness crab (*Cancer magister*) from the same bay were 25 µg/g, probably, in part, because it is a predator of *C. amurensis*.

13.9   TROPHIC TRANSFER

Prologue. Se is always efficiently transferred from one trophic level to the next. Biomagnification between trophic levels is probably more the rule than the exception. Because of the difference in the amount of Se bioaccumulated by different prey, some types of predators are exposed to much more Se than others. This is one of several factors that cause the poor correlation between bioaccumulation in predators and

Table 13.4 *Se concentrations in five common invertebrates from an uncontaminated rocky intertidal zone at Jervis Bay, Australia. Animals were sampled at six different times. These results represent the grand mean of all samplings. The variability among species, even in uncontaminated circumstances, is typical of Se in invertebrates; although the absolute differences among species would increase if contamination increased. The predators with the highest concentrations appear to be predators of predaceous invertebrates (and perhaps fish) and predators of bivalves; although more needs to be known about this*

Species	Grand mean Se concentration (µg/g tissue dw)	Variability (95% confidence interval)
Herbivorous gastropods		
Nerita atramentosa	0.11	0.03
Bembicium nanum	0.64	0.10
Austrocochlea constricta	1.13	0.12
Carnivorous gastropod		
Morula marginalba	3.03	0.42
Oyster		
Ostrea angasi	1.27	0.14

Source: From Baldwin and Maher (1997).

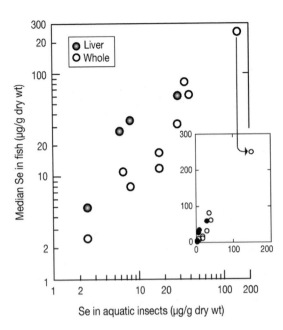

Fig. 13.5 Correlation between median Se concentrations in predaceous fish and a dominant species of insect (prey) from four evaporation ponds in the San Joaquin Valley receiving irrigation drainage. (Data from Skorupa, 1998.)

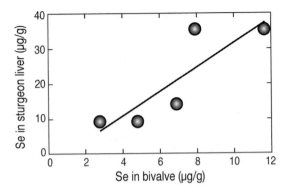

Fig. 13.6 Se concentrations in bivalves (*Corbicula fluminea*, aka *Potamocorbula amurensis*) and their predator (white sturgeon, *Acipenser transmontanus*) as concentrations in bivalves changed over time. (After Linville *et al.*, 2002.)

dissolved Se concentrations. Assessments of risks from Se must consider the complexities of trophic transfer.

As in the case of invertebrates, dietary bioaccumulation (trophic transfer) is unquestionably the pathway of Se exposure in fish and birds. If feeding relationships are simple or accurately identified, concentrations in predator and prey are related. Such relationships are relatively strong across habitats, suggesting that Se concentrations in prey are good indicators of Se exposure to the most vulnerable organisms in a habitat: predators (Fig. 13.5). One result is that Se bioaccumulation differs among predators in the same habitat reflecting differences in bioaccumulation among prey species.

Traditional experiments or analyses of Se in food webs have reported an absence of biomagnification, when all predators and all prey have been considered together (without differentiating dietary choices: Saiki and Lowe, 1987; Besser *et al.*, 1993; Jarman *et al.*, 1996). But when exact feeding relationships are identified and each prey is compared to its specific predator, biomagnification can be observed (Figs. 13.6 and 13.7). Careful studies of this sort show that the ratio between concentrations in predator and concentrations in prey range from 1.0 to 2.0 (T. Presser and S. N. Luoma, unpublished data), although more data of this specificity are needed. Thus, it is likely that biomagnification is the rule, rather than the exception, for Se.

Speciation, transformation, physiological dynamics and food choice thus combine to determine Se exposure of predators. Phase transformation and prey bioaccumulation are driven by biogeochemistry and physiology. That influence is propagated up the food web. For example, stilt, the wading bird *Himantopus mexicanus*, averaged about the same Se concentration (20 to 30 µg/g Se in eggs) at Chevron Marsh in San Francisco Bay as at Kesterson Reservoir (25 to 37 µg/g in eggs). But the water in Chevron Marsh contained about 10% the concentration of Se found at Kesterson (20 versus 300 µg/L at maximum) (Skorupa, 1998). Transfer of Se from water to particulates and on to aquatic invertebrates was greater at Chevron Marsh, and the effects of that enhanced bioavailability were passed on to the stilt. Selenate was the original form at Kesterson, and the lower bioavailability was reflected in the food web; selenite was probably the predominant form in dissolved phase at Chevron Marsh.

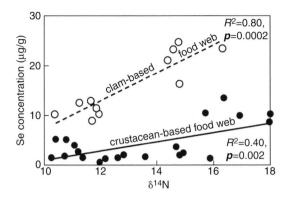

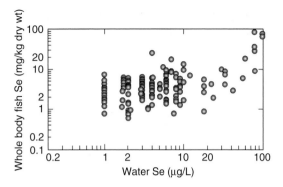

Fig. 13.7 Se concentrations in San Francisco Bay fauna as a function of trophic level, as identified by stable isotope analysis ($\delta^{15}N$). Stewart *et al.* (2004) sampled a broad array of predator and prey fauna from Grizzly Bay, California, USA in the autumn of 1999. They used stable isotopes of nitrogen to quantitatively define trophic level. $\delta^{15}N$ is accumulated preferentially to other N isotopes when one organism eats another, so it proportionately increases (higher $\delta^{15}N$) up food chains. A higher $\delta^{15}N$ determined in the tissues of an organism signifies a higher trophic level, if all else is carefully controlled. They also used general ecological knowledge of the animals to verify predator–prey feeding relationships. The correlation between $\delta^{15}N$ and Se concentrations was not strong among all data; but when food chains were carefully separated, biomagnification was evident. In particular, Stewart *et al.* (2004) separated a bivalve-based (clam) food web from a zooplankton-based (crustacean) food web in San Francisco Bay. Much higher Se concentrations were found in sturgeon (whose stomachs were full of bivalves) than in striped bass (which ate zooplankton). Both individual food webs biomagnified Se. The difference between the food webs occurred because initial differences in prey Se concentrations were propagated up each food web to create two correlational relationships. Predators at the top of the bivalve food web (sturgeon) will suffer considerably higher Se exposures than predators at the top of the zooplankton food web (striped bass). In contrast, Jarman *et al.* (1996) characterised isotopes and Se in a Pacific Ocean food web. They did not observe Se-build up from lower to higher trophic levels (no biomagnification); but they compared an array of feeding relationships without separation into known food chains. (After Stewart *et al.*, 2004.)

Fig. 13.8 Se concentrations in a variety of fish species from rivers in the western USA, and Se concentrations in water at the same locations. Data taken from the National Irrigation Drainage study for lentic (standing water) sites as reported by Brix *et al.* (2005). There is no correlation between bioaccumulation and dissolved Se except at the most extreme Se concentrations. Brix *et al.* (2005) termed this a 'hockey stick' relationship. (After Brix *et al.*, 2005.)

The result of the influences of food selection, efficient trophic transfer and ultimate dependence upon speciation is that the correlation is universally poor between dissolved Se concentrations and Se concentrations in fish and birds, if diverse habitats are compared. For example, this relationship was highly variable among areas of Se-enriched irrigation in the western USA (Brix *et al.*, 2005) (Fig. 13.8). If only lentic (standing water) habitats were considered a weak relationship appeared at the highest concentrations only (a 'hockey-stick' shaped relationship). Among rivers (lotic habitats) no relationship was evident. The differences between habitats could reflect differences in opportunities for recycling and hence speciation between lentic and lotic environments, or differences in the length of food webs. A site-specific risk assessment for a system like a river must not assume that the same bioavailability will apply in backwaters, in in-stream or oxbow lakes, or in all the habitats downstream (Lemly, 1999; Orr *et al.*, 2005).

The complexities that influence trophic transfer of Se have important management implications. First and most important, assessments of risk and regulatory policies that depend upon dissolved Se concentrations are unlikely to be effective. Instances of either overestimation or underestimation of risk, depending upon the circumstance, are likely from any choice of a generic dissolved water quality guideline. Supporting evidence like concentrations of Se in prey might facilitate prediction of predator exposure better than concentrations in water. It follows that invertebrate species might be used to biomonitor threats to specific predators. Resident species like bivalves, polychaetes, amphipods or zooplankton are practical to collect and can be present in areas of Se contamination (Chapter 8; Phillips and Rainbow, 1994). The direct threat from each prey species will be to its specific predator, however, not to birds and fish in general.

Large predator species are the species of concern but are much more difficult to monitor frequently and in large enough numbers to detect differences in Se exposure with much statistical power. Knowledge of which predators bioaccumulate (or are likely to bioaccumulate) the most Se is critical to assessing ecological risks.

Box 13.7 Selenium toxicity in livestock

'Blind staggers' disease is a disease in livestock that results from acute consumption of plants high in Se (Martin and Gerlach, 1972). It is characterised by impaired vision, aimless wandering behaviour, reduced consumption of food and water, and paralysis. 'Alkali disease' is a disease in livestock resulting from chronic consumption of high levels of Se. It is characterised by hair loss, deformation and sloughing of the hooves, erosion of the joints of the bones, anaemia, and effects on the heart, kidney, and liver. Both are associated with consuming plants of the genera *Astragalus*, *Xylorhiza*, *Oonopsis* and *Stanleya* that hyperaccumulate Se from Se-rich soils.

13.10   TOXICITY

Prologue. Se toxicity is manifested as adverse effects on the development of young life stages or reduced reproductive success (teratogenesis, or birth deformities, and poor hatching success of eggs are classic signs of Se poisoning).

Field evidence shows that effects of Se on reproduction can result in extirpation of predator populations, with no evidence of dead adults. A universal relationship between concentrations of Se in predator tissues (especially eggs or reproductive tissue) and onset of Se toxicity is a more realistic expectation for Se than for most metals, although differences in vulnerability and exposure among predators add some uncertainty.

13.10.1 Signs of Se toxicity

Prologue. The window between sufficiency and toxicity is smaller for Se than for any other trace element. Se expresses its toxicity biochemically by substituting for sulphur in important enzymes and proteins and by excessive lipid peroxidation. Toxicity is expressed as deformed young and failure of eggs to hatch.

Confusion about Se toxicity partly stems from its role as an essential micronutrient in animal nutrition. Biochemically, Se is necessary for the functioning of glutathione peroxidase, which protects cell membranes from damage resulting from the production of oxidants or peroxides that are produced naturally during metabolism. Symptoms of Se deficiency occur in humans, livestock and wildlife where soils are deficient in Se. Epidemiological evidence also suggests that Se is an anti-carcinogen in humans, although that is probably not expressed until it reaches levels near or at toxicity (Spallholz, 1997).

The toxicity of Se was noted in the 1930s in the form of 'alkali disease' or 'blind staggers' in horses and cattle living in areas with Se-enriched soils in western North America (Box 13.7). Signs of Se toxicosis have since been thoroughly studied. Because they are relatively distinct, some are diagnostic of Se toxicity in wildlife.

When Se substitutes for sulphur in proteins, it disfigures the protein, typically disturbing normal development of tissues (this is especially true of hard tissues that are protein rich: Spallholz and Hoffman, 2002). It may also generate hydrogen selenide in tissues, which is toxic to liver cells.

Most important, the difference is very narrow between the intake required for minimum physiological needs for Se, and the intake that is toxic. For example, Hilton *et al.* (1980) showed that dietary Se levels between 0.15 and 0.38 μg/g in dry feed were necessary to prevent Se deficiency symptoms in the diet of rainbow trout (*Oncorhynchus mykiss*).

They also found chronic dietary toxicity (reduced growth) in the trout at 13 μg/g dry feed (only 30 times higher than sufficiency requirements). This difference is smaller than for other trace metals.

In fish and birds, the earliest signs of toxicity occur via the pathway: mother to egg to developing young life stages. Elevated Se from the diet of females is deposited in the liver, where it can be transferred to the egg yolk precursor protein, vitellogenin (Se substitutes for sulphur in the vitellogenin). Deformities (teratogenesis) or death of the developing organism is induced when the yolk is consumed during development, and the selenoproteins disrupt normal protein functions.

In birds the first sign of toxicity is failure of the eggs to hatch. Occurrence of deformities first appears at concentrations higher than those which begin to inhibit hatching. In fish, failure of eggs to hatch occurs at the same or higher concentrations than signs of Se effects on development. The signs of developmental toxicity include oedema (larvae with distended abdomens), bulging or protruding eyes (exophthalmus), cataracts (white coating on eyes), curvature of spine, misshapen heads and mouths, and distorted caudal or missing fin rays.

13.10.2 Detecting toxicity: Se bioaccumulation in target organisms

Prologue. Selenium concentrations bioaccumulated in fish or bird tissues can be correlated (as a measure of dose) with signs of toxicity. Concentrations in reproductive tissues show the best correlations.

Because the relationship between adverse effects of Se and dissolved concentrations is so variable, it is necessary to find other measures to relate dose (Se exposure) and response. Bioaccumulated Se is used in fish and birds as related to the frequency of malformations, survival or growth retardation in offspring. These relationships are less variable than for most trace metals, partly because detoxification of bioaccumulated Se seems to be weak. The relevant experiments also must involve realistic exposure scenarios (dietary exposures), and realistic endpoints (reproductive effects). The relationships

have also been validated in field studies (e.g. Skorupa, 1998). The choice of tissue in the animal in which to determine bioaccumulation is an important consideration, because different tissues concentrate Se to a different degree (at the same dose). For fish, the US Environmental Protection Agency recommends whole body analysis, largely for practical purposes. But a choice that correlates more closely with effects is probably reproductive tissue, in particular eggs or ovaries, although those can be difficult to sample in natural waters for many fish.

Heinz (1996) concluded that, for birds, Se concentrations in eggs were better predictors of adverse effects on reproduction than were concentrations in the liver. Two measures of developmental effects in birds are commonly employed: teratogenicity or deformities, and chick mortality. The latter is a combined effect of teratogenesis, hatchability and chick survival on net bird productivity (Fairbrother *et al.*, 1999). Chick mortality, or hatchability alone, is sensitive to Se at about one-half to one-quarter the concentration that induces deformities (two to four times more sensitive).

13.10.3 Differences in sensitivity

Prologue. Fish and bird species differ in their susceptibility to extirpation in Se contaminated habitats. Differences in feeding habit are probably the largest source of such differences, although some differences in inherent sensitivity to Se also occur.

The relationship between bioaccumulated Se concentrations and effects of Se also varies somewhat among species. In the studies of toxicity in Belews Lake (Box 13.8), there was a progressive increase in deformities in sensitive fish species as Se concentrations increased from 5 to 30 μg/g Se whole body dw. At 40 to 50 μg/g Se, 50% to 80% deformities were found in larvae of the more sensitive species. However, when all species (sensitive and insensitive) were considered together, it was the *range* in prevalence of deformities that increased as Se increased to 5 to 30 μg/g Se whole body dw. At high end of the range, high rates of deformity were seen

Box 13.8 Belews Lake

The most detailed study of Se toxicity in nature is that from Belews Lake, North Carolina, USA (Lemly, 1985, 1993a, 1996, 2002). Belews Lake was contaminated in the 1970s by Se in waste water from an ash impoundment from a coal-fired power plant. Impacts on the resident fish community were first described in 1983. To test the route of uptake, Se concentrations were determined in bluegills (*Lepomis macrochirus*) and largemouth bass (*Micropterus salmoides*) from the lake. These were then compared with concentrations in those species when experimentally exposed to the same concentration and form of Se in water alone. The concentration factor from the experiments was too low to explain the field observations, demonstrating the importance of diet as the exposure route. This was corroborated by the observation that Se was bioaccumulated differently among fish with different feeding habits (Lemly, 1985), following the ranking: piscivores (bass and perch) > omnivores > planktivores. This was also the order in which fish disappeared from the contaminated lake and then recovered when the contamination was eliminated.

Lemly's studies showed that selenite was the primary form of input to Belews Lake, and that signs of toxicity were present when dissolved selenite concentrations were less than 5 µg/L. He showed contamination of sediment and invertebrates. Signs of chronic Se poisoning were present in fish that were not extirpated. These included organ dysfunction, altered blood chemistry, and cellular pathology in liver, kidney, heart and ovary. The most obvious effects, however, were reproductive failure due to egg mortality, and teratogenic deformities of the spine, head, mouth and fins. Lemly described Se poisoning in fish as 'invisible', because the primary point of impact is the egg, which receives Se from the female's diet and stores it until hatching; and then biochemical functions are disrupted in the developing chick. Teratogenic deformities and, sometimes, death follow. Adult fish can survive and appear healthy despite the fact that extensive reproductive failure is occurring. But usually, a population that is not reproducing will disappear. Field surveys in the lake showed that 19 of the 20 fish species in Belews Lake were eliminated in the most severely contaminated parts of the lake by the 1980s. All piscivores and omnivores succumbed to the poisoning in the lake, as expected from the bioaccumulation (exposure) data. A few lower trophic level fish survived.

Se-laden waste water inputs to the lake were eliminated in 1986. Concentrations in water, sediment and animals had declined by 1996. Some fish species were successfully reproducing in 1992, although deformities were common and predators were scarce. By 1996 predaceous largemouth bass has reappeared and were common in the lake, although deformed fry of all species were still observed. The recovery of the fish community after elimination of the Se inputs erased any doubts about Se being the original cause of the disappearance of fish populations. Lemly rated the lake as moderately stressed from Se in 1996, due to recycling of the legacy of Se from sediments. Table 13.5 shows changes in Se concentrations that accompanied these changes in biological communities.

Table 13.5 *Concentrations of Se (µg/g, except in water µg/L) in various media in Belews Lake in 1986, and in 1996, 10 years after removal of wastes*

Ecosystem component	Se concentrations[a]	
	1986	1996
Water	5–20	<1
Sediment	4–12	1–4
Invertebrates	15–57	2–5
Fish eggs		
Carp (*Cyprinus carpio*)	45–119	5–18
Bluegill (*Lepomis macrochirus*)	40–133	3–20
Largemouth bass (*Micropterus salmoides*)	77–159	4–16
Mosquitofish (*Gambusia affinis*)	63–120	5–20

[a]Concentrations in a control lake (High Rock Lake) were water and sediment <1 µg/g, invertebrates <2 µg/g, and all fish eggs <3 µg/g.
Source: From Lemly (1996).

in some species and low rates were seen in others. The differences among species led to a variety of interpretations. For example, USEPA (2002c) initially suggested that a chronic value of 44.6 μg/g Se in whole fish would be protective. That was based upon the rapid rise in deformities among all species. At 40 to 50 μg/g Se, 20% to 30% of all young from all species were deformed. From the same data, Lemly (1996) concluded 5 to 10 μg/g Se was the protective value. The lowest value was based upon a 6% increase in deformities in the more sensitive species. Differences in the protection goal (protect sensitive species versus protect the average species) was the cause of the disparity in protective choices.

Predator species differ in their exposure to Se because of their differences in feeding (and because prey species vary widely in Se concentrations). But differences among predators in sensitivity to Se also occur when exposures are similar (Table 13.6). Dose–response curves, sometimes based upon a restricted range of data, suggest at least twofold differences among some species in the relationship of tissue concentrations to observations of effects (Skorupa, 1998) (Table 13.6).

Differences in trace metal sensitivity among species are typically caused by different mechanisms of detoxification (Chapter 9), but there is not strong evidence for a variety of detoxification mechanisms for Se in animals. Recently Se was found (in association with Hg) in 'crystalline particles' in the tissues of birds (cormorant) and mammals (Nigro and Leonzio, 1996). Although conjugation of this type is used for detoxification of other trace metals, the physiological significance of the Se-rich conjugates is unclear. Se is not otherwise detoxified by conjugation with inorganic granules or metallothioneins.

Differences in reproductive strategy among species might cause differences in the sensitivity of development of young. Researchers speculate that the young from species with eggs that are large and rich in yolk would be more susceptible to Se effects. The young of species that use more yolk to support development could be subjected to a greater Se load, for example. Vitellogenin may also differ in its characteristics among species. If its sulphur content differed, its ability to hold Se might also differ. The timing of vitellogenesis, yolk utilisation and organ development is another factor that could vary among species (R. Linville and J. Skorupa, personal communication). None of these has been adequately explored, however. In short, the differences in sensitivity that add so much complexity to understanding effects of other metals relative to bioaccumulated concentrations are not as extreme with Se, but need to be better understood.

13.10.4 Interaction of Se with other contaminants

Prologue. Interactions of Se with other elements, in particular Hg, are known, but not well understood. The problem of element interactions adds some uncertainty to interpretation of Se bioaccumulation and effects; but it is probably a more important consideration for Hg than for Se.

It is quite likely that Se interacts with other elements in ways that might affect toxicity or bioaccumulation. Unfortunately, such relationships seem to differ in different circumstances. In the open ocean, strong correlations exist between Se and Hg concentrations in the tissues of birds, mammals

Table 13.6 *Dose–response data from Tulare Lake, California, USA where a variety of habitats with different Se contamination occur in close proximity, and several species co-occur. The values are different effect concentrations (EC) in eggs at which non-viable chicks of stilt and avocet were found (Skorupa, 1998). There was about a two-fold difference in sensitivity between the species*

	Stilt (*Himantopus mexicanus*)	Avocet (*Recurvirostra americana*)
	μg/g Se	μg/g Se
EC–1%	14	41
EC–10%	37	74
EC–50%	58	105

and predators like tuna, although ratios differ among species (e.g. Leonzio *et al.*, 1992; Kim *et al.*, 1996). It is thought that one role of the interaction might be to reduce the toxicity of Hg (Magos and Webb, 1980; Pelletier, 1986). But in the oceans, concentrations and bioavailabilities of Se and Hg are relatively consistent across large areas. In other types of environments, the Se:Hg relationship is less clear. For example, a significant inverse relationship between Hg and Se was found in perch and walleye in lakes of the Sudbury region; just the opposite of the relationship seen in ocean predators (Chen and Belzile, 2001). Nor did Se and Hg 'move together through the ecosystem' in experimental lakes dosed with Se in Canada (Rudd *et al.*, 1980). When Canadian and Swedish lakes were dosed with Se, Hg bioaccumulation in fish was reduced (Rudd *et al.*, 1980). This has led people to suggest that Se additions to lakes might ameliorate Hg contamination of fish, as long as Se toxicity does not ensue (Rudd *et al.*, 1980) (Box 13.9).

In the laboratory, very high Se concentrations in diet will protect against methyl mercury toxicity from spiked food in rats (Ralston *et al.*, 2007), but it is not clear how consistently this translates to other animals. Methyl mercury exposure in solution does not affect Se uptake in mussels (*Mytilus edulis*: Pelletier, 1988). Nor were any protective effects observed when shrimp (*Pandalus borealis*) were fed mussel tissue spiked with methyl mercury and Se (Roleau *et al.*, 1992). So the reasons for the amelioration of effects of one by the other must have to do with internal detoxification mechanisms, but it is not clear what those are and when they are or are not effective. Given the inconsistent results between experiments, the safest conclusion at present is that no consistent conceptual model fits all the data. If the effect is that subtle, then its implications in contaminated field situations should probably be assumed to be minor unless shown otherwise (Pelletier, 1986).

Antagonistic interactions between Se and As are also indicated, as are interactions of Se with Cd (Magos and Webb, 1980; Bjerregaard, 1985) and Ag (Pelletier, 1986). All these observations occur at high concentrations; so again, the implications for a field contaminated site are not clear.

Box 13.9 Disappearance of perch but not pike from Swedish lakes

A series of 11 lakes, including Lake Öltertjärn, in Sweden (Paulsson and Lundbergh, 1991) was dosed with selenite in the mid-1980s in an attempt to mitigate Hg contamination of edible fish. After higher dosing in the first year (about 5 µg/L), a second year of dosing resulted in a water concentration of 1 to 2 µg/L Se. Se concentrations in pike (*Esox lucius*) increased from an average of 1.3 µg/g Se dw tissue to 4.6 µg/g Se after the second year, with no evidence of catastrophic population declines. In contrast, Se concentrations in perch (*Perca fluviatilis*) averaged 0.8 to 2.0 µg/g tissue before dosing. The mean concentration after one year of dosing was 23 µg/g Se, in the same system where little bioaccumulation was observed in pike. The perch had disappeared from four of the lakes by the second year, and were rare in a fifth lake. Perch survival was found only in lakes where water exchange was rapid. Although the authors did not conclusively address whether Se was the cause of extirpations, other authors suggest that effects were consistent with Se effects on reproduction (Skorupa, 1998). Paulsson and Lundbergh (1991) suggested that if Se were to be used in such treatments, dosing should not exceed 2 µg/L. A somewhat overlooked aspect of the outcomes was that different species of predaceous fish bioaccumulated different concentrations of Se. Effects were then as expected from internal exposures, probably because of the limited detoxification differences among species as is typical of Se. Any inherent physiological differences in susceptibility were small enough not to affect conclusions. The differences in exposure presumably occurred because the two fish occupied different food web niches.

13.11   EFFECTS

Prologue. For Se, a number of relatively well-documented case studies demonstrate the links between toxicity and adverse ecological effects. The case studies link observations indicative of Se toxicity in the development process to disappearance of some species (extirpation). More species are extirpated as the contamination becomes more severe. The challenges in evaluating ecological effects (species extirpations) from Se contamination include: (1) determining that Se is sufficiently bioavailable to extirpate species; (2) determining which species should be affected and which might not; and (3) determining if the absence of potentially vulnerable species is attributable to Se or to other factors.

It is a well-established principle in population biology that populations subjected to high enough reductions in survival of their young will slowly dwindle in size. Extirpation is inevitable if the recruitment of young into the population slows to rates below the natural rate of mortality (Chapter 11). Some of the difficulties in detecting or demonstrating Se effects may be mitigated because the signs of reproductive toxicity (teratogenesis and reduced hatching rates) are particularly distinct.

Differences in trophic transfer and differences in inherent sensitivity to Se toxicity both suggest that Se will not affect every species similarly; especially if species are parts of different food webs. Thus, it should be expected, in an affected system, that some populations will disappear, while others may not, or may be less affected (as in Belews Lake: Box 13.8). The populations with the greatest exposure must be considered the most at risk, as a first-order assumption (i.e. exceptions will occur). For example, percentages of nests with embryo toxicity at Kesterson were eared grebe and coot (63% to 64%) > mallard ducks (23% to 46%) > stilt (22% to 24%). The order of Se exposure, as measured by mean Se concentrations in eggs, was the same (Ohlendorf *et al.*, 1986). Exposure to Se was also the best indicator of species' vulnerability in Swedish lakes spiked with Se to eliminate Hg toxicity (Box 13.9).

Demonstration of ecological effects from Se in any field situation is a formidable challenge, even where signs of toxicity are distinct. As was discussed in Chapter 11, one of the challenges is obtaining a sufficiently robust sample size to detect individual effects or reduced recruitment. Another complexity is that there are circumstances where reproductive toxicity may be detectable, but

the population is not extirpated (e.g. Box 13.10). This can occur if population-level mechanisms provide compensation for the reduced juvenile survival. Population models suggest that cutthroat trout (*Oncorhynchus clarki*) populations would not decline until Se concentrations reached 7 μg/g dw (based on toxicity tests) and then the extirpation may not occur for decades (van Kirk and Hill, 2007). The cause of the eventual extinction of the population was the inability of the trout to achieve high survival during favourable years of fry production. Even where local reproduction fails, adults may be present, if recruitment is enhanced by immigration into the habitat from elsewhere. This is a particularly important consideration for fish in rivers, where immigrants can be constantly entering a contaminated system from uncontaminated tributaries (Skorupa, 1998; Hamilton, 1999). Immigration also adds to the variability in signs of Se contamination. Recently arrived adults from tributaries may have no previous history of exposure, so immigrants may show low tissue concentrations and infrequent signs of toxicity.

13.12   CONCLUSIONS

Arguably as much may be known about basic aspects of Se chemistry and biology as for any other trace metal. Repeatedly, studies show that the total dissolved concentration of Se, alone, is not a good predictor of adverse biological effects in aquatic habitats. This is because Se bioavailability can change dramatically with different environmental circumstances. As with all trace metals, chemical speciation influences bioavailability. But Se is different from most trace metals. The effect of speciation on transformation of dissolved Se to particulate phases is more important than its direct effect on uptake by animals. Transformation is important because different processes result in different particulate forms, and different concentrations in the particles. Both influence bioaccumulation at the first step in the food web, and the Se bioaccumulated by those species is propagated up the food webs to predators. It is in predators that the earliest signs of Se toxicity are usually observed. Toxicity can

Box 13.10 Se toxicity in streams affected by Canadian coal mining: high concentrations – low effects?

In the 1990s elevated Se was discovered in the waters of otherwise pristine streams in western Canada, downstream from coal mining activities: the Elk River Basin, British Columbia and Luscar Creek/McLeod River, Alberta. Several studies illustrated the complexity of determining if there was toxicity from Se in a field environment with high inputs of what was probably selenate. Se concentrations in water in these streams were high enough to expect some effects (maxima about 30 µg/L), based on experience in other water bodies; but the nature of the effects was more difficult to ascertain than might be expected.

Effects on two birds in the Elk River area (American dippers, *Cinclus mexicanus*, and spotted sandpipers, *Actitis macularia*) were compared to local reference areas. There was a high incidence of sandpiper nests in which the whole clutch failed to hatch, and hatchability was significantly reduced in the presence of contamination compared to the reference area. It was also notable that the authors were unable to locate sandpiper nests in the watershed of the most contaminated area. For dippers, effects on hatchability were not statistically significant, although just barely so (*p* < 0.056). No gross deformities were found in embryos of eggs at either site. Kennedy *et al.* (2000) studied effects on reproduction in cutthroat trout (*Oncorhynchus clarki lewisi*) in the same watershed. Twenty female fish from a reference site and 17 from a contaminated site were stripped of their eggs and the eggs were raised to determine hatchability and frequency of deformities. The tissues of the trout from the mine-affected Fording River were more contaminated than at a reference (no coal mining) lake. The authors found no statistically significant linear correlation between Se concentrations in the egg masses of the different mothers and the occurrence of seven measures of reproductive toxicity. Kennedy *et al.* (2000) concluded that there were no adverse effects on cutthroat trout in the Elk River. A re-evaluation of the data by other authors, however, has suggested that more careful approaches and a different statistical analysis might

have led to a different conclusion (Hamilton and Palace, 2001).

Holm *et al.* (2005) used a similar approach near uranium and coal mines in Alberta, Canada. Concentrations of Se in both rainbow trout (*O. mykiss*) and brook trout (*Salvelinus fontinalis*) exceeded protective tissue criteria; but the authors noted that such criteria were derived with warm water fish. They compared egg development from fish collected in both Se-contaminated and reference streams. They stripped eggs from fewer females than Kennedy *et al.* (2000), but they used more sensitive methodologies for detecting deformities and abnormalities (more sensitive than gross abnormalities). A significant correlation was found between Se in eggs and craniofacial defects, skeletal deformities and oedema in rainbow trout. They proposed that 15% of the population would be affected at egg concentrations of 8.8 to 10.5 µg/g ww (35 to 44 µg/g dw). Brook trout had significant elevation of some types of skeletal deformities but no dose response could be derived. Holm *et al.* (2005) concluded that there was evidence of toxicity in rainbow and brook trout from the Se generated by coal mining. Again, reanalysis of these data by other authors raised questions, including about the small number of fish employed (MacDonald and Chapman, 2007). Adult populations of both trout and birds from these two sites showed no outward evidence of adverse effects.

The uncertainties in all studies were rigorously examined and vigorously debated (Hamilton and Palace, 2001; MacDonald and Chapman, 2007). Small details in the approach and analysis seemed to affect conclusions in these complex circumstances. None of the studies evaluated Se speciation, transformation, Se in prey species and, especially, the specific feeding relationships of the animals being compared. Partly because the dialogue was mostly restricted to toxicology, explanations for differences among species and among studies were difficult to resolve. It was clear, however, that Se toxicity can affect reproduction without affecting adults, and that it will affect some species but not others. Conclusions about the concentration threshold where effects appeared were more controversial (Holm *et al.*, 2005; Chapman, 2007), as were conclusions about significance to populations (MacDonald and Chapman, 2007).

be linked to either Se concentrations in food (prey) or Se bioaccumulation in the predator itself.

Se toxicity strikes some predators more so than others; often those that eat the invertebrates that accumulate the most Se. So the broad effect of Se is to eliminate the predators experiencing the highest dietary exposure, changing the structure of communities and, ultimately, biodiversity. A number of case studies and experiments now support this conceptual model of Se behaviour.

Clearly, direct evaluations of effects, without collaborative evidence, must be undertaken with caution in assessing risks from Se. As always, a body of evidence is desirable to avoid either underestimating or overestimating risks. In Chapter 17 we will discuss protocols for accumulating evidence across boundaries to derive protective values for specific circumstances. But part of the controversy about Se still rests with whether or not effects significant to populations are occurring where contamination is known. If observations are focused on predator populations, five lines of evidence might be particularly valuable addressing the effect/no effect question.

(1) The presence of Se contamination in the environment and in the species of interest (if they are present), at concentrations indicative of possible effects is the first sign of potential risk.
(2) The presence of substantial Se contamination in the specific prey of the species that is absent or stressed can be important corroborative evidence if predators are rare or absent.
(3) The presence of Se-specific indications of reproductive stress in the species that are present is a strong line of evidence if observations are positive, but is vulnerable to Type II error if signs of stress are absent or weak.
(4) The presence of the potentially affected population in less contaminated subhabitats along with its absence where Se contamination is most severe can be a convincing line of evidence.

(5) If the contamination can be eradicated, recovery of the population is the ultimate evidence that the contamination was the cause of the absence.

Despite a strong state of knowledge, diverging viewpoints and interpretations characterise two decades of advocacy debate about Se risks and risk management (e.g. Hamilton and Palace, 2001; Chapman, 2007). As we will see in Chapters 17, some regulations to manage Se contamination are based upon the flawed assumption that dissolved Se is a source of toxicity to animals; while others are based upon field evidence that includes food web transfer. The two types of regulations diverge more than regulations for any other single chemical. New efforts are under way in Canada and the USA to redesign Se regulations around a concept more consistent with the state of knowledge.

Suggested reading

Chapman, P.M. (2007). Selenium thresholds for fish from cold freshwaters. *Human and Ecological Risk Assessment*, **13**, 20–24.

Hamilton, S.J. (1999). Hypothesis of historical effects from selenium on endangered fish in the Colorado River Basin. *Human and Ecological Risk Assessment*, **5**, 1153–1180.

Luoma, S.N., Johns, C., Fisher, N.S., Steinberg, N.A., Oremland, R.S. and Reinfelder, J. (1992). Determination of selenium bioavailability to a benthic bivalve from particulate and solute pathways. *Environmental Science and Technology*, **26**, 485–491.

Presser, T.S. (1994). The Kesterson effect. *Environmental Management*, **18**, 437–454.

Skorupa, J.P. (1998). Selenium poisoning of fish and wildlife in nature: lessons from twelve real-world examples. In *Environmental Chemistry of Selenium*, ed. W.T. Frankenberger Jr. and R.A. Engberg. New York: Marcel Dekker, pp. 315–354.

Stewart, A.R., Luoma, S.N., Schlekat, C.E., Dobin, M.A. and Hieb, K.A. (2004). Food web pathway determines how selenium affects aquatic ecosystems. *Environmental Science and Technology*, **38**, 4519–4526.

14 • Organometals: tributyl tin and methyl mercury

14.1   INTRODUCTION

Prologue. Methyl mercury (MeHg) is perhaps the most dangerous and the most complex of the metals. Management of risks remains a challenge. Tributyl tin (TBT) represents a contrasting case study illustrating the power of a successful science–policy collaboration.

The organometals are often excluded from discussions of the fate and effects of trace metals because of claims that they are unique. However, many attributes of organometals like tributyl tin and methyl mercury fall within the range of behaviours of other trace metals, if geochemistry, toxicology, biology, ecology and risk management are all considered. In fact a discussion of metal science is incomplete without considering such compounds. Episodes of human and environmental toxicity with such chemicals are powerful reminders of the possible results of mismanaging trace metal risks. Successes in management are lessons about the power of appropriate science-aided policies.

Box 14.1 Definitions

alkyl mercury: mercury bound to an alkyl group such as methyl (CH_3), ethyl (C_2H_5) or propyl (C_3H_7). Examples therefore include methyl mercury. Synthetic mercury-based fungicides, typically applied to seeds, often contained alkyl mercury (for example 2-methoxymethylmercury (C_3H_7ClHgO)), and were banned in the 1970s after being linked to episodes of accidental human poisoning and massive disappearances of seed-eating birds.

antifoulant: a chemical agent added to paint to prevent encrustation by aquatic fouling organisms such as algae, barnacles or mussels.

hydrophobic: hydrophobic compounds do not dissolve easily in water, and are usually non-polar. Examples are oils and other long hydrocarbons. The trace organic contaminants posing the greatest ecological risks are typically neutral hydrophobic compounds. Their toxicity partly stems from their ability to penetrate hydrophobic lipid membranes that surround living cells.

imposex: the imposition of male sexual characters on to females, as exemplified by the effect of tributyl tin on many marine gastropod molluscs.

methylation: the transfer of a methyl (CH_3) group from an organic compound to, for example, a metal ion, as in the case of methyl mercury.

methyl mercury: CH_3Hg (strictly CH_3Hg^+ in solution); mercury bound to a methyl group; a very toxic form of mercury.

organometal: a combination of a metal and an organic entity covalently bonded (at least partly); the organic entity is usually a small alkyl group; examples are methyl mercury and tributyl tin.

tributyl tin: $(C_4H_9)_3SnH$, a synthetic organotin compound, formulated to kill fouling organisms that attach to the hulls of boats.

Tributyl tin (TBT) and methyl mercury (MeHg) are emphasised in this chapter. Behaviours of Hg and MeHg are especially complex, from its global cycle, to speciation/methylation to the mode of its adverse effects. MeHg presents some of the greatest ecological and human health risks of any metal. Effective management of MeHg is especially important, but complexity and its high toxicity represent barriers to achieving that goal. But MeHg is the only metal for which at least some lateral holistic management approaches are in place in some jurisdictions.

TBT represents a unique story for metal contamination. Its release into the environment has a relatively short history, especially compared to contaminants like Hg and issues like mining. Strong evidence about adverse environmental effects of TBT was developed from toxicology, geochemistry, biology and field observation. It included evidence of commercial harm and extermination of non-commercial species. A political debate followed that pitted environmental groups and one special interest, the fishing industry, against larger, typically more powerful industrial interests from the paint and shipping industries. Fishing interests gained at least a partial victory. TBT policies were built from all lines of evidence; a good example of lateral management. The partial risk management solution also may have limited some side effects of the ban while allowing environmental recovery to begin.

14.2   TRIBUTYL TIN

Prologue. TBT is perhaps the most effective antifoulant paint ever applied internationally.

Accumulation of marine organisms on the bottoms of vessels increases drag and fuel consumption at sea, with consequences for economics and emissions. This problem was recognised by the Romans, 2000 years ago, who sheathed their hulls with sheets of lead or copper to prevent fouling (by killing fouling organisms). Modern antifouling methods involve incorporating toxins into paints applied to boat bottoms. The paints gradually release the antifoulant as the paint wears away.

Clark and Steritt (1988) estimated that costs to the shipping industry were US$1 billion per year as a result of marine fouling in 1975.

Tributyl tin compounds, when they were incorporated into ship-bottom paints, set the standard for effective elimination of encrusting organisms. They are not only very toxic to fouling organisms, but they also release slowly. Thus repainting is only necessary once in about 5 years, as compared to about once per year for the earlier standard copper-based paints (Champ, 2000).

14.2.1   Sources and uses

Prologue. Tens of thousands of tons of TBT were released into the environment from boat bottoms before its ban in the 1970s.

Tributyl tin exists in various formulations. It is a broad-spectrum biocide, with low mammalian toxicity. This facilitates its use as a fungicide, bactericide, insecticide and preservative in textiles, paper, leather, electrical equipment and plastics (Clark and Steritt, 1988). The greatest environmental concerns arise from one formulation, n-tributyl tin (TBT) which was used throughout the world as an antifoulant incorporated into paint for the bottom of boats. Langston (1995) cited world production of organotins in 1994 at 35000 tons per year.

Use of TBT as an antifoulant began to spread in the 1960s (Huggett et al., 1992) and grew through the 1970s in the shipping industry and with small boat owners (Santillo et al., 2001). The compound was registered by the US Environmental Protection Agency as safe for use in 1978. By the late 1970s, 80% of the world's commercial fleet of vessels was using TBT (Ruiz, 2004). It was also widely used as an antifoulant on nets, piers, buoys, cooling towers and other nautical devices. In all cases, TBT acts as it is leached from the surfaces where it is applied. Ruiz (2004) estimated that the 50000 ships using one port in northwest Spain lost about 1.64 tons of TBT per year. Such leaching of TBT from boat bottoms exposes non-target organisms as well as fouling organisms to the toxin.

Between 1976 and 1981, repeated failures of larval settlement and proliferation of chambers in oyster shells (balling) led to a near collapse of the commercial oyster fishery (*Crassostrea gigas*) on the Atlantic coast of France (Alzieu, 2000). Financial losses were estimated at US$150 million between 1977 and 1983. Once TBT was suspected to be the cause, France banned use of TBT on vessels smaller than 25 m (largely pleasure craft). The first controls on TBT use were introduced in Britain in 1985, then expanded in 1987. In 1988 the US Congress passed the Organotin Paint Control Act to limit the use of TBT there. Canada, Australia and New Zealand introduced similar bans on the use of TBT on small vessels in 1989.

14.2.2   Concentrations and geochemistry

Prologue. Like other metals that do not occur in great abundance (e.g. Ag, Cd) the concentrations of TBT in contaminated natural waters do not seem high in absolute terms (0.001 to 1 µg/L); but they are significant because of the extreme toxicity of the compound to invertbrates. Although it can ultimately be broken down or metabolised by bacteria, TBT accumulates in sediments and is retained after contamination recedes. Once inputs were banned, sediments provided a record of the legacy of TBT usage, and became a source of TBT.

Tributyl tin, like all metals, distributes among all major compartments in the aquatic environment: water, sediment and biota. Determination of environmental concentrations was initially impeded by an absence of analytical techniques. In the earliest studies, G.W. Bryan and colleagues extracted tissues with concentrated HCl, then with hexane to remove all forms of organotin. They further extracted with sodium hydroxide to separate tributyl tin from dibutyl tin. They dried and ashed the extracts and analysed for Sn, then corrected for the differences in weight. Later methods involved extraction with hexane or other organic solvents and analysis by gas chromatography. Because concentrations characteristic of natural waters were extremely low (0.001 µg/L: USEPA, 1997), extraction

of large volumes of water was usually necessary to reach the detection limits of the instruments involved. This resulted in problems in at least some data with both contamination and intereferences. Extraction and instrumentation methods were also developed to separate tributyl tin from dibutyl tin, but those differences in speciation will not be considered here.

As a result of the evolving methodologies, TBT concentrations are reported using different conventions, typically ng Sn/unit weight or volume (g or mL) or ng TBT/g or mL (Meador, 2000). To convert from ng Sn in TBT to ng TBT, it is necessary to multiply by 2.44, since 1 ng of Sn would be contained in 2.44 ng of TBT ($(C_4H_9)_3SnH$) (Meador, 2000).

Clark and Steritt (1988) reported the highest recorded water column concentrations of TBT as 1.89 µg/L, but cited more frequent maximum concentrations to be 0.1 to 0.2 µg/L (100 to 200 ng/L). More recent surveys show concentrations ranging from 1 to 500 ng/L in the waters of harbours and marinas around the world. Concentrations as high as 1.0 µg/L may have been common in marinas in southwest England (Bryan and Gibbs, 1991) during the times of greatest contamination. Concentrations above 0.100 µg/L found in the more open waters of estuaries were related to boating activity (highest concentrations in the estuaries with the most yachting facilities). Concentrations were lower in areas frequented by commercial shipping than in areas with large numbers of small boats (Bryan and Gibbs, 1991). Seasonally, concentrations were highest in summer, during the period of greatest boating activity. Comparable data are available from North America, Australia and Europe.

Maximum levels in sediments in harbours and marinas in various countries show concentrations in the range of hundreds of parts per billion to low parts per million (e.g. 0.2 to >1.0 µg/g dw), although values as high as 10 µg/g are not unprecedented (Meador, 2000).

The processes controlling the partitioning of TBT between water and sediment reflect the complex nature of a combination of organic compound and metal. It can be argued that TBT partitions between water and sediment like a neutral hydrophobic organic compound; concentrations in pore waters are driven by an inherent property: TBT solubility in octanol (K_{ow}: Di Toro *et al.*, 1992). But the K_{ow} is also affected by pH because TBT is ionisable, like an inorganic metal (Meador, 2000). Partitioning in any specific circumstance is also influenced by organic carbon concentrations of the particulate material. Absolute concentrations of TBT in particulate material increase as the organic carbon content of the particulate material increases. Equilibrium partitioning predicts, from that relationship, a partition coefficient relative to organic carbon concentrations (K_{oc}) of 32 000 or $10^{4.5}$ (Meador, 2000). Total TBT partitioning coefficients (K_d) from natural waters vary from 500 to 10 000; the corresponding K_{oc} are 1 to 5×10^5. The high end of these values is about 10 times higher than predicted from equilibrium partitioning (Meador, 2000). This is a common observation when comparing field- and laboratory-derived partitioning. Natural waters provide an infinite sink of the dissolved constituent that is not depleted and/or does not reach equilibrium in the same way as in an enclosed laboratory vessel (Meador, 2000).

The persistence of TBT where it accumulates in sediments is a subject with significant policy implications. Unlike metals, which do not degrade, this compound is subject to breakdown into its less toxic parts. Inorganic tin (Sn), the end product of breakdown, is much less toxic than TBT. Biodegradation, rather than chemical degradation, is the important process in TBT breakdown in natural waters. Early studies suggested a half-life for TBT of 16 weeks in bacteria-rich sediments (half of the compound would degrade every 16 weeks). Intermediate breakdown products (dibutyl tin or DBT) were also found in sediments contaminated with TBT; evidence that such breakdown does occur in nature. Such findings suggested that biotic and abiotic processes would 'result in eventual removal of this compound from aquatic systems' (Clark and Steritt, 1988); and 'TBT is not persistent in the environment, suggesting cautious use may be acceptable' (Laughlin and Linden, 1987).

Later studies were not so optimistic. Some found little or no degradation in sediment; others suggested half-lives on the order of months to years

(Meador, 2000). Field evidence supports the latter. After legislative restrictions reduced TBT inputs to natural waters, concentrations in water declined considerably, but sediment contamination remained a problem.

14.2.3   Bioaccumulation

Prologue. TBT is strongly bioaccumulated by aquatic organisms, with concentrations in animals 1000 to 30000 higher than concentrations in water, and 5 to 50 times higher than concentrations in sediments. Bioaccumulation also varies widely among invertebrate species. Many species are able to bioaccumulate TBT far above the expected thermodynamic maximum predicted from their lipid content and the total organic content (TOC) in local sediment. Differences in ability to metabolise TBT, different biodynamics, lipid content and environmental factors drive TBT uptake and cause large interspecific variability in TBT bioaccumulation. These differences appear to partly influence which species are most vulnerable to TBT toxicity in contaminated environments.

A fundamental characteristic of TBT is the great difference in bioaccumulation among species from the same location (Fig. 14.1), and similarly large differences in sensitivity of biota. Explaining these observations is important to understanding the ecological effects of the chemical. If TBT bioaccumulation followed principles established for hydrophobic organic compounds (equilibrium partitioning theory), then the relationship between the bioconcentration factor (BCF or the ratio [TBT_{tissue}]/[TBT_{water}]) and the lipid solubility (expressed as K_{ow} or the octanol–water partition coefficient) of the compound should be strong. Bioaccumulation should be a function of lipid content of the organism (Box 14.2) and exposure route should not be important.

In contrast, metals theory (biodynamics) suggests that bioaccumulation is a function of the sum of uptake rates from each environmental phase or compartment, which are also metal-, species-, environment-, and exposure route-dependent (Chapter 7; Box 14.2). Aspects of TBT bioaccumulation are

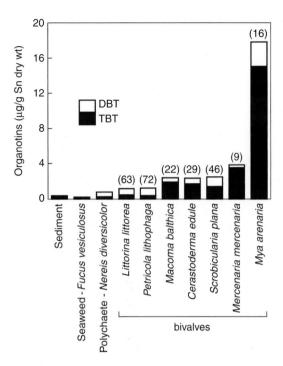

Fig. 14.1 TBT concentrations in sediment, seaweed and eight invertebrates collected from the same mudflat of a contaminated harbour. Differences among species are large, as with all metals. Langston (1995) cites a range in BCF of 2000 to 50000 among estuarine organisms collected from the same site at the same time. Figures in parentheses on bivalve histogram indicate the percentage (DBT/DBT+TBT). One factor very important in determining differences in bioaccumulation among species is the different abilities to metabolise TBT into its breakdown products. One outcome of TBT metabolism is the build-up of dibutyl tin (DBT). Bivalves have widely differing abilities to metabolise TBT as shown here by higher DBT percentages in some bivalves than others. Bivalves like *Mya arenaria* and *Mercenaria mercenaria* accumulate very high TBT concentrations because they do not break down the compound effectively (Langston, 1996).

have the greatest sensitivity to TBT, indicating a lack of significant detoxified storage of accumulated TBT. Whatever the controlling processes, the BCFs for TBT from natural waters vary from 1000 for the freshwater zooplankton *Daphnia magna*, to 900 000 for a freshwater bivalve (Meador, 2000). BSAF (bioaccumulated concentration/sediment concentration ratios) can range from at least 5 to 50 among species. Molluscs have the highest bioaccumulation factors, with particularly high concentrations found in some species of bivalves and snails (e.g. Fig. 14.1).

TBT bioaccumulation is also affected by metabolic breakdown of the toxin, another characteristic unique to organic chemicals. Fish and some crustaceans (like decapod crabs) have an active system for enzymatically breaking TBT down to metabolites that can be excreted (Lee, 1991). This, of course, reduces the steady state concentration bioaccumulated and reduces the risk of significant TBT accumulation. On the other hand, this breakdown system is relatively inactive in molluscs where TBT metabolites may bind to cellular proteins that inhibit the breakdown system (Lee, 1991). The result is accentuated bioaccumulation in molluscs via slower elimination (rate constants of loss are 0.01 to 0.03: Langston and Burt, 1991). When TBT is metabolised in invertebrates, one of the breakdown products is the simpler, and less toxic form, dibutyl tin (DBT). The ratio of TBT/DBT can vary widely even among mollusc species, reflecting differences in capabilities for TBT metabolism (Langston, 1995) (Fig. 14.1).

Trophic transfer of TBT is not well known, but it is probably most significant along some invertebrate food chains. Invertebrate predators, like predaceous neogastropod *Nucella* and *Thais* species (Bryan and Gibbs, 1991), appear to be the organisms most affected by the chemical; food was the primary source of their TBT exposure.

14.2.4   Sublethal effects

Prologue. TBT is a slow-acting poison that causes effects that are sublethal to individual adults, but eventually extirpates populations by preventing

consistent with both processes. The very species-specific nature of the bioaccumulation of TBT can only be explained by biodynamics. But the chemical processes that drive bioavailability are consistent with equilibrium partitioning (Meador, 2000; Box 14.2). Toxicity is also highly variable among species. Those species with the highest bioaccumulation

Box 14.2 Bioaccumulation and toxicity of TBT: biodynamics or equilibrium partitioning?

Traditionally, bioaccumulation of TBT has been assumed to occur via equilibrium partitioning. Data from individual species are consistent with this concept. For example, the ratio of bioaccumulated TBT ($[TBT_{tissue}]$) to sediment-bound TBT ($[TBT_{sediment}]$) within the amphipod *Rhepoxynius abronius* and the polychaete *Armandia brevis* is strongly related to the total organic carbon (TOC) content of the sediments. There is also evidence that the lipid content of the organism is influential; for example, in internal partitioning of the TBT to different organs (Meador, 2000). These factors can be accounted for in a BSAF (typically derived for organic contaminants) where

$$BAF = [TBT_{tissue}]/[TBT_{sediment}]$$

and

$$BSAF = ([TBT_{tissue}]/f_{lipid})/([TBT_{sediment}]/f_{TOC})$$

where bioaccumulated TBT is normalised to the lipid content of the organism (f_{lipid}) and TBT in sediment is normalised to content of total organic carbon content of the sediment (f_{TOC}).

For neutral hydrophobic organic compounds the theoretical maximum BSAF is 1 although values of 2 to 4 are observed in the field. Nevertheless, the relationships among these factors are relatively constant. But all measures of TBT bioaccumulation show a degree of variability among species (BCF, BAF, BSAF) inconsistent with a bioaccumulation process controlled only by equilibrium partitioning. For example, BCFs range from 1 000 to 900 000, and BAFs range from 2 to 100. This degree of variability is not expected from any influence of lipid content. Bivalves can have BSAFs as high as 40; theoretically impossible if only lipid and TOC controls are considered.

When the BAF versus TOC relationship is compared among the two invertebrate species above, it is clear that the bioaccumulation is species-specific, even if bioaccumulation in each species correlates with TOC

in sediment (Fig. 14.2). The differences between species is consistent with differences in specific biodynamic constants. For example, the BCF for the amphipod *Eohaustorius washingtonianus* is 15 times higher than that for the amphipod *Rhepoxynius abronius* after a 10 day exposure to dissolved TBT, and is predicted to be 85 times higher at steady state (Meador, 2000). These differences are consistent with rate constants of loss which are 4.5 times slower in *E. washingtonianus*. Uptake rates are also thought to be substantially faster in *E. washingtonianus* (Meador, 2000). Thus the interspecies differences are consistent with biodynamics.

The differences in bioaccumulation among species are important because they partly explain the differences in sensitivity to TBT among species, if minimal influence from detoxified storage is assumed. Meador (2000) showed that differences in LC_{50}s among species vary with the inverse of the BCF. In the field molluscs tend to be the organisms most affected by TBT and they have the highest BCFs (in the field).

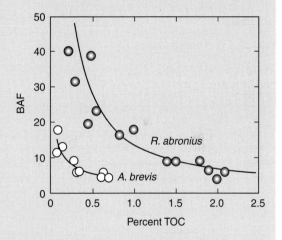

Fig. 14.2 Accumulation of TBT (quoted as BAF: bioaccumulation factor) as a function of total organic carbon in sediments for a polychaete, *Armandia brevis* and an amphipod, *Rhepoxynius abronius*. (After Meador, 2000.)

reproduction and recruitment of young. The most sensitive sublethal responses to TBT exposure are growth inhibition, inhibition of larval settlement, shell chambering in oysters, and, especially, imposex in prosobranch gastropod molluscs (aquatic snails).

Adverse effects from TBT are documented in at least 118 species worldwide (Huet et al., 2004). Perhaps one of the most notorious morphological responses to a trace metal contaminant is the induction of imposex by TBT exposure in gastropod molluscs (predaceous

Box 14.3 TBT and imposex

Imposex is the imposition of male sexual characters onto females, including a penis and a vas deferens. Imposex causes a loss of fertility, and ultimately mortality, in females of the dogwhelk *Nucella lapillus* and other predaceous gastropods common on rocky shores (e.g. *Thais* spp.). One measure of the intensity of imposex in a dogwhelk population was the Relative Penis Size (RPS) index, calculated as

([mean female penis length^3/mean male penis length^3] × 100)

(Gibbs *et al.*, 1988). Another index, the Vas Deferens Sequence (VDS) index is based on six stages of the development of imposex in a female dogwhelk (Gibbs and Bryan, 1987), as follows:

(1) The growth of a proximal section of vas deferens close to the genital papilla.
(2) The commencement of penis development, with formation of a ridge behind the right tentacle.

(3) The formation of a small penis; development of the distal section of vas deferens from the base of this penis.
(4) The joining of sections of the vas deferens; at this stage, the penis size in females approaches that of males.
(5) The proliferation of the vas deferens, which overgrows the genital papilla and sterilises the female.
(6) The accumulation in the capsule gland of compressed aborted egg capsules that cannot be expelled, forming a brown mass.

Females with advanced imposex are permanently sterile; recovery does not appear possible. Mortality also may occur in severely affected females, apparently as an effect of the rupture of the distended capsule gland (Gibbs *et al.*, 1986). The VDS index describes the average stage of imposex in the females of a dogwhelk population. When the VDS index exceeds 4, the presence of sterile females much reduces the reproductive capacity of the population (Table 14.1). The consequence is elimination of populations, as evidenced in southern Britain during the 1980s (Langston *et al.*, 1990).

Table 14.1 *Summary of the effects of exposure to TBT on the female reproductive system of the gastropod mollusc, the dogwhelk* Nucella lapillus

TBT in water (ng Sn/L)	RPS Index (%)	VDS Index	Effect on reproductive system
<0.5	<5	<4	Breeding normal; development of penis and vas deferens
1–2	40+	4–5+	Breeding capacity retained by some females; others sterilised by oviduct blockage; aborted capsules in capsule gland
3–5	90+	5	Virtually all females sterilised; oogenesis apparently normal
10+	90+	5	Oogenesis suppressed; oocytes resorbed; spermatogenesis initiated
20	90+	5	Testis developed to variable extent; vesicula seminalis with mature sperm in most affected animals
100	90+	5	Sperm-ingesting gland undeveloped in some 'females'

Source: After Gibbs *et al.* (1988).

aquatic snails). In particular the masculinisation of the female (imposex) (Box 14.3) is well known in the dogwhelk *Nucella lapillus* (Bryan *et al.*, 1986), the result of endocrine disruption by the TBT (Matthiessen and Gibbs, 1998; Meador, 2000). *Nucella lapillus* is a predator common on temperate rocky shores, extending into estuaries; and, as a predator, is a keystone species with an important role in affecting the structure of intertidal rocky shore communities (Spence *et al.*, 1990). Concentrations of organotins as low as

Table 14.2 *Median toxicity of TBT to sensitive species: conclusions derived from long-term studies and field observations*

Response	Water (μg/L)	Sediment (μg/g dw)	Tissue concentrations (μg/g dw)
Mortality (median)	0.2	0.1–0.2 (amphipods)	30–115
Sublethal (imposex in gastropods)	0.001		0.5
Elimination of populations	0.005 (gastropods)	0.70 (gastropods) <0.30 (bivalves)	
Sublethal (larval settlement in some species of oyster; growth in other species)	0.002		3.0
Protective criteria (USEPA)	0.001		

Source: The table represents a compilation of data from Bryan and Gibbs, 1991; Langston, 1995; USEPA, 1997; Meador, 2000.

50 ng Sn/g in tissues are associated with imposex in *N. lapillus* (Langston *et al.*, 1990). Imposex can reduce populations dramatically (Spence *et al.*, 1990).

14.2.5   Toxicity testing

Prologue. Traditional toxicity testing pointed out that TBT was a hazard but did not capture its extreme toxicity. The slow-acting nature of the toxin and its very specific effects on reproduction, of course, are not detectable in acute toxicity tests. As a result the chemical was registered for use and effects went undetected until field observations indicated that a problem was occurring.

Because some animals, particularly vertebrates, metabolise and degrade TBT, this compound is not as hazardous for such animals as it is for invertebrates with lesser degradation capabilities. For those animals with minimal degradation capabilities, TBT causes environmental damage at concentrations lower than recorded for almost any other marine pollutant.

Early studies showed that mortality responses to TBT have been reported at concentrations that vary by more than three orders of magnitude among species: 0.1 to 200 μg/L (Laughlin and Linden,

1987); although most the acute toxicity testing responses reported by Hugget *et al.* (1992) were in a much higher range (200 to 2000 ng/L). The relative insensitivity seen in such acute tests, and the low toxicity to mammals, were the likely reasons that the US Environmental Protection Agency licensed TBT for use in 1988.

Chronic toxicity sufficient to eliminate populations occurs at much lower concentrations than acute toxicity. Reduction of growth in 60 day tests with mussels (*Mytilus edulis*) occurs at dissolved concentrations of 20 ng/L; and in field tests with oysters (*Crassostrea gigas*) at water concentrations of 48 ng/L (Meador 2000). Meador (2000) reported that the median concentration for chronic toxicity amongst a variety of species subjected to traditional chronic, sublethal testing was 0.22 μg/L (Meador, 2000). More important, reproductive damage in the neogastropod *N. lapillus* (Box 14.3) begins at about 0.001 μg/L (1 ng/L) (Bryan and Gibbs, 1991). Populations begin to disappear when females are sterilised at water concentrations of 5 ng/L in natural waters (Langston, 1995) (Table 14.2). Impacts on France's commercial oyster facilities were thought to start at 2 ng/L because *Crassostrea gigas* is sensitive to TBT. In the 1980s it became clear that the concentrations at which TBT was toxic to sensitive species were commonly exceeded by TBT concentrations

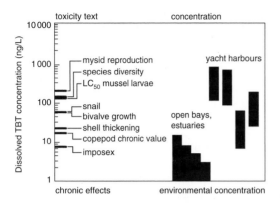

Fig. 14.3 Concentrations of TBT in water at which different sublethal effects occur on different species, compared to concentrations of dissolved TBT observed in open bays or estuaries and yacht harbours (after Huggett *et al.*, 1992). Data are from both laboratory and field observations. Note that imposex, the most sensitive response to TBT, occurs at concentrations common in contaminated environments. The validity of the suggestion of toxicity was verified by widespread observations of imposex and/or disappearance of sensitive species from many such environments.

in harbours, marinas and some more open bays and estuaries (Huggett *et al.*, 1992) (Fig. 14.3). These types of observations drove the final USEPA ambient water quality criteria of 1 ng/L (USEPA, 1997) (Table 14.2).

14.2.6 Ecological observations from natural waters

Prologue. 'Direct evidence of affected animals in the environment rather than extrapolations from laboratory bioassays and field chemistry data resulted in the European regulations. There is some question whether any TBT restrictions, in the United States or abroad, would have occurred had the biological effect as reported in Europe not been so easily observable.' (Huggett *et al.*, 1992)

French biologists were the first to notice abnormalities and a significant decrease in the abundance of Pacific oysters *C. gigas* inhabiting Arcachon Bay and other localities on the Atlantic coast of France in the late 1970s. Rare shell malformations affected 80% to 100% of the surviving animals. TBT was only

suspected to be the cause, however. Nevertheless, TBT was banned in France in 1982 because of the commercial impact. A significant decrease in TBT concentrations in French waters followed within several years, as well as some recovery of oyster production.

Shortly after the French discovery, marine biologist Peter Gibbs noticed an anomaly in female dogwhelks *Nucella lapillus* that he had never seen in all his years of exploring British shores: a 'male' penis. He showed these structures to his colleague Geoff Bryan (also of the Laboratory of the Marine Biological Association of the United Kingdom at Plymouth) who had long studied metals and their effects on marine animals (G.W. Bryan, personal communication). Bryan remembered studies by an American expert on whelk and other snail reproduction, B.S. Smith (1971, 1980), who had found similar abnormalities in the mudsnail *Nassarius obsoletus* in proximity to harbours and marinas in 1981. Smith (1981) felt that antifouling paints on boats might be the cause, but had no direct evidence. Gibbs, Bryan and, later, colleague Bill Langston began a series of studies that combined field observations, laboratory tests and process studies, investigating the relationship between TBT and imposex. Within a few years they, and others, established an unequivocal body of evidence showing that TBT was decimating the intertidal invertebrate communities in the British Isles (Box 14.4). Later studies showed that this was a global problem, wherever small boats were in use. As Huggett *et al.* (1992) noted above, it was this careful combination of scientific approaches, not just extrapolation from toxicity tests or observations alone, that built the body of evidence supporting a largely worldwide risk management action.

14.2.7 Risk management

Prologue. The risk management approach chosen for TBT was a ban that only applied to vessels less than 25 m long. Imposition of the international ban was slowed by arguments citing the cost-effectiveness of TBT paint and the lack of obvious replacements. Total bans on TBT were (and continue to be) proposed but have only recently been agreed to by the necessary fraction of shipping interests.

Box 14.4 Science supporting TBT effects

The sequence of studies that demonstrated the biological impacts of TBT followed the lines of evidence necessary for evidencing effects of metals from a body of work (Chapter 11). The work of Bryan and colleagues started with observations that populations of the dogwhelk *Nucella lapillus* had declined dramatically between the mid-1970s and 1985 in an estuary (the Fal) which Bryan had been studying for years (Bryan *et al.*, 1986). Bryan also verified that the surviving dogwhelks in the estuary had a high frequency of imposex. They described the abnormalities in careful drawings and developed an indexing system by which they could objectively quantify the seriousness of the imposex (Gibbs and Bryan, 1986, 1987). They demonstrated how the abnormality could cause sterility in females and suggested that severe cases could also cause premature death of the female. They also used new (at the time) analytical capabilities to establish that the animals with imposex in the field contained high concentrations of hexane-extractable organotin.

Bryan, Gibbs and Langston conducted experiments to demonstrate that TBT could cause imposex. Laboratory experiments were first conducted by introducing the TBT via a rod painted with antifouling paint (Bryan *et al.*, 1986). They showed that leaching of 0.02 µg/L of Sn as TBT into the water resulted in accumulation of 1 µg Sn as TBT/g dw in the dogwhelks and development of imposex after 4 months (with severity increasing over time).

They also transplanted unaffected animals (from a cleaner estuary) to two contaminated sites (0.04 to 0.35 µg Sn/L, probably as TBT, in the water: Bryan *et al.*, 1986). At both sites they saw an increase in both incidence and severity of imposex in the transplanted animals within a period of 2 to 6 months. They later established that blockage of the oviduct and sterilisation occurred only after 12 months exposure to 0.107 µg Sn as TBT/L in the laboratory, and after 18 months in adults transplanted to a rocky shore exposed to 0.028 µg Sn as TBT/L. Growing juveniles were more sensitive, and could be sterilised by exposure to <0.005 µg Sn as TBT/L. When they allowed animals to lose their burden of TBT by exposure to an uncontaminated system, they found no evidence of remission of the imposex. Thus, with experiments in both the laboratory and the field, they showed that the presence of TBT in concentrations typical of contaminated waters can result in induction of imposex if exposures were of sufficient duration. But it is notable that the time period necessary for expression of the toxicity was longer than typical exposures in traditional tests.

They then used their index and biomonitoring capabilities in a survey of about 70 estuaries and harbours around the southwest peninsula of England (Bryan *et al.*, 1986). They noted that the incidence and severity in one estuary near their laboratory had increased between 1969 and 1985. They found in the survey that the incidence of imposex was widespread and severity correlated with bioaccumulated TBT concentrations, but not with concentrations of other metal contaminants. Dogwhelks in all water bodies were affected to some degree, although the severity of the toxicity differed. They compared an index of abundance with earlier surveys and showed that many of these populations were in decline or animals were absent entirely (about 26 of 34 populations). The stressed populations were characterised by moderate to high degrees of imposex, contained relatively fewer females, few juveniles and a general scarcity of laid egg capsules. All these were signs of reduced reproductive capacity that would reduce recruitment of young into the population and result in a decline in numbers.

Later work by this group and others expanded the effects studies to other species, and to shipping lanes outside confined harbours. The ecological implications of extirpating a keystone predator from an intertidal shore environment were also discussed (Spence *et al.*, 1990).

Early concerns over TBT paints focused on leisure craft which are often moored in large numbers in estuaries, harbours and marinas. Water exchange with the sea in such areas is (intentionally) limited and contaminants can accumulate. In the 1980s water concentrations in such environments were often >1 µg/L, much higher than concentrations that cause reproductive toxicity (Langston, 1995) (Fig. 14.3) National legislation and international directives hence banned use of TBT-based paints on small boats (<25 m) in most of western Europe, North America, Australasia and South Africa, in addition to freshwater lakes in Switzerland, Germany and Austria. Regulations in the USA and

Europe have also restricted release rates of TBT from paints, and eliminated the use of free TBT in paint (Champ, 2000).

Water quality criteria were also developed for TBT. The original US Environmental Protection Agency water quality criterion for TBT in 1997 was 0.01 µg/L (Meador, 2002). Because of concern about continuing TBT inputs from commercial shipping, the standard was reconsidered in 2002. USEPA (2002a) concluded:

- TBT is an immunosuppressing agent and an endocrine disruptor;
- TBT biomagnifies through the food chain and has been found in tissues of marine mammals;
- TBT causes adverse reproductive and developmental effects in aquatic organisms at very low concentrations;
- TBT degrades much more slowly in sediment than earlier studies had indicated, and is likely to persist in sediments at concentrations which cause adverse biological effects.

USEPA (2002a) based their conclusion that the standard should be lowered on the following lines of evidence:

- the traditional endpoints of adverse effects on survival, growth and reproduction as demonstrated in numerous laboratory studies;
- the endocrine disrupting capability of TBT as observed in the production of imposex in field studies;
- the bioaccumulation of TBT in commercially and recreationally important freshwater and saltwater species;
- the vulnerability of an important commercial organism to a prevalent pathogen that was made even more vulnerable by prior exposure to TBT.

For these reasons, the criterion to protect saltwater aquatic life from chronic toxic effects was set at 0.001 µg/L (1 ng/L). The inclusion of non-traditional evidence (experimental data other than standardised toxicity tests; sublethal field observations) is notable compared to the way standards are set for many other metals, at least by the US Environmental Protection Agency (Chapter 15). In addition to the Aquatic Water Quality Criterion, Meador (2002) suggested a sediment quality criterion of 0.12 µg/g dw at 2% TOC (half that at 1% TOC).

Organotin is still used as an antifoulant on commercial and military vessels. It is estimated that such vessels account for 90% of the market for antifouling paints and two-thirds of these contain TBT (Langston, 1995). The justification for retaining these uses was that such vessels traverse the open seas where dilution is much greater. But they do spend time in harbours and dry docks. Langston (1995) reported TBT concentrations of 0.100 µg/L in 1994 in a commercial maintenance facility in Southampton, UK, a value equivalent to that in marinas in the 1980s. But the number of facilities with this degree of contamination is certainly much smaller than it once was.

In 2001 it was clear that imposex (although not necessarily population damage) was being observed in snails from shipping lanes in open waters (e.g. Ide et al., 1997). The International Maritime Organisation (IMO) adopted the International Convention on the Control of Harmful Antifouling Systems on Ships (AFS Convention) as a result. Once in force the AFS Convention would ban the application on all ships of TBT-based antifouling paints. The AFS Convention was designed to enter into force 1 year after 25 nations representing at least 25% of the world's shipping tonnage ratified it. In early 2007, 23 states representing 17.06% of the world's shipping tonnage had ratified the AFS Convention. Fifteen jurisdictions representing 10.5% of the world's shipping tonnage had not ratified. Germany, the UK and the USA have not ratified. Australia, France and the Scandinavian countries have. In October 2007, Panama signed the agreement and the required proportion of world shipping was reached, suggesting the global ban of TBT on all shipping will go into effect in 2008.

The European Union has banned the application of TBT-based paints on EU-flagged vessels and, as of 1 January 2008, it will be an offence for any ship visiting an EU port to have TBT present on its hull. But the European TBT ban will only be effective if it is properly enforced in EU ports. That will require new protocols and testing with entirely new procedures (Seas at Risk, 2007).

14.2.8   Recovery

Prologue. Incidence rates of imposex are declining and populations of locally extirpated species are beginning to reappear after the partial ban of TBT. Full recovery has not occurred, however.

Because the signs of TBT toxicity are relatively easy to observe, studies of its effects continued throughout the world after the ban. Within a decade, signs of TBT toxicity in molluscs, once evident in Europe, North America, Australia and Asia, began to disappear. For example, Evans *et al.* (1994) reported evidence of recovery from signs of imposex in dogwhelks (*Nucella lapillus*) in the waters of the Isle of Cumbrae in Scotland as early as 1993. The numbers of dogwhelks on the shore also had increased. At a site where dogwhelks were absent in 1988, they were abundant in 1992.

TBT concentrations in waters of marinas, harbours and estuaries recorded the most obvious declines. Concentrations in estuaries of southern England provide a good example of trends (Langston, 1996) (Fig. 14.4). Concentrations at one site there (Lymington), for example, declined from 0.200 to 0.007 µg/L TBT between 1988 and 1994. However the rate of decline slowed in the 1990s. Predictions from early data for the Hamble estuary (south England), for example, were that recovery would be complete by 1995 (concentrations below the criterion of 0.002 µg/L) (Fig. 14.4). However, data from 1992 to 1994 suggested that this concentration would not be reached until 2011 (Langston, 1996). Recovery from metal contamination often follows such trends, reflecting continuing inputs from the legacy of contamination in the sediments as well as continued inputs from other sources than those controlled.

The restrictions on TBT use appeared to reverse damage to oyster stocks in the north Atlantic. Recruitment of larvae and production of marketable shellfish had returned in France in 1996, although some shell chambering was still evident in some bays (Langston, 1996). A lower incidence of imposex and reproductive impairment in dogwhelks was one of the first signs of the recovery in the UK and France (Fig. 14.5). Recolonisation

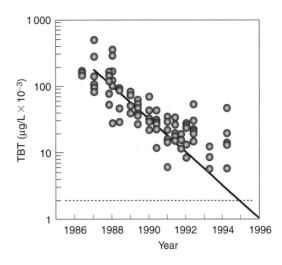

Fig. 14.4 TBT concentrations in water in several estuaries from south England, between 1986 and 1994, after restrictions on the use of TBT-based paints on leisure vessels (after Langston, 1996). The dashed horizontal line represents the 0.002 µg/L protective criterion level. Trend line represents predictions of future concentrations based upon all the data. Between 1990 and 1994 the decline slowed considerably, however, extending the time before contamination reached the protective criterion level.

of areas previously devoid of *N. lapillus* had occurred in some areas by 1996, but was slowed in others by the inherently slow dispersal of this species which lacks a planktonic larval phase. The irreversibility of imposex was also a factor because adults themselves had to die before they could be replaced by animals capable of reproduction (Langston, 1995). Completely unaffected *N. lapillus* were difficult to find, even in 1996, in southwest England (Langston, 1996), and some of the places with the heaviest boat traffic still did not have viable dogwhelk populations.

An increasing number of studies illustrate the characteristics of the recovery (Box 14.5):

- an early reduction in severity of symptoms;
- a slower improvement in the frequency of symptoms (reduced percentage of female gastropods with signs of imposex);
- recovery most evident in areas frequented primarily by pleasure craft and in coastal areas

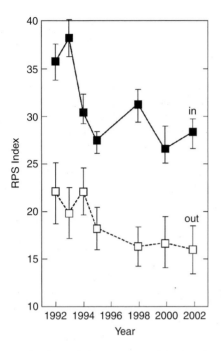

Fig. 14.5 The change between 1992 and 2002 in the relative penis size (RPS) in female dogwhelks (*Nucella lapillus*) in (in) and around (out) the Bay of Brest (France). RPS is a measure of imposex induced by TBT exposure. (After Huet *et al.*, 2004.)

suffering from secondary contamination (contaminants transported in from bays, harbours and estuaries);

• slowest recovery, or no sign of recovery, in hotspots (or shipping lanes) where commercial traffic remained heavy or where scrapings from boat bottoms had accumulated.

14.2.9   Was the partial ban adequate?

Prologue. 'The human-aided spread of species beyond their natural range is a significant form of global change and a major threat to biodiversity' (Nehring, 2005). Fouling on ship hulls is major vector by which exotic species invade habitats where they previously did not occur. When TBT was in use, ballast water was a greater source of such invasive organisms than ship hulls. Transport of invasive species on ship hulls now has returned as the primary

vector where TBT paints are not in use. A ban of TBT use on larger vessels could exacerbate the invasive species problem, more so than a ban on use on smaller vessels. This has been an unanticipated trade-off of the choice of risk management strategies.

One side effect of banning TBT was that it became easier for marine organisms to be transported from one place to another, at least on small vessels. Thus the ban increased the risk of invasions of harbours by exotic (non-native) species (Nehring, 2001). Fouling organisms were a major component of invasive species before the use of TBT. The effectiveness of TBT caused a shift in the major vector of introduction to ballast water, because so many fewer organisms were carried on the hulls of ships. In a 1992 to 1996 survey of 186 vessels, Gollasch (2002) found that hulls, once again, were the major source of transport of invading species. Ships lacking TBT protection had a higher diversity of species attached to their hulls than ships that were coated with TBT paint. Minchin and Gollasch (2003) summarised the converging events that have led to invasive species becoming a problem of increasing concern:

> ships today are generally larger, and faster, and have a high frequency of port visits thereby increasing the number of spawning opportunities, perhaps with a larger inoculum size. With trade expansion, new trading routes, political events and changes in climate, new pathways for invasion will emerge. Greater controls on industrial discharges, improved treatments of urban wastes and better management of waste runoff into rivers as well as a phasing out of organotin antifoulants will mean a reduced toxicity in port regions. This may enable a smaller inoculum to colonize by creating opportunities for establishment not present in the previous 25 years.

International shipping represents the most important vector of introduction for aquatic alien species to harbours, estuaries and coastal waters throughout the world (Nehring, 2005). Thus, it might be argued that had the ban on TBT been extended to large ships, which travel much farther with

Box 14.5 Studies documenting partial recoveries from TBT contamination

Several studies from around the world have documented recoveries from TBT damage in predatory gastropods, which were the most heavily affected. Recoveries of *Thais* species provide an example from Australia. By 1999, imposex had declined at 11 sites near Perth, Western Australia. But, 100% of females showed signs of imposex at sites close to commercial harbours and docks. Approximately 30% of these had reached the stage of effective sterility (Reitsema *et al.*, 2003). It was concluded that recovery was greatest in areas dominated by the use of smaller pleasure craft, while sites associated with commercial shipping activities showed little change. Poorly managed slipways where vessel hulls are scraped down and resprayed in open air conditions were also a likely point source of TBT.

Gibson and Wilson (2003) found no decline in the frequency of imposex in Sydney Harbour, Australia (which has both pleasure craft and commercial vessels) but a consistent decline in severity. On the open coast outside the harbour they found a decline in frequency of imposex. They also cited 'hotspots' as a concern.

In England the bivalve *Scrobicularia plana* has returned in some abundance to estuaries of southwest England from which it had disappeared in the 1980s (S. N. Luoma and P. S. Rainbow, personal observation). These small estuaries are primarily used by pleasure craft.

Huet *et al.* (2004) conducted a 10 year study at 56 locations in northwest Brittany (France), beginning in 1992. They assessed the level of imposex in the *N. lapillus* populations, primarily in the rich fishing waters in and around the Bay of Brest, France. The Bay of Brest was highly polluted by TBT, and its contamination affected nearby coastal waters in 1992. At three locations within the Bay of Brest, new populations of *N. lapillus* appeared after 1998. At other stations, numbers of individual dogwhelks increased and the percentage of less severely affected females also increased (Fig. 14.5). Nevertheless, imposex in *N. lapillus* populations remained prevalent inside the Bay; no population was free of symptoms. Even if the percentages of sterility decreased, sterility is still common. Because the Bay of Brest has a great deal of military and commercial traffic, reformulations of antifouling paints with slower leaching rates may have contributed to the improvements more than the ban of uses on leisure craft.

great frequency than leisure craft, this side effect could have been exacerbated. This hypothesis will presumably be tested beginning in 2008.

The point is not that the ban on TBT was a poor policy choice. Some ban was, in fact, essential. But managing every environmental risk comes with trade-offs. Risk management for TBT differed from the complete ban imposed for chemicals like DDT (Chapter 1; Box 1.2). Despite arguments to the contrary (e.g. Champ, 2000) both bans were ultimately based upon a strong, interdisciplinary and convincing body of scientific evidence. There was minimal uncertainty that these were serious hazards and that the risks were leading to actual effects on populations of important species. Aggressive action was justified.

The reaction in the case of DDT was a ban on all uses; the maximum application of precaution. An important trade-off of the complete ban was an increase in malaria in people from poor tropical regions. But, closer study shows that the most

dispersive uses of DDT were agricultural; not those used in controlling vectors that were threats to human health. The complete ban of the use of DDT resulted in some loss of a benefit that might have cost human lives (Box 1.2). In 2002, the World Health Organisation (United Nations) approved use of indoor spraying for DDT.

The reaction to the TBT threat was a partial ban (a more step-by-step approach than the ban on DDT). It appears that adverse ecological effects are at least partly being reversed, although the full story is not yet in. The partial ban may have avoided worsening the unanticipated side effect of increased invasions by exotic species, although more study of the trade-offs with continuing TBT use on international shipping is also necessary. Advocates of a total ban on TBT, as proposed by the IMO, cite high levels of TBT in commercial ports and continued observations of imposex in shipping lanes of open waters. Nehring (2000) showed that populations of prosobranch gastropod snails in the North Sea (outside ports) are declining and cited TBT as the most likely explanation. On the other hand, proponents of TBT use argue that there has not been a full discussion of the ecological risks and other costs of antifouling alternatives; nor has full consideration been given to the economic benefits of TBT use (Champ, 2000). In any case, complex policy trade-offs can still exist in the modern world, even when the science is relatively clear.

14.2.10 Conclusions: tributyl tin

Tributyl tin is one of the most effective antifoulants ever applied to the bottoms of boats. It is extremely toxic to invertebrates which comprise most fouling organisms. It is relatively non-toxic to vertebrates because of their ability to metabolise TBT into non-toxic forms. TBT is also economically attractive as an antifoulant because it does not need to be reapplied for several years. The same toxicity that made it a good antifoulant also made it a species-specific environmental hazard.

A combination of experimentation and field observation identified the characteristics of TBT toxicity and unambiguously demonstrated that it was the cause of readily observable declines in important invertebrate species, including commercially grown oysters. They showed that TBT is especially strongly bioaccumulated by those invertebrates that lack the ability to metabolise it; in particular molluscs. TBT concentrates in sediments and is primarily bioaccumulated from the diet of affected organisms. Some of the first effects were observed in invertebrate predators, further suggesting an important role for trophic transfer. Bioaccumulation in vulnerable species exceeds that predicted by equilibrium partitioning; but does follow the expectations from a biodynamically driven process. Thus this toxin has characteristics of both an organic contaminant and a metal contaminant. TBT is an endocrine disrupter; thus its toxicity is manifested by effects on the reproductive system. Adults do not die unless the contamination is extreme but their ability to reproduce is eliminated. Therefore the population disappears within a few generations. The ecological risks from TBT were not detectable from toxicity tests used for the registration process. Field observations led to the discovery of its ecological impacts (similar to events with organochlorine pesticides).

Perhaps because effects were so obvious and the scientific case for cause and effect was so strong, TBT use on small boats was banned within a decade of the discovery of its effects. That ban was effective in reducing the frequency of reproductive effects throughout the world, although it did not eliminate the problem. The ban appeared to have the side effect of reinvigorating invasions of harbours by exotic fouling species. In 2008, an international ban on TBT usage will take effect. It is likely that this will control all adverse effects from the compound, if it is enforceable. Monitoring of invasive species should be an important requisite for large harbours heretofore protected by TBT usage on large ships.

14.3   METHYL MERCURY

14.3.1   Introduction

Methylmercury is a potent toxin, bioaccumulated and concentrated through the aquatic food chain, placing at risk people, throughout the globe and across the socioeconomic spectrum, who consume predatory fish or for whom fish is a dietary mainstay. Methylmercury developmental neurotoxicity has constituted the basis for risk assessments and public health policies. Despite gaps in our knowledge on new bioindicators of exposure, factors that influence MeHg uptake and toxicity, toxicokinetics, neurologic and cardiovascular effects in adult populations, and the nutritional benefits and risks from the large number of marine and freshwater fish and fish-eating species, the panel concluded that to preserve human health, all efforts need to be made to reduce and eliminate sources of exposure.

Mergler *et al.* (2007)

The properties of mercury were described by Hippocrates in 400 BC. The metal has been mined in Almadén, Spain for 27 centuries (Clarkson, 2002), but its hazard potential and ecological risks have been recognised for nearly as long. The greatest ecological and health risks from mercury come from methyl mercury (MeHg), which is a neurotoxin that readily crosses biological membranes, can accumulate to harmful concentrations in exposed organisms, and biomagnifies to concentrations of toxicological concern in aquatic food webs.

Of all the metals Hg is, arguably, the most hazardous to human health. Its potency as a neurotoxin was recognised in the person of the 'mad hatter' in *Alice in Wonderland* who represented mercury-poisoned workers in the hat trade. Good top hats at the time in Britain were made from beaver fur, but cheaper ones used furs such as rabbit instead. With the cheaper furs, it was necessary to brush a solution of mercurous nitrate onto the fur to roughen the fibres so that they would mat more easily. Beaver fur had natural serrated edges that made this process unnecessary, but the cost and scarcity of beaver meant that the alternative furs had to be used, and hatters consequently breathed in toxic Hg fumes. Musicians even speculate that Beethoven's deafness may have originated from wearing such hats. Death and disease resulting from the industrial discharge of mercury waste products into Minamata Bay, Japan were discussed in Chapter 1. Although two laboratory technicians died of poisoning creating the first synthetic, short chain, organic (alkyl) mercury compounds in the 1860s, cereal crops were treated with the compound to destroy fungus until outbreaks of poisoning and ecological destruction of seed-eating birds made its dangers clear (Clarkson, 2002). Most notably an outbreak in rural Iraq in 1971 to 1972 poisoned as many as 40 000 people when bread was made directly from Hg-treated grain intended for use as seed.

MeHg is generated by bacteria in natural waters; and that form is difficult to control. Discovery of high concentrations of Hg in fish from lakes in Sweden and Canada and in the Great Lakes, USA led to the closure of fisheries in the 1970s. The developing foetus and children continue to be at risk, particularly among people who eat large quantities of fish; whether they be wealthy urban consumers eating ocean fish like tuna or swordfish (Hightower and Moore, 2003), or indigenous people who depend upon lakes and rivers for their subsistence. Those risks were known in the 1970s but have been rediscovered in the last decade (Mergler *et al.*, 2007). MeHg is also considered to be the metal posing the greatest ecological hazard, in particular threatening charismatic predators, especially fish-eating wildlife (Heinz and Hoffman, 1998). That will be the focus of the second part of this chapter.

Understanding the chemistry, biology and ecology of MeHg is critical to managing this metal. Its chemistry leads to a sizeable atmospheric presence, making Hg a global problem. Like TBT, its worst effects are detectable only from the study of wildlife and from observations from natural waters. Traditional single-species dissolved tests do not address the mechanisms defining MeHg's hazard and risks.

14.3.2 Sources of mercury

Prologue. Conclusions of an expert panel on source attribution of mercury (Lindberg *et al.*, 2007):

- North American and European emissions of Hg are declining while those of Asia and Africa are increasing, but the latter changes are less well known.
- The overall global increase in Hg deposition (from atmosphere to Earth) since the Industrial Revolution is 3 ± 1 times higher than natural deposition rates.
- Atmospheric deposition of Hg contains a significant fraction from anthropogenic sources, even at the most remote locations.
- There has been no discernible net change in the size of the atmospheric pool of Hg in the northern hemisphere since the mid-1970s because of the complexity of global emission trends.
- Decreases or increases in local emissions result in decreased or increased contamination in the nearby environments and food webs.
- Depositional processes and their impacts are greatly influenced by the biogeochemical speciation of Hg.

Coherent management policies for a contaminant ultimately depend upon identification of sources. For Hg this is a complex question. Hg is a waste product from specific human activities. But understanding sources also requires understanding the global cycle of Hg, biogeochemical speciation of the element and transport between different compartments of the environment. The mass of input has significance at the local, regional and global scale. Each scale is important in a different way to the impacts of the contaminant. The form of Hg also has much to do with its fate and effects.

14.3.2.1 The global mercury cycle

Prologue. The greatest masses of mercury occur as elemental Hg (Hg^0) and reactive species of divalent Hg (Hg(II)). The natural mercury cycle is dominated by large reservoirs of Hg(II) in soils and in the oceans. Smaller, rapidly exchanging reservoirs also occur in the atmosphere, in the form of mostly Hg^0; and in surface waters, in the form of reactive Hg(II). Anthropogenic inputs have increased Hg fluxes to the atmosphere and increased the size of the atmospheric reservoir by two to four times (generally assumed to be threefold). Atmospheric Hg is especially important because the Hg that accumulates there is ultimately deposited in remote locations, far from sources of input.

The global cycle is ultimately driven by the fundamental characteristics of the different species of Hg. Elemental Hg^0 is the only metal that is a liquid at room temperature, and also occurs as a vapour or gas. It has little tendency to dissolve in water (Morel et al., 1998). This results in rapid volatilisation or flux of Hg^0 out of water into air. Approximately 95% of total mercury in the atmosphere is in the elemental state, where it is slowly oxidised to the mercuric Hg(II) form, and ultimately redeposited. In water the dominant forms are reactive species, Hg(II), which like other metals can be associated with complexes like dissolved organic material, chlorides, hydroxides, or even dissolved sulphides. Speciation processes are driven by geochemistry and dependent upon the characteristics of the water body.

At the concentrations typical of natural waters, reactive Hg(II) can be reduced to Hg^0 by either microbes or by geochemical processes enhanced by sunlight, under certain conditions (Morel et al., 1998). The result of the reduction process is loss of Hg^0 into the gas form (ultimately into the atmosphere). But Hg(II) can be stabilised by formation of strong chloro-complexes or complexes with dissolved organic matter, avoiding reduction to Hg^0. The concentration of Hg(II) in a water body thus depends upon inputs, speciation, pH and the balance between oxidation and reduction.

The atmosphere, surface soils and both surface and deep ocean waters are the major compartments in the global Hg cycle (Chapter 4; Mason et al., 1994). The masses of Hg in the atmosphere and the surface of the oceans track changes in inputs within a few years, because Hg cycles rapidly through these compartments. Changes in the masses of Hg in the deep oceans and soils occur very slowly. For examples, declines in anthropogenic Hg inputs could take one or two centuries to be reflected in deep ocean waters. Terrestrially deposited anthropogenic Hg is probably even more resistant to change (Lindberg et al., 2007).

Natural inputs of Hg to the atmosphere include emission of Hg from the land and the sea. Wind entrainment of dust particles from land is a source, especially from areas with Hg-rich soils. Volcanic eruptions, forest fires and biogenic emissions are other natural sources. When natural deposits of Hg from deep in the earth are exposed by mining, a small potential natural source becomes a large anthropogenic source. Other anthropogenic sources include metal production, pulp and paper facilities that use Hg in a chlor-alkali process, and waste handling and treatment facilities (e.g. incinerators especially). The energy sector, especially combustion of coal, peat and wood for energy, is one of the most important anthropogenic sources.

Mercury is typically emitted into the air from anthropogenic source as gaseous elemental Hg^0 and divalent, reactive Hg(II). Hg^0 retains its gaseous form and remains in the atmosphere for long periods (Lindberg et al., 2007), creating a reservoir that spreads globally as air masses move. The elemental vapour form in the atmosphere exchanges with the oceans and the surface soils of the land via slow oxidation of Hg^0 to Hg(II) (see cycle in

Chapter 4). This source of deposition is spread evenly over the Earth with the most important ramifications occurring in remote areas with no local Hg sources. In contrast, Hg(II) becomes associated with particulates and usually deposits relatively rapidly out of the atmosphere, but in more concentrated deposits.

Anthropogenic activities have increased Hg fluxes into the atmosphere. Estimates are that 3.57 million tons of anthropogenic Hg are maintained in the atmosphere, compared to a natural reservoir (pre-human) of 1.6 million tons. Experts are thus relatively certain that two-thirds of the Hg in the atmosphere is of human origin. Humans have added 25 million tons of Hg to the natural reservoir of 302 million tons in the oceans (data from Lindberg *et al.*, 2007).

One of the results of the global spread of Hg^0 is that lakes, ocean waters and soils far from any Hg source can be Hg contaminated under appropriate conditions. The effects of the enriched atmospheric reservoir of Hg are felt in northern Canada, Arctic and Antarctic seas, and especially the open oceans. Concentrations of Hg in fish and other predators in such locations can exceed the safety standard for human consumption, even if there are no known Hg sources for long distances (Fitzgerald *et al.*, 1998).

14.3.2.2 Historical mercury sources
Prologue. Hundreds of thousands of tons of Hg have been emitted to the environment by human activities over the last five centuries, mostly from mining for silver and gold.

In contrast to TBT, the history of human contamination of the Earth with Hg is a long one. Silver and gold have long been precious commodities to humans. The silver market has been dominated by sources from North and South America since 1570. The availability of these precious metals stems from mercury amalgamation as a method for extracting them from relatively low-grade ores. When ores are passed over a bed of Hg, the Ag and Au react chemically with the Hg and are retained while the remainder of the ore material is discarded. In the crudest methods, the Hg is collected and volatilised, leaving behind the precious metals.

Estimates from the eighteenth century are that 1 kg of Hg was lost to the environment for every 6 to 8 kg of ore that was processed, or 1.5 kg of Hg was lost for every 1 kg of Ag produced. Efficiencies improved in the nineteenth century with an estimated loss of about 1 kg Hg for every ton of Au ore (Nriagu, 1994).

For example, between 1850 and 1900, the dominant sources of Hg in the world were mines in the Americas. About 90% of that Hg was used in the recovery of Au and Ag (Nriagu, 1994). Nriagu (1994) estimated that average atmospheric emission rates of Hg from North America at that time were 780 tons/yr. The result was discharge to the environment of somewhere between 245 and 2540 tons of Hg per year. This wide dissemination of Hg is reflected in data from cores collected from long-standing glaciers. One such example is an ice core from Fremont Glacier in Wyoming, USA. The increase in atmospheric Hg was sufficient during the period between 1850 and 1900 to create a notable peak in Hg concentrations in ice deposited at that time (Schuster *et al.*, 2002) (Fig. 14. 6). The peak reflects at least a regional influence of the mining on Hg^0 in the atmosphere, if not a global influence.

In all, estimates are that 257 000 tons of Hg were lost to the environment from the entire period of historical mercury mining in the Americas (Nriagu, 1994). Large Hg mines elsewhere report similar inputs. For example, the Indrija mercury mine in Slovenia, in use since 1490, produced 107 000 tons of Hg over its five century history.

14.3.2.3 Modern mercury sources
Prologue. Modern inputs of Hg from the traditionally industrialised nations are declining, but inputs from rapidly growing economies increasing. Thus, global trends in the mass of Hg in the biosphere are not clear or at least not unidirectional. Future trends will be influenced by factors such as trends in the nature of gold mining and the degree of mercury removal from the wastes of coal-fired power generators.

Modern sources of Hg are relatively well known (Lindberg *et al.*, 2007). Despite activities in the Americas and Europe over the last century, source

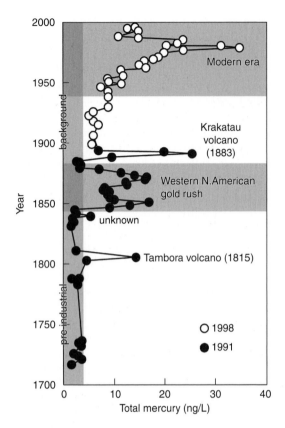

Fig. 14.6 Hg concentrations in dated cores (1991, 1998) from the Upper Fremont Glacier in Wyoming, western USA. Trends reflect effects from volcanoes, the 1850 to 1885 gold rush in the western USA (where Hg was used to extract Au from the ore), and the growth in modern discharges of Hg until the 1990s and the subsequent regional decline in usage. (After Schuster *et al.*, 2002.)

inputs from North America and Europe are now declining. Future increases in inputs with new technologies (e.g. nanotechnologies) are possible but cannot be forecast quantitatively. Data from ice cores, lake sediments and ocean waters show a downward trend in areas dominated by inputs from North America and Europe, beginning in the early 1990s (Fig. 14.6). On average, global gaseous Hg concentrations in the atmosphere are not declining, however (Lindberg *et al.*, 2007). Emissions of Hg from dynamic economies such as China, India, Korea and others are increasing, although there is only a limited amount of quantitative data.

Hg inputs to the atmosphere from Asia overall are estimated at 1460 tons per year (Renner, 2005). Asia contributed 56% of worldwide Hg emissions in 1995, compared to 30% in 1990. The forms in the atmosphere suggest that much of the Hg emitted from Asia is redeposited in the nearby ocean (and probably also onto the soils in that continent) (Renner, 2005). Trends in economic development suggest that these globally significant Hg releases will continue to increase.

China is now regarded as the largest anthropogenic source of Hg. Total Hg consumed in China was >900 tons in 2000 (Jiang *et al.*, 2006). Total Hg emissions from all anthropogenic sources in China increased at an average annual rate of 2.9% during the period 1995 to 2003, reaching 696 ($\pm$307) tons in 2003 (Wu *et al.*, 2006). As a comparison, average atmospheric discharges of Hg from the USA were 260 to 600 tons per year in the 1990s and 120 tons in 2004 (Johnson, 2003). China is also rich in Hg deposits suitable for mining, having the third largest deposits in the world. Large-scale mining operations release 0.79 g Hg/g Au produced (Jiang *et al.*, 2006). But very small mining by artisans is also traditional, and releases much more Hg than the large operations (about 15 g Hg/g Au).

Coal combustion is an important source of Hg in Asia. China consumed 1531 million tons of coal (28% of world consumption) in 2000, and this is expected to double to 3037 million tons of coal by 2020. Total Hg emission from coal combustion in China in 1995 was about 303 t including 213.8 t into the atmosphere and 89.1 t into ash and cinder (Wang *et al.*, 2000). Hg emission from power generation was growing 5.9% annually in 2003 (Wu *et al.*, 2006).

Just as in China, coal combustion is expected to grow rapidly throughout the world in the decades ahead. Traditional coal-fired boilers emit 64.0% to 78.2% of the (average) 220 µg/g Hg in coal. But new technologies can remove Hg more efficiently. Future trends in Hg emissions are dependent upon political and economic decisions about whether to employ those technologies. Policy debates about such controls (Box 14.6) are occurring throughout the world; outcomes in one region are likely to affect decisions in others.

Box 14.6 *Modern sources: policy*

Rates of Hg input to the atmosphere from North America and Europe have decreased over the last 20 years as authorities have progressively set standards for Hg emission from various sources. But concerns remain about accumulation of Hg in hotspots near sources of emission, as well as the global contribution.

The successes in controlling Hg emissions in the USA stem from the Clean Air Act of 1990 wherein emitters of hazardous substances were required to install 'maximum achievable control technology (MACT)'. Rapid inclusion of better technology resulted in relatively rapid declines in emission. Coal-fired power plants are the last unregulated source of Hg air emissions in the USA. They were given a temporary release from the 1990 law, but that release never ended. The coal plants emit about 48 tons of Hg per year, about one-third of total US emissions in 2004. Coal produces 53% of US electricity (Johnson, 2003, 2004, 2005). That contribution will undoubtedly grow in the decades ahead. Many coal-fired power plants are more than 40 years old and have limited pollution abatement equipment for Hg. Hg capture varies widely among these plants: from 10% to 90%. But technology exists to consistently capture 90%, with investment in activated carbon scrubbing.

Controlling coal plant Hg emissions became controversial in 2004 when the Bush administration changed the status of Hg so it was not covered by the MACT law. The administration suggested that Hg emissions could be controlled secondarily by improving controls over other air pollutants. On 15 March 2005, the Environmental Protection Agency issued the Clean Air Mercury Rule in which they imposed a cap on overall Hg emissions at about the level it was thought could be achieved by the secondary controls. Then they set up a trading system for Hg credits that, it was proposed, would be a theoretical incentive to install the most efficient controls. The costs of such an approach would be relatively modest. The cap proposed for the trading was 34 tons. The Environmental Protection Agency administrator suggested that emissions would be reduced to 15 tons by 2018 following this approach.

Opponents say that a 90% reduction in emissions (down to <5 tons) could be achieved within a few years by imposing the MACT approach, and that trading credits has no place in the management of a neurotoxin. They also suggest that the cap proposed by the administration, under which Hg would be traded, is so large it will create no incentive for a viable trading scheme. They argue that the recent actions would allow existing hotspots of Hg contamination to persist. Perhaps more important, the failure to directly address Hg emissions could set a technological precedent for nations that have not yet imposed controls. China, for example, is building power plants at an extremely rapid rate. Their decisions about technologies will affect Hg emissions for decades into the future, with important implications for the global Hg cycle. Coal is an upcoming source of energy everywhere in the world; and the installation of modern technology is the only way to control Hg contamination as that energy source grows.

The European Union is proposing an international initiative to phase out Hg production (Stokstad, 2005). The EU would permanently close a major mine in Spain and ban all exports from the EU, the world's largest modern source of Hg, starting in 2011. The Bush administration in 2004, in contrast, favoured a voluntary programme to control Hg emissions internationally.

It can be difficult to determine the influences of the globally complex emission trends on Hg contamination. One reason is that regional influences are difficult to separate from global influences at any particular sampling location. For example, the suspected increase in Hg emissions from Asia have not been detectable in rain or gaseous Hg in North America or Europe, suggesting little response yet in global atmospheric Hg (Lindberg *et al.*, 2007). Hg concentrations in the upper ocean near Bermuda declined between 1979 and 1999 to 2000. But this probably reflects trends in the North America atmosphere rather than the global atmosphere (Mason and Gill, 2005; analogous to Pb (Chapter 5). Atmospheric concentrations of Hg at what was thought to be a remote site at Rörvik, Sweden showed increasing trends in the mid-1980s, then declined between 1987 and 1997. It is now recognised that this site was recording emissions from eastern Europe before and after the break-up of the Soviet

Union (Lindberg *et al.*, 2007). As noted above, the overall conclusion is that global atmospheric concentrations of Hg are not changing because new inputs are balancing declines in traditional inputs.

It is important to differentiate the global cycle and the local cycles of Hg in terms of where impacts occur and their characteristics. Reducing or increasing smaller Hg emissions might seem irrelevant on a global scale because of the large reservoirs, slow transport rates and slow exchange rates in the global Hg budget. But regional and local impacts can be important. Although half of the Hg emitted by anthropogenic sources spreads considerable distances, the rest contributes to hotspots of local or regional contamination. With increases in Hg emissions from Asia, perhaps the greatest ecological and human health concern is an increase of Hg contamination in hotspots near the sources of emission (Jiang *et al.*, 2006). Increasing overall emissions of a metal contaminant are usually indicative of a high frequency and great intensity of local hotspots.

Watersheds are a source of Hg to freshwater environments via runoff. In the USA, about 0.3 to 30 $\mu g/m^2/$ yr is deposited into watershed soils and local water bodies (USEPA, 1997). The characteristics of the watershed determine the release of Hg to the aquatic environment. The size and topography of the watershed, land cover and land use are all important (Munthe *et al.*, 2007). Soils can hold onto Hg and delay recovery of water bodies receiving their runoff. But if erosion is accelerated (e.g. by clearing of forests), Hg inputs to associated water bodies are accelerated. For example, lakes with logged watersheds in Canada have higher Hg concentrations in predatory fish than lakes whose watersheds have not been logged (Garcia and Carignan, 2000).

14.3.3   Mercury methylation

Methylation is the 'bridge between the seemingly incongruent observations that inorganic Hg is the dominant form released to the environment, and MeHg is the dominant form of Hg found in edible fish' (Krabbenhoft *et al.*, 1999).

Prologue. MeHg is formed primarily by sulphate reducing bacteria in anoxic sediments. It is a toxin of primary concern in aquatic environments, although the other forms play major roles in the Hg cycle. In freshwaters, mercury methylation is greatest in watersheds with significant wetland density, organic sediments, and with low surface water pH. The influence of wetland abundance on MeHg is broadly important because human activities have strong positive and negative influences on wetlands.

The most toxic form of Hg is methylated or methyl Hg (MeHg). Methylation is defined as the transfer of a methyl (CH_3) group from an organic compound to the metal ion. This reaction is not favoured by chemical thermodynamics but can be enhanced (catalysed) when microorganisms create carbon-to-metal bonds that are stable in water because they are partly covalent. Hydrolysis reactions that might break the bond are very slow (Morel *et al.*, 1998). Thus the source of MeHg is a natural process. But when human activities add inorganic mercury to the environment, the most important outcome is generation of higher concentrations of MeHg.

Several lines of evidence show that specific microbes that reduce sulphate (or Fe) are responsible for most Hg methylation in natural waters. They methylate Hg in culture and when molybdate, a specific inhibitor of sulphate-reducing bacteria, is added to the culture Hg methylation is inhibited (Morel *et al.*, 1998). Methylation rates correlate in time and space with the abundance and activity of sulphate-reducing bacteria, although in full-salinity seawater, sulphate reduction may not be especially efficient. The balance between bacterial formation of MeHg and bacterial MeHg breakdown (demethylation) determines net MeHg inputs to the environment. Demethylation is accomplished by a variety of mechanisms, the best-studied being in microbes that have a gene (*mer* operon) that allows them to detoxify the metal (Marvin-DiPasquale *et al.*, 2000). Detoxification processes include reduction of MeHg to Hg^0, formation of Hg sulphide or extracellular oxidation.

Complex processes control the net methylation of Hg in rivers, lakes, estuaries and the oceans. In freshwaters, estuaries and coastal zones, methylation occurs in sediments. The source of MeHg to ocean fish is less certain. Rates of net MeHg formation from different types of sediments vary widely. The concentration of inorganic Hg is important but so are environmental conditions (Munthe *et al.*, 2007). Influential environmental factors include sulphur, pH, organic matter, Fe and the type of bacteria present. Net methylation

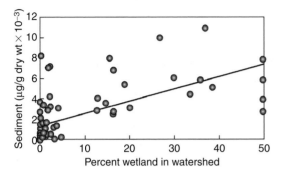

Fig. 14.7 Relationship between sediment methyl Hg concentrations in rivers from across North America and percent wetlands covering the watershed upstream of the sample in those rivers. $R^2 = 0.41$. Watersheds with more wetlands produce more MeHg. (After Krabbenhoft *et al.*, 1999.)

is favoured under anoxic conditions. Wetland sediments, which have zones of anoxia near the surface and tend to be organically rich, generate MeHg more quickly than other types of sediments. The area of wetland in a watershed, indeed, is the single most important basin-scale factor controlling MeHg production (Krabbenhoft *et al.*, 1999) (Fig. 14.7). Artificial wetland ecosystems are often constructed to replace those destroyed by development; but if the 'raw materials' for the artificial wetland include Hg contaminated water or sediment, MeHg loads to the water body can be increased (Box 14.7). Warm, shallow, organically rich lake sediments are also important zones of methylation. When such environments are created by damming and flooding, the result is generation of MeHg (Box 14.7).

14.3.4 Concentrations

Prologue. Understanding the range of total Hg and MeHg concentrations typical of uncontaminated and contaminated waters is an essential perspective for the

Box 14.7 Processes influencing MeHg occurrence and transport in freshwaters

Several studies show that the amount of MeHg input to a water body or exported by a watershed is tied to the abundance of wetlands or the construction of shallow reservoirs. In general, a hydrological unit with more wetlands may produce, be affected by and export more MeHg than a hydrological unit with very few wetlands. Wiener and Shields (2000) showed that Hg-contaminated sediments in two reservoirs in Massachusetts contributed much less MeHg to the river system they occupied than did downstream wetlands into which the sediments deposited. Transport of the contaminated sediments into the wetlands greatly increased the MeHg concentrations exported by that river. St Louis *et al.* (1994) showed that yields of MeHg were 26 to 79 times higher from wetland portions of watersheds than from upland areas or lakes in the Experimental Lakes area of Canada. Catchments and lakes were sites of MeHg retention and demethylation in general, while wetlands

were sources of net methylation. A strong correlation is also found between MeHg in the sediments of a river and percent wetland in the sub-basin (Krabbenhoft *et al.*, 1999) (Fig. 14.7). Dissolved MeHg shows a weaker relationship with abundance of wetlands, presumably because of the complex processes that influence sediment–water partitioning (K_d). Although complex processes affect how exact this generalisation is, it raises a flag of caution about restoring and adding wetlands in a Hg-contaminated hydrological unit.

Newly flooded lakes also generate MeHg. Kelly *et al.* (1997) studied a boreal forestland that was experimentally flooded to a shallow depth. Methyl mercury production increased 39-fold after flooding, with resultant increased contamination of plants, lower trophic level invertebrates and fish. The authors suggested that the total area of such land that is flooded should be minimised (shallow flooding is most conducive to MeHg formation), and that flooding of organically rich wetlands created the greatest risks for MeHg formation.

management of Hg. Hg concentrations are always low compared to those of many other metals; so extremely rigorous methodologies are necessary to produce acceptable data. MeHg concentrations are 1% to 11% of total Hg concentrations but not necessarily related to total Hg concentrations, because of influences from the nature of the watershed, the original source of the Hg and other physicochemical processes.

Concentrations of MeHg and total Hg in fresh and marine waters are much lower than once believed (Wiener and Spry, 1996). Before 1980 many attempts to quantify concentrations of Hg in surface waters were hampered by large measurement errors. Ultra-clean, contamination-free protocols for collection, handling and analysis of water samples, along with Hg-specific techniques for concentrating extremely low concentrations of Hg and MeHg from large volumes of filtered water, were necessary to understand true Hg concentrations in water. Determinations require special digestion and storing protocols (to prevent formation and volatilisation of Hg^0). Hg is detected by oxidising all forms to Hg(II), then reducing it to Hg^0 in a closed chamber, and passing the gaseous vapour past a light source from which absorption is determined. This method is entirely different from that used for other trace elements; thus a unique sample is required. Development of this methodology allowed determination of Hg in sediments and organisms in the 1970s. But ultra-clean protocols for water analysis were not introduced until the late 1980s (e.g. Bloom, 1989). It should be recognised that some of the downward trends in concentrations of aqueous Hg reported in the literature reflect changes in methodologies (Spry and Wiener, 1991), rather than meaningful environmental trends.

Reliable Hg and MeHg concentrations are available from an increasing number of environments, in organisms, sediments, water, the atmosphere and unique materials like ice cores. These advances allow generalisations to be made about the ranges of concentrations that might be expected in different circumstances (Box 14.8, Table 14.3). Hg occurs in uncontaminated natural waters in concentrations from 0.1 to 1 ng/L (μg/L $\times$ 10^{-3}), although MeHg-enriched food webs are found associated with dissolved concentrations as low as 0.04 ng/L. Concentrations in contaminated waters rarely exceed 10 ng/L, except in mining districts. Urbanised estuaries provide a good example of typical Hg contamination found in large water bodies subjected to both direct anthropogenic inputs and inputs from the contaminated atmosphere (Box 14.8).

Particulate or sedimentary Hg concentrations in undisturbed sediments are also low compared to other metals: typically <0.1 μg/g dw. Concentrations follow trends in inputs but absolute concentrations in areas of non-mining contamination have a relatively narrow range compared to some metals. For example, sediments from Wisconsin lakes were more than twice as contaminated in 1989 as in the period before atmospheric emissions in North America (Rada et al., 1989). Concentrations were between 0.04 and 0.07 μg/g dw in sediments deposited before anthropogenic enrichment and increased to 0.09 to 0.24 μg/g dw in 1989. Concentrations (in sediment cores) in San Francisco Bay increase from 0.04 μg/g dw before 1840 to 0.3 to 0.6 μg/g dw, regionally, in 1990, with hotspots containing up to 3 μg/g dw (Hornberger et al., 1999). Sediment cores from lakes also reflected rapid declines in Hg deposition in the 1990s in the USA and Europe (Engstrom and Swain, 1997). But the range of change is small.

MeHg was 1% to 11% of dissolved Hg (Krabbenhoft et al., 1999) in a survey of 61 US freshwater sites with different land uses. Thus, MeHg concentrations of 0.1 to 0.5 ng/L are a strong signal of contaminated waters; and concentrations of >5 ng/g dw signal contaminated sediments. MeHg concentrations are at least partly a function of total Hg in sediment in moderately contaminated sediments. But at high total Hg concentrations in sediments (>1.0 μg/g dw), the proportion of MeHg often declines (Krabbenhoft et al., 1999). Some mining districts have very high concentrations of both dissolved MeHg and sedimentary MeHg (e.g. the ancient Almadén mining district of Spain). But the efficiency of MeHg formation is typically <2% in such environments.

Box 14.8 Mercury dynamics in urbanised Chesapeake Bay

Mason *et al.* (1999) characterised Hg and MeHg dynamics in Chesapeake Bay, an urbanised estuary in Virginia/Maryland, USA, and Bloom *et al.* (1999) characterised Lavaca Bay, Texas, USA, an industrially contaminated, urbanised estuary. The contrast indicates the differences that might be expected between urban/atmospheric contamination and industrial contamination with Hg.

Concentrations in water in both systems are in the range of ng/L; nevertheless, both water bodies are enriched in MeHg and Hg compared to undisturbed environments. In the Chesapeake Bay system, Hg concentrations in particulate material were 0.56 and 0.48 µg/g dw in the two most contaminated locations (e.g. Baltimore Harbor) but otherwise ranged from 0.06 to 0.16 µg/g dw. Hg concentrations in Lavaca Bay were higher than in Chesapeake Bay (0.12 to 0.79 µg/g dw). The concentrations of MeHg in sediments of the two waters were similar, however. Chesapeake Bay sediments contained between 0.4 and 1.6 ng/g dw MeHg, and concentrations in Lavaca Bay sediments were 0.8 to 1.0 ng/g dw. Clearly, the percentage MeHg formation in Lavaca Bay was the lower of the two. In Chesapeake Bay, typical partitioning of MeHg concentrations between particulate and dissolved phases were described by a K_d of 7×10^3 to 1×10^4 in

the Pautaxent arm of the Bay (Benoit *et al.*, 1998). The K_d for MeHg were $10^{2.7 \pm 0.43}$ in Lavaca Bay. Again, the higher K_d for MeHg in the Chesapeake Bay represents a higher methylation rate. MeHg ranged from <1% to 5% of total Hg in both the water and sediment of Chesapeake Bay. Those data were not available for Lavaca Bay. But in San Francisco Bay (another urbanised estuary, but subject to mining inputs) it is usually <1% with the highest values being 2.2%.

About half of the Hg input to Chesapeake Bay was via atmospheric inputs. Inputs were presumed to be dominated by industrial discharge in Lavaca Bay, especially near the industrial outfall. Mason *et al.* (1999) also found that Chesapeake Bay trapped inorganic Hg coming in from the rivers. Concentrations decreased from the river mouths to sea. But more MeHg was leaving the Bay than entering it. Thus MeHg was being formed by microbial activity within the estuary and it was a source for the ocean. About 2% of Hg inputs per year were methylated. Mason *et al.* (1999) also found that the soils of the Chesapeake Bay watershed contained most of the Hg deposited from the atmosphere. The heightened atmospheric input due to human activities was being stored in these soils and was slowly leaking into the coastal zone. Whatever changes occurred in releases of Hg by human activities, the soils would continue to be a source for the estuary for a significant period of time.

Waters and sediments from mining areas can contain exceptional concentrations of total Hg. Artisanal mining (small independent miners working outside the scope of normal practices – often illegally) is of special concern in developing parts of the world, where human exposure are often thought to be problematic (Box 14.9).

14.3.5 Bioaccumulation

Prologue. The basic question about MeHg bioaccumulation is how do concentrations of <1 ng/L in water yield concentrations six orders of magnitude higher (mg/kg) in fish (Morel *et al.*, 1998). Biomagnification occurs when MeHg crosses rapidly into cells and is lost slowly. The mechanism of uptake

(via MeHg-cysteine) and redistribution in the body accentuates the process. Particularly sinister forms of toxicity occur when MeHg-cysteine moves through the body unrecognised as a toxin, including to sites where other toxicants are excluded. Differences in food sources and differences in biodynamics explain differences in how invertebrates and vertebrate animals manage Hg.

The same processes that influence bioaccumulation of all metals control the characteristics of Hg and MeHg bioaccumulation (Chapter 7). Those factors differentiate accumulation of inorganic Hg and MeHg and define which organisms are at greatest risk. The nature of the MeHg that crosses membranes and accumulates in cells is also fundamental to the mechanism by which MeHg exerts its toxicity.

Table 14.3 *Dissolved and particulate Hg and MeHg concentration ranges in different settings*

	Dissolved Hg (µg/L × 10^{-3})	Particulate Hg (µg/g dw)	Dissolved MeHg (µg/L × 10^{-3})	Particulate MeHg (µg/g dw)	Reference
Freshwater					Jiang *et al.*, 2006
Uncontaminated	0.3–1.0	0.05–0.17			Horvat *et al.*, 1999
Low alkalinity lakes	0.9–6.5		<0.01–0.4		Spry and Wiener, 1991
Streams: 61 in 26 watersheds	3.3–11[a]	0.034–0.22[a]	0.01–0.48[a]	0.001–0.003[a]	Krabbenhoft *et al.*, 1999
Lake Ontario	0.7				Wiener & Spry, 1996
Lake Erie	1.8				Wiener & Spry, 1996
Contaminated	2–10		0.4–2.0		Wiener & Spry, 1996
Mining affected					
China	3.2–331	24–343			Jiang *et al.*, 2006
14 sites USA	84	0.80	0.10	0.002	Krabbenhoft *et al.*, 1999
Almadén, Spain	6–11 200	0.5–800	0.04–30	0.003–0.082	Higueras *et al.*, 2006
Lahonten Reservoir, USA	0.8–56	30–600	0.1–3.1	0.002–0.014	Bonzongo *et al.*, 1996
Coastal waters					
Near shore	0.5–2.0	0.04			Horvat *et al.*, 1999
Offshore	0.2–0.4				Horvat *et al.*, 1999
Contaminated	0.6–2.3	0.1–2.1			Horvat *et al.*, 1999
Gulf of Trieste	0.5–4.9				Horvat *et al.*, 1999
Chesapeake Bay	6.2 ± 5.8[b]	0.06–1.0	0.05–0.2	0.002–0.010	Mason *et al.*, 1999
San Francisco Delta	0.1–0.5	0.1–0.4	<0.1–0.3	0.001–0.008	Krabbenhoeft *et al.*, 1999

[a]Mean for land uses: background, agriculture, forest, urban, mining.
[b]Average ± standard deviation.

14.3.5.1 Invertebrate bioaccumulation

Prologue. Invertebrates are of interest in Hg bioaccumulation because they act as a vector for Hg transfer to higher trophic levels. Invertebrates typically bioaccumulate lower proportions of MeHg than do fish. The proportion of MeHg varies among species but this is driven by differences in Hg forms within different food sources (in contrast to Se). Uptake of MeHg itself does not vary greatly among invertebrate species. Most differences are consistent with the biodynamics of Hg(II) and MeHg uptake.

Concentrations of total Hg vary five- to ten-fold among invertebrate species collected from the same environment (Francesconi and Lenanton, 1992) (Table 14.4). In field surveys, bivalves are among the most efficient invertebrates at accumulating total Hg (Langston, 1982). Langston (1982) also noted that a significant relationship occurred between sediment concentrations and accumulated concentrations in deposit-feeding bivalves only when sediment concentrations were normalised to the percentage of organic material (Fig. 14.8). A high organic content in the sediment (the food supply for the deposit feeding bivalve *Scrobicularia plana*) had a negative effect on the accumulation of total Hg by this bivalve although the mechanism for that influence is not clear.

MeHg concentrations do not vary as widely among species as does inorganic Hg bioaccumulation; but

Box 14.9 Artisanal Hg mining

The most extreme concentrations of Hg are often found associated with Hg mining (Table 14.3). In the case of artisan mining, illegal miners (usually) use the traditional amalgamation methods to remove Ag and Au from ores. The ore is washed over a bed of Hg; the silver and gold amalgamate or attach to the Hg; then the Hg is volatilised and precious metals captured (Nriagu, 1994). This was the dominant type of mining in the Sierra Nevada during the latter stages of the California Gold Rush which began in 1850. Hg was mined on the east side of California's Central Valley, then transported across the valley to be used for Au mining in the mountains on the west side. Entire mountains containing ore potentially rich in Au were washed into sluices containing the Hg (termed hydraulic mining). The sediments released from hydraulic mining were dumped into rivers and streams and spread over northern California watersheds between 1850 and 1880. The result was Hg contamination over much of northern California that remains today. Pure Hg^0 can be found in some places where Au mining was active. Concentrations of Hg in stream sediments are 1 to 10 µg/g dw outside such pockets. In San Francisco Bay, 100 to 250 miles from the hydraulic gold mines, the regional background sediment concentration of Hg is 0.3 to 0.6 µg/g dw, with deposits as high as 3.0 µg/g dw occurring within the sediments near mouths of creeks with Hg mines in their watershed.

A new global gold rush started in 2001 when gold prices rose from \$260 to \$725 per ounce. An artisan mining industry arose in developing countries in response to the price increase, with the traditional amalgamation methods of extracting the precious element. Suppliers of the necessary Hg include US Hg recyclers as well as chemical plants and plastics manufacturers for which Hg is a waste product. In the Agusan River basin of eastern Mindanao, the Philippines, water concentrations of 2906 µg/L and sediment concentrations of >20 µg/g dw Hg are reported 15 to 20 km downstream of mining activities (Appleton *et al.*, 1999); although these concentrations seem extreme compared to many. Dissolved Hg concentrations in waters from the Xiaoquinling Region in China, with a long history of artisanal mining, were 110 to 3100 ng/L averaging 740 ng/L (very high compared to Table 14.3) (Jiang *et al.*, 2006).

Artisanal gold mining is also important in Latin America. The total number of miners was probably between 200000 and 400000 within the Brazilian Amazon in the late 1990s. Half as many are thought to exist in Colombia and Ecuador. For all Latin American countries combined, the estimate is between 543000 and 1039000 (Malm, 1998). Since the 1980s, Brazil has ranked first in South American Au production with annual production from 100 to 200 tonnes per year. Nearly 90% came from informal mining or *garimpos* (Malm, 1998). Malm (1998) estimated that this production would correspond to about 2000 to 3000 tons of Hg released into the Brazilian Amazon environment during the present gold rush. Hg concentrations in fish in such regions exceed health standards for consumption, and Hg concentrations in the hair of local people indicate occupational exposure to this neurotoxin (Appleton *et al.*, 1999). Contamination is expanded when Hg is carried into water bodies, like reservoirs (Palheta and Taylor, 1995) from which fish are consumed (Akagi *et al.*, 1995).

Table 14.4 *Total Hg and MeHg concentrations in different groups of invertebrates from Princess Royal Harbour in Australia*

Organism	Total Hg (µg/g dw)	MeHg (µg/g dw)	Average MeHg (%)
Crustaceans	0.21–0.47	0.14–0.43	75
Gastropods	0.11–0.39	0.05–0.29	57
Bivalves	0.38–1.2	0.08–0.30	24
Polychaetes	0.14–0.41	0.06–0.13	36
Echinoderms	0.06–0.20	0.01–0.09	35

Source: Francesconi and Lenanton (1992).

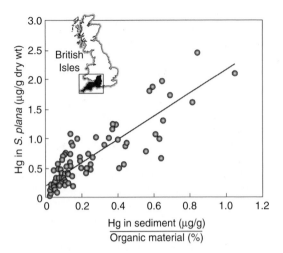

Fig. 14.8 The relationship between Hg concentrations in the deposit-feeding bivalve *Scrobicularia plana* from the estuaries of southwest England, and the ratio of Hg/organic material in fine-grained sediments, the food supply. More organic material in sediments appeared to lower Hg bioavailability. (After Langston, 1982.)

the proportion of MeHg does vary. Invertebrates are potentially exposed to Hg(II) and MeHg from the dissolved phase, from food and from sediments, although dietary uptake usually dominates (Tsui and Wang, 2004). The differences in the proportion of MeHg among invertebrate species appear to reflect differences in choice of food accentuated by differences in biodynamics (uptake and loss) between inorganic Hg and MeHg. Invertebrates, typically, have a lower proportion of MeHg compared to fish. In a survey of a marine bay in Australia, 91% to >95% of the Hg in six species of fish was MeHg; only 24% to 36% of the Hg in three groups of invertebrates was MeHg (Francesconi and Lenanton, 1992) (Table 14.4). Physiological uptake and loss do not differ greatly between methyl and inorganic forms in invertebrates, but inorganic Hg occurs in much higher concentrations in water and in their food compared to MeHg (Box 14.10). Thus inorganic Hg uptake is favoured. The exception was the 75% MeHg in several species of crustaceans (Table 14.4). In the phytoplankton food of many such species, inorganic Hg is bound to membranes, while MeHg occurs in cytoplasm (Mason *et al.*, 1995). Cytosolic metals are the form of

metals bioavailable to at least crustaceans like copepods (Reinfelder and Fisher, 1991). This would accentuate preferential accumulation of MeHg in the species that use this specific food source. In contrast to invertebrates, fish lose MeHg much more slowly than they do inorganic Hg; thus bioaccumulation of MeHg is favoured and total bioaccumulation is enhanced compared to invertebrates (Wiener and Spry, 1996).

14.3.5.2 Fish bioaccumulation
Prologue. Biodynamics, dietary exposure and trophic transfer are dominant processes defining risks to fish and to human health from fish consumption. For fish, high assimilation efficiencies from food, and slow rate constants of loss (Chapter 7) yield high bioaccumulation of MeHg, and allow the preferential accumulation of the organic form of the metal. Biodynamics also is the basic cause of food web biomagnification in predators. Rate constants of loss of <0.001/d make it possible that a fish can progressively accumulate MeHg over a very long time, contributing to a positive correlation between size or age and MeHg concentration.

Bioaccumulation of MeHg in fish is a long-standing concern. Surveys of Hg concentrations in fish across many environments followed the development of reliable analytical methodologies for tissues. Widespread Hg contamination was identified in the 1970s, especially in some kinds of fish or fish-eating predators. Johnels *et al.* (1967) characterised extremely high concentrations of Hg in pike (*Esox lucius*) from Swedish lakes. Further studies of those lakes (Jernelov and Lann, 1971) showed that MeHg increased in concentration at each trophic level through the lake food webs. Most of the Hg was in the form of MeHg. In 1990 Håkanson *et al.* (1990) calculated that there were about 10 300 Swedish lakes with >1 μg Hg/g dry wt in fish (the Swedish blacklisting limit). High concentrations of MeHg have also repeatedly been shown in top fish predators from the oceans (e.g. tuna and swordfish: Tollefson and Cordle, 1986).

Thus, the stage was set four decades ago for understanding the most serious impacts of Hg contamination: effects on upper trophic level

Box 14.10 Why do invertebrates typically have a low proportion of MeHg in their bodies?

Specific studies with bivalves (the mussel *Mytilus edulis*: Gagnon and Fisher, 1997), polychaete worms (*Nereis succinea*: Wang *et al.*, 1998), copepods (Lawson and Mason, 1998), amphipods (Mason *et al.*, 1995) and freshwater zooplankton (*Daphnia magna*: Tsui and Wang, 2004) illustrate some common characteristics of biodynamics among the invertebrates. These are of interest because of the contrast to bioaccumulation in upper trophic level organisms (the predators of the invertebrates). Commonalities in Hg bioaccumulation by invertebrates include:

- *MeHg concentrations in food and solution are much lower than inorganic Hg concentrations.* As noted above MeHg is typically <10% of total Hg in all environmental matrices.
- *MeHg uptake rates from solution by invertebrates are higher than inorganic Hg uptake rates, but the differences are relatively small.* For example, the K_u for MeHg in *Daphnia magna* was 0.46 compared to the K_u for inorganic Hg of 0.35 L/g/d (Tsui and Wang, 2004).
- *Assimilation efficiencies vary with food types but are higher for MeHg than for inorganic Hg.* For example, in mussels, the AE for Hg(II) from sedimentary food was only 1% to 9%, but for MeHg it was >30% up to 87% (Gagnon and Fisher, 1997). AEs were 2.0 to 2.8 times higher for MeHg than for inorganic Hg in copepods and amphipods (Lawson and Mason, 1998). In the *N. succinea* AEs for inorganic Hg were 7% to 30% and

were unaffected by sediment composition including the presence of sulphides, whereas AEs of $CH_3Hg(II)$ ranged between 43% and 83%. Similarly, in *D. magna* AE of inorganic Hg varied with food type from 9% to 47%, and AE of MeHg was 64% to 97%.

- *Diet is the dominant source of both inorganic Hg and MeHg* in most cases, although there are circumstances where uptake from the dissolved phase can be important (Wang *et al.*, 1998; Tsui and Wang, 2004).
- *MeHg is lost more slowly than inorganic Hg, but the differences are not especially large.* Replicate experiments with inorganic Hg in *N. succinea* found rate constants of loss of 0.027/d and 0.031/d compared to a rate constant of 0.014/d for MeHg. Rate constants of loss for *D. magna* were the same for inorganic Hg and MeHg: 0.045 to 0.061 for inorganic Hg and 0.050 to 0.063 for MeHg, respectively. This is the greatest single difference between invertebrates and vertebrate predators. In the vertebrates, loss rate constants are 10 times or more slower than in invertebrates.
- *Biodynamic modelling predicted that, under conditions typical of coastal sediment environments,* MeHg accumulation would contribute only 5% to 17% of total Hg accumulation in polychaetes. This is because the higher assimilation efficiencies and slightly slower loss rates for MeHg do not compensate for the very much lower MeHg concentrations in food and water. On the other hand, MeHg was usually the dominant form in *Daphnia*, as noted above; because all bioavailable Hg is MeHg.

predators and threats to human health from consumption of such organisms for food. Interest in Hg since 1967 has been cyclic, however. There was a wane of funding between the late 1970s and the early 1990s. The rediscovery of high Hg concentrations in some of our most common sources of fish protein (e.g. tuna and swordfish) in the twenty-first century has reinvigorated research on the subject.

The factors that link anthropogenic Hg inputs to MeHg contamination of fish and other upper trophic level animals are complex. Biological factors are clearly important. The biodynamics of Hg in fish is of long-standing interest because fish are the primary vectors of human exposure. MeHg

uptake by fish is almost entirely from food (Morel *et al.*, 1998); and MeHg is assimilated much more efficiently from food than inorganic Hg. The slow rate constant of MeHg loss is also an important factor in the high percentage of MeHg in fish, the elevated bioaccumulation and food web magnification. In general, the consensus is that the half time of loss of MeHg in large fish is 1 or more years. In a comparison of different data, Trudel and Rasmussen (1997) found that the rate constant of loss for MeHg averaged 0.003/d in fish weighing ≤1 g and declined to <0.001/d in fish weighing more than 100 g. Wiener and Spry (1996) cited the most careful studies showing similar rate constants, yielding half retention times of 200 to 600 days in rainbow trout (*Oncorhynchus mykiss*), mosquitofish (*Gambusia affinis*) and northern pike (*Esox lucius*). Experiments conducted for less than 90 days did not capture the very slowly exchanging pool of MeHg in fish (Trudel and Rasmussen, 1997), even though this is the pool most likely to dominate in nature.

Because fish preferentially 'hang on' to MeHg more effectively than inorganic Hg, they selectively concentrate MeHg progressively over time, even though the Hg in their food is usually >50% inorganic. The high trophic transfer potential of MeHg, the ratio of AE/rate constant of loss (Chapter 7) is the cause of the biomagnification seen in natural waters. The trophic transfer potential reflects the basic mechanisms driving MeHg uptake. For example, sophisticated instrumentation shows that all Hg in fish occurs as a strongly bound sulphide complex, presumably with the amino acid cysteine (Harris *et al.*, 2003) or the tripeptide glutathione which contains cysteine. When one organism eats another, uptake of the complex occurs because the complex of MeHg–cysteine mimics another amino acid (L-methionine). The organometal and the amino acid are moved across the gut epithelial membrane on very abundant methionine transporter sites (Clarkson, 2002) (Box 14.11). Once taken up, there is a dynamic redistribution of MeHg among tissues and organs largely because MeHg–cysteine mimics the essential compound (Clarkson, 2002).

Some tissues (particularly the central nervous system) detoxify MeHg by demethylation. This creates inorganic Hg which the central nervous system cannot excrete (Box 14.11), allowing accumulation and eventual toxicity, but from inorganic Hg not MeHg itself (Clarkson, 2002).

The most common tests of MeHg toxicity in the aquatic environment use waterborne MeHg which occurs as a hydroxide or chloride complex. These forms are toxic, but the mode of transport across the epithelial cell membrane is different and the potency is less than in the case of the MeHg–cysteine. Thus traditional dissolved, single-species toxicity tests cannot address the major mechanisms driving exposure. They therefore underestimate toxicity and do not address the most sinister effects of MeHg. Transport across the

gut epithelium membrane as an amino acid mimic also explains the extremely high assimilation of MeHg when one organism eats another. This also might explain the strong correlation observed between MeHg concentrations and $\delta^{15}N$ (a proxy for trophic position – based on the enrichment of ^{15}N relative to ^{14}N in protein of predators relative to their prey) in food webs (Campbell *et al.*, 2005).

14.3.6   Food web contamination

Prologue. The organisms most likely to accumulate the highest concentrations of Hg are the birds, fish and mammals that occupy the highest trophic levels. Bio-availability of MeHg to the food web and the characteristics of the food web itself influence the degree of contamination in the top predators. Also important are the mass of Hg input, the form of the input, and environmental characteristics. The latter include hydrological and ecological properties that affect Hg delivery to zones of methylation as well as biogeochemical controls on methylation.

Mercury is the most hazardous of the metals because note only is MeHg highly toxic, but it accumulates to ever higher concentrations with increasing trophic level in food webs. The regulatory focus on Hg is motivated largely by the health risks to humans consuming contaminated food. Human exposure to MeHg is almost wholly due to consumption of fish (Wiener and Spry, 1996). Even though nearly all the total Hg in fish is MeHg, fish do not methylate Hg.

14.3.6.1 Relationship of Hg loading to food web contamination

Ecosystems that have low methylation efficiency may exhibit low or moderate bioaccumulation even under high Hg loads. . . .On the other hand, significant bioaccumulation of methylmercury can result even when very low Hg loading rates exist if methylation efficiency is high.

Krabbenhoft *et al.* (1999)

Prologue. When local and regional Hg emissions increase or decrease, concentrations of Hg in

Box 14.11 Uptake of MeHg

Conflicting models characterise uptake rates of MeHg. The traditional concept in ecotoxicology was derived from studies of organic chemicals. Because MeHg has some lipid solubility, it was assumed that MeHg diffused across membranes, analogously to what was thought to be the mode of uptake for organic chemicals. Mason *et al.* (1995), however, noted that inorganic Hg complexes (e.g. common zero charge chloro-complex of Hg) were nearly as lipid soluble as MeHg, yet MeHg was taken up faster. They also noted that MeHg in fish resides in protein rather than in fat tissues, characteristic of a lipophobic chemical not a lipophilic entity like the most dangerous organic chemicals. Finally, the lipid solubility of MeHg is not comparable to organic chemicals concentrated as effectively as MeHg (e.g. PCBs or other lipophilic contaminants). Despite these contradictions, citation of the lipid solubility as the source of efficient transport of MeHg into organisms remains common.

The alternative model was suggested by Clarkson (2002). In humans, MeHg is assimilated with >90% efficiency when fish are the food source; the same is true of fish-eating fish. Within cells MeHg occurs primarily as a complex with sulphur-bearing amino acids, specifically L-cysteine (Harris *et al.*, 2003). Structurally, the MeHg–L-csyteine complex is similar to the large neutral amino acid L-methionine. Specific transporters for L-methionine exist in all membranes, particularly in alimentary tract epithelia. Cell membranes mistake the MeHg–cysteine complex for that amino acid and co-transport MeHg and cysteine together. Inside the cell, the complex retains its form and can be incorported into larger molecules containing cysteine: glutathione (a tripeptide made of cysteine, glutamic acid and glycine) or other cysteine-rich proteins; or it can be moved around in the body. MeHg crosses the barrier between the blood and brain, and between a mother and her foetus in humans and in fish (Hammerschmidt *et al.*, 1999). Glutathione–MeHg can be removed from the brain, but that occurs slowly so it accumulates. Some is demethylated to inorganic Hg (Clarkson, 2002) for which there is no removal mechanism and no detoxification mechanism (again, in both humans and fish: Gonzalez *et al.*, 2005). It is actually the inorganic Hg inside the central nervous system which is the source of strong central nervous system toxicity. The central nervous system of young organisms (e.g. human infants) are particularly vulnerable to adverse effects when Hg accumulates. Cells appear to stop dividing and migrating, and damage occurs throughout the brain. In adults damage tends to be more localised in the brain. Thus MeHg toxicity, as it occurs in nature, can only be determined from dietary toxicity study using foods that have transformed MeHg into MeHg–cysteine.

receiving waters may respond accordingly, but complex factors will affect the response. The magnitude and timing of the response of individual environments to Hg inputs depend upon the form of Hg in the inputs as well as a complex of physical, chemical and biological factors in the environment.

It is widely observed that when experimental systems are subjected to increased Hg loading, they respond rapidly with increased MeHg levels in the food web (Munthe *et al.*, 2007). This was demonstrated when large mesocosms in a boreal lake in Canada were spiked with different concentrations of inorganic Hg (Orihel *et al.*, 2007). MeHg bioaccumulated in zooplankton, benthic invertebrates and fish within weeks and the degree of build-up was related to the original loading of inorganic Hg.

More Hg loading correlated with more MeHg in the food web. This suggests a simple relationship between loadings and food web response.

Indeed, within an environment, regulation of contaminant input can sometimes result in simple changes in contamination of the food web. For example, in 1995, the US Environmental Protection Agency ruled that all municipal and medical incinerators must reduce Hg emissions to the atmosphere by 90% to 94%. After emissions from a local incinerator met that target in Florida, news stories reported that Hg concentrations in fish and wading birds in the Florida Everglades declined by 75% at some locations.

In nature, the relationships between Hg inputs and MeHg contamination of food webs are typically more complex than this, however. For example, spatial patterns of Hg input were linked to food web contamination across 1500 lakes in Scandinavia (Munthe *et al.*, 2007). But there was substantial variability among environments. Similarly, no simple relationship was found between MeHg in organisms and Hg inputs across 831 surface waters in the USA (Krabbenhoft *et al.*, 1999). Sources of variability were proximity to emission sources, nature of landscape (wetlands, forest cover), and water level manipulation in reservoirs. Trends in contamination in ocean fish are also difficult to decipher. Part of the problem may be the complexity of inputs; inputs from some continents are certainly declining, but not all sources (see earlier section). It is also difficult to capture the pre-contamination era by comparative analyses because of the very long history of Hg use by humans. Data compared between the earliest Hg analyses on ocean fish (about the early 1970s) and the present (Kraepiel *et al.*, 2003) show little change. Comparisons of modern specimens and museum specimens of fish go back further than the 1970s (Miller *et al.*, 1972) but are also equivocal. However, pre-1900 specimens of bird feathers (from some but not all species) appeared to contain less Hg than specimens captured in the early 2000s (Thompson *et al.*, 1992).

Physicochemical characteristics of different environments affect the net rate of MeHg generation and thus affect the relationship between loading and food web contamination. Wetlands, shallow water habitats and organically rich environments all have accelerated rates of MeHg generation, as noted above. Thus, as with Se, the entire hydrological unit should be considered in evaluating influences of Hg input. Watersheds where Hg can be transported into hotspots for methylation will be most vulnerable to food web contamination (Munthe *et al.*, 2007).

Physical processes are especially important in lakes and reservoirs. For example, MeHg may accumulate in the anoxic bottom waters of lakes. If the lake mixes (as often occurs at some seasons) a pulse of MeHg may enter the biologically rich surface layers (Munthe *et al.*, 2007). In reservoirs, the ratio of the flooded area to the volume of the reservoir influences methylation. Upstream inputs are also important (Johnston *et al.*, 1991). Newly constructed reservoirs may experience increased methylation efficiency without increasing the Hg load (Bodaly *et al.*, 1997). After flooding, the methylation efficiency of many reservoirs shows a strong upward trend in net MeHg production.

The form of Hg input to aquatic environments and the geochemical characteristics of the environment affect the production of MeHg and the bioavailability of the MeHg that is produced. Geochemical characteristics differ widely among water bodies. They also control speciation of all metals, including Hg and MeHg. Speciation affects the entry of MeHg into primary producers at the base of food webs and into invertebrates via uptake from water in complex ways. For example, dissolved organic material (DOM) may reduce MeHg association with sediments and increase its mobility, but its effects on bioavailability remain controversial. Munthe *et al.* (2007) noted both the uncertainties and the importance of this discussion: 'The overall role of DOM in transport, methylation, and bioaccumulation is complex, with competing influences, but central to the physical and biogeochemical behaviour of Hg in watershed.'

The form of Hg in sediments has a particularly strong influence on MeHg formation. In particular, a lower proportion of MeHg appears to occur in the most contaminated sediments (Krabbenhoft *et al.*, 1999; Munthe *et al.*, 2007); whether this reflects a single process or differences in source is not well known (mining or industrial releases of Hg seem to yield a lower proportion of MeHg than atmospheric inputs). But the lower proportion does not necessarily mean a lower concentration of MeHg. For example, 250 tons of Hg were released by an industrial operation into East Fork Poplar Creek, Tennessee, USA in the 1950s and 1960s. Total Hg concentrations in sediments increased 100- to 1000-fold compared to undisturbed sediments. But MeHg concentrations in the sediments increased by only about 10-fold (reviewed by Munthe *et al.*, 2007).

14.3.6.2 Food web characteristics influence contamination

Prologue. The level of MeHg contamination in a food web, as a whole, is controlled by Hg inputs and the factors that influence the net percentage of MeHg formed. But food web characteristics (including trophic pathways and food chain length) determine the concentration (degree of contamination), especially at the highest trophic levels. The number of trophic levels between predators and prey determines the degree of contamination of the predators.

Ecological factors like the nature of the food web play a role in the degree of Hg contamination. Enrichment begins when phytoplankton accumulate concentrations of both inorganic Hg and MeHg that are orders of magnitude higher than occur in solution. The greatest increase in Hg concentration occurs at this step (Mason *et al.*, 1995). Bioconcentration factors from water to seston are often 10^5 to 10^6 (Wiener *et al.*, 2007). Zooplankton assimilate more MeHg than Hg when they eat phytoplankton and thus begins the discrimination process against inorganic Hg. In general, the average proportion of MeHg over total Hg increases 'from about 10% in the water column to 15% in phytoplankton, to 30% in zooplankton, and to 95% in fish' (Morel *et al.*, 1998). In natural waters this is manifested by a correlation of MeHg concentrations in prey with MeHg in their predators (Fig. 14.9).

Trophic position determines the concentration in each species, from algae to top predators; with consumers and predators always having higher MeHg concentrations than their food. The length of a food chain is particularly important: the higher the trophic level, the higher the concentration of MeHg. For example, the presence of planktivores, such as lake herring (*Coregonus artedii*), rainbow smelt (*Osemerus mordax*) or mysid crustaceans, in lakes typically increases the number of trophic levels. Where planktivores are present, higher Hg concentrations occur in top predators than is the case where the planktivores are absent (and food chains are shorter) (Box 14.12). For example, MeHg concentrations in lake trout (*Salvelinus namaycush*) vary by 100-fold (0.03 to 3.96 µg/g dw) between lakes in the St Lawrence River watershed (Cabana and Rasmussen, 1994), which have similar Hg inputs (only from the atmosphere). Where pelagic forage fish and the crustacean *Mysis relicta* were present, adding to food web length,

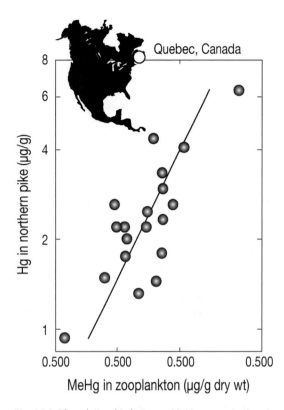

Fig. 14.9 The relationship between MeHg concentrations in the fish, northern pike *Esox lucius*, from a variety of boreal Canadian lakes and the MeHg concentrations in zooplankton from the fish's food web. MeHg concentrations in predators are related to MeHg concentrations in their prey. (After Garcia and Carignan, 2000.)

Hg concentrations were about 3.6-fold higher than in lakes with shorter food chains. Similarly, Hg in northern pike (*Esox lucius*) in a Finnish lake lacking forage fish were one-quarter of those in a similar lake with forage fish (Wiener and Spry, 1996).

MeHg inputs are crucial in determining the positioning of the relationship between exposure and bioaccumulation of MeHg. The relationship linking trophic level and Hg concentrations in the food web is shifted upwards if Hg inputs and net MeHg generation are greater in one water body than other (Fig. 14.11). But the relationship among trophic levels is maintained (R. Stewart *et al.*, unpublished observations), which appears to be the case for many food webs regardless of productivity, latitude or salinity (Campbell *et al.*, 2005).

Box 14.12 Biomagnification of Hg

The highest concentrations of Hg are commonly found in those fishes occupying the top trophic level (e.g. Bodaly and Fudge, 1993; Kidd *et al.*, 1995). As noted above biomagnification of MeHg through food webs occurs partly because assimilation efficiencies range from 60% to 95%, with the highest AE occurring in piscivores (fish-eating animals); and partly because loss rates of MeHg from fish are extremely slow. But progressive accumulation of MeHg is also linked to the feeding habits of the different organisms. Determining feeding habits can be challenging, however. Conventional methods involve determining the gut content of fishes from various species and then drawing a generalisation about feeding. Fishes are opportunistic feeders, however, whose diets (and trophic levels) often change as they grow or with environmental conditions. Their diet can also differ somewhat among individuals of the same species.

Stable isotope ratios of nitrogen ($\delta^{15}N/^{14}N$ or $\delta^{15}N$) provide a means of studying average feeding habits of individual fish, as a complement to conventional methods (Kidd *et al.*, 1995). The heavier isotope of nitrogen increases an average of 3 parts per thousand (‰) from prey to predator due to the preferential excretion of the lighter isotope through metabolic processes. The $\delta^{15}N$ at any point in time in a fish, for example, reflects what trophic level that fish has been feeding from in its recent past. $\delta^{15}N$ can therefore be used as a continuous measure of trophic behaviour.

Kidd *et al.* (1995) demonstrated the efficient trophic transfer of Hg by comparing six lakes in northern Ontario, Canada. They collected eight species of fish from the lakes (all species were not present in all lakes) and determined Hg concentration and $\delta^{15}N$ from samples of each. Nearly all Hg in such species can be assumed to be MeHg. A significant relationship was seen between Hg concentrations and $\delta^{15}N$ among species within individual lakes. But the position of each

species was different in the different lakes depending upon its specific place in the food web of each lake. Figure 14.10 shows the general tendencies in the relationship between $\delta^{15}N$ and Hg concentration in each species of fish averaged among lakes (each point represents the Hg concentration in a species averaged among all the lakes where it was present). This ignores both the effect of different food web lengths (and does not include a correction for 'baseline' $\delta^{15}N$). But it shows that the powerful influence on MeHg bioaccumulation of the averaged (general) trophic position of each species of fish.

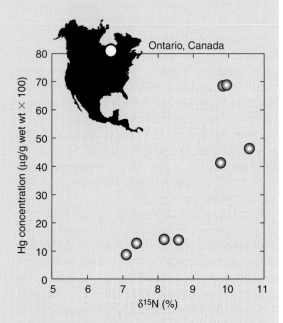

Fig. 14.10 Mean MeHg concentrations in each of eight different species of fish collected from six different lakes in northwestern Ontario, Canada. Bioaccumulation of MeHg in predators is related to the general feeding habits that determine the average trophic position (represented by $\delta^{15}N$) of the species. (Data from Kidd *et al.*, 1995.)

14.3.7 Ecological effects

Prologue. Both field and laboratory evidence illustrate the high toxicity of MeHg. Some organisms (e.g. some marine mammals) accumulate very high concentrations of Hg but are not affected. Detoxification

of MeHg via complexation with Se occurs in the liver of some marine mammals, presumably before the MeHg gets to other organ systems. Field observations suggest that other organisms, like some fish, are very sensitive to MeHg. Toxicity is expressed when a mother is exposed to MeHg via her diet and the

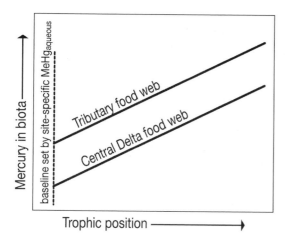

Fig. 14.11 A conceptual schematic derived from data from two different MeHg-enriched habitats in the San Francisco Bay-Delta watershed. The relationship between MeHg in different species and trophic position (determined by $\delta^{15}N$) followed a similar slope in the two habitats. But average MeHg concentrations in the waters of the 'tributary food web' were higher than average MeHg concentrations in the waters of the 'Central Delta food web' raising the y-intercept of the relationship in the former case. (Concept from R. Stewart, US Geological Survey, personal communication.)

MeHg is transported into the egg, adversely affecting development of the embryo. Such effects occur at concentrations that are 1% to 10% of the concentrations defined as toxic by traditional dissolved toxicity testing. Effect levels in the tissues of sensitive fish species like walleye *Sander vitreus vitreus* (a lake predator) can be as low as 7.5 μg/g dw (1.5 μg/g wet wt.).

Neurotoxicity is an important aspect of MeHg toxicity in birds and mammals as it is in humans. Neurological disorders were observed in the 1950s in predatory birds from Sweden (Clarkson, 2002). Evidence emerged linking these anomalies to consumption of prey that were feeding on Hg-treated grain in Swedish fields. In fish, long-term dietary exposure to MeHg showed loss of coordination, inability to feed and diminished responsiveness (Wiener and Spry, 1996). Such damage is especially important in the developing embryo and in young, where damage to a single cell can proliferate to much greater effects.

Traditional metal detoxification mechanisms apply to inorganic Hg, but fewer defenses seem to exist for MeHg. Demethylation does occur in the liver of at least some marine organisms. Total Hg is 92% to 98% MeHg in the muscle and skin of marine mammals like beluga whales (*Delphinapterus eucas*), narwhal (*Monodon monoceros*) and ringed seals (*Pusa hispida*). But in the livers of the same animals only 3% to 12% of the Hg is MeHg (Wageman *et al.*, 1998). It is thought that an association with Se demethylates the Hg. Se builds up in livers of many marine animals because they are top predators and feed from a Se-rich food web. Indeed, there is a strong statistical correlation between Hg and Se concentrations in the livers of open ocean mammal species (e.g. Currant *et al.*, 1996). The demethylation by formation of Hg–Se in mammalian livers contrasts to what happens in the brain when Hg builds up. Perhaps the lack of detoxification in the brain occurs because Se does not build up in the brain, and that detoxification mechanism is not available in the central nervous system.

A variety of laboratory data exists on concentrations of Hg associated with toxicity. Three factors define informative data for fish (Wiener and Spry, 1996):

- the sensitive mode of action is internal (e.g. in the brain or central nervous system), not on external organs exposed to Hg, like the gill;
- toxic effects result from dietary exposure, both because of the insidious form of MeHg accumulated and because accumulation is so efficient;
- the route of uptake ultimately does not affect where the toxin accumulates, because MeHg (as MeHg–cysteine) is redistributed among tissues and organs so pervasively.

Experiments with dissolved Hg and MeHg show effects at concentrations above the range typical of contaminated waters, except in the case of extreme mining contamination. The lowest concentration at which effects of dissolved MeHg on behaviour occurred was 200 ng/L, and the lowest concentration at which reproduction was affected was 100 ng/L (compare to Table 14.3). Wiener and Spry (1996) suggested that concentration in tissues might be a better basis of comparison, recognising confounding influences of interspecific differences in sensitivity, rates of accumulation and exposure time. They suggested that the average fish with Hg concentrations of 35 to 75 µg/g dw in the brain was probably suffering potentially lethal effects. For example, Wiener and Spry (1996) suggested that 25 µg/g dw in either brain or whole body of brook trout (*Salvelinus fontinalis*) indicated a likelihood of effects; whereas 15 µg/g dw was a likely no effect level. In walleye (*Sander vitreus vitreus*), another sensitive species, 7.5 µg/g dw appeared to be the no effect level. Diminished avoidance of predators occurred in mosquitofish (*Gambusia affinis*) at 3.5 to 25 µg/g dw.

Dietary exposure studies show that perhaps the most sensitive effect of MeHg on fish populations would be reduced reproductive success resulting from the toxicity of maternally derived MeHg to embryonic and larval life stages (Wiener and Spry, 1996). Reproduction is affected when MeHg accumulated by the mother is passed into the yolk of the egg (Fig. 14.12). Subsequent exposure of the growing embryo occurs as the MeHg–cysteine

disrupts cell proliferation. Reproductive failures begin at concentrations in eggs that are 1% to 10% of the concentrations that adversely affect adults. Thus Hg concentrations in eggs are the most sensitive measure of exposure (and the best indicator of potential effects). Embryonic mortality appeared to coincide with Hg concentrations of 0.35 to 0.5 µg/g dw in fertilised eggs; a concentration within six- to ninefold of that in the eggs of fish from water bodies without direct Hg inputs other than the atmosphere (Lake Ontario: Wiener and Spry, 1996). Among freshwater fish, walleye (*Sander vitreus vitreus*) populations could be affected in many natural waters at present rates of input.

Similarly, at concentrations that cause no measurable adverse effects in adult birds, impairment can occur to egg fertility, hatchling survival and overall reproductive success. As early as 1979, Heinz (1979) showed that a MeHg concentration of 2.5 µg/g dw in the diet of mallard ducks (*Anas platyrhynchos*) decreased reproductive success and altered behaviour of young ducklings. Hatchability declines in the eggs of ring necked pheasants (*Phasianus colchicus*) (Fimreite, 1971) and mallard eggs (Heinz and Hoffman, 2003) at concentrations in the egg of 2.5 to about 5.0 µg/g dw. Such concentrations are within the range observed in Hg-contaminated water bodies (Box 14.13)

14.3.8 Managing risks from methyl mercury

If the primary route of human exposure to MeHg is fish consumption, the most effective means of protecting public health is the issuance of fish consumption advisories. These advisories are designed to inform the general population, recreational and subsistence fishers, pregnant women, nursing mothers and children that unacceptable concentrations of MeHg (or other contaminants) are found in local fish and wildlife. The advisories include recommendations to limit or avoid consumption of certain fish and wildlife species from specific waterbodies or, in some cases, from specific waterbody types (e.g., all lakes). In the USA, the Environmental Protection Agency issues advisories for non-commercial fish, and the Food and Drug

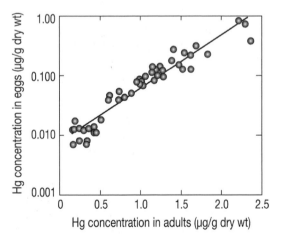

Fig. 14.12 The relationship between Hg concentrations in eggs and in the adult whole body of yellow perch *Perca flavescens* from lakes in northern Wisconsin (after Hammerschmidt *et al.*, 1999). Selected MeHg analyses suggested that MeHg was a very high proportion of the Hg in both. The data show that MeHg taken up by a reproductively mature female is passed to her eggs.

Box 14.13 Hg effects on an endangered marsh bird species

New techniques can sometimes create breakthroughs in understanding the effects of metals in nature and in the laboratory. Heinz and Hoffman (2003) developed a method for injecting MeHg into the eggs of birds that provides a basis for tying nature to experiments. The ease with which experiments can be conducted allowed them to compare sensitivity of developing bird embryos among a variety of species. Sensitivity varied widely, but one of the more sensitive species was an endangered wetlands species: the California clapper rail (*Rallus longirostris obsoletus*).

The clapper rail is an obligate salt marsh inhabitant of tidal marshes in San Francisco Bay. Only 1040 to 1264 individuals are thought to remain (Schwarzbach *et al.*, 2006). Loss of habitat is the first-order cause of the decline of this species. As much as 90% of suitable marsh around the Bay has been lost (Nichols *et al.*, 1986). Elimination of other stressors is essential to preserving the remnant populations however. When stressors were compared in a study of six marshes over four breeding seasons, the reproductive potential of the clapper rail was found to be 'much reduced over natural potential' (Schwarzbach *et al.*, 2006). Predation by rats, snakes and foxes appeared to eliminate one-third of the eggs of the species. Nest surveys showed clutch sizes (number of eggs per nest, presumably after predation) of about seven eggs per nest. The more alarming observation was that clapper rails produced on average 1.9 to 2.5 young out of those seven eggs.

Only 45% of the nests successfully hatched at least one egg; a value less than comparable species elsewhere. Deformities, embryo haemorrhaging and embryo malposition were also noted in many eggs. In attempting to define a cause for these observations, Schwarzbach *et al.* (2006) eliminated flooding, organochlorine pesticides, and other trace elements, including Se as causative factors. One stressor common to all marshes (reduced potential occurred in all marshes) was Hg contamination. They could only sample non-viable eggs (so as not to affect survival of this endangered species), but they found Hg concentrations varying from 0.55 to 12.6 µg/g dw in 59 eggs. Concentrations of Hg typical of clapper rail eggs from southern California (where Hg contamination is less pervasive) were <0.11 µg/g dw. MeHg was 95% of the Hg in the eggs. The highest MeHg concentrations occurred in four marshes in South San Francisco Bay. South Bay receives runoff from what was once one of the largest Hg mines in the world (now abandoned). Of the failed eggs collected in South Bay, 50% had concentrations exceeding the 2.5 µg/g dw threshold above which adverse effects on hatchability occur in pheasant, and 25% were above the 3.0 µg/g dw threshold for mallard ducks. Given the sensitivity of clapper rails compared to these surrogate test species, Scharzbach *et al.* (2006) concluded that only Hg contamination could explain the reduced reproductive potential of the species. They suggested that strategies for recovery of this species would have to involve more than just creating new habitat. Reducing Hg exposure in that habitat was essential.

Administration issues advisories or regulates commercially sold products.

Advisories and regulations usually relate to top predators. The highest concentrations of Hg in ocean fish are in swordfish (*Xiphias gladius*), marlin (e.g. species of *Makaira*) and tuna (e.g. species of *Thunnus*) (especially the larger tuna). These animals feed at the top of one of the longest food webs on Earth and live a long time. Even in the absence of local sources of Hg, atmospheric inputs are sufficient to contaminate these sources of human nutrition. Top predators from estuaries, lakes, streams and rivers can also be contaminated, in contaminated waters where net production of MeHg is

elevated. Examples include pike (*Esox lucius*), walleye (*Sander vitreus vitreus*), lake trout (*Salvelinus namaycush*), striped bass (*Morone saxatilis*) and other fish that feed at the top of the food web.

The major concern is exposure of children to the MeHg in fish. When mothers eat such fish, MeHg is passed onto the developing child. Young children sometimes also eat large quantitities of the fish (e.g. canned tuna). People who are heavy fish consumers, especially of tuna or swordfish, can exhibit signs of Hg toxicity (Hightower and Moore, 2003). Exposure of pregnant women and children is particularly worrisome. As noted above, once taken up, MeHg spreads throughout the developing brain

Box 14.14 Defining safe concentrations of MeHg in food

Fish advisories, of course, require defining what an 'unsafe' concentration is in fish tissue; and as noted above that can vary among jurisdictions. The ingredients in defining risk are concentration in fish tissue (µg MeHg/g fish), times consumption rate by the person (g fish/kg person/d). This is compared to the dose thought to result in health risks to the human foetus as determined from toxicity tests and observations. For methyl mercury that maximally recommended dose is 0.1 µg MeHg/kg body wt per day. The US Environmental Protection Agency assumes that the average person eats 17.5 g per day and from that calculates what degree of contamination is acceptable. The concentration in fish tissue that should not be exceeded based on a total fish and shellfish consumption-weighted rate of 17.5 g fish per day (about 1 pound or two meals per month) is 0.3 µg/g wet wt or 1.5 µg/g dry wt.

of children and can cause damage during cell development.

Estimates of safe concentrations of MeHg in food (Box 14.14) vary among jurisdictions from 0.1 µg/g wet wt (or about 0.5 µg/g dry wt, in some European countries and Japan), to 0.3 µg/g wet wt (or about 1.5 µg/g dry wt: the US Environmental Protection Agency advisory), to the US Food and Drug Administration's limit of 0.5 µg/g wet wt (or about 2.5 µg/g dry wt). In the USA 45 states have fish consumption 'advisories' because of Hg levels, making up 1782 advisories for Hg contamination in specific fish or specific localities (Krabbenhoft et al., 1999). Over 1000 individual water body segments were identified by the states as specifically having Hg contamination. In addition, over 3900 water body segments were identified as impaired due to contamination by metals, which may include Hg. The National Research Council (from the National Academies of Science, USA) estimated that as many as 60 000 children may be born each year with neurological deficits as a result of Hg exposure in the womb (NRC, 2000).

The advisories also appear to be effective. Diminished consumption of dark meat fish, canned tuna and white meat fish was observed after a recent national Hg advisory was widely publicised. These decreases resulted in a reduction in total fish consumption of approximately 1.4 servings per month (95% confidence limits 0.7, 2.0) from December 2000 to April 2001, with ongoing declines past this period (Oken et al., 2003). Advisories do not tell the whole story, however. Consumption of fish also has great benefits to human health (because fish is high in proteins that protect people from heart disease). There is presently substantial controversy about the balance between the risks from Hg contamination and the benefits from eating fish (Cohen et al., 2005). Moderation on all counts, and special attention to vulnerable groups of people, seem to be commonalities amongst all advocates in the debate.

The preferred approach for relating a concentration of MeHg in fish tissue to a concentration of Hg in ambient water (and thus derive a water quality standard) is to derive a site-specific BAF based on water and fish collected in the water body of concern (Box 14.15). Site-specific BAFs incorporate the net effects of the biological and environmental conditions that can affect bioaccumulation at a particular location. Alternatively, bioaccumulation models can be used; or in the absence of data to use these approaches, the US Environmental Protection Agency PA provides a generic BAF for each trophic level. Assumptions also must be made about the proportion of total Hg that is MeHg, since most analyses are of total Hg.

The MeHg criterion is the only fish tissue criterion issued by the Environmental Protection Agency (all other criteria are based upon water concentrations: Chapter 15). Historically, the ambient water quality criterion for protection of aquatic life was 12 ng/L total Hg (USEPA, 1984). This value was derived from the desired tissue residue value divided by a universal bioconcentration factor. More recently, many states have adopted the Agency's 1997 criteria recommendations of 50 ng/L total Hg for human health protection from the consumption of both water and organisms, and 51 ng/L total Hg

Box 14.15 A site-specific MeHg water quality standard

Hope (2003) developed a site-specific MeHg criterion for the Willamette River basin, Oregon, USA. He used the US Environmental Protection Agency fish tissue criterion of 0.3 μg MeHg/g wet wt (1.5 μg MeHg/g dw). He estimated a biomagnification factor for each trophic level (1 to 5) using a bioaccumulation model. And he used a probability function from empirical data for the basin to estimate the fraction of total Hg that was MeHg. The distribution of dissolved MeHg to total Hg in the basin was 0.20 ± 0.07. The 50th percentile bioaccumulation factor for trophic levels 3 and 4 ranged from 1.9×10^6 to 7.0×10^6. He calculated that 'for the northern pikeminnow *Ptychocheirus oregonensis* (trophic level 4), when total Hg water concentration exceeds 2.6 ng/L, a 95% probability exists that the concentration in an individual fish will exceed the fish tissue criterion'. The value is on the higher end of Hg contamination in natural waters. The probability falls to 50% when the water concentration is about 0.4 ng/L; a value common in uncontaminated waters. A 95% probability exists that tissue exceedances will occur in 95% of the population when water concentrations exceed 9.5 ng/L; again, this value is typical of the highest levels of contamination observed outside of adjacency to an Hg mine. These values pass the simplest test of justifiable standards, although they have not been validated in other systems. The advantage of forecasting several probabilities is that it provides choices for decision makers with regard to how conservative they wish to be.

for human health protection from the consumption of organisms only (USEPA, 2001a). These values were derived using toxicological and exposure input values current at the time of its publication, including a bioconcentration factor. These values have not been validated against field data; but do represent extreme contamination. They are not particularly consistent with the effects levels developed from the body of evidence in most of this chapter. Effects at environmental concentrations lower than this seem quite likely.

14.3.9 Conclusions: methyl mercury

Much is known about Hg in the environment. There is also much left to learn, especially with regard to forecasting the implications of Hg contamination in different settings. In particular, translating the knowledge about MeHg to environmental standards is, at the best, inconsistent.

Reasonable estimates exist for the various fluxes in the global Hg budget; and it is recognised that atmospheric contamination is responsible for MeHg contamination in surprisingly remote regions. Mass balance budgets also exist for particular water bodies. The chemical and biological processes that control both local and global Hg budgets are challenging to quantify, however, because of their complexities and the very low concentrations involved.

The entry of MeHg into the base of the food web and its subsequent transfer in the lowest trophic levels are not as well studied as MeHg in fish. Nevertheless, it is established that the concentration of MeHg in all trophic levels is strongly correlated with its supply from methylating environments. Food web processes then determine the concentrations in the organisms at the top of the food web. Geochemical factors influence methylation but seem to have less influence on MeHg uptake by animals than is the case with many metals. Consumption of animals at the top of the food web provides the greatest risks to human health.

The adverse effects of MeHg in both humans and fish are probably the result of transport of the MeHg–cysteine complex, which is not recognised by the body as a toxicant. The most serious effects occur when MeHg–cysteine is broken down into inorganic Hg in the central nervous system or when MeHg–cysteine crosses the barrier between a mother and her developing embryo. Some animals appear to be able to detoxify at least some of the MeHg burden by complexation with Se in the liver.

But adverse effects on the most sensitive species of fish and birds appear likely at contamination levels that occur in at least some modern waters.

Traditional dissolved metal toxicity testing does not address the mechanisms by which the insidious toxicity of MeHg is manifested; criteria derived from such tests are unlikely to be protective of nature. The US Environmental Protection Agency has developed criteria for acceptable concentrations of Hg in fish tissues to protect human health. This is the beginning of what we have described as a more lateral approach to management and one that is likely to lead to improved efficiency and effectiveness in the regulations. Protection from ecological risks will ultimately require a similar type of approach (Chapter 17).

Suggested reading

Bryan, G.W., Gibbs, P.E., Hummerstone, L.G. and Burt, G. R. (1986). The decline of the gastropod *Nucella lapillus* around South-West England: evidence for the effect of tributyltin from antifouling paints. *Journal of the Marine Biological Association of the United Kingdom*, **66**, 611–640.

Clarkson, T.W. (2002). The three modern faces of mercury. *Environmental Health Perspectives*, **110**, 11–23.

Huggett, R.J., Unger, M.A., Seligman, P.F. and Valkirs, A.O. (1992). The marine biocide tributyltin: assessing and managing environmental risks. *Environmental Science and Technology*, **26**, 232–237.

Lindberg, S., Bullock, R., Ebinghaus, R., Engstrom, D., Feng, X., Fitzgerald, W., Pirrone, N., Prestbo, E. and Seigneur, C. (2007). A synthesis of progress and uncertainties in attributing the sources of mercury in deposition. *Ambio*, **36**, 19–33.

Morel, F.M.M., Kraepiel, A.M. and Amyot, M. (1998). The chemical cycle and bioaccumulation of mercury. *Annual Review of Ecology and Systematics*, **29**, 543–566.

Munthe, J., Bodaly, R.A., Branfireun, B.A., Driscoll, C.T., Gilmour, C.G., Harris, R., Horvat, M., Lucotte, M. and Malm, O. (2007). Recovery of mercury-contaminated fisheries. *Ambio*, **36**, 33–44.

15 • Hazard rankings and water quality guidelines

15.1 INTRODUCTION

Prologue. Quantitative definitions of acceptable ambient water quality for metal contamination will be of increasing importance in the future in risk management. This chapter considers the science underlying risk assessment and derivation of ambient water quality criteria, guidelines or standards. We discuss the biological and geochemical uncertainties that influence the choice of criteria and the outcomes of risk assessments, as well as the methodologies available to address at least some of these. Criteria from around the world are compared, and some generalisations are drawn about their effectiveness.

In Chapter 2 we discussed the different scales at which metal contamination is regulated. At the global scale, hazard analysis ranks the potential for environmental disruption by each metal. At the national and international scale, hazardous metals are regulated via water quality criteria, standards or guidelines (Box 15.1). At the local scale, national criteria can be converted to local, enforceable standards. Site-specific objectives (in the USA) or watershed analyses (as directed by the European Union's Water Framework Directive) are other ways to account for local conditions. Risk assessments are employed at the global, national and local scales to evaluate threats. Many of the same uncertainties apply to risk assessment as to criteria derivation.

An important question is whether the existing approaches to managing metal contamination 'work' (Chapman, 1991). It is well documented that metal contamination has declined in the recent decade, at least for some metals in some places. The declines are primarily responses to regulations on effluents from urban and industrial point sources, improvements in mining practices and remediation, and continual implementation of improved technologies by dischargers (best management practices) where mandated by government or corporate policy. Modern contamination, however, also requires addressing dispersed sources, geographical areas where government control of contamination is not strong, and new technologies (e.g. metal-based nanotechnology). Modern approaches to managing risks from such sources include risk assessment and regulatory approaches that are comprehensive in addressing the various sources of contaminant input in a watershed. The Total Maximum Daily Load (TMDL) process (Chapter 4) in the USA involves not only determining inputs from all sources in the watershed, but also regulating those based upon a limit on the ambient concentration in the water body of interest (defined by ambient water quality criteria). The Hg regulations in the USA even involve trading Hg 'credits' among sources, a system that only works if a tight cap is set on allowable concentrations in the environment (i.e. a protective but realistic ambient environmental standard). The Water Framework Directive, and its marine/estuarine equivalent in Europe, also relies upon ambient environmental standards to classify water bodies on the basis of chemical contamination. Risk assessments frequently depend upon or suggest ambient standards. Thus it can be argued that ambient environmental standards were not the most important factor in the successes to date with managing metal contamination; but they will be of increasing importance in the future (Ford, 2001). It is important to get them 'right'. In this chapter we examine the science underlying these important environmental

Box 15.1 Definitions

acute-to-chronic ratio (ACR): ratio between acute toxicity and chronic toxicity.

ambient criterion or environmental standard: criterion defining the concentration of a contaminant in natural waters above which environmental damage is expected.

criterion continuous concentration: USEPA terminology describing a criterion developed from chronic toxicity data (or acute data with an application factor to correct to chronic data) meant to define the 4 day average concentration not to be exceeded.

criterion maximum concentration: USEPA terminology describing a criterion developed from acute toxicity data defining the one hour average not to be exceeded (limiting episodic or pulse exposures).

hazard identification: hazard identification (described in Chapter 1) determines the potential for adverse effects based on a substance's fundamental properties (Floyd, 2006).

lowest observed effect concentration (LOEC): lowest concentration at which a detectable adverse response to a metal is observed.

no observed effect concentration (NOEC): highest concentration at which no effect is observed.

precautionary: action to avoid the potential adverse impact of the release of hazardous substances before

scientific research has convincingly shown a causal link between these substances and adverse impacts.

predicted environmental concentration (PEC): concentration of metal observed or predicted in a natural water for comparison to a predicted no effect level.

predicted no effect concentration (PNEC): highest concentration at which no effect is predicted.

Total Maximum Daily Load (TMDL): calculated maximum amount of a pollutant that a water body can receive and still meet water quality standards, and an allocation of that amount to the pollutant's sources. The TMDL process is defined by USEPA as the process whereby an allocation of loadings from each source is achieved.

Water Effects Ratio (WER): ratio of the toxicity of a metal in the water from a specific site, compared to the toxicity of the same metal in standard laboratory water.

water quality criterion (synonym: *water quality objective*; also used: *water quality guideline*): numerical concentration or narrative statement, the goal of which is to support and maintain a designated water use (e.g. protection of aquatic life). A guideline may be more flexible than a criterion or objective. Guidelines (and sometimes criteria) are not necessarily legally binding.

water quality standard: objective that is a legally enforceable environmental regulation; often developed from criteria.

standards. We evaluate the protectiveness of a few standards as examples, and discuss opportunities to manage risks in a more effective manner.

The choice of a numerical ambient environmental standard can be greatly influenced by the availability of data and by how geochemical and biological uncertainties are addressed. Some ambient criteria for trace metals appear well justified on the basis of relevance to natural waters and comparisons with modern scientific knowledge of metal effects. But others are difficult to justify. Many experts assume that metal criteria are uniformly over-protective, because of safeguards built

into standardised toxicity testing (Di Toro, 2003). We show that examples exist of apparent over-protection, usually because geochemical uncertainties are not addressed. But examples also exist of likely under-protection, usually because of biological uncertainties in the data used to derive the criteria.

15.2   HAZARD IDENTIFICATION

Prologue. Persistence, bioaccumulation and toxicity are the typical criteria defining the potential hazard from a chemical (Chapter 2), but they are difficult to apply unambiguously to trace metals. Persistence is a characteristic of all metals. Simplistic characterisations of bioaccumulation can be misleading. Toxicity determinations emphasise data from acute, dissolved exposures, and outcomes that are influenced by how the test is conducted. Hazards from metals that are efficiently transferred via food and food webs are mischaracterised by the traditional approaches. Potential for trophic transfer is a valuable, but underutilised inherent property that differs among trace metals and greatly affects potential for risks. The order of hazard rankings is likely to change if potential for trophic transfer is added to the traditional measures. Se, Cd, Zn and perhaps other elements would be ranked as greater hazards if trophic transfer were to be considered.

Managing environmental risks from chemical contamination involves regulating thousands of chemicals and wastes. The goal is to control harmful chemical releases to the environment, with emphasis on those chemicals that can pose long-term problems when released. Most jurisdictions prefer source reduction and recycling methods over waste treatment and disposal methods for managing risks. That requires determining what chemicals are most hazardous and restricting the volume of their releases (see REACH in Chapter 2, Box 2.3). The scientific process that guides this effort is termed *hazard identification.*

The potential hazards of trace metals are ranked both among metals and relative to other chemicals. Persistence, bioaccumulation, and aquatic toxicity (PBT) are used in the international ranking system for hazard classification of chemical substances (Fig. 15.1). *Persistent* (P) chemicals do not readily break down in the environment. For the purposes of hazard assessment, *bioaccumulation* (B) is defined by whether a chemical is easily metabolised and whether it can accumulate in human or ecological food chains. *Toxicity* (T) is defined from dose–response determinations. A fourth category included in some analyses is potential for long-range transport (LRT) (Mackay *et al.*, 2003), or potential for dispersal. It is assumed that chemicals that exhibit these characteristics will present a long-term toxic threat to human

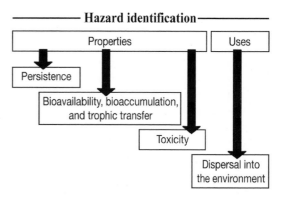

Hazard identification

Fig. 15.1 Components of hazard identification for trace metals.

health and the environment, if released even in small quantities.

15.2.1 Persistence

Prologue. Metals are inherently persistent, so this property does not differentiate their hazard. Bioavailability is relevant to risk but not to the determination of hazard on the basis of persistence, because metal forms change when environmental conditions change.

Chemicals that are persistent take a long time to break down to non-toxic forms once they are released into the environment. From the perspective of national or international rankings, most observers agree that all metals are equally persistent. As elements, their degradation rate constants are zero. This categorisation does not allow differentiation or ranking of hazard among trace metals.

However, trace metals occur in different forms in the environment, some of which are not bioavailable or toxic. This led to consideration of whether conversion to less bioavailable forms reflects reduced persistence (the equivalent of breaking down). There is now a consensus that most environmental forms of metals can transform if environmental conditions change. Thus bioavailability status is not a permanent or inherent property of a metal; and not an indicator of persistence.

15.2.2 Bioaccumulation and alternatives

Prologue. Bioaccumulation was included in hazard identification for organic chemicals because of concerns about those that increase in concentration at each trophic step in a food web; but this concept is not typically applied to hazard analysis for metals. For metals, bioaccumulation has been used as an indicator of chronic toxicity on the assumption that toxicity potential was proportional to accumulated concentration. But objective measures of metal-specific bioaccumulation (e.g. bioconcentration or bioaccumulation factors) are problematic to apply.

The criteria traditionally used to define bioaccumulation potential for organic chemicals yield highly variable, inconsistent values for metals. Traditionally, bioaccumulation is determined in standardised tests, where organisms are exposed to a metal in water. A bioconcentration factor (BCF) is calculated to quantify bioaccumulation from water.

$$BCF = \text{whole body concentration} \,/$$
$$\text{exposure mediadissolved concentration}$$

A bioaccumulation factor (BAF) can also be used. The BAF is formally defined as a measure of accumulation from water and diet together; and formally designated as derived from a combination of dissolved and tissue-bound metal data from a field site.

$$BAF = \text{whole body concentration} \,/$$
$$\text{a defined environmental concentration}$$

In practice, BAFs and BCFs are sometimes used interchangeably creating confusion in the literature. It is not uncommon that values cited as BAFs may include data from simplistic dissolved laboratory exposures. This is important because BCFs for trace metals in laboratory tests are consistently lower (often by orders of magnitude) than BAFs determined from reliable field data (see Se example in Box 15.2). Yet field-based BAF data have had minimal impact in hazard identification, either because of lack of data or because of challenges with interpretation.

The threshold BCF above which a substance is deemed a hazard is between 500 and 5000, depending on the jurisdiction (McGeer et al., 2003). To define whether a metal BCF falls within that range, a small range of values should characterise each metal. Unfortunately, BCF or BAF for each metal is an uncertain measure of bioaccumulation potential.

Firstly, BCFs or BAFs for one metal can vary by orders of magnitude among different biological species (Chapters 6 and 13). The value is also affected by the environmental conditions under which the test is conducted and/or the biological characteristics of the species. The theoretical basis for applying a BCF or BAF does not consider influential factors like species-specific biodynamics, essentiality, natural background, homeostasis and detoxification.

Secondly, the toxic threat of a metal can vary among species at the same BCF. Some aquatic organisms store metals in detoxified forms, such as in inorganic granules, and these in particular accumulate very high concentrations that are not toxic (Chapter 7). High concentrations of the same metal in another species may be highly toxic if a higher proportion of accumulated metal is not in detoxified form. The degree to which the BCF is related to the inherent threat of the metal is thus widely different among organisms.

Thirdly, BCFs are negatively correlated with concentration in the animal's tissues in many circumstances (McGeer et al., 2003) (Fig. 15.2). Background concentrations, essentiality and biodynamics contribute to the non-linearity of this relationship.

Variability in BAF can be reduced by relating concentrations in a specific consumer to its specific diet, for example, but broader uses are usually of limited value. A recent workshop of experts concluded that commonly recognised methods for characterising metal-specific bioaccumulation potential (e.g. BCF) are ambiguous. Direct assessments are the best basis for rankings that consider chronic toxicity or trophic transfer potential (Adams and Chapman, 2006).

15.2.3 Toxicity

Prologue. Toxicity can be ranked for trace metals from the large dissolved toxicity testing database.

Box 15.2 Ranking hazards from a trophically transferred trace metal: Se

Rankings of the hazard of Se, compared to other trace metals, varies widely depending upon how the analysis of bioaccumulation and toxicity are conducted. Dissolved Se has extremely slow uptake rates by animals (Chapter 13). That means it has very low toxicity to animals in dissolved toxicity tests. It also means that, in natural waters, dissolved Se does not play a significant direct role in the pathway of Se exposure. For example, the BCF of Se in dissolved exposures for all animals is $<1 \times 10^3$. Fish from San Francisco Bay that eat bivalves, like sturgeon (*Acipenser transmontanus*) and Sacramento splittail (*Pogonichthys macrolepidotus*), have Se BAFs of $\geq 1 \times 10^5$. Thus field observations differ by two orders of magnitude from laboratory tests.

The Toxics Release Inventory program in the USA uses a benchmark BCF of 1000 to classify a compound as 'trophically transferred' and a value of 5000 to classify a substance as 'highly trophically transferred' (USEPA, 1999a). The experimental BCF for Se based upon dissolved exposures (<1000) would rank it as of the lowest hazard. A BAF based upon environmental observations, however, would show that Se exceeds the 'highly trophically transferred' benchmark by two- to 10-fold (highest hazard). Clearly the bioaccumulation hazard potential is influenced by how it is determined.

In nature dissolved Se is transformed by primary producers to particulate Se, most forms of which are of high trophic bioavailability. This is the primary route by which Se enters food webs (Chapter 13). Particulate concentrations are 10^2 to 10^5 higher than the dissolved concentrations. Particulate Se is efficiently assimilated from particulate diets by invertebrates. The hazard of Se is then further amplified by its effective transfer from invertebrates to their predators through food webs. Laboratory experiments and field observations both show that higher trophic level birds and fish are the organisms most at risk from Se toxicity, because of this food web transfer. No trace metal is as effectively transferred through food webs in natural waters, but so inefficiently taken up from solution in traditional experiments.

If dietary exposure is considered, Se concentrations in prey organisms begin to elicit toxic effects in their predators at tissue concentrations that can be <5 times above natural background. This narrow window between essentiality and toxicity is another aspect of the hazard of Se. Few trace metals have such a small margin of error for managing the contaminant. The potential ecological hazard of Se is verified by clear examples in nature where trophic transfer of Se contamination has resulted in the collapse of populations (extirpation) of fish and birds (Chapter 13; Box 13.8).

But the rankings are influenced to some degree by the choice of species and test conditions. Nevertheless there is a general consensus about the rank of dissolved metal toxicities.

In hazard classification for organic chemicals, acute dissolved toxicity is often used to identify the potential for hazard in the short term, while persistence and bioaccumulation are used as surrogates for chronic hazard potential as determined by chronic toxicity tests. Thus, in practice, the ranking rests upon toxicity; and in nearly all cases, dissolved toxicity.

Intuitively, it seems that acute dissolved toxicity should be a relatively unambiguous way to accomplish a toxicity ranking. In Chapters 10 and 11, however, we saw that the concentration at which a trace metal is toxic is greatly affected by the way

toxicity is determined. If it is determined over very short periods of time, for example, comparability among species becomes an issue (exposure time compared to generation time varies widely among species). If dietary toxicity is important, the ranking from dissolved tests can be misleading (Box 15.2). Rankings may also vary between chronic and acute toxicity.

Table 15.1 shows traditional hazard rankings of nine metals from the website of the NGO Environmental Defense. The total hazard rankings include both human health and ecological risk. The ecological hazard uses traditional approaches, and is an amalgam of conclusions reached by three slightly different methodologies (two US universities and USEPA). All three approaches employ toxicology databases to draw conclusions; and so dissolved toxicity is a primary driving factor in the

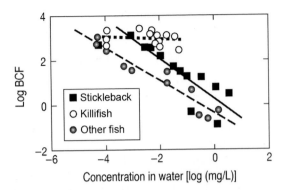

Fig. 15.2 Bioconcentration factors (BCF) for Cd as a function of dissolved Cd exposures in stickleback (*Gasterosteus aculeatus*), killifish (*Fundulus heteroclitus*) and 'other fish' for which bioassays data exist (after McGeer et al., 2003). The authors found that BCFs for Cd, Cu, Hg, Ni, Pb and Zn decrease as a function of exposure concentration. Bioaccumulated concentrations increased as a function of exposure in all studies, but the ratio of the two (BCF) was negatively related to exposure. This could be a result of natural background concentrations, physiological regulation of essential metals (not applicable for Cd), the way the exposures were conducted or the physiology of uptake. The important point is that no simple generic BCF can be derived that characterises bioaccumulation of all species, in all circumstances for any metal.

Table 15.1 *Relative rankings of the environmental hazard from different trace metals. Numbers are unitless and meaningful only in comparison to one another. Se and Hg would rank highest in all the categories if trophic transfer and effects on predators were appropriately considered*

Trace metal	Ecological risk	Environmental hazard	Total hazard
Hg	9	140	29
Cd	9	140	33
Pb	9	140	33
Se	7	85	22
Cu	7	100	29
Zn	7	140	11
Ni	7	140	32
Cr	7	160	33

Source: Data from Environmental Defense (http://www.scorecard.org/chemical-profiles/index.tcl) using USEPA, Purdue University and University of Tennessee ranking methods for ecological and human health hazards from trace metals.

ecological ranking. Most analyses arrive at the same general order of ranking on this traditional basis. Hg, Cd and Pb are ranked most ecologically hazardous; Se is not ranked highly. The greatest total hazard potential is attributed to Cd, Pb, Ni and Cr, with Hg and Se well down the list.

The European Commission, in a similar exercise, established List I, most dangerous substances, and List II, substances of concern. The metals on List I are Hg and Cd. The metals on List II are Pb, Cr, Zn, Cu, Ni and V. Se is not listed as a dangerous substance. While there are many similarities in these lists, the fact that there are differences reflects the complexities in deciding which metals are most hazardous.

15.2.4 Trophic transfer

Prologue. Trophic transfer potential varies among trace metals, and is an inherent property of each metal. Usually a good test of whether the choice of methods is justified is by comparison to circumstances in natural waters. Contradictions occur between predictions of toxicity in traditional tests and observations of metal toxicity in natural waters, many of which are explained by trophic transfer. Adding trophic transfer to hazard identification is a way to make the hazard identification process more 'lateral' (crossing the boundary between toxicology, biology and ecology), and more holistic.

Greater impacts from organic chemicals occur in large, long-lived, higher trophic level animals than would be expected on the basis of toxicity testing. That is why efficiency of transfer through food webs is a consideration for organic chemicals. At present, hazard identification protocols for trace metals do not consider food web transfer. This is apparently a carry-over from the now disproven assumption (Chapters 6, 10 and 11) that dietary transfer of metals is unimportant. As we have seen earlier, a strong body of evidence shows that significant metal bioaccumulation does occur via diet and through food webs. This trait is different

Table 15.2 *An alternative approach for rankings hazard for several trace metals if the potential for dietary uptake, trophic transfer, and toxicity are added considerations. If dietary toxicity is additive with dissolved toxicity, then trace metals for which there is great potential of exposure via that route will be a greater hazard. Trophic transfer in this case means propensity to be transferred to upper trophic levels, where the greatest effects will occur. This is an effect additive to dietary bioaccumulation because it involves magnified exposure in upper trophic levels. The scoring system is 1 = low to 3 = high risk. The values were determined by professional judgement, and are only meant as an example of how such factors could be considered*

	Dissolved toxicity	Dietary exposure	Trophic transfer	Total score
Hg	3	3	3	9
Cd	2	3	2	7
Se	1	3	3	7
Ag[a]	3	2	1	6
Cu	3	2	1	6
Ni	2	2?	2?	6
Zn	2	2	2	6
Pb	1	1	1	3
Cr	1	1	1	3

[a]Seawater.

among metals and among animal species; and is a factor in determining the potential hazard to animals.

As best is known, the trophic transfer potential of trace metals approximately follows the order Hg > Se > Cd ≥ Zn ≥ Ag > Cu > Pb > Cr, with Ni unknown (Wang, 2002). Table 15.2 shows hazard rankings if trophic transfer potential and dietary bioaccumulation are considered. The most important conclusions from Table 15.2 agree with what we know about trace metal effects in nature. Hg is unquestionably the most hazardous of the metals. But Se and Cd are also of high hazard, if they are mobilised in bioavailable form. Only consideration of trophic transfer would result in such a ranking for Se, yet it seems justified from what we know about its effects in nature (Box 15.2; Chapter 13).

15.3 PROTOCOLS: AMBIENT CRITERIA AND RISK ASSESSMENT

Prologue. Ambient water quality criteria and risk assessments are the primary tools for managing metal contamination in the water column.

Ambient water quality criteria define the (dissolved metal) concentration in natural waters above which ecological damage can be expected. Single, numerical standards are the dominant approach (Chapters 2 and 3). Risk assessment is a more flexible approach, but can also be focused on a simple numerical definition of the concentration at which ecological damage is expected from contamination.

15.3.1 Ambient water quality criteria

It is the mark of an instructed mind to rest easy with the degree of precision which the nature of the subject permits, and not seek an exactness where only an approximation of the truth is possible.

Aristotle, from Chapman (1991)

Water quality criteria attempt to set the limits that assure that no adverse effects on the use of water by humans and organisms (Adams and Rowland, 2003; Chapter 2). Criteria take several forms: (1) numerical values applied to all waters; (2) criteria modified to reflect site-specific conditions; (3) narrative criteria that explain the goal rather than quantify it (e.g. no further deterioration); and (4) criteria derived using scientifically defensible methods other than those formally specified by the ruling jurisdiction. Numerical values are the most common approach for managing water quality at the national level.

Formal guidelines for deriving numerical criteria for aquatic life differ in some details among jurisdictions. But most assume that a single numeric will protect an array of different environments, sometimes at the geographical scale of continents. Exceedance of the designated value is expected to result in ecological damage. Thus a numerical criterion is a single concentration that represents the 'bright line' between adverse effect and no effect. Numerical criteria are also consistent with the regulatory demand for 'simple measures' that can be applied 'in a timely manner' (Chapter 2). They are easily written into law, and are simple to enforce. The critical question is whether they are suitable for the future; and what the alternatives are.

The problems with the 'one size fits all' numerical standards are well known (Di Toro, 2003). Most are derived from single-species toxicity testing data (Foran, 1993) which have important assumptions and limitations in extrapolating to nature (Chapter 10). The simplest numerical criteria implicitly assume that a safety (or 'application') factor can be derived to account for differences between laboratory toxicity test results and field conditions. As methods become more advanced, it is expected that these somewhat arbitrary factors will become less influential. Even if those assumptions held, however, a science-based approach inherently produces a range of values, not absolute numbers (Chapman, 1991). For example, the coefficients of variation of the LC_{50}s varied by 31% to 74% when Anderson and Norberg-King (1991) conducted 10 replicate toxicity tests, with each of two metals (Cu and Cr) and four species (*Ceriodaphnia*, fathead

minnow, echinoderm fertilisation and bivalve embryos). Yet from such tests a single protective number must be derived. The uncertainties that this introduces are not important if levels of contamination are far in excess of protective guidelines; but become important where high stakes decisions rest on small deviations above or below a criterion level.

The contradiction between the legal demands for an exact number and the scientific supply of a range of numbers is a long-standing source of controversy in managing contamination. Preston (1979) argued that a viable alternative was a consistent process, not one number, as means of managing contamination (Chapter 2). Chapman (1991) argued for use of the term 'guidelines', defined as a framework to guide management instead of a single number. The USEPA Science Advisory Board (e.g. USEPA, 1995a), and experts in other fora (Reilly *et al.*, 2003) repeatedly suggest that a more flexible approach to managing risks would be more compatible with the state of knowledge (e.g. Box 15.3).

Recognition of the limitations of numerical criteria results in modifications in how they are employed. In some jurisdictions (e.g. Canadian Environmental

Box 15.3 Adding flexibility to management of risk from metals: a proposed scheme

Scientists from government, industry and academia (Reilly *et al.*, 2003; summarised by Di Toro, 2003) recently defined three different types of criteria that might be employed in the optimal scheme for managing metal contamination. National criteria (Type 1) were to be protective, but not necessarily predictive of environmental effects from metals (i.e. they were precautionary, protecting aquatic life under most circumstances). These criteria would be the least modified for local conditions. Type 2 criteria would include modifications to consider local environmental conditions and local ecology. Thus they would include methodologies to modify the Type 1 criteria for considerations such as bioavailability and biologically unique attributes of a water body. Type 3 criteria were fully integrated risk-based, site-specific assessments. Multi-media data from the specific site of interest were necessary for Type 3 criteria. They would be justified where unusual or difficult problems

affected determination of concentrations at which effects might be expected (e.g. if bioavailability was difficult to determine from standardised methods; suspicions existed about over- or under-protection, or focus was needed on a specific population). Presumably, Type 3 criteria would be easiest to justify where the stakes were high.

The workshop suggested that a multi-tier regulatory system would

- clarify the level of protection expected,
- reduce uncertainties where it is important to do so,
- improve the efficiency and effectiveness of regulations,
- improve risk communication.

This scheme has not been implemented in risk management by any regulatory jurisdiction, nor is it typically considered in risk assessment. But it is typical of ongoing advice from advisory boards and the scientific community in general: greater flexibility would allow better accounting for complexities when managing risks from metal contaminants.

Quality Guidelines) national criteria are not legally binding, but are guides for local, state, provincial or regional (legally binding) 'standards'. But the local standards are still numerical standards. European 'Directives' are more narrative than numerical. The Water Framework Directive, for example, sets a target of 'progressive reduction' for emissions of the 33 priority substances (including the trace metals As, Cd, Cr, Cu, Hg, Ni, Pb, Zn and tributyl tin). But legally binding numerical criteria are retained in member states. Site-specific objectives, similar to Type 3 criteria in Box 15.3, are a common way to modify national criteria in high stakes situations in the USA; but again the objectives are numerical.

15.3.2   How are criteria derived?

Prologue. Regulatory bodies provide formal guidance about what constitutes acceptable scientific data for derivation of criteria, guidelines and standards (Foran, 1993). That guidance largely restricts data to standardised toxicity tests, with an emphasis on acute tests. Correction factors are used to account for the most important assumptions behind those tests.

The administrative process whereby water quality criteria are derived is usually determined by formal administrative guidelines (e.g. the US process is pictured in Fig. 15.3). Derivation of the numerical criteria is the part of the process that leans most heavily upon science. Differences exist in terminologies, practice, and even goals among jurisdictions; but most derivation procedures have more fundamental similarities than they do differences (see comparison of procedures in Canada, New Zealand, the Netherlands and the USA in Reilly *et al.*, 2003).

The original objective was to use chronic or partial life cycle toxicity test data in criteria derivation, supplemented by 'good scientific judgment' (Adams and Rowland, 2003). It was assumed that chronic tests more closely represented how adverse effects might be manifested in nature. But chronic studies, and especially life cycle tests, proved expensive, time-consuming and complicated to

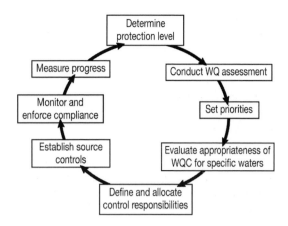

Fig. 15.3 The administrative process that governs the USEPA water quality based approach to managing metals is depicted (after Keating, 2003). The process is followed carefully through the step of establishing source controls. Compliance is better monitored in some systems than in others. In practice, the least emphasis is given to careful evaluations of progress, and no step is included that explicitly calls for learning from experience and using feedback to re-evaluate protection levels.

conduct. Thus, despite decades of toxicity testing, there are many metals and species for which chronic data sets are not robust (e.g. Cd in marine waters: USEPA, 2001c). But the simpler acute toxicity tests exist for a number of species under many conditions. Therefore it is typical to rely mainly on acute toxicity data and apply 'application factors', to adjust acute toxicity values to what would be expected from chronic exposures (e.g. Brix *et al.*, 2001). The criteria are usually based on the lowest concentration of a substance that affects the test organisms (Lowest Observable Effect Concentration, LOEC). What constitutes the 'lowest concentration' may vary (e.g. lowest 5th percentile).

The science that is acceptable for use in the derivation is defined by how well the test design follows a formally defined toxicity testing protocol (e.g. USEPA, 2003). In some approaches, a minimum number of species is designated for a suitable data set (e.g. Box 15.4). In other jurisdictions the data must include specific types of species. In Germany for example, toxicity studies for freshwater criteria must include data from primary

> **Box 15.4** Protocols for using toxicity tests to develop criteria
>
> Protocols for metals sometimes seem to be a complex way to preserve the 'simplicity' of numerical criteria. For example, the minimum database and requirements for criteria development by USEPA include:
>
> - toxicity data for the most sensitive life stage from a minimum of 1 species in each of 8 taxonomic families;
> - calculate mean acute values for each species and for each genus, and rank them from most to least sensitive;
>
> - use the lowest 4 genus mean values to calculate a final acute value (FAV), or final chronic value (FCV);
> - using the lowest 4 genera to interpolate (or extrapolate) the 5th percentile value (a value low enough to protect 95% of species represented by the test organisms);
> - the criterion represents the FAV or FCV divided by 2;
> - if insufficient chronic toxicity data are available, then the FAV is divided by an acute-to-chronic ratio to obtain the FCV (Keating, 2003).

producers (e.g. the green alga *Scenedesmus subspicatus*), primary consumers (e.g. the crustacean *Daphnia magna*), secondary consumers (e.g. fish) and reducers (e.g. the bacterium *Pseudomonas putida*). In the USA, acceptable tests must include sufficient data to determine a dose–response curve, under restricted water chemistry conditions, in the absence of feeding (e.g. USEPA, 2003).

Conceptually, defining acceptable science by the protocol employed seems a feasible way to assure quality. And it seems a logical way to sort a large, complex information base. In practice, however, it restricts consideration to databases based upon standardised toxicity testing and focuses on data, not understanding. Peer-reviewed literature is often excluded (L. Kaputska, personal communication, 2006, SETAC Asia-Pacific) on the basis of protocol, not creativity of the question or whether the outcome adds basic knowledge. For example, derivation of the Cd criteria for the USA (USEPA, 2001c) explicitly excluded hundreds of studies, including work that would have fundamentally changed the choice of the marine criteria (see later discussion).

15.3.3   Ecological risk assessment

Prologue. Risk assessments, as defined in Chapter 2 (Box 2.2) may use ambient water quality criteria, or may complement ambient criteria. They can be used to develop hazard rankings and protective criteria at the national scale.

A risk assessment comprises hazard characterisation, exposure assessment, risk characterisation and risk assessment (for definitions see Chapter 2, Box 2.2; Suter, 2006). The scientific requirements are different for the different steps. *Hazard characterisation* requires a dose–response curve for the metal. *Exposure assessment* involves:

- estimation of emissions to environmental compartments,
- determination of the fate of such emissions in the environment,
- derivation of predicted environmental concentrations (PEC),
- estimation of uptake by fauna (including exposure to humans).

In its simplest form, risk characterisation is defined by a risk quotient (RQ) that is derived from the comparison of the predicted (or observed) environmental concentrations (PEC) with the predicted no effect concentrations (PNEC) for the corresponding environmental compartment (e.g. dissolved metal):

$$RQ = PEC/PNEC.$$

Different terminologies are used in different jurisdictions to achieve the same end. For example, the USEPA estimates risk using a hazard quotient (HQ) where an estimate of the exposure concentration is divided by a measure of the concentration at which toxicity is predicted (reference exposure):

$$HQ = Site\ exposure/Reference\ exposure.$$

The RQ and HQ approaches both compare site exposure levels to literature-based concentrations believed to cause no or minimal toxic effects. Formally, in the US approach, HQ values do not account for site-specific factors, although other methods are available that can (see below). Effects are not expected when the ratio <1, while values >1 indicate that effects may occur. Where values >1, it has to be decided whether further information and/or testing are required to clarify the concern, whether to refine the risk assessment, or whether to act to reduce the risks.

Reliance on the PNEC or reference exposure in a risk assessment restricts data to toxicity tests. But risk assessment can be flexible. For example, exposure and effects data can be based on probability distributions rather than point estimates. Potential risk is then expressed as a probability distribution rather than a single value (e.g. Brix et al., 2001). The risk assessment is further strengthened if other data are added to supplement the toxicity data. Local (e.g. geochemical) influences can be incorporated into the PEC to account for bioavailability; or bioavailability can be incorporated into the PNEC, if testing data are available across different geochemical conditions. It is less common to include bodies of knowledge that include observations from nature, like those used to understand metal effects in Chapter 11, however.

15.4   UNCERTAINTY

Prologue. The extrapolation of toxicity testing data to protective measures and risk assessment requires some important assumptions. New methods address many of these, but the net effect of incorporating those methods awaits validation.

Because all risk assessments are predictive, they are inherently uncertain (Chapter 2). Risk assessment must always include characterisation of the implications of these attendant uncertainties (Suter, 2006). This is different from justifications for ambient criteria, which rarely express assumptions and uncertainties. Unfortunately, there is not a uniform scientific consensus on how to resolve the cumulative uncertainties that influence choices of criteria and outcomes of risk assessments (Bodar et al., 2005), but methods are rapidly evolving to address individual uncertainties.

However, it is possible to clarify uncertainties. Attention must also be paid to maintaining a balance, so as not to lean unrealistically toward either over- or under-protection as result of giving unbalanced weight to different uncertainties. For example, some of the assumptions of toxicity tests are biological: short exposures, the use of surrogate species, and the contention that only dissolved toxicity is important in natural waters. If not corrected for, these assumptions would mostly lead to under-protection of the environment. Others are geochemical: test conditions are designed to maximise toxicity. If not corrected for, these would mostly lead to over-protection.

Historically, large corrections or 'assessment or application factors' were applied to correct criteria for cumulative uncertainties. New protocols increasingly take careful account of geochemical limitations. They address some, but not all, of the biological simplifications; and in many cases eliminate application factors. This raises questions about balance if biological corrections are not as complete as geochemical corrections; i.e. the historical 'balance-of-errors' may be lost.

15.4.1   Geochemical uncertainties

Prologue. Risk assessments and ambient criteria both involve comparing ambient metal concentrations (PEC) to a predicted no effect concentration (PNEC or reference exposure). Assumptions about geochemical conditions are instrumental in the interpretation of both.

15.4.1.1 Total metal versus dissolved metal
Prologue. Total metal concentrations are higher than dissolved concentrations; but dissolved concentrations are, unequivocally, the best indication of environmental metal concentrations for criteria and risk assessment.

Determinations of ambient metal concentrations or PEC were originally based on total metal concentrations in water (unfiltered). The purpose was to take a precautionary approach to environmental concentrations, because it was well known that total metal concentrations are always at least equal to, and usually higher than, dissolved concentrations (separated with a 0.45 µm filter). Total metals were also easier to detect, and reliable data could be obtained without using cumbersome ultra-clean methodologies.

The problem is that total metal concentrations are not indicative of anthropogenic input or adverse effects, because of their dependency on particulate abundance in the sample (Chapter 5). Most trace metals have a particulate/dissolved (K_d) distribution $>10^3$; thus a high proportion of the total metal concentration will depend upon the amount of suspended of sediment in the water sample.

A total metal PEC is also not comparable to a typical toxicity test derived PNEC. Tests of dissolved toxicity are conducted in particle-free water explicitly for the purpose of eliminating the ambiguities caused by metal association with particles.

Modern criteria almost all use dissolved metal concentrations as their measure of environmental metal concentrations. Moving to dissolved metal concentrations as the environmental exposure indicator was an essential change, based upon solid scientific evidence. But it puts more pressure on getting right the other aspects of the protocols in criteria making.

15.4.1.2 Hardness and water effects ratio

Prologue. Hardness corrections and empirically determined 'water effects ratios' are simplified approaches to correct criteria and risk assessments for local speciation. Neither addresses metal bioavailability as effectively as the Free Ion Activity Model, but their use is widely accepted in the regulatory community as a step forward.

Another precautionary aspect of early criteria was the requirement to conduct toxicity tests in water with highly soluble, completely dissociated metal salts. The water was to be low in dissolved organic substances, particulate matter and major ions. The result was tests that maximised dissolved bioavailability.

Metal speciation strongly influences dissolved bioavailability and toxicity in freshwater and in marine waters, and it varies widely from place to place (Chapters 5 and 7). Hardness, salinity, specific ion levels, pH, alkalinity, inorganic complexes and dissolved organic matter can all be important. The Free Ion Activity Model (FIAM) was the original formulation defining effects of dissolved metal speciation on metal toxicity (Sunda and Guillard, 1976; Anderson and Morel, 1977; Chapters 5 and 7). Despite its direct application to marine natural waters, the free ion approach was never adopted, as such, by any regulatory agency. Free metal ion determinations (modelled or analysed), rarely, if ever, appear in risk assessments. The free metal ion approach was the harbinger of later approaches, but it remains largely in the domain of scientific study.

The first sign that the regulatory community recognised the importance of water chemistry was in the 1980s when European institutions and the USEPA developed ambient water quality criteria for metals that were different in freshwater and marine waters (Niyogi and Wood, 2004). This was an implicit recognition that high ionic strength influenced dissolved trace metal toxicity.

The next step was correcting freshwater data for 'hardness', as determined from calcium carbonate concentration. Logarithmic regressions, based on correlations between hardness and increasing LC_{50}s for different species, are now used to generate site-specific adjustments (using the hardness of local receiving water) of criteria or standards in most jurisdictions. Criteria concentrations are presented at different hardness concentrations (see Cd criteria: Box 15.10 below) to national audiences.

The Water Effects Ratio (WER) is an empirical approach that corrects for speciation issues caused by dissolved organic material, in addition to hardness and ionic strength influences. WER protocols involve side-by-side toxicity tests in the site-specific water and in hardness-matched 'clean' laboratory water. The WER is defined as the ratio of the toxicity of a metal in the water from a site, compared to the toxicity of the same metal in standard laboratory water:

$$WER = LC_{50}\text{site water}/LC_{50}\text{laboratory water.}$$

A site-specific adjustment can be determined by multiplying the national ambient water quality criterion value by the WER,

$$\text{site-specific criterion} = WER * AWQC.$$

At a minimum, WER determinations are usually repeated seasonally or during different hydrological or hydrodynamic regimes. A 'final' WER (fWER: Welsh *et al.*, 2000) is the mean of several WER determinations and can summarise the effects of speciation on toxicity at a site (Welsh *et al.*, 2000).

Diamond *et al.* (1997) concluded that seasonal effects on water quality and stream flow both influenced metal toxicity in a Pennsylvanian stream (USA). According to their WER, Cu was at least five times less toxic at some times of year than at others. In a mine-impacted creek, Welsh *et al.* (2000) reported ratios (site/criteria) that varied from 3.5 to 15.9 over time. Sixteen different determinations of a WER for San Francisco Bay, using larvae of the marine mussel *Mytilus edulis*, resulted in WERs ranging from about 2 to about 5.2.

WERs are typically run using acute toxicity tests because biological conditions are easiest to control. Implicitly it is usually assumed that an acute WER can be extrapolated to a chronic exposure. But Welsh *et al.* (2000) cited two different studies showing that a geometric mean acute WER for Cu was greater than the geometric mean from a chronic test with *Ceriodaphnia* and with fathead minnow (3.68_{acute} and 6.27_{acute} versus $2.85_{chronic}$ and $3.58_{chronic}$ respectively). They concluded that such results indicated a risk of under-protection in modifying chronic criteria using acute WER.

Hardness and WER corrections address an important assumption in the standardised toxicity tests: that the clean water toxicity tests represent conditions in natural waters. Neither is as theoretically robust nor as flexible as the models and tools typically used by practising geochemists; but most practitioners agree that they have provided some value as an interim tool for incorporating influences of geochemistry into risk assessment and criteria development.

15.4.1.3 Biotic Ligand Model (BLM)

Prologue. The Biotic Ligand Model (BLM) is a mechanistic approach to considering metal bioavailability from solution. In the regulator arena it is used to incorporate site-specific conditions into interpretations of toxicity, as a more sophisticated and flexible substitute for the hardness correction and the WER. It uses equilibrium geochemical modelling to calculate metal speciation, and adds an experimentally characterised biotic ligand to the model as a competitor for metal binding. Using the biotic ligand as a receptor for metal uptake allows the model to consider competition between the free metal ion and other naturally occurring cations (e.g., Ca^{2+}, Na^+, Mg^{2+}, H^+: Meyer *et al.*, 1999). An advantage of the BLM is that it shifts the emphasis from the exposure solution to the biological receptor. Concentrations on the bioreceptor (e.g. fish gill, invertebrate whole body) can then be correlated against toxicity. The method also has important uncertainties:

- organic complexation adds uncertainty to the speciation modelling. The most common outcome is under-prediction of bioavailability, unless a downward correction is made for organic complexation;
- uptake properties of the biotic ligand have only been characterised for a very few species;
- the use of acute toxicity data for the correlations require corrections to apply outcomes to chronic, sublethal or dietary toxicity;
- detoxification processes are not considered.

Theoretically, the Biotic Ligand Model (BLM) is an offshoot of the Free Ion Activity Model. The BLM predicts metal toxicity on the basis of calculated free ion activities, but it adds competitive effects of major ions and pH. The latter factors are important considerations in freshwaters (Meyer *et al.*, 1999). In terms of empirically derived applications, the BLM has been described as a way of doing a WER test in a computer, with a saving of time, effort and money (Niyogi and Wood, 2004).

There are several components to the Biotic Ligand Model:

- Characterisation of metal speciation.
- Characterisation of processes controlling uptake rates in the species of interest; these define the characteristics of the 'biotic ligand'.
- Characterisation of the outcome of the competition between ligands in the water, ligands on the membrane and other influential factors; i.e. how does water chemistry affect accumulation on the biotic ligand, expressed in terms of nM metal/g organism?
- Correlation of the accumulated metal (concentration on the ligand or LA_{50}) with percentage survival in a toxicity test. The emphasis has been on acute toxicity tests, to date. Presumably this is because of the difficulty of testing a geochemically complex methodology against a biologically complex methodology (chronic or life cycle tests) where variables other than geochemistry might be important.

The BLM concept has some important advantages:

(1) It quantitatively addresses the question of bioavailability from solution.
(2) Geochemically, it is mechanistically based, and thus much more flexible and robust than are the empirical approaches for evaluating influences of bioavailability.
(3) Biologically, it considers relevant uptake mechanisms, at least in theory. It assumes that toxicity stems from the basic, quantifiable properties that determine how metals are taken up by a species (e.g. fish gill properties).
(4) It has an explicit conceptual model for toxicity that is testable and can be validated, modified and improved upon. It assumes that toxicity will occur when a specified proportion of a functional site (the biotic ligand) is captured by a metal.

The BLM concept also has limitations and needs for further (or alternative) development. The uncertainties with the BLM are both geochemical and biological.

The most important geochemical uncertainty is how to quantify metal complexation with organic material in a general way, the long-standing weak link in speciation modelling. It is difficult to build into a model (by nature models are simplifications) all the complexities that govern organic complexation. Natural organic matter from different sources tends to bind metals differently, and affect toxicity differently (Ryan et al., 2004). This is especially important for metals like Cu and Pb. It is critical for environments high in organic material and for low concentrations of metals (as typically occur in nature: Cheng and Allen, 2006). Recent advances have improved upon these short-comings (WHAM VI: Tipping, 1994). Nevertheless, when computing Cu complexation, for example, there remains a tendency to over-predict Cu complexation, and thus under-predict bioavailability.

Cheng and Allen (2006) characterised Zn complexation with natural organic matter from (only) three freshwater sources. They compared the Zn binding characteristics to those predicted by WHAM VI. They found that the Zn complexation characteristics of surface water did not vary with source in this example. Nevertheless at Zn concentrations lower than 65 µg/L, and/or at pH $= 8$, the model over-predicted complexation, and under-predicted bioavailable Zn.

One approach used to 'correct' organic complexation for experimental (or local) conditions is to fit the predicted toxicity to experimental observations of toxicity (Box 15.5). Very different waters usually require their own fitting exercise. It is not uncommon in modelling to determine unknowns by 'fitting'. But where that is done, the number of uncertain fitting parameters should be small. Independent validation of the model forecasts is also critical (Niyogi and Wood, 2004); independent meaning comparisons are between two fundamentally different methodological approaches. BLM formulations derived by fitting to one toxicity testing data set have been compared to toxicity tests with the same species but in an array of different waters (e.g. Villavicencio et al., 2005). No published studies yet, however, test BLM forecast in natural waters to actual ecological change in those waters.

The biological coefficients that define the biotic ligand in the BLM have only been developed for a few species (Box 15.6). Thus the model commonly uses characteristics from one species (the fathead

Box 15.5 Organic complexation and the Biotic Ligand Model

Different approaches are used to account for the complexity of modelling metal–organic complexes. The difficulty lies in using a single set of constants for a variety of waters with a variety of organic matter. Operational methods to characterise the organic matter at each site are available, but no such method has yet proven fully reliable across a variety of circumstances. There are also uncertainties about what metal fractions are bioavailable (Campbell *et al.*, 2002). These problems are addressed in the BLM by calibrating the model to the toxicity data. The values for the organic complex are adjusted until the predicted toxicity (from experimental water) fits the toxicity observed in experiments with that water. Of course, other parameters must be held constant with each fitting exercise. The results usually involve assuming that only a fraction of the organic matter is active in complexing the metal. It seems unlikely that model formulations derived from such fitting exercises could be generalised to a wide array of circumstances (e.g. waters with different kinds of organic matter) without conducting toxicity tests in those waters and rerunning the fitting exercise. The errors are small compared to past efforts, but would be significant if the BLM were used to implement decisions about exceedances of numerical criteria.

minnow *Pimephales promelas*: Playle *et al.*, 1993) to determine bioaccumulation in other species. It is either assumed that uptake characteristics are similar among species, or coefficients are adjusted by fitting to species-specific toxicity testing data. Determination of biological uptake parameters by fitting toxicity curves is a second source of uncertainty. Correcting for this uncertainty requires making assumptions about the value that will correct for the uncertainties regarding organic complexation since both parameters cannot be derived by fitting. The mechanistic meaning becomes uncertain when multiple fitting parameters are employed in such models (Villavicencio *et al.*, 2005). Box 15.6 describes the details of how biological considerations influence BLM construction and use.

The BLM was initially developed as an indicator of acute toxicity. Acute toxicity of metals is rarely relevant to assessing risks from metals in aquatic environments. But fitting exercises with chronic toxicity data are just as feasible as fitting exercises with acute data; and are becoming more common (De Schamphelaere and Janssen, 2004; De Schamphelaere *et al.*, 2005). Both, of course, are subject to the same limitations as criteria development (variable acute-to-chronic ratios and a deficiency of reliable chronic toxicity data).

Equilibrium-based concepts assume that the only factor of importance in linking metal exposure and effects is concentration at the ligand. This contrasts to the dynamic conceptual model, described in earlier chapters: toxicity occurs when metal intake rates exceed the combination of detoxification and loss rates (Chapters 7 and 9). In its present state, the BLM offers no option for determining influx rates from diet, rates of loss or information about detoxification.

When the biological values are held constant, and the data are fitted to a species-specific toxicity test, the BLM is a construct primarily designed to quantify the correlation between toxicity and speciation in solution. An important question raised by such an approach is how widely results can be extrapolated (especially if data are to be used for development of ambient criteria). BLM formulations derived by fitting to one toxicity testing data set were compared with good success to toxicity tests with the same species but in an array of different waters (e.g. Villavicencio *et al.*, 2005) (Fig. 15.4). Toxicity was predicted over large ranges of water quality within a factor of two.

To further establish its credibility, the BLM construct must include:

- biological constants from a variety of organisms,
- methods to improve incorporation of influential characteristics of organic matter,
- relationships to effects data more sensitive than acute toxicity tests,
- consideration of dietary exposure,
- further validation against observations in nature or complex toxicity tests.

Box 15.6 Biological considerations in the Biotic Ligand Model

The Biotic Ligand Model applies the basic construct of speciation models to accumulation at the biotic ligand:

$$[M_iL_b] = K * [L_b] * [M_i^+]$$

where M_iL_b is the metal concentration associated with the biotic ligand, K is the log stability constant defining the affinity or strength of metal binding to the ligand (in the organism), L_b is the number of biotic ligand sites available and M_i^+ is the bioavailable metal as defined by the speciation model. Definition of M_i^+ includes definition of competition constants for the major ions, using the correlation with LC_{50}s. The LA_{50} is the $[M_iL_b]$ at which 50% toxicity occurs. The terms of this equation were developed with fathead minnow (Playle et al., 1993) and with rainbow trout. These same constants are typically employed in most studies. The LA_{50} for the species and waters of interest is determined from the fit to the toxicity data. Thus, for any speciation regime, it is possible to calculate $[M_iL_b]$ and compare it to the LA_{50} to determine risk for acute toxicity ($[M_iL_b]/LA_{50} > 1$ = toxic conditions). Conversely, the total metal concentration in solution can be calculated, at which acute toxicity will be observed in any speciation regime (that value will vary with speciation, of course).

For example, the approach to developing a daphnid BLM is to 'recalibrate' the fish model by adjusting the LA_{50} (downward) (e.g. De Schamphelaere et al., 2005).

Modellers use K and L_b from the fish model and adjust $[M_iL_b]$ until the predicted toxicity correlates with the toxicity testing data (Santore et al., 2001). Of course, the modeller could also obtain a correlation with the toxicity data by holding $[M_iL_b]$ constant and adjusting K until a successful fit was found. Neither case provides mechanistic information about the biological traits affecting toxicity. This approach to the Biotic Ligand Model has been expanded to a number of metals, several fish species and various species of daphnid cladocerans.

Villavicencio et al. (2005) compared BLM-calculated water quality criteria for Cu to the USEPA approach to determining water quality criteria (WQC). The USEPA acute freshwater WQC for Cu was calculated (using the latest protocols) as a function of water hardness:

$$WQC = \exp\{0.9422 \cdot [\ln(H) - 1.7]\}$$

where H is the hardness (as mg $CaCO_3$/L). Villavicencio et al. (2005) used, for their comparison, acute toxicity experiments with the zooplankton *Daphnia magna*, and tested nearly 50 waters in Chile. They found that BLM-predicted criteria were not statistically different from criteria suggested by USEPA in low DOC waters, both based upon short-term exposures. In high DOC waters, the USEPA approach did not account for organic complexation, so the limits suggested from the BLM in these waters were much higher than the suggestions from USEPA. The latter result was consistent with general theoretical expectations.

The BLM offers a scientifically derived basis for adjusting clean water testing for site-specific geochemical conditions. It is unquestionably an improvement over empirical hardness corrections and WER studies. But the BLM does not correct for the biological uncertainties inherent in toxicity testing data; so many of the limitations of the traditional criteria development approach are retained. Dietary exposure, extrapolation across species and extrapolation from single species laboratory tests to ecologically complex conditions in nature remain unaddressed.

Caveats and uncertainties do not mean that new methodologies, like the BLM, will not be useful in some aspects of risk assessment and metal regulation. The difficulty that managers and risk assessors must face is when, where and how to incorporate such new technologies. The BLM is certainly a more advanced approach than was traditional. But it seems doubtful it is enough of an advance to justify eliminating application factors from criteria before the full suite of uncertainties are better accounted for.

Box 15.7 provides one of the few independent validation exercises with the BLM. It shows that BLM predictions agree well with toxicity tests. But the BLM under-predicts toxicity when it is tested against outcomes of a totally different methodology (mesocosm toxicity testing that includes

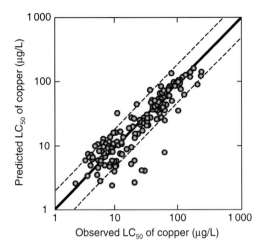

Fig. 15.4 Concentrations of Cu that are predicted to be toxic by different versions of the BLM are correlated against observed LC_{50}s for three species of *Daphnia* in a variety of natural waters (from Chile) and synthetic waters (constructed to represent a wide range of conditions). The LA_{50}s for the models were derived by fitting model outputs to toxicity data in previous experiments with *Daphnia magna*. Biological parameters were those derived for fathead minnow *Pimephales promelas*. (After Villavicencio *et al.*, 2005.)

feeding animals). Such independent tests are the true measure of a methodology. Many additional tests of this concept will follow in the years ahead. Methodologically independent comparisons will ultimately demonstrate when and where to best employ this new tool.

15.4.2 Biological uncertainties

15.4.2.1 Acute-to-chronic ratios
Prologue. Chronic toxicity almost always occurs at lower concentrations than acute toxicity; but fewer data exist for chronic toxicity. Ratios between acute and chronic toxicity concentrations (ACR) are used to adjust acute toxicity data to chronic toxicity; even though, for each metal, ACR can vary by more than two orders of magnitude. It is critical to better understand species-specific metal-sensitivities in long-term exposures, especially for those species that are potential indicators of metal effects.

Acute toxicity tests use extremely short exposures, for practical purposes, and those exposures are extended in chronic tests. Much more data exist for acute toxicity than for chronic or life cycle tests. For example, derivation of a freshwater criterion for Cr used acute toxicity from 13 species: six invertebrates and seven fish. On the other hand, chronic toxicity data for Cr exist for only three freshwater fish and no invertebrates. Acute toxicity data for Cr in marine environments exist for 20 species. But the only acceptable chronic data were life cycle studies for the polychaete *Neanthes arenaceodentata*, and the mysid crustacean *Mysidopsis bahia* (USEPA, 1980).

Where limited chronic data are available, acute data are often used for determining ecological risks from metals (Brix *et al.*, 2001; USEPA, 2001c, 2003). The acute data are corrected for chronic exposures using an *application factor* (AF) or *acute-to-chronic ratio* (ACR). In USEPA terminology, the ratio is used to calculate a Final Acute Value or FAV. Traditionally the USEPA (1998) Framework for Ecological Risk Assessment called for a correction factor of 1 (no correction) if the toxicity data were derived from field studies. USEPA divides chronic no-effect concentrations by 10 to estimate risk because 'chronic' toxicity tests usually are not full life cycle tests. USEPA uses a factor of 100 for acute toxicity data and 1000 for a single test of acute toxicity.

As the body of chronic testing data grew, chemical-specific averages were developed for the correction factor. The acute-to-chronic ratio was calculated as:

$$ACR = \text{species mean acute value}/$$
$$\text{species mean chronic value}$$

where the acute effect concentration was divided by the average of the No Observed Effect Concentration (NOEC) and the Lowest Observed Effect Concentration (LOEC) from a chronic toxicity test, using data from two or more similar species. Kenaga (1982) found that the ratios among a variety of chemicals and species varied from 1 to 18000. The mean for the central 93% of the data was 25.

The variability for metals alone was not quite that great. In deriving an ambient water quality

Box 15.7 Secondary ecological effects of Cd in a stream mesocosm: validation of the BLM

Brooks *et al.* (2004) dosed stream mesocosms with either 15 or 138 μg/L Cd during a 10 day study. The organisms for the mesocosms were recruited onto artificial substrates in local streams from the region, and two replicates were conducted for each treatment. The stream community was dominated by the snail *Physa* sp. on day 0 (>50% of individual invertebrates). At the higher Cd concentration, the abundance of the snails declined dramatically by day 10. Two other species increased in abundance. Chironomid (midge) larvae became abundant, and biomass of periphyton increased greatly (Fig. 15.5). Increases in periphyton appeared to be a direct response to reduced grazing on the plants as snails disappeared. In metal-contaminated rivers, increased abundance of periphyton to nuisance proportions is a common occurrence, although it is usually unclear whether nutrient or metal pollution is the cause. The mesocosm results demonstrated that metals can cause such an effect. Increased chironomid abundance appeared to be partly linked to a greater metal tolerance in this species. But increased chironomid abundance is also often associated with increased periphyton abundance, as decaying periphyton provide a nutrient source for the midge larvae.

The authors also conducted single-species toxicity tests using the water from the stream and a surrogate test species, the daphnid *Ceriodaphnia dubia*. They predicted toxicity to *C. dubia* should occur at 280 μg/L. Thus toxicity to *C. dubia* would not be expected at the mesocosm exposure of 138 μg/L, based upon Cd speciation predicted from the BLM modelling approach. And tests showed that the higher Cd concentration, indeed, was not toxic to *C. dubia*. Thus dissolved toxicity testing and speciation modelling were internally consistent with one another. But the single-species tests did not predict the final outcomes of the full mesocosm test: disappearance of the snails, the growth of periphyton and growth of midge larvae at 138 μg/L. These all happened at bioavailable Cd concentrations lower than that predicted to be toxic by the BLM. The BLM underpredicted toxicity by twofold or more.

Typical of multi-species tests, this study relied upon only a few replicates and considered only two exposure concentrations. But it did capture biological complexities that single-species testing cannot address.

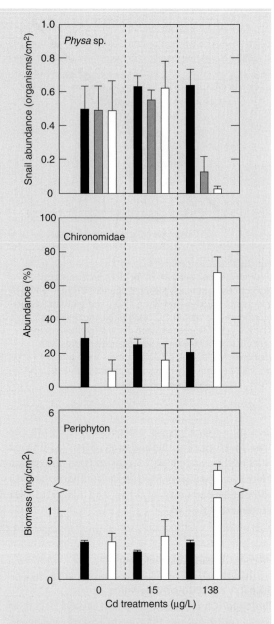

Fig. 15.5 Snail (*Physa* sp.), chironomid midge larva abundance and periphyton biomass in replicates of stream mesocosms treated with different concentrations of Cd. Cd treatments are average observed concentrations of Cd in stream water (after Brooks *et al.*, 2004). Single species tests predicted an EC_{50} of 280 μg/L in these waters, using BLM calculations; but obvious effects were observed in the 138 μg/L treatment.

As we noted in Chapter 10, complex mesocosm tests are ideal for uncovering fundamental effects on processes that cannot be observed in single-species tests; even though they are not ideal for dose–response studies. The independent test of the BLM predictions (do they predict ecological change?) is the standard that any new method should meet to be used as a risk assessment or regulatory tool. The BLM output agreed with the measure it was designed to predict (dissolved toxicity) but it under-predicted the more ecologically complex outcome of the overall test.

criterion for Cd, USEPA (2001c) reported that ACRs ranged from 0.90 for the chinook salmon (*Oncorhynchus tshawytscha*) to 434 for flagfish (*Jordanella floridae*), with other values scattered throughout this range. Brix *et al.* (2001) compared ACRs for Cu among nine species of freshwater lake species. They found a range of 2.8 to 31.7. The ACR was higher where the acute toxicity value was higher. This was because differences in chronic sensitivities between taxonomic groups were narrower than the differences in acute toxicity.

The ACR range reported by USEPA (2001c) for Zn, based on comparisons of a variety of test species was 0.7 to 41; the average was 2. Hickey and Golding (2002) compared acute and chronic toxicity for Zn among tests with a metal-sensitive species, the mayfly larva *Deleatidium* sp. This was a species that they suggested might be useful as an indicator of metal effects in New Zealand streams. There was a high chronic sensitivity for this species and a relatively low acute sensitivity. Thus the Zn ACR was 66 to 184, much higher than the average for Zn in the standard toxicity testing database.

It is not surprising that ACRs are highly variable and difficult to generalise with any certainty. They reflect a diversity of testing approaches, as well as the species-specific biological variability characteristic of responses to all metals (Chapters 7 to 11). Nevertheless some correction is essential if only acute toxicity data are used to derive a predicted no effect concentration, an ambient criterion or an estimate of risk. Acute toxicity tests, themselves, cannot realistically address risks in nature (see Chapter 10).

15.4.2.2 Most sensitive species

Prologue. Toxicity testing protocols originally assumed that the data would include the most sensitive species

typical of nature; and data from those species would protect others. If databases are robust, species sensitivity distributions are a method for reducing statistical uncertainties in determining the level at which a metal is likely to affect 95% of species. Whatever the statistical approach, the reliability of the protective measure is dependent upon whether or not species among the most sensitive species from nature are included in the data set.

Decades of toxicity testing have allowed databases for many metals to grow to tens of species, especially for acute toxicity. To determine the concentration that would be protective, *species sensitivity distributions* (Brix *et al.*, 2001), or statistical extrapolation methods (Bodar *et al.*, 2005), are employed to take advantage of the large toxicity testing data sets. Species sensitivity distributions plot the effect concentration for different species against the cumulative rank of the value among the species (Fig. 15.6a). Risk is estimated as the percentage of species expected to be affected in an aquatic community above any dissolved concentration. Criteria for protection can be derived by calculating the concentration required to protect a given percentage of species. For example, that percentage is defined by USEPA as 95% of species.

An example is the use of the database for Cd by USEPA (2001c) to derive the Cd aquatic criterion. The derivation found 'acceptable data' on the acute effects of Cd in freshwater for 43 species of invertebrates and 27 species of fish. A robust species sensitivity distribution was generated from these data (Fig. 15.6a). Fewer acceptable chronic toxicity tests were available (21 species, including seven invertebrates and 14 fishes) but still enough to derive a species sensitivity distribution (Fig. 15.6b). USEPA requires at least 10 toxicity values for different

(a)

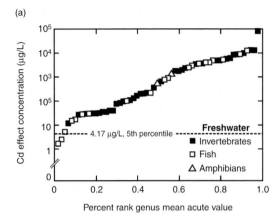

(b)

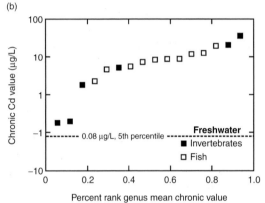

Fig. 15.6 (a) The concentrations of Cd causing acute toxicity ranked for different genera of invertebrates, fish and amphibians. This species sensitivity distribution was used to derive a freshwater final acute value for Cd, the concentration at which toxicity was observed for the most sensitive 5% of taxa (4.17 µg/L at 50 mg/L hardness) (after USEPA, 2001c). (b) The species sensitivity curve based upon freshwater chronic toxicity tests. Sensitivity at the 5th percentile was calculated to occur at 0.08 µg/L at 50 mg/L hardness; but the paucity of data from sensitive species added uncertainty to the choice of a 5th percentile criterion. (After USEPA, 2001c.)

species covering at least eight taxonomic groups before a species sensitivity distribution can be employed, recognising that power increases as the number of species increases. The use of the statistical approach reduces uncertainties in the derived value. The alternative is to derive the criterion from the experiment showing the lowest long-term No

Observed Effect Concentration (NOEC). Dependence upon a single test raised uncertainties. The distribution approach exploits the statistical power of many experiments to interpolate a final toxicity value from the entire distribution.

Where insufficient data exist, correction factors are employed. For example, to make up for the absence on data on chronic toxicity for Cr (see above), chronic sensitivity was estimated from the 5th percentile of the acute distribution. Then an ACR is applied (Brix *et al.*, 2001). Data on the ratio were available for two marine animals, a polychaete (for which the ratio was 120) and a mysid crustacean (the ratio was 15). A compromise ACR of 72 was chosen to calculate the Criterion Continuous Concentration (CCC) for Cr. Use of the average ACR instead of the higher value probably results in a less protective chronic criterion than would occur if enough data existed to use a species sensitivity distribution; and a more protective criterion than if the lowest value was chosen. The ultimate chronic criterion is more influenced by policy than science in such an instance. It is easy to see how either over-protection or under-protection are likely to occur, and controversy to ensue in difficult circumstances, when data are as uncertain as this.

Success of the species sensitivity methodology is very dependent upon whether the organisms most sensitive to a particular metal are included in the testing data (Forbes and Calow, 2002). For Cr, an amphipod crustacean, *Gammarus pseudolimnaeus*, was the most sensitive species in the acute tests, with an LC_{50} value about one-fiftieth of the next lower acute value. But so few species have been tested that it is unknown if there are other species similar or more sensitive to Cr under these conditions. Species senstivities are species-specific, metal-specific and poorly known in advance of testing or from a mechanistic basis. By nature, routine toxicity testing also tends to employ hardy surrogates. Thus, it is not predictable in advance what species should be tested to determine the most sensitive response. It is also likely, if the available data set is small, that the most sensitive species from nature will not be included (Chapter 10). The occurrence of only one sensitive species in the tests (as above for Cr) may add as much uncertainty as reassurance that the criterion is reasonable. Such uncertainties have led to accumulation of ever larger toxicity testing data sets (to improve statistics). A more lateral approach, on the other hand, might be to expand of definition of 'acceptable data'.

A criterion that is the lowest value in any study for any effect can also be irrelevant if highly sensitive species do not reside in a community or habitat. For example, freshwater Cd criteria are based upon the high Cd sensitivity of some species of cladoceran zooplankton in soft water lakes. But this criterion may overstate sensitivities of stream environments. There the invertebrate community is dominated by insect larval communities that seem to be less sensitive to dissolved Cd than cladocerans (USEPA, 2001c). The cladoceran-driven toxicity standard may again be appropriate for larger rivers, however, where highly sensitive freshwater mussels are present (e.g. USEPA, 2001c), if corrected for hardness. Divergences between test species and the realities of communities in nature are inevitable when acceptable data are restricted to laboratory derived toxicity tests following rigid protocols. Evaluations of uncertainties do not consider such uncertainties.

Thinking across the laboratory–field boundary (i.e. adding some 'eco' to the application of 'ecotoxicology') about the water body where a criterion is being applied can help address uncertainties about extrapolation among species, even if the definition of acceptable data is not expanded. For example, Brix et al. (2001) used separate Cu sensitivity distributions for different types of organisms typical of a lake environment. By comparing those distributions against a normal distribution of two Cu concentrations typical of lakes (expressed as a probability distribution), they predicted what species were most likely to disappear and the implications of that disappearance for the food web of the lake (Box 15.8). They could therefore derive a protective value from the species sensitivity distribution that was relevant to the most sensitive group of species from that habitat. Risk assessments and criteria derivation protocols could improve their credibility by including considerations that address chemical and ecological attributes of the environment of interest, even if it is done in the most general sense.

15.4.3   Uncertainties deriving from the unique attributes of metals

Prologue. Background concentrations, physiological regulation and essentiality are often cited as factors confounding interpretations of metal toxicity. All are influential in exceptional circumstances more than in general; enough is known to readily account for such influences where they occur.

Some of the unique attributes of metals, compared to other contaminants, add unique biological uncertainties to risk assessment and risk management. Firstly, natural background concentrations exist for all metals (Chapter 5) and differ widely among trace metals. There are some circumstances where natural concentrations of metals may be toxic, and, in fact, may influence the natural community that is present, although they are probably rare. Two examples are Cu in low dissolved organic carbon (DOC) open ocean waters (Brand et al., 1986) and the soils around serpentine rock formations (where Ni concentrations are extremely high and unique plant communities are known). In general, a numerical criterion value that is less than the background concentration is likely to be either unenforceable (one cannot regulate natural inputs) or overprotective. A bioaccumulation (or toxicity) test that adds less metal than occurs naturally also will not result in a response.

Essentiality, or the requirement for a trace metal for metabolic function, is also a consideration. Toxicity tests that work with metal concentrations lower than the level required for essentiality could mistake metal insufficiency for metal toxicity (De Schamphelaere et al., 2005). The links between concentrations in a water body and essentiality are not known, however. From an evolutionary point of view, it is probably rare that natural concentrations are below requirements. The window between essentiality and toxicity also differs among metals. For Se that window is extremely narrow. Small increases in Se exposure can have great effects on some species, emphasising the challenge of managing that hazardous trace metal. The window seems broader for Cu and Zn, at least for most animals.

Box 15.8 Determination of effects of Cu risks using a species sensitivity curve

Brix *et al.* (2001) used species sensitivity distributions (SSD) and probabilistic risk assessment to evaluate Cu toxicity for the different taxonomic groups expected in a hypothetical lake environment (Fig. 15.7). Dissolved Cu distributions were assumed to be normally distributed with a mean of 5 μg/L and a standard deviation of 3 μg/L. For the predicted effects concentration they used acute toxicity data from 87 species. They used the acute-to-chronic ratio (ACR) to convert this to chronic toxicity due to a paucity of chronic toxicity data in the acceptable databases. The ACR was derived from nine studies as cited above. Species sensitivity curves were derived separately for several different taxonomic groups of organisms (e.g. cladoceran zooplankton, insects, fish). Cladocerans were clearly the most sensitive taxonomic group, and insects the least sensitive (Fig 15.7). Assuming the normally distributed Cu concentrations reflected a probability distribution, the analysis predicted that risks to warm-water fish and invertebrates other than cladocerans would be low to negligible 85% of the time. The estimated risk to cladocerans was much higher. Up to 50% of the cladoceran species would be expected to be affected 20% of the time and 25% of the cladoceran species affected 45% of the time. Applying the analysis to a food web typical of a lake, Brix *et al.* (2001) noted that zooplankton, like the cladocerans, were the greatest source of carbon for planktivorous fish. Although Cu posed no direct risk to the fish, it was possible that disappearance of cladoceran species could affect planktivorous fish biomass. Reduced biomass of planktivorous fish could present a risk for their predators, the piscivorous fish. Piscivorous fish were an important recreational asset in the lake. Thus risk to them, even though indirect, was considered unacceptable. Thus application of even rudimentary and generic ecological knowledge, used with toxicity tests, yielded an analysis with realistic applications to the real world. These are examples of small and feasible steps toward lateral thinking that should be a necessary part of risk analyses and criteria development.

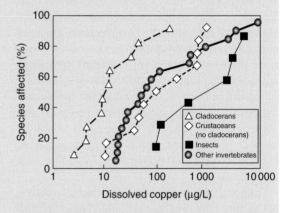

Fig. 15.7 The percentage of test species (cladocerans, crustaceans other than cladocerans, insects and other invertebrates) suffering acute toxicity at different dissolved Cu concentrations. The elevated sensitivity of cladocerans is demonstrated, focusing risk implications on this group of animals. (After Brix *et al.*, 2001.)

Risk assessment is a flexible tool. Thus state-of-the-science approaches can adjust for at least some of the uncertainties stemming from the unique nature of metals and the challenges in testing their toxicity. Box 15.9 gives a detailed example of a large-scale risk assessment for Zn that employs several such approaches. The example also illustrates the policy choices that must be made in the adjustments.

The Zn risk assessment procedures emphasise that uncertainties stemming from the state of the science can be addressed logically. But addressing uncertainties involved more than just science; policy choices were necessary in deriving a guideline. The policy choices that were made fell within the accepted bounds of what is known about metals. Validating the risk assessment, at least in terms of its practical regulatory implications (do the number and location of non-compliances make sense?) was also a useful and manageable exercise. Validation in terms of non-standardised toxicological or field data would be a next step in expanding the lateral approach to deriving an internationally acceptable standard (e.g. Chapter 11).

Box 15.9 Zn risk assessment and uncertainty

The new chemical policies of the European Union (REACH, or Registration, Evaluation, Authorisation of Chemicals) will require risk assessments for many of the high production volume chemicals, in order to determine if there is a need for risk reduction measures. Many of these chemicals are metals or metal products. Bodar *et al.* (2005) reported on the risk assessment for Zn (one of the first). They argued that risk assessments must be built on scientific principles; but that pragmatic considerations were also necessary to determine how uncertainties were dealt with during the decision process (Bodar *et al.*, 2005). So while the risk assessment was originally developed by experts from member states, non-governmental organisations (NGOs) and industry, it also involved consensus choices from a multi-stakeholder process. Bodar *et al.* (2005) described examples of considerations that had fundamental implications for the risk characterisation.

Firstly, Zn is present at a natural background concentration in all environmental compartments. To account for this factor, both the PEC and the PNEC are determined on the basis of the amount of Zn that was added to the toxicity test (rather than the total concentration of exposure). The use of both an added predicted environmental concentration (PEC_{add}) and an added predicted no effect concentration ($PNEC_{add}$) was thought to balance environmental and regulatory implications. For example, assume a $PNEC_{add}$ water of 7.8 µg/L is applied to waters with either a Zn background concentration of 5 µg/L or one of 20 µg/L. From the environmental point of view, there would be different environmental risks where the final Zn concentration was about 13 µg/L Zn (in the case of the 5 µg/L background) as compared to about 28 µg/L Zn (in the case of 20 µg/L background). But from the regulatory point of view, a discharger should only have to regulate their own contribution to that environmental risk. The implicit policy choice in this case was to use the equitable management approach in the risk assessment.

Because Zn is an essential element it was considered that organisms, including humans, have a minimum requirement for Zn that supplies their needs, and a maximum concentration above which Zn is toxic. The range between the minimum and maximum is often called the 'window of essentiality' (Hopkin, 1989). The policy choice was the assumption of no risk if discharges resulted in concentrations lower than that necessary to supply essential needs.

Chemical and biological processes that affect transformation and speciation of zinc were considered in derivation of the PEC_{add}. But the risk assessment also assumed that only the dissolved fraction is potentially bioavailable for aquatic organisms. To correct for bioavailability, they considered influences of pH, hardness, and dissolved organic carbon (DOC) using the Biotic Ligand Model (BLM). Different species respond differently to changes in the bioavailability of Zn. To account for this uncertainty, the Zn risk assessment used toxicity data ($PNEC_{add}$) for three aquatic species (an alga, a species of *Daphnia* and a fish) and compared these to a bioavailable PEC_{add} from local water conditions. Corrections were applied to the PEC to account for bioavailability; depending upon local conditions PEC was multiplied by 0.4 to 1.0.

The large amount of Zn chronic ecotoxicity data allowed the use of a sensitivity curve with the chronic data itself to derive a $PNEC_{add}$. No toxicity data extended into the 5th percentile (Fig. 15.8) so a value was extrapolated from the curve of 15.5 µg/L; then an application factor (a twofold multiplier) was used to compensate for the remaining uncertainties, yielding a $PNEC_{add}$ of 7.8 µg/L. The choice of a risk level remains controversial. The authors defended the policy choice

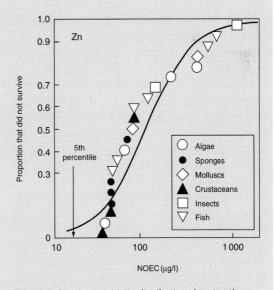

Fig. 15.8 Species sensitivity distribution showing the chronic toxicity of Zn (NOEC) versus the cumulative frequency of tested species that did not survive for all available aquatic data. The no effect concentration for the 5th percentile of species was 15.5 µg/L.

to extrapolate beyond the data then apply a further correction by citing field studies which pointed toward effects at or (slightly) below the 5th percentile. A NOEC-based approach would have derived a toxicity value of 17 µg/L. If a traditional assessment factor of 10 was applied to the NOEC, it would have resulted in a $PNEC_{add}$ of 1.7 µg/L. That value was deemed too low because it was less than many natural background determinations of dissolved Zn (also the essentiality argument).

The Zn risk assessment concluded that the safe level was 7.8 µg/L in typical surface waters and 3.1 µg/L in soft water (<25 mg/L Ca, presumably as $CaCO_3$). In comparison the current USEPA chronic ambient water quality criteria for Zn is 120 µg/L in freshwater and 81 µg/L in saltwater (USEPA, 2002c).

The chosen value for Zn risk in the risk assessment was compared to concentrations in Western European waters to determine the number of non-compliances under existing conditions (PEC/PNEC ratios >1). Zn production facilities and processing plants (e.g., electro-galvanising, brass production and die-casting) were usually associated with ratios >1. At the regional scale, large German rivers, waters in the Flanders and Walloon provinces of Belgium and some rivers in France were areas where such ratios were most common. The differences, in general, did not exceed 3. The average over larger regions was <1. Monitoring data were lacking for several heavily industrialised regions in a number of southern European countries.

Because of the complexity of the considerations, collection of field data, the addition of experiments to test uncertainties and the large database of toxicity testing data analysed, the Zn risk assessment took 8 years to complete. Controversy continues about risks from Zn; and the reality is that the controversy is probably irresolvable scientifically within the constraints of the data considered. Part of the problem was reliance on a toxicity-based species-sensitivity curve that could not resolve low-level toxicity unambiguously. Although field studies were cited, no thorough analysis of them was included in the published article.

15.5   VALIDATING AMBIENT WATER QUALITY CRITERIA

Prologue. The prevailing perception is that ambient water quality criteria are uniformly over-protective for metals (Di Toro, 2003). In fact examples of under-protection are probably as common as is over-protection. Incoherence is the major problem. Differences between ambient water quality criteria for the same metal can also be found within and among jurisdictions. A systematic approach is needed to determine the causes and degree of variability in protectiveness and international disharmony.

The USEPA first released ambient water quality criteria for metals in the USA in 1986. Revised criteria were released beginning in 2001 (USEPA, 2002c). The greatest difference between the earlier and later versions is the addition of some methodologies to correct for uncertainties. As predicted by Cairns and Mount (1990), the later criteria also have smaller and fewer application factors. Internationally, jurisdictions have followed similar protocols and timelines (e.g. Reilly *et al.*, 2003).

The critical questions are whether the existing criteria have kept up with the state of knowledge about metal effects in nature and how well they serve as useful risk management tools. This question is of increasing importance as risk management moves toward approaches that rely increasingly on such guidelines. The advantages and disadvantages of the universal numerical approach to criteria were discussed earlier, but we did not directly address whether specific criteria are over- or under-protective compared to what we know about natural waters. Surprisingly no body of work directly addresses such questions.

15.5.1   How comparable are criteria from different jurisdictions?

Prologue. Protocols for deriving environmental metal standards are similar in many ways, internationally. But the standards themselves vary much more widely among jurisdictions than suggested from the strict approaches to their application, illustrating the difficult of adapting generic toxicity testing to region-specific conditions. The most effective change might be to develop new protocols to derive the criteria.

One test of the validity might be comparisons among values from different jurisdictions. If consistent protocols and consistent databases are available, criteria should be generally similar wherever such approaches are followed. Variability in the same criterion might reflect influential uncertainties and remaining data gaps.

Table 15.3 is one set of such comparisons. Protocols for deriving environmental standards have many similarities, internationally. And criteria for each metal are usually (but not always) within the same order of magnitude. But most vary among jurisdictions by at least twofold and more commonly by more. Examples can be found where criteria for one metal can differ by 10 to 50 times (e.g. Hg in marine waters, Se in marine waters, Ag comparing marine and freshwaters, Pb in both marine and freshwaters, and Ni in general).

Such variability is at least partly driven by the challenges explained by Tong *et al.* (1999) in deriving interim Cu marine criteria for ASEAN (the Association of Southeast Asian Nations). After an impressive review of the toxicity testing literature and how it applied to tropical species, they noted that there were few chronic toxicity data available with sensitive tropical species. Therefore criteria development would require an application factor. They wrote:

> Using the lowest LOEC from a chronic study (77 µg/L for reproduction for *Mysidopsis bahia*) and a safety factor of 0.1 would result in a value of 7.7 µg/L. However, this copper concentration was not considered protective enough as it exceeded the EC_{50} value for at least one study on bivalve larval development (from non-tropical regions). There was one acute-to-chronic ratio (ACR = 3.3) available, also for the warm water mysid shrimp *M. bahia*. If the most sensitive LC_{50} from an acute study (48 hour EC_{50} for *Crassostrea commercialis* larval development) was divided by this ACR value, the result was 0.9 µg/L. If this EC_{50} value was multiplied by a universal application factor of 0.03 then the resultant criterion would be 0.03 µg/L.

But the standard application factors resulted in a value below natural baselines for the area (0.03 to 0.4 µg/L). They continued:

> Thus, using either of these values as an interim criterion value appears anomalous . . . Further work is necessary to generate sufficient chronic toxicity and ACR data for copper to tropical marine life to develop an ASEAN water quality criterion. In the meantime, a value of 2.9 µg/L, which is the value set by the USEPA as a 4 day average continuous concentration, is recommended as a conservative but not unrealistic interim ASEAN marine environmental quality criterion for copper.

The implication is that toxicity testing data seem to be available internationally, but single-discipline protocols and region-specific data are inadequate to adjust standards for region-specific conditions. An improved protocol for derivation might be as valuable as more data, especially if these data are dominated by the historical approaches.

Se criteria represent the ultimate in incoherence (see Chapter 17); the result of the historical failure to consider trophic transfer in toxicity testing. Some criteria use field data to derive freshwater criteria, but some still use dissolved toxicity tests. Guidelines for Se range from none (Europe), to 150 µg/L for some marine waters (USA) to 1 µg/L for some freshwaters (Canada).

15.5.2 Are environmental metal standards congruent with expected metal concentrations in contaminated waters?

Prologue. Widely accepted protocols do not exist for validating whether environmental standards for metals are justifiable. Comparisons of standards to concentrations in natural waters show rare instances where standards are lower than natural background concentrations, but many more instances where standards are higher than would ever be expected in a contaminated water, if modern analytical protocols are employed.

Table 15.3 *Some examples of dissolved metal concentrations set as protective criteria for freshwater and seawater in several jurisdictions across the world. All values are µg/l. For perspective, the criteria values are compared to metal concentrations in undisturbed waters and large water bodies affected by human activities (as defined in Chapter 5 unless otherwise cited)*

Jurisdiction	Ag	Cd	Cu	Hg	Ni	Pb	Se	Zn
USA								
Freshwater	1.9	0.25	9.0	0.77	52	2.5	5.0	
Marine	3.2	8.8	3.1	0.94	8.2	8.1	71.0	
Canada								
Freshwater	0.1	0.012	2.0–4.0	0.1	25–150	4–16	1.0	
Marine	1.5	0.17	2.0	0.02		2.0	2.0	
ANZEC[a]								
Freshwater		0.2–2.0	2.0–5.0			1.0–5.0		5–50
Marine		2.0	2.0			5.0		50
Europe (UK)								
Marine		5.0	5	0.3	30	25	None	40
Disturbed coastal waters	0.02–0.10	0.03–0.20	0.5–5.0		0.4–4.2	0.02–0.27	0.3–1.0	0.6–5.1
Undisturbed coastal waters	0.0006–0.004	0.003–0.08	0.04–0.40	0.0001–0.002	0.27–0.32	0.0006–0.038	0.04–0.19	0.003–0.61

[a]ANZEC is the Australia–New Zealand Environment Council.

Source: Data from USEPA (2002c).

Ideally the concentrations of metals that constitute water quality guidelines or criteria should reflect concentrations associated with ecological change in field studies. Although effects are known in the field (Chapter 11), dose–response is not well known. At a minimum, however, a comparison of criteria against high-quality analytical data from natural waters seems an important requirement to validate whether a criterion is justifiable (the approach used by Bodar et al., 2005).

For example, criteria concentrations should not fall within the range of concentrations typical of uncontaminated waters. Such criteria would be both unenforceable and over-protective in most circumstances. None of the examples in Table 15.3 is lower than metal concentrations in undisturbed waters. However, there are a few that are near the higher concentrations seen in such waters. The lowest Cd standard among the examples in Table 15.3 is the 0.012 μg/L (12 ng/L) Canadian fresh-water guideline for Cd (Environment Canada, 2002). This criterion was derived from Cd toxicity tests with cladocerans and effects observed in soft water lakes. Table 5.4 shows that a typical range of Cd concentrations in rivers with human activities in their watershed is 0.014 to 0.045 μg/L. The implication is that any river on which there is some human activity has a risk of ecological disturbance by Cd. Cd concentrations near the 0.012 μg/L standard are also seen in harder water lakes typically thought to be not heavily disturbed (e.g. 0.09 μg/L in the interior Great Lakes: Nriagu et al., 1996; Table 5.4). Thus this seems a case of a standard that will be difficult to enforce. It is consistent with toxicity to the most sensitive known species of zooplankton, but not with effects on field populations of similar species if population mechanisms compensate for toxicity to individuals (Marshall and Mellinger, 1980). Similarly, for Se, the Canadian standard of 1 μg/L is below concentrations in rivers draining Se-rich geology.

The most surprising observation in Table 15.3 is that some standards are an order of magnitude or more higher than the top 5th percentile of dissolved metal data from disturbed water bodies. The criteria concentrations for Ag, Cd and Pb in marine waters are unlikely ever to be seen except in effluents. Most criteria for Hg and Ni are similarly suspect. Se in marine waters is comparable to the most problematic drainage (Chapter 17).

What does it mean when criteria are orders of magnitude higher than are likely to occur in a contaminated water body? In practical terms it means that no regulatory actions based only upon ambient criteria are likely anywhere they are in force. It also means that the ceilings set from the criteria for Total Maximum Daily Loading management (TMDLs) could allow substantial degradation of water bodies. The implication is that there are no effects in nature from metals at present, or in the known past, anywhere but the very most extreme circumstances. Such a conclusion does not seem consistent with observations in Chapter 11. These criteria are protective in toxicity testing conditions but their relevance to nature seems worth further investigation.

15.5.3   When should corrections for geochemical conditions be applied?

Prologue. Applying geochemical corrections to standards that already understate important biological influences is likely to be problematic.

Methodologies to correct for metal speciation in dissolved toxicity tests are more advanced and more sophisticated than methodologies to correct for biological assumptions. Geochemical corrections are undoubtedly useful in preventing over-protection in those circumstances where criteria concentrations are near typical concentrations in nature, and geochemical/biological processes reduce metal bioavailability. Cu is a well-known example where over-protection is a possibility, because of its strong affinity for organic complexation. Speciation might also lead to over-protection in the case of the freshwater Cd criteria, in some circumstances.

However, bioavailability corrections are more questionable where criteria concentrations appear higher than observed in contaminated waters. Use of geochemical corrections in these circumstances will raise criteria concentrations even higher. These are examples where correcting for geochemistry alone seems to disrupt a balance of uncertainties.

It appears to be difficult to determine when to apply well-established advances in one science before other advances have occurred, unless some sort of correction is employed.

15.5.4   Were all data adequately considered?

Prologue. A lateral approach might use effects data from complex tests, dietary exposures and from natural waters to test the validity of criteria derived from dose–response data. A systematic protocol is necessary to determine what data should be employed and how to employ them.

A validation exercise might consider whether data exist that are of high quality and informative, but perhaps outside the scope of the rigidly defined protocols that determine data acceptable for use in derivation of criteria. A more holistic or lateral approach to validation would consider more than simply a reanalysis of toxicity testing data. Nature and her responses to metals are the arena in which criteria are designed to work. Therefore nature should be included in the analysis. Several specific considerations might comprise a systematic analysis:

- Was chronic toxicity adequately addressed?
- Were there enough data using species sensitive to the metal?
- At what concentrations do adverse effects occur in mesocosm studies, complex multi-species tests or life cycle tests? Such experiments rarely include full dose–response data, but many are suitable to evaluate the validity of criterion chosen from a dose–response test.
- Are the criteria justifiable based upon field observations where adverse effects of the metal are observed? Field data, like mesocosm data, rarely result in dose–response data, but can be useful in ground-truthing such results. Cause-effect can be an issue, but a growing number of case studies and methodologies are available that minimise such uncertainties (Chapter 11).
- How comparable to the criteria are data on dietary toxicity, dietary exposure or mesocosms where organisms fed during the experiments?

Was dietary toxicity considered for those metals where it is likely to be especially important? Again, dose–response data are less important if validation, not derivation, is the goal.

15.6   INFLUENCE OF ASSUMPTIONS ON CRITERIA DERIVATION: CADMIUM

Prologue. The Cd criteria by USEPA (2001c, 2006) provide an example of holistic validation of water quality criteria. Freshwater Cd criteria appear to be well justified; even though all biological assumptions were not considered and application factors are not used. On the other hand, salt water criteria seem poorly justified. For example, the marine standard of 8.8 µg/L Cd is more than an order of magnitude too high to protect either sensitive phytoplankton or feeding copepods in marine and estuarine environments. Influential uncertainties include data availability/acceptability, failure to include non-standardised studies relevant to nature and lack of consideration of dietary exposure. If these types of inconsistencies are widespread in the existing trace metal criteria or guidelines, a new generation of criteria is needed.

The Cd criteria released by USEPA in 2006 (USEPA, 2006) were derived from a USEPA analysis in 2001 (details in Box 15.10). The freshwater analysis included a fivefold correction factor for geochemistry. It concluded that dissolved Cd should not exceed dissolved concentrations of 0.08, 0.16 and 0.32 µg/L respectively, at hardnesses of 50, 100 and 200 mg/L as $CaCO_3$ (the 4 day average). The 2006 Cd criterion (CCC) was 0.25 µg Cd/L at 100 mg Ca/L. The USEPA (2001c) also derived an allowable pulse or 1 hour average dissolved exposure of 2.1, 4.8 and 11 µg/L at the hardnesses of 50, 100 and 200 mg/L, although these are probably less relevant to nature, as discussed earlier.

The freshwater Cd chronic criteria were derived from a robust chronic toxicity data set, strongly influenced by the inclusion of a sensitive group of species, cladoceran zooplankton. The criteria are consistent with observations of effects on zooplankton populations in metal impacted, soft water lakes, as observed in mesocosms and in

Box 15.10 Derivation of ambient criteria for Cd

Freshwater standards

To derive a Cd criterion for freshwater species, USEPA (2001c) used species sensitivity distributions for both acute and chronic toxicity data. For the acute data they used a data set of 50 genus mean acute values in the computation of the final acute value; selecting the four lowest values as closest to the 5th percentile (protect 95% of species). They lowered the derived Final Acute Value calculated from the distribution to protect rainbow trout (*Oncorhynchus mykiss*) and concluded that 4.296 µg/L was protective from acute effects at a hardness of 50 mg/L as CaCO$_3$. Brown trout (*Salmo trutta*) and striped bass (*Morone saxatilis*) had mean acute values lower than this; thus the 2006 acute standard was lowered to 2.0 µg/L.

Chronic toxicity was observed at less than the Final Acute Value for many species, with toxicity commonly observed between 1 and 5 µg/L Cd. Few of the acceptable studies considered concentrations less than 1 µg/L, however; so effects are poorly quantified in this lower range (still an order of magnitude above the highest Cd concentrations in disturbed waters). The most sensitive species in chronic tests were cladoceran zooplankton (*Daphnia magna*: chronic value <0.1 µg/L); although freshwater amphipods (*Hyalella* sp.) also showed chronic toxicity at <1 µg/L. The chronic data set was not as robust as the acute data set. But USEPA (2001c) concluded that ACRs were too variable to use. Thus the criterion was interpolated from the distribution of the chronic data (Fig. 15.6b). They concluded that:

> except possibly where a locally important species is very sensitive, freshwater aquatic organisms and their uses should not be affected unacceptably if the four-day average concentration of dissolved cadmium does not exceed 0.08, 0.16 and 0.32 µg/L respectively, at hardnesses of 50, 100, and 200 mg/L as CaCO$_3$ (the four-day average).

The freshwater water value compared well to Cd effects on zooplankton observed in mesocosms in freshwater lakes (where cladocerans are an important part of the community) (Chapters 8 and 11). Although 0.08 to 0.32 µg Cd/L represent low concentrations from the standpoint of toxicity tests, this range is high compared to concentrations of Cd in uncontaminated lake waters (0.0006 to 0.009 µg/L in the Great Lakes, for example: Chapter 5; Nriagu *et al.*, 1996) and comparable to concentrations in contaminated lakes where species

of zooplankton disappear. Thus, because the Cd toxicity testing data set for freshwaters included highly sensitive species, USEPA forecast toxicities appear credible for lakes (concentrations are >10× above 'background'; multiple species or field test results concur within twofold).

Marine standards

Many studies verify that dissolved Cd should be less bioavailable and less toxic in salt water than in freshwater (Chapters 5 and 7). A salt water Final Acute Value of 80.55 µg/L was derived from species sensitivities of 50 species. This was lowered using an application factor of two to obtain a short-term (acute) exposure limit of 40 µg/L. Very few chronic toxicity tests were acceptable. Therefore USEPA divided the final acute value by the mean ACR of 9.106, an average from comparable acute and chronic tests with one set of species (a zooplankton mysid crustacean). USEPA (2001c) concluded:

> invertebrates should not be affected unacceptably if the four-day average dissolved concentration of Cd does not exceed 8.8 µg/L more than once every three years on the average and if the one-hour average dissolved concentration does not exceed 40 µg/L more than once every three years on the average. However, the limited data suggest that the acute toxicity of Cd is salinity-dependent; therefore the one-hour average concentration might be under-protective at low salinities and overprotective at high salinities.

This seawater criterion concentration is 100-fold higher than the criteria for soft water; much greater than any known effect of salinity on Cd bioavailability. From a geochemical point of view, the salt water criterion is suspicious.

Concentrations of Cd in undisturbed seawater are ≤0.01 µg/L, although in upwelled seawater they can be 0.1 µg/L. Values of 9 µg/L are essentially unheard of, if the analytical chemistry is properly conducted; even the worst effluents rarely contain 40 µg/L. Data have existed for 20 years that suggest that Cd is toxic in marine waters at about the same concentration as it is toxic in freshwater. Brand *et al.* (1986) showed, with very high quality protocols, that Cd toxicity to some species of phytoplankton begins at about 0.1 µg/L, if Cd is not organically complexed. These data were deemed unacceptable for the USEPA analysis because of the buffering approach used to hold free ion concentrations constant (Chapter 11). Even though geochemists view this work as the most credible study of Cd toxicity to oceanic flora, it was not influential in the USEPA analysis.

More important, Cd exposures in the marine environment are typically >90% via food, because uptake from water by animals is relatively slow and Cd is readily assimilated from food (e.g. Wang, 2001). But the understanding of Cd toxicity from diet is just beginning. In the one available study, Hook and Fisher (2001a) found the acute LC_{50} of dissolved Cd to estuarine zooplankton copepods was 50 µg/L. But when they fed the copepods phytoplankton cells exposed to 0.5 µg/L dissolved Cd, they found a toxic effect in the zooplankton that ate the phytoplankton. The Cd concentration in the phytoplankton was 6.4 µg Cd/g. Egg production in the copepods eating those algae decreased by 50%. Reproductive toxicity to zooplankton at 0.5 µg/L is consistent with the concentration that causes toxicity to phytoplankton in the sea and to

zooplankton in freshwaters. Like the freshwater criterion, this is a relatively high concentration of Cd compared to uncontaminated seawater; more like what would be found in heavily contaminated marine or estuarine waters.

An important issue in the derivation of standards is the choice of acceptable data. Protocols for choosing data for the Cd analysis eliminated hundreds of studies (cited but not used: USEPA, 2001c). The acceptable data adhered rigidly to a single type of data: traditional toxicity testing dominated by acute toxicity data sets with strong dose–response curves. Some of the reasons for exclusion seemed legitimate, especially if each study is considered independently. But it is problematic that two extremely important studies, considered classic works by any scientific standard, were excluded from the analysis. As we suggested in Chapters 3 and 11, a holistic understanding of the body of evidence about a metal and its effects can lead to a different answer about risks than do traditional toxicity tests alone.

nature (Marshall and Mellinger, 1980; Yan et al., 2004; Chapter 11). The criteria would probably protect against recurrence of such effects in contaminated lakes. These criteria also make sense from a consideration of natural background in lakes; they are well above such concentrations.

Soft water lake ecosystems appear to be highly vulnerable to Cd, however. Thus downward correction for the lack of major ions in such waters is especially important in protection. Corrections for other geochemical factors (organic complexation) could be quite important in making such criteria applicable to other lake environments. It is also possible that the freshwater criterion could be overprotective for rivers, as noted above where aquatic insect larvae and fish like trout may not be as sensitive to Cd as cladocerans (USEPA, 2001c). The nature of the biological community is as important as water chemistry (Brix et al., 2001) in such evaluations.

Compared to freshwater, the marine criteria were derived from a particularly weak chronic toxicity data set; not enough data are available to assure that sensitive marine species are tested. Acute to chronic ratios were deemed too variable for use in freshwater criteria derivation, but were the basis for the marine standard because

the acute toxicity data set was robust compared to the chronic toxicity data deemed acceptable (Box 15.10, Fig. 15.6). Dietary exposure also differs in importance between freshwater and seawater habitats. Dissolved Cd is an important source of exposure in freshwater, because the free ion dominates speciation (Chapter 5). In high chloride marine waters, however, Cd speciation is dominated by low bioavailability chloro-complexes. Dissolved Cd uptake rates by animals are reduced and direct exposure via solution is relatively unimportant. Thus Cd toxicity appears lower in marine than in freshwaters in traditional tests. However in marine waters, dietary exposure is dominant, analogous to the case for Se (Wang, 2001). Dietary toxicity tests, however, were excluded from consideration in developing the marine standard.

The marine standard of 8 µg/L is orders of magnitude higher than concentrations of Cd typical of marine surface waters, even where human activities are quite influential (Tables 5.4 and 15.3). Geochemically corrected toxicity tests on phytoplankton (Brand et al., 1986) and toxicity tests (Hook and Fisher, 2002) both suggest that Cd is toxic in seawater at dissolved concentrations roughly similar to toxicity levels in freshwater (Text Box 15.10). The marine criterion fails most of our requirements for a justifiable regulatory measure, despite the use of correction factors. Clearly more careful study of Cd toxicity in marine environments is needed to establish the risks of this metal in these types of waters.

Cd is an example of how inconsistencies can occur in managing metal contamination. Lack of standardised data in crucial areas and the use of protocols that restrict what data can be considered in a derivation are important problems. In combination these can result in criteria that make sense from a toxicological point of view, but not from a view closer to nature.

Other similar cases exist. Ag forms a bioavailable neutral apolar chloro-complex in brackish and marine environments (Chapters 7 and 11) and is especially problematic in these types of environments (Luoma et al., 1995). Dietary exposure studies and field studies (Hornberger et al., 2000; Hook and Fisher, 2002) suggest that adverse effects from Ag can occur in coastal waters at about 0.1 µg/L. This is approximately the concentration found in San Francisco Bay when adverse effects were occurring there (Chapter 11). Ag concentrations in uncontaminated brackish waters are considerably lower than this (0.004 to 0.006 µg/L in San Francisco Bay: Smith and Flegal, 1993; Squire et al., 2002). A criterion of >1 µg/L Ag reflects a concentration that has never been observed in disturbed estuarine waters, even during times of heavy contamination and adverse biological effects. It is unlikely to be either protective or relevant in those environments. On the other hand, site-specific analysis of Cu in South San Francisco Bay raised the Cu criterion in that system from 3.1 to about 8 µg/L (USEPA, 2004). Careful geochemical studies (Donat et al., 1994; Buck and Bruland, 2005), and field monitoring both suggest that this change was well justified (Chapter 11).

One would assume that the framework of administrative process in which derivation of numerical criteria is embedded (Fig. 15.3) would have exposed such uncertainties. The USA framework calls for periodic evaluation of regulations, and their modification as experience grows. Protocols for such evaluations, beyond compliance monitoring, do not exist, however. Compliance monitoring (do dissolved metals exceed the criteria?) are not informative if

criteria are too high. A next-generation approach to protecting water quality is needed. Experts agree that evolution away from numerical, one-size-fits-all, standards would be constructive. Aspects of that change might include:

- Protocols for making adjustments, beyond only geochemistry, for different circumstances (Box 15.3).
- Goals and trajectory-orientated guidelines might be more flexible than numerical standards.
- Protocols for employing multiple lines of evidence beyond toxicity testing to at least test the validity of proposed guidelines would be an important step forward.
- Studying outcomes from watershed-based evaluations might facilitate more holistic approaches to validate guidelines (the TMDL and Water Framework Directive).
- Values to describe the thresholds where metals begin to exert adverse effects will always be needed. But ranges of values might reflect the realities of the science (Chapman, 1991) more effectively than single numerics that reflect mostly the needs of a one-dimensional law.

15.7 CONCLUSIONS

Traditional approaches to determining hazard rankings for trace metals are problematic because they have drifted away from the original purpose for recognising the dangers of bioaccumulation across trophic levels (trophic transfer potential was the original purpose for including bioaccumulation of organic contaminants). This has happened because of overreliance on dissolved toxicity testing, an approach that increasingly appears to have reached the point of diminishing returns when used alone to derive risk management approaches in at least some circumstances and may be irrelevant in circumstances (e.g. Se, Cd in marine waters) where exposure route is dominated by dietary uptake.

Work needs to continue on addressing the assumptions and limitations of standardised toxicity testing. But unbalanced incorporation of single scientific advances may sometimes be less constructive than cautious, balanced progress toward implementing change holistically. Political decisions about application factors will continue to be necessary until the full suite of uncertainties is narrowed, as long as standardised toxicity tests dominate 'acceptable' data. Rigorous adherence to only toxicological data from one paradigm is at the heart of many of the least justifiable decisions in the choice of risks and criteria. The next generation approach must include recognition of the benefits from an approach to managing metal contamination that uses knowledge that crosses the boundaries of geochemistry, toxicology, biology, field observation and experiment. Geochemistry, ecology and toxicology are increasingly adept at explaining nature when used together. But departure from the protocol of the standardised test is necessary for such progress. Demands for integrated, lateral consideration of bodies of knowledge could constructively be added to the protocol for defining sound science. With trace metals the problem is not too many chemicals to regulate; it is too little understanding about a manageable number of chemicals. A shift in emphasis from standardised testing to deeper understanding of both broad processes and specific circumstances might benefit both risk assessment and risk management.

Recognition that water, sediments and biota interact in every contaminated situation is also essential. Dissolved ambient water quality standards in isolation leave important uncertainties unaddressed. They are doomed to contentious, unresolvable disagreements where stakes are high. Harmonised standards (Chapter 17) and/or integrated consideration of multiple standards and observations must be fitted into the legal system for assuring protection from metal contamination.

Attention must also be paid to maintaining a balance, so as not to lean unrealistically toward either over- or under-protection when weighting influences of uncertainties. For example, some of the assumptions of toxicity tests are biological: short exposures, the use of surrogate species, and the contention that only dissolved toxicity is important in natural waters. If not corrected for, these assumptions would mostly lead to under-protection of the environment. Others are geochemical: test conditions are designed to maximise toxicity. If not corrected for, these would mostly lead to over-protection.

Historically, large 'assessment or application factors' were applied to correct criteria for cumulative uncertainties. The methodologies incorporated into the more recent criteria proposals increasingly take careful account of geochemical limitations. They address some, but not all, of the biological simplifications. If corrections for the latter are not as robust as corrections for the former, the historical 'balance of errors' may be lost. Thus, choice of application factors is still a critical decision in deriving criteria. Similarly a risk assessment that corrects for geochemical uncertainties is not likely to be in balance unless accompanied by a correction for biological uncertainties.

There is too little literature that systematically evaluates the effectiveness of criteria by converging geochemistry, dietary toxicity testing and bioaccumulation, biological or ecological data from natural waters. There is no widely accepted methodology for evaluating the justification for a criterion. Accountability seems a critical need in the development of the next generation of risk assessments and risk management approaches.

Suggested reading

Brand, L.E., Sunda, W.G. and Guillard, R.R. L. (1986). Reduction of marine phytoplankton reproduction rates by copper and cadmium. *Journal of Experimental Marine Biology and Ecology*, **96**, 225–250.

Chapman, P.M. (1991). Environmental quality criteria: what type should we be developing? *Environmental Science and Technology*, **25**, 1352–1359.

Clements, W.H. (2000). Integrating effects of contaminants across levels of biological organization: an overview. *Journal of Aquatic Ecosystem Stress and Recovery*, **7**, 113–116.

Moffett, J.W., Brand, L.E., Croot, P.L. and Barbeau, K.A. (1997). Cu speciation and cyanobacterial distribution in harbors subject to anthropogenic Cu inputs. *Limnology and Oceanography*, **42**, 789–799.

Niyogi, S. and Wood, C.M. (2004). Biotic Ligand Model, a flexible tool for developing site-specific water quality guidelines for metals. *Environmental Science and Technology*, **38**, 6177–6192.

Wang, W.-X. (2002). Interactions of trace metals and different marine food chains. *Marine Ecology Progress Series*, **243**, 295–309.

16 • Sediment quality guidelines

16.1 INTRODUCTION

Prologue. Sediments progressively accumulate metal contaminants in response to metal inputs and retain a legacy of contamination after inputs recede. Sediment contamination presents an ecological risk that must be managed in all metal-contaminated environments.

Sediments are the largest reservoir of metal in a contaminated aquatic environment (Chapter 6). Sediments are also an ecologically important component of the aquatic habitat (Luoma, 1983, 1989). Therefore, wherever metal contamination occurs, management of contaminated sediments is important. Most risk assessments consider total metal concentrations in sediment as linked to sediment toxicity. But the importance of adding consideration of metal bioavailability is well recognised. The challenges in understanding bioavailability have slowed (or even halted) progress (NRC, 2003). A major barrier to lateral management is that sediment contamination is managed separately from contamination of the water column and management of risks to aquatic life other than the benthos. Recognition that sediment contamination is inextricably linked to the degree of contamination in all other components of the aquatic environment is necessary for the next steps forward. A related barrier is an inadequate appreciation of the roles of vertical zonation and areal heterogeneity in the sedimentary environments of natural waters. A lateral view of the living sedimentary environment offers opportunities to better understand the risks that metals pose in aquatic environments; and thus to better manage those risks.

16.2 THE SETTING

Prologue. Equilibration of metals among sediments, water and living organisms is driven by the complex geochemistry of the sediments. Oxidised and reduced zones in the sediments affect metal form. These zones affect and are affected by the organisms living in the sediment. Disturbances in the structure and function of benthic communities can be observed in the presence of metal contamination in sediment, but dose–response and cause–effect are complex.

In previous chapters we have established the setting that makes management of metal contamination in sediments an important but complex issue (see especially Chapters 6 and 11).

The concentrated repository of metals in aquatic sediments does not exist in isolation. Equilibration, via physical, biological and biogeochemical processes, occurs among pore waters within the sediments, the water column and the living components of the sediment. The benthic community of the aquatic environment lies at the base of a larger food web that occupies benthic, pelagic (water column) and riparian habitats. Predators of the benthos are exposed to metals accumulated by the many forms of life living within sediments. But the physical and chemical processes that govern metal equilibration among sediments, water and living organisms are variable among circumstances. Thus distributions are variable and site-specific.

Metals are distributed, biogeochemically, between the surfaces of particles and pore waters as defined by a partition coefficient (K_d) (Chapter 6). K_d is influenced by dissolved speciation, redox potential of the sediment, the nature of the

Box 16.1 Definitions

Apparent Effects Threshold (AET): sediment concentration of a metal above which toxic effects always occur.

acid volatile sulphide (AVS): the sulphide released when a sediment is extracted with dilute hydrochloric acid (HCl) and the simultaneously released metal (SEM) is the metal released into the acid by that extraction.

benthos or benthic community: organisms (plants and animals) that live at the bottom of a water body; including organisms living on or within the sediments.

Effects Range Low (ERL): marine and estuarine sediment concentration of metal below which toxic effects risks are considered minimal. When data are assembled in ascending concentration against effects, the ERL is defined by the 10th percentile.

Effects Range Median (ERM): marine and estuarine sediment concentration of metal above which toxic effects often might be expected. The ERM is defined as the median of the ascending concentration data set, at which effects are observed.

in situ: a Latin term meaning 'in place' or not removed; used here to signify toxicity tests conducted in the natural setting (i.e. in the sediments of a water body).

Probable Effects Level (PEL): usually freshwater sediment concentration of metal below which toxic effects risks are considered minimal. When data are assembled in ascending concentration against effects, the PEL is defined by the 15th percentile of the data.

SEM: see *acid volatile sulphide*.

Threshold Effects Level (TEL): usually freshwater sediment concentration of metal above which toxic effects often might be expected.

particulate material and microbiologically driven processes. Influential sedimentary components include Fe oxides, Mn oxides, organic coatings, sulphides, organic detritus and organisms (from bacteria to macrofauna) living within the sediments.

The balance between the influx of oxygen into sediments and consumption of oxygen by the living component of sediments influences a vertical zonation of oxidised and reduced layers in sediments. These redox processes influence metal form, biological activity and the nature of the associated food web. Grain size is another trait of sediments that has important biogeochemical and biological influence.

Sediments are inhabited by diverse communities of microorganisms, meiofauna and macroorganisms (the benthic community) in freshwater, estuarine and marine environments. Experiments and field studies show that metal contamination in sediments can be toxic to these organisms. In general, metal bioaccumulation (and toxicity) increases throughout a benthic community as the degree of contamination increases, but the variability of concentration and response is large

(Chapters 8 and 11). The complexity in these relationships is related to both the geochemical complexity of the sedimentary environment and the biological complexity in the relationship of organisms to their environment.

The exposure of organisms in benthic communities to the sedimentary environment, and its associated metal contamination, is influenced by life history and functional ecology. Animals that feed on sedimentary deposits are more directly exposed to local contamination than animals that feed from suspension, for example. The organism's relationship to redox conditions (Chapter 6) in the sediments is also important. Oxic conditions are extraordinarily important to macrofauna and meiofauna (Rhoads and Boyer, 1983). Epibenthic species live on the surface of the sediments in direct contact with oxidised waters. Some animals (e.g. most meiofauna) live exclusively within surface oxidised layers. Others live deeper in the sediment column but use burrows or siphons to draw oxygenated water (e.g. from above the surface of the sediment) across their gills. Most also possess mechanisms that allow feeding upon the more organically rich material from the surface layers. Some members of the benthos also migrate from oxic to anoxic sediments or from the sediments to the water column on a regular basis.

Exposure routes of benthic organisms include direct ingestion of detritus, isolated food particles, prey or non-living particles coated with Fe oxides, Mn oxides or organic material. Organisms are also exposed to metals via pore waters, waters inside defined burrows, or oxic overlying waters. Metal exposure in the benthic community is affected by biological traits including food selection, feeding rate, filtration rate and digestive processing in addition to the approaches used to acquire oxygen (e.g. burrow or tube irrigation). Many benthic animals depend at least partly upon the microflora and microfauna or meiofauna of the sediment for their food.

The benthic community can modify the geochemical nature of the sediments by physically mixing (bioturbation) sediments of different character and by affecting oxygen influx (and the depth of the interface between reducing and oxidising conditions). Many organisms that live within sediment create a microenvironment that influences the geochemical nature of their relationship to the sediments and the water column.

Metal-specific, geochemical, biological and physical complexities, like those above, are the reason that it is complicated to determine metal toxicity in sediments and the concentrations related to some predicted frequency or intensity of adverse biological effects (NRC, 2003; Batley et al., 2005).

16.3   TOXICITY

Prologue. It is well known that there is a link between metal contamination of sediments and toxicity. But quantifying and generalising about dose response is especially challenging. Sediment toxicity can be empirically tested by spiking sediments or testing sediment collected from nature. Standardised testing protocols yield repeatable results, but are insensitive

to many of the nuances in the way toxicity is expressed in nature.

It is unequivocal that contaminated sediments can be toxic to aquatic life (Chapman, 1989b). Sediment toxicity is readily demonstrated when clean sediments are 'spiked' with trace metals, or when benthic organisms are exposed to highly contaminated sediments from nature (Kemp and Swartz, 1988). Toxicity disappears in a dose-responsive manner when contaminated sediments are diluted with uncontaminated sediments or contamination,recedes in nature (Hornberger et al., 2000). Sediment contamination is often accompanied by damaged communities, missing species and stressed survivors (Chapter 11). The bioassays that demonstrate such toxicity are more realistic than dissolved toxicity tests, because they include two major components of the environment (water and sediment). Nevertheless, they require simplifications and assumptions that deviate from the realities of nature. Thus determining the quantitative link between chemical concentration and ecological risk using toxicity testing alone is a challenge.

16.3.1   Bioassays

Prologue. Bioassays are important tools for understanding the processes that affect sediment toxicity. But sediment sampling, transport, storage and/or manipulation affect the validity of extrapolations from bioassays to field conditions, as do biological considerations. If uncertainties are not recognised and addressed in the assessment process, then erroneous conclusions may result (Batley et al., 2005).

The toxicity of sediments is commonly evaluated via sediment bioassays. Sediment bioassays are both more complex and more realistic than standardised dissolved toxicity tests because they add an additional concentrated repository of metals. Single-species tests are the least expensive and time-consuming method to evaluate the responses of biota to contaminated sediment and its associated pore and overlying waters, under controlled conditions.

Ambient or whole sediment toxicity tests involve using the bioassay as a sensor. A sediment is collected from a field location and a chosen response in a selected species is used to characterise its toxicity. Such tests empirically address the question of whether contaminants that occur in the sediments (even in mixtures) can cause detectable adverse effects over a chosen exposure period (Chapman, 1989b). The tests, of course, reflect effects from all the contaminants in the sediment, acting in concert.

Some study designs compare responses from test sites to responses from a reference (uncontaminated) site (Long et al., 1990). Comparisons of responses along a contaminant gradient in nature are possible, as are comparisons of responses from different locations within a system. Dose–response

can also be tested by diluting contaminated sediments from nature with an uncontaminated sediment (Swartz et al., 1985). Ambient bioassays circumvent the difficulty of predicting bioavailability from geochemical analyses; and address multi-contaminant problems empirically (with data from a direct test). However, the results are specific to the aliquot of sediment tested. Extrapolation of results is limited. Nor do ambient tests answer the question of why the sediment is toxic.

Bioassays with spiked sediments can be used to determine dose–response or investigate mechanistic questions (Kemp and Swartz, 1988). Sediment bioassays can address sediment toxicity at any level of biological organisation. But the single-species, whole organism bioassay is the most common approach (Box 16.2). In the early sediment bioassays, the recommended test species were mostly

---

Box 16.2 Single-species sediment bioassays

A variety of macrofauna or meiofauna are employed in whole sediment bioassays. Choice of species is governed by availability, sensitivity to contaminants and tolerance to test conditions. Compatibility with the environmental conditions, such as salinity, grain size, presence of other toxins like ammonia (Spies, 1989) can affect survival.

Amphipod crustaceans are some of the most commonly employed organisms in sediment toxicity testing in the estuarine/marine environment. The US Environmental Protection Agency has established guidelines for conducting sediment toxicity tests (Burton et al., 2003). Species for which standardised protocols exist include a variety of marine and freshwater amphipods, polychaetes (especially *Neanthes arenaceodentata)*, decapod crustaceans (for example, *Palaemonetes pugio)*, white urchin (*Lytechinus pictus*) and bivalve mollusc larvae from species like *Mercenaria mercenaria*.

Protocols have similarities to those for chronic dissolved bioassays. Amphipod (*Hyalella azteca*) tests start with juvenile animals and continue up to 29 days until reproductive maturation (Ingersoll et al., 1996). Tests with midge larvae (e.g. *Chironomus tentans*) start

with first instar larvae (<24 hours old) and continue up to 29 days through adult emergence (Ingersoll and Nelson, 1990). Most protocols recommend collection of sediments from nature from the surficial 2 to 5 cm at the sediment–water interface, then mixing or homogenisation of the sediments to aid reproducibility. Replicates of the sediment sample are essential for each test, as is removal of native fauna by sieving. Test animals must be acclimated in similar sediments for several days, and the number of animals per test must be adequate (e.g. about 20 animals for each test). The choice of static or flow-through conditions is important because the overlying water can influence results. Longer-term tests, in which survival and growth are measured as animals mature, replicate effects relevant to nature more sensitively than do the shorter-term or acute tests (Ingersoll et al., 1998). Yet most tests restrict the time of exposure to 10 to 28 days for practical reasons (Melzian, 1990). Controls include clean sediment from the animals' habitat, sediment with the same particle size distribution, or a positive response control employing a chemical and sediment with known toxicity. Bioaccumulation and survival are the most reproducible endpoints, but ability to bury and emergence from sediment are also typical responses.

Growth and gonad production can be studied in longer experiments (Thompson *et al.*, 1989). Life history endpoints have been employed in studies with cultured species (Scott and Redmond, 1989; Pesch *et al.*, 1991).

Results from standardised tests can be robust. Swartz *et al.* (1982) showed that five replicate tests were 75% certain to detect a mean survival change of 15% using the *Rhepoxynius abronius* test. Inter-laboratory calibrations show greater variability. Five laboratories distinguished 'toxic' from 'non-toxic' in a similar manner in several sediments from nature (Mearns *et al.*, 1986), but differences occurred in the ranking of the toxic sediments. Chapman (1991) suggested that a 50% coefficient of variation was common in assessments of dose–response from the same sediment. Burton *et al.* (1996) found a smaller range of variation when different test conditions were compared among seven laboratories (13% to 39% coefficient of variation) with tests of the freshwater organisms *H. azteca* and *C. tentans*.

adult bivalves, polychaetes and fish. These test organisms were generally insensitive to contaminated sediments in the types of short-term exposure employed. Shorter-lived smaller species, and/or longer-term chronic endpoints (behaviour, growth, reproduction, larval settlement) proved more sensitive.

Choice of species for single-species whole sediment tests must also consider the match between the requirements of the organism and test conditions. A principal criterion in choosing a test species is that it represents an important group in uncontaminated environments and is among the most pollution sensitive of benthos. Some of the best-developed sediment bioassays use highly sensitive amphipods like *Rhepoxynius abronius* or *Eohaustorius estuarius* in marine environments (e.g. Lamberson *et al.*, 1992; Hunt *et al.*, 2001) or *Hyalella azteca* in freshwater situations (Ingersoll *et al.*, 1996); although tests for several other sensitive organisms are also well developed (Box 16.2).

Because sediments add an additional and more complex matrix to toxicity testing, results are more sensitive to test conditions than is the case with dissolved toxicity testing. Sediments sometimes generate toxic metabolites, such as sulphide or ammonia (Spies, 1989), that can themselves cause toxicity and lead to false positive predictions (predictions of contaminant effects when none exists: Spies, 1989). Species also differ widely in ecological, physiological and life history characteristics that affect exposure to and survival in sediments. For example, some species are sensitive to the characteristics of the sediment, like grain size. *Rhepoxynius abronius* turned out to have a narrow range of salinity and particle size tolerance. As a result, protocols were developed with *E. estuarius*, which tolerates a broader salinity range and a broader range of particle types than *R. abronius*. However, *E. estuarius* is more tolerant to Cd than *R. abronius*, i.e. it is not the 'most sensitive species' to all toxicants. Thus fundamental understanding of ecological requirements is a greater challenge for sediment toxicity testing than for dissolved testing. Use of a battery of established surrogate species can be important (Box 16.2).

A toxicity test that is designed for widespread use usually necessitates simplifying assumptions that make the test manageable, repeatable and pragmatic (Chapter 10). Standardisation improves repeatability but the assumptions limit extrapolation.

16.3.2   Complex sediment toxicity tests

Prologue. Complex toxicity testing procedures more closely simulate conditions in nature and are amenable to studying sublethal measures, like growth rates, reproduction or effects over multiple generations. Organisms are typically more responsive to contamination *in situ*, than when sediments are taken to the laboratory for testing. In general, laboratory assays may be attractive from pragmatic and convenience perspectives, but the increased risks of underestimating toxicity must be carefully considered (Ringwood and Keppler, 2002).

An alternative to standardised laboratory testing of sediments is exposing toxicity test organisms *in situ* or within (on) the sediments of the water body of interest. Typically, organisms are caged in direct contact with intact sediments. This approach reduces sampling related artifacts and better simulates natural exposures. *In situ* bioassays with fish have a long history. Sophisticated methods for testing macroinvertebrates in sediments (e.g., bivalve larvae, cladocerans, amphipods and midges) are more recently developed (Ringwood and Keppler, 2002; Burton *et al.*, 2003).

In situ protocols must demonstrate survival of control organism for 10 to 28 day exposure periods. This requires test organisms that can survive handling stress and fluctuating field conditions. A major challenge is distinguishing between effects from metal contamination and effects from moving, caging or responding to fluctuation in the habitat conditions. Careful, extensive study of the organism in uncontaminated conditions can separate these influences (Ringwood and Keppler, 2002) (Box 16.3). Variability in the response variable must also be understood. Stochastic variability is more problematic than dynamically stable variability (e.g. seasonal cycles). For example, Olsen *et al.* (2001) transplanted fourth instar *Chironomus riparius* larvae for 48 hours at 13 uncontaminated river sites across southeast England. Activities of two enzymatic biomarkers varied almost twofold across the sites, with statistically significant differences detectable between sites (Olsen *et al.*, 2001). They concluded that biomarker results must be treated with caution because natural variability in responses can occur even in the absence of toxicant exposure.

Comparisons of *in situ* tests with tests of the same sediments in the laboratory report mixed results (Box 16.3). Tests in the field often show greater toxicity, but the cause is not always clear. Environmental factors such as human activities in the water body, storms, temperature or inputs of unmeasured toxicants are potential external causes of toxicity (Kater *et al.*, 2001). Some of these can be controlled by selectively choosing optimal times of year for the test. Controls must be used to eliminate influences of the field chamber itself, and designs are available that separate influences of overlying water from influences of sediment (although the two can also be interrelated).

Ringwood and Keppler (2002) illustrated a successful approach. They conducted preliminary studies that determined the influences of environment alone at a number of uncontaminated sites, then compared those results to 13 contaminated sites of similar character in Charleston Harbor, USA (Box 16.3). They suggested that laboratory tests often falsely conclude no toxicity. False positive outcomes in their study (predicting toxicity when none was likely) occurred at a similar rate in field and laboratory tests (about 17%). But the rate of false negatives (false predictions of no toxicity) was much higher in the laboratory tests (69%).

On the other hand, the opposite results were observed in tests of toxicity to an estuarine amphipod using sediments heavily spiked with Cd (273 to 1900 µg/g Cd dw), then transplanted to both laboratory and the field (DeWitt *et al.*, 1999) (Box 16.3). Based upon pore water toxicity, LC_{50}s occurred at similar concentrations in the laboratory (50.4 mg/L Cd) as in the field (51.6 mg/L Cd). The extreme concentrations and artificial conditions that favoured extremely high pore water metal concentrations may affect conclusions in tests of this type.

As discussed in Chapter 10, mesocosms are used to test questions that single species bioassays cannot. For example, important secondary ecological effects on communities may only be detectable using multiple species in a mesocosm, *in situ* (Bonsdorff *et al.*, 1990). Because mesocosm studies can be costly, labour intensive, difficult to control and difficult to replicate, they are rarely used in routine toxicity testing. Thus mesocosms are better suited to testing process questions than to determining dose–response or replicating nature (Cairns, 1983).

One example of a mesocosm approach used with sediments is *in situ* colonisation by local fauna when sediments spiked with metals are placed in open containers in nature. Hare *et al.* (1994) compared the lake species that settled in precontaminated (Cd) trays of sediment with those that settled in uncontaminated trays. Settlement

Box 16.3 *In situ* versus laboratory tests of sediment toxicity

Responses of juvenile bivalves (newly metamorphosed 'seed clams' of the species *Mercenaria mercenaria*) were compared between tests of sediments taken into the laboratory and the same sediments tested *in situ* (Ringwood and Keppler, 2002). Juvenile clams were deployed for 1 week at a variety of degraded and non-degraded reference sites in Charleston Harbor, South Carolina, USA (13 contaminated sites; 16 uncontaminated sites). Mortality and a sublethal endpoint, clam growth rate, were used to compare reference and degraded sites. Newly metamorphosed bivalves exhibit rapid growth, so effects on growth can be detected in a relatively short time-frame. The juvenile clams are infaunal, crawling through the sediments and feeding at the surface–water interface. Clam growth was not significantly biased by sediment type, so the test was applicable across sandy as well as muddy habitats. Sites were classified *a priori* as reference or degraded, based on chemical analyses of contaminant levels. Growth rates of field-deployed clams at reference sites tended to be higher than growth rates in the laboratory; apparently an effect from holding sediments in the laboratory.

Significant adverse effects on either mortality or growth were detectable in laboratory assays at 4 of the 13 potentially degraded sites. But 9 of the 13 sites were toxic in the field, including most of the sites that were toxic under laboratory conditions. The results strongly supported the conclusion that laboratory assays do not detect potential sediment toxicity at all degraded sites. A major issue in comparing laboratory and *in situ* assays is whether the greater toxicity in the field is a response to chemical contamination or to habitat differences, such as salinity regimes. To understand influences of variability in environmental conditions, Ringwood and Keppler (2002) conducted an extensive study of the 16 reference sites over time. They showed that growth rates in the absence of

contamination were affected by such factors as pH, dissolved oxygen and salinity. In interpreting data from the contaminated sites they paired comparisons between reference and contaminated sites at similar salinities; and used regressions between toxic responses and pH and/or dissolved oxygen from the reference sites to subtract out the influence of those factors. Thus their final conclusions attributed differences in growth, *in situ*, among sites to contamination in sediments with a minimum of ambiguity (Fig. 16.1).

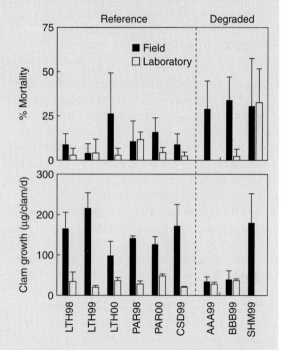

Fig. 16.1 Percent mortality and growth in juvenile hard clams (*Mercenaria mercenaria*) at high salinity sites that were either degraded (identified as contaminated by chemical analyses) or non-contaminated (reference) were compared between a 7 day laboratory bioassay and an *in situ* deployment. (After Ringwood and Keppler, 2002.)

of one sediment-dwelling species (the midge larva *Chironomus salinarius*) was affected by elevated Cd exposures. Other species were not influenced by the Cd contamination. Interestingly, many of the species that colonised the trays were primarily exposed to Cd via the water column, not the

contaminated sediment in their habitat. In most cases, their life history traits involved migration out of the sediments, thus limiting contact with sedimentary food and pore waters (Warren *et al.*, 1998). Colonisation also varied among habitats (Hare, 1995). Negligible colonisation occurred in deep lake (profundal) habitats, and faster colonisation occurred in littoral (nearshore) environments. Surface-dwelling species and migrating species also colonised the trays most quickly (within months).

16.3.3 Geochemical uncertainties unique to sediment toxicity tests

> At the microenvironment level there are approximately one million bacteria per gram of sediment that, over the space of millimeters occupy critical niches for metabolizing – cycling specific organic compounds, organic acids, nitrogen, sulfur, methane and hydrogen.
>
> Burton *et al.* (2003)

Prologue. The results of sediment bioassays are affected by manipulation of the sediments during collection, storage and experimentation. Standardised handling procedures for sediments have been published, but it is difficult to avoid changes in the sediments that can influence extrapolation of results, even if such procedures are followed. Immediate removal (and testing) of pore waters is one way to bypass some of the effects from collection, storage and handling; but this approach has its own difficulties.

Outcomes of sediment bioassays are least uncertain when the goals are simple and purely empirical. But bioassays have serious limitations when the goals become more ambitious (Chapter 9; Chapman and Long, 1983; Chapman, 1989b). One of the perceived advantages of laboratory assays is that if adverse effects occur, they can be ascribed to contaminants, because all other confounding variables have been controlled. In reality, the issues with collecting, handling and preparing sediments for experimentation, as well as choice of species and test conditions, create enough uncertainties that either underestimation or overestimation of potential toxicity must be seriously considered for different circumstances (Ringwood and Keppler, 2002).

16.3.3.1 Sediment collection
Prologue. Most sediment collection procedures disrupt the zonation and fluxes that characterise a natural sediment. Homogenising the sediments,

restricting collection to visibly oxidised layers or reducing the entire sediment after collection then allowing zonation to develop are alternative approaches to addressing the issue. The last two choices are most likely to produce results most applicable to nature.

Sediments, in nature, are open, dynamic, structured biogeochemical systems (Chapter 6). A typical natural sediment deposit exists in contact with an oxidised water column, and the sediment is composed of oxic sediments on top of anoxic sediments. Fluxes of chemical constituents occur between all compartments. The gradients in this dynamic system are important geochemically and biologically. Sample collection is inevitably disruptive of this environment, as is extended sample storage and manipulation of sediments for a bioassay. If storage or manipulation alter geochemistry, they will affect exposure routes and either enhance (by producing toxic products like ammonia or hydrogen sulphide) or inhibit (by affecting metal chemistry) toxicity (Burton *et al.*, 2003).

For example, merely constraining the sediment can create a 'bottle effect' that disrupts oxygen input, affects fluxes and quickly changes zonation (Westerlund *et al.*, 1986). Most recommendations for whole sediment bioassays suggest collecting surface sediments to a fixed depth (e.g. 2 to 5 cm: ASTM 2003). Given the variability in the depth of the oxic zone, a sample taken from the surficial 3 cm of one sediment may contain a very different proportion of oxidised and reduced sediments than a sample of similar depth from another sediment (Chapter 6). The result will be concentrations of metal reactive components such as Fe or Mn oxides and sulphides that reflect the collecting procedure rather than nature.

A standard procedure is to address redox complexity in a collected sediment is to homogenise the sediment before experimentation. This reduces internal variability and 'controls' differences among treatments. But mixing reduced sediments with oxic sediments, or oxidising a mixture of sediments will create a maze of reactions whose outcome will be dependent upon the initial portions of reactants, the availability of oxygen and other oxidants to the mixture, and time. New

Box 16.4 Handling effects

Simpson and Batley (2003) showed that metal partitioning between particles and pore water in sediments containing anoxic material (Fe(II)-rich pore waters) was altered as sediments underwent preparations for sediment toxicity testing:

> Experiments with Zn-contaminated estuarine sediments demonstrated large and often unpredictable changes to metal partitioning during sediment storage, removal of organisms, and homogenization before testing. Small modifications to conditions, such as aeration of overlying waters, caused large changes to the metal partitioning. Disturbances caused by sediment collection required many weeks for re-establishment of equilibrium. Bioturbation by benthic organisms led to oxidation of pore-water Fe(II) and lower Zn fluxes because of the formation of Fe hydroxide precipitates that adsorb pore-water Zn. For five weeks after the addition of organisms to sediments, Zn fluxes increased slowly as the organisms established themselves in the sediments, indicating that the establishment of equilibrium was not rapid.

products (such as Fe or Mn oxides) are rapidly formed during such a process and existing products will change. Pore water concentrations and gradients can disappear, re-form and redistribute according to the specific conditions in each container over the course of a 30 day experiment (Thomson *et al.*, 1980). The end result is to add uncertainty about extrapolation to nature (Box 16.4).

Uncertainties associated with sampling sediments of different redox characteristics can be reduced by restricting samples to the surficial visually oxidised layer of surface sediments (Luoma and Bryan, 1978, 1979). Visual interpretation of oxidising conditions is an operational approach, but it nevertheless greatly reduces variability compared to sediments that clearly include mixed redox layers. Published 'standards' for handling sediments for bioassays do not address reducing/oxidising conditions (USEPA, 2001b; ASTM, 2003), unfortunately. Systematic studies that compare

toxicity responses in sediment sampled by different procedures or from different redox conditions are important requisites.

Addition of supplemental food is another important consideration in a sediment toxicity test. Many test species do not exclusively obtain their nutrition from ingesting just sediment particles; so an additional source of food is often essential. But standardised protocols that add food do not equilibrate it with the contaminated test media. In contrast, all components of the environment are typically contaminated in nature. Studies show that adding food equilibrated with the metal contaminant increases metal exposure in polychaetes (Lee *et al.*, 2001). Exposures in bioassays with unequilibrated supplemental food underestimate exposures that would occur in nature in those sediments.

Storage of sediments is another important source of bias. Most investigators agree that storage time changes toxicity. This is not surprising because storage is likely to change at least redox conditions. Storage in a sealed contamination can eliminate oxygen, for example, and result in a completely anoxic sediment. Published recommendations are maximum storage from 5 to 7 days under refrigeration (Swartz, 1987). In general it is probably best to test sediments as soon as possible after collection. Unavoidable storage should be at 4 °C in the dark. An alternative is to intentionally reduce all the sediment, then allow vertical rezonation to develop in the experimental chamber before conducting the test. Lee *et al.* (2000a, b, c) used this approach to simulate exposures of different benthos to metals in San Francisco Bay sediments. The redox gradients that developed in the sediments were similar to the types of gradients observed in the field, suggesting that exposures of organisms were also more similar than if sediments were stored in their original condition then homogenised.

Most sediment collection, handling or storage procedures also change or eliminate meiofauna, microflora like benthic diatoms, and the bacterial flora of sediments. These are important food sources for some species (including some typical test species). Thus the food source of test species from stored sediments is likely to be different from that in nature.

Intuitively, it seems that the unique complexities of testing sediments themselves could be alleviated by presentation of metal removed from the sediment in dissolved form. Solutions are less complex systems to test than are sediments and easier to control. Changes in reducing/oxidising conditions and the build-up of metabolites occur more slowly in the absence of sediment. Exposures can be kept short and the effects of extraneous factors such as feeding can be minimised. The effects of natural sediment factors such as grain size also can be eliminated by removing contaminants from the sediments before testing (Carr *et al.*, 1989; Chapman, 1989b). This is also an attractive choice for test organisms that are difficult to work with in a sediment (e.g. many larvae and embryos). Data can be comparable to dissolved bioassay tests like those used to derive water quality criteria (Chapter 15).

Centrifugation, squeezing, filtration or a combination of the above are used to obtain pore waters. To preserve the original character of the waters, a nitrogen atmosphere must be rigorously maintained during the removal process, assuming that the sediment sample contains at least some anoxic sediment. For elutriate preparation, standardised procedures call for a slurry of sediment and oxidised seawater (1:4), vigorous mixing and then settling (for 1 hour in the most common method). The elutriate is either decanted or filtered, or the sediment is not removed. The greatest problem with these methods is precipitation of particulate, highly metal-reactive Fe oxides and Mn oxides. These reactions occur within seconds when the waters associated with anoxic sediments contact oxygen (Troup *et al.*, 1974; Simpson and Batley, 2003). Trace metals co-precipitate with the Fe oxides and are stripped from solution, undoubtedly affecting toxicity. Similarly, if investigators filter pore waters, a loss of metal toxicity is likely as metals associated with colloidal or particulate precipitates are removed, unless strict adherence to oxygen-free conditions is maintained.

Unfortunately, macrofauna require oxygen for a suitable test environment. Procedures that mix reduced sediments with oxidised seawater will inevitably precipitate the oxides. Exposure change as metals move from dissolved to particulate. Thus, bioassays that remove pore waters from anoxic sediments or elute reduced sediments are likely to reduce dissolved metal concentrations and change routes of exposure. In short-term tests the result is an underestimate of toxicity.

16.3.4   Appropriate uses of sediment bioassays

Prologue. Single-species whole sediment bioassays, when used alone, are most effective at screening natural sediments for obvious problems, especially where multiple contaminants are present, and in empirically testing specific process questions. Large data sets are useful to constrain high and low probability risks. Single-species whole sediment bioassays should be interpreted cautiously in site-specific assessments or in drawing fine distinctions between levels of risk at different locations.

Whole sediment bioassays can be (but are not always) concordant with other lines of evidence about toxic effects. Concordance is most common when comparing sites with extreme contamination to sites that lack of contamination. Discordance is more common in the intermediate contamination ranges where uncertainties about risk are greatest (Box 16.5). A body of evidence that cuts across disciplines is usually necessary to reduce the probability of false negatives and false positives at any individual site unless contamination is extreme (Chapter 11).

16.4   METAL CONCENTRATIONS IN SEDIMENTS: LINKS TO BIOACCUMULATION AND ECOLOGICAL RISK

Prologue. Disagreements over the use of bioaccumulation information, historically, inhibited a consensual view of the state of knowledge about risks from metal-contaminated sediments, and thus slowed the development of lateral risk management.

Box 16.5 Chemical concentrations, sediment toxicity and community structure: two examples

(a) Rhine–Meuse Delta

In 1992 to 1995, sediment toxicity, sediment contaminant concentrations, the nature of the sediments and ecological processes were studied at 106 sites in the enclosed Rhine–Meuse delta, a freshwater environment in the Netherlands (Peeters *et al.*, 2001). Multivariate statistical relationships among the variables were analysed. The variation in species abundance data was partitioned into a spatial and an environmental component using multivariate techniques (canonical correspondence analysis). Pore water bioassays were conducted with the cladoceran *Daphnia magna*, and sediment bioassays were conducted with the midge *Chironomus riparius*. There was a very weak relationship between species composition in the benthic community and bioassays results: 1.9% of the variation in resident macroinvertebrate fauna was explained by the bioassay responses. Fewer taxon abundances were related to results from the pore water bioassays than from the whole sediment assay, consistent with the likelihood of geochemical interferences in the pore water methodology. Distributions of some of the resident macroinvertebrate species showed a significant relationship with the responses in the *Chironomus riparius* bioassay. But various species of *Chironomus* were present in the delta, and their distribution patterns showed no significant associations with the results of the *C. riparius* bioassay.

Elevated contaminant concentrations were significantly associated with differences in the macroinvertebrate food web structure. The composition of the benthic community was also influenced by grain size in the sediment and habitat variables associated with geographical zone. Based upon the multivariate statistics, the authors concluded that sediment contamination affected sediment-feeding taxa but not taxa with other feeding modes. The macroinvertebrate field surveys and the laboratory bioassays yielded different types of information. Effects on a specific component of the benthic community could not be predicted from the toxicity tests. They speculated that manipulations of the sediment before the test, sample storage and the pollutant tolerance of many macroinvertebrates in the lake contributed to the disconnect between bioassay results and observations from nature.

(b) San Francisco Bay

Hunt *et al.* (2001) evaluated 111 sites in San Francisco Bay using ecological surveys, chemical analyses and bioassays. Preliminary studies showed that pore water bioassays with sea urchin larvae were of questionable reliability because of chemical reactions during and after pore water removal. But sediment bioassays were successful with the estuarine amphipod *Eohaustorius estuarius*. Ecological data were quantified by a 'relative benthic index' which took into account general pollution tolerance of resident fauna. No toxicity was observed at 54 of the 111 sites. Both chemical concentration and negative results in bioassay toxicity tests appeared to identify these sites as having the lowest probability of risks; although the design did not allow testing for false negatives. Nine locations in three extremely contaminated coves showed effects in toxicity tests, chemical concentrations above established thresholds (see later discussion) and benthic communities indicative of severe pollution (in one case all fauna were absent). Thus, all three methodologies were able to identify extreme effects accompanied by extreme contamination. Among the remaining 48 sites, results in general were a complex mix from the three measures. Weight of evidence was calculated to estimate contamination effects at these sites, but the validity of the different tests in this moderate contamination, moderate toxicity or disturbed ecology range was uncertain.

Contradictions between bioaccumulation observations and toxicity experiments slowed the development of consensus about ecological risks from contaminated sediments.

Appreciation of the strengths and weaknesses of different approaches to linking concentration and risk from contaminated sediments is a key to unravelling the difficult choices about managing those risks. A consensus on risks from metals in sediments was confounded for a long time by disagreements about the appropriate approach to use in evaluating exposure and toxicity. There were questions about whether metals bound to the particulate material are directly available to organisms (e.g. Meyer *et al.*, 2005); with the suggestion that pore waters were the only important route (or at least indicator) of exposure (Di Toro *et al.*, 1991). It was assumed that a simple distribution coefficient (K_d) and a simple ratio of exposure between organisms and sediments (BAF or bioaccumulation factors) could be used to define exposure. Bioaccumulation was rarely used to help interpret exposures. Many publications assumed that no metal was available via direct ingestion of sediment or the route of exposure was assumed irrelevant. As stated by Di Toro *et al.* (1991): 'It is assumed that the organism receives an equivalent exposure from a water-only exposure or from any equilibrated phase; either from pore water via respiration; from sediment carbon via ingestion; or from a mixture of the routes.' The basic assumption was that the total concentration of metal in sediment did not matter; the variable driving toxicity was the distribution of metal between particle and water, and exposure was limited to pore waters (this is termed the *equilibrium partitioning* theory of bioavailability). Bioaccumulation and biomonitoring were also disregarded in traditional assessments of risks from contaminated sediments (e.g. the 'sediment quality triad' of Chapman and Long, 1983). It was stated that bioaccumulation is a phenomenon not a process (Chapman, 1989b), and many authors saw bioaccumulation data as only marginally useful. Presumably these authors meant that not all bioaccumulated metal is directly toxic. Indeed, a complex interplay among geochemistry, biology and metal-specific attributes make the relationship between bioaccumulation and toxicity variable or complex (e.g. Pb in Hare and Tessier, 1998). But, the concept widely accepted elsewhere, that internal accumulation of metals is an informative first step in the cascade of toxicity processes (Chapter 7), has not been widely appreciated in studies of sediment toxicity. Thus, historically, it was convenient to assume that direct bioavailability of particulate metal was unimportant and that either toxicity tests or equilibrium partitioning calculations were predictive of risk. Toxicity tests could be used to empirically evaluate specific circumstances (e.g. dredge

spoil disposal) and prediction of risk could be derived from pore water concentrations and the dissolved toxicity testing database (Chapters 10 and 16; Di Toro *et al.*, 1991).

16.4.1. How do metal concentrations in sediment relate to risk: bioaccumulation?

Prologue. Significant relationships between bioaccumulated metal and metal concentration in sediments suggest that exposure and concentrations can be broadly related. This is most common if the data range includes extreme and/or uncontaminated sediments. Relationships are not necessarily simple in narrowly specific circumstances, across narrow data ranges or intermediate values.

As discussed in earlier chapters, it is recognised that bioaccumulation is relatively easy to determine, unambiguous to interpret (where appropriately used) and contaminant-specific (Chapter 7). But linking the bioavailability inferred from bioaccumulation in an organism in nature to concentrations in sediments proved challenging to unravel. A common approach was to assess statistical linkages between bioaccumulated metal (in benthos that ingest particulate material) and sedimentary metal concentrations. Significant correlations are often observed between bioaccumulated metal and concentrations in sediments where data are restricted to simple gradients or geochemically similar sediments. Usually (but not always) significant correlations are found over broad concentration ranges among dissimilar sediments (Bryan, 1985) (Box 16.6). But weak relationships, or no relationship, are observed over small data ranges especially if sediments differ widely in character. The linkage between bioaccumulation and metal concentration in sediments is therefore a (first-order) broad one with the degree of variability reflecting influences on metal bioavailability.

Regressions between sedimentary concentrations of metals and bioaccumulated metal concentrations do not address whether exposure to

Box 16.6 Total metal concentrations in sediments versus bioaccumulated metal in large data sets containing dissimilar sediments

Large data sets have compared metal concentrations in sediments and three species of biomonitors (two bivalves and a polychaete) from the geochemically diverse estuaries of southwest England (Luoma and Bryan, 1978, 1981; Bryan, 1985; Bryan et al., 1985; Bryan and Langston, 1992). The relationships between bioaccumulated metal concentrations in benthic species and metal concentrations in sediments were derived from multiple sites in each of 17 English estuaries (Luoma and Bryan, 1981). The strength of the relationship varied among metals and among species (Table 16.1; see examples of figures in Chapter 8). Although correlations were insignificant in a few instances (e.g. Cd in the polychaete *Nereis diversicolor*; Cu in the bivalve *Scrobicularia plana*), sediment metal concentrations explained about 30% to 80% of the variance in bioaccumulation in 8 of 11 comparisons (Table 16.1).

These studies were carefully designed to minimise confounding influences:

- Concentrations of metals were determined in surface sediments, judged to be oxidised by appearance.
- Elimination of grain size biases by sieving made concentrations in sediments comparable across diverse settings.
- Obvious sources of bias such as large redox differences or dilution of tissue concentrations by reproductive tissue were carefully controlled.
- The estuaries included a wide range of physical and biogeochemical conditions. This maximised influences of geochemistry but minimised co-variances among geochemical factors.

- Data ranges for all elements were over several orders of magnitude. A wide range of data maximises the likelihood of uncovering broad relationships.

It was clear that concentrations of metals in sediments were, more often than not, a first-order factor in determining metal bioavailability to these benthic species, but not the only factor. The differences among metals reflected the importance of geochemical/biological factors on metal bioavailability from sediments. There was more variance among metals within a species than there was among species for each metal.

Table 16.1 *The percentage of variance in bioaccumulation of Ag, Cd, Cu and Pb explained by variation in sedimentary concentrations of these metals (1N HCl extractions). Data are from a broad range of estuaries in the United Kingdom*

	Ag (%)	Cd (%)	Cu (%)	Pb (%)
Nereis diversicolor (polychaete)	41–62	0–16	72	54
Scrobicularia plana (bivalve)	49	29	0	50
Macoma balthica (bivalve)	69	21	10	47

Source: From regressions published in Luoma and Bryan (1981), Bryan et al. (1985), Bryan and Langston (1992); reviewed in Luoma and Fisher (1997).

metals is directly related to metal on the particulate material or whether exposure is via water equilibrated with the sediment. Under similar conditions a correlation between bioaccumulation and one component of the environment can occur, even though another component is the source of the exposure. For example, a significant correlation was found between dissolved free Cd ion activities predicted from sediment characteristics and Cd concentrations bioaccumulated by an aquatic insect larva, a freshwater mussel and an amphipod in the lakes of Quebec (Tessier et al., 1984; Amyot et al., 1994). But experimental studies showed that dietary uptake was overwhelmingly the source of exposure for the aquatic insect larva in these systems (Munger and Hare, 1997). The correlations in nature presumably reflected a relatively invariant linkage between dissolved and particulate metal across these broadly similar waters.

16.4.2   Route of exposure

Prologue. Experimental evidence unequivocally shows that direct uptake of particulate metals is instrumental in determining metal risks from sediments (pore waters are not the only route of exposure). The existing body of modelling and mechanistic studies with deposit feeding animals shows that sediment should be considered as a direct source for metal accumulation. The body of evidence shows that:

- Uptake via ingestion and via water is additive.
- Every metal can be assimilated by some organism from nearly every type of particulate material. The degree of assimilation is influenced by a combination of geochemical form, the traits of the biological species and the trace metal.
- The concentration of metal bioaccumulated via ingestion of sediment usually, but not always, exceeds the concentration accumulated from water.
- Sediment/water partitioning is crucial in determining the relative importance of the two routes.

The relative importance of the different contaminant exposure pathways (overlying waters, pore waters, sediment, suspended detritus) has only recently been well understood (Batley et al., 2005). But it is an influential factor in understanding metal exposures from sediments. The exchange between particles and water creates difficulties in experimentally separating uptake from direct ingestion of sedimentary material from uptake via pore waters. These difficulties were addressed in early experiments by physically separating exposure routes in the experimental treatments. For example, Luoma and Jenne (1976, 1977) placed one set of experimental animals (Macoma petalum) in dialysis bags, which separated them from feeding on sediments. The 'controls' were starved, but exposed to only metal in water within the experiment. A second set of experimental animals was placed within sediments and allowed to feed. There was unequivocally greater uptake of metals in the animals that fed on sedimentary particulates radiolabelled with ^{109}Cd, ^{110m}Ag and ^{65}Zn than in organisms separated from the particulate material.

Versions of this design were used repeatedly (reviewed by Luoma, 1989). Results varied with regard to which route was more important for uptake. For example, Borgmann and Norwood (1995) found dissolved Pb to be more important than Pb in sediment as a source for the freshwater amphipod Hyalella azteca. Luoma and Jenne (1977) concluded that 75% to 89% of uptake Cd, Ag and Zn by M. petalum was from ingestion of sediment. Timmermans et al. (1992) concluded that Cd uptake by predatory water mites (Limnesia maculata) and caddisfly larvae (Mystacides longicornis) was from food while Zn uptake was from water.

To some extent, such results were highly dependent upon details of the experimental design. But it was unequivocal that direct uptake of metal from particulate metal *can* occur. When uptake occurred from ingestion of sediment, it was also clear that uptake from sediment and water was additive (Luoma and Fisher, 1997).

More recent studies have used biodynamic modelling (Chapter 7; Wang et al., 1996; Luoma and Fisher, 1997; Griscom et al., 2002) to disaggregate quantitatively the different contributions of ingested sediment, overlying water and pore waters for particular species. This quantitative approach has eliminated many uncertainties about routes of exposure and allowed some important generalisations.

High concentrations of metals in sediments combined with high ingestion rates by benthic animals result in a very large contribution of ingested sediment to bioaccumulation (sometimes >90%) in many common sediment-ingesting species and for most metals (Wang and Fisher, 1999; Griscom et al., 2002) (Box 16.7). Ranges of assimilation efficiency vary widely among species and among metals. Some metals are more likely to be assimilated efficiently (Zn) than others (Cr); while assimilation of yet other metals is highly variable (Ag, Cd). For example, for Zn the range among benthos is 12% to 86%; for Cd 13% to 44%; for Cr 0.7% to 12.5%; for Ag 11% to 50% (Wang and Fisher, 1999; Wang et al., 1999; Fan and Wang, 2001; Griscom et al., 2002; Ke and Wang, 2002; Yan and Wang, 2002; King et al., 2005).

When evaluating exposure route the biological traits of the experimental animal must be

Box 16.7 Determination of exposure routes

Biodynamic studies with sediments quantitatively address the relative importance of exposure routes, metal form and food types in determining bioaccumulation (Griscom *et al.*, 2002; King *et al.*, 2005; Chapter 7). Griscom *et al.* (2002) showed that ingested sediment was the major source for Cd and Ag bioaccumulation by the clam *Macoma petalum* (as *M. balthica*) under all realistic environmental conditions (Fig. 16.2). Across the range of feasible values for major variables in the model, the Ag bioaccumulated from food varied from 60% to 90% of total bioaccumulation. Over the same range Cd accumulation from food varied from 40% to 80% of the total bioaccumulation. But Co accumulation was only from the dissolved phase.

King *et al.* (2005) carried out a similar study with a species of bivalve (*Tellina deltoidalis*) from the same family. They also showed that ingestion of sediment dominated Cd and Cu uptake under conditions typical of sediments, but that shifts in feeding rate or partitioning between water and sediment could be crucial in determining the source of bioaccumulation.

Wang *et al.* (1999) showed that >98% of the Cd, Co, Se and Zn in a polychaete (*Nereis succinea*) originated from sediment ingestion, due to the high ingestion rates of these animals and the low uptake rate of metals from the dissolved phase (pore water or overlying water). For Ag, which is accumulated from solution faster than the other metals, approximately 65% to 95% of uptake originated from ingestion of sediments. Similarly, for *N. succinea*, Wang *et al.* (1998) showed that >70% of Hg(II) accumulation was derived from sediment ingestion. Uptake from the dissolved phase and sediment ingestion were equally important for MeHg accumulation, but depended upon partitioning. Yan and Wang (2002) showed that bioaccumulation of Cd, Cr and Zn in the sipunculid worm (*Sipunculus nudus*) is 'largely dominated by sediment ingestion due to the low uptake from the solute phase as well as the high metal concentrations in the sediment'.

Mechanistically, the digestive environment of deposit feeders includes enzymes, surfactants and other chemicals capable of effectively solubilising sedimentary food substrates (Mayer *et al.*, 1996) and efficiently releasing amino acids (Mayer *et al.*, 1996). Contaminant metals bound to sediments are, thus,

subject to considerable solubilisation during passage of the sediments through the digestive systems of deposit feeders. Complexation by the dissolved amino acids is an important desorption process (Chen and Mayer 1998a). Amino acid-bound metals are also directly available for co-transport across the gut membrane (Chapters 7 and 14). Gut retention time (GRT) influences the degree of amino acid and metal solubilisation (Chen and Mayer, 1999a), mechanistically explaining why gut retention time is one of the most important physiological parameters explaining metal assimilation from particulates (Decho and Luoma, 1994). Up to 63% of Cd was extracted by the digestive fluids of *S. nudus* (Yan and Wang, 2002) whereas <4% of Cr was extracted. There was a significant relationship between the metal AE and the metal extraction by gut juice for Cd. There was no significant relationship for Cr and Zn, however.

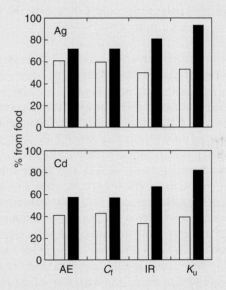

Fig. 16.2 The proportion of metal bioaccumulation in the bivalve *Macoma petalum* that is attributable to uptake from directly ingesting sediment, under scenarios with the lowest value (open bars) and the highest value (closed bars) for each of four parameters crucial to the calculation: AE (assimilation efficiency), C_f (concentration in food), IR (ingestion rate) and K_u (rate constant of uptake from water). (After Griscom *et al.*, 2002.)

considered. The experimental organisms must naturally ingest sediments or have external permeable surfaces in close contact with sediments. As noted earlier, lake species that live on the surface of the sediments (*Hyalella azteca*) or that migrate in and out of the sediment periodically (*Chaoborus punctipennis*) are less influenced by contaminated sediment than species that live within the contaminated material (Warren *et al.*, 1998). Taxa that do not feed on the sediment but construct burrows across the oxic surface and into deeper anoxic sediment show a moderate response to contamination (a number of insect larval species).

16.4.2.1 Influence of sediment–water partitioning (K_d)

Prologue. Many experimental conditions enrich pore waters much more than they enrich particulate metal, resulting in a shift of routes of uptake toward dissolved metal. Where K_d values do not mimic nature, routes of exposure and sources of toxicity will not mimic nature.

A factor that contributes to contradictions in the literature is that metal speciation in water, geochemical conditions in sediment, contact time between metal and sediment and other aspects of experimental design will modify conclusions about uptake route by affecting partitioning of the metal between water and sediment (see discussion below; Rainbow, 1995a; Luoma and Fisher, 1997; Wang and Fisher, 1999; Wang *et al.*, 1999; Griscom *et al.*, 2000; Fan and Wang, 2001). If partition coefficients (K_d) are low compared to nature that means metal concentrations in water are higher relative to sedimentary concentrations than occurs naturally. A design that results in strong partitioning to water will, of course, show greater influence of the dissolved vector on uptake. To obtain a relevant answer about route of exposure, partitioning has to be relevant.

Although the biodynamic studies make a convincing case for the importance of ingested sediments in risk evaluations, another body of work, usually with toxicity tests, reached the opposite conclusion (e.g. Swartz *et al.*, 1985; Di Toro *et al.*, 1992). Early biodynamic studies showed that the relative importance of uptake routes was highly dependent upon the partitioning of metals between water and particulates (Wang *et al* 2006). Models that assume a K_d typical of nature show a dominance of dietary exposure and predictions agree well with observations of bioaccumulation in nature (Luoma and Rainbow, 2005). If experimental conditions enhance pore water concentrations more than sediment concentrations, the importance of the dissolved vector of uptake increases proportionally. Partitioning between sediments and pore waters is affected by sediment properties (Simpson and King, 2005) and approach to spiking. If sediments are washed after they were labelled, then placed into clean seawater the partition coefficient is the result only of the slow exchange from the sediment back to water and K_d is more similar to nature than if the original spiking solution is used in the experiment (Luoma and Jenne, 1977). In an experimental setting, contact time is also influential (Lee *et al.*, 2004). The K_d is lower if a relatively short spiking period is employed (less than 30 days). Two examples of how such choices affect results are shown in Box 16.8

> **Box 16.8** Food versus water depends upon the ratio of concentrations
>
> Manipulation of sediments in the laboratory can result in great changes in partitioning of metals between sediments and pore waters (Simpson and Batley, 2003). A very early study by Kemp and Swartz (1988) illustrated the importance of partitioning between particles and pore waters on determining ecological risks from Cd-contaminated sediments. They showed that Cd bioavailability was predictable from pore water concentrations (and consistent with dissolved bioassay toxicities). However, the experimental conditions created a ratio of Cd concentration in pore waters to those in sediments much higher than typical of nature. In spiking the sediments, pore water concentrations of Cd were enhanced by 1000 to 10000 times over the concentrations found in contaminated estuarine waters, while concentrations in food were enhanced by 16- to 72-fold compared to uncontaminated

sediments. Luoma and Fisher (1997) applied biodynamic modelling to those conditions. They predicted the same conclusion as the toxicity experiment: 99% of Cd bioaccumulation by *Macoma nasuta* uptake should be from pore water under the experimental conditions in the original *toxicity test* (Luoma and Fisher, 1997). Adjusting the partition coefficient to that typical of a moderately contaminated estuary predicted that both food and water would be sources of Cd to the bivalve.

Lee *et al.* (2004) showed a similar outcome experimentally. They used a 10 day toxicity test with *Leptocheirus plumulosus* to evaluate Zn toxicity in sediments that were spiked, then aged for 5 to 95 days. In 95 days pore water Zn decreased 11- to 23-fold, the partitioning coefficient increased 10- to 20-fold and the Zn LC_{50} increased as pore water Zn concentrations declined (the short-term toxicity tests also primarily test dissolved toxicity). The authors concluded that short equilibration times for spiked sediments will accentuate metal partitioning toward the dissolved phase and shift the pathway for metal exposure toward the dissolved phase.

Thus, experiments with low partitioning coefficients do not refute the hypothesis that sediments are not a direct source for contaminant bioaccumulation under the conditions that occur in nature. The conditions of exposure determine the relative importance of pathways. If experiments do not mimic partitioning conditions typical of nature, they will not yield conclusions that can be accurately extrapolated to nature.

16.5   METAL BIOAVAILABILITY: VARIABILITY IN THE RELATIONSHIP BETWEEN CONCENTRATION AND TOXICITY

Prologue. Toxicity and bioaccumulation of metal contaminants are influenced by sediment geochemistry and the biology of the organisms involved. Geochemical processes may cause bioaccumulation or toxicity to vary widely (Luoma and Bryan, 1978; Luoma, 1989; Di Toro *et al.*, 1991).

A 2003 report from the US National Academies of Sciences (Ehlers and Luthy, 2003; NRC, 2003) defined bioavailability processes as 'the individual physical, chemical and biological interactions that determine the exposure of organisms to sediment bound chemicals' (Fig. 16.3). Research on risks from metal contaminated sediment has repeatedly demonstrated the importance of such processes. It is clear that toxic concentrations determined with one sediment will not be precisely predictive of toxic concentrations in all sediments. However, it is also important to recognise that differences in bioavailability may not affect risk assessment or risk management more than other processes in some cases and may in others (Ehlers and Luthy, 2003). For example, if concentrations are extreme, or cumulative effects from multiple contaminants are likely, bioavailability of a single contaminant is not necessarily the most important consideration.

The geochemical form of trace metals in pore water and in sediments is an important determinant of bioavailability, as is the functional ecology of the species, which determines what forms and concentrations of the metal a species is exposed to (Rainbow, 1995a; Luoma and Fisher, 1997; Wang and Fisher, 1999; Wang *et al.*, 1999; Griscom *et al.*, 2000). Differences in the way that experiments address these processes affect conclusions about bioavailability (NRC, 2003). The NRC report (Ehlers and Luthy, 2003) defined several processes that influence whether a metal in sediment becomes bioavailable and toxic. These include:

- Physical, chemical and biochemical phenomena that bind, unbind, expose or solubilise a metal.
- Movement of a released metal to the membrane of an organism (or vice versa).
- Movement of metals still bound to solid phases.
- Movement from the external environment through a physiological barrier and into the living system.
- Fate after passage through the membrane (detoxification, movement to a metabolically active site).

A number of physical, geochemical and biological processes influences the bioavailability of trace metals, but the most important are:

- differences in bioavailability between oxidised and reduced sediments;
- the form of metal in oxidised sediment: balances among Fe oxides, Mn oxides, living or non-living

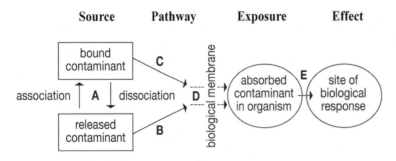

Source   Pathway   Exposure   Effect

Fig. 16.3 Processes affecting metal bioavailability from sediment, put in the risk assessment context of source, pathway, exposure and effect. (After Ehlers and Luthy, 2003; NRC, 2003.)

organic materials, or minor abundances of sulphide are influential, and those influences differ among metals and species (Harvey and Luoma, 1984; Griscom *et al.*, 2000; Ke and Wang, 2002);

- the abundance of organic material in the sediment (detritus, living material and/or prey);
- how different species interact with their environment;
- the accessibility of the deposited metal or likelihood of dispersal, redistribution or transformation.

Bioavailability is also, to some degree, a credibility or risk communication issue. Searching for reduced bioavailability can be viewed negatively as an excuse to 'do nothing' in a contaminated environment (Ehlers and Luthy, 2003). Or it can be viewed positively as a way to either improve the accuracy of risk assessment or justify risk management goals that might leave contaminants in place but still protect the environment (and human health). A great deal is usually at stake in evaluating bioavailability. And, the issue is made even more contentious by its complexity.

16.5.1   Anoxic sediments

Prologue. The geochemical process with the single greatest influence on metal form and sediment–pore water partitioning in sediments is redox condition and the associated presence of sulphides. For bioavailability the crucial questions are:

- how much sulphide is present in the (usually oxidised) microenvironment where most organisms intersect their environmental milieu;
- how to quantify sulphide in that microenvironment;
- whether pore water exposure dominates uptake for species exposed to sediments containing sulphide;
- how much sulphide-bound metal is assimilated by ingestion of sediment.

There is great momentum to widely apply the SEM–AVS theory to risk assessment and risk management, but implementation protocols do not yet exist that explain how to address some of the important questions or limit the influence of uncertainties.

Changes in metal form are dramatic at the interface between oxidised sediments and reduced sediments in a vertical column of sediment (below the layer of adequate oxygen penetration: Chapter 6). A large body of literature shows that acid-extractable sulphide (termed acid volatile sulphide or AVS) is a major reactive constituent in anoxic sediments. Sulphide binds strongly with metals and inhibits metal exchange with pore waters. The result is greatly reduced metal concentrations in the pore water in the zone of anoxia. When such sediments are homogenised and tested, the reduced pore water concentrations correspond to declines in toxicity (Ankley *et al.*, 1996). No acute toxicity in 10 day whole sediment toxicity tests is typically observed when the molar concentration of the acid volatile sulphide (AVS) is greater than the molar concentration of the simultaneously extracted metals (SEM) released from a sediment sample during the AVS extraction (Di Toro *et al.*, 1991) (Box 16.9).

Box 16.9 Effect of SEM and AVS on Cd and Ni toxicity

Di Toro *et al.* (1992) is the classic example of the early demonstrations of sulphide controlling toxicity in whole sediment bioassays. The experiments were short-term (10 day) whole sediment acute toxicity tests using amphipods (*Ampelisca abdita* and *Rhepoxynius hudsoni*), oligochaetes (*Lumbriculus variegatus*) and snails (*Helisoma* sp.). Freshwater and marine sediments with different AVS concentrations were either spiked with Cd or Ni, or collected from Cd- and Ni-contaminated sites.

Dose–response relationships varied widely among the nine different sediments for each species and each metal (Fig. 16.4a). For each sediment, an SEM (extractable metal)/AVS (extractable sulphide) ratio was calculated. When mortality was plotted versus SEM/AVS, the scattered data from the separate experiments fell within a single dose–response relationship (Fig. 16.4b). No significant mortality occurred compared to controls if the molar concentration of acid volatile sulphide (AVS) in the sediment was greater than the molar concentration of simultaneously extracted Cd and/or Ni, with some variability around the breakpoint. Where AVS concentrations exceeded metal concentrations, it appeared that all metal was associated with sulphide and none was available to exert toxic effects. As metal concentrations began to exceed AVS concentrations, toxicity increased. Experiments using this approach consistently show that 'the AVS concentration of a sediment established the boundary below which these metals cease to exhibit acute toxicity in freshwater and marine sediments' (Di Toro *et al.*, 1992). While the strong association of metals and sulphides was well known, the reactive nature of those sulphides when the metal was added was an important observation.

The whole sediment bioassay paradigm was repeatedly used after these initial discoveries to refine the SEM–AVS concept and demonstrate its applicability to other metals with strong sulphide forms (e.g. Ag, Cu, Pb, Zn). Several generalisations were developed from this body of work:

- In order to apply the relationship it was necessary to sum molar concentrations of all extractable metals that might form sulphides and compare them to molar concentrations of AVS (Di Toro *et al.*, 1992);
- The difference between the summed SEM and AVS (SEM–AVS) was a more powerful approach than the

ratio of SEM/AVS (Hansen *et al.*, 1996), except at low total metal concentrations in sediments where SEM > AVS does not result in toxicity (Burton *et al.*, 2005).
- Toxicity in these experiments correlated strongly with pore water concentrations of metals, leading to the conclusion that 'it should be possible to infer bioavailability based upon pore water metal concentrations' (Ankley *et al.*, 1996). Equilibrium partitioning was therefore assumed to be the basis of the toxicity (analogous to organic contaminants: Di Toro *et al.*, 1992).
- If pore water concentrations of metals are determined by the partitioning of metals from sediments as influenced by AVS concentrations, then the toxicity of the sediment can be assessed by comparing pore water concentrations predicted by an EP calculation to a selected set of whole water toxicity data (Di Toro *et al.*, 1991).

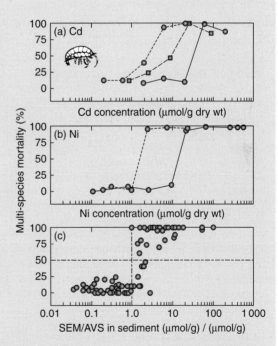

Fig. 16.4(a, b) Amphipod mortality after 10 days in different sediments spiked with Cd and Ni at a range of concentrations. (c) Amphipod mortality in nine sediments (including those above) plotted against the molar concentrations of Cd plus Ni normalised to molar concentrations of acid volatile sulphide. (After Di Toro *et al.*, 1992.)

The body of literature relating sediment toxicity to SEM–AVS in whole sediment bioassays is large (e.g. Box 16.9). The conclusions that sulphides control pore water metal concentrations and pore waters concentrations control toxicity are elegantly demonstrated and consistent across several sediment dwelling species and several trace metals.

Box 16.10 emphasises the interplay of geochemical and biological complexities in determining the influences of metal bioavailability. Lee *et al.* (2000a, b, c) conducted a complex experiment that tested the influence of both of these on Cd, Ni and Zn bioaccumulation (Box 16.11).

In situ bioassays in the field can show both consistencies (Burton *et al.*, 2003) and inconsistencies (Warren *et al.*, 1998) with the toxicity tests from the laboratory (Box 16.10).

Two factors important in experiments that address metal bioavailability in general are the role of the surface oxidised layer as both a source of food and the zone of exposure for many organisms, and the species-specific traits that influence how organisms experience the sediment environment (feeding depth, burrow depth, irrigation of burrows, digestive processing and food selection). Details concerning the influence of both on metal exposure are considered in only a few studies (Charbonneau and Hare, 1998; Warren *et al.*, 1998). Integrated models defining metal bioavailability in/from sediments are needed (Ankley, 1996) and feasible (Griscom *et al.*, 2002), considering exposure and therefore metal uptake from both sediments and pore (or overlying) water, defined for the species of interest that would experience that environment.

16.5.2   Oxidised sediments

Prologue. The single most important component of the sedimentary environment to which animals are exposed is the oxidised zone. Careful sampling of

---

Box 16.10 The SEM/AVS approach in the field

Burton *et al.* (2005) spiked 2 cm deep trays with homogenised anoxic sediments with no Zn, 400 µg/g Zn dw or 4000 µg/g Zn dw. After 10 days equilibration, they moved these trays to streams with different ecological and hydrological characteristics and held them for 11 to 37 weeks. Zinc adversely affected colonisation by benthic animals where physical processes created conditions that oxidised AVS over time and SEM/AVS ratios increased to values >1 (Fig. 16.5). Where AVS was not oxidised, sediments were not toxic. For example, no difference in colonisation was observed between reference trays and trays containing sediment with 913 µg/g Zn dw (at the end of the experiment) that had been transplanted to Ankeveen Lake (the Netherlands). Sulphides were retained in these sediments throughout the study period and the SEM/AVS ratio in this sediment was always <1.

Colonisation was significantly disturbed in trays with 175 µg/g Zn dw that had been transplanted to a hydrologically active stream (the Pallanza in Italy) wherein oxidation of the AVS resulted in SEM/AVS ratios of 43. Results were most complex from a slow moving stream in the Netherlands. As AVS disappeared from the sediments over time, the SEM/AVS ratio rose from 1.6 to between 2.3 and 2.9. Strong effects on colonisation were observed only in the later samplings (perhaps reflecting stronger oxidation in the surface layers than was measured in the bulk sediments).

Hare *et al.* (2001) and Warren *et al.* (1998), as described in Section 16.3.2, also studied colonisation of trays of homogenised sediments with different SEM–AVS conditions, but in lake environments. They found SEM–AVS to be influential for some species and not for others. Warren *et al.* (1998) found a Cd–AVS effect only in the case of sediment ingesting species like the chironomid insect larva *Chironomus* sp; whereas the source of Cd for many of the lake species was the water column. Changes in sediment chemistry after homogenised sediments were placed in the field were influential in both experiments; and these changes were crucial to the results of both. Strong oxidation occurred in Quebec lakes as evidenced by vertical gradients in pore water Cd that were typical of many natural sediments (Fig. 16.6). Concentrations of Cd in pore waters at the surface of these sediments were often similar to overlying water. Presumably this

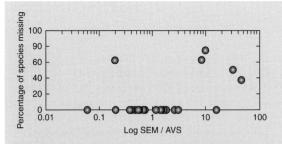

Fig. 16.5 Percentage disruption of colonisation by benthic macroinvertebrates in four streams and lakes as a function of SEM/AVS where Zn was spiked into sediments transplanted in defaunated trays into those environments. (After Burton *et al.*, 2005.)

profile, typical of many natural sediments, contributed to the lower Cd exposure of epibenthic species. Among the experiments with different hydrological regimes, Zn was bioavailable where oxidation occurred at the sediment surface, but not where sediments retained their largely anoxic character (Burton *et al.*, 2005). Thus Warren *et al.* (1998) used biological differences to test the influence of sediment oxidation, while Burton *et al.* (2005) used hydrological regime. Both verified the reduced bioavailability of sulphide-associated metal, and both demonstrated the labile nature of that association and the complex ways it is experienced by different fauna.

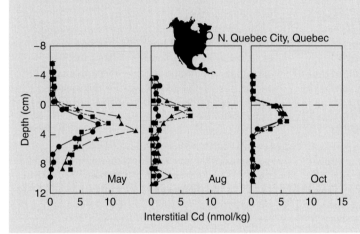

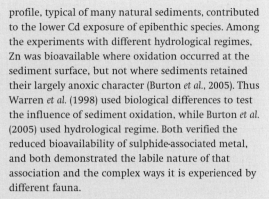

Fig. 16.6 Depth profiles of pore water Cd concentrations after several weeks in trays of homogenised sediment placed in the sediments of a lake in Quebec, Canada (after Warren *et al.*, 1998). The oxidised zone is characterised by higher pore water Cd concentrations than the deeper anoxic zone, but concentrations of Cd in pore water near the sediment surface are usually similar to overlying water due to diffusion out of the sediments. The depth of the enriched pore water concentrations changes with season.

more oxidised layers of sediments provides correlative evidence that points to factors that affect bioavailability of cationic metals from sediments in nature. But no simple operational procedure exists for directly removing or isolating bioavailable fractions from oxidised sediments, despite decades of research in this area (Luoma, 1989).

Oxidised sediments are usually dominant at the zone in sediments where most living species intersect their environment (because all benthic macrofauna require oxygen to survive). A number of early studies evaluated relationships between metal chemistry of oxidised sediment and metal exposure of biota, usually emphasising bioaccumulation. The statistical relationship between metal concentration in sediment dwelling animals and concentrations in sediments was usually improved if sediment geochemistry was considered. To see such influences, however, it was necessary to use careful sampling and data from multiple environments across a broad range of contamination. Some studies focused sediment collection on the surficial layer of sediment, as a way of focusing on the environmental interface experienced by most (Luoma and Bryan, 1978, 1982; Langston, 1980). Some used sampled or quantified specific phases in the oxidised layers that influence metal form (Tessier and Campbell, 1987; Belzile *et al.*, 1989b). From these observational studies empirically derived generalisations developed about relationships between bioaccumulation and indicators of metal form in oxidised sediments (Box 16.12).

Box 16.11 Metal bioavailability reflects the interplay of biology and labile redox chemistry

Lee *et al.* (2000a, b, c) studied bioaccumulation of Cd, Ni and Zn by four species of invertebrates that behave differently. They spiked a range of metal concentrations into sediments with one concentration of AVS in one set of experiments. This is a typical design with AVS studies, but Lee *et al.* (2000a) noted that SEM and the SEM/AVS ratio co-vary in this approach. Thus they ran a second simultaneous set in which they used one metal concentration and various AVS concentrations (no co-variance). They allowed vertical zonation to develop in the experimental sediments, creating a situation like nature where the animals had access to oxidised sediments and were more likely to feed. They also employed a relatively narrow range (10×) of concentrations, typical of different levels of contamination found in nature. Pore water concentrations, AVS and SEM were determined from the vertical layer from which each animal fed for the comparisons with bioaccumulation – surface sediments in the case of two bivalves, *Corbula amurensis* and *Macoma petalum*, and deep layers for the polychaete *Heteromastus filiformis*. From this complicated design they showed that metal bioaccumulation typically followed SEM alone (Fig. 16.7a, b). AVS strongly controlled pore water metal concentrations, vertically and among treatments. Metal bioaccumulation followed pore water concentrations where SEM

co-varied with SEM–AVS (Fig. 16.7c, d). But where SEM did not co-vary with SEM–AVS, there was no relationship between bioaccumulation and pore water metal concentrations, in general (Fig. 16.7e, f). Overall they found a poor association between bioaccumulation of metal and pore water metal concentrations in 10 of the 12 metal–animal combinations (3 metals, 4 animals). Thus SEM–AVS did not appear to control metal bioaccumulation when conditions similar to nature were carefully established, dietary exposure was encouraged, extreme concentrations were avoided (thus avoiding extreme pore water loading atypical of nature), and co-variances were carefully avoided. Experimental design, dietary exposure and ecological traits (functional ecology) were all important to the conclusions.

These experiments addressed bioaccumulation not toxicity; although it would seem that the two should be linked. More important the experimental design used a new approach to address what appeared to be a large body of elegant and convincing toxicity data that had largely been tested by approaches that had many similarities (homogenised sediment bioassays with large concentration gradients). The results of Lee *et al.* (2000a) were controversial because they questioned what was fast becoming conventional wisdom when they concluded 'important uncertainties remain in the application of the AVS-equilibrium concept as . . . a regulatory tool . . .'.

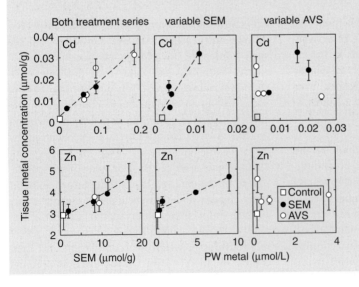

Fig. 16.7 Accumulation of Cd and Zn by the bivalve *Macoma petalum* as a function of Cd or Zn concentration in sediments with differing combinations of SEM and AVS, and as a function of pore water metal concentrations (controlled by AVS) in a treatment series where AVS was varied independently of SEM. (After Lee *et al.*, 2000a, b, c.)

Box 16.12 Bioavailability to estuarine benthos of metals from oxidised sediments: field and laboratory synthesis

Arsenic, lead and chromium

Concentrations of these metals in surficial sediment layers (visually the oxidised layer) were compared to concentrations in the deposit-feeding bivalve *Scrobicularia plana* from 27 estuaries in England and Wales (Luoma and Bryan, 1978; Langston, 1980). Concentrations of all metals in sediment ranged widely (e.g. As from 2 µg/g dw to 2500 µg/g dw). Fe oxides appeared to control the form of these three metals in the sediment; consistent with theoretical knowledge of their surface chemistry (Luoma and Bryan, 1978). Concentrations in *S. plana* correlated more significantly with the metal to Fe ratio in sediments than with metal concentration alone (e.g. Fig. 8.3); suggesting that high concentrations of reactive oxidic Fe inhibited bioavailability in surface sediments. More recently, laboratory experiments showed that Fe oxide binds Cr strongly and reduces assimilation efficiency in at least some bivalve clams (3% AE in *Macoma petalum*: Decho and Luoma, 1994; 13% in *Ruditapes philippinarum*: Fan and Wang, 2001). Bioavailability of Cr can be enhanced somewhat in organic-rich conditions (Decho and Luoma, 1991; Lee and Luoma, 1998; Chong and Wang, 2000).

Silver

Across the group of 17 southwest England estuaries, Ag concentrations in the deposit-feeding tellinid bivalves *Macoma balthica* and *Scrobicularia plana* correlated strongly with Ag concentrations extracted by 1 N HCl from sediments. Ag concentrations in sediments and the tellinid *Macoma petalum* from San Francisco Bay fell on the same relationship as that for *M. balthica* from southwest England (Luoma *et al.*, 1995). Berry *et al.* (1999) showed a similar correlation between HCl-extractable Ag and toxicity from Ag-spiked sediments in experiments with the deposit-feeding amphipod *Ampelisca abdita*. Mechanistically, the relationship with this simple extraction technique probably reflects the insolubility of AgS (silver sulphide) in HCl. The fraction of Ag that exists as silver sulphide is not bioavailable. Griscom *et al.* (2000) showed that assimilation efficiencies of two bivalves from anoxic sediments were lower for Ag than for any other metal. Ag assimilation from anoxic sediment was 4.6 ± 2% by the mussel *Mytilus edulis* and 11 ± 1%

by the clam *Macoma petalum*. Other forms of sedimentary Ag showed a range of AE of 9% to 38% for *M. edulis* and 13% to 35% for *M. petalum* (i.e. many forms are quite bioavailable). 'Other forms' were abundant in every sediment from the southwest England estuaries, however.

Cadmium and zinc

Most evidence suggests that these two metals are distributed in complex ways among forms in oxidised sediments (Luoma and Bryan, 1981) and consistently show complex relationships between concentrations in sediment and bioaccumulation by benthos. Diet is the primary route of exposure for these metals in most studies and assimilation efficiencies for both elements vary widely among different types of sediments.

Anoxic sediment

Cd and Zn are highly available to *M. edulis* from anoxic sediments (AE 35 ± 6% and 32 ± 4%, respectively); but less available to *M. petalum* (Griscom *et al.*, 2000). Griscom *et al.* (2000) showed that this reflected oxidising conditions in the digestive tract of *M. edulis*, and greater solubility of CdS and ZnS than the sulphide form of Ag. Chong and Wang (2000) also concluded that Cd and Zn would be taken up from anoxic sediment by the clam *Ruditapes philippinarum*, the mussel *Perna viridis* and the polychaete *Nereis succinea* (Wang *et al.*, 1999). But AE was higher (about 3.1× for Cd; about 1.4× for Zn) from oxic than anoxic sediment. For species and conditions where pore waters dominate uptake, AVS could play an important role. But for species that experience primarily oxidised sediment or assimilate metal from ingested anoxic sediment, that relationship will not be simple.

Fe oxides

In oxidised sediment, Fe oxides, and perhaps humic materials, reduce Cd and Zn assimilation (Decho and Luoma, 1994; Fan and Wang, 2001).

Mixed forms of Cd and Zn in oxidised sediments

AE also varied widely across different forms of metals found in oxidised sediments in most organisms studied (e.g. 10% to 41% for *M. edulis*, 10% to 23% for *M. petalum*: Griscom *et al.*, 2000).

Living material or biologically productive sediments

An extremely important factor for both Cd and Zn appears to be the presence of living material in the sediment (Decho and Luoma, 1996; Lee and Luoma, 1998; Chong and Wang, 2000; King *et al.*, 2005). AE for these metals from organic-rich or plant-enriched sediments are always higher than AE from less productive sediments (Table 16.2). Both metals are assimilated quite efficiently from algae or algae-enriched sediment. Thus during periods of plant growth on sediments or in the water column, in eutrophic conditions, or in species that selectively feed on algae (Ke and Wang, 2002), bioavailability of Cd and Zn could be enhanced in natural waters.

Copper

Quantitative evaluations of factors affecting the bioavailability of Cu are just beginning (e.g. Croteau *et al.*, 2004). Results to date suggest, as for other metals,

that diet is the predominant exposure route for benthos, especially where the water column contains dissolved organic material (which will reduce the bioavailability of dissolved Cu: disproportionately: Croteau and Luoma 2005). The geochemistry of Cu suggests that a complex mix of factors will also dominate bioavailability, but questions remain. Bioaccumulation, and hence uptake, of Cu followed sediment concentrations in some species (the polychaete *Nereis diversicolor*) but not others (the bivalve *Scrobicularia plana*) in southwest England. Some types of anoxic conditions appeared to enhance Cu bioavailability, to at least tellinid bivalves, in some locations (Luoma and Bryan, 1982); but in general equimolar concentrations of extractable sulphide will reduce the bioavailability of Cu (Ankley *et al.*, 1993).

Table 16.2 *Assimilation efficiencies (AE%) from different types of sediments and from pure cultures of benthic and planktonic microalgae by two species of tellinid bivalves (*Macoma petalum *and* Tellina deltoidalis*) in different studies. Adding organic materials, especially algal food, resulted in increased assimilation efficiency. Metals were assimilated from anoxic sediments, but at lower efficiencies than from oxic sediments*

Metal	Species	AE(%)	Comment	Reference
Ag	*Macoma petalum*	18 ± 4	Sediment – oxic organic-poor (SFB)	Griscom *et al.*, 2000
	Macoma petalum	21 ± 4	Sediment – oxic organic-rich (Flax Pond, Long Island)	Griscom *et al.*, 2000
	Macoma petalum	11 ± 2	Sediment – anoxic organic-rich (Flax Pond, Long Island)	Griscom *et al.*, 2000
	Macoma petalum	39 ± 3	Algae (*Thalassiosira pseudonana*)	Griscom *et al.*, 2000
	Macoma petalum	38 ± 1	Algae (*Isochrysis galbana*)	Reinfelder *et al.*, 1997
	Macoma petalum	49 ± 1	Algae (*Thalassiosira pseudonana*)	Reinfelder *et al.*, 1997
Cd	*Macoma petalum*	9.7 ± 3	Sediment – oxic organic-poor (SFB)	Griscom *et al.*, 2000
	Macoma petalum	9 ± 2	Sediment – anoxic organic-rich (Flax Pond, Long Island)	Griscom *et al.*, 2000
	Macoma petalum	8.2 ± 2	Organic-poor sediment with benthic microalgae (SFB)	Lee and Luoma, 1998
	Macoma petalum	23 ± 2	Sediment – oxic organic-rich	Griscom *et al.*, 2000
	Tellina deltoidalis	28 ± 2	Sediment – sub-oxic, organic-rich	King *et al.*, 2005
	Macoma petalum	32 ± 2	Benthic microalgae (SFB)	Lee and Luoma, 1998
	Macoma petalum	13 ± 2	Spring bloom particles (SFB)	Lee and Luoma, 1998
	Macoma petalum	51 ± 4	Algae (*Thalassiosira pseudonana*)	Griscom *et al.*, 2000
	Macoma petalum	69 ± 2	Algae (*Isochrysis galbana*)	Reinfelder *et al.*, 1997
	Macoma petalum	88 ± 10	Algae (*Thalassiosira pseudonana*)	Reinfelder *et al.*, 1997
	Tellina deltoidalis	73 ± 3	Algae	King *et al.*, 2005
Cu	*Tellina deltoidalis*	30 ± 2	Sediment – sub-oxic, organic-rich	King *et al.*, 2005
	Tellina deltoidalis	49 ± 4	Algae	King *et al.*, 2005

As a result of the strong binding of Cu to organic material, organic forms of Cu are observed in natural sediments even when sulphides are abundant (e.g. Chen and Mayer, 1999a). CuS is insoluble (like Ag_2S) in HCl (Allen *et al.*, 1993), so the SEM found in almost every sediment represents some form not free to bind with AVS (perhaps organically complexed in particles). But, unlike Ag, simple relationships are not typically found between HCl-soluble Cu and either bioaccumulation or toxicity, although normalisation of Cu/TOC is being promoted for use in setting sediment quality guidelines (Di Toro *et al.*, 2005);

the empirical evidence that organic complexation reduces Cu bioavailability from sediments is, however, not clear. Chen and Mayer (1999a) extracted Cu from sediment with digestive fluids of two deposit feeders and one suspension feeder. The digestive fluids extracted less Cu than did HCl and the extractable Cu declined as sulphide concentrations in the sediment increased. Chen and Meyer (1999a) concluded that 'subsurface deposit feeders feeding on anoxic sediments may be exposed to less Cu than their surface-feeding counterparts in Cu-contaminated environments'.

Soil scientists and aquatic scientists have long sought the 'silver bullet' (i.e. simple protocol) that would identify the bioavailable fraction of metals in any sediment or soil (NRC, 2003); with limited success. That fraction probably lies within an acid-soluble fraction of metal (Luoma, 1989). Thus an HCl extraction is informative, although not predictive (except perhaps for Ag). Normalisation of the HCl-extractable metal (aka SEM) in oxidised sediments improves relationships, and adding AVS to such normalisations (again on surface layers of sediment) might be a valuable addition for organisms whose exposures are influenced by redox processes. Despite decades of field evidence, surface sediment sampling and normalisations are not mentioned in regulatory or risk assessment protocols. Apparently this is because of perceived challenges in collecting oxidised sediments (Batley *et al.*, 2005). It also may reflect another example of the dichotomies that occur between observational science and experimental science in management of metal contamination.

16.6   PHYSICAL PROCESSES

Prologue. The physical fate of sediments is as important as sediment toxicity in defining risks. Potentially mobile deposits spread risks more widely than deposits that are inaccessible or unlikely to move.

Recognition of the role of physical processes in controlling access of organisms to contaminated sediments is as critical to successful management (or evaluation) of risks from contamination as is the level of contamination itself. If sediments are inaccessible to organisms (e.g. buried), risks are greatly reduced. Their toxicity is an important consideration only if the sediments are re-exposed by dredging or erosion. Sediments that are constantly remobilised, or dispersed from hotspots, pose great risks.

In a river, seasonal changes in flow are large and greatly affect sediment transport. Major floods can disperse uncontained wastes across vast areas. The mass of sediment transported by a river increases exponentially as discharge increases; so very high river discharges move much more sediment than lower flows. Contaminated deposits that might move during floods pose much greater risks than deposits that are more isolated. Metals that are deposited on floodplains are later mobilised (Chapter 12) extending the time and expanding the spatial scale of contamination.

It is often assumed that wastes will lie where they are deposited (e.g. in dump sites) in bays, estuaries and oceans. But, ultimately, metals are redistributed by tidal forces in nearly all such systems. Hydrodynamics can create patches of contamination where none previously existed, and spread contamination over unexpectedly wide areas (Heiss *et al.*, 1996) (Box 16.13). Tidal currents can move sediment contamination landward (upstream) in estuaries as well as seaward from a point of input. On the other hand, deposits buried by natural sedimentation are eventually of reduced

Box 16.13 Physical processes can spread risks from contaminated sediments and can change through time

San Francisco Bay

A very large Pb smelter discharged Pb into the atmosphere and the waters of San Francisco Bay between 1870 and 1970. This created a strong hotspot of contaminated sediment in the Bay adjacent to the discharge (Hornberger et al., 2000). The history of contamination from this facility was reconstructed from a sediment core in 1990. It appeared that the local contamination was rapidly buried at this site after the smelter shutdown. The accumulation rate of sediment was about 3 cm per year; enough to bury the contamination deeper than the 30 cm depth of bioturbation within 10 years. But continual resuspension also appeared to redistribute much of the Pb from this hotspot before the contamination was buried. Pb from the smelter was the major source of regional Pb contamination in sediments of the entire Bay in 2002, more than 50 years after the smelter had closed (Ritson et al., 1999). The unique isotopic signature of the smelter Pb was evident in sediments and in water everywhere in the Bay. Over time, however, the dynamics of the estuarine system have changed. Upstream dams on the rivers coming into the Bay have greatly reduced the sediment load of the rivers. Geological equilibrium has been disturbed and the sediments of the estuary have begun eroding

rather than depositing. Thus the buried contamination legacy of San Francisco Bay is gradually being re-exposed because of another, seemingly unrelated, human activity (Foxgrover et al., 2004).

Mecklenberg Bay

Events, rather than day-to-day processes, can have a great influence on the fate of contaminated sediments. For example, storms in coastal zones, bays and estuaries can mobilise large pulses of contaminated sediments from accessible hotspots, and transport them long distances. Wave-driven resuspension during gale events in metal-contaminated Mecklenberg Bay (Germany) mobilised sediments to 10 cm sediment depth, at a water depth of 23 m. Wave events during one gale were estimated to have mobilised and transported 650 kg of Zn from a fly ash dump site (Kernsten et al., 2005). The consequence was a net erosion of the particulate contaminant load from the hotspot, and redistribution into the wider Mecklenburg Bay. Using a typical gale recurrence rate, Kernsten et al. (2005) estimated that 4.3 tonnes of Zn were annually spread over an area of 360 km^2 by storms. The authors suggest that periods of accumulation in hotspots, and in their depositional zones, are interrupted by periods of erosion, resuspension and transportation. Ultimately the progressive transport of wastes in this way explained 50% to 80% of the contamination deposited in the deep Baltic Sea.

risk. Physical changes that result in mobilisation of sediment-bound metals from a deposit cause also changes in character of the deposit (e.g. oxidising previously reduced sediments) adding uncertainty to characterisations of risk based upon sediment chemistry.

16.7   MANAGEMENT OF SEDIMENT CONTAMINATION

Prologue. The choices of approach in attempting to create consistent guidelines for managing contaminated sediment include: (a) define risk by chemical contamination (not bioavailability); (b) define by sediment bioassays (difficult to extrapolate beyond test); (c) consider bioavailability; (d) use a mix of methods

or a consensus value. Approaches to sediment quality guidelines differ widely across jurisdictions. There is not yet a global consensus on the best approach to protect aquatic environments from the risks from metal contamination in sediments.

Across the world most jurisdictions share concerns about sediment management (Babut et al., 2005). There is general agreement about the nature of the problem. But there is not a consensus among jurisdictions about preferred methodologies for risk assessment or risk management. The basic approaches include (Babut et al., 2005):

- no specific concentration targets (USEPA);
- direct toxicity testing of each sediment (dredge spoils in the USA);

- derivation of concentration limits based upon consideration of some reference (Flanders in Belgium);
- 'effects-based' regulations based upon total concentration (NOAA, USA);
- regulations that add bioavailability as a consideration (the Netherlands);
- a tiered combination of several approaches (Australia and New Zealand).

Confusion in this arena is exemplified in the USA. After years of dialogue about possible criteria, USEPA, in 2002, published a collection of technical procedure documents to provide users with information for deriving their own values to assess sediment contamination (Engler *et al.*, 2005). Thus the USA has no formal nationwide sediment quality guidelines. The USA does have testing requirements for disposal of dredged sediments, derived by the US Army Corps of Engineers. Sediment quality guidelines were developed by another agency (National Oceanic and Atmospheric Administration); and different states use different combination of all of the above. Some states promulgate use of SEM–AVS methodology to include consideration of bioavailability. European Union member states also differ in their approaches, from a concentration-based approach to consideration of bioavailability (Babut *et al.*, 2005). Tiered approaches expand intensity as the degree of the problem (and the amount at stake) is clarified (Melzian, 1990). Australia and New Zealand recently adopted the tiered combination approach.

Part of the incoherence among jurisdictions is a lack of consensus about the science. Different scientific approaches to deriving guidelines result in great differences in the concentration suggested as protective (Fig. 16.8). Disagreement about the relative importance of different uncertainties is common (see also NRC, 2003; Wenning *et al.*, 2005).

16.7.1 Toxicity testing and sediment quality management

Prologue. Toxicity or bioaccumulation testing is used to manage the fate of discrete batches of sediment

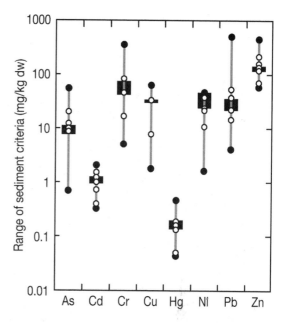

Fig. 16.8 Comparison of the range of metal concentrations recommended as low range sediment quality guidelines (protective values, trigger points, or the concentration above which effects are increasingly likely) suggested by different scientific approaches. Black squares represent guidelines defined as Effects Range Low (ERL) and Threshold Effects Level (TEL). (After Babut *et al.*, 2005.)

taken at a well-defined place and time, and/or evaluate risks in particular circumstances. Results are difficult to extrapolate more widely. Toxicity testing is a practical compromise in such circumstance, if not especially informative about potential effects from metals on benthic communities in natural waters.

Initial attempts to manage sediment contamination in the USA came from recognition that harbours were filling with contaminated sediments and regulations were needed to guide disposal of those sediments. Guidelines were published as early as 1977 for case-by-case evaluation of the toxicity of dredge materials to be disposed in the oceans (summarised in USEPA, 1992). An 'integrated' approach was recommended. In the original tiered approach (Melzian, 1990) chemical analyses were first employed to screen for the occurrence of contamination. A succession of bioassays was then employed on sediments where contamination was deemed potentially problematic. The elutriate test is one example of a bioassay used for assessing sediments for ocean disposal (Engler *et al.*, 2005). Sediments are mixed with water (shaking a sediment sample with four times its volume of seawater) to simulate release of metals that might occur during water column disposal of dredged material. Metal concentrations are analysed in the water after filtration (or centrifugation) to determine if ambient water quality guidelines are exceeded. Alternatively, toxicity tests are conducted with the elutriated water. Unfortunately, elutriate tests are considered the least reliable of more modern toxicity testing approaches (Ingersoll *et al.*, 1997). Nevertheless they remain the guideline for ocean disposal of dredged material in many international contexts (e.g. Environment Australia, 2002). Toxicity and bioaccumulation testing protocols were also published in a 1991 manual for marine sediments and a 1998 manual for inland testing (freshwaters), to complement elutriate testing (Engler *et al.*, 2005). Direct toxicity testing is particularly suited to testing and regulating individual batches of sediment, even though results are not particularly relevant to actual ecological effects from the sediment disposal.

16.7.2   Managing sediments on the basis of metal concentrations

Prologue. Management of risks from metal contaminated sediments can also involve numeric concentration limits, analogous to ambient water quality guidelines or criteria, termed 'sediment quality guidelines'. Providing a justifiable basis for the guideline is the challenge. Uncertainties affect the precision of every approach, but optimism about their use abounds:

> Though the scientific underpinnings of the different SQG approaches vary widely, none of the approaches appear to be intrinsically flawed. All approaches are grounded in concepts that, viewed in isolation, are sound.
>
> Wenning *et al.* (2005)

The earliest attempts to manage metal contamination in sediments established concentration limits. In the USA early limits for Hg (1 μg/g Hg), Pb (50 μg/g Pb) and Zn (50 μg/g Zn) were established in 1971 (Engler *et al.*, 2005). These proved infeasible to implement. The Zn values were below concentrations in many background sediments and bioavailability issues were constantly raised. This is the fourth decade of a search for ways to better screen, evaluate and manage contaminated sediments. Although the techniques may not be 'inherently flawed', important uncertainties continue to slow the development of consensus about best practice.

Numerical sediment quality guidelines identify the concentrations of chemicals that are thought to cause or substantially contribute to adverse effects on sediment-dwelling organisms. Chemical analyses have definite advantages for use in managing contaminated sediments. Effective analytical methodologies are available for determining total concentrations of trace metals in sediments, and interpretation of concentration data is relatively straightforward. Nearly all existing environmental standards take the form of chemical-specific concentrations (Chapman, 1989b), so a precedent exists for applying that concept to sediments. Therefore, chemical analyses are recommended as an essential part of any evaluation of sediment toxicity. The biggest challenge is establishing risk levels; and different jurisdictions do this in different ways (Babut *et al.*, 2005).

16.7.3   Guidelines defined relative to some reference concentration

Prologue. Some guidelines attempt to define risks by exceedances of either the undisturbed or the existing regional background concentrations. The challenge is choice of the degree to which exceedance is acceptable.

In Europe, the reference condition approach is or was widely practised (e.g. by Flanders in Belgium, and in France, Germany and Italy: Babut *et al.*, 2005). For example, in Flanders an extensive effort was made to define reference values from sediments in locations where benthic communities

Table 16.3 *Sediment metal concentrations definitive of effects as identified by various approaches (see USEPA, 2006)*

Metal	Babut et al., 2005 Reference value[a]	Long et al., 1995 ERL (SW)[b]	ERM (SW)[b]	Riba et al., 2004 Range of uncertainty	Barrick et al., 1989 AET	USEPA, USGS and NOAA TEL (FW/SW)[b]	PEC (FW)[b]
As	11	8	85	27–213	700	6/7	33
Cd	0.4	1.2	9.6	0.5–1.0	9.6	0.6/0.7	5.0
Cr	17	81	370		270	37/52	111
Cu	8	34	270	209–979	1300	36/19	149
Pb	14	47	218	260–270	660	35/30	128
Hg		0.2	0.7	0.5–1.47	2.1	0.2/0.1	1.1
Ni	11	21	52			18/16	49
Ag		1	3.7		6.1	0.7	1.8
Zn	67	150	410	513–1,310	1600	123/124	459

[a]Approach used in Flanders (Belgium), from Babut et al., 2005.
[b]FW and SW mean freshwater and seawater respectively; FW/SW means both. For other abbreviations, see Box 16.1.

appeared to be undisturbed. These reference values are used as the guidelines. Any exceedance triggers action (Table 16.3). This approach is the most precautionary if the reference is defined as truly undisturbed environment. But it also is problematic to implement. Human activities are consistently accompanied by metal contamination in sediments. Thus exceedances will occur wherever there is human activity (which is almost everywhere). It seems that a guideline so cautious that it does not prioritise problems could create the equivalent of no guideline, because it cannot be enforced.

A related approach is to define the reference as a regional concentration. The implicit policy goal in such an instance is to prevent further degradation. For example, in Germany the lowest threshold for marine waters is established by the prevailing concentrations in the Wadden Sea between 1982 and 1992. The second or action level is defined as five times the reference value. Probably the choice of enrichment is more political than scientific; but it creates a condition that can be enforced even if the numeric choice is not particularly well justified.

16.7.4 Empirically based guidelines

Prologue. Empirically based guidelines are increasingly being used for many other applications in risk assessment and risk management (Babut et al., 2005). They seem to provide a justifiable definition of the range above which effects are likely and below which effects are not likely, despite some important limitations. The uncertainties in the empirical guidelines have their greatest effects when the levels are developed from a narrow database or for a use that requires precision.

Some sediment quality guidelines relate the concentrations of contaminants in sediment to some predicted frequency or intensity of biological effects (Batley et al., 2005). Justifying the guideline on the basis of some measure of biological impact eliminates one of the strongest sources of controversy about the use of total metal concentrations alone. These 'effects-based guidelines' are in practice in Australia/New Zealand, Hong Kong, France and the Netherlands. The UK is developing such standards.

Box 16.14 Empirically based guidelines

These methods choose those natural sediments for which adverse biological responses have been observed and try to relate the probability of a response to a metal concentration in those sediments. Any data on 'effects' are acceptable for use (acute, chronic or field data; spiked sediments or sediments contaminated in nature). The most common source of data is acute toxicity tests on whole, fine-grained sediments from field sites (Batley et al., 2005). The tests include a wide range of organisms (e.g. from marine environments: amphipods, sea urchins, mysids, polychaetes, shrimp, bivalves and bacteria). Sediment quality guidelines are typically a concentration on a dry weight basis. But the same approach can be used to develop guidelines on metal concentrations normalised to AVS (Long et al., 2006).

After assembling data for a contaminant in order of ascending concentration, the 10th percentile (ERL: Long et al., 1998a) or 15th percentile (PEL: MacDonald et al., 1996) is defined as the concentration below which effects are not expected. The value above which effects are often expected is determined by the median (ERM) or 50th percentile concentration (TEL) at which effects are observed. The effects level for an AET is the concentration above which effects always occur. ERMs and ERLs classify toxic and non-toxic samples similar to paired PELs and TELs, as might be expected. ERMs and TELs minimise Type I error (false positives) and ERLs and PELs minimise Type II error (false negatives), especially compared to AETs.

One of the challenges with the empirical approach is that most sediments contain multiple contaminants, but the indices relate to single contaminants. To address this question, mean sediment quality guidelines (SQG) quotients (mSQGQ) were proposed (e.g. Long et al., 2006). The quotients are determined by dividing the concentrations of each chemical in a contaminated sediment by its respective guideline value (determined from the individual contaminant data sets). Then the arithmetic mean of the ratios is derived. The result is a single, unitless, effect-based index of the relative degree of contamination in the sediment overall. Long et al. (2006) claim that this represents the likelihood that a sediment sample would be toxic to sediment-dwelling organisms.

The mean index is not sensitive to a situation where one contaminant occurs in very high concentrations and others are in very low concentration. It will reduce the first value by the others. This contradicts nature, where the lack of other contaminants will not reduce the toxicity of a single contaminant. This was addressed by testing a wide variety of sediments and determining the quotient at which the onset of toxicity occurred (rather than just assuming that quotient was 1). Ingersoll (2001) found that a consistent increase in toxicity occurred at a mean quotient >0.5, for example. The sums of quotients would better address the issue of unbalanced toxicity (e.g. Clements, 2000). But sums are subject to variability if different numbers of chemicals were measured in different samples. So the mean approach is recommended as a general guideline (Ingersoll, 2001; Long et al., 2006).

The effects-based approaches fall into three groups:

- empirically based guidelines;
- mechanistically based (or theoretical) guidelines that consider bioavailability;
- consensus guidelines.

Empirically based guidelines are derived by statistically aggregating results from studies that include both determination metal concentrations in sediments collected from the field and measures of toxicity ('effects') in those same sediments (Box 16.14). The goal is to determine what proportion of the studies show toxicity at different total metal concentrations in the sediments. For any individual trace metal, the data are assembled in order of ascending concentration at which some effect was observed. A statistical method is used to define concentrations below which effects rarely occur and a second concentration above which effects are often observed (Table 16.3, from Long et al., 1998a). Effects are defined mostly by toxicity tests, but any data are acceptable if an effects measure is accompanied by a concentration (Box 16.14).

One approach uses only data from marine/ estuarine sediments to define an ERL or *Effects Range Low* below which risks are minimal, and an ERM or *Effects Range Median* above which effects often might be expected. An alternative methodology uses a similar methodology, with freshwater sediments, to define a TEL or *Threshold Effects Level* and a PEL or *Probable Effects Level*, respectively (MacDonald *et al.*, 1996). A third approach defines the concentration above which effects always occur: an *AET* or *Apparent Effects Threshold* (Barrick *et al.*, 1989).

Validation of the empirical approach against community structure data was undertaken by Long (2000). He showed general correspondence between exceedance of ERM and disturbance of community structure in several sites from a USEPA monitoring data set (Fig. 16.9).

There are important limits to the empirical guideline approach. A primary one is that the method does not address bioavailability. It also puts extremely different endpoints for toxicity in the same comparison. And it ignores route of exposure. All these factors add imprecision. However, as we observed earlier, concentrations, toxicity tests and simple measures of change in communities are all most effective at defining risks

at the extremes of contamination. A great strength of the empirical effects approach (implicitly) is that it uses a statistically acceptable method to define those extremes. In that case, the imprecisions resulting from the uncertainties are less important than if the guidelines were used as numeric standards. The statistical power of the more established approaches is also an advantage; hundreds of data sets are available. As a result TELs and ERLs agree rather closely (Table 16.3) for most metals, as do TECs and ERMs. This internal consistency is impressive. But data from a single water body can yield results that differ widely than those derived from larger databases (Riba *et al.*, 2004; Table 16.3), the result of lower statistical power of the local data. Because of the limits, Chapman *et al.* (1999) suggested that the term 'sediment quality values' was the best description for the empirical guidelines. In Australia and New Zealand, the lower threshold is explicitly defined as a 'trigger value' to imply that further action (presumably study) is needed; while the upper values are used as informative of where effects might be more likely (Babut *et al.*, 2005). A practical advantage is that the empirical methods are simple to apply and their limited goals are simple to understand.

16.7.5 Mechanistically based guidelines

Prologue. Unambiguously determining bioavailability from sediments raises important challenges that have not been resolved (NRC, 2003). If policy specifies use of bioavailability tests (e.g. SEM–AVS methodologies) challenges in implementing the standards are likely. Inconsistent results and even regulatory chaos could result if data from individual sites do not recognise the complexity of natural sediments and their resident biota. If the SEM–AVS protocols are employed in nature, they should at least restrict sampling to zones of sediments where the most organisms intersect their environment (zones that are mostly oxidised).

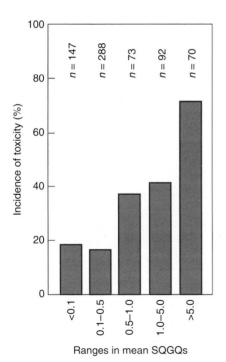

Fig. 16.9 Percentage incidence of amphipod toxicity over five ranges of sediment quality guideline quotients (SQGQs). (After Long *et al.*, 2006.)

Mechanistically based guidelines add consideration of bioavailability to the concentration approach. To date, the only mechanistic guidelines recommended for risk assessment and risk management have their roots in equilibrium partitioning theory (Batley *et al.*, 2005). Concentrations of metals in the sediment are related to corresponding concentrations in pore waters, and sediment guidelines are derived by choosing an effect concentration from the dissolved toxicity database (e.g. ambient water quality guidelines: Chapter 15). This guideline is then compared to the concentration in the sediment from nature, as determined through the SEM–AVS methodology.

The mechanistic approach is attractive because of the experimental support for the concept. But it has not yet been demonstrated that applying the approach adds value from a regulatory point of view. Long *et al.* (1998b) compared concentrations

causing effects in all experiments that used SEM/AVS or SEM/AVS*OC (organic carbon) protocols, and found that there were few differences from empirical guidelines developed from total metal concentrations. Similarly Babut *et al.* (2005) concluded that where bioavailability was considered by the AVS protocols, it was unclear 'whether the additional measurements actually help refine the assessment'. There is not yet enough experience to determine if, when, where and how frequent contradictory results might be expected (as in the experiments of Warren *et al.*, 1998; Lee *et al.*, 2000a).

Any implementation of AVS–SEM methodology (Box 16.15) might benefit from considering in advance:

Box 16.15 Applying AVS–SEM methodologies in nature

The reactive sulphide in sediments is typically measured for toxicological studies by collecting a sediment sample from a selected (explicitly or implicitly) segment of the sediment column. The sample is taken to the laboratory and stored until a subsample can be immersed in a cold solution of dilute (1N) hydrochloric acid (HCl). The acid extracts Fe and other metals from their solid phase sulphide forms into the acid, and releases the sulphide as a gas (hence the term acid volatile sulphides). Released concentrations of both sulphide and metals are then measured analytically.

Geochemists recognise that HCl extracts a complex and variable mix of sulphide forms from the sediment: it is not necessarily 100% efficient or specific for amorphous Fe monosulphide (Rickard and Morse, 2005). Thus practitioners term the extraction 'operational' in its definition of reactive sulphide (Allen *et al.*, 1993). That is, the extraction shows the tendency of the sediment to generate reactive sulphide; but it is not precisely restricted to reactive sulphide.
Sampling natural sediments is also a consideration.

(1) The depth at which oxidised conditions change to reducing conditions will be different from sediment to sediment, and can vary at the same location over time (Chapter 6). Van den Berg *et al.* (1998)

concluded: 'Due to the increase in AVS concentrations with increasing sediment depth, it is obvious that SEM/AVS ratios of homogenized sediment samples (e.g., 0–10 cm) are not suited to assess the potential heavy metal toxicity in the surface sediments.' They also noted that the surface layer is the connection of most life forms to their environment.
(2) Sampling of the visually oxidised layer of sediment on the surface may be a consistent method for addressing bioavailability, especially given the importance of surface sediments to organisms. This also is an operational approach, because the boundary between oxidised and reduced sediments is not necessarily distinct or visibly predictable (Morse and Rickard, 2004). But it will constrain sampling toward the most relevant type of sediment, biologically. A surface sample might be accompanied by a clearly reduced subsurface sample, to obtain some measure of the gradient in the sediment.
(3) Multiple spatial replicates are essential to verify the repeatability of the observations. In some systems, AVS can be variable from one location to another. In the Brisbane River, Queensland, Australia, AVS varied from 0.3 to 22.6 µM/g dw along eight transects (Mackey and Mackay, 1996). Variation is much smaller in surface sediments (S.N. Luoma *et al.*, unpublished data).

(4) Seasonal variability is also likely in most systems. Leonard *et al.* (1993) found that AVS varied over two orders of magnitude in monthly measurements for 16 months in three Minnesota, USA lakes.
(5) Bulk samples have limited sensitivities to microenvironments created by at least some (if not many) of the species that comprise the benthos. In this respect, a surface sample would be environmentally precautionary.
(6) Resuspension of bed sediment results in the rapid decrease in acid volatile sulphide (AVS) as

determined by the oxidation of Fe monosulphide phases. During 8 hour sediment resuspension experiments, AVS decreased and SEM(Cu) increased from 0.1 to about 2 µM/g, presumably due to oxidation of CuS, while the SEM measured for the other metals remained constant (Simpson *et al.*, 1998). Again, surface sediments best characterise the fraction of the sediment bed that is most likely to undergo resuspension and redeposition.

- How to sample the vertical zonation of sediments, given its site-specific character.
- To which habitats the determinations of AVS are most relevant. For example, most members of the benthos create microenvironments that are lost in a bulk sediment measurement of AVS.
- How to consider possible transformation of reduced metals. Neither the surface layers of sediments nor suspended sediments will stay reduced if exposed to an oxidised water column. Conclusions about reduced bioavailability based upon evaluations of the sediment bed are not applicable to metals that exchange between the bed and the water column (at the least).
- What animals are being protected. Differences in functional ecology and dietary exposure, especially, affect how different organisms are exposed to metals in their sedimentary environment.

16.7.6 Lateral management: multiple lines of evidence

Prologue. All methodologies devised to date are operational and have trade-offs, especially when considering bioavailability (NRC, 2003). Weight-of-evidence approaches, multiple lines of evidence and adaptive management are necessary ingredients in managing an issue in such an uncertain environment.

Any single approach to deriving sediment quality guidelines is likely to be variably sensitive and relatively imprecise, except at the extremes. It is possible to aggregate guidelines or conclusions

developed from different methods (e.g. Table 16.3). This can be done narrowly or broadly. In the narrow case central tendency guidelines were developed from different approaches to the empirical guideline concept (Ingersoll, 2001; Batley *et al.*, 2005). Simpson and King (2005) took an entirely new approach in addressing how 'the bioavailability of metals in sediments is dependent on (i) speciation (e.g. metal binding with sulphide, Fe hydroxide, organic matter), (ii) sediment–water partitioning relationships (e.g. K_d) ([Sediment-Cu]/ [Water-Cu], with concentrations expressed in mg/ kg and mg/L respectively), (iii) organism physiology (uptake rates from waters, assimilation efficiencies from particulates) and (iv) organism feeding and other behaviour (feeding selectivity, burrow irrigation). They showed that rates of uptake from multiple routes (sediment and water) helped explain differences in organism sensitivities to contaminated sediments in short-term acute toxicity experiments. The more explanatory approach would seem to provide a better bridge to ecological assessments; i.e. a basis for predicting which species might or might not survive sedimentary contamination.

The sediment quality triad was the earliest attempt at a broader approach to using multiple lines of evidence at field sites (Chapman and Long, 1983; Chapters 10 and 11). The original triad included sediment metal concentrations, bioassays and community data. More advanced approaches employ tests that better consider *in situ* conditions, recognising the limits of bringing sediments into the laboratory. Burton *et al.* (2005) cited instances from employment of the triad where survival in 10 to 28 day sediment toxicity tests correlated with (or 'general good agreement' was found with) abundance of species similar to the test species, species richness, colonisation of artificial substrates or other measures of community structure.

Combining measures that have their individual limitations does not necessarily always ease concerns about specific risk management decisions (Wenning et al., 2005). Part of the problem is that evaluations of complex issues, like metal bioavailability from sediments, are not necessarily amenable to solution with a batch of simple tests (no matter how many tests are added). Systematically continuing to develop the body of evidence and understanding is the long-term approach most likely to broadly resolve the bioavailability dilemma (NRC, 2003). Locally, solutions can develop more rapidly if explanatory experiments can be paired with well-designed ecological studies (Chapter 11). Such work takes time and resources, but if enough is at stake and uncertainty about effects is great, investment in understanding the nature of geochemical, toxicological, biological and ecological circumstance is much more constructive (and cost-effective) than unresolvable advocacy debates, political deadlock and/or endless litigation.

Given the potential benefits of explicit consideration of bioavailability but the uncertainties of all approaches, adaptive management seems the best way forward. Ehlers and Luthy (2003) described this as 'pilot studies with different tools and models'. Such studies might involve experimenting with a lateral management approach in a single water body or watershed. A test might compare two approaches to regulating sedimentary contamination (e.g. with and without bioavailability for example) in the same water body, in terms of environmental protection and social acceptability for example. A well-designed and systematic series of such studies might help authorities find the path between theory and practice and help refine a common approach for when and how to incorporate bioavailability into risk management.

16.8   CONCLUSIONS

Unified protocols are clearly needed to address uncertainties about risks from sediment contamination with trace metals. In no area of study is it more essential that the traditional approaches to metal contamination science be merged. In no field is incoherence in risk assessment more evident. Management of ecological risks from metal contaminated sediment is also an ongoing focus of intense debate because the economic and natural resources at stake are large (NRC, 2003; Wenning et al., 2005). Complex combinations of different types of contamination, complicated environmental processes, different mixtures of receptors, and conflicting social values add to the challenge. Most of all, there is a lack of scientific consensus on the best approaches to meet the needs perceived of either the science or the policy community. The difficulties are evidenced by the variable and inconsistent management approaches used across the world (Babut et al., 2005).

Environment-specific, metal-specific and species-specific traits interact to determine actual exposures to metals of each species from each site. Geochemical uncertainties are one source of the slow growth of knowledge in this field. Analytical techniques and models are not as advanced as in determining metal speciation in solution (NRC, 2003). Most methods for determining specific forms of metals in sediments are imprecise and unselective (i.e. operational) and, at best, point toward tendencies rather than specific forms. Models of metal distributions among forms in sediments need improvement. Linking geochemical and biodynamic models is within reach, however. Comparisons between biodynamic forecasts and observations of bioaccumulation in nature could be used to resolve the basic dilemma about the degree to which different organisms are exposed to reduced versus oxidised sediments, for example. That of course would require toxicologists accepting that bioaccumulation is a useful measure of exposure from sediments; a simple step that as yet seems difficult to attain in this contentious field. Biological uncertainties are also a direction that might yield to modelling forecasts. Species-specific behaviours, feeding choices and digestive environments could be addressed with models, comparing outcomes to nature. Biomonitors, biodynamics and incorporation of species- (or function-)specific sensitivities are underutilised technical approaches that work in other contexts but are rarely applied to understanding risks from contaminated sediments.

Just using the existence of metal contamination as a trigger for risk management seems like a simple precautionary way to address sediment quality (the reference approach). But the goal needs to be explicit (e.g. hotspots should not exceed existing regional levels of contamination?) or it seems the outcomes will be difficult to enforce. Sediment bioassays are another approach that is simple to implement in a regulatory context. But bioassay data alone are relevant only to specific sediment samples, predictive guidelines are not possible and it should not be assumed that the outcomes of the tests reflect ecological responses. The use of empirical guidelines to provide context in scientific studies or ambient monitoring is not particularly controversial (interpret trends, interpret monitoring results, help choose sites for detailed studies, establish reference conditions, rank contaminated areas, compare to more detailed analyses). The statistically derived thresholds in the empirical approach also seem well suited for constraining the concentration range where uncertainty about risks is greatest. Their use in regulatory actions like formally establishing a mandatory trigger level for clean up action or in defining clean-up limits will always be more controversial, however.

Bioavailability is the key consideration in identifying site-specific risks from contaminated sediments. If the situation is such that bioavailability corrections will make a detectable difference in the outcome, then addressing the issue is justified. The advantage is that clean-up costs could be reduced, efforts prioritised to areas where risks are greatest, and effective (e.g. *in situ*) remediation might be implemented in environments where contamination was difficult to treat or most likely to have the greatest adverse effects. But it is essential to recognise that the bioavailability dilemma is not amenable to a simple resolution with a universal measure. A credible resolution, at best, requires a systematic body of evidence that probably includes both field observations and relevant experimental data, as well as consideration of the sociological implications.

Predictive tools, adaptively monitoring outcomes as remediation proceeds, and validation of proposed risk management approaches are necessary parts of a lateral risk management approach. An experimental approach to risk management is not attractive to most policy makers, but may be the best way to sort through the choices. The limits and uncertainties of various approaches to addressing contaminated sediments continue to be the focus of contentious debates about exact values for each guideline. It is possible that this debate cannot be resolved on purely scientific grounds; a protocol rather than a guideline may be more constructive. Adaptive management in the form of pilot risk management exercises may be a way forward in the face of such dilemmas.

Suggested reading

Batley, G.E., Stahl, R.G., Babut, M.P., Bott, T.L., Clark, J.R., Field, L.J., Ho, K.T., Mount, D.R., Swartz, R.C. and Tessier, A. (2005). Scientific underpinnings of sediment quality guidelines. In Use of Sediment Quality Guidelines and Related Tools for the Assessment of Contaminated Sediments, ed. R.J. Wenning, G.E. Batley, C.G. Ingersoll and D.W. Moore. Pensacola, FL: SETAC Press, pp. 39–120.

Di Toro, D.M., Mahony, J.D., Hansen, D.J., Scott, K.J., Carlson, A.R. and Ankley, G.T. (1992). Acid volatile sulfide predicts the acute toxicity of cadmium and nickel in sediments. *Environmental Science and Technology*, **26**, 96–101.

Lee, B.-G., Griscom, S.B., Lee, J.-S., Choi, H.J., Koh, C.-H., Luoma, S.N. and Fisher, N.S. (2000). Influences of dietary uptake and reactive sulfides on metal bioavailability from sediments. *Science*, **287**, 282–284.

Long, E.R., Ingersoll, C.G. and MacDonald, D.D. (2006). Calculation and uses of mean sediment quality guideline quotients: a critical review. *Environmental Science and Technology*, **40**, 1726–1736.

Luoma, S.N. (1989). Can we determine the biological availability of sediment-bound trace elements? *Hydrobiologia*, **176/177**, 379–396.

Ringwood, A.H. and Keppler, C.J. (2002). Comparative *in situ* and laboratory sediment bioassays with juvenile *Mercenaria mercenaria*. *Environmental Toxicology and Chemistry*, **21**, 1651–1657.

17 • Harmonising approaches to managing metal contamination: integrative and weight-of-evidence approaches

17.1 INTRODUCTION

Prologue. Existing approaches to managing metal contamination treat water, sediments and food webs as separate entities in a contaminated water body. A coherent process for integrating or harmonising these different types of data is an essential ingredient in risk assessment and, ultimately, lateral management of risks (see also Preston, 1979). Integrative measures implicitly include the influence of processes linking inputs to environmental concentrations and to environmental effects. Holistic, weight-of-evidence or lateral approaches interpret multiple measures (lines of evidence) to derive conclusions about exposure and effect.

There are technical limitations to every individual approach used in ecological risk assessment or risk management (Kennedy *et al.*, 2003). The philosophy that simple approaches are essential to risk assessment and risk management came from policies designed to address organic contaminants. Metals are many fewer in number than organic contaminants, and have unique complexities that limit the suitability of simplifying assumptions. Integrative tools, integrative assessments and/or multiple lines of evidence (holistic management) are therefore essential in addressing many of the ecological risks from metal contamination (Burton *et al.*, 2003).

Weaknesses exist in the traditional management of metal contamination as evidenced by incoherence among jurisdictions in water quality criteria and examples of over- and under-prediction (Chapters 15, 16 and 17). Se is a specific example (Box 17.2; Fig. 17.1). Existing standards for Se in the USA have been under review since 1998 with no resolution. The large differences between ambient water quality

criteria across the world reflect contradictions between results from 'acceptable' standardised toxicity tests and observations of Se toxicity in nature.

Box 17.1 Definitions

Bioaccumulated Metal Guideline (BMG): measure of toxic exposure based on bioaccumulated metal concentrations. BMGs can be expressed as total or fractional bioaccumulated concentrations. BMGs vary among species in most circumstances and within species under some circumstances, particularly varying total metal uptake rate. The degree of variability differs among trace metals.

Dietary Metal Guideline (DMG): concentration (total or in a specified fraction) in the diet of a consumer species that correlates with the onset of adverse effects in the consumer.

holistic management or integrated risk assessment: use of multiple indicators of contamination to determine environmental concentrations (PEC) and multiple measures to derive PNEC.

integrative criteria: single measures that integrate processes linking environmental concentrations to environmental effects.

PEC: Predicted (or measured) environmental concentrations.

PNEC: Predicted low or no effect concentrations.

weight of evidence (WOE) approach: process of combining information from multiple lines of evidence.

Box 17.2 Ambient water quality standards for Se

In deriving the salt water (marine) criterion for Se, USEPA employed their large database of toxicity tests. They used a species sensitivity curve with 17 saltwater and 18 freshwater species to derive an acute toxicity standard. The range of mean acute values was 255 to 17350 µg/L Se for selenite. To protect aquatic environments from short-term (1 hour) exposures to Se, USEPA suggests a value of 185 µg/L Se for freshwater (USEPA, 2004).

In the 1980s it was found that the toxicity testing data, even when corrected for uncertainties, were under-protective compared to field observations of Se toxicity in lakes. Therefore freshwater criteria were derived from experience at Se contaminated sites in the field; primarily Martin's Lake in Texas and Belews Lake in North Carolina (see Box 13.8 for the Belews Lake story). Numerous fish species were extirpated from these lakes at Se concentrations far below those predicted to be toxic in toxicity tests. Dietary exposure was the cause of the toxicity (Lemly, 1993a). The freshwater standard (USEPA, 1987) reflected this field experience, concluding that concentrations of Se greater than 5 µg/L Se caused toxicity in areas of the lake where fish toxicity was observed. Later, another US agency (US Fish and Wildlife Service, USFWS) found reduced hatching success in birds in the Central Valley of California at 2 µg/L Se in local waters. USFWS promulgates that as a standard. These data also form the basis of the freshwater standards in other jurisdictions (e.g. Canada and Australia).

The contradictions between the field-based criterion and the toxicity testing criterion was reflected in a draft document released for public comment, that addressed using fish tissue concentrations of Se to assess chronic risks (USEPA, 2002c). The authors concluded: 'freshwater aquatic life should be protected if the concentrations of selenium in whole-body fish tissue does not exceed 7.9 µg/g dw and if the short-term average concentration of Se dissolved in water seldom exceeds 185 µg/L' (USEPA, 2002c). The contradiction between a fish criterion that is typical of contaminated waters and a dissolved criterion more typical of Kesterson Reservoir (Chapter 13) seems incongruous if natural waters are at all considered. But the possible outcome of such thinking was more alarming. For example, the State Water Quality Control Board in California marine water quality objective is 15 mg/L Se. This is based upon the toxicity of dissolved total Se, divided by a safety factor of 10. Correction for

speciation was part of the protocol. The speciation correction would lead, for example, to a protective guideline of 58 mg/L Se for marine water body like Newport Bay, USA. The Acute Water Quality Criterion for Se adopted by USEPA in 2004, when corrected for speciation, (USEPA, 2004) would allow >150 mg/L Se in the coastal waters of northern California. This led the US Bureau of Reclamation to suggest that irrigation drainage from the Central Valley of California (the water that caused the Kesterson Reservoir disaster: Box 13.2) could be safely disposed of in the ocean.

It is probably safe to say that the concentrations of Se now allowed in the marine environment will never occur in anything but untreated effluents and heavily contaminated irrigation drainage waters (Table 17.1:

Table 17.1 *A comparison of dissolved Se concentrations in a gradient from extreme contamination in the watershed (irrigation drainage) to the coastal waters off San Francisco Bay, compared to ambient water quality criteria for Se from around the world. Signs of adverse effects of Se were severe in Kesterson Reservoir but are also seen or suspected to a lesser degree in all systems through upper San Francisco Bay (Chapter 13)*

	Dissolved Se concentrations (µg/L)
Water body	
Kesterson Reservoir – entrance pond	350
Kesterson Reservoir – terminal evaporation pond	14
Constructed Treatment Wetland	5–30
San Joaquin River: 1986 to 1997	0.8–4.8
Upper San Francisco Bay	0.2–0.5
Pacific Ocean	0.08
Ambient criteria (1996)	
European Commission	Not a chemical of concern
Freshwater – Canada	1
– USA	5
– Australia	5
Marine Criterion – California	15
– Australia	80
USEPA Acute Criterion for Californian coastal waters	>150[a]

[a]Corrected for speciation for coastal California.

Chapter 13). Extreme ecological effects occur at these concentrations, where they are found (e.g. Kesterson National Wildlife Refuge, California, USA: Chapter 13). Risks from Se toxicity exist in in San Francisco Bay at concentrations orders of magnitude below these concentrations (Presser and Luoma, 2006). The remarkable difference between freshwater and marine standards occurred because there are no field data acceptable to USEPA from marine or estuarine environments. Although authors cite sulphate competition with selenate in marine waters, most Se is not selenate in estuaries or the oceans. And there is no evidence to support the >10-fold difference in ecological sensitivity among such systems (e.g. Luoma *et al.*, 1992; Luoma and Presser, 2000). The paralysis reflected in nearly a decade of debate over revision of such criteria is partly a result of confusion about how to correct the discrepancies between experience from nature and toxicity tests.

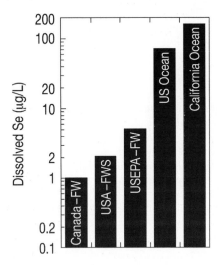

Fig. 17.1 Se standards in different jurisdictions. There is no guideline for managing Se in Europe, nor is it listed as a hazardous chemical.

Criteria or guidelines chosen on the basis of field observations (the freshwater Se standards of 1 to 5 µg/L in Canada and the USA) are not consistent with criteria chosen on the basis of dissolved toxicity (the criterion of 71 µg/L for marine waters in the USA; the decision that Se is not an important hazard in Europe). Incoherence is the result wherever such contradictions in the scientific literature are not resolved in an integrative fashion.

Traditional criteria also regulate water, sediments and food webs independently; whereas in nature they all interact (Kennedy *et al.*, 2003). The cost is uncertainty and contradictions (Di Toro, 2003). Recognition of the strengths and limits to individual approaches (water quality criteria, sediment criteria, wildlife criteria) can improve their applications. But, ultimately, uncertainties are best reduced by addressing how components of the environment are linked with integrative measures and weight of evidence analyses.

Figure 17.2 is a depiction of the chain of linkages among processes that lead to adverse ecological effects from metals, derived from the conceptual model in Chapter 2. Lower-order approaches to managing risks from contamination manage mass loading or concentration. These are the simplest, and the most precautionary management tools to implement. But they also have the greatest uncertainties, and thus are the most likely to over-protect or under-protect an environment. As one proceeds through the processes that govern effects, each higher-order process integrates the influences of processes that precede it in the conceptual model, but require assumptions about or corrections for the processes above it. For example, ambient water quality criteria integrate the influences of metal loads and hydrology. But assumptions or simplistic corrections for influential geochemical and biological processes are necessary to manage toxicity (Fig. 17.2a).

17.2   INTEGRATIVE MANAGEMENT APPROACHES

Prologue. Lower-order tools for managing ecological risks from metals require multiple assumptions. The most integrative measures require fewer assumptions, but issues with detection and causation can limit

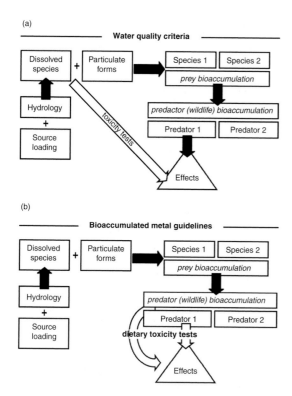

Fig. 17.2 Every management approach makes assumptions about the linkages to toxicity. Higher-order approaches skip fewer linkages; lower-order approaches skip the most links and are usually the most uncertain. One way that metal contamination is managed is to depend on water concentrations (e.g. water quality criteria) as shown in Fig. 17.2a. These make assumptions about partitioning, speciation, bioavailability and biological/ecological processes. In contrast, the use of bioaccumulated metal concentration (Fig. 17.2b) would integrate processes that influence bioaccumulation, but does not quantify the toxic influence of those processes.

their usefulness when implemented in the absence of other data.

Integrative measures or *environmental standards* are single measures that integrate processes linking environmental concentrations to environmental effects. Bioaccumulated metal concentrations represent a more integrative measure than mass loading or water concentrations (Fig. 17.2b), for example. The influences of concentration in water, geochemistry, biology and food web interactions

are reflected in the bioaccumulated concentration in an organism; assumptions about those influences are not necessary. The attraction of the more integrative measures of risk is that fewer links in the conceptual model are skipped. Bioaccumulated metal is not necessarily toxic, however. Therefore the processes affecting the link between bioaccumulation and toxicity require assumptions, corrections or understanding to best assess and manage risks (Chapman, 1995). The other challenges with integrative measures lie in detecting the response and interpreting its cause. Several examples of prospective integrative measures are discussed below with their strengths and weaknesses. All have caveats; but many integrative measures reduce important uncertainties overall.

17.2.1 Dietary Metal Guidelines (DMG)

Prologue. Metal concentrations in prey species strongly influence metal exposure of specific predators (e.g. Chapters 7 and 13). Thus it should be possible to derive a guideline for a metal concentration in prey species (Dietary Metal Guideline, DMG), assuming dietary exposure is potentially a cause of toxicity. The threshold dietary concentration causing a toxic effect in a consumer ($PNEC_{prey}$) can be determined in dose–response experiments in which diet is the source of the dose. Monitoring metal concentrations (PEC_{prey}) or some metal fraction concentration in invertebrates or other prey would be a pragmatic and effective way to determine bioavailable environmental exposure. Although there is still much to learn, implementation of a DMG is feasible if ecological (e.g. food web) processes are carefully considered, and the methodologies developed for biomonitors employed (Chapter 8).

A *Dietary Metal Guideline (DMG)* is derived from the metal concentration in a prey organism that causes (and thus signals) the onset of toxicity in a consumer. Derivation and implementation of a DMG can be similar to that for a water criterion. The assumed route of exposure is different (diet versus solution) and many of the same caveats apply (e.g. metal form affects the relationship between dose and response); although the order of magnitude of the uncertainty is probably lower. Arithmetically, risk from dietary metal exposure might be approximated by a risk quotient (RQ) where

$$RQ = PEC_{prey} / PNEC_{prey}$$

where PEC_{prey} is the metal concentration in a prey species in the water body and $PNEC_{prey}$ is the highest concentration in the prey that causes no effect on its predator. RQ < 1 if $PEC_{prey} < PNEC_{prey}$ and no adverse effects are anticipated. The $PNEC_{prey}$ is the Dietary Metal Guideline (DMG).

Experimental determination of the $PNEC_{prey}$, of course, must involve dietary toxicity experiments. The literature on dietary toxicity to date is much smaller than the literature on dissolved toxicity; and there is still much to learn. Endpoints must be considered carefully in dietary exposure experiments, for example. Acute toxicity may not be a relevant mode of action when exposure is from diet. Other factors often interfere before direct mortality is observed. Thus the short-term exposure paradigm that dominates dissolved exposures does not yield the same kind of results when diet is the source of exposure. One of the reasons is that dietary exposure can affect or inhibit gut processes. Animals avoid ingesting heavily contaminated food, modify the way they digest it (Decho and Luoma, 1996) or lose gut functions when exposed (Woodward et al., 1995; Maltby et al., 2002). Metals taken up from diet also may be accumulated or processed differently by the consumer than when metal exposure is via solution (e.g. Hook and Fisher, 2002). For example, metals accumulated in the liver of the consumer are likely to associate with reproductive products manufactured in the same organ (vitellogenins in fish liver, for example). Finally, there may be some uncertainties associated with determining a $PNEC_{prey}$ from surrogate food sources. Much more needs to be known about the degree of variability associated with toxicity from different food sources.

The biomonitoring literature illustrates that prey organisms are practical to sample in statistically valid numbers and better satisfy the criteria for effective biomonitoring than do larger predator organisms (Chapter 8). This is a major advantage of a DMG compared to other integrative-type approaches to assessing or managing risks. However, not all prey species are suitable choices for determining the $PEC_{prey.}$. If the species chosen for monitoring is being affected in a contaminated habitat, it will be either rare, absent or represented by adults that migrated in from elsewhere. Thus a more tolerant organism is a more reliable biomonitor than a more sensitive organism (Chapter 8). Animals that regulate body concentrations of Zn (Chapter 7) are not good biomonitors (Chapter 8).

Bioavailability of the metal from the prey to the predator may be a consideration in development of a $PNEC_{prey}$. With some metals there are differences in metal bioavailability among prey types. For example, Nunez-Nogueira et al. (2006) showed that Cd AEs by a penaeid prawn (Penaeus indicus) were 39% to 50% from an alga and 64% to 83% from squid muscle. The approximately twofold variation was significant, although not comparable to the much larger variability driven by speciation in dissolved metal exposures. On the other hand, assimilation efficiencies were the same (40% to 70%) for Zn whether this omnivore was fed muscle from a squid (Loligo vulgaris) soaked in radiolabelled solution after dissection, or a macrophytic alga that had accumulated labelled metal from the radiolabelled solution over several days.

Biochemical fractionation of metals within the prey may partly account for differences in assimilation from prey species (Chapter 7). Reinfelder and Fisher (1991) first proposed that the metal within the cytoplasm of phytoplankton (released as soluble metal when the cells were broken) was bioavailable, but metal in other fractions was not. Metal in the cytoplasm correlated strongly with differences in assimilation efficiency by zooplankton when a variety of metals were compared (Fig. 17.3). But the digestive processes of consumers or predators also influence what fraction of metal is bioavailable. Cytoplasmic concentrations were not consistent predictors of bioavailability from phytoplankton in animals with complicated digestive processes, like bivalves (Decho and Luoma, 1991). Fractions other than cytoplasmic metal are potentially bioavailable. Predictable fractions of cytoplasmic and organelle-associated metal (termed TAM

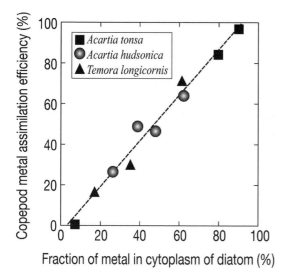

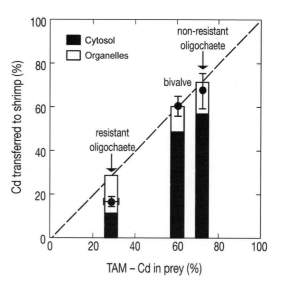

Fig. 17.3 Assimilation efficiency (AE, percent) of ingested trace metals by zooplankton (copepods) fed algal cells (the phytoplanktonic diatom *Thalassiosira pseudonana*) as a function of the cytoplasmic fraction of those metals in the diatom. Each data point for a species is the average AE for a different metal. (After Reinfelder and Fisher, 1991.)

Fig. 17.4 Assimilation efficiency (filled circles) of Cd by shrimps as a function of the percentage of Cd as cytosol and organelle-associated metal together constituting TAM (histograms) in the food: *Palaemonetes pugio* fed Cd-tolerant and non-tolerant oligochaetes (*Limnodrilus hoffmeisteri*), and *Palaemon macrodactylus* fed bivalves (*Corbula amurensis*) labelled with ^{109}Cd. Both the soluble forms of Cd in the prey and a proportion of the Cd associated with organelles appeared to be bioavailable. (After Wallace and Lopez, 1996; Wallace and Luoma, 2003.)

or trophically available metal: Chapter 7) appeared to be transferred to a shrimp consumer when two types of prey were considered Wallace and Lopez, 1996; Wallace and Luoma, 2003) (Fig. 17.4; see also Box 17.3). Neogastropod predators with very strong digestive systems appear to assimilate metals from an even wider variety of biochemical fractions in the prey including the metal-rich granule fraction (Cheung and Wang, 2005; Cheung et al., 2005; Rainbow et al., 2007). The implication is that prey species with an accumulated metal content that has been stored predominantly in a relatively digestion-resistant form such as an inorganic pyrophosphate granule, would be associated with an absolute DMG value that is much higher than a DMG associated with a prey organism that does not store accumulated metal detoxified in an inert insoluble form.

Prey species can vary widely in their total metal concentrations. Any generalised PEC$_{prey}$ might best be determined from prey with higher, rather than lower, bioaccumulated metal concentrations. For example, the animals most threatened by Se are predators that feed on prey that bioaccumulate

Se efficiently (Stewart et al., 2004; Chapter 13, Fig. 13.6). If zooplankton were sampled in San Francisco Bay to compare to a Se DMG, little likelihood of risk would be found (Fig. 17.7). If bivalves were sampled, a high probability of reproductive damage in predators would probably be found (PEC/PNEC > 1). This actually reflects the reality of conditions in the environment, where predators of bivalves are at more risk than predators of zooplankton (Linville et al., 2002; Stewart et al., 2004).

Implementation of a DMG should take trace metal-specific influences into account. Methyl mercury is passed most efficiently to consumers that feed on phytoplankton because MeHg accumulates in the cytosol of phytoplankton and inorganic Hg accumulates in an unavailable form on the surface of the cell (Chapter 14). Thus, phytoplankton-based food webs are a good choice for comparisons to a

Box 17.3 Effect of pre-exposure and detoxification on the passage of metals from prey to predator

Rainbow *et al.* (2006) fed polychaetes (*Nereis diversicolor*) to the shrimp *Palaemonetes varians*. The prey were from two UK estuaries with different levels of metal contamination. The hypothesis tested was that polychaetes with a history of metal exposure would have greater proportions of accumulated metal in insoluble detoxified form, and bioavailability of dietary metal to the consumer would be reduced.

The degree of effect differed among metals, however. The full range of assimilation efficiencies for Ag in 14 different experiments was 11% to 50%. About threefold less Ag was assimilated by the shrimp when the worms were from the contaminated estuary (mean AEs were 13% from prey worms from the contaminated water body versus 41% from the uncontaminated water body). Both Cd and Zn were assimilated more efficiently than Ag but had a smaller range of AEs (Fig. 17.5). Their bioavailability to the predator appeared to be less affected by the exposure history of the prey than was Ag. When all metals were considered together, a greater fraction of metal-rich granules (the insoluble detoxified fraction) in the prey usually resulted in reduced bioavailability in the predator (Fig. 17.6). Thus detoxification processes in the prey appear to influence the degree to which metal is passed from prey to predator; but the bioavailable fraction was not fully predictable by the operational fractionation protocol.

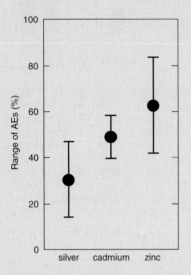

Fig. 17.5 The full range of assimilation efficiencies of Ag, Cd and Zn by the shrimp *Palaemonetes varians* fed on the polychaete *Nereis diversicolor* from the uncontaminated Blackwater estuary and the metal-contaminated estuary Restronguet Creek, after radiolabelling of prey from solution or sediment (from Rainbow *et al.*, 2006). The variability in bioavailability from the prey was substantial, but small compared to variability in bioavailability due to speciation in solution or form in sediment (e.g. sulphide versus non-sulphide forms).

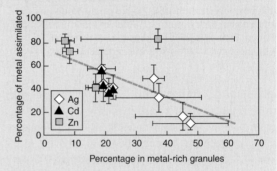

Fig. 17.6 Assimilation efficiencies of Ag, Cd and Zn by the shrimp *Palaemonetes varians* fed the radioactively labelled polychaetes (*Nereis diversicolor*) from estuaries with different histories of metal exposure, as a function of the percentage of the metal in metal-rich granules within the prey. Some bioavailability of granule-bound Zn was evident, but in no case was AE greater than 20% when the percentage of accumulated metal in the prey bound in granules was greater than 40%. (After Rainbow *et al.*, 2006.)

MeHg DMG. But bivalves accumulate total Hg most efficiently, overall, among invertebrates; thus they might be the best choice for an Hg DMG. Detoxification in prey species appears influential in the trophic transfer of Cd and Zn (Box 17.3); thus the PNEC$_{prey}$ could be quite species specific. A solution might be to conduct the dietary toxicity test with a specific prey as food, and define a DMG that is specific for that prey type.

At least for these trophically transferred metals, the threat from a prey species is to its specific predator, not to predatory animals in general. If risk assessment or risk management identifies the higher trophic level species of greatest interest

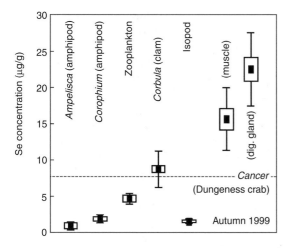

Fig. 17.7 Se concentrations in different species of invertebrates observed in autumn, 1999 in Suisun Bay, within San Francisco Bay. The annual mean Se concentrations in different invertebrates from the same habitat in San Francisco Bay varied from 2 μg/g Se prey dw for zooplankton to 11 μg/g Se prey dw for a bivalve. In the winter, concentrations in the bivalve increased to a December mean of 15 μg/g Se dw. This difference spans the threshold for the onset of Se effects in predators. In the summer concentrations dropped to 5 μg/g Se dw. The choice of invertebrate is important in monitoring dietary exposure.

for protection, then the correct choice for determination of PEC_{prey} is the prey of that species. Exact feeding relationships can be narrowed by gut content and by stable isotope analyses, if the questions are important enough (these are not expensive techniques: Stewart *et al.*, 2004).

17.2.2 Bioaccumulated Metal Guidelines (BMG)

Prologue. The total bioaccumulated metal concentration in an organism is informative about exposure in the field. High or low field concentrations (or ranges of concentrations) are known for many species and are at least a partial indicator of risk. For metals that are efficiently transferred through food webs, bioaccumulated metal concentrations may be essential in evaluating risks from contamination, because other measures can lead to completely erroneous answers. But it should also be recognised that toxicity occurs

when a metabolically active fraction of metal accumulates faster than it can be detoxified or excreted. This process may not be predictable from the total body burden in the organism. For metals that are detoxified and retained (e.g. Ag, As, Cd, Cu, Pb or Zn in many invertebrates), the bioaccumulated concentration at which toxicity occurs will vary widely among species and can vary within species. For trace metals that are not detoxified to a significant or varying extent and retained, the variability among species is more manageable (e.g. Se, TBT, MeHg).

Criteria based upon metal bioaccumulated in organisms were originally termed Tissue Residue Criteria. The term *tissue residue* is used in the regulatory context for concentrations of organic contaminants bioaccumulated into either whole bodies or specific tissues of organisms. This terminology is problematic for metals because whole body measures are common, 'tissue' is not an accurate descriptor, and 'residue' is not a descriptor commonly employed for bioaccumulated trace metals. Thus we will use a terminology more appropriate for trace metals: Bioaccumulated Metal Guideline (BMG).

Just because a metal bioaccumulates to elevated levels in the tissues of an organism does not mean it is exerting a toxic effect. That is why regulatory uses of metal bioaccumulation are controversial (Chapman, 1995). But the bioaccumulated metal concentration is usually an effective indicator of bioavailable metal exposure (not in the case of regulators: Chapter 7). It reflects an integrated influence of exposure concentration(s), metal-specific behaviour, environmental processes (loads, geochemical form, physics of the system) and the biology of the organism (e.g. weak or strong bioaccumulator: Chapter 7). Bioaccumulated metal also takes dietary exposure into account, reducing another important source of uncertainty about exposures.

A guideline based upon bioaccumulated metal concentrations must link bioaccumulation and toxicity, however. Unfortunately, it is unreasonable to expect that a field 'critical body concentration' exists for every metal, beyond which toxicity ensues (Borgmann *et al.*, 1991). The processes that link toxicity and bioaccumulation are more complicated than that (Chapters 7 and 9). Just as the dissolved concentration at which each metal is toxic varies among species and among circumstances, so does the internal concentration at which toxicity occurs. Physiologically, metal bioaccumulation is a function of uptake rate from all sources minus loss rates (Chapter 7). Metal toxicity is influenced by the rate of accumulation of the internal dose (net uptake rate); but it is also influenced by the rate of detoxification of that internal dose (Rainbow, 2002; Vijver et al., 2004). Toxicity occurs when uptake rate exceeds the sum of loss rate and detoxification rate. This balance among rates determines the accumulation of the metabolically available fraction (MAF) of metal (that which is not detoxified and associates with sites active in an essential aspect of life) in the organism. A 'true' BMG should reflect the accumulation of this MAF, or accumulation of metal at the site(s) of action.

Differences are common among species in the internal concentration at which toxicity occurs. But the degree of difference varies among trace metals. Bioaccumulation differences are the result of large interspecific differences in uptake rates (from food and water) and loss rates (Chapter 7). In addition, detoxification mechanisms and capabilities differ among species. Thus metabolically available concentrations are driven by species-specific differences between the total rate of metal uptake and the combined rates of detoxification and excretion. There is no a priori reason to expect that species which are strong accumulators of total metal are necessarily more tolerant or more sensitive to metal exposure; the differences in accumulation of the MAF must first be understood (e.g. Cain and Luoma, 1999).

Toxicity might also occur at different body concentrations within the same species. For example, under very high exposure conditions (and therefore very high uptake rates), toxicity can occur at low bioaccumulated concentrations if detoxification capabilities are quickly overwhelmed or saturated by extremely high influx rates. Similarly, if the detoxification rate falls relative to the uptake rate as metal bioaccumulates, the MAF will also reach its critical level (toxicity) at lower and lower total bioaccumulated concentrations. For this

reason, experiments that include extreme exposures do not yield consistent critical body burdens even within a species (Jarvinen and Ankley, 1999).

The sometimes weak linkage between bioaccumulation and toxicity thus suggests that application of a BMG as a measure of risk will be complex in many, but not necessarily all, circumstances.

17.2.2.1 Inappropriate applications of a BMG

Prologue. It will be complex to implement a BMG for metals that are detoxified and stored in high concentrations. A BMG for one species will probably not be applicable to others. Within sample variability (among individuals) is also higher for bioaccumulated metal than for replicated samples of dissolved metals, for example. This may influence the precision with which a guideline can be enforced; but uncertainties are not greater than those imposed by the uncertainties in toxicity testing.

It is well known that some species accumulate extremely high concentrations of some metals, but detoxify them (e.g. Zn in barnacles: Chapter 7). Thus what is a high accumulated concentration of Zn that might be used as a BMG in a barnacle would not be a reasonable BMG for other species like a mussel (Chapters 7 and 8). Toxicity in mussels occurs at much lower bioaccumulated Zn concentrations that those ever seen in a barnacle (Chapters 7 and 8). Thus it is inappropriate to expect one BMG to cover species that differ so greatly in the biodynamics governing their bioaccumulation of Zn.

For the same reasons, a single critical body burden for the onset of toxicity would not be expected for metals that are detoxified and retained as they are accumulated, because species differ so widely in how they employ this process. This situation is likely to be common for Ag, Cd, Cu, Pb and Zn (examples of metals for which granule formation has been documented).

For example, field data show how interspecific differences in biological detoxification result in very different internal concentrations of Ag at the point of toxicity (Figs. 17.8 and 17.9). A relationship between Ag exposure and inability to produce mature gametes was seen in two different field studies with two different species of bivalves

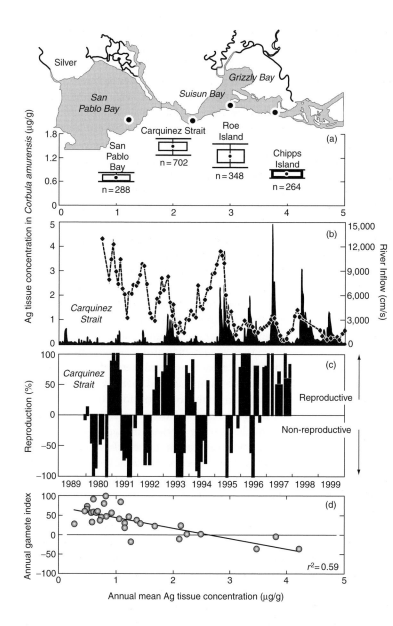

Fig. 17.8 (a) Ag concentrations in the bivalve *Corbula amurensis* as a function of location in San Francisco Bay, showing average enrichment of Ag concentrations in Carquinez Strait. (b) Changes in Ag concentrations in *C. amurensis* between 1990 and 1999, showing the overall decline in bioaccumulated Ag (and seasonal fluctuations related to river inflows). (c) Percentages of 10 animals that were reproductively mature (reproductive) and not mature (non-reproductive) in most months between 1990 and 1997, showing a higher frequency of non-maturity in years when Ag contamination was greatest. (d) An index of the annual average reproductive index at all sites, determined by subtracting the percentage immature from the percentage mature in each month and averaging for each year, as a function of annual mean Ag concentrations in the bivalve. The relationship illustrated the decline in proportion of animals that reached reproductive maturity when Ag contamination was greatest. Cross-cutting spatial and temporal sampling eliminated the likelihood that other sources of stress explained the change of reproductive maturity when Ag concentrations were high. An annual average of 50% of the animals reached reproductive maturity at 2 to 3 μg/g Ag dw. (After Brown *et al.*, 2003.)

(Hornberger *et al.*, 2000; Brown *et al.*, 2003). In both cases, Ag contamination from a local source dissipated as a source was removed. Adverse effects on reproduction were observed in both case studies when metal exposures were high, but declined as the clean-up progressed. In Fig. 17.9, 50% of *Macoma petalum* (as *M. balthica*) did not produce mature gametes when tissue concentrations were 45 μg/g Ag dw. The same proportion of non-reproductive

animals occurred at the very different tissue concentration of 3 μg/g Ag dw in the other bivalve, *Corbula amurensis*, in the second study (Fig. 17.8; Brown *et al.*, 2003). The higher tolerance of *M. petalum* to accumulated Ag probably occurred because this bivalve produced granules in its tissues that detoxify Ag. *Corbula amurensis* does not detoxify metals via granule production to the extent that *M. petalum* does (Wallace *et al.*, 2003), so toxic effects

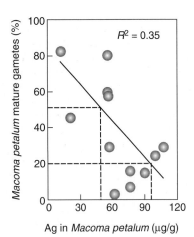

Fig. 17.9 The annual average percentage of the bivalve *Macoma petalum* that reached reproductive maturity between 1980 and 1992 as a function of Ag concentrations in the bivalve (after Hornberger *et al.*, 2000). Bioaccumulation and sections from reproductive tissue were collected monthly through multiple years when a period of Ag contamination was followed by recovery as Ag was eliminated from waste inputs. Ag concentrations were high during periods of input and reproduction was inhibited. Concentrations were low after recovery and reproductive capabilities returned. The percentage of animals with mature gametes was reduced to 20% at 90 µg/g Ag dw and to 50% at 45 µg/g Ag; a much higher threshold than seen in *C. amurensis* (Fig. 17.8).

are seen at much lower total accumulated body concentrations. A single BMG for Ag based upon bioaccumulation in either of the bivalve species would be inappropriate for the other species. Furthermore, a precautionary BMG of 3 µg Ag/g dw would protect *C. amurensis,* but would not be enforceable broadly, because these concentrations occur in *M. petalum* in uncontaminated (and thus unaffected) environments.

Furthermore, some metal concentrations in certain tissues (and in some cases whole bodies) are regulated by some animals to a relatively constant concentration irrespective of metal availabilities in the environment (Chapter 7). At least some arthropods regulate Zn in this way; a few species regulate Cu and some species lose metals like Cd and Se very rapidly. Similarly, Cd, Cu, Ag, Zn or Pb do not necessarily accumulate in proportion to exposure

(or with detectable efficiency) in the muscles of fish. The portal vein carries blood directly from the digestive tract to the liver in fish. Detoxification mechanisms appear to trap these metals in the liver before they can pass to the muscle. Concentrations in liver may be related to exposure, but concentrations of these metals in muscle are not (Wiener and Giesy, 1979). Nor are muscle concentrations indicative of risks for toxicity. If regulation of bioaccumulated metal concentration occurs (at body or tissue level), changes in bioaccumulated metal concentration (at body or tissue level) might not be detectable at the onset of toxicity. If regulation breaks down (and bioaccumulated concentration rises: Chapter 7) before toxicity ensues, however, then a BMG (higher than the usual regulated concentration) might be possible for such a species. But there is only a narrow window of exposure before toxicity ensues and the animal disappears in those circumstances.

Route of uptake may also affect accumulation of the MAF of bioaccumulated metal. Hook and Fisher (2002), for example, found that dissolved Ag accumulated to high concentrations in zooplankton without causing toxicity; primarily because it accumulated adsorbed to the exterior of the animals. An internal MAF accumulated slowly during dietary exposure of the zooplankton, however. Ag from the dietary source correspondingly exerted effects on reproduction at much lower (internally accumulated) body concentrations than did dissolved Ag (higher proportion of total body concentration not internalised). Thus a whole tissue BMG might understate or overstate risks for some metals and some species.

Precision also is an issue in implementing a bioaccumulated metal guideline. Variability among species and variability within a species are typical of bioaccumulated metal concentrations. Sample size for each species is a practical consideration. Statistical power is weak if only a few individuals are analysed (e.g. <10). Characterisation of a PEC thus must include consideration of intraspecific variability as well as interspecific variability. Differences in guidelines that lie within the range of variability will not be separable when monitoring compliance (e.g. Box 17.4).

Box 17.4 Precision in selection of a BMG

The only specific BMG in regulatory use is the USEPA proposal of a chronic tissue residue criterion for whole fish of 7.8 μg Se/g dw. A controversy developed between government institutions when US Fish and Wildlife Service advocated a fish criterion of 5.8 μg Se/g dw. DeForest *et al.* (1999) disagreed with both, criticising the use of the data used in developing the criteria. These authors used a selected set of toxicological studies that were consistent with USEPA methodology to suggest a 6 μg Se/g dw criterion for cold water anadromous fish and a 9 μg Se/g dw criterion for warm water fish. In fact, the differences between all these suggestions are remarkably small especially compared to some of the differences in existing water quality criteria.

To illustrate this concept, Fig. 17.10 shows Se concentrations in four important resource and predator species in San Francisco Bay. The variability among species and the variability among samples within a species (in some cases) is greater than the variability between the proposed guidelines. Species-specific experimental studies suggest the concentrations of Se in sturgeon (*Acipenser transmontanus*) from the Bay could inhibit successful reproduction (Linville, 2006); concentrations in Dungeness crab (*Cancer magister*) are also high enough to be of concern. Either standard would protect these species. Deformed young Sacramento splittail (*Pogonichthys macrolepidotus*) have been found in Suisun Marsh (Stewart *et al.*, 2004), a finding consistent with experimental studies that suggest deformities should be expected at the concentrations found in these fish (Teh *et al.*, 2004). No indications of adverse effects in striped bass (*Morone saxatilis*) have been documented.

All the above choices of BMG for Se, it appears, would provide a sensitive, but realistic, warning level, for vulnerable species. Issues such as choice of species to implement the standard, number of organisms sampled, constraining sampling to a single habitat or time are more influential in evaluating compliance than is the choice of either exact value for the BMG for Se.

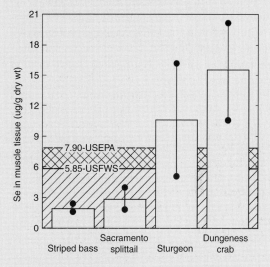

Fig. 17.10 Concentrations of Se in muscle tissue of four species in northern San Francisco Bay in October 1999 (data from Stewart *et al.*, 2004). Horizontal lines are bioaccumulated metal guidelines proposed by different regulatory agencies. Variability in the Se concentration within species and among species is much greater than between the different proposed guidelines, although it was the latter difference that was the focus of debate about the guideline.

17.2.2.2 Appropriate applications of a BMG

Prologue. BMGs can be implemented if means can be found to account for the uptake of the metabolically available fraction of metals or for trace metals that are not detoxified and strongly accumulated in that form. Calibration of bioaccumulation in a tolerant biomonitors against effects in sensitive species may offer a powerful alternative approach to the BMG concept.

Even where metals are detoxified to some degree, a BMG may reduce uncertainties compared to traditional approaches. For example, bioaccumulated concentrations of Cd were determined at mortality after a 4 to 10 week exposure to dissolved Cd in the freshwater amphipod *Hyalella azteca* (Borgmann and Norwood, 1995; Borgmann *et al.*, 2004). The experiments were conducted in geochemically different waters, so the LC_{50}, based upon dissolved concentrations, varied >35-fold. But the bioaccumulated concentration at lethality varied only threefold. There was not one critical body concentration at which the amphipods exhibited toxicity, but the variability in the bioaccumulated concentration associated with lethality was less than the variability associated with changes in speciation.

Biological and ecological considerations are important in those circumstances where a BMG is feasible. In defining environmental concentrations for a water quality criterion, water samples are straightforward to collect and interpret. High-quality analyses and speciation are the most difficult issues. In implementing a BMG, analysis problems are less formidable, but the choice of species and interpretation of the bioaccumulated concentration require care.

Implementation of a BMG for some trace metals could require consideration of metabolically available metal accumulation, rather than total metal accumulation. The rate-based toxicity concept is at the core of determining how and where a bioaccumulated metal concentration might be employed as an integrative measure of risk. The model is similar to the Biotic Ligand Model (BLM) (Chapter 15) in the recognition that metal accumulation at the site of action causes adverse effects, not total metal accumulation. In the case of the BLM, the site of action is considered to be transport sites on the gill. Metal bioaccumulation from the diet, metal loss and metal detoxification are not accounted for in the BLM, but they could be in an appropriately defined BMG.

It is possible to identify operationally a fraction of metal that is metabolically available and assess rates of accumulation versus rates of loss and detoxification. Only a few studies have attempted to validate such schemes against toxicity, however. Cu associated with high and low molecular weight proteins, after removal of metallothionein-like proteins, was correlated with reduced growth in crab larvae (Sanders and Jenkins, 1984). Accumulation of 1 to 2 $\mu g/g$ Cd dw in soluble protein was associated with a significant loss of the ability of the shrimp *Palaemonetes pugio* to capture its prey (Wallace *et al.*, 2000). In this case, Cd accumulated more rapidly in soluble non-metallothionein protein than in metallothioneins as Cd exposures increased. A more detailed operational scheme was also proposed in which the bioactive fraction of metal (termed the metal sensitive fraction) was

purported to include non-metallothionein, soluble protein plus metal associated with organelles (Wallace *et al.*, 2000). Accumulation rates in those fractions have not been correlated with or calibrated in toxicity tests, however. Cain and Luoma (1999) also showed that it is feasible to monitor a fraction of metal in different species and compare bioaccumulation in the different fractions compared to a measure of exposure like total tissue or cytosolic metal concentration.

It might also be possible to model metabolically available metal accumulation from quantification of critical physiological processes. For example, recent studies have compared Cd dissolved uptake rates, loss rates and detoxification capability among species of aquatic insect larvae (Martin *et al.*, 2007). A model was constructed that estimated steady state accumulation of metabolically available Cd at an exposure concentration of 1 $\mu g/L$ Cd. Based upon physiology alone, both bioaccumulation and the concentration of the MAF can differ by 1000-fold among species in the same environmental circumstances. The ranking of potential sensitivity is determined by the proportion of the bioaccumulated body burden that is not detoxified, not by the total bioaccumulated concentration alone. The advantage of this technique is that the ranking can be calibrated against ecological change in a metal contaminated environment. Mayfly species with the greatest tendency to accumulate Cd in the MAF are the same species that tend to disappear from metal-contaminated streams.

Not all trace metals are detoxified and stored or accumulated to high concentrations. Se is an example of a trace metal for which a relatively consistent relationship is found between bioaccumulated concentrations in various organs, eggs or young life stages of the receptor organism, and measurable toxic responses. It would appear, therefore, that bioaccumulated Se is either not detoxified, or that there is a consistent relationship between concentrations of total accumulated Se and concentrations of Se in metabolically available form at all accumulated Se concentrations. Granule formation and conjugation by metallothioneins are not known for Se. In the vertebrate liver, Se attaches to vitellogenin, which eventually ends up in the egg. Se–vitellogenin interferes with successful development of the egg to an embryo and/or causes deformities in the embryo. When the contaminant occurs in high enough concentrations in the egg or developing organism (strictly in metabolically available form as opposed to total concentration, although the two seem to be directly related in the case of Se), it disrupts development. Se concentrations in eggs are therefore thought to be the most consistent criteria by which to judge risks to various species of birds. The effectiveness of judging risks to fish and birds varies with the tissue used, but concentrations in eggs and juveniles are the best measures.

There are differences between the bioaccumulated concentration of Se that signifies toxicity among bird species, but those differences are only about two- to fourfold (Box 17.5). The variability in a BMG for Se thus appears to be manageable, if relegated to appropriate tissues in appropriate species.

A range of values (Chapman, 1995) might be a better indicator of Se risk than an absolute value if there is variability among species in the bioaccumulated concentration linked to toxicity. A risk guideline based upon a range of values would implicitly quantify the uncertainties about the choices of risk because the boundaries of uncertainty are clarified. Box 17.6 presents an estimate of four risk levels for Se for fish and birds, determined from dietary toxicity tests and experience in natural waters. A political choice balancing precaution and potential over-protection will be necessary in deciding specific actions from such ranges. But the scientific credibility of the approach would be improved.

The range of variability in the bioaccumulated concentration at the onset of toxicity has not been systematically studied for most metals. We could speculate that detoxified storage mechanisms may be rare for forms of metals to which organisms have not been exposed in their evolutionary history. In those cases, interspecific variability could be low. For example, a BMG seems reasonably feasible for the synthetic organo-metal tributyl tin. Other metals may be difficult for organisms to detoxify and/or are reproductive toxins (e.g. Se and MeHg).

For these a BMG based upon concentrations in reproductive tissues might be feasible.

For conditions where the balance between loss/detoxification processes and uptake rates is not disrupted during different exposures of the same species, a species-specific BMG might be possible. In Fig. 17.8, bioaccumulated Ag correlated strongly with increasing reproductive failure within each bivalve species. The onset of toxicity correlated with a particular threshold uptake rate which presumably overcame a combination of detoxification and excretion. Accumulated concentration was the indicator, if not the proximate cause, of toxicity within but not between species.

Another alternative is to calibrate a tolerant biomonitor against responses in sensitive species. A requirement of a biomonitor (Chapter 8) is that it should be a species that is sufficiently tolerant to survive exposure to high availabilities of the metal and ideally it should be a strong bioaccumulator. Bioaccumulated metal concentrations in a suitable biomonitor species might be feasible to calibrate with ecological stress in the more sensitive aspects of a community. For example, heptageniid mayflies in rocky bottom rivers are both the strongest bioaccumulators of metabolically available Cu and Cd in streams (Martin *et al.*, 2007) and the most sensitive of the local benthic species to those metals. If high bioaccumulated metal concentrations are found in mayflies in a stream, and detoxification capabilities are known, it is feasible to obtain an indication of risk (Chapter 11). But a major issue is that physiologically sensitive species will probably not be abundant or widespread along a metal gradient.

Hydropsychid caddisflies are widespread in metal-contaminated streams and are also excellent biomonitors of metals like Cd and Cu (Cain *et al.*, 1992; Gower *et al.*, 1994). The bioaccumulated concentration at which hydropsychid caddisfly larvae are stressed is not the same as the bioaccumulated concentration at which heptageniid mayfly larvae are stressed. But there is a correlation between the bioaccumulated concentration in hydropsychids and the abundance of heptageniid mayflies (Maret *et al.*, 2003) (Fig. 17.11). The concentration in *Hydropsyche* sp. at the onset of the decline in

Box 17.5 Differences in bioaccumulated concentrations of Se that correspond with the onset of toxicity (or are non-toxic) in several species

USEPA's chronic Se toxicity value (concentration in whole fish tissue) derived from 'acceptable tests' with chinook salmon (*Oncorhynchus tshawytscha*), rainbow trout (*Oncorhynchus mykiss*) and bluegill (*Lepomis macrochirus*) range from only 9 to 16 µg/g tissue dw. But the chronic value for fathead minnow (*Pimephales promelas*) in USEPA 'acceptable experiments' is 51.4 µg/g tissue dw, if food avoidance is excluded as an endpoint (if not, 6 µg/g tissue dw is the chronic value). Lemly (1993a, 1996) compared effects thresholds from different species of fish from the episode of Se toxicity in Belews Lake. 'Critical body concentrations' were similar enough to develop recommendations for bioaccumulated Se standards that were much less uncertain than 100-fold difference in Se water quality criteria (an indicator of the uncertainty of that

approach). Differences in the body concentration of Se that corresponded to a toxic response in fish varied about twofold among the few species for which a BMG was tested (Table 17.2).

Similarly, among wading birds, avocets (*Recurvirostris avosetta*) tolerate twice the body burden of Se as stilts (*Himantopus himantopus*); and both are more tolerant than mallard ducks (*Anas platyrhynchos*). But, mallards are exposed to less Se than the other species, because of their diet. Most bird toxicity tests use mallards as the surrogate species because of their sensitivity.

Instances of very high Se concentrations occur in a few top predators, where exposure to Se contamination is unlikely. Giant petrel (*Macronectes* sp.) from presumably pristine arctic and antarctic habitats have the highest Se concentrations known for any birds in the wild. Their tissue concentrations (101 to 136 µg Se/g dw: Gonzalez-Solis et al., 2002) are associated with reproductive failure in other species; but there is no evidence that these concentrations are the result of

Table 17.2 *Relationship between tissue concentrations of Se and the onset of adverse effects in a variety of fish species, from a synthesis of field studies (Lemly, 1993a, 1996). An overall effects threshold was recommended by integrating all studies; in this case by interpolation to a No Effects Level from the results of all the studies. The purpose of the table is to illustrate that despite some variability in the onset of impairment among species, justifiable guidelines can be derived. It is also important to note that tissue residue concentrations at which effects occur differ among organs. Both the degree of interspecies variability and the differences among organs must be considered in setting standards and in evaluating risk*

	Tissue	Concentration at earliest effect (µg/g Se tissue dw)	Effect studied	Number of published studies
Species				
Five species[a]	Whole animal	6–24	Mortality,[b] reproduction, growth	12
Bluegill	Liver	29–34	Mortality,[b] reproduction	2
Striped bass, Bluegill	Muscle	14–20	Mortality,[b] reproduction	3
Bluegill	Ovaries	18–34	Reproduction	4
Lemly (1996)	Whole body	4	Effects threshold	
recommendation	Liver	12	Effects threshold	
	Muscle	8	Effects threshold	
	Ovary/egg	10	Effects threshold	

[a]Bluegill (*Lepomis macrochirus*), striped bass (*Morone saxatilis*), rainbow trout (*Oncorhynchus mykiss*), chinook salmon (*Oncorhynchus tshawytscha*), fathead minnow (*Pimephales promelas*).
[b]Juvenile or larval mortality.

anthropogenic sources or that petrel populations are adversely affected. High Se concentrations occur in muscle and feathers of oystercatchers (*Haematopus ostralegus*), from the Wadden Sea, the Netherlands (Goede, 1993). Where Se concentrations in muscle tissues are high, Se in the eggs are below toxicity thresholds. The mechanism of tolerance that the oystercatcher uses is the avoidance of transfer of Se to the eggs. At least some species, therefore, have evolved mechanisms to tolerate high Se in their tissues. Data from surrogate or sensitive species cannot be extrapolated to effects in these species.

Box 17.6 Ranges for BMG and DMG for Se

Although relationships of diet, fish tissues and bird tissues with toxicity have been studied in the field and laboratory, there is not a simple consensus among toxicologists on a precise value for dietary or bioaccumulated Se concentrations that would be protective. However, the frequencies and intensity of effect within a relatively narrow range of exposures are relatively consistent within the body of literature. For the purposes of this discussion we will define four categories of risk. R1 to R4 will be used to designate ascending risk (highest risk at R4). We give the range of concentrations for each measure that various authors cite as supported by the literature. All data reported here are presented as dw, assuming ww values were corrected for 80% water.

R1: No adverse effects likely
DMG All prey: ≤3 μg Se/g dw in prey
BMG Fish: <3 μg Se/g dw, whole body
BMG Birds: <4 μg Se/g dw

R2: Some effects on individuals detectable. Effects on sensitive populations possible but uncertain
Deformities, reproductive difficulties and mortality of larvae were observed at the low end of this range, but not in all circumstances. The probability of toxic effects, in at least some species, increases toward the higher end of the range. Experiments are least ambiguous toward the higher end of the range. Proposals for criteria fall within this range.
DMG All prey: 4–10 μg Se/g dw
BMG Fish tissue: 4–10 μg Se/g dw
BMG Bird eggs: 6–15 μg Se/g dw
Some specific guidelines suggested include:

- 2–5 μg/g dw diet of fish (mainly based on 50% mortality at 5.1 μg/g dw diet in a laboratory study

of winter stress in juvenile bluegill (*Lepomis macrochirus*): Lemly, 1993a);
- 3.6 to 5.7 μg/g dw diet in birds (based on 4.87 μg/g dw diet in birds, 10% effect level in mallards, hatchability; confidence intervals 3.56 to 5.74 μg/g dw: Ohlendorf, 2003);
- 4–6 μg/g dw whole-body fish tissue (based on 4.5 μg/g dw fish tissue whole body 10% effect level (the 40–50% mortality is the 5.8 μg/g dw whole body level) mortality in winter stressed bluegill: Lemly, 1993a);
- 6 μg/g dw bird egg (based on 3% hatchability effect level in sensitive species, stilts (*Himantopus himantopus*): Skorupa, 1998, 1999);
- 6.4–16.5 μg/g dw in birds eggs (based on 12.5 μg/g dw in eggs, 10% effect level in mallards (*Anas platyrhynchos*), hatchability, confidence intervals 6.4 to 16.5 μg/g dw: Ohlendorf, 2003).

R3: Deformities in some species and reproductive failure in sensitive species
Risk of eventual extirpation of a sensitive population is high. But correlations between tissue concentration and effects can be highly variable in some circumstances; probably because of differences in sensitivity among individuals and species. Population dynamics are crucial in probability of extirpation. Long-lived, low-fecundity populations that are exposed to high concentrations in their food web could be lost (e.g. sturgeon Acipenser transmontanus) *in San Francisco Bay)*
DMG All prey: 10–20 μg Se/g dw
BMG Fish tissues: 10–40 μg Se/g dw
BMG Bird eggs: 20–40 μg Se/g dw egg

R4: Expect evidence of gross deformities and examples of reproductive failure
Probability of multiple populations collapsing increases above these concentrations.
DMG All prey: >20 μg Se/g dw
BMG Fish tissues: >40 μg Se/g dw
BMG Bird eggs: >40 μg Se/g dw

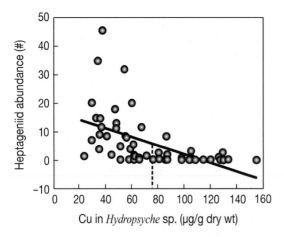

Fig. 17.11 The accumulation of high body concentrations of Cu by the caddisfly larvae *Hydropsyche* sp. is associated with a decrease in the sympatric abundance of heptageniid mayfly larvae in Cu-rich streams. A high accumulated Cu concentration in the Cu-tolerant *Hydropsyche* can therefore be used as an indirect Bioaccumulated Metal Guideline (BMG) indicating the presence of high enough local bioavailability of Cu to have an ecotoxicological effect on the more Cu-sensitive heptageniid mayfly larvae. (After D. J. Cain *et al.*, unpublished observations.)

abundance of heptageniid mayflies might be an appropriate BMG; a version of a surrogate species approach. An appropriate surrogate biomonitor in lakes might be the widespread insect larvae of the genus *Chaoborus* (Hare and Tessier, 1998); in marine and estuarine environments it might be one or more of the biomonitors for which atypically high bioaccumulated concentrations have been identified in Chapter 8. Further testing of such concepts is required.

17.3   WILDLIFE CRITERIA

Prologue. Wildlife criteria are based upon dietary toxicity studies with birds and fish. They include rate-based toxicity assessments and species-specific attributes. Existing protocols are weakened by reliance on BAFs and assumptions of 100% assimilation efficiency of metals from food.

Wildlife criteria were meant to represent a biologically justifiable approach to protecting a variety of

upper trophic level species using the dietary toxicity data from a few species. Although they were primarily developed for organic chemicals, they have been used to manage Hg in specific environments (e.g. the Great Lakes, USA: USEPA, 1995b). An advantage over traditional toxicity tests is that they specifically address dietary exposure. Few such criteria have been implemented, however.

Wildlife criteria assume that the important factors in an animal's exposure are daily intake rate of food and metal concentration in food. Unfortunately the existing approaches usually assume that assimilation efficiency is the same across all foods (100%) and that rate constants of loss are the same for all wildlife. The latter two assumptions will lead to serious over-protection errors in such criteria for most metals. But it is quite feasible to correct those errors with relatively simple experiments (Chapter 7). For the criteria the dietary concentration associated with the onset of toxicity (often termed the reference dose or RfD) is calculated as a rate of contaminant ingestion (e.g. µg metal ingested/d). Toxicity occurs when the rate of metal ingestion overwhelms the ability of the predator to eliminate and/or detoxify the trace metal. To make the determination species-specific food ingestion rates and body weights are considered for the species of interest (Box 17.7).

The advantages of wildlife criteria are their relevance to nature and their use of species-specific biological knowledge when it is available. The latter is also a disadvantage if ecological data are difficult to obtain for some of the species of greatest interest; so surrogates or universal values must be used. Interestingly application factors are rarely applied, probably because the dietary toxicity tests have fewer uncertainties than the dissolved tests other criteria are based upon.

17.4   HOLISTIC OR LATERAL APPROACHES TO MANAGING METAL CONTAMINATION: WEIGHT OF EVIDENCE

Prologue. Use of multiple lines of evidence reduces uncertainties, especially if the lines of evidence include

Box 17.7 Wildlife criteria: Se

Wildlife criteria define the concentrations of metal in food and the rate of trace metal intake that is associated with the onset of toxicity in a feeding animal. To determine if the rate of ingestion in a particular situation exceeds the dose that would cause toxicity (i.e. the predator is at risk), food ingestion rates, body weights and Se concentrations in food are considered.

$$([Se_{diet}] * \text{mass of food ingested per day})/$$
$$\text{body weight} = \text{dose}$$

or

$$(\mu g\ Se/g\ food * g\ food/d)/g\ \text{body weight}$$
$$= \mu g\ Se/g\ \text{body weight}/d$$

Toxicologists have developed tables describing mass of food ingested per day and typical body masses for different wildlife species. These values can be inserted into the equation to yield a dose at any concentration of Se in diet. This dose is then compared to the reference dose (dose at which toxicity occurs) to evaluate risk.

The critical choices in applying wildlife criteria are the dietary reference dose against which to compare the outcome of the equation, the Se concentration in food and the species-specific biological data. What is not always recognised is that the uncertainty in selecting values for ingestion rate (which is difficult to evaluate experimentally) and Se concentration (as noted above this is prey-dependent in natural waters) are probably as great or greater than the uncertainty in selecting a reference dose for comparison.

USEPA (1993) defined the toxicity threshold for wildlife as ingestion of 200 to 250 μg Se/g body wt/d, using the level that causes growth inhibition in small mammals. The no effects level (NOEL) for Se in food, ranged from 4 to 25 μg Se/g dw. Natural foods were rarely used in their experiments. USEPA (1993) concluded that the NOEL for reproductive effects in birds is 0.078 μg Se/g body weight/d.

These values can be converted to a species-specific acceptable dose (dose that should not be exceeded) by rearranging the first two equations, using species-specific biological values.

$$RfD_{species} = (NOEL * FI_{species})/(BW_{species})$$

where NOEL represents the dietary threshold for toxicity as determined in a surrogate species in μg Se/g$_{food}$, FI represents mass of food ingested per day by the specific species in g/d, and BW$_{species}$ is the body weight of the specific species (in g). RfD is the maximum dose that species can ingest per day and avoid toxicity.

Wildlife criteria, in some approaches, are converted to water criteria, by using a bioaccumulation factor (BAF). For a universal criterion, this approach assumes that the same BAF applies across all species and all environments (many speciation regimes); an assumption that will add orders of magnitude uncertainty to any conclusions (Chapter 15).

---

observational as well as investigative data. Some formal guidelines exist as to how to employ multiple lines of evidence. Ultimately, it involves recognising the linkages among dissolved, particulate bound and bioaccumulated metal concentrations. Integrative risk assessments use weight of evidence from experiments, toxicity tests, geochemistry and field observations.

Holistic management uses multiple indicators of contamination to determine environmental concentrations (PEC); optimally, effect concentrations (e.g. PNEC) are considered from several media (water, sediment, bioaccumulated concentrations). *Integrated risk assessments* consider data from toxicity tests, environmental chemistry, biology and field observation. Conclusions about risk are derived from the integrated understanding and weight of evidence (WOE) (Box 17.8).

Policy decisions derived from a single approach are plagued by uncertainties. If the circumstance is complex and/or enough is at stake, it is more likely that multiple studies and multiple lines of evidence represent the shortest road to a body of evidence that can justify a trustworthy policy. Toxicology, geochemistry and biological/ecological observations are powerful ingredients in combination in a data set; much more powerful than any one set of data alone. Implementation of holistic management is impeded to the extent that regulators and risk assessors avoid use of field evidence,

Box 17.8 What is weight of evidence (WOE)?

The weight-of-evidence approach (WOE) is defined as the process of combining information from multiple lines of evidence. Applications include assessments of impairment, prioritisation of site contamination, or decision making on criteria or other management actions. The WOE process can help determine the extent of pollution, or its ecological significance. It can contribute to defining the optimal remediation approach or the urgency of corrective actions. The WOE process is often used within the context of ecological risk assessments.

The lines of evidence employed can include (but are not restricted to):

- chemical determination of compliance with criteria, standards and guidelines, as well as comparison to reference conditions;
- site chemistry to characterise speciation and other processes affecting bioavailability;
- comparisons of laboratory-based toxicity responses of surrogate organisms (e.g. survival, growth, reproduction) between test site samples and reference sites and controls;

- comparisons of bioaccumulated metal concentrations in local biota with the same species from reference sites or literature values;
- *in situ* toxicity and bioaccumulation experiments;
- behavioural or biomarker responses;
- comparisons of indigenous populations between test and reference sites;
- comparisons of habitat conditions and stressors that might affect ecological relationships;
- comparisons of ecosystem function;
- evaluations of model predictions of fate, exposure and/or effects to a gradient, reference sites or literature values.

Effective weight-of-evidence evaluations must include both observations from natural waters and experiments. Stakeholder concerns, availability of data or resources for further study all differ among situations. Environments vary widely in character, available expertise is always limited and abilities or facilities to execute a plan may also vary. The choices of lines of evidence (LOE) are affected by all these considerations. The Sediment Quality Triad (Long and Chapman, 1985) and the consensus-based approach are examples of published approaches providing formal guidance on applying weight of evidence (Burton *et al.*, 2002).

understanding of biological/ecological processes and complex (e.g. *in situ;* mesocosm) toxicity test data.

In reality, every body of evidence has its own emphasis among the possible approaches. The individual lines of evidence retain their assumptions, strengths and limits, which must be taken into account. Weight of evidence is a systematic approach to incorporating the combination of available knowledge into the decision at hand. If the methods all yield similar conclusions, confidence in the conclusion is greatly increased. If different methods yield different conclusions, it is more constructive to identify the basis of the discrepancy than to debate which data are the most correct (Chapter 18). Dialogue about why interpretations differ will identify studies that might address discrepancies.

Formal processes are available for reducing different lines of evidence to numerical values or otherwise quantitatively weight evidence. But, ultimately, the process of weighting the lines of evidence incorporates judgements about the quality, extent and

congruence of the information in each line of evidence. An effective weighting process is transparent, systematic, logical, justifiable, and multidiscipline. It considers all the evidence fairly and explicitly identifies and balances the important uncertainties (Burton *et al.*, 2002). One temptation is to use formal processes to reduce multiple lines of evidence to a single numeric describing the situation. Experienced observers suggest too much information is lost in such exercises; and they can be misleading (Burton *et al.*, 2002).

17.4.1   Lateral management

Prologue. An integrated evaluation of multiple criteria may make risk management strategies more holistic. But integration can be complex. Extrapolating back to the water criteria as the final policy, for example, retains at least some of the original uncertainties.

17.4.1.1 Site-specific criteria developed from added lines of evidence

Site-specific water quality objectives represent a formal alternative means of overriding national water quality criteria at a specific location, at least in the USA. For example, the California Regional Water Quality Control Board adopted a policy for implementing toxics standards that concluded that site-specific objectives (SSOs) are appropriate, if (a) a criterion or objective is not achieved in the receiving water, and (b) there is a demonstration that the discharger cannot be assured of achieving the criterion or objective and/or effluent limitation through reasonable treatment, source control and pollution prevention measures. Site-specific objectives were deemed necessary for a waterbody if (a) the current objectives are not being consistently met despite reasonable treatment, source control or pollution prevention measures; (b) the chemical features of water body or source reduce the toxicity and bioavailability of the contaminant below that of the standardised toxicity test-based value; (c) an impairment assessment conducted for the specific system demonstrates that the established water quality objectives could be relaxed while still fully protecting beneficial uses; and (d) further reductions in loading will be difficult and costly and will not provide corresponding water quality improvements. European Directives allow similar caveats in implementation of contaminant regulations, although the terminologies are different.

Multiple lines of evidence are typically employed to help make water quality guidelines more site specific (e.g. DeForest *et al.*, 1999; Fairbrother *et al.*, 1999). For example, Brix *et al.* (2005) recommended a technique for developing a site-specific ambient water quality Se criterion using dissolved Se and bioaccumulated concentrations (Box 17.9). They cited several advantages to their approach: limited data collection, two measures of compliance, use of a broad data set to interpret local data and accounting for variability. A major disadvantage was the perceived need to extrapolate the bioaccumulated standard back to a water standard. The weak correlation between bioaccumulated and dissolved Se in the field data added to uncertainties about the new standard. The method also did not address the question of which species might be most threatened.

Extrapolation of a bioaccumulated metal concentration back to a water standard remains the desired approach by the regulatory community, because of the precedent for protecting aquatic environments with ambient water quality guidelines. But extrapolation from a tissue concentration to a water concentration requires assumptions about speciation, transformation and bioaccumulation. Thus setting a water standard from tissue concentrations could add back some of the uncertainties

---

Box 17.9 Development of a site-specific water quality standard using minimal local data collection

Brix *et al.* (2005) suggested a three-tiered approach to determine what data were necessary for a site-specific standard:

(1) If Se concentrations in water were <5 µg/L Se (the ambient chronic water quality criterion for Se in freshwater), then no further action was necessary;

(2) If Se concentrations were greater than 5 µg/L Se but bioaccumulated concentrations (fish and birds, respectively) were less than 7.9 and 13 µg Se/g dw, then again, the site was in compliance and no further analysis was necessary;

(3) If both the water and tissue criteria were exceeded then the statistical method was used to calculate what the water criterion should be. For example, if fish tissues at the site were 64 ± 1 µg Se/g dw, and water concentrations were 1 µg/L Se, then site-specific criteria would have to be well below 1 µg/L Se to bring fish tissues down to 7.9 µg Se/g dw. If, on the other hand, fish tissues had a mean of 10 µg Se/g dw at dissolved concentrations of 64 µg/L Se, then a dissolved standard not much below 64 µg/L Se would be allowed (not much change necessary to reach 7.9 µg Se/g dw). When variability in the fish data was great, the site-specific criterion was adjusted downward. A more careful approach was called for when even a few high values for bioaccumulated Se were observed.

of the traditional water standard unless care is taken to avoid that.

The added lines of evidence in development of a site-specific objective often use metal speciation to account for effects of speciation on bioavailability. For example, a site-specific Cu criterion was developed for South San Francisco Bay (USEPA, 2003) in conjunction with a programme of pollution prevention measures. The scientific data formally used to justify the site-specific objective were a Water Effects Ratio study (Chapter 16), comparing toxicity in site water to toxicity in standardised laboratory waters. That study showed that Cu toxicity was 0.3 to 0.5 of toxicity projected in standardised tests. As a result, the ambient water quality objective was raised from 3.1 μg/L Cu to 6.9 μg/L Cu. The WER approach alone is a typical unilateral justification for a site-specific objective. But because it relies upon toxicity testing alone, it would not have resulted in a scientifically convincing policy in South Bay. However, other data were also available that all pointed to low Cu bioavailability in South San Francisco Bay. The multiple lines of evidence from those studies illustrate how a convincing body of evidence might be built:

- More modern and more sophisticated tools suggested Cu in the Bay waters was strongly complexed. Labile Cu concentrations were determined with advanced techniques and showed that free Cu ion concentrations formed an unusually low proportion of total Cu in this water body (Donat et al., 1994; Buck and Bruland, 2005).
- The estimates of free ion Cu were orders of magnitude below the free ion activities that caused toxicity to the most sensitive types of marine algae in continuous culture toxicity tests (Brand et al., 1986; Chapter 11) and in field studies using related techniques (Moffet et al., 1997; Chapter 11).
- Model forecasts showed that Cu^{2+} would remain well below toxic levels even if concentrations rose to the level stipulated in the new objective.
- The low bioavailability of Cu was further verified by a lack of Cu removal during phytoplankton blooms in the water body; other metals (Cd and Ni) were taken up by phytoplankton and removed from solution during the bloom (Luoma et al., 1998).

- Biomonitoring with bivalves showed that Cu concentrations had declined dramatically from the highly contaminated conditions that had been documented in the 1980s. Cu concentrations in bivalves were as low as found in very slightly contaminated environments when the new objective was implemented (Moon et al., 2005).

In this case a combination of experiment, sophisticated geochemistry, realistic toxicity tests results and relevant field observations (biomonitors and toxicity) supported the policy decision.

17.4.1.2 Using weight of evidence from a water body to establish a regulation

Prologue. The data least utilised in developing criteria are comprehensive data from natural waters. Dose–response studies in the field are possible but require large data sets from a variety of sites. The range of exposures is more difficult to control in the field, so precision in identifying the point of toxicity can be a source of controversy. Attributing adverse effects solely to the toxicant (attributing cause and effect) can also add uncertainty. Nevertheless, observations from, and experiments in, natural waters can be an effective way to address the realities of nature and test the validity of existing regulations.

Field data bring reality to evaluations of risk for trace metals. But validating or deriving criteria based upon field evidence is rejected by some as a line of evidence. Chapman (1995) states: 'field studies can never validate laboratory studies since there is no certainty that effects observed (or not) in the field result from mechanisms, effects or phenomena observed in the laboratory and correlation is not necessarily related to cause and effect'. While this statement is usually true for individual studies, development of an appropriate body of evidence can greatly reduce uncertainties (e.g. Chapter 11).

Other commonly cited reasons for not using (or collecting) data from natural waters include:

- High costs.
- Exposure-response data can be obtained but are less systematic from a body of field studies than from broadly designed toxicity tests. Interpolation or

extrapolation to determine 'no effect' levels can appear less certain than in experiments where exposures can be planned.

- Confounding circumstances are more difficult to control, than in a planned experiment (see Chapter 11).
- Detecting effects at low levels is also statistically challenging. Early field assessments of Se effects on birds (in Kesterson Reservoir) left no doubts about the cause of the deformities observed. But the data were all for very high exposures. The available data thus represented a combination of extreme effects and no effects (from uncontaminated environments). The spread of data was too coarse to predict with confidence the threshold at which toxicity would occur (Skorupa and Ohlendorf, 1991).
- Comparisons among multiple data sets can be complex. In the Kesterson experience, there was also an uneven embryo sampling effort among studies; different bird species were evaluated in different studies, and the chemical environments were diverse. Later, federally funded studies assembled random data from >400 bird (stilt *Himantopus himantopus*) nests across the areas affected by Se-rich irrigation drainage in the American west. Skorupa (1998) derived a threshold for toxicity from these field data, relating effects to Se concentrations in eggs.
- Questions always arise about consistency, data quality and influences of other factors in large, statistically valid studies (Fairbrother *et al.*, 1999); but these uncertainties are not any greater than the uncertainties of the assumptions associated with use of toxicity testing data alone.
- A set of field studies may involve only one speciation regime, so extrapolation to the regimes of other waters raises uncertainties.

These are balanced by advantages of the field approach:

- Natural waters are the systems that risk assessment and criteria are designed to protect; so observations from natural waters are essential to validating criteria.
- Field studies have uncertainties, but the outcome from a appropriately designed effort can be uniquely useful as a first-order demonstration of levels of risk.
- The expense of conducting the field study is rarely weighed against the cost of a poor policy decision or a political stalemate (Cairns, 1986); it is unclear that the former outweighs the latter, especially where stakes are high.
- Multiple lines of evidence, especially if centred around known mechanisms whereby adverse effects are manifested, can reduce uncertainties about cause and effect.
- It is also unclear that uncertainties in careful field studies (Chapter 11) are greater than the uncertainties in toxicity testing.

The best use of comprehensive data from natural waters, of course, is as a complement to toxicity testing and geochemistry. Although the effort and the scale of the study necessary to derive a field-based criterion can be large, the result can be great reduction in uncertainties; especially when interpretations are combined with results from toxicity testing. This does not mean that controversies are eliminated, necessarily (e.g. Fairbrother *et al.*, 1999; Skorupa, 1999). But the combination of data greatly reduces the range of possible criteria and validates that the range in dispute is within the bounds of what is important in nature. Ultimately, a policy decision about a criterion value will be made; but that decision is much more likely to be valid if based upon a lateral evaluation of multiple lines of evidence.

Frameworks for weight-of-evidence analyses state that both observational and investigative data are essential for a comprehensive analysis (Burton *et al.*, 2002). Chapman (1995) concluded his discussion of validation with the conclusion that verification of toxicity results must be based upon a burden of evidence not simple comparisons. It is increasingly clear that the most competent evaluations of risk give equal weight to field observations and experimental data.

Field evidence has been used in regulation where toxicity data are obviously at odds with observations from nature. As noted elsewhere, field evidence was influential in the development of the freshwater Se criteria (Box 17.10). Most of the components of a useful field study were present in the Se fieldwork. Multiple lines of evidence

Box 17.10 Derivation of the freshwater ambient Se criteria

The 1986 USA Chronic Continuous Criterion for freshwater is 5 μg/L Se (USEPA, 1987); the same standard is used in Australia. These criteria were based, mostly, upon field studies at Belews and Martins Lake (Lemly, 1985, 1993a, 1997, 2002; USEPA, 1987). The study at Belews Lake, itself, is described in detail in Box 13.8. These field observations were used in developing the regulation because they clearly contradicted expectations from toxicity tests, as noted earlier. But there were several characteristics of the Belews Lake study that made it adequate for use in developing criteria.

- Clear diagnosis that Se was the cause of toxicity in the lake, based upon knowledge of inputs, comparisons to similar lakes and, eventually, partial recovery of the community once Se inputs were eliminated.
- Chemical concentrations, speciation, concentrations in sediments, bioaccumulation in invertebrates and predators, teratogenesis and changes in the fish community were all documented in Belews Lake, and compared to similar systems uncontaminated by Se.
- Demonstration that survival in dissolved Se toxicity tests did not give an accurate ranking of the sensitivity of fishes in the lake. The fish most tolerant to dissolved Se toxicity were some of the first to disappear from the lake.
- Experiments that showed that 'diet represents the primary route by which wildlife are exposed to Se in nature' and only dietary exposure could explain the Belews Lake results (Lemly, 1985). A plausible mechanism was therefore shown to explain why the observations from the lake differed from expectations of toxicity tests.
- Belews Lake included habitats with a variety of Se concentrations and a variety of effects. Because both were carefully documented in repeated studies, it was possible to at least estimate dose response.
- Recovery of populations in the lake also allowed reassessment of the validity of choices for protective values (Lemly, 1997).

Despite the careful study, the situation was sufficiently complex that varying proposals for ambient criteria were possible. USEPA concluded from that 5 μg/L Se was the concentration in a portion of the lake where no chronic effects were observed. The US Fish and Wildlife Service and the National Marine Fisheries Service suggested that 5 μg/L Se would be protective in only 50% to 70% of circumstances and that a 1 to 2 μg/L dissolved total Se standard was necessary to be 95% protective (Skorupa, 1998). But there were too few data from this one site to resolve the discourse with science alone.

The source of Se in Belews Lake was primarily selenite, and the 5 μg/L Se did not differentiate between selenite and selenate. Some authors claimed that the criteria based on Belews Lake would be greatly over-protective for environments where selenate is the form of Se release. Subsequent criteria (e.g. USEPA criteria for Great Lakes: USEPA, 1995b) accounted for the abundance of selenite, selenate and organo-selenium with a formula. However, only dissolved toxicity data were available to derive the formula. Therefore this approach was not amenable to using what was learned from the field experience.

Further details of this case study are presented in Box 13.8.

indicated that toxicity was occurring in more than one case. Evidence came from both birds and fish in different studies. It was clear that Se was the cause both at the time of each study and from evidence of recovery. And enough data were available to estimate dose–response (Skorupa, 1998; Lemly, 2002). The range of criteria in debate from these studies was, in the end, relatively small (2 to 5 μg/L). High-quality data from natural waters are increasingly available; and formal methods exist for conclusively linking cause and effect (Forbes and Calow, 2002; Chapter 18). *A priori* exclusion of field data, therefore, does not seem constructive in either risk assessment or criteria development.

17.4.1.3 Schemes for weighting lines of evidence in holistic criteria

Prologue. Schemes exist to numerically weight lines of evidence in order to facilitate risk decisions. The greatest danger is elimination of information in the numerical schemes, leading to oversimplified and therefore potentially misleading conclusions.

Table 17.3 *Lines of evidence in risk characterisation levels for Se in various media, after the hazard clasification of Lemly (1996) and the risk forecasting approaches of Luoma and Presser (2000). The ranges are those classified as R1 to R4 in Box 17.6*

	R1	R2	R3	R4
Water (µg/L Se)	Not clear	1–2	2–5	>5
Particulate or sediment (µg/g Se dw)	<1.5	1.5–3	2–4	>4
Prey species: invertebrate or forage fish (µg/g Se dw)	≤3	4–10	10–20	>20
Upper trophic level fish (µg/g Se dw)	≤3	4–10	10–40	>40
Birds (µg/g Se dw)	<4	6–15	20–40	>40
Signs of toxicity in the field	No toxicity likely	Detectable toxic signs are possible	Detectable toxicity is likely	Signs of toxicity and population effects detectable

Protocols exist for weighting lines of evidence when combining criteria from different parts of the environment (Burton *et al.*, 2002). Lemly (1998, 1999, 2002) proposed a protocol for Se risk assessment that incorporated measurements from several ecosystem components: water, sediment, benthic invertebrates, fish eggs or bird eggs. For each he devised five levels of risk based upon concentrations at the site. Each component was given a numerical score based upon its risk category (1 to 5). The sum of the values for each component gave a single number to characterise risk for the water body. Table 17.3 uses risk categories based upon the ranges of values in Box 17.6 to illustrate the concept. A range of Se concentrations is shown for each component arranged by category of risk (R1 being lowest risk; R4 being greatest). A narrative defines the category of risk in each category, based upon existing knowledge. The use of ranges in this example, rather than absolute numbers (proposed by Lemly, 1999), is a means of recognising the uncertainties in the existing state of knowledge.

The advantage of such a risk characterisation is that it is a systematic way to address ecosystem contamination, using all lines of evidence. Consideration of all lines of evidence improves the credibility of the analysis. By itself, however, use of such a set of numerics to evaluate an ecosystem is not sufficient. Narrative interpretations are necessary to address crucial questions like prioritisation

of data gathering, explanations of the choices of categories, contradictions among lines of evidence and attributes of the site itself are essential for a credible risk analysis.

17.4.2 Integrative risk assessment

Prologue. Uncertainties are reduced, reflected in a narrowing of possible policy choices, as more lines of evidence are integrated into a risk assessment. But overreliance on unilateral lines of evidence and/ or inadequate representation of the environmental context can weaken interpretations and confuse risk assessments and criteria derivation.

By formal definition ecological risk assessment integrates multiple lines of evidence, from all sources (Suter, 2006). But in reality, analyses of ecological risk from metals often rely heavily upon the traditional and standardised approaches approved by regulatory agencies and less upon understanding the water body involved. The former often require controversial techniques for extrapolation of laboratory data to field situations, like application factors and acute-to-chronic ratios (Chapter 15). Although it is recognised that the best substitute for a correction factor is further study, expanded risk approaches also can be overly simplistic. Some studies termed integrative risk assessments and weight-of-evidence studies retain uncertainties if the emphasis is on 'acceptable' toxicity testing data. Uncertainties are reduced somewhat if risk protocols are used to add in chemical analysis, and/or simple measures of one or more resident communities (Chapman, 1995). But such approaches are generally most sensitive only to extreme effects (Chapter 11). For example, a tool proposed early on for ecotoxicology-based integrative risk assessment was the sediment quality triad (Long and Chapman, 1985; Chapman, 1990). Chapman (1990) defined the triad as including synoptic measurements of chemistry to measure contamination, bioassays to measure toxicity, and a one-time *in situ* biological assessment to measure effects such as benthic community structure.

As traditionally implemented, the triad is vastly superior to toxicity testing alone, but low to moderately robust for detecting adverse effects in nature. It is least robust where circumstances are complex (Burton *et al.*, 2002). The limited robustness stems from the use of one-time snapshots to quantify toxicity, concentrations and community structure in an environment where those all three are expressed in complex ways. The triad concept, however, does not exclude more elaborate approaches. These can include *in situ* toxicity tests or mesocosm tests, evaluations of environmental conditions beyond chemical concentrations of contaminant, evaluation of speciation and/or 'biomonitoring' to determine bioavailable metal concentrations (Chapter 8). As the body of evidence grows, and the knowledge of environmental context grows, the results become more robust (Chapman *et al.*, 1997).

The Clark Fork Ecological Risk Assessment, conducted by USEPA (1999b) is an example of a toxicology-based weight-of-evidence approach that expands triad methodologies (Box 17.11).

The Clark Fork ecological risk assessment is an instructive example of the complications and suspicions about non-traditional lines of evidence that can arise when science and politics become intertwined, and the stakes are high. This risk assessment was probably as 'state-of-the-art' as any conducted in the USA to date; but it also reflected the deficiencies of the 'art'. What seemed like a relatively straightforward analysis of the effects of historical mining became a complex, contentious and even convoluted situation. Settlement of one set of litigation (hundreds of millions of dollars to go to remediation) occurred after 15 years of controversy, while other aspects of the litigation continue. The cost was more than a decade of delay before serious remediation began.

One of the difficulties the Clark Fork risk assessors faced was the perceived necessity of reaching a single, simple conclusion about risks from metals in the river. An alternative approach less commonly employed in risk assessment is identification of reasonable scenarios, and development of quantitative forecasts for each scenario. If the scenarios are built from a feasible range of conditions, uncertainties can be characterised by the range of outcomes in the forecasts. The goal is to give policy makers a range of outcomes that are possible under a range of reasonable possibilities (most of which involve choosing different levels of risk). From this they can determine the risks from various management strategies. For example, Luoma and Presser (2000) forecast effects on predators under different Se loading scenarios to San Francisco Bay from a discharge canal proposed to remove irrigation drainage from the San Joaquin Valley (Chapter 13). They systematically followed a conceptual model for fate and effects of Se to develop scenarios and forecast effects:

- Different projections for possible Se loadings to the Bay were derived from historical proposals. Outcomes were shown for choices ranging from greatly reducing salts in the valley soils with minimal treatment of the effluent, to less salt removals with greater treatment of effluents.
- River discharges during three different hydrological regimes were combined with the different Se loads to forecast a range of possible dissolved Se concentrations at the head of the Bay.
- Three alternative (but feasible) speciation and particulate transformation scenarios were used to forecast particulate concentration and form. Policy makers could choose between the risk that speciation would retain the characteristics

Box 17.11 Clark Fork Ecological Risk Assessment

The risk assessment for the mine-impacted Clark Fork River Superfund site (Chapter 12) had available an unusual wealth of environmental and toxicity data relevant to local conditions (Chapter 11). The risk assessment considered data on exposure and effects on fish, benthic macroinvertebrates, algae, terrestrial plants, terrestrial animals and soil organisms. Multiple lines of evidence were defined by three approaches:

(1) Predictive approach
This approach used site exposure concentrations (PEC) defined by dissolved metal concentrations. These were compared to literature based exposure levels that were believed to cause no or minimal toxic effects (hazard quotient (HQ) or PNEC: Chapters 2 and 15). The assessment found that only three out of 232 dissolved metal samples from the river had chronic HQ values (PEC/PNEC) for Cu between 1.1 and 1.3, and one sample had an HQ value of 2.8. No other metal exceeded criteria. This line of evidence suggested little risk of chronic toxicity.

(2) Site-specific toxicity studies
A second approach was to measure the response of receptors that are exposed *in situ*.

- Invertebrates were exposed to water from the Clark Fork River that was either spiked with metals or used in its ambient condition (Water Effects Ratio (WER) as defined in Chapter 15). The WERs were variable when conducted at different times and places. Typical WERs showed that Cu in the river was 3.5- to 17-fold less toxic than in standardised test water, and Zn was 1- to 7-fold less toxic (Welsh *et al.*, 2000).
- Historical data from *in situ* toxicity tests were also considered. They showed toxicity when juvenile trout were held at contaminated field sites for 1 month (Chapter 12). Metal concentrations were elevated but no metal consistently exceeded any criteria where the toxicity occurred.

This line of evidence suggested toxicity was occurring despite the reduced bioavailability of Cu and the lack of criteria exceedances. The risk assessors were left to conclude that undetected pulses of Cu were responsible for the toxicity.

(3) Direct observations of receptor diversity and abundance
Field observations were used or studied including:

- Fish populations and fish community structure. As noted in Chapter 13, the fish community in

the Clark Fork River is impaired. It is of low diversity, and there is a low abundance of surviving species compared to carefully selected reference sites.
- Benthic community structure. Otherwise common species, known to be sensitive to metals, were missing from the benthic community, although broad indices were less indicative of disturbance.

The risk assessors were extremely thorough in the data they reported from the watershed but, in their conclusions, they were generally distrustful of the links between metals and these lines of evidence. They were also distrustful of studies of exposure via the dietary pathway, and the assessment raised doubt about bioavailability from contaminated sediments (linkages between sediment, dissolved and biological contamination were not considered important in the end). They noted that many factors could have affected the nature of the existing community. They cited habitat suitability, availability of food, predator pressure, natural population cycles, and meteorological conditions as examples. It was interesting that none of these alternative explanations was subjected to the same degree of scrutiny that metals were; nevertheless they were carefully considered in conclusions like the following:

> even though it is possible that some of the differences observed may be attributable to habitat factors that were not perfectly matched between site and reference segments, the consistency of the patterns across sites and across time is strong evidence that some of the difference is attributable to metals exposure.

The analysis included some integration of the multiple lines of evidence supporting important effects from metals, but were sceptical, as was clear above, about the studies that excluded alternative explanations of the impaired fish (Box 11.4). The end result was an extremely cautious statement, perhaps reflecting the high stakes (hundreds of millions of dollars; intense political pressure):

> it is unknown to what degree chronic stress contributes to population-level effects such as the decrease in standing fish population. Based on the available data, it is concluded that acute exposures to pulses or other high-concentration events are more important than chronic stresses in causing population-level effects. . . . It is also concluded that decreases in fish populations may be due in part to other (non-metal) stressors.

of the river versus the possibility that speciation would behave as it had historically and change as Se entered and was retained in the estuarine environment.

- A biodynamic model was used to forecast bioaccumulation from each particulate form/concentration into invertebrates. Bivalves were the invertebrates of choice because they bioaccumulate Se more effectively than any other prey species in the Bay. Forecasts were conducted for different assimilation efficiencies that might accompany the different types of Se transformation.
- A regression from the field data was used to link bioaccumulated Se in bivalves to bioaccumulated Se in their specific predators: sturgeon (*Acipenser transmontanus*) and the migratory waterfowl surf scoter (*Melanitta perspicillata*).
- The likelihood of toxicity was evaluated from the bioaccumulated concentration in the predators. These were later followed up by experiments that tested dietary toxicity in two species of native fish that were bivalve predators in the Bay: sturgeon (*A. transmontanus*) and Sacramento splittail (*Pogonichthys macrolepidotus*) (Teh *et al.*, 2004; Linville, 2006).
- The validity of the model was tested by developing calculations for existing conditions in the Bay and comparing the output to field observations. In each case, the agreements were reasonable.
- Overall risks were determined by comparison of model forecasts to proposed guidelines for each environmental component.

The model calculations were sensitive to hydrology, the choice of partitioning coefficient (K_d) and the choice of invertebrate species. Sensitivity analyses showed that ecological effects from Se were most likely during the low flow season and dry years. Risks were greater if the K_d was that typical of the existing estuary, as compared to the K_d presently typical of the incoming river. And risks to the bivalve food web were much greater than risks to zooplankton-based food web. Figure 17.12 shows outputs for three loading scenarios using speciation and K_d typical of the Bay in 2000.

17.5   PRIORITISING

Prologue. Schemes for prioritising data collection are necessary to help develop a systematic integrative assessment.

Traditional water quality criteria have a proven track record and are relatively simple to implement because high-quality water analyses are feasible and relatively inexpensive. But their assumptions add uncertainties and inhibit their coherence, consistency and effectiveness (Chapter 15). Available case studies suggest that the more integrative the assessment and the more lines of evidence considered (including field data), the more likely a justifiable (and sometimes even more timely) risk management policy. If interpretations from natural waters converge with experimental data, policy choices can be greatly narrowed. Less contentiousness may also be possible.

The weight-of-evidence and modelling approaches also have the advantage of allowing consideration of more than one combinations of possible outcomes (which is usually the reality of the situation). The major weakness is that such approaches are data/knowledge intensive. Credibility and likelihood of success are improved by prioritisation of data collection. Every individual approach for evaluating risk and/or managing trace metals has different costs, different benefits, different types of uncertainties and different degrees of uncertainty (Table 17.4). These all must go into decisions about priorities. The choice of a weight-of-evidence strategy should be developed from the management goals for the situation.

17.6   CONCLUSIONS

The difficulty of interpreting risk from traditionally derived single criteria is accentuated by the unique nature of different locations and the unique influences of newly known processes like dietary exposure and trophic transfer. Universal criteria developed from effects under worst-case conditions do not necessarily reflect the likelihood

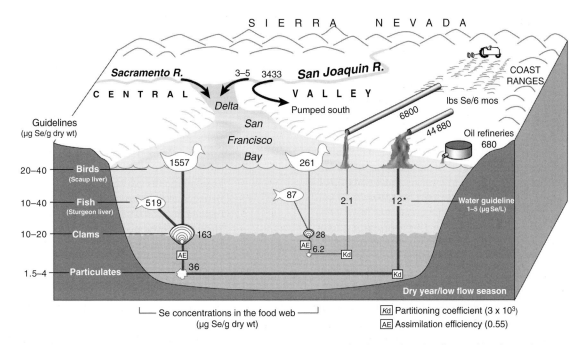

Fig. 17.12 Estimation of Se concentrations in water, sediment, bioaccumulated Se and trophically transferred Se in the environment of San Francisco Bay. In this figure, different scenarios for Se inputs to San Francisco Bay undergo speciation and transformation (K_d), typical of existing conditions in the Bay, to yield particulate metal concentrations. Biodynamic modelling employing assimilation efficiencies typical of a mix of Se forms in food was used to determine bioaccumulation by bivalves at each loading. Trophic transfer to predators was determined by extrapolating regressions characterising historical data. Forecast concentrations in each media (water, particulates, clams, fish and birds) were compared to guidelines for each that are available in the scientific literature.

of effects at the more common sites where complex conditions and less-than-extreme contamination dominate. The Clark Fork Ecological Risk Assessment, and the deadlock in Se policies in the USA in the 1990s and 2000s, are examples of the policy deadlock that can occur when scientific uncertainties are compounded by suspicions of multiple lines of evidence.

The result of the above is that at least some of the criteria or guidelines for managing ecological risks from metal contamination are problematic. The scientific knowledge is now available to begin to build policies that are more effective, transparent and efficient. Implementing those polices will benefit from greater consideration of interdisciplinary interpretations and fuller consideration of the body of available evidence about

risks for each metal (harmonisation). More use of approaches that integrate processes, when appropriate (e.g. dietary or bioaccumulative criteria), can, at least, provide opportunities to validate risk management decisions. Better validation is a form of continued learning from experience that is critical to better risk management in the future. Case studies demonstrate that interdisciplinary, cross-boundary validation can narrow the range of justifiable alternatives and thereby lead to informed risk analysis and improved risk management. Consciously beginning the process of thinking and managing across the traditional boundaries that have separated metal science is the first requirement toward making management of ecological risks from metals the most effective it can be.

Table 17.4 *Some advantages and disadvantages of different lines of evidence*

Line of evidence	Advantages	Uncertainties
Concentration in water, sediment from water body	Reality Relatively simple Standard methods	No cause–effect Not directly related to toxicity
Laboratory toxicity tests	Measure of effect Widely used Standardised methods	Poor connection to reality Not a measure of causation Unrealistic exposure conditions
Bioaccumulated Metal Concentration (BMG)	Measure of exposure and bioavailability Illustrates food web processes	Not a measure of effect Regulation effect in some cases Detoxification affects results Implementation challenges
Critical Body Residue	Chemical causality	Theory not validated No standard method Serious data deficiency
Dietary exposure (PEC$_{prey}$) as related to dietary toxicity of the prey to a predator (DMG aka PNEC$_{prey}$)	Realistic Measure of effect Statistically valid collections Considers bioavailability	Dose–response not well developed for many metals
Resident biota (ecology)	Measure of effect Public interest Standard methods	High variability Cause–effect difficult Confounding stressors
Presence/absence of indicator species	Simpler than full community Diagnostic if accompanied by BMG	Metal-sensitive indicators not known for all environments
Field toxicity (e.g. *in situ* tests)	More realistic than laboratory Address natural stressors	No measure of causation Complex exposures affect interpretations
Effects observed in field	Realistic Directly relevant Public interest	Power of detection can be low Large sample size for dose–response Response must be specific
Biomarkers (*in situ*)	Sensitive response Rapidly measured May be stressor-specific	Power of detection

Source: Partially after Burton *et al.* (2002).

Suggested reading

Brix, K.V., Toll, J.E., Tear, L.M., DeForest, D.K. and Adams, W.J. (2005). Setting site-specific water-quality standards by using tissue residue thresholds and bioaccumulation data. II. Calculating site-specific selenium water-quality standards for protecting fish and birds. *Environmental Toxicology and Chemistry*, **24**, 231–237.

Brown, C.L., Parchaso, F., Thompson, J.K. and Luoma, S.N. (2003). Assessing toxicant effects in a complex estuary: a case study of effects of silver on reproduction in the bivalve, *Potamocorbula amurensis*, in San Francisco Bay. *Human and Ecological Risk Assessment*, **9**, 95–119.

Buck, K.N. and Bruland, K.W. (2005). Copper speciation in San Francisco Bay: a novel approach using multiple analytical windows. *Marine Chemistry*, **96**, 185–198.

Fairbrother, A., Brix, K.V., Toll, J.E., McKay, S. and Adams, W.J. (1999). Egg selenium concentrations as predictors of avian toxicity. *Human and Ecological Risk Assessment*, **5**, 1229–1253.

Lemly, A.D. (1998). A position paper on selenium in ecotoxicology: a procedure for deriving site-specific water quality criteria. *Ecotoxicology and Environmental Safety*, **39**, 1–9.

Skorupa, J.P. (1999). Beware missing data and undernourished statistical models: comment on Fairbrother *et al.*'s critical evaluation. *Human and Ecological Risk Assessment*, **5**, 1255–1262.

18 • Conclusions: Science and policy

18.1 INTRODUCTION

Prologue. Management of environmental issues (like metal contamination) requires ongoing interaction among scientists, managers and the public. Scientists need mechanisms to communicate research and management results. Understanding of scientific, management and social issues, and the ability to communicate with scientists, managers and the public are important ingredients.

Education about environmental issues often involves learning the fundamentals of the science (s) or the substance of the policy issues. It is more rare that the interaction between science and policy is explicitly discussed or taught. All experts agree that effective environmental management must be responsive to science; although science need not (or cannot) be prescriptive of policy (Schneider, 2002). Thus the relationship between science and policy is a delicately balanced, but necessary, ingredient in addressing complex environmental problems (e.g. Dietz *et al.*, 2003).

There is also a broad consensus that the interaction between environmental policy and environmental science needs to be improved (Dietz *et al.*, 2003; Houck, 2003). The problems with the interaction are well diagnosed (Coglianese and Marchant, 2003; Houck, 2003), but there is surprisingly little discussion about the ingredients of an effective interaction. In fact, from a policy perspective, there are great differences in how 'sound science' is defined and applied, or how adaptive management is implemented (Lee, 1999).

Managing metal contamination in aquatic environments requires that both the science community and the policy community (Box 18.2) be aware of not only the state of the science, but also the basic attributes of a science–policy interaction. Here, we incorporate some conclusions about the state of the science of metal contamination into a discussion of factors that influence how that science is communicated. We also consider in some detail an important role that managers and policy makers alike often must undertake: construction of a science agenda that is relevant to their policy problems, as well as productive and credible.

18.2 STATE OF KNOWLEDGE

Fully integrative risk assessments and lateral risk management are rare at the present state of the science. This is partly because of cross-boundary scepticism; but probably more related to a state of knowledge that was heretofore inadequate for integration. But, as we have shown in the preceding chapters, that state of knowledge has grown greatly. Enough is now known to address integration aggressively.

In every chapter we have shown important advances in our knowledge of trace metals as

---

Box 18.1 Definitions

policy: A plan or course of action, by a government, or business, for example, intended to influence and determine decisions.

science: The observation, identification, description, experimental investigation and theoretical explanation of phenomena.

Box 18.2 The policy community and the science community

We will define two communities that interact in this discourse; members of both may have scientific training. *The policy community* includes stakeholders such as non-governmental groups (NGOs, usually representing environmental interests), resource users (e.g. industry) and the affected or interested public. An important part of the policy community consists of personnel from governmental agencies with regulatory responsibilities. These can include scientists working as resource managers and their technical advisors. These individuals implement policies and advise upper-level managers who devise and recommend new policy (the policy makers). Influential stakeholders are also policy makers, in the sense that they also recommend policy options. The final group consists of politicians, the decision makers. In democracies they make the final broad decisions on policy, but not necessarily the detailed decisions that determine how policies will be implemented.

The *science community* is the diverse group of practising scientists. Practising scientists are those who conduct original studies in the field or laboratory, develop new methodologies or provide new interpretations of existing data. These can include academics (usually from universities), scientists from government organisations without regulatory responsibilities, scientists from the private sector and some stakeholder scientists.

contaminants. There are uncertainties in all areas, but they are greatly reduced compared to the past. The important questions have long been known for managing metal contamination, as evidenced by early papers in the field; most of those questions still frame what is important to know. There is at least an implicit consensus on a common conceptual model for the processes that control contamination. A systematic application of that model in formulating risk analysis could help standardise the risk assessment process. Sources of each metal are also relatively well known and quantitative biogeochemical cycles are available for many trace metals. It is clear that, for the most hazardous

trace metals, human inputs to the biosphere equal or exceed natural inputs. The global risk from metals lies in their ubiquitous uses. Risks are often the sum of many confined hotspots of effects, although regional-scale (Se irrigation drainage, particulate Hg and aerosol Pb in the atmosphere, unconstrained mine wastes) and global-scale (Hg^0) risks are also known.

Several classifications and definitions of metals (and metalloids) are available on a geochemical basis. For the sake of simplicity, we argue that trace metals follow a continuum of hydro-, geo- bio-, eco-characteristics that allows their classification by that single term. Great advances have occurred in determining trace metal concentrations in natural waters. The range of possible dissolved concentrations is now well known for many types of waters and many water bodies; although more data from many specific locations are still needed. It is also possible to define speciation with much greater reliability than in the past. Speciation models have greatly reduced uncertainties, especially if some empirical corrections are included. Just as important, analytical techniques have developed that allow powerful, operational characterisations of labile metal. The latter (e.g. diffusive gradients in thin films or DGT technique) are being increasingly incorporated into studies of different water bodies. Direct determination of important aspects of some metal species is almost routinely possible for some trace metals (e.g. Se forms and MeHg).

It is well known that metals transform from dissolved to particulate state such that most of the contamination ends up in sediments or suspended on particles in the water column. Transformation is affected by speciation in solution and the nature of the particulate material. Thus the ratio of particulate to dissolved metal (K_d) can vary widely. Traditional approaches to modelling the transformation process once assumed that there would be a narrow range of K_ds for each metal. It is now known that the variation is orders of magnitude. However, analytical capabilities are such that K_ds can be determined empirically in a given environment. Some generalisations based upon environmental conditions are also possible.

Transformation is important because the concentration of metal associated with particulate material ultimately affects contamination in all compartments of the environment. Case studies from nature show that contamination can be retained in sediments long after water column inputs have declined. Hydrological dispersion of contaminated sediments is also possible. Methodologies, however, are well developed for site-specific determination of metals. The most important caveat for interpreting such data is accountability for particle size, but several protocols to correct for such biases are available. If sediment data are not corrected for particle size, it is possible to seriously underestimate the degree of contamination in an environment with coarse sediments.

Bioavailability remains one of the critical processes determining fate and effects of metals. Metals are bioavailable, to at least some degree, from water, sediments and organisms used by animals as food. Important recent advances allow quantification of metal uptake from diet, and show that it is a dominant source of metal exposure in many circumstances. The long-standing assumption that trophic transfer of metals is unimportant is disproven. It is now well established that the degree of bioavailability depends upon speciation in solution, the chemistry of the sediments and the intra-organismal partitioning of metals in prey species. One important uncertainty stems from questions about whether anoxic sediments are important in metal exposure, since most organisms require contact with an oxygenated environment. Using metal bioavailability in regulating metal contamination in sediments thus remains contentious. The precision of operational methods to determine the bioavailable metal fraction in prey species could also be improved. Nevertheless the uncertainties about metal bioavailability are much constrained compared to the past.

Biodynamic models are available that allow unification of the complexities that control bioaccumulation. Models that forecast metal-, environment-, species- and route-specific influences on bioaccumulation seem to predict bioaccumulation with about two- to fourfold accuracy, if some data from the environment and species of interest are available. These models have strong potential applications to management of metal risks.

Links between bioaccumulation and toxicity are increasingly known. If rates of uptake from all routes exceed rates of detoxification and excretion, toxicity will ultimately ensue. Knowledge of detoxification rates, or as a surrogate the detoxified fraction of metal, can facilitate understanding of the sensitivity of a species, if accompanied by knowledge of the physiological constants governing bioaccumulation (net uptake rates from water and food for relevant circumstances and rate constant of loss). Sensitive species are those that have limited detoxification capabilities accompanied by traits that allow substantial bioaccumulation. A few studies have even demonstrated that these are the species that disappear from contaminated environments. Growth in the understanding of how bioaccumulation and toxicity are linked mechanistically might offer opportunities to put the large but underutilised database on biomonitoring to use. Biomonitors are known from nearly every type of aquatic environment, and data are available from around the world to calibrate bioaccumulation in these species to the degree of contamination. The next step could be calibration to toxic responses in other species. For example, exercises could use stress data from the sub-organismal level, *in situ* toxicity tests with whole organisms, or presence/abundance of sensitive species.

Large databases now exist that demonstrate thresholds of metal toxicity in short-term, dissolved metal exposures for tens of species. Smaller, but sometimes robust, data sets also exist for traditional definitions of chronic toxicity for some metals but not all. Some very sensitive protocols for determining reproductive toxicity (e.g. in *Daphnia magna* or other zooplankton) are available; and some very sophisticated mesocosm and *in situ* approaches allow study of complex responses to metals. The Biotic Ligand Model is an example of a new approach that ties quantitative determination of metal speciation with traditional toxicity testing. Modelling, in this case, is an unquestionable improvement over the traditional adjustments of environmental standards via 'hardness corrections' or 'water effects ratios'. What is missing are robust validation studies that address the degree to which the BLM provides effective protective measures for natural waters; whether as corrected acute toxicity tests or as BLM-linked chronic tests. We have emphasised the weaknesses of the toxicity testing approach (no dietary exposure, limited exposure duration, test species chosen on the basis of convenience rather than a mechanistic expectation of sensitivity). We submit that these biological limitations will probably have to be addressed if any toxicity-testing-based concept is to efficiently and effectively protect nature better than the last generation of risk evaluation approaches. The challenge could be viewed as a null hypotheses that has yet to be refuted by independent validation tests.

A growing number of case studies demonstrate that the feasibility of metal toxicity, as shown in the laboratory, can be demonstrated in nature, especially where a robust body of data is available. An important outcome of this body of work applies to the long-standing recognition that species differ in their sensitivity to metal contamination. Statistical approaches are available to objectively sort such data from the toxicity testing database, thus narrowing traditional criteria development to the data (or species) most sensitive to dissolved exposures. The combination of mechanistic and field data, however, were most useful in demonstrating which species are most sensitive and most tolerant in nature. Understanding the traits that make species vulnerable to metal contamination offers new opportunities for uncovering metal-specific signs of stress in the field. This, in turn, will facilitate demonstration of cause and effect in field studies. Signs of responses to metals (biomarkers) at lower levels of organisation will be an essential set of indicators as well (lysosomal destabilisation, metallothionein synthesis, signs of oxidative stress), as long as it is appreciated that these are only manifested in the populations that can survive the metal stress.

These are a few of the scientific principles that seem ready for application to risk management.

An important unaddressed question is how we make the connection between scientific advances and applications to policy. That is a 'science' in its own right.

18.3 WHAT IS A SCIENCE–POLICY INTERACTION?

Prologue. Scientific input is essential in creating policies for dealing with risks from metal contamination, but science, alone, cannot dictate policy. Scientific uncertainties define the need for an explicit science–policy interaction. The need for interaction must be recognised by all interested parties if it is to succeed. The science–policy interface will benefit from explicit goals (e.g. nurturing public trust) and mechanics that address those goals.

The same basic questions come up over and over again in managing metal contamination. By how much must pollutant loads be reduced in order to limit ecological risks to an acceptable level? What concentrations are safe in different components of natural waters? How do we assure and validate that the criteria recommended for the local water body are neither over- nor under-protective? How do we assure and validate that the regulations are actually protecting the environment and are not a waste of money? The answer to every one of these questions results in a policy that guides the actions (e.g. risk management) of the jurisdiction asking the question. The use of scientific knowledge to develop that policy is the science–policy interaction. As we have seen in earlier chapters, the science underlying such questions is complex. Every approach involves uncertainties and trade-offs. Multiple lines of scientific evidence are usually necessary to draw defensible interpretations. The credibility of the institution responsible for managing the pollutant issue depends upon the policy chosen; credibility grows if the answers to the questions above prove effective when they are implemented. Credibility is lost if policies ultimately prove, or are perceived as, ineffective.

Justifiable policy is the goal of any institution (public or private) in such circumstances. For metal contamination this means a policy that protects the environment, is defensible economically and engenders trust from all interested parties and the public in general. This is a tall order in environmental science. At all scales and for all institutions, it is expected that environmental policies will derive from scientific knowledge. But science rarely points clearly to a single answer; and it is not unusual that scientific knowledge is incomplete. The result is that tough decisions must be made despite technical uncertainty, complexity and substantial constraints, as well as conflicting human values and interests (Dietz *et al.*, 2003). These decisions define much of environmental policy.

Although policy choices need to incorporate the most advanced scientific knowledge, they also must consider societal factors, politics and administrative process. It is not feasible for policy action to wait for all the scientific knowledge to develop; partly because there is, inherently, no end to learning. And we must accept that the status quo is a form of action too. Action almost always must proceed despite at least some, and sometimes many, uncertainties. If scientific knowledge is plagued by disagreements and uncertainties, the policy maker or risk manager must be aware of what those are, and make a judgement.

How we deal with scientific uncertainty greatly defines the interaction between environmental policy and environmental science. We discussed above the rapid growth in the state of knowledge; nevertheless metals science is young, with much left to learn. Uncertainties remain. How we communicate those uncertainties while leaving room to incorporate the new science in a new generation of policies is a major challenge.

Does uncertainty about technical matters mean that science is just too difficult to deal with in the policy arena? Should we pay little attention to science when disagreements appear, and go on to make the decisions based upon political influence? As we have discussed, some argue that the solution is to impose extreme versions of the precautionary principle, on the assumption that we will never know enough to develop policy from science. They conclude that values deserve the most emphasis in deriving policies and risk

management strategies. Others argue that no policy regulating metal contamination is justifiable until adverse effects are unequivocally demonstrated. This limits policy actions to using only that science which is the most certain. Such policies have long been the basis of under-protection of environmental values. On the other hand, we also must be sure that reliance on precaution does not mean that the analysis of complex issues and the details of solutions are left to intuition. Environmental issues like metal contamination are too complex for intuitive solutions to necessarily be good solutions; nor is that necessary at the rapidly advancing state of our knowledge.

We assume here that under-protection of the environment is politically unacceptable in the modern world; there are too many instances in our history that show that metals can cause loss of species and ecosystem simplification, sometimes over broad areas. Policies that are not sensitive to environmental disruption are costly and ultimately not cost effective. However ignoring science in the name of environmental precaution is just as unacceptable. Policies that overreact to contamination can be environmentally ineffective, costly in terms of trust and in terms of economics. Tributyl tin policies are an excellent example of the fine line that must be walked to resolve one problem with a minimum exacerbating of another (Chapter 14).

The balance between precaution and positive proof of harm is an active interface where the dialogue between science–policy continues to grow. Ideally, all parties in such an interaction seek a common understanding about what is known or unknown about the technical questions. And new knowledge is continually injected into the process. But the ideal requires many ingredients to be successful.

History shows that lack of clarity in explaining what is known and unknown about an issue can destroy public trust, if and when surprises occur (Löfstedt, 2005). Poor technical justifications for decisions can result in serious failures, either in costs of compliance or in protecting the environment. Such failures also affect trust and can derail other justifiable management options.

So uncertainties are a given in complex metal issues, but clarification is usually a better option than ignoring uncertainty or assuming that it can be resolved.

18.4 DEVELOPING A SCIENCE DIALOGUE USEFUL TO POLICY

Prologue. A policy-neutral dialogue might constantly re-evaluate the status of knowledge about an issue. But the goal is not to provide simple policy solutions. It is to inform interested parties about their choices. The benefit is that policy-neutral dialogue usually exposes marginal ideas; the cost is that the selection of a policy is left to the policy and decision makers. But a neutral dialogue can help establish trust among interested parties and with the public. It is a way of managing conflict such that scientific disagreement and uncertainty do not unnecessarily disrupt a delicate policy making environment. An alternative is an advocacy dialogue. In this case, the final conclusion is advocacy of a specific position. The degree of scientific analysis involved in the choice varies. Some argue that advocacy is every scientist's responsibility. An alternative and common argument is that advocacy disrupts trust in the scientific enterprise. The nature of dialogue with the policy community is a fundamental choice for every individual. Each has important implications.

The issues associated with metal contamination are not unique among environmental issues. But it is increasingly critical that we find an effective structure to guide the role of science and natural/ social in environmental policy. Nowhere in the world is this done as well as it needs to be. Despite the common call for 'science-based management', scepticism is often expressed about what or how science can contribute to management of environmental issues (Houck, 2003).

The individual ingredients that we describe for science–policy interaction are known. But explicitly choosing goals and a combination of ingredients to foster the interaction is less discussed. It is argued that sustainable solutions in complex circumstances depend upon adapting management actions to new knowledge (Walters, 1997; McDaniels and Gregory, 2004). Sustainable solutions in such circumstance also may require maintaining fragile political coalitions. The nature of the dialogue between the science community and the policy community is crucial to all of these.

18.4.1 Goals

Prologue. It is constructive to set broad goals for the dialogue between the science and policy community in specific circumstances, so that all parties understand what the dialogue is trying to achieve. Policy-neutral goals might include (a) enhancing public trust in the process, (b) sustaining a constructive dialogue about new science relevant to the policy, (c) avoiding policy gridlock or (d) designing a policy flexible enough to respond to new knowledge. The goal of advocacy is to promote a specific policy.

An important starting point for any science–policy interaction, whether it is policy-neutral or advocacy, is recognition that the interaction must be explicitly created and nurtured. It will not just happen because the affected parties think it should. And, like every complex endeavour, the interaction will benefit from all parties understanding why they are getting together: explicit goals that are specific to the circumstance (English et al., 1999). Once the goals are established they will guide the mechanics necessary to shape the interaction appropriately. For example, the two most important goals in including scientists in the policy process are (a) to assure that policy is informed, and (b) to assure that the public trusts that the policy is scientifically justified. Informed (justifiable) policy builds from defensible interpretations of the existing scientific knowledge. Public trust is established by mechanisms that assure openness and honesty, especially about uncertainties. A minimum goal of a science–policy interaction is to define publicly the state of knowledge with sufficient clarity and credibility that policy proposals that are counter to that knowledge are recognised as such. The more open the discussion of the state of knowledge and its uncertainties, the greater the likelihood of enhancing public trust. This is the most fundamental rule in dealing with scientific uncertainty.

A constructive, ongoing scientific dialogue centred around the most important policy issues is an effective way for management to obtain scientific input and follow the advancing science it is supporting. Focusing the dialogue on a specific policy demands a different kind of science input than the typical conference. Experts from several disciplines are assembled to discuss how the aspect of science they know best applies to that specific policy problem; without necessarily recommending solutions. The application of science to the policy problem, not the latest advances in the science, should be the focus for each expert.

The site-specific water quality criteria developed for Cu in South San Francisco Bay, California (USA) was an example of how basic science influenced policy in a constructive way (Chapter 17). The policy for a Cu criterion for South Bay was discussed for more than a decade among all parties. Specific scientific studies were supported as the discussion ensued. The new studies from independent experts eventually showed that concentrations of dissolved organic matter were sufficient to reduce Cu bioavailability to levels orders of magnitude below those that caused harm in other environments. Sufficient communication about the declining Cu contamination and the low bioavailability occurred that ultimately a local Cu criterion was developed and accepted by all parties. The new criterion relaxed by twofold the existing standard for South Bay alone (Chapter 17). This was accompanied by the requirement for ongoing monitoring to assure that conditions did not degrade. The dialogue took patience, but a very contentious issue was resolved to a general level of satisfaction with both costs and environmental protection.

The difficult issues surrounding management of Hg contamination provide another example of how the balance between scientific information and political choices can fluctuate. Science is not always foremost in the policy decision. For example, California (USA) is developing a strategy for managing Hg in its watersheds. Hg is

a statewide problem because of historical Hg and Au mining which spread Hg through much of the northern half of the state. Scientists originally diagnosed the Hg problem, but policy makers must decide how to address this diffuse and widespread problem. After a long process, a recent decision was made to require that all sources of Hg to waterways in the state be reduced by a fixed percentage, whether they are large or small. Scientific determinations of mass balances suggested that cutting small sources (in this case waste treatment facilities) would be expensive but would be largely irrelevant to environmental concentrations of Hg. In this case the policy left the perception that the political need to show action was more important than the basic goal of reducing Hg in the environment.

Policy gridlock is the least attractive outcome of a science–policy dialogue. Gridlock will slow the development of new knowledge and adversely affect management, and does not allow advance of policy beyond the status quo. Gridlock also destroys public trust. Although scientific uncertainties are often cited as a cause of gridlock (Houck, 2003), in fact this is determined by how conflict about the uncertainties is managed. In the USA decisions about new environmental standards for Se have been stalled since 1996, for example. International standards for Se vary extremely widely (Chapter 17). Mostly this stems from inconsistent recognition of the growing understanding of Se effects on upper trophic levels (Chapters 13, 15 and 17). The scientific dialogue about Se was often addressed in forums designed to determine 'who is right'. It could be argued that this contentious approach to scientific dialogue perpetuated the inability to make decisions despite the availability of a relatively robust science.

A contrast to Se policy was the strong action taken to develop an international policy for tributyl tin (TBT). Strong scientific evidence of adverse effects on important aquatic resource species, and natural communities of animals, led to an internationally consistent ban on TBT, preventing its use as an antifoulant on the bottoms of small boats (Chapter 14). A clear body of scientific evidence identified the problem and unequivocally showed both the cause and significance of the problem. Communication was clear and unequivocal. An international dialogue brought recognition of the problem to the highest political levels. Coherent policy followed.

Policies that leave room for flexibility or adaptability are the solution where scientific knowledge is incomplete. Scientists can help policy makers find ways to achieve such flexibility. Flexible policies recognise that surprises are inevitable, unexpected crises are likely to arise and policy will probably have to adjust to new knowledge. But they also require data, so that implementers can accumulate knowledge about their effectiveness. Creating a system that is flexible enough to respond to what is learned as experience is accumulated is another goal of incorporating scientific expertise into the policy activities. Table 18.1 gives some other examples of generic goals for a science–policy interaction.

Finally, it is important to recognise that all policies do not come only from government. The legacies of mineral mining create issues with metal contamination across the world. Some (but not all) mining corporations are developing (or have in

Table 18.1 *Examples of goals for a science–policy interaction*

- Establish a credible programme of relevant scientific studies
- Build common ownership of knowledge among academics, government personnel, stakeholders and policy makers/political leaders by developing and sustaining an active and constructive scientific discourse among all parties
- Avoid advocacy debates and otherwise manage the conflict that stems from disagreements about scientific uncertainties
- Establish trust via open proceedings centred upon defining the existing state of knowledge from:
  - converging lines of evidence
  - identifying key uncertainties and points of technical disagreement
  - plans or advice for next steps
- Marginalise misinformation by exposure in the open proceedings

place) operable sustainable development policies that are making a difference in terms of environmental effect. These include policies for limiting contamination and minimising its dispersion as well as explicit plans for post-mining remediation. Industrial globalisation has allowed international consistency in such policies, and pushed improved environmental practices into remote regions where governments have traditionally been ineffective. Independent evaluation of the effectiveness of these policies has not begun, but could provide a heretofore underexplored interface where science and policy dialogue could grow (Chapter 12).

18.4.2   Trust

Prologue. Public trust is not automatically enhanced by including scientists in the policy dialogue. The science must be impartial and competent, and it must effectively address the policy in question. Internal trust among scientists from different backgrounds and institutions is also important. Trust develops from constructive interactions among scientists with different missions; and distrust grows when professionals with different assignments rarely communicate.

Many policy issues cannot be resolved in an immediate or simple way. But sustaining the trust of the public during the process of addressing a complex issue is probably as important as resolving the issue itself. Mistrust grows when the public is misinformed. In science as in politics, where there is not a clear answer, the public at least expects an open debate. The greatest failures of science occur when results are distorted to satisfy politicians (for an example see Medvedev, 1969). Failure is also likely if studies are designed to prove that preconceptions are correct (this is called normative science: Lackey, 2001). It is tempting for any institution to try to control scientific findings *a priori*, or to deny or avoid communicating findings that might seem politically unfavourable, when stakes are high. But history shows that distorting results is not a solution in the long term. Inevitably, effective policy suffers, and the trust of the public erodes. Thus, managing scientific input into policy also requires decisions about peer review, whether or not to allow public access to scientific discussions or how to balance representation on advisory committees. Public trust will be affected by each of these decisions. The importance of building trust should be an early consideration in making such decisions.

Internal trust and a sense of common ownership among individuals (including scientists) from different institutions can also aid policy solutions. But missions and cultures differ widely among the technically orientated parties associated with environmental policy. Academics, agencies and stakeholders typically know best their own studies and interpretations. But they may not be especially familiar with the details of work outside their 'silos' of interaction. Suspicions and potential for conflict increase when poorly linked groups carry different perceptions of the state of understanding about issues (Adams *et al.*, 2003). For example, it is common for regulatory scientists to suspect that academic work is irrelevant to policy or management needs. Academics suspect the quality of unpublished data and interpretations. Suspicions exist about interpretations from people paid by entities that were either regulating or have a stake in the process. Where suspicions prevail, constructive dialogue can break down. Different groups try to convince each other that their opinion is best supported by the evidence; often without really understanding what the alternative evidence or interpretation is.

What science to do, how to do it, who should do it and who should be involved in communicating the state of knowledge are examples of choices that influence trust. Building appreciation of the contributions from different professionals is the way to build trust and expand the credibility of the science enterprise as a whole.

18.4.3   Scientific uncertainty

Prologue. Explicitly considering how to deal with uncertainties is a critical ingredient in effective environmental policy making. If trust is a goal of the interaction, then ignoring uncertainty is not an acceptable approach. Advocacy may be essential in situations where all parties are not aware of critical aspects of knowledge. Its role in scientific dialogue or in science–policy dialogue can come at a cost, however.

Choosing how to deal with uncertainty is a key ingredient in policy and integral to the science–policy interaction. As we noted above, ignoring uncertainties does not engender trust from anyone. Nevertheless, there are many reasons that uncertainties are ignored in a policy discussion. Admitting uncertainties can be perceived as weakening the case of an organisation that has decided upon a course of action. It might not be attractive to parties who are wary of litigation. It might be threatening to experts who are recommending policy on the basis of their professional judgement. And it is usually avoided by advocates who are promoting a policy choice. It is also argued that uncertainty confuses the public, and is best left to the experts. Yet uncertainties are a part of every aspect of the science that guides our management of risks from metal contamination. Of course, some uncertainties are more influential than others. The consequences of some will be minor and the consequences of others large. Nothing destroys the trust of the public more effectively than being misinformed about uncertainty, especially uncertainties with large consequences only to have the unexpected consequences appear (the European experience with 'mad cow disease' is an example: Brown, 2004).

As noted above, acceptance that all solutions have uncertainties is the best protection from conspiracy theories, charges of incompetence and/or perceptions of political underhandedness. Resolving scientific uncertainty is not necessarily a realistic expectation. But in a complicated endeavour, discussion and appreciation of uncertainty can be constructive. In the face of uncertainty, scientists who cannot agree on an interpretation can often agree on why they disagree about that interpretation. Surprisingly, such a discussion can often reframe issues in constructive ways and even yield new insights and solutions. Perhaps counter-intuitively, the best way to expose poorly substantiated conjecture is a neutral discussion of the uncertainties that every option presents. Three decades ago, when interest in the environment was just beginning, it was true that knowledge of metal issues was very thin. Today, uncertainties exist, but so does a wealth of experience and new knowledge. Constructive discussion of existing knowledge and its uncertainties is a fruitful avenue for exploring and framing the options for policy making.

At present there seems to be general public and professional trust in most of the existing risk management strategies for metals; probably stemming from the successful reduction of contamination that has occurred where regulations have been in place. However, validation of criteria against observations of metal effects in natural waters is very rare, partly because of the complexity of the sciences involved and partly because validation is not a serious requirement of most existing policies. In a way, this is an example of ignoring uncertainty. International harmonisation of such criteria via validation by studies of natural waters seems a necessary goal for the next generation of policies (Chapter 17). Openly addressing widely recognised uncertainties in the data used to derive most criteria might be politically complex at first, but will ultimately be essential. Nevertheless, neither the science nor policy communities have given priority to validation, even though accountability ultimately seems inevitable.

Advocacy is commonly and effectively used in litigation. Each side advocates its best case about the resolution of an uncertainty and a judge decides the outcome. A 'yes–no' decision is the only possible outcome. At the beginning of the twenty-first century advocacy has become a very popular mode of public discourse about sensitive issues. The analogy between litigation and policy discourse is not a good one, however. Yes–no solutions are often unnecessary in environmental policy. Where uncertainties are important, some of the most satisfactory outcomes can result from incremental steps along a path where consensus is possible. Jurisdictions differ in their tendencies to litigate. But it is a universal observation that litigation and advocacy do not help in clarifying converging lines of evidence where they occur. Advocacy can also give undeserved weight to poorly supported conjecture, especially if the judging body does not have detailed technical knowledge of the subject. Nor does advocacy assist in identifying the critical stumbling blocks to understanding (advocates rarely address uncertainties in their own argument). Some argue that advocacy weakens the credibility of the scientists involved in the dialogue (Schneider and Ingram, 1997) and it presents no clear message for the policy maker.

Advocacy debates can also harden the boundaries between disciplines. For example there is an ongoing and heated debated about whether biomarkers represent acceptable measures of metal effects. The result is policies that typically exclude biomarker data, even where such data bring knowledge about how the contaminant affects the aquatic environment and at what concentration. Similarly, dietary exposure to metals was explained away for decades, despite existing data to the contrary; only for it to be found now that it is crucial piece of determining what and how metals affect animals in nature. It seems that thinking across these old boundaries will be necessary to achieve the exacting environmental standards that new policies and policy approaches require to be effective. It is not clear that advocacy debates are the most effective route to that end.

Relying on professional judgement to choose the best policy is a subtle and more accepted form of advocacy. The risk lies in the choice of the professional. Professional judgement is heavily dependent upon the biases of the individual professional involved. Nor is consensus among experts always the best guide for determining how an uncertainty will ultimately be resolved. Professional judgement can be useful as advice but risky as a strict guide for policy. In a letter to *Science* magazine, Guzelian and Guzelian (2004) wrote: 'when authority based consensus is mistaken for conclusive evidence, uncertainties are shrouded and errors are likely'. The flaw in the approach is the expectation the 'best science' can be reliably and consistently defined by one (or a small group's) opinion.

18.5   THE MECHANICS OF A USEFUL SCIENCE–POLICY INTERACTION

Prologue. The traditional ways of communicating science can all be effective in establishing a communication channel with the policy community. However, it is very important how the communication is framed and implemented. Mechanisms that build a sense of ownership in the communication by all stakeholders will increase the likelihood that traditional boundaries will be penetrated.

Publications, conferences, advisory panels, workshops, white papers and consensus opinions are all traditional ways to communicate science for use in policy. An effective interaction requires that proceedings are organised so that both communities are included as full partners. Including representatives of both science and policy communities wherever technical issues are a consideration provides one opportunity. Conferences can provide a forum where both policy and science can be discussed in presentations and posters. Working groups, workshops and other proceedings can be explicitly designed to present information to higher-level policy makers or decision makers; or to update the technical experts of the policy community to share the latest findings with academics. Blue-ribbon panels of outside, independent experts can be used to review programmes or provide ongoing advice. The latter can be particularly valuable in that an impartial, independent panel of outside experts can provide fresh, novel insights free of the inherent prejudices of the policy making organisation (Freudenburg, 1999). And the panels can learn about and help legitimate the actions of the policy makers.

Traditionally the above activities might occur once, or once in a while. In the most complicated circumstances, an active science–policy dialogue is required that involves ongoing contact. Table 18.2 lists some suggestions for sustaining proceedings that help the connection between policy and science. Carpenter (2002) describes the sequence in an ongoing dialogue referring to long, challenging experiences of scientists and managers managing fisheries:

Initially the debates were intense, and it was difficult for the players to find acceptable compromises. Gridlock prevented changes in policy. Gridlock was broken by massive crashes in fish numbers (a catastrophe of some sort). As the players gained understanding of their . . . problem, it became easier to reach consensus on experimental policies. Stock declines became smaller and less frequent.

Table 18.2 *Some guidelines for organising technical proceedings when the goal is a policy-neutral dialogue between the science community and the policy community. These are designed to provide assessments of the scientific state of knowledge for the policy community*

- **Goal: Expand the sense of joint ownership and build a common view of the state of scientific knowledge**
  - Proceedings open to all interested parties;
  - Proceedings organised by representatives of the science and policy community together;
  - Introductory presentations by representatives of the policy community frame the policy questions and interests;
  - Scientific assessments are summaries designed for the non-expert;
  - Include panels that discuss the intersection between science and policy.

- **Goal: Clarify important uncertainties, clarify where evidence converges, and marginalise poorly substantiated conjecture.**
  - Impartial analyses systematically lay out conceptual models, explain technical data and interpretations, articulate trade-offs and openly admit uncertainties;
  - Separate factual issues from value judgements;
  - Where differences in interpretations exist, the focus can be on why the interpretations differ (what are the most important uncertainties) rather than who is right;
  - Identify important uncertainties; make clear where uncertainties have a minor impact, or are overstated;
  - Do not try to reach consensus on the best interpretations of inconclusive data.

A more complicated issue is the nature of the communication. A great challenge in creating a constructive science–policy dialogue is connecting the science and policy cultures. The science culture is usually uncomfortable with simple answers, but the policy community ultimately must produce them. Policy happens frenetically with many deadlines. Science happens more methodically. And the languages of the two can be quite different. Table 18.3 presents several models for the ways that the science and policy communities communicate, all of which are influenced by these differences. Whatever the model, an important goal is for the science and policy communities to learn about each other's issues before catastrophe strikes. Creating a stream of information through multiple proceedings over time is critical to sustain the connection. If an emergency or surprise does occur, the system is in place to engage science effectively and work collectively toward the best policy response. Any one of the choices for a scientist's role can, of course, be legitimate for an individual. There are many important roles in the social debate over environmental issues: advocacy, consensus building, pure science, marginal engagement in policy making, and promotion of values. But, if the goal is to effectively engage science in a policy making situation, the choices must be viewed in a different light; weighing the advantages and disadvantages for that specific goal in that specific circumstance.

18.6   CONSTRUCTING A SCIENCE AGENDA

Prologue. Many environmental managers must make choices about funding new science; yet there is little formal training to prepare for such policy decisions. Questions about timeliness and unpredictability of science work against it. But experience shows that a relevant body of knowledge is the only substitute for litigation and/or contentious dialogue. The latter can go on for years in the absence of a growing body of knowledge. Policy frenetically demands short-term decision points, but few of those terminate the issue. Science methodically accumulates knowledge as those decision points come and go. The effective timeline for the two often ends up being quite similar.

Successful scientific input in any high stakes environmental issue requires a continual infusion of new knowledge. One of the most important jobs of a manager in such situations is to sponsor development of a base of scientific knowledge that is useful to advances in the policy issue of interest. Yet, like other aspects of the science–policy interaction, the mechanics of creating a credible science agenda are rarely discussed. Some basic elements in building a body of science are introduced below.

The crucial questions in a policy situation are what science should be done, who should do it and what are the expectations? Relevance is the primary expectation from policy makers. Scientists must be aware of the questions most important to the policy community. To be relevant to its stakeholders, a programme of study must derive from the questions that policy makers and politicians are asking. The policy community usually has representatives who are generally familiar with the conventional knowledge about technical issues. They often have a long history of debate about these issues. So their questions can be insightful statements about uncertainties, and where those uncertainties intersect with political issues. The policy community has the task of articulating these questions clearly; the science community's task is to listen carefully.

The challenge for scientists (sometimes unrecognised) is to develop tractable science questions from the policy questions. The policy community defines what issues are addressed; but the science community, not the policy community, is most qualified to dictate how they are addressed. For example, a basic question from policy makers for metal contamination is what concentrations in the environment cause ecological degradation? This question has been deconstructed into science questions about how to test toxicity, how to evaluate bioavailability, what is the nature of ecological degradation, etc. As is clear from this example, it is rare that a policy question can be unequivocally addressed and resolved in one scientific study. Ultimately a body of work provides what solutions are possible for such complex questions. In some of our case studies, metal problems were studied

Table 18.3 *Models for science–policy communication*

Trickle-down

Definition: Scientists passively engage in policy discourse. Assumes that if good science (usually defined by peer review) is conducted, published and presented in traditional technical outlets, its influence will ultimately show up in policy.

Advantages: Clear separation of roles; usually identifies uncertainties; indirect engagement reduces political conflict.

Disadvantages: Policy message and boundaries of knowledge often not clear. Will the policy community find the appropriate papers, read them and understand their relevance? No means to clarify scientific disagreements about interpretations, or implications of uncertainties.

Scientific management

Definition: The policy community assumes that the best policies are a direct outgrowth from 'good science'; the policy community expects science to dictate policy.

Advantages: Clear message if there are no major uncertainties; joint ownership of decision.

Disadvantages: Roles confused: if scientists dictate policy, values, politics and process may be underrepresented. Who defines the 'good science'? Uncertainties can cloud the best scientific choice. Conflict or confusion result, if uncertainties are common or interpretations of data differ (common case).

Advocacy debate

Definition: Pits advocates with different points of view against each other so that the conveners can judge what policy choice is best supported by science.

Advantages: Appropriate roles retained. Relative strengths of the best arguments are examined.

Disadvantages: Little co-ownership of ideas; poor perspective on uncertainties; stimulates conflict. Deadlock is a likely outcome if uncertainties are large and disagreements are strong. Disagreement becomes the message; credibility damaged.

Professional judgement

Definition: Policy follows best judgement on policy from one or more experienced experts.

Advantages: Clear message.

Disadvantages: Narrow view of boundaries of knowledge and uncertainties. Can worsen conflict if uncertainties are large. Confuses roles (policy community abandons responsibility for value judgements).

No science

Definition: Values of the decision maker should dictate the policy. Science has little to offer beyond conventional knowledge held by all interested parties.

Advantages: Simple; keeps roles distinctly separated.

Disadvantages: Can create conflict if technical issues are ignored. Can lead to serious technical errors. No co-ownership. Critical uncertainties likely to be missed. Loss of public trust in policy makers if mistakes occur.

Policy-neutral dialogue

Definition: Science–policy dialogue identifies state of knowledge without drawing final judgement as to best policy choice.

Advantages: Can minimise unnecessary conflict. Not 'who is right', but why do we disagree? What is next? Clearly defined roles, with appropriate role for all. Wide participation. Clarifies and addresses uncertainties; and what is known. Thorough analysis clarifies and isolates 'outlier' concepts.

Disadvantages: Does not directly deliver a policy recommendation.

for decades without a policy solution moving forward. Usually this is because direct policy questions were difficult to address unambiguously and approaches that involved learning about the environmental context of the question were not thought to be policy relevant. Eventually many of these hurdles were overcome as a body of work developed; but that body included ecosystem, biological and geochemical perspectives as well as purportedly directly relevant studies.

Even if the science is relevant and the approach is tractable, an effective science–policy partnership requires sustained support from influential leaders and stakeholders (from most sides). But political support also creates challenging expectations. One common expectation is timeliness. In fact, most in the policy community feel that timelines on policy decision are often too short to allow scientific understanding to develop. The timeliness issue can be an ongoing source of disconnect.

To some extent, the timeliness disconnect is exacerbated in a science programme when it is managed to react to ever-changing policies emergencies. Constant redirection slows the process of scientific discovery. Progress is not accelerated by abrupt changes of direction or artificially short timelines. Inevitably, individual policy timelines always seems short. But the prospective policy maker also might ask: is this issue a simple, one-time circumstance or long-term in nature and inherently complex? Will I be making this decision, then leaving the issue to manage itself, or will follow-up management of the issue require a continuing series of implementation decisions? How will I learn from the experiences I gain in implementing a metal policy? Shouldn't both the decision and implementation timelines govern the choice of science? Experience shows that the actual evolution, development and implementation of policy is usually an ongoing situation. Decisions break down into pieces and decision making usually stretches over years in a complicated situation (and metal issues can be very complicated). In these circumstances there is always time for scientific knowledge to develop. New knowledge may not develop quickly enough to affect the first decision, but if new science is not funded

while that first decision process is occurring, there will be no science later as the solutions evolve and are implemented. Applied science must address the issue at hand and cannot proceed in a completely open-ended, aimless wander. But a timeliness disconnect is not necessary if both the science and policy community appreciate their different needs within the largest context of the issue.

Many applied programmes also choose, as a first step, to constrain how their scientific investment will be spent. This is usually the reaction to an austere budget combined with the expectations cited above. 'We do not want to pay for research' is an example of a common initial constraint. The message behind this is usually that 'basic' research is not desirable, where 'basic' is defined by the expectation of either no clear application to the policy issue, or no possibility of timeliness. But narrowing the approaches to a problem, does not necessarily yield quicker growth of relevant understanding. It is important to recognise that multiple approaches to a question may be necessary to find that crucial breakthrough (Table 18.4); creative approaches come from narrowing proposals but within an atmosphere willing to engage a wide possibility of fruitful avenues. There are trade-offs with regard to timing, control and likelihood of success from individual proposals. An effective goal might be a coherent science strategy, within the allotted budget, instead of predetermined constraints on what will work.

For example, the policy community often provides the funding so a desire for firm control of the science agenda, in terms of questions and approaches, is not surprising. The assumption is usually that limiting science to that which promises short-term or predictable outcomes, as perceived by the policy maker, is the best assurance of useful products. Indeed, studies promising to be of immediate relevance to broad policy questions are the easiest for the policy community to justify to their leaders, when activities are described. And, there are some opportunities in all fields to make quick, relevant advances. But the wise manager recognises that rapid advances are much less common than the steady accumulation of evidence within a strategic framework tailored to the

Table 18.4 *Approaches to developing new science: trade-offs*

Science to fill immediate policy needs

Definition: A small number of scientific studies with short time horizons that directly address and quickly resolve the most crucial policy questions.

Advantages: Relevance is direct. Policy community is in control and can demand what it sees is most important. In theory, no time is lost to marginally relevant detail.

Disadvantages: Policy questions are usually not tractable scientifically. Very few policy questions are so simple that one study will result in resolution of the issue. Time and money can result in promises unkept and little gain. Controversy results if outcomes are spun to meet expectations.

Best approach: Seek 'low-hanging fruit'; i.e. questions where technology, talent and state of knowledge are ready for rapid advance. Peer review can help define whether proposals will meet this objective.

Occurrence and distribution

Definition: Describe distributions and concentrations of metal contamination, and describe the systems of interest.

Advantages: Relatively straightforward. Establishes a baseline for future observations and monitoring. First chance to uncover surprises. Often may build from existing data.

Disadvantages: Requires infrastructure for field work. Is not explanatory (why concentrations are what they are). Often perceived to not meet immediate policy needs, but will build base for future policy.

Best approach: Understanding existing conditions is essential; but often taken for granted. Studies should be sensitive to the scale of the anticipated problems, but need not be intense within those scales (to start).

Process science

Definition: Scientific studies that describe and/or explain the environmental processes relevant to an issue (e.g. metal fate, cycling, environmental interactions and effects – local or generic).

Advantages: A body of such studies will build a robust understanding of the issue; process understanding may apply to multiple issues. Policy built on robust process understanding is the least controversial and the most likely to be trusted.

Disadvantages: Takes time to develop. Projects must be carefully selected for relevance and potential effectiveness; but should include multiple approaches. Can be difficult to understand relevance to issues, so an ongoing dialogue is crucial to interpret outcomes for policy community.

Best approach: Set goals that allow building a body of crucial knowledge, over time. Combine peer review and policy review to assure potential to succeed and relevance. Allow time for knowledge to grow.

Monitoring/assessment

Definition: Keeping track of key variables or performance measures, and, ideally, a few critical system variables, over a long period of time.

Advantage: The only way to track performance of policy and trends in key indicators of environmental welfare.

Disadvantages: Takes time to develop trends. Failure to allow for ongoing interpretation and reporting of data is the single greatest flaw. Monitoring is also easy to over design; in such cases the proposal can be too difficult or too expensive to sustain over the necessary time period to detect and follow trends. It is challenging to design a cost-effective monitoring programme.

Best approach: Begin now. Keep it simple but as explanatory ('process-based') as possible. Know the budget and practical limits; build within those. Build in interpretation. Do not underestimate the design challenge.

Adaptive management experiments

Definition: Treat a management action as an experiment; designing a manipulation of the system to allow rigorous study of the implications of a management action.

Advantages: A formal way to determine, rigorously, the implications of a management action. Where successful, experiments have proven very fruitful and uncovered important surprises.

Disadvantages: Many biophysical, institutional, regulatory and scientific barriers make successful manipulations rare. Feedback will be slow.

Best approach: Make sure that the budget and time allowed are adequate for the scale of the experiment. Begin slowly. Proceed carefully. Make sure that institutional regulations and politics are appropriate to allow the experiment before investing heavily in the science.

question at hand. It is rare that one mandated, policy-oriented study should suddenly uncover what decades of work had not. Proposals to quickly resolve intractable policy questions have a long history in science. The most common outcome is wasted resources spent chasing wishful thinking.

But how does the policy maker sort the great ideas out? One rule is that a relevant question must be accompanied by a thorough proposal detailing how it will achieve the promised outcome. Careful, technical, review by experts in the exact field of the proposal (asking the question: 'can this be done?') combined with policy review by those with a broader view (asking: 'is this a good question?') will address both relevance and whether a proposal is likely to deliver what it promises. If a study that promises to suddenly solve a policy dilemma is deemed likely to succeed, it should certainly be funded; but first it must pass that practical litmus test of whether it can be done.

Monitoring is usually justifiable for the policy community. Long-term monitoring is very tractable for metals, because reliable, relatively inexpensive, analytical capabilities are widely available for most elements. Designing cost-effective monitoring is the greatest challenge. The second is assuring that the monitoring data do not languish uninterpreted. Designing a cost-effective monitoring programme is a major challenge, but an effective monitoring programme can provide invaluable results as time goes on. Modelling is also moderately easy to justify in an applied programme, in that models offer the opportunity to explore options without large investments in experimental or field data. These advantages are likely only if the model is already or nearly developed. Model development, itself, can be as expensive and time-consuming as basic research. Again there are well-established guidelines for effective modelling that are detailed in many publications (e.g. Boesch, 2006).

(1) Models, via their assumptions and inputs, create their own reality; none simulates nature perfectly. Their outcomes are descriptions of how the world will behave within the bounds of that reality.

(2) Models are better used to generate insights and questions than to predict in absolute terms. Overexpectations from predictions are the largest source of modelling 'failures'.

(3) Experienced modellers suggest not to overinvest in one modelling approach, especially if it involves a large, complex model that will, theoretically, solve all problems. As models become larger and more complex it becomes more and more difficult to understand the underlying assumptions. Multiple modelling approaches are the investments most likely to yield useful answers in the long run; and often many smaller models are more powerful than one large one (Boesch, 2006).

Descriptions of problems in nature and studies of ecosystem processes are crucial to risk management decisions. Unambiguous observation was how our most serious problems with contaminants were uncovered. But observation should be supplemented by key mechanistic studies. Policy develops quickly from science when scientists unequivocally demonstrate the mechanisms that explain a phenomenon. Investment in research to explain why something is observed may take time to yield results, but the probability of beneficial outcomes increases the longer the programme goes on.

Inherent to all of the approaches above is recognition that one discipline should not control the strategic science agenda. That theme is prevalent throughout this book, and carries over to choosing a science strategy. Metals science developed from several disciplinary approaches. Representative studies from each, and, especially, studies that integrate disciplinary approaches will eventually result in an optimally robust view of problems and solutions. Optimally one should invest in science the way that investment experts suggest managing money: diversify the portfolio. In large programmes, it can be useful to shift some emphasis within the investment to studies with the greatest chance of addressing an aspect of an immediate policy question; but investing all resources in such studies is usually not wise given their low probability of outcomes that will stand the test of time. Process studies are important because difficult uncertainties will never be resolved unless they are persistently pursued; but a complete programme of process study will yield few results in the short term. Monitoring and modelling are essential to determine successes and failures and provide feedback to management decisions. But their success rests on continually improving knowledge of processes in nature.

The most difficult question asked by policy makers is 'can science deliver'? The history of work on metal contamination shows that science can deliver especially in a proactive, fairly supported mode where the expectations are realistic. The proactive science agenda is more likely to have a true impact than a reactive agenda. An effective science agenda is constantly new knowledge about some aspect relevant to the issue at hand. If the problem is large and complex, the agenda may not deliver each piece of desired new knowledge as it is prescribed. There are many examples where accelerated funding for science facilitated influenced policy. But it is also common that the new knowledge does not immediately resolve the original policy question. This is not because such knowledge is not relevant. But individual findings of science are not necessarily predictable in advance, especially if the question is complex. Resolution of one question often leads to another. What is most common is that new knowledge reframes critical aspects of the debate, or raises questions about the original premise. The most important outcomes may clarify management alternatives as time goes on, and either legitimise the original decisions (e.g. management of lead to avoid human health impacts) or raise questions about their legitimacy (e.g. US water quality criteria for divalent metals). The complexity of ways that science does deliver adds to the challenge of satisfying expectations.

18.7   CONCLUSIONS

Most environmental issues, including the metal issues that we focus on in this book, are ongoing in the sense that many decisions must be made over long periods of time to implement the chosen strategy. An effective science–policy interaction is not something that happens once, results in a decision and goes away. It is an interface at which multiple audiences intersect frequently over time, with each audience playing its appropriate role in influencing each decision and its management. A constant input of new knowledge is critical to the viability of the interface. The dialogue about that new knowledge builds from technical communication among scientists, to discussions of technical issues that included scientists and the policy community, to participation of representatives of the science viewpoint in policy discussions.

We have incorporated a number of case studies that show how the principles in this book can play out in the management of risks from different trace metals: metal wastes in mining, Se, Hg, TBT, and the cationic metals as managed by ambient criteria or sediment quality criteria. In each case lateral interpretations that cross barriers between approaches could narrow uncertainties, even though every methodology will inevitably retain some uncertainty. For example, lateral management recognises that toxicity from one medium (water, sediment) never acts in isolation from the other processes that determine metal fate, exposure and effects. It facilitates use of multiple lines of evidence, to draw conclusions about protective values, with a priority given to observations from nature when contradictions develop between evidence from different types of studies. Thus it seems well designed to facilitate learning from management experience.

In order to incorporate lateral approaches into risk assessment/management, biology and ecology must become as prominent as toxicology and geochemistry. Holistic analyses of data that cut across disciplines must become as important as rigorously identifying the ingredients of a valid toxicity test. The final decision about a policy always involves a political choice. But, ultimately that choice is most justifiable where it takes advantage of convergences of evidence and is explicit about uncertainties. We have come a long distance in the science underlying management of ecological risks from metals. The growth of that knowledge sets the stage for an era in which much more can be accomplished. But it also sets the stage for a new era of management. Environmental contamination with metals will be an ongoing issue as long as is conceivable. Proactively advancing the goal of ever more efficient and effective management of that contamination should be top agenda item for both science and policy in perpetuity.

Suggested reading

Adams, W.M., Brockington, D., Dyson, J. and Vira, B. (2003). Managing tragedies: understanding conflict over common pool resources. *Science*, **302**, 1915–1916.

Dietz, T., Ostrom, E. and Stern, P.C. (2003). The struggle to govern the commons. *Science*, **302**, 1907–1911.

Carpenter, S.R. (2002). Ecological futures: building and ecology of the long now. *Ecology*, **83**, 2069–2083.

Lee, K. N. (1999). Appraising adaptive management. *Conservation Ecology*, **3**, 3. http://www.conservecol.org/vol3/iss2/art3

References

Abu Saba, K.E. and Flegal, A.R. (1995). Chromium in San Francisco Bay: superposition of geochemical processes causes complex spatial distributions of redox species. *Marine Chemistry*, **49**, 189–199.

Abu-Saba, K.E. and Flegal, A.R. (1997). Temporally variable freshwater sources of dissolved chromium to the San Francisco Bay Estuary. *Environmental Science and Technology*, **31**, 3455–3460.

Adams, S.M. (2003). Establishing causality between environmental stressors and effects on aquatic ecosystems. *Human and Ecological Risk Assessment*, **9**, 17–35.

Adams, S.M., Bevelhimer, M.S., Greeley, J., Levine, D.A. and Teh, S.J. (1999). Ecological Risk Assessment in a large river-reservoir. VI. Bioindicators of fish population health. *Environmental Toxicology and Chemistry*, **18**, 628–640.

Adams, W.J. and Chapman, P.M., eds. (2006). *Assessing the Hazard of Metals and Inorganic Metal Substances in Aquatic and Terrestrial Systems*. Boca Raton, FL: CRC Press.

Adams, W.J. and Rowland, C.D. (2003). Aquatic toxicology test methods. In *Handbook of Ecotoxicology*, ed. D.J. Hoffman, B.A. Raffner, G.A. Burton Jr and J. Cairns Jr. Boca Raton, FL: CRC Press, pp. 19–45.

Adams, W.M., Brockington, D., Dyson, J. and Vira, B. (2003). Managing tragedies: understanding conflict over common pool resources. *Science*, **302**, 1915–1916.

Adams, W., Brix, K.V., Cothern, K.A., *et al.* (1998). Assessment of selenium food chain transfer and critical exposure factors for avian wildlife species: need for site-specific data. In *Environmental Toxicology and Risk Assessment*, ASTM Special Technical Publication No.133, ed. E.E. Little, A.J. DeLonay and B.M. Greenberg. Washington, DC: American Society for Testing and Materials, pp. 312–342.

Ahsanullah, M. and Florence, T.M. (1984). Toxicity of copper to the marine amphipod *Allorchestes compressa* in the presence of water- and lipid-soluble ligands. *Marine Biology*, **84**, 41–45.

Akagi, H., Malm, O., Kinjo, Y., *et al.* (1995). Methylmercury pollution in the Amazon, Brazil. *Science of the Total Environment*, **175**, 85–95.

Akberali, H.B. and Black, J.E. (1980). Behavioural responses of the bivalve *Scrobicularia plana* (Da Costa) subjected to short-term copper (Cu-II) concentrations. *Marine Environmental Research*, **4**, 97–107.

Akkeson, B. (1983). Methods for assessing the effects of chemicals on reproduction in marine worms. In *Methods for Assessing Effects of Chemicals on Reproductive Functions*, ed. V.B. Voulk and P.J. Sheehan. Chichester, UK: John Wiley, pp. 459–470.

Allen, H.E., Fu, G. and Deng, B. (1993). Analysis of acid-volatile sulfide (AVS) and simultaneously extracted metals (SEM) for the estimation of potential toxicity in aquatic sediments. *Environmental Toxicology and Chemistry*, **12**, 1441–1453.

Allen, J.I. and Moore, M.N. (2004). Environmental prognostics: is the current use of biomarkers appropriate for environmental risk evaluation? *Marine Environmental Research*, **58**, 227–232.

Aller, R.C. (1978). Experimental studies of changes produced by deposit feeders on pore water, sediment, and overlying water chemistry. *American Journal of Science*, **278**, 1185–1234.

Aller, R.C. (1980). Relationships of tube-dwelling benthos with sediment and overlying water chemistry. In Marine Benthic Dynamics, 11th Belle W. Baruch Symposium in Marine Science, Georgetown, SC, April 1979, ed. K.R. Tenore and B.C. Coull. *Belle W. Baruch Library in Marine Science*, 11, 285–308.

Al-Mohanna, S.Y. and Nott, J.A. (1987). R-cells and the digestive cycle in *Penaeus semisulcatus* (Crustacea: Decapoda). *Marine Biology*, **95**, 129–137.

Al-Mohanna, S.Y. and Nott, J.A. (1989). Functional cytology of the hepatopancreas of *Penaeus semisulcatus* (Crustacea: Decapoda) during the moult cycle. *Marine Biology*, **101**, 535–544.

Alongi, D.M., Tirendi, F. and Robertson, A.L. (1991). Vertical profiles of copper in sediments from the Fly

Delta and Gulf of Papua (Papua New Guinea). *Marine Pollution Bulletin*, **22**, 253–255.

Alzieu, C. (2000). Impact of tributyltin in invertebrates. *Ecotoxicology*, **9**, 71–76.

Amiard, J.-C. and Amiard-Triquet, C. (1993). Zinc. In *Handbook of Hazardous Materials*, ed. M.W. Corn. San Diego, CA Academic Press, pp. 733–744.

Amiard, J.-C., Amiard-Triquet, C., Barka, S., Pellerin, J. and Rainbow, P. S. (2006). Metallothioneins in aquatic invertebrates: their role in metal detoxification and their use as biomarkers. *Aquatic Toxicology*, **76**, 160–202.

Amiard, J.-C., Amiard-Triquet, C., Berthet, B. and Métayer, C. (1986). Contribution to the ecotoxicological study of cadmium, lead, copper and zinc in the mussel *Mytilus edulis*. I. Field study. *Marine Biology*, **90**, 425–431.

Amiard, J.-C., Amiard-Triquet, C., Berthet, B. and Métayer, C. (1987). Comparative study of the patterns of bioaccumulation of essential (Cu, Zn) and non-essential (Cd, Pb) trace metals in various estuarine and coastal organisms. *Journal of Experimental Marine Biology and Ecology*, **106**, 73–89.

Amiard-Triquet, C. and Amiard, J.-C. (1998). Influence of ecological factors on accumulation of metal mixtures. In *Metal Metabolism in Aquatic Environments*, ed. W. J. Langston and M. Bebianno. London: Chapman and Hall, pp. 351–386.

Amiard-Triquet, C., Burgeot, T. and Claisse, D. (1999). Biomonitoring of the marine environmental quality: experiences of the French National Observation Network (RNO) and the development of biomarkers. *Océanis*, **25**, 651–684.

Amiard-Triquet, C., Jeantet, A.-Y. and Berthet, B. (1993). Metal transfer in marine food chains: bioaccumulation and toxicity. *Acta Biologica Hungarica*, **44**, 387–409.

Amyot, M., Pinel-Alloul, B. and Campbell, P. G. C. (1994). Abiotic and seasonal factors influencing trace metal levels (Cd, Cu, Ni, Pb, and Zn) in the freshwater amphipod *Gammarus fasciatus* in two fluvial lakes of the St. Lawrence River. *Canadian Journal of Fisheries and Aquatic Sciences*, **51**, 2003–2016.

Amyot, M., Pinel-Alloul, B., Campbell, P. G. C. and Désy, J. C. (1996). Total metal burdens in the freshwater amphipod *Gammarus fasciatus*: contribution of various body parts and influence of gut contents. *Freshwater Biology*, **35**, 363–373.

Anderson, D. M. and Morel, F. M. M. (1977). Copper sensitivity of *Gonyaulax tamarensis*. *Limnology and Oceanography*, **23**, 284–293.

Anderson, D. T. (1994). *Barnacles: Structure, Function, Development and Evolution*. London: Chapman and Hall.

Anderson, S. L. and Norberg-King, T. J. (1991). Precision of short-term chronic toxicity tests in the real world. *Environmental Toxicology and Chemistry*, **10**, 143–145.

Ankley, G. T. (1996). Evaluation of metal/acid-volatile sulfide relationships in the prediction of metal bioaccumulation by benthic macroinvertebrates. *Environmental Toxicology and Chemistry*, **15**, 2138–2146.

Ankley, G. T., Di Toro, D. M., Hansen, D. J. and Berry, W. J. (1996). Technical basis and proposal for deriving sediment quality criteria for metals. *Environmental Toxicology and Chemistry*, **15**, 2056–2066.

Ankley, G. T., Mattson, V. R., Leonard, E. N. and West, C. W. (1993). Predicting the acute toxicity of copper in freshwater sediments: evaluation of the role of acid-volatile sulfide. *Environmental Toxicology and Chemistry*, **12**, 315–320.

Apitz, S. E., Elliott, M., Fountain, M. and Galloway, T. S. (2006). European environmental management: moving to an ecosystem approach. *Integrated Environmental Assessment and Management*, **2**, 80–85.

Appleton, J. D., Williams, T. J., Breward, N., *et al.* (1999). Mercury contamination associated with artisanal gold mining on the island of Mindanao, the Philippines. *Science of the Total Environment*, **228**, 95–109.

Apte, S. C., Howard, A. G., Morris, R. J. and McCartney, M. J. (1986). Arsenic, antimony and selenium speciation during a spring phytoplankton bloom in a closed experimental ecosystem. *Marine Chemistry*, **20**, 119–130.

Arnot, J. A., Mackay, D., Webster, E. and Southwood, J. M. (2006). Screening level risk assessment model for chemical fate and effects in the environment. *Environmental Science and Technology*, **40**, 2316–2323.

ASTM (American Society for Testing and Materials) (2001). Standard terminology relating to biological effects and environmental fate. E 943-000. In *Annual Book of ASTM Standards*, Vol. 11.05, Philadelphia, PA: American Society for Testing and Materials. pp. 246-248.

ASTM (American Society for Testing and Materials) (2003). *Standard Guide for Designing Biological Tests with Sediments*, ASTM Designation: E 1367-03. Philadelphia, PA: American Society for Testing and Materials.

Atkinson, R. L. (1987). *Copper and Copper Mining*, Shire Album 201. Haverfordwest, UK: Shire Publications.

Au, D. W. T. (2004). The application of histo-cytopathological biomarkers in marine pollution monitoring: a review. *Marine Pollution Bulletin*, **48**, 817–834.

Averett, R. C. (1961). *Macro-invertebrates of the Clark Fork River, Montana*. Helena, MT: Montana Board of Health.

Axtmann, E. V. and Luoma, S. N. (1991). Large-scale distribution of metal contamination in the fine-grained sediments of the Clark Fork River, Montana, U.S.A. *Applied Geochemistry*, **6**, 75–88.

Ayling, G.M. (1974). Uptake of cadmium, zinc, copper, lead and chromium in the Pacific oyster, *Crassostrea gigas*, grown in the Tamar River, Tasmania. *Water Research*, **8**, 729–738.

Babut, M.P., Ahlf, W., Batley, G.E., *et al.* (2005). International overview of sediment quality guidelines. In *Use of Sediment Quality Guidelines and Related Tools for the Assessment of Contaminated Sediments*, ed. R.J. Wenning, G.E. Batley, C.G. Ingersoll and D.W. Moore. Pensacola, FL: SETAC Press, pp. 345–382.

Baines, S.B., Fisher, N.S., Doblin, M.A. and Cutter, G.A. (2001). Uptake of dissolved organic selenides by marine phytoplankton. *Limnology and Oceanography*, **46**, 1936–1944.

Baines, S.B., Fisher, N.S., Doblin, M.A., Cutter, G.A. and Cole, B. (2004). Light dependence of Se uptake by phytoplankton and implications for predicting Se incorporation into food webs. *Limnology and Oceanography*, **49**, 566–578.

Baines, S.B., Fisher, N.S. and Stewart, A.R. (2002). Assimilation and retention of selenium and other trace elements from crustacean food by juvenile striped bass (*Morone saxatilis*). *Limnology and Oceanography*, **47**, 646–655.

Baird, D.J., Barber, I., Bradley, M., Calow, P. and Soares, A.M.V.M. (1989). The *Daphnia* bioassay: a critique. *Hydrobiologia*, **188/9**, 403–406.

Baldwin, S. and Maher, W. (1997). Spatial and temporal variation of selenium concentration in five species of intertidal mollusks from Jervis Bay, Australia. *Marine Environmental Research*, **44**, 243–263.

Ball, A.L., Borgmann, U. and Dixon, G. (2006). Toxicity of a cadmium-contaminated diet to *Hyallela azteca*. *Environmental Toxicology and Chemistry*, **25**, 2526–2532.

Banaoui, A., Chiffoleau, J.-F., Moukrim, A., *et al.* (2004). Trace metal distribution in the mussel *Perna perna* along the Moroccan coast. *Marine Pollution Bulletin*, **48**, 378–402.

Barata, C., Markich, S.J. and Baird, D.J. (2002). The relative importance of water and food as cadmium sources to *Daphnia magna*. *Aquatic Toxicology*, **61**, 143–154.

Barnett, B.E. and Ashcroft, C.R. (1985). Heavy metals in *Fucus vesiculosus* in the Humber estuary. *Environmental Pollution B*, **9**, 193–213.

Barrick, R., Beller, H., Becker, S. and Ginn, T. (1989). Use of the apparent effects threshold approach (AET) in classifying contaminated sediments. In *Contaminated Marine Sediments: Assessment and Remediation*, ed. Committee on Contaminated Marine

Sediments, NRC. Washington, DC: National Academy Press, pp. 64–77.

Batley, G.E., Stahl, R.G., Babut, M.P., *et al.* (2005). Scientific underpinnings of sediment quality guidelines. In *Use of Sediment Quality Guidelines and Related Tools for the Assessment of Contaminated Sediments*, ed. R.J. Wenning, G.E. Batley, C.G. Ingersoll and D.W. Moore. Pensacola, FL: SETAC Press, pp. 39–120.

Baudrimont, M., Andrès, S., Metivaud, J., *et al.* (1999). Field transplantation of the freshwater bivalve *Corbicula fluminea* along a polymetallic contamination gradient (River Lot, France): metallothionein response to metal exposure. *Environmental Toxicology and Chemistry*, **18**, 2472–2477.

Bayne, B.L., Addison, R.F., Capuzzo, J.M., *et al.* (1988a). An overview of the GEEP workshop. *Marine Ecology Progress Series*, **46**, 235–243.

Bayne, B.L., Brown, D.A., Burns, K., *et al.* (1985). *The Effects of Stress and Pollution on Marine Animals*. New York: Praeger.

Bayne, B.L., Clarke, K.R. and Gray, J.S. (1988b). Biological effects of pollution: results of a practical workshop. *Marine Ecology Progress Series*, **46**, 1–278.

Belzile, N., Lecomte, P. and Tessier, A. (1989a). Testing readsorption of trace elements during partial chemical extractions of bottom sediments. *Environmental Science and Technology*, **23**, 1015–1020.

Belzile, N., Vitre, R.R.D. and Tessier, A. (1989b). *In situ* collection of diagenetic iron and manganese oxyhydroxides from natural sediments. *Nature*, **340**, 376–377.

Benayoun, G., Fowler, S.W. and Oregioni, B. (1974). Flux of cadmium through euphausiids. *Marine Biology*, **27**, 205–212.

Bendell-Young, L.I., Harvey, H.H. and Young, J.F. (1986). Accumulation of cadmium by white suckers (*Catostomus commersoni*) in relation to fish growth and lake acidification. *Canadian Journal of Fisheries and Aquatic Sciences*, **43**, 806–811.

Bengtsson, B.-E. (1979). Biological variables, especially skeletal deformities in fish, for monitoring marine pollution. *Philosophical Transactions of the Royal Society of London B*, **286**, 457–464.

Bengtsson, B.-E. and Larsson, A. (1986). Vertebral deformities and physiological effects in fourhorn sculpin (*Myxocephalus quadricornis*) after long-term exposure to a simulated heavy metal-containing effluent. *Aquatic Toxicology*, **9**, 215–229.

Benner, S.G., Smart, E.W. and Moore, J.N. (1995). Metal behaviour during surface–groundwater interaction, Silver Bow Creek, Montana. *Environmental Science and Technology*, **29**, 1789–1795.

Benoit, J. M., Gilmour, C. C., Mason, R. P., Riedel, G. S. and Riedel, G. F. (1998). Behaviour of mercury in the Patuxent River estuary. *Biogeochemistry*, **40**, 249–265.

Berry, C. (2006). The precautionary principle – more sorry than safe. *Science in Parliament*, **63**, 18–19.

Berry, W. J., Cantwell, M. G., Edwards, P. A., Serbst, J. R. and Hansen, D. J. (1999). Predicting toxicity of sediments spiked with silver. *Environmental Toxicology and Chemistry*, **18**, 40–48.

Berthet, B., Mouneyrac, C., Amiard, J.-C., *et al.* (2003). Accumulation and soluble binding of cadmium, copper, and zinc in the polychaete *Hediste diversicolor* from coastal sites with different trace metal bioavailabilities. *Archives of Environmental Contamination and Toxicology*, **45**, 468–478.

Besser, J. M., Canfield, T. J. and La Point, T. W. (1993). Bioaccumulation of organic and inorganic selenium in a laboratory food chain. *Environmental Toxicology and Chemistry*, **12**, 57–72.

Bielmyer, G. K., Grosell, M. and Brix, K. V. (2006). Toxicity of silver, zinc, copper, and nickel to the copepod *Acartia tonsa* exposed via a phytoplankton diet. *Environmental Science and Technology*, **40**, 2063–2068.

Biggs, R. B., DeMoss, T. B., Carter, M. M. and Beasley, E. L. (1989). Susceptibility of U. S. estuaries to pollution. *Reviews in Aquatic Science*, **1**, 40–51.

Binz, P. A. and Kägi, J. H. R. (1999). Metallothionein: molecular evolution and classification. In *Metallothionein IV*, ed. C. Klaassen. Basel, Switzerland: Birkhäuser, pp. 7–13.

Birch, G. F. (2003). A test of normalization methods for marine sediment, including a new post-extraction normalization (PEN) technique. *Hydrobiologia*, **492**, 5–13.

Bjerregaard, P. (1985). Effect of selenium on cadmium uptake in the shore crab *Carcinus maenas* (L.). *Aquatic Toxicology*, **7**, 177–189.

Bjerregaard, P. and Vislie, T. (1986). Effects of copper on ion- and osmoregulation in the shore crab *Carcinus maenas* (L.). *Marine Biology*, **91**, 69–76.

Blackmore, G. (2000). Field evidence of metal transfer from invertebrate prey to an intertidal predator, *Thais clavigera* (Gastropoda: Muricidae). *Estuarine, Coastal and Shelf Science*, **51**, 127–139.

Blackmore, G. and Wang, W.-X. (2002). Uptake and efflux of Cd and Zn by the green mussel *Perna viridis* after metal preexposure. *Environmental Science and Technology*, **36**, 989–995.

Blaise, C., Gagné, F. and Pellerin, J. (2003). Bivalve population status and biomarker responses in *Mya arenaria* clams (Saguenay Fjord, Québec, Canada). *Fresenius Environmental Bulletin*, **12**, 956–960.

Blanck, H., Wängberg, S.-A. and Molander, S. (1988). Pollution-induced community tolerance: a new ecotoxicological tool. In *Functional Testing of Aquatic Biota for Estimating Hazards of Chemicals ASTM 988*, ed. J. Cairns Jr and J. R. Pratt. Philadelphia, PA: American Society for Testing and Materials, pp. 219–230.

Bloesch, J. (1995). Mechanisms, measurements and importance of sediment resuspension in lakes. *Marine and Freshwater Research*, **46**, 295–304.

Bloom, N. S. (1989). Determination of picogram levels of methylmercury by aqueous phase ethylation, followed by cryogenic gas chromatography with cold vapour atomic absorption spectrophotometry. *Canadian Journal of Fisheries and Aquatic Sciences*, **46**, 1131–1140.

Bloom, N. S., Gill, G. A., Cappellino, S., *et al.* (1999). Speciation and cycling of mercury in Lavaca Bay, Texas, sediments. *Environmental Science and Technology*, **33**, 7–13.

Blowes, D. (2002). Tracking hexavalent Cr in groundwater. *Science*, **5562**, 2024–2025.

Blum, D. J. W. and Speece, R. E. (1990). Determining chemical toxicity to aquatic species. *Environmental Science and Technology*, **24**, 284b–290b.

Boalch, R., Chan, S. and Taylor, D. (1981). Seasonal variation in the trace metal content of *Mytilus edulis*. *Marine Pollution Bulletin*, **12**, 276–280.

Bodaly, R. A. and Fudge, R. J. P. (1993). Uptake of mercury by fish in an experimental boreal reservoir. *Archives of Environmental Contamination and Toxicology*, **37**, 103–109.

Bodaly, R. A., St Louis, V. L., Paterson, M. J., *et al.* (1997). Bioaccumulation of mercury in the aquatic food chain in newly flooded areas. *Metal Ions in Biological Systems*, **34**, 259–287.

Bodar, C. W. M., Pronk, M. E. J. and Sijm, D. T. H. M. (2005). The European Union risk assessment on zinc and zinc compounds: the process and the facts. *Integrated Environmental Assessment and Management*, **1**, 301–319.

Boesch, D. F. (2006). Scientific requirements for ecosystem-based management in the restoration of Chesapeake Bay and coastal Louisiana. *Ecological Engineering*, **26**, 6–26.

Bonsdorff, E., Bakke, T. and Pedersen, A. (1990). Colonization of amphipods and polychaetes to sediments experimentally exposed to oil hydrocarbons. *Marine Pollution Bulletin*, **21**, 355–358.

Bonzongo, J.-C., Heim, K. J., Chen, Y., *et al.* (1996). Mercury pathways in the Carons River–Lahanton Reservoir system, Nevada, USA. *Environmental Toxicology and Chemistry*, **15**, 677–683.

Bopp, F. and Biggs, R. B. (1981). Metals in estuarine sediments: factor analysis and its environmental significance. *Science*, **214**, 441–443.

Bordin, G., McCourt J. and Rodríguez, A. (1992). Trace metals in the marine bivalve *Macoma balthica* in the Westerschelde Estuary (The Netherlands). I. Analysis of total copper, cadmium, zinc and iron concentrations: locational and seasonal variations. *Science of the Total Environment*, **127**, 255–280.

Borgmann, U. (1998). A mechanistic model of copper accumulation in *Hyalella azteca*. *Science of the Total Environment*, **219**, 137–145.

Borgmann, U. and Norwood, W. P. (1995). Kinetics of excess (above background) copper and zinc in *Hyalella azteca* and their relationship to chronic toxicity. *Canadian Journal of Fisheries and Aquatic Sciences*, **52**, 864–874.

Borgmann, U., Norwood, W. P. and Babirad, I. M. (1991). Relationship between chronic toxicity and bioaccumulation of cadmium in *Hyalella azteca*. *Canadian Journal of Fisheries and Aquatic Sciences*, **48**, 1055–1060.

Borgmann, U., Norwood, W. P. and Dixon, D. G. (2004). Re-evaluation of metal bioaccumulation and chronic toxicity in *Hyalella azteca* using saturation curves and the biotic ligand model. *Environmental Pollution*, **131**, 469–484.

Boult, S., Collins, D. N., White, K. N. and Curtis, C. D. (1994). Metal transport in a stream polluted by acid mine drainage: the Afon Goch, Anglesey, UK. *Environmental Pollution*, **84**, 279–284.

Bourassa, A. (2004). The environmental impact of producing nonrenewable resources: the case for mining. In *Life-Cycle Assessment of Metals: Issues and Research Directions*, ed. A. Dubreuil. Pensacola, FL: SETAC Press, pp. 132–137.

Bourgoin, B. P., Risk, M. J. and Aitken, A. E. (1991). Factors controlling lead availability to the deposit-feeding bivalve *Macoma balthica* in sulphide-rich oxic sediments. *Estuarine, Coastal and Shelf Science*, **32**, 625–632.

Boyden, C. R. (1974). Trace element content and body size in molluscs. *Nature*, **251**, 311–314.

Boyden, C. R. (1977). Effect of size on metal content of shellfish. *Journal of the Marine Biological Association, of the United Kingdom*, **57**, 675–714.

Boyden, C. R. and Phillips, D. J. H. (1981). Seasonal variation and inherent variability of trace elements in oysters and their implications for indicator studies. *Marine Ecology Progress Series*, **5**, 29–40.

Boyle, E. A. (1979). Copper in natural waters. In *Copper in the Environment*, Part 1, *Ecological Cycling*, ed. J. O. Nriagu. New York: John Wiley, pp. 77–88.

Boyle, E. A., Chapnick S. D., Shen G. T. and Bacon, M. (1986). Temporal variability of lead in the western North Atlantic. *Journal of Geophysical Research*, **91**, 8573–8593.

Boyle, E. A., Edmond, J. M. and Sholkovitz, E. R. (1977). The mechanism of iron removal in estuaries. *Geochimica et Cosmochimica Acta*, **41**, 1313–1324.

Boyle, E. A., Huested, S. S. and Grant, B. (1982). The chemical mass balance of the Amazon Plume. II. Copper, nickel, and cadmium. *Deep Sea Research*, **29**, 1355–1366.

Brand, L. E., Sunda, W. G. and Guillard, R. R. L. (1986). Reduction of marine phytoplankton reproduction rates by copper and cadmium. *Journal of Experimental Marine Biology and Ecology*, **96**, 225–250.

Bremner, I. (1991). Metallothionein and copper metabolism in liver. In *Methods in Enzymology*, vol. **205**, *Metallobiochemistry*, Part B, *Metallothionein and Related Molecules*, ed. J. F. Riordan and B. L. Vallee. San Diego, CA: Academic Press, pp. 584–591.

Brick, C. M. and Moore, J. N. (1996). Diel variation of trace metals in the Upper Clark Fork River, Montana. *Environmental Science and Technology*, **30**, 1953–1960.

Bright, D. A. and Ellis, D. V. (1989). Aspects of histology in *Macoma carlottensis* (Bivalvia: Tellinidae) and *in situ* histopathology related to mine-tailings discharge. *Journal of the Marine Biological Association of the United Kingdom*, **69**, 447–464.

Brix, K. V., DeForest, D. K. and Adams, W. J. (2001). Assessing acute and chronic copper risks to freshwater aquatic life using species sensitivity distributions for different taxonomic groups. *Environmental Toxicology and Chemistry*, **20**, 1846–1856.

Brix, K. V., Toll, J. E., Tear, L. M., DeForest, D. K. and Adams, W. J. (2005). Setting site-specific water-quality standards by using tissue residue thresholds and bioaccumulation data. II. Calculating site-specific selenium water-quality standards for protecting fish and birds. *Environmental Toxicology and Chemistry*, **24**, 231–237.

Brook, E. J. and Moore, J. N. (1988). Particle size and chemical control of As, Cd, Cu, Fe, Mn, Ni, Pb, and Zn in bed sediment from the Clark Fork River, Montana (U. S.A.). *Science of the Total Environment*, **76**, 247–266.

Brooke, L. T., Ankley, G. T., Call, D. J. and Cook, P. M. (1996). Gut content weight and clearance rate for three species of freshwater invertebrates. *Environmental Toxicology and Chemistry*, **15**, 223–228.

Brooks, B. W., Stanley, J. K., White, J. C., *et al.* (2004). Laboratory and field responses to cadmium: an experimental study in effluent-dominated stream mesocosms. *Environmental Toxicology and Chemistry*, **23**, 1057–1064.

Brown, B. E. (1982). The form and function of metal-containing 'granules' in invertebrate tissues. *Biological Reviews*, **57**, 621–667.

Brown, C. L., and Luoma, S. N. (1995). Use of the euryhaline bivalve *Potamocorbula amurensis* as a biosentinel species to assess trace metal contamination in San Francisco Bay. *Marine Ecology Progress Series*, **124**, 129–142.

Brown, C. L., Parchaso, F., Thompson, J. K. and Luoma, S. N. (2003). Assessing toxicant effects in a complex estuary: a case study of effects of silver on reproduction in the bivalve, *Potamocorbula amurensis*, in San Francisco Bay. *Human and Ecological Risk Assessment*, **9**, 95–119.

Brown, D. A. and Parsons, T. R. (1978). Relationship between cytoplasmic distribution of mercury and toxic effects to zooplankton and chum salmon (*Oncorhynchus keta*) exposed to mercury in a controlled ecosystem. *Journal of the Fisheries Research Board of Canada*, **35**, 880–884.

Brown, P. (2004). Mad-cow disease in cattle and human beings. *American Scientist*, **92**, 334–341.

Bruland, K. W. (1983). Trace elements in sea-water. In *Chemical Oceanography*, vol. 8, ed. J. P. Riley and R. Chester. London: Academic Press, pp. 157–220.

Bruland, K. W. (1989). Oceanic zinc speciation: complexation of zinc by natural organic ligands in the central north Pacific. *Limnology and Oceanography*, **34**, 267–285.

Bryan, G. W. (1971). The effects of heavy metals (other than mercury) on marine and estuarine organisms. *Proceedings of the Royal Society of London B*, **177**, 389–410.

Bryan, G. W. (1984) Pollution due to heavy metals and their compounds. In *Marine Ecology*, vol. 5, Part 3, ed. O. Kinne. London: John Wiley, pp. 1289–1430.

Bryan, G. W. (1985). Bioavailability and effects of heavy metals in marine deposits. In *Wastes in the Ocean: Near Shore Waste Disposal*, ed. I. Duedorf, B. Perk and D. Kester, London: John Wiley, pp. 41–61.

Bryan, G. W. and Gibbs, P. E. (1983). Heavy metals in the Fal Estuary, Cornwall: a study of long-term contamination by mining waste and its effects on estuarine organisms. *Occasional Publications of the Marine Biological Association of the United Kingdom*, **2**, 1–112.

Bryan, G. W. and Gibbs, P. E. (1987). Polychaetes as indicators of heavy-metal availability in marine deposits. In *Oceanic Processes in Marine Pollution*, vol. 1, ed. J. M. Capuzzo and D. R. Kester. Malabar, FL: Robert E. Kreiger, pp. 37–49.

Bryan, G. W. and Gibbs, P. E. (1991). Impact of low concentrations of tributyltin (TBT) on marine organisms: a review: In *Metal Ecotoxicology: Concepts and Applications*, ed. M. C. Newman and A. W. McIntosh. Ann Arbor, MI: Lewis Publishers, pp. 323–361.

Bryan, G. W. and Hummerstone, L. G. (1971). Adaptation of the polychaete *Nereis diversicolor* to estuarine sediments containing high concentrations of heavy metals. I. General observations and adaptation to copper. *Journal of the Marine Biological Association, of the United Kingdom*, **51**, 845–863.

Bryan, G. W. and Hummerstone, L. G. (1973a). Brown seaweed as an indicator of heavy metals in estuaries in south-west England. *Journal of the Marine Biological Association of the United Kingdom*, **53**, 705–720.

Bryan, G. W. and Hummerstone, L. G. (1973b). Adaptation of the polychaete *Nereis diversicolor* to estuarine sediments containing high concentrations of zinc and cadmium. *Journal of the Marine Biological Association of the United Kingdom*, **53**, 839–857.

Bryan, G. W. and Hummerstone, L. G. (1977). Indicators of heavy-metal contamination in the Looe estuary (Cornwall) with particular regard to silver and lead. *Journal of the Marine Biological Association of the United Kingdom*, **57**, 75–92.

Bryan, G. W. and Langston, W. J. (1992). Bioavailability, accumulation and effects of heavy metals in sediments with special reference to United Kingdom estuaries: a review. *Environmental Pollution*, **76**, 89–131.

Bryan, G. W., Gibbs, P. E., Hummerstone, L. G. and Burt, G. R. (1986). The decline of the gastropod *Nucella lapillus* around South-West England: evidence for the effect of tributyltin from antifouling paints. *Journal of the Marine Biological Association of the United Kingdom*, **66**, 611–640.

Bryan, G. W., Langston, W. J. and Hummerstone, L. G. (1980). The use of biological indicators of heavy metal contamination in estuaries. *Occasional Publications of the Marine Biological Association of the United Kingdom*, **1**, 1–73.

Bryan, G. W., Langston, W. J., Hummerstone, L. G. and Burt, G. R. (1985). A guide to the assessment of heavy-metal contamination in estuaries. *Occasional Publications of the Marine Biological Association of the United Kingdom*, **4**, 1–92.

Bryan, G. W., Langston, W. J., Hummerstone, L. G., Burt, G. R. and Ho, Y. B. (1983). An assessment of the gastropod, *Littorina littorea*, as an indicator of heavy-metal contamination in United Kingdom estuaries. *Journal of the Marine Biological Association of the United Kingdom*, **63**, 327–345.

Buchwalter, D. B. and Luoma, S. N. (2005). Differences in dissolved cadmium and zinc uptake among stream insects: mechanistic explanations. *Environmental Science and Technology*, **39**, 498–504.

Buck, K. N. and Bruland, K. W. (2005). Copper speciation in San Francisco Bay: a novel approach using multiple analytical windows. *Marine Chemistry*, **96**, 185–198.

Burd, B. (2002). Evaluation of mine tailings effects on a benthic marine infaunal community over 29 years. *Marine Environmental Research*, **53**, 481–519.

Burkhardt, L. P., Cook, P. M. and Mount, D. R. (2003). The relationship of bioaccumulative chemicals in water and sediment to residues in fish: a visualization approach. *Environmental Toxicology and Chemistry*, **22**, 2822–2830.

Burton, G. A., Chapman, P. M. and Smith, E. P. (2002). Weight of evidence approaches for assessing ecosystem impairment. *Human and Ecological Risk Assessment*, **8**, 1657–1673.

Burton, G. A., Denton, D. L., Ho, K. and Ireland, D. S. (2003). Sediment toxicity testing: issues and methods. In *Handbook of Ecotoxicology*, ed. D. J. Hoffman, B. A. Raffner, G. A. Burton Jr and J. Cairns Jr. Boca Raton, FL: CRC Press, pp. 111–150.

Burton, G. A., Nguyen, L. T. H., Janssen, C., *et al.* (2005). Field validation of sediment zinc toxicity. *Environmental Toxicology and Chemistry*, **24**, 541–553.

Burton, G. A., Norberg-King, T. J., Ingersoll, C. G., *et al.* (1996). Interlaboratory study of precision: *Hyalella azteca* and *Chironomus tentans* freshwater sediment toxicity assays. *Environmental Toxicology and Chemistry*, **15**, 1335–1343.

Bury, N. R. and Wood, C. M. (1999). Mechanism of branchial apical silver uptake by rainbow trout is via the proton-coupled Na$^+$ channel. *American Journal of Physiology*, **277**, R1385–R1391.

Bury, N. R., Walker, P. A. and Glover, C. N. (2003). Nutritive metal uptake in teleost fish. *Journal of Experimental Biology*, **206**, 11–23.

Byrne, R. H., Kump, L. R. and Cantrell, K. J. (1988). The influence of temperature and pH on trace metal speciation in seawater. *Marine Chemistry*, **25**, 163–181.

Cabana, G. and Rasmussen, J. B. (1994). Modelling food chain structure and contaminant bioaccumulation using stable nitrogen isotopes. *Nature*, **372**, 255–257.

Cade, B. S. and Noon, B. R. (2003). A gentle introduction to quantile regression for ecologists. *Frontiers of Ecology and Environment*, **1**, 412–421.

Cain, D. J. and Luoma, S. N. (1985). Copper and silver accumulation in transplanted and resident clams (*Macoma balthica*) in south San Francisco Bay. *Marine Environmental Research*, **15**, 115–135.

Cain, D. J. and Luoma, S. N. (1990). Influence of seasonal growth, age, and environmental exposure on Cu and Ag in a bivalve indicator, *Macoma balthica*, in San Francisco Bay. *Marine Ecology Progress Series*, **60**, 45–55.

Cain, D. J. and Luoma, S. N. (1991). Benthic insects as indicators of large-scale trace metal contamination in the Clark Fork River, Montana. In *US Geological Survey Toxic Substances Hydrology Program*, Proceedings of the Technical Meeting, Monterey, Ca, ed. G. E. Mallard and D. A. Aronson. *Geological Survey Water Resources Investigations Report* **91**-4034, 525–529.

Cain, D. J. and Luoma, S. N. (1999). Metal exposures to native populations of the caddisfly *Hydropsyche* (Trichoptera: Hydropsychidae) determined from cytosolic and whole body metal concentrations. *Hydrobiologia*, **386**, 103–117.

Cain, D. J., Buchwalter, D. B. and Luoma, S. N. (2006). Influence of metal exposure history on the bioaccumulation and subcellular distribution of aqueous cadmium in the insect *Hydropsyche californica*. *Environmental Toxicology and Chemistry*, **25**, 1042–1049.

Cain, D. J., Luoma, S. N. and Axtmann, E. V. (1995). Influence of gut content in immature aquatic insects on assessments of environmental metal contamination. *Canadian Journal of Fisheries and Aquatic Sciences*, **52**, 2736–2746.

Cain, D. J., Luoma, S. N., Carter, J. L. and Fend, S. V. (1992). Aquatic insects as bioindicators of trace element contamination in cobble-bottom rivers and streams. *Canadian Journal of Fisheries and Aquatic Sciences*, **49**, 2141–2154.

Cain, D. J., Luoma, S. N. and Wallace, W. G. (2004). Linking metal bioaccumulation of aquatic insects to their distribution patterns in a mining-impacted river. *Environmental Toxicology and Chemistry*, **23**, 1463–1473.

Cairns, J. (1983). Are single species toxicity tests alone adequate for estimating environmental hazard? *Hydrobiologia*, **100**, 47–57.

Cairns, J. (1986). What is meant by validation of predictions based on laboratory toxicity tests? *Hydrobiologia*, **137**, 271–278.

Cairns, J. (1989). Foreword. *Hydrobiologia*, **188/9**, 1–5.

Cairns, J. (1993). Environmental science and resource management in the 21st century: scientific perspective. *Environmental Toxicology and Chemistry*, **12**, 1321–1329.

Cairns, J. (2003). The emergence and future of ecotoxicology. In *Fundamentals of Ecotoxicology*,

2nd edn., ed. M. C. Newman and M. A. Unger. Boca Raton, FL: Lewis Publishers, pp. 8–9.

Cairns, J. and Mount, D. I. (1990). Aquatic toxicology. *Environmental Science and Technology*, **24**, 154–161.

Cairns, J., Pratt, J. R., Neiderlehner, R. R., and McCormick, P. V. (1986). A simple, cost-effective multispecies toxicity test using organisms with a cosmopolitan distribution. *Environmental Monitoring and Assessment*, **6**, 207–220.

Calabrese, A., MacInnes, J. R., Nelson, D. A., Greig, R. A. and Yevich, P. P. (1984). Effects of long-term exposure to silver or copper on growth, bioaccumulation and histopathology in the blue mussel *Mytilus edulis*. *Marine Environmental Research*, **11**, 253–274.

Calmano, W., Forstner, U. and Hong, H. (1994). Mobilization and scavenging of heavy metals following resuspension of anoxic sediments from the Elbe River. *American Chemical Society Symposium Series*, **550**, 298–321.

Calow, P. (1992). The three Rs of ecotoxicology. *Functional Ecology*, **6**, 617–619.

Calow, P. and Sibly, R. M. (1990). A physiological basis of population processes: ecotoxicological implications. *Functional Ecology*, **4**, 283–288.

Calow, P., Sibly, R. M. and Forbes, V. E. (1997). Risk assessment on the basis of simplified population dynamics scenarios. *Environmenal Toxicology and Chemistry*, **16**, 1983–1994.

Campbell, L. M., Norstrom, R. J., Hobson, K. A., *et al.* (2005). Mercury and other trace elements in a pelagic Arctic marine food web (Northwater Polyna, Baffin Bay). *Science of the Total Environment*, **351/2**, 247–263.

Campbell, P. G. C. (1995). Interaction between trace metals and aquatic organisms: a critique of the free-ion activity model. In *Metal Speciation and Aquatic Systems*, ed. A. Tessier and D. R. Turner. New York: John Wiley, pp. 45–102.

Campbell, P. G. C. and Stokes, P. M. (1985). Acidification and toxicity of metals to aquatic biota. *Canadian Journal of Fisheries and Aquatic Sciences*, **42**, 2034–2049.

Campbell, P. G. C., Errecalde, O., Fortin, C., Hiriart-Baer, V. P. and Vigneault, B. (2002). Metal bioavailability to phytoplankton: applicability of the biotic ligand model. *Comparative Biochemistry and Physiology C*, **133**, 189–206.

Campbell, P. G. C., Hontela, A., Rasmussen, J. B., *et al.* (2003). Differentiating between direct (physiological) and food-chain mediated (bioenergetic) effects on fish in metal-impacted lakes. *Human and Ecological Risk Assessment*, **9**, 847–866.

Camus, L., Pampanin, D. M., Volpato, E., *et al.* (2004). Total oxyradical scavenging capacity responses in *Mytilus*

galloprovincialis transplanted into Venice lagoon (Italy) to measure the biological impact of anthropogenic activities. *Marine Pollution Bulletin*, **49**, 801–808.

Cantillo, A. Y. (1998). Comparison of results of Mussel Watch programs of the United States and France with worldwide Mussel Watch studies. *Marine Pollution Bulletin*, **36**, 712–717.

Cao, Y., Bark, A. W. and Williams, P. (1996). Measuring the responses of macroinvertebrate communities to water pollution: a comparison of multivariate approaches, biotic and diversity indices. *Hydrobiologia*, **341**, 1–19.

Carlisle, D. M. and Clements, W. H. (2003). Growth and secondary production of aquatic insects along a gradient of Zn contamination in Rocky Mountain streams. *Journal of the North American Benthological Society*, **22**, 582–597.

Carlisle, D. M. and Clements, W. H. (2005). Leaf litter breakdown, microbial respiration and shredder production in metal-polluted streams. *Freshwater Biology*, **50**, 380–390.

Carpenter, S. R. (2002). Ecological futures: building and ecology of the long now. *Ecology*, **83**, 2069–2083.

Carr, R. S., Williams, J. W. and Fragata, C. T. B. (1989). Development and evaluation of a novel marine sediment pore water toxicity test with the polychaete *Dinophilus gyrociliatus*. *Environmental Toxicology and Chemistry*, **8**, 533–543.

Carson, R. (1962). *Silent Spring*. Boston, MA: Houghton Mifflin.

Castro J. M. and Moore, J. N. (2000). Pit lakes: their characteristics and the potential for their remediation. *Environmental Geology*, **39**, 1254–1260.

Cavaletto, M., Ghezzi, A., Burlando, B., *et al.* (2002). Effect of hydrogen peroxide on antioxidant enzymes and metallothionein level in the digestive gland of *Mytilus galloprovincialis*. *Comparative Biochemistry and Physiology C*, **131**, 447–455.

Chadwick, J. W., Canton, S. P. and Dent, R. L. (1986). Recovery of benthic invertebrate communities in Silver Bow Creek, Montana, following improved metal mine wastewater treatment. *Water, Air and Soil Pollution*, **28**, 427–438.

Champ, M. A. (2000). A review of organotin regulatory strategies, pending actions, related costs and benefits. *Science of the Total Environment*, **258**, 21–71.

Chan, H. M., Bjerregaard, P., Rainbow, P. S. and Depledge, M. H. (1992). Uptake of zinc and cadmium by two populations of shore crabs *Carcinus maenas* at different salinities. *Marine Ecology Progress Series*, **86**, 91–97.

Chapman, P.M. (1986). Sediment quality criteria from the sediment quality triad: an example. *Environmental Toxicology and Chemistry*, **5**, 957–964.

Chapman, P.M. (1989a). A bioassay by any other name might not smell the same. *Environmental Toxicology and Chemistry*, **8**, 551.

Chapman, P.M. (1989b). Current approaches to developing sediment quality criteria. *Environmental Toxicology and Chemistry*, **8**, 589–599.

Chapman, P.M. (1990). The sediment quality triad approach to determining pollution-induced degradation. *Science of the Total Environment*, **97/8**, 815–825.

Chapman, P.M. (1991). Environmental quality criteria: what type should we be developing? *Environmental Science and Technology*, **25**, 1352–1359.

Chapman, P.M. (1995). Extrapolating laboratory toxicity tests to the field. *Environmental Toxicology and Chemistry*, **14**, 927–930.

Chapman, P.M. (1997). Is bioaccumulation useful for predicting impacts? *Marine Pollution Bulletin*, **34**, 282–283.

Chapman, P.M. (2007). Selenium thresholds for fish from cold freshwaters. *Human and Ecological Risk Assessment*, **13**, 20–24.

Chapman, P.M. and Long, E.R. (1983). The use of bioassays as a part of a comprehensive approach to marine pollution assessment. *Marine Pollution Bulletin*, **14**, 81–84.

Chapman, P.M., Anderson, B., Carr, S., et al. (1997). General guidelines for using the sediment quality triad. *Marine Pollution Bulletin*, **34**, 368–372.

Chapman, P.M., Caldwell, R.S. and Chapman, P.F. (1996). A warning: NOECs are inappropriate for regulatory use. *Environmental Toxicology and Chemistry*, **15**, 77–79.

Chapman, P.M., Dexter, R.N. and Long, E.R. (1987). Synoptic measures of sediment contamination, toxicity and infaunal community composition (the Sediment Quality Triad) in San Francisco Bay. *Marine Ecology Progress Series*, **37**, 75–96.

Chapman, P.M., Wang, F., Adams, W.J. and Green, A. (1999). Appropriate applications of sediment quality values for metals and metalloids. *Environmental Science and Technology*, **33**, 3937–3941.

Charbonneau P. and Hare, L. (1998) Burrowing behaviour and biogenic structures of mud-dwelling insects. *Journal of the North American Benthological Society*, **17**, 239–249.

Chen, C.Y., Stemberger, R.S., Klaue, B., et al. (2000). Accumulation of heavy metals in food web components across a gradient of lakes. *Limnology and Oceanography*, **45**, 1525–1536.

Chen, Y.-W. and Belzile, N. (2001). Antagonistic effect of selenium on mercury assimilation by fish populations near Sudbury metal smelters? *Limnology and Oceanography*, **46**, 1814–1818.

Chen, Z. and Mayer, L.M. (1998a). Mechanisms of Cu solubilization during deposit feeding. *Environmental Science and Technology*, **32**, 770–775.

Chen, Z. and Mayer, L.M. (1998b). Digestive proteases of the lugworm, *Arenicola marina*, inhibited by Cu from contaminated sediments. *Environmental Toxicology and Chemistry*, **17**, 433–438.

Chen, Z. and Mayer, L.M. (1999a). Assessment of sedimentary Cu availability: a comparison of biomimetic and AVS approaches. *Environmental Science and Technology*, **33**, 650–652.

Chen, Z. and Mayer, L.M. (1999b). Sedimentary metal bioavailability determined by the digestive constraints of marine deposit feeders: gut retention time and dissolved amino acids. *Marine Ecology Progress Series*, **176**, 139–151.

Chen, Z., Mayer, L.M., Quetel, C., et al. (2000). High concentrations of complexed metals in the guts of deposit feeders. *Limnology and Oceanography*, **45**, 1358–1367.

Chen, Z., Mayer, L.M., Weston, D.P., Bock, M.J. and Jumars, P.A. (2002). Inhibition of digestive enzyme activities by copper in the guts of various marine benthic invertebrates. *Environmental Toxicology and Chemistry*, **21**, 1243–1248.

Cheng, T. and Allen, H.E. (2006). Comparison of zinc complexation properties of dissolved natural organic matter from different surface waters. *Journal of Environmental Management*, **80**, 222–226.

Chester, R. and Hughes, M.J. (1967). A chemical technique for the separation of ferromanganese minerals, carbonate minerals and adsorbed trace elements from pelagic sediments. *Chemical Geology*, **2**, 249–262.

Cheung, M.S. and Wang, W.-X. (2005). Influences of subcellular compartmentalization in different prey on the transfer of metals to a predatory gastropod from different prey. *Marine Ecology Progress Series*, **286**, 155–166.

Cheung, M.S., Fok, E.M.W., Ng, T.Y.-T., Yen, Y.-F. and Wang, W.-X. (2005). Subcellular cadmium distribution, accumulation, and toxicity in a predatory gastropod, *Thais clavigera*, fed different prey. *Environmental Toxicology and Chemistry*, **25**, 174–181.

Chiffoleau, J.-F., Auger, D., Roux, N., Rozuel, E. and Santini, A. (2005). Distribution of silver in mussels and oysters along the French coasts: data from the national monitoring program. *Marine Pollution Bulletin*, **50**, 1719–1723.

Chong, K. and Wang, W.-X. (2000). Assimilation of cadmium, chromium, and zinc by the green mussel

Perna viridis and the clam *Ruditapes philippinarum*. *Environmental Toxicology and Chemistry*, **19**, 1660–1667.

Chong, K. and Wang, W.-X. (2001). Comparative studies on the biokinetics of Cd, Cr, and Zn in the green mussel *Perna viridis* and the Manila clam *Ruditapes philipinnarum*. *Environmental Pollution*, **115**, 107–121.

Clark, E. A. and Steritt, R. N. (1988). The fate of tributyl tin in the aquatic environment. *Environmental Science and Technology*, **22**, 600–603.

Clarke, K. R. and Ainsworth, M. (1993). A method of linking multivariate community structure to environmental variables. *Marine Ecology Progress Series*, **92**, 205–219.

Clarkson, T. W. (2002). The three modern faces of mercury. *Environmental Health Perspectives*, **110**, 11–23.

Clearwater, S. J., Farag, A. and Meyer, J. S. (2002). Bioavailability and toxicity of dietborne copper and zinc to fish. *Comparative Biochemistry and Physiology C*, **132**, 269–313.

Clements, W. H. (2000). Integrating effects of contaminants across levels of biological organization: an overview. *Journal of Aquatic Ecosystem Stress and Recovery*, **7**, 113–116.

Clements, W. H. (2004). Small-scale experiments support causal relationships between metal contamination and macroinvertebrate community responses. *Ecological Applications*, **14**, 954–957.

Clements, W. H. and Newman, M. C. (2002). *Community Ecotoxicology*. Chichester, UK: John Wiley.

Clements, W. H., Carlisle, D. M., Courtney, L. A. and Harrahy, E. A. (2002). Integrating observational and experimental approaches to demonstrate causation in stream biomonitoring studies. *Environmental Toxicology and Chemistry*, **21**, 1138–1146.

Clements, W. H., Carlisle, D. M., Lazorchak, J. M. and Johnson, P. C. (2000). Heavy metals structure benthic communities in Colorado Mountain streams. *Ecological Applications*, **10**, 626–638.

Clements, W. H., Cherry, D. S. and Cairns, J. Jr (1988). Structural alterations in aquatic insect communities exposed to copper in laboratory streams. *Environmental Toxicology and Chemistry*, **7**, 715–725.

Clifton, R. J. and Hamilton, E. I. (1979). Lead-210 chronology in relation to levels of elements in dated sediment core profiles. *Estuarine and Coastal Marine Science*, **8**, 250–269.

Coakley, J. P. and Poulton, D. J. (1993). Source-related classification of St. Lawrence Estuary sediments based upon spatial distribution of adsorbed contaminants. *Estuaries*, **16**, 873–876.

Coale, K. H. and Bruland, K. W. (1990). Spatial and temporal variability in copper complexation in the North Pacific. *Deep-Sea Research*, **47**, 317–336.

Coglianese, C. and Marchant, G. E. (2003). *Shifting Sands: The Limits of Science in Setting Risk Standards*, KSG Working Paper No. RWP03-036. Available online at http://ssrn.com/abstract=443080.

Cohen, J. T., Bellinger, D., Connor, W., *et al.* (2005). A quantitative risk–benefit analysis of changes in population fish consumption. *American Journal of Preventive Medicine*, **29**, 325–333.

Connell, D., Lam, P., Richardson, B. and Wu., R. (1999). *Introduction to Ecotoxicology*. Oxford, UK: Blackwell Science.

Connell, J. H. (1985). The consequences of variation in initial settlement versus post-settlement mortality in rocky intertidal communities. *Journal of Experimental Marine Biology and Ecology*, **93**, 11–45.

Conrad, E. M. and Ahearn, G. A. (2005). 3H-I-histidine ^{65}Zn^{2+} are cotransported by a dipeptide transport system in intestine of lobster *Homarus americanus*. *Journal of Experimental Biology*, **208**, 287–296.

Cooke, T. D. and Bruland, K. W. (1987). Aquatic chemistry of selenium: evidence of biomethylation. *Environmental Science and Technology*, **21**, 1214–1219.

Cooper, D. C. and Morse, J. W. (1998). Extractability of metal sulfide minerals in acidic solutions: application to environmental studies of trace metal contamination within anoxic sediments. *Environmental Science and Technology*, **32**, 1076–1078.

Corner, E. D. S. and Sparrow, B. W. (1957). The modes of action of toxic agents. II. Factors influencing the toxicities of mercury compounds to certain Crustacea. *Journal of the Marine Biological Association of the United Kingdom*, **36**, 459–472.

Cossa, D., Bourget, E. and Piuze, J. (1979). Sexual maturation as a source of variation in the relationship between cadmium concentration and body weight of *Mytilus edulis* L. *Marine Pollution Bulletin*, **10**, 174–176.

Costa, F. O., Neuparth, T., Correia, A. D. and Costa, M. H. (2005). Multi-level assessment of chronic toxicity of estuarine sediments with the amphipod *Gammarus locusta*. II. Organism and population-level endpoints. *Marine Environmental Research*, **60**, 93–110.

Couillard, Y., Campbell, P. G. C., Pellerin-Massicotte, J. and Auclair, J. C. (1995b). Field transplantation of a freshwater bivalve, *Pyganodon grandis*, across a metal contamination gradient. II. Metallothionein response to Cd and Zn exposure, evidence for cytotoxicity, and links to effects at higher levels of biological organisation. *Canadian Journal of Fisheries and Aquatic Sciences*, **52**, 703–715.

Couillard, Y., Campbell, P. G. C., Tessier, A., Pellerin-Massicotte, J. and Auclair, J. C. (1995a). Field transplantation of a freshwater bivalve *Pyganodon grandis*, across a metal contamination gradient. I. Temporal changes in metallothionein and metal (Cd, Cu and Zn) concentrations in soft tissues. *Canadian Journal of Fisheries and Aquatic Sciences*, **52**, 690–702.

Courtney, L. A. and Clements, W. H. (2002). Assessing the influence of water quality and substratum quality on benthic macroinvertebrate communities in a metal-polluted stream: an experimental approach. *Freshwater Biology*, **47**, 1766–1778.

Couture, P. and Kumar, P. R. (2003). Impairment of metabolic capacities in copper and calcium contaminated wild yellow perch (*Perca flavescens*). *Aquatic Toxicology*, **64**, 107–120.

Croteau, M. N. and Luoma, S. N. (2005). Delineating copper accumulation pathways for the freshwater bivalve *Corbicula* using stable copper isotopes. *Environmental Toxicology and Chemistry*, **24**, 2871–2878.

Croteau, M. N., Hare, L. and Tessier, A. (1998). Refining and testing a trace metal biomonitor (*Chaoborus*) in highly acidic lakes. *Environmental Science and Technology*, **32**, 1348–1353.

Croteau, M. N., Luoma, S. N. and Stewart, A. R. (2005). Trophic transfer of metals along freshwater food webs: evidence of cadmium biomagnification in nature. *Limnology and Oceanography*, **50**, 1511–1519.

Croteau, M. N., Luoma, S. N., Topping, B. R. and Lopez, C. B. (2004). Stable metal isotopes reveal copper accumulation and loss dynamics in the freshwater bivalve *Corbicula*. *Environmental Science and Technology*, **38**, 5002–5009.

Cullen, J. T., Lane, T. W., Morel, F. M. M. and Sherrell, R. M. (1999). Modulation of cadmium uptake in phytoplankton by seawater CO_2 concentration. *Nature*, **402**, 165–167.

Cumbie, P. M. and Van Horn, S. L. (1978). Selenium accumulation associated with fish mortality and reproductive failure. *Proceedings of the Annual Conference of the Southeastern Association of Fish and Wildlife Agencies*, **32**, 612–624.

Currant, F., Navarro, M. and Amiard, J.-C. (1996). Mercury in pilot whales: possible limits to the detoxification process. *Science of the Total Environment*, **186**, 95–104.

Curtis, T. M., Williamson, R. and Depledge, M. H. (2001). The initial mode of action of copper on the cardiac physiology of the blue mussel, *Mytilus edulis*. *Aquatic Toxicology*, **52**, 29–38.

Cutshall, N. (1974). Turnover of zinc-65 in oysters. *Health Physics*, **26**, 327–331.

Cutter, G. A. (1982). Selenium in reducing waters. *Science*, **217**, 829–831.

Cutter, G. A. (1989). The estuarine behavior of selenium in San Francisco Bay. *Estuarine, Coastal and Shelf Science*, **28**, 13–34.

Cutter, G. A. and Bruland, K. W. (1984). The marine biogeochemistry of selenium: a re-evaluation. *Limnology and Oceanography*, **29**, 1179–1192.

Cutter, G. A. and Cutter, L. S. (2004). Selenium biogeochemistry in the San Francisco Bay estuary: changes in water column behavior. *Estuarine, Coastal and Shelf Science*, **61**, 463–476.

Cutter, G. A. and San Diego-McGlone, M. L. C. (1990). Temporal variability of selenium fluxes in San Francisco Bay. *Science of the Total Environment*, **97/8**, 235–250.

Dai, M.-H. and Martin, J.-M. (1995). First data on trace metal level and behaviour in two major Arctic river-estuarine systems (Ob and Yenisey) and in the adjacent Kara Sea, Russia. *Earth and Planetary Science Letters*, **131**, 127–141.

Dallinger, R., Berger, B., Hunziker, P. and Kägi, J. H. R. (1997). Metallothionein in snail Cd and Cu metabolism. *Nature*, **388**, 237–238.

Daskalakis, K. D. and O'Connor, T. P. (1995). Distribution of chemical concentrations in US coastal and estuarine sediment. *Marine Environmental Research*, **40**, 381–398.

Davenport, J. and Manley, A. (1978). The detection of heightened sea-water copper concentrations by the mussel *Mytilus edulis*. *Journal of the Marine Biological Association of the United Kingdom*, **58**, 843–850.

De Coen, W. M. I. and Janssen, C. R. (1997), The use of biomarkers in *Daphnia magna* toxicity testing. IV. Cellular Energy Allocation: a new biomarker to assess the energy budget of toxicant-stressed *Daphnia* populations. *Journal of Aquatic Stress and Recovery*, **6**, 43–55.

De Coen, W. M. I. and Janssen, C. R. (2003). The missing biomarker link: relationships between effects on the cellular energy allocation biomarker of toxicant-stressed *Daphnia magna* and corresponding population parameters. *Environmental Toxicology and Chemistry*, **22**, 1632–1641.

De Coen, W., Robbens, J. and Janssen, C. (2006). Ecological impact assessment of metallurgic effluents using in situ biomarker assays. *Environmental Pollution*, **141**, 283–294.

de Kock, W. C. and Kramer, K. J. M. (1994). Active biomonitoring (ABM) by translocation of bivalve molluscs. In *Biomonitoring of Coastal Waters and Estuaries*, ed. K. J. M. Kramer. Boca Raton, FL: CRC Press, pp. 51–84.

De Schamphelaere, K. A. C. and Janssen, C. R. (2004). Development and field validation of a biotic ligand model predicting chronic copper toxicity to *Daphnia magna*. *Environmental Toxicology and Chemistry*, **23**, 1365–1375.

De Schamphelaere, K. A. C., Lofts, S. and Janssen, C. R. (2005). Bioavailability models for predicting acute and chronic toxicity of zinc to algae, daphnids, and fish in natural surface waters. *Environmental Toxicology and Chemistry*, **24**, 1190–1197.

De Wolf, H., Van den Broeck, H., Qadah, D., Backeljau, T. and Blust, R. (2005). Temporal trends in soft tissue metal levels in the periwinkle *Littorina littorea* along the Scheldt estuary, The Netherlands, *Marine Pollution Bulletin*, **50**, 463–484.

Dearfield, K. L., Bender, E. S., Kravits, M., *et al.* (2005). Ecological risk assessment issues identified during the U. S. Environmental Protection Agency's examination of risk assessment practices. *Integrated Environmental Assessment and Management*, **1**, 73–81.

Decho, A. W. and Luoma, S. N. (1991). Time-courses in the retention of food material in the bivalves *Potamocorbula amurensis* and *Macoma balthica*: significance to the absorption of carbon and chromium. *Marine Ecology Progress Series*, **78**, 303–314.

Decho, A. W. and Luoma, S. N. (1994). Humic and fulvic acids: sink or source in the availability of metals to the marine bivalves *Macoma balthica* and *Potamocorbula amurensis*? *Marine Ecology Progress Series*, **108**, 133–145.

Decho, A. W. and Luoma, S. N. (1996). Flexible digestion strategies and trace metal assimilation in marine bivalves. *Limnology and Oceanography*, **41**, 568–572.

DeForest, D. K., Brix, K. V., and Adams, W. J. (1999). Critical review of proposed residue-based selenium toxicity thresholds for freshwater fish. *Human and Ecological Risk Assessment*, **5**, 1187–1228.

Delemos, J. L., Bostick, B. C., Renshaw, C. E., Sturup, S. and Feng, X. (2006). Landfill-stimulated iron reduction and arsenic release at the Coakley Superfund Site (NH). *Environmental Science and Technology*, **40**, 67–73.

Depledge, M. H. (1989a). Re-evaluation of metabolic requirements for copper and zinc in decapod crustaceans. *Marine Environmental Research*, **27**, 115–126.

Depledge, M. H. (1989b). The rational basis for detection of the early effects of marine pollutants using physiological indicators. *Ambio*, **18**, 301–302.

Depledge, M. H. (1990). New approaches in ecotoxicology: can inter-individual physiological variability be used

as a tool to investigate pollution effects? *Ambio*, **19**, 251–252.

Depledge, M. H. (1994a). Genotypic toxicity: implications for individuals and populations. *Environmental Health Perspectives*, **102**, 101–104.

Depledge, M. H. (1994b). Series foreword. In *Ecotoxicology in Theory and Practice*, ed. V. E. Forbes and T. L. Forbes. London: Chapman and Hall, pp. vii–ix.

Depledge, M. H. (1996). Genetic ecotoxicology: an overview. *Journal of Experimental Marine Biology and Ecology*, **200**, 57–66.

Depledge, M. H. and Andersen, B. B. (1990). A computer-aided physiological monitoring system for continuous long-term recording of cardiac activity in selected invertebrates. *Comparative Biochemistry and Physiology A*, **96**, 473–477.

Depledge, M. H. and Bjerregaard, P. (1989). Haemolymph protein composition and copper levels in decapod crustaceans. *Helgoländer wissenschaftliche Meeresuntersuchungen*, **43**, 207–223.

Depledge, M. H. and Fossi, M. C. (1994). The role of biomarkers in environmental assessment. II. Invertebrates. *Ecotoxicology*, **3**, 161–172.

Depledge, M. H., Aagaard, A. and Györkös, P. (1995). Assessment of trace metal toxicity using molecular, physiological and behavioural biomarkers. *Marine Pollution Bulletin*, **31**, 19–27.

Depledge, M. H., Amaral-Mendel, J. J., Daniel, B., *et al.* (1993). The conceptual basis of the biomarker approach. In *Biomarker Research and Application in the Assessment of Environmental Health*, ed. D. B. Peakall and R. L. Shugart. Berlin, Germany: Springer-Verlag, pp. 15–29.

Dethlefsen, V. (1984). Diseases in North Sea fishes. *Helgoländer wissenschaftliche Meeresuntersuchungen*, **37**, 353–374.

DeWitt, T. H., Hickey, C. W., Morrisey, D. J., *et al.* (1999). Do amphipods have the same concentration response to contaminated sediment *in situ* as *in vitro*? *Environmental Toxicology and Chemistry*, **18**, 1026–1037.

Di Toro, D. M. (2003). Executive summary. In *Reevaluation of the State of the Science for Water-Quality Criteria Development*, ed. M. C. Reiley, W. A. Stubblefield, W. J. Adams, D. M. Di Toro, P. V. Hodson, R. J. Erickson and F. J. Keating, Pensacola, FL: SETAC Press, pp. xxi–xxv.

Di Toro, D. M., Mahony, J. D., Hansen, D. J., *et al.* (1992). Acid volatile sulfide predicts the acute toxicity of cadmium and nickel in sediments. *Environmental Science and Technology*, **26**, 96–101.

Di Toro, D. M., McGrath, J. A., Hansen, D. J., *et al.* (2005). Predicting sediment metal toxicity using a sediment biotic ligand model: methodology and initial application. *Environmental Toxicology and Chemistry*, **24**, 2410–2427.

Di Toro, D. M., Zarba, C. S., Hansen, D. J., *et al.* (1991). Technical basis for establishing sediment quality criteria for nonionic organic chemicals using equilibrium partitioning. *Environmental Toxicology and Chemistry*, **10**, 1541–1583.

Diamond, J. M., Gerardi, C., Leppo, E. and Miorelli, T. (1997). Using a water-effect ratio aproach to establish effects of an effluent-influenced stream on copper toxicity to the fathead minnow. *Environmental Toxicology and Chemistry*, **16**, 1480–1487.

Diamond, M. L. (1995). Application of a mass balance model to assess in-place arsenic pollution. *Environmental Science and Technology*, **29**, 29–42.

Dickson, K. L., Waller, W. T., Kennedy, J. H. and Ammann, L. P. (1992). Assessing the relationship between ambient toxicity and instream biological response. *Environmental Toxicology and Chemistry*, **11**, 1907–1322.

Dietz, T., Ostrom, E. and Stern, P. C. (2003). The struggle to govern the commons. *Science*, **302**, 1907–1911.

Dines, H. G. (1956). *The Metalliferous Mining Region of South-West England*. London: Her Majesty's Stationery Office.

Distel, D. L. (1998). Evolution of chemoautotrophic endosymbioses in bivalves. *BioScience*, **48**, 277–286.

Doblin, M. A., Baines, S. B., Cutter, L. S. and Cutter, G. A. (2006). Sources and biogeochemical cycling of particulate selenium in the San Francisco Bay estuary. *Estuarine, Coastal and Shelf Science*, **67**, 681–694.

Dodge, K. A., Hornberger, M. I. and Lavgine, I. R. (2003). Water-quality, bed-sediment and biological data (October 2001 through September 2002) and statistical summaries of data for streams in the Upper Clark Fork Basin, Montana. *US Geological Survey Open File Report*, 203–356.

Dollar, N. L., Souch, C. J., Filippelli, G. M. and Mastalerz, M. (2001). Chemical fractionation of metals in wetland sediments: Indiana Dunes National Lakeshore. *Environmental Science and Technology*, **35**, 3608–3615.

Donat, J. R. and Bruland, K. W. (1995). Trace elements in the oceans. In *Trace Elements in Natural Waters*, ed. B. Salbu and E. Steinnes. Boca Raton, FL: CRC Press, pp. 247–281.

Donat, J. R., Lao, K. A. and Bruland, K. W. (1994). Speciation of dissolved copper and nickel in South San Francisco Bay: a multi-method approach. *Analytica Chimica Acta*, **284**, 547–571.

Dondero, F., Dagnino, A., Jonsson, H., *et al.* (2006). Assessing the occurrence of a stress syndrome in mussels (*Mytilus edulis*) using a combined/gene expression approach. *Aquatic Toxicology*, **78**, 13–24.

Dorward-King, E. J. (2007). Extractive development and the valuation of biodiversity and community needs in Madagascar. In *Valuation of Ecological Resources*, ed. R. Stahl. Boca Raton FL: CRC Press, pp. 187–194.

Dwane, G. C. and Tipping, E. (1998). Testing a humic speciation model by titration of copper-amended natural waters. *Environment International*, **24**, 609–616.

Eggleton, J. and Thomas, K. V. (2004). A review of factors affecting the release and bioavailability of contaminants during sediment disturbance events. *Environment International*, **30**, 973–980.

Ehlers, L. J. and Luthy, R. G. (2003). Contaminant bioavailability in soil and sediment. *Environmental Science and Technology*, **37**, 296A–302A.

Eisler, R. (1981). *Trace Metal Concentrations in Marine Organisms*. New York: Pergamon.

Elbaz-Poulichet, F., Garnier, J.-M., Guan, D. M., Martin, J.-M. and Thomas, A. J. (1996). The conservative behaviour of trace metals (Cd, Cu, Ni and Pb) and As in the surface plume of stratified estuaries: example of the Rhône River (France). *Estuarine, Coastal and Shelf Science*, **42**, 289–310.

Encarta (2001). Mining. Available online at http://encarta. msn.com/encyclopedia_761575410/Mining.html

Engler, R. M., Long, E. R., Swartz, R. C., *et al.* (2005). Chronology of the development of sediment quality assessment methods in North America. In *Use of Sediment Quality Guidelines and Related Tools for the Assessment of Contaminated Sediments*, ed. R. J. Wenning, G. E. Batley, C. G. Ingersoll and D. W. Moore. Pensacola, FL: SETAC Press, pp. 311–344.

English, M. R., Dale, V. H., Van Riper-Geibig, C. and Ramsey, W. H. (1999). Overview. In *Tools to Aid Environmental Decision Makers*, ed. V. H. Dale and M. R. English. New York: Springer-Verlag, pp. 1–32.

Engstrom, D. R. and Swain, E. B. (1997). Recent declines in atmospheric mercury deposition in the upper midwest. *Environmental Science and Technology*, **31**, 60–67.

Environment Australia (2002). *National Ocean Disposal Guidelines for Dredged Materials*. Canberra, ACT: Commonwealth of Australia. Available online at www.environment.gov.au/coats/pollution/dumping/guidelines/pubs/guidelines.pdf

Environment Canada (2002). *Canadian Environmental Quality Guidelines*. Ottawa, ON: Canadian Council of Ministers of the Environment.

Erel, Y. and Patterson, C. C. (1994). Leakage of industrial lead into the hydrocycle. *Geochimica et Cosmochimica Acta*, **58**, 3289–3296.

Errécalde, O. and Campbell, P. G. C. (2000). Cadmium and zinc bioavailability to *Selenastrum capricornutum*

(Chlorophyceae): accidental metal uptake and toxicity in the presence of citrate. *Journal of Phycology*, **36**, 473–483.

Etxeberria, M., Sastre, I., Cajaraville, M. P. and Marigómez, I. (1994). Digestive lysosome enlargement induced by experimental exposure to metals (Cu, Cd, and Zn) in mussels collected from a zinc-polluted site. *Archives of Environmental Contamination and Toxicology*, **27**, 338–345.

European Food Safety Authority (2006). *Opinion of the PPR Panel on the Scientific Principles in the Assessment and Guidance Provided in the Area of Environmental Fate, Exposure, Ecotoxicology, and Residues between 2003 and 2006.* Availale online at www.efsa.europa.eu/en/science/ppr/ppr_opinions/1512.html

Evans, S. M., Hawkins, S. T., Porter, J. and Samsoir, A. M. (1994). Recovery of dogwhelk populations on the Isle of Cumbrae, Scotland following legislation limiting the use of TBT as an antifoulant. *Marine Pollution Bulletin*, **28**, 15–17.

Fairbairn, D. W., Olive, P. L. and O'Neill, K. L. (1995). The comet assay: a comprehensive review. *Mutation Research*, **339**, 37–59.

Fairbrother, A., Brix, K. V., Toll, J. E., McKay, S. and Adams, W. J. (1999). Egg selenium concentrations as predictors of avian toxicity. *Human and Ecological Risk Assessment*, **5**, 1229–1253.

Fan, T. W-M. and Higashi, R. M. (1998). Biochemical fate of selenium in microphytes: natural bioremediation by volatilization and sedimentation in aquatic environments. In *Environmental Chemistry of Selenium*, ed. W. T. Frankenberger Jr and R. A. Engberg. New York: Marcel Dekker, pp. 545–564.

Fan, W. and Wang, W.-X. (2001). Sediment geochemical controls on Cd, Cr and Zn assimilation by the clam *Ruditapes philippinarum*. *Environmental Toxicology and Chemistry*, **20**, 2309–2317.

Farag, A. M., Boese, C. J., Woodward, D. F. and Bergman, H. L. (1994). Physiological changes and tissue accumulation in rainbow trout exposed to food-borne and water-borne metals. *Environmental Toxicology and Chemistry*, **13**, 2021–2029.

Farag, A. M., Skaar, D., Nimick, D. A., MacConnell, E. and Hogstrand, C. (2003). Characterizing aquatic health using salmonid mortality, physiology, and biomass estimates in streams with elevated concentrations of arsenic, cadmium, copper, lead, and zinc in the Boulder River watershed, Montana. *Transactions of the American Fisheries Society*, **132**, 450–467.

Farag, A. M., Stansbury, M. A., Hogstrand, C., MacConnell, E. and Bergman, H. L. (1995). The physiological impairment of free-ranging brown trout exposed to metals in the Clark Fork River, Montana. *Canadian Journal of Fisheries and Aquatic Sciences*, **52**, 2038–2050.

Farag, A. M., Woodward, D. F., Brumbaugh, W., *et al.* (1999). Dietary effects of metals-contaminated invertebrates from the Coeur d'Alene river, Idaho, on cutthroat trout. *Transactions of the American Fisheries Society*, **128**, 578–592.

Farrington, J. W., Goldberg, E. D., Risebrough, R. W., Martin, J. H. and Bowen, V. T. (1983). U. S. "Mussel Watch" 1976–1978: an overview of the trace-metal, DDE, PCB, hydrocarbon, and artificial radionuclide data. *Environmental Science and Technology*, **17**, 490–496.

Feng, H., Cochran, J. K. and Hirschberg, D. J. (1999). ^{234}Th and ^{7}Be as tracers for transport and dynamics of suspended particles in a partially mixed estuary. *Geochimica et Cosmochimica Acta*, **63**, 2487–2505.

Fialkowski, W. and Rainbow, P. S. (2006). The discriminatory power of two biomonitors of trace metal bioavailabilities in freshwater streams. *Water Research*, **40**, 1805–1810.

Fialkowski, W., Fialkowska, E., Smith, B. D. and Rainbow, P. S. (2003a). Biomonitoring survey of trace metal pollution in streams of a catchment draining a zinc and lead mining area of Upper Silesia, Poland using the amphipod *Gammarus fossarum*. *International Review of Hydrobiology*, **88**, 187–200.

Fialkowski, W., Klonowska-Olejnik, M., Smith, B. D. and Rainbow, P. S. (2003b). Mayfly larvae (*Baetis rhodani* and *B. vernus*) as biomonitors of trace metal pollution in streams of a catchment draining a zinc and lead mining area of Upper Silesia, Poland. *Environmental Pollution*, **121**, 253–267.

Fialkowski, W., Rainbow, P. S., Fialkowska, E. and Smith, B. D. (2000). Biomonitoring of trace metals along the Baltic coast of Poland using the sandhopper *Talitrus saltator* (Montagu) (Crustacea: Amphipoda). *Ophelia*, **52**, 183–192.

Fialkowski, W., Rainbow, P. S., Smith, B. D. and Zmudzinski, L. (2003c). Seasonal variation in trace metal concentrations in three talitrid amphipods from the Gulf of Gdansk, Poland. *Journal of Experimental Marine Biology and Ecology*, **288**, 81–93.

Fimreite, N. (1971). Effects of dietary methylmercury on ring-necked pheasants. *Canadian Wildlife Service Occasional Paper*, **9**, 1–399.

Fisher, N. S. and Reinfelder, J. R. (1995). The trophic transfer of metals in marine systems. In *Metal Speciation and Bioavailability in Aquatic Systems*, ed. A. Tessier and D. Turner. Chichester, UK: John Wiley, pp. 363–406.

Fisher, N. S., Teyssié, J.-L., Fowler, S. W. and Wang, W.-X. (1996). Accumulation and retention of metals in mussels from food and water: a comparison under field and laboratory conditions. *Environmental Science and Technology*, **30**, 3232–3242.

Fitzgerald, W. F., Engstron, D. R., Mason, R. P. and Nater, E. A. (1998). The case for atmospheric mercury contamination in remote areas. *Environmental Science and Technology*, **32**, 1–7.

Fleeger, J. W., Carman, K. R. and Nisbet, R. M. (2003). Indirect effects of contaminants in aquatic ecosystems. *Science of the Total Environment*, **317**, 207–233.

Flegal, A. R. and Coale, K. H. (1989). Trends in lead concentrations in major U. S. rivers and their relation to historical changes in gasoline-lead consumption. *Journal of the American Water Resources Association*, **25**, 1279–1281.

Flegal, A. R., Nriagu, J. O., Nemeyer, S. and Coale, K. H. (2005). Isotopic tracers of lead contamination in the Great Lakes. *Nature*, **339**, 455–458.

Flegal, A. R., Smith, G. J., Gill, G. A., Sanudo-Wilhelmy, S. A. and Anderson, L. C. D. (1991). Dissolved trace element cycles in the San Francisco Bay estuary. *Marine Chemistry*, **36**, 329–363.

Flegal, R. and Sanudo-Wilhelmy, S. A. (1993). Comparable levels of trace metal contamination in two semi-enclosed embayments: San Diego Bay and South San Francisco Bay. *Environmental Science and Technology*, **27**, 1934–1936.

Flik, G., Verbost, P. M. and Wendelaar Bonga, S. E. (1995). Calcium transport processes in fish. In *Cellular and Molecular Approaches to Fish Ionic Regulation*, ed. C. M. Wood and T. J. Shuttleworth. San Diego, CA: Academic Press, pp. 317–342.

Florence, T. M. and Stauber, J. L. (1986). Toxicity of copper complexes to the marine diatom *Nitzschia closterium*. *Aquatic Toxicology*, **8**, 11–26.

Floyd, P. (2006). Future perspectives in risk assessment of chemicals issues in environmental science and technology. In *Chemicals in the Environment: Assessing and Managing Risk*, ed. R. M. Harrison and R. E. Hester. London: Royal Society of Chemistry, pp. 41–63.

Flury, M., Frankenburger, W. T. and Jury, W. A. (1997). Long-term depletion of selenium from Kesterson dewatered sediments. *Science of the Total Environment*, **198**, 259–270.

Fones, G. R., Davison, W. and Hamilton-Taylor, J. (2004). The fine-scale remobilization of metals in the surface sediment of the North-East Atlantic. *Continental Shelf Research*, **24**, 1485–1504.

Foran, J. A. (1993). *Regulating Toxic Substances in Surface Waters*. Boca Raton, FL: Lewis Publishers.

Forbes, V. E. and Calow, P. (2002). Species sensitivity distributions revisited: a critical appraisal. *Human and Ecological Risk Assessment*, **8**, 473–492.

Forbes, V. E. and Calow, P. (2004). Systematic approach to weight of evidence in sediment quality assessment: challenges and opportunities. *Aquatic Ecosystem Health and Management*, **7**, 339–350.

Forbes, V. E. and Depledge, M. H. (1992). Predicting the population response to pollution: the significance of sex. *Functional Ecology*, **6**, 376–381.

Forbes, V. E. and Forbes, T. L. (1994). *Ecotoxicology in Theory and Practice*. London: Chapman and Hall.

Ford, L. (2001). Development of chronic Aquatic Water Quality Criteria and Standards for silver. *Water Environment Research*, **73**, 248–253.

Forrow, D. M. and Maltby, L. (2000). Toward a mechanistic understanding of contaminant-induced changes in detritus processing in streams: direct and indirect effects on detritivore feeding. *Environmental Toxicology and Chemistry*, **19**, 2100–2106.

Fortin, C. and Campbell, P. G. C. (2001). Thiosulphate enhances silver uptake by a green alga: role of anion transporters in metal uptake. *Environmental Science and Technology*, **35**, 2214–2218.

Foster, P., Hunt, D. T. E. and Morris, A. W. (1978). Metals in an acid mine stream and estuary. *Science of the Total Environment*, **9**, 75–86.

Fowler, S. W. and Benayoun, G. (1976). Influence of environmental factors on selenium flux in two marine invertebrates. *Marine Biology*, **37**, 59–68.

Foxgrover, A. W., Higgins, S. A., Ingraca, M. K., Jaffe, B. E. and Smith, R. E. (2004). Deposition, erosion, and bathymetric change in South San Francisco bay: 1858–1983. *US Geological Survey Open File Report* 2004–1192.

Francesconi, K. A. and Lenanton, R. C. J. (1992). Mercury contamination in a semi-enclosed marine embayment: organic and inorganic mercury content of biota, and factors influencing mercury levels in fish. *Marine Environmental Research*, **33**, 189–212.

Freedman, B. and Hutchinson, T. C. (1980). Pollutant input from the atmosphere and accumulation in soils and vegetation near a nickel-copper smelter at Sudbury, Ontario, Canada. *Canadian Journal of Botany*, **58**, 108–132.

Frenzilli, G., Nigro, M., Scarcelli, V., Gorbi, S. and, Regoli, F. (2001). DNA integrity and total oxyradical scavenging capacity in the Mediterranean mussel, *Mytilus galloprovincialis*: a field study in a highly eutrophicated coastal lagoon. *Aquatic Toxicology*, **53**, 19–32.

Freudenburg, W. R. (1999). Tools for understanding the socioeconomic and political settings for environmental decision making. In *Tools to Aid Environmental Decision Makers*, ed. V. H. Dale and M. R. English. New York: Springer-Verlag, pp. 94–130.

Fuller, C. C. and Davis, J. A. (1989). Influence of coupling of sorption and photosynthetic processes on trace element cycles in natural waters. *Nature*, **340**, 52–55.

Funes, V., Alhama, J., Navas, J. I., López-Barea, J. and Peinado, J. (2006). Ecotoxicological effects of metal pollution in two mollusc species from the Spanish South Atlantic littoral. *Environmental Pollution*, **139**, 214–223.

Gagnon, C. and Fisher, N. S. (1997). Bioavailability of sediment-bound methyl and inorganic mercury to a marine bivalve. *Environmental Science and Technology*, **31**, 993–998.

Galay Burgos, M. and Rainbow, P. S. (1998). Uptake, accumulation and excretion by *Corophium volutator* (Crustacea: Amphipoda) of zinc, cadmium and cobalt added to sewage sludge. *Estuarine, Coastal and Shelf Science*, **47**, 603–620.

Galloway, T. S., Brown, R. J., Browne, M. A., *et al.* (2004). A multibiomarker approach to environmental assessment. *Environmental Science and Technology*, **38**, 1723–1731.

Galloway, T. S., Brown, R. J., Browne, M. A., *et al.* (2006). The ECOMAN project: a novel approach to defining sustainable ecosystem function. *Marine Pollution Bulletin*, **53**, 186–194.

Gammons, C. H., Shope, C. L. and Duaime, T. E. (2005). A 24 h investigation of the hydrogeochemistry of baseflow and stormwater in an urban area impacted by mining: Butte, Montana. *Hydrological Processes*, **19**, 2737–2753.

Garcia, E. and Carignan, R. (2000). Mercury concentrations in northern pike (*Esox lucius*) from boreal lakes with logged, burned, or undisturbed catchments. *Canadian Journal of Fisheries and Aquatic Sciences*, **57**, 129–135.

Gault, N. F. S., Tolland, E. L. C. and Parker, J. G. (1983). Spatial and temporal trends in heavy metal concentrations in mussels from Northern Ireland coastal waters. *Marine Biology*, **77**, 307–316.

Geffard, A., Amiard, J.-C. and Amiard-Triquet, C. (2002). Use of metallothionein in gills from oysters (*Crassostrea gigas*) as a biomarker: seasonal and intersite fluctuations. *Biomarkers*, **7**, 123–137.

Geffard, A., Amiard-Triquet, C. and Amiard, J.-C. (2005). Do seasonal changes interfere with metallothionein induction by metals in mussels? *Ecotoxicology and Environmental Safety*, **61**, 209–220.

Geffard, A., Amiard-Triquet, C., Amiard, J.-C. and Mouneyrac, C. (2001). Temporal variations of metallothionein and metal concentrations in the digestive glands of oysters *Crassostrea gigas* from a clean and a metal-rich site. *Biomarkers*, **6**, 91–107.

Geffard, A., Jeantet, A. Y., Amiard, J. C., *et al.* (2004). Comparative study of metal handling strategies in bivalves *Mytilus edulis* and *Crassostrea gigas*: a multidisciplinary approach. *Journal of the Marine Biological Association of the United Kingdom*, **84**, 641–650.

George, S. G. (1983). Heavy metal detoxification in the mussel *Mytilus edulis*: composition of Cd-containing kidney granules (tertiary lysosomes). *Comparative Biochemistry and Physiology C*, **76**, 53–57.

George, S. G., Carpene, E., Coombs, T. L., Overnell, J. and Youngson, A. (1979). Characterisation of cadmium-binding proteins from mussels, *Mytilus edulis* (L.), exposed to cadmium. *Biochimicae t Biophysica Acta*, **580**, 225–233.

George, S. G., Pirie, B. J. S. and Coombs, T. L. (1976). The kinetics of accumulation and excretion of ferric hydroxide in *Mytilus edulis* (L.) and its distribution in the tissues. *Journal of Experimental Marine Biology and Ecology*, **23**, 71–84.

Geret, F., Jouan, A., Turpin, V., Bebianno, M. J. and Cosson, R. P. (2002). Influence of metal exposure on metallothionein synthesis and lipid peroxidation in two bivalve mollusks: the oyster (*Crassostrea gigas*) and the mussel (*Mytilus edulis*). *Aquatic Living Resources*, **15**, 61–66.

Gerhardt, A. (1993). Review of impact of heavy metals on stream invertebrates with special emphasis on acid conditions. *Water, Air and Soil Pollution*, **66**, 289–314.

Ghazi, P. (2003). Unearthing controversy at the OK Tedi mine. *World Resources Institute Features*, **1**, 1–5.

Giambérini, L. and Cajaraville, M. P. (2005). Lysosomal responses in the digestive gland of the freshwater mussel, *Dreissena polymorpha*, experimentally exposed to cadmium. *Environmental Research*, **98**, 210–214.

Gibbs, P. E. and Bryan, G. W. (1986). Reproductive failure in populations of the dog-whelk, *Nucella lapillus*, caused by imposex induced by tributyltin from antifouling paints. *Journal of the Marine Biological Association of the United Kingdom*, **66**, 767–777.

Gibbs, P. E. and Bryan, G. W. (1987). TBT paints and the demise of the dogwhelk, *Nucella lapillus* (Gastropoda). In *Oceans '87 Proceedings*, vol. **4**, *International Organotin Symposium*. New York: Institute of Electrical and Electronic Engineers, pp. 1482–1487.

Gibbs, P. E., Pascoe, P. L. and Burt, G. R. (1988). Sex change in the female dog-whelk *Nucella lapillus*, induced by tributyltin from antifouling paints. *Journal of the Marine Biological Association of the United Kingdom*, **68**, 715–731.

Gibson, C. P. and Wilson, S. P. (2003). Imposex still evident in eastern Australia 10 years after tributyltin restrictions. *Marine Environmental Research*, **55**, 101–112.

Giguère, A., Couillard, Y., Campbell, P. G. C., *et al.* (2003). Steady-state distribution of cadmium and copper among metallothionein and other cytosolic ligands in bivalves living along a polymetallic gradient. *Aquatic Toxicology*, **64**, 185–200.

Gill, G. A., Bloom, N. S., Cappellino, S., *et al.* (1999). Sediment–water fluxes of mercury in Lavaca Bay, Texas. *Environmental Science and Technology*, **33**, 663–669.

Girvin, D. C., Hodgson, A. T. and Panietz, M. H. (1975). *Assessment of Trace Metal and Chlorinated Hydrocarbon Contamination in Selected San Francisco Bay Shellfish*, Lawrence Berkeley Laboratory Publication No. UCID3778. Berkeley, CA: University of California.

Glover, C. N. and Hogstrand, C. (2002). Amino acid modulation of *in vivo* intestinal zinc absorption in freshwater rainbow trout. *Journal of Experimental Biology*, **205**, 151–158.

Gobeil, C., Rondeau, B. and Beaudin, L. (2005). Contribution of municipal effluents to metal fluxes in the St. Lawrence River. *Environmental Science and Technology*, **39**, 456–464.

Goede, A. A. (1993). Selenium in eggs and parental blood of a Dutch marine wader. *Archives of Environmental Contamination and Toxicology*, **25**, 79–84.

Goedkoop, M. and Dubreuil, A. (2004). Overview (metal mining). In *Life-Cycle Assessment of Metals: Issues and Research Directions*, ed. A. Dubreuil. Pensacola, FL: SETAC Press, pp. 103–108.

Goldberg, E. D. (1954). Marine geochemistry. I. Chemical scavengers of the sea. *Journal of Geology*, **62**, 249–265.

Goldberg, E. D., Koide, M., Hodge, V., Flegal, A. R. and Martin, J. (1983). U. S. Mussel Watch: 1977–1978 results on trace metals and radionuclides. *Estuarine, Coastal and Shelf Science*, **16**, 69–93.

Gold-Bouchot, G., Sima-Alvarez, R., Zapate-Perez, O. and Guemez-Ricalde, J. (1995). Histopathological effects of petroleum hydrocarbons and heavy metals on the american oyster (*Crassostrea virginica*) from Tabasco, Mexico. *Marine Pollution Bulletin*, **31**, 439–445.

Gollasch, S. (2002). The importance of ship hull fouling as a vector of species introductions into the North Sea. *Biofouling*, **18**, 105–121.

Gonzalez, P., Dominique, Y., Massabuau, J. C., Boudou, A. and Bourdineaud, J. P. (2005). Comparative effects of dietary methylmercury on gene expression in liver, skeletal muscle, and brain of the zebrafish. *Environmental Science and Technology*, **39**, 3972–3980.

Gonzalez-Solis, J., Sanpera, C. and Ruiz, X. (2002). Metals and selenium as bioindicators of geographic and trophic segregation in giant petrels *Maronectes* spp. *Marine Ecology Progress Series*, **244**, 257–264.

Goodyear, K. L. and McNeill, S. (1999). Bioaccumulation of heavy metals by aquatic macro-invertebrates of different feeding guilds: a review. *Science of the Total Environment*, **229**, 1–19.

Gordon, M., Knauer, G. A. and Martin, J. H. (1980). *Mytilus californianus* as a bioindicator of trace metal pollution: variability and statistical considerations. *Marine Pollution Bulletin*, **11**, 195–198.

Gordon, R. B., Graedel, T. E., Bertram, M., *et al.* (2003). The characterization of technological zinc cycles. *Resources, Conservation and Recycling*, **39**, 107–135.

Gorinstein, S., Arancibia-Avila, P., Moncheva, S., *et al.* (2006). Changes in mussel *Mytilus galloprovincialis* protein profile as a reaction of water pollution. *Environment International*, **32**, 95–100.

Gorski, J. and Nugegoda, D. (2006). Sublethal toxicity of trace metals to larvae of the blacklip abalone, *Haliotis rubra*. *Environmental Toxicology and Chemistry*, **25**, 1360–1367.

Gower, A. M. and Darlington, S. T. (1990). Relationships between copper concentrations in larvae of *Plectrocnemia conspersa* (Curtis) (Trichoptera) in mine drainage streams. *Environmental Pollution*, **65**, 115–168.

Gower, A. M., Myers, G., Kent, M. and Foulkes, M. E. (1994). Relationships between macroinvertebrate communities and environmental variables in metal-contaminated streams in south-west England. *Freshwater Biology*, **32**, 199–221.

Grant, A., Hateley J. G. and Jones, N. V. (1989). Mapping the ecological impact of heavy metals in the estuarine polychaete *Nereis diversicolor* using inherited metal tolerance. *Marine Pollution Bulletin*, **20**, 235–238.

Gray, J. S. (1979). Pollution induced changes in populations. *Philosophical Transactions of the Royal Society of London B*, **286**, 545–561.

Griscom, S. B., Fisher, N. S. and Luoma, S. N. (2000). Geochemical influence on assimilation of sediment-bound metals in clams and mussels. *Environmental Science and Technology*, **34**, 91–99.

Griscom, S. B., Fisher, N. S. and Luoma, S. N. (2002). Kinetic modeling of Ag, Cd and Co bioaccumulation in the clam *Macoma balthica*: quantifying dietary and dissolved sources. *Marine Ecology Progress Series*, **240**, 127–141.

Grosbois, C.A., Horowitz, A.J., Smith, J.J. and Elrick, K.A. (2001). The effect of mining and related activities on the sediment–trace element geochemistry of Lake Coeur d'Alene, Idaho, USA. III. Downstream effects: the Spokane River Basin. *Hydrological Processes*, **15**, 855–875.

Grosbois, C.A., Horowitz, A.J., Smith, J.J. and Elrick, K.A. (2002). The effect of mining and related activities on the sediment trace element geochemistry of the Spokane River Basin, Washington, USA. *Geochemistry: Exploration, Environment, Analysis*, **2**, 131–142.

Guerlet, E., Ledy, K. and Giambérini, L. (2006). Field application of a set of cellular biomarkers in the digestive gland of the freshwater snail *Radix peregra* (Gastropoda, Pulmonata). *Aquatic Toxicology*, **77**, 19–32.

Guieu, C., Martin, J.-M., Tankere, S.P.C., *et al.* (1998). On trace metal geochemistry in the Danube River and western Black Sea. *Estuarine, Coastal and Shelf Science*, **47**, 471–485.

Guthrie, J.W., Hassan, N.M., Salam, M.S.A., *et al.* (2005). Complexation of Ni, Cu, Zn, and Cd by DOC in some metal-impacted freshwater lakes: a comparison of approaches using electrochemical determination of free-metal-ion and labile complexes and a computer speciation model, WHAM V and VI. *Analytica Chimica Acta*, **528**, 205–218.

Guttman, S.I. (1994). Population genetic structure and ecotoxicology. *Environmental Health Perspectives*, **102** (Suppl. 12), 97–100.

Guzelian, P.S. and Guzelian, C.P. (2004). Authority-based explanation. *Science*, **303**, 1466–1469.

Håkanson, L., Andersson, T. and Nilsson, N. (1990). Mercury in fish in Swedish lakes: linkages to domestic and European sources of emission. *Water, Air and Soil Pollution*, **50**, 171–191.

Hamilton, S.J. (1999). Hypothesis of historical effects from selenium on endangered fish in the Colorado River Basin. *Human and Ecological Risk Assessment*, **5**, 1153–1180.

Hamilton, S.J. and Palace, V.P. (2001). Assessment of selenium effects in lotic ecosystems. *Ecotoxicology and Environmental Safety*, **50**, 161–166.

Hammerschmidt, C.R., Wiener, J.G., Frazier, B.E. and Rada, R.G. (1999). Methylmercury content of eggs in yellow perch related to maternal exposure in four Wisconsin lakes. *Environmental Science and Technology*, **33**, 999–1003.

Han, B.-C. and Hung, T.-C. (1990). Green oysters caused by copper pollution on the Taiwan coast. *Environmental Pollution*, **65**, 347–362.

Handy, R.D., McGeer, J.C., Allen, H.E., *et al.* (2005). Toxic effects of dietary metals: laboratory studies. In *Toxicity of Dietborne Metals to Aquatic Organisms*, ed. J.S. Meyer, W.J. Adams, K.V. Brix, S.N. Luoma, D.R. Mount, W.A. Stubblefield and C.M. Wood, Pensacola, FL: SETAC Press, pp. 59–102.

Hansen, D.J., Berry, W.J., Mahony, J.D., *et al.* (1996). Predicting the toxicity of metal-contaminated field sediments using interstitial concentration of metals and acid-volatile sulfide normalizations. *Environmental Toxicology and Chemistry*, **15**, 2080–2094.

Hansen, D., Duda, P.J., Zayed, A.D.L. and Terry, N. (1998). Selenium removal by constructed wetlands: role of biological volatilization. *Environmental Science and Technology*, **32**, 591–597.

Hansen, J.A., Lipton, J., Welsh, P.G., Cacela, D. and MacConnell, B. (2004). Reduced growth of rainbow trout (*Oncorhynchus mykiss*) fed a live invertebrate diet pre-exposed to metal-contaminated sediments. *Environmental Toxicology and Chemistry*, **23**, 1902–1911.

Hansen, J.I., Mustafa, T. and Depledge, M.H. (1992a). Mechanisms of copper toxicity in the shore crab, *Carcinus maenas* (L.). I. Effects on Na,K-ATPase activity, haemolymph electrolyte concentrations and tissue water contents. *Marine Biology*, **114**, 253–257.

Hansen, J.I., Mustafa, T. and Depledge, M.H. (1992b). Mechanisms of copper toxicity in the shore crab, *Carcinus maenas* (L.). II. Effects on key metabolic enzymes, metabolites and energy charge potential. *Marine Biology*, **114**, 259–264.

Haq, F., Mahoney, M. and Koropatnick, J. (2003). Signaling events for metallothionein induction. *Mutation Research*, **533**, 211–226.

Harada, M. (1995). Minamata disease: methylmercury poisoning in Japan caused by environmental pollution. *Critical Reviews in Toxicology*, **25**, 1–24.

Hare, L.A. (1992). Aquatic insects and trace metals: bioavailability, bioaccumulation and toxicity. *Critical Reviews in Toxicology*, **22**, 327–369.

Hare, L.A. (1995). Sediment colonization by littoral and profundal insects. *Journal of the North American Benthological Society*, **14**, 315–323.

Hare, L.A. and Campbell, P.G.C. (1992). Temporal variations of trace metals in aquatic insects. *Freshwater Biology*, **27**, 13–27.

Hare, L.A. and Tessier, A. (1996). Predicting animal cadmium concentrations in lakes. *Nature*, **380**, 430–432.

Hare, L.A. and Tessier, A. (1998). The aquatic insect *Chaoborus* as a biomonitor of trace metals in lakes. *Limnology and Oceanography*, **43**, 1850–1859.

Hare, L.A., Campbell, P.G.C., Tessier, A. and Belzile, N. (1989). Gut sediments in a burrowing mayfly (Ephemeroptera, *Hexagenia limbata*): their contribution to animal trace element burdens, their removal, and the efficacy of a correction for their presence. *Canadian Journal of Fisheries and Aquatic Sciences*, **46**, 451–456.

Hare, L. A., Carignan, R. and Herta-Diaz, M. A. (1994). A field study of metal toxicity and accumulation by benthic invertebrates: implications for the acid-volatile sulfide (AVS) model. *Limnology and Oceanography*, **39**, 1653–1668.

Hare, L. A., Tessier, A. and Borgmann, U. (2003). Metal sources for freshwater invertebrates: pertinence for risk assessment. *Human and Ecological Risk Assessment*, **9**, 779–793.

Hare, L. A., Tessier, A. and Campbell, P. G. C. (1991). Trace element distributions in aquatic insects: variations among genera, elements and lakes. *Canadian Journal of Fisheries and Aquatic Sciences*, **48**, 1481–1491.

Hare, L. A., Tessier, A. and Warren, L. (2001). Cadmium accumulation by invertebrates living at the sediment–water interface. *Environmental Toxicology and Chemistry*, **20**, 880–889.

Harkins, W. D. and Swain, R. E. (1907). The determination of arsenic and other solid constituents of smelter smoke, with a study of the effects of high stacks and large condensing flues. *Journal of the American Chemical Society*, **29**, 970–979.

Harkins, W. D. and Swain, R. E. (1908). The chronic arsenical poisoning of herbivorous animals. *Journal of the American Chemical Society*, **30**, 928–946.

Harley, M. B. (1950). Occurrence of a filter feeding mechanism in the polychaete *Nereis diversicolor*. *Nature*, **165**, 734–735.

Harris, H. H., Pickering, I. J. and George, G. N. (2003). The chemical form of mercury in fish. *Science*, **301**, 1203.

Harvey, R. W. and Luoma, S. N. (1984). The role of bacterial exopolymer and suspended bacteria in the nutrition of the deposit-feeding clam *Macoma balthica*. *Journal of Marine Research*, **42**, 957–968.

Hateley, J. G., Grant, A. and Jones, N. V. (1989). Heavy metal tolerance in estuarine populations of *Nereis diversicolor*. In *Reproduction, Genetics and Distribution of Marine Organisms*, ed. J. S. Ryland and P. A. Tyler. Fredensborg, Denmark: Olsen and Olsen, pp. 379–385.

Hawkins, A. J. S., Rusin, J., Bayne, B. L. and Day, A. J. (1989). The metabolic/physiological basis of genotype-dependent mortality during copper exposure in *Mytilus edulis*. *Marine Environmental Research*, **28**, 253–257.

Heinz, G. H. (1979). Methylmercury: reproductive and behavioral effects on three generations of mallard ducks. *Journal of Wildlife Management*, **43**, 394–401.

Heinz, G. H. (1996). Selenium in birds. In *Environmental Contaminants in Wildlife*, ed. W. N. Beyer, G. H. Heinz and A. W. Redmon-Norwood. New York: Lewis Publishers, pp. 447–458.

Heinz, G. H. and Hoffman, D. J. (1998). Methylmercury chloride and selenomethionine interactions on health and reproduction in mallards. *Environmental Toxicology and Chemistry*, **17**, 139–145.

Heinz, G. H. and Hoffman, D. J. (2003). Embryotoxic thresholds of mercury: estimates from individual mallard eggs. *Archives of Environmental Contamination and Toxicology*, **44**, 257–264.

Heiss, S., Flicker, S., Hamilton, D. A., *et al.* (1996). Osmium isotopes and silver as tracers of anthropogenic metals in sediments from Massachusetts and Cape Cod bays. *Geochimica et Cosmochimica Acta*, **60**, 2753–2763.

Helgen, S. O. and Moore, J. N. (1996). Natural background determination and impact quantification in trace metal-contaminated river sediments. *Environmental Science and Technology*, **30**, 129–135.

Hettler, J., Irion, G. and Lehmann, E. (1997). Environmental impact of mining waste disposal on a tropical lowland river system: a case study on the Ok Tedi Mine, Papua New Guinea. *Mineralium Deposita*, **32**, 280–291.

Hickey, C. W. and Clements, W. H. (1998). Effects of heavy metals on benthic macroinvertebrate communities in New Zealand streams. *Environmental Toxicology and Chemistry*, **17**, 2338–2346.

Hickey, C. W. and Golding, L. A. (2002). Response of macroinvertebrates to copper and zinc in a stream mesocosm. *Environmental Toxicology and Chemistry*, **21**, 1854–1863.

Hickey, J. J. and Anderson, D. W. (1968). Chlorinated hydrocarbons and eggshell changes in raptorial and fish-eating birds. *Science*, **162**, 271–273.

Hightower, J. M. and Moore, D. (2003). Mercury levels in high-end consumers of fish. *Environmental Health Perspectives*, **111**, 604–608.

Higueras, P., Oyarzun, R., Lillo, J., *et al.* (2006). The Almadén district (Spain): anatomy of one of the world's largest Hg-contaminated sites. *Science of the Total Environment*, **356**, 112–124.

Hillman, T. W., Chapman, D. W., Hardin, T. S., Jensen, S. E. and Platts, W. S. (1995). Assessment of injury to fish populations: Clark Fork River NPL sites, Montana. Appendix G in *Aquatic Resources Injury Assessment Report, Upper Clark Fork River Basin*, ed. J. Lipton, D. Beltman, H. Bergman, D. Chapman, T. Hillman, M. Kerr, J. Moore, D. Woodward. Helena, MT: State of Montana Natural Resource Damage Litigation Program.

Hilton, J. W., Hodson, P. V. and Slinger, S. J. (1980). The requirement and toxicity of selenium in rainbow trout (*Salmo gairdneri*). *Journal of Nutrition*, **110**, 2527–2535.

Hogstrand, C., Wilson, R. W., Polgar, D. and Wood, C. M. (1994). Effects of zinc on the kinetics of branchial calcium uptake in freshwater rainbow trout during adaptation to waterborne zinc. *Journal of Experimental Biology*, **186**, 55–73.

Holling, C. S. (1996). Surprise for science, resilience for ecosystems and incentives for people. *Ecological Applications*, **6**, 733–735.

Holm, J., Palace, V., Siwik, P., *et al.* (2005). Developmental effects of bioaccumulated selenium in eggs and larvae of two salmonid species. *Environmental Toxicology and Chemistry*, **24**, 2373–2381.

Hook, S. E. and Fisher, N. S. (2001a). Reproductive toxicity of metals in calanoid copepods. *Marine Biology*, **138**, 1131–1140.

Hook, S. E. and Fisher, N. S. (2001b). Sublethal effects of silver in zooplankton: importance of exposure pathways and implications for toxicity testing. *Environmental Toxicology and Chemistry*, **20**, 568–574.

Hook, S. E. and Fisher, N. S. (2002). Relating the reproductive toxicity to five ingested metals in calanoid copepods with sulfur affinity. *Marine Environmental Research*, **53**, 161–174.

Hope, B. (2003). A basin-specific aquatic food web biomagnification model for estimation of mercury target levels. *Environmental Toxicology and Chemistry*, **22**, 2525–2537.

Hopkin, S. P. (1989). *Ecophysiology of Metals in Terrestrial Invertebrates*. Barking, UK: Elsevier Applied Science.

Hornberger, M. I., Lambing, J. H., Luoma, S. N. and Axtmann, E. V. (1997). Spatial and temporal trends of trace metals in surface water, bed sediment, and biota of the upper Clark Fork basin, Montana. *US Geological Survey Open File Report* **1997–669**.

Hornberger, M. I., Luoma, S. N., Cain, D. J., *et al.* (2000). Linkage of bioaccumulation and biological effects to changes in pollutant loads in South San Francisco Bay. *Environmental Science and Technology*, **34**, 2401–2409.

Hornberger, M. I., Luoma, S. N., van Geen, A., Fuller, C. and Anima, R. (1999). Historical trends of trace metals in the sediments of San Francisco Bay, California. *Marine Chemistry*, **64**, 39–65.

Horowitz, A. J. (1991). *A Primer on Trace Metal-Sediment Chemistry*, 2nd edn. Ann Arbor, MI: Lewis Publishers.

Horowitz, A. J., Elrick, K. A. and Cook, R. B. (1993). Effect of mining and related activities on the sediment trace element geochemistry of Lake Coeur D'Alene, Idaho, USA. I. Surface sediments. *Hydrological Processes*, **7**, 403–423.

Horowitz, A. J., Elrick, K. A., Robbins, J. A. and Cook, R. B. (1995). Effect of mining and related activities on the sediment trace element geochemistry of Lake Coeur D'Alene, Idaho, USA. II. Subsurface sediments. *Hydrological Processes*, **9**, 35–54.

Horowitz, A. J., Elrick, K. A. and Smith, J. J. (2001). Estimating suspended sediment and trace element fluxes in large river basins: methodological considerations as applied to the NASQAN programme. *Hydrological Processes*, **15**, 1107–1132.

Horvat, M., Covelli, S., Faganeli, J., *et al.* (1999). Mercury in contaminated coastal environments: a case study: the Gulf of Trieste. *Science of the Total Environment*, **237/8**, 43–56.

Houck, O. (2003). Tales from a troubled marriage: science and law in environmental policy. *Science*, **302**, 1926–1929.

Howarth, R. W., Giblin, A., Gale, J., Peterson, B. J. and Luther, G. W. (1983). Reduced sulfur compounds of a New England salt marsh. In *Symposium on Environmental Biogeochemistry*, ed. R. O. Halberg. *Ecological Bulletin*, **35**, 135–152.

Huerta-Diaz, M. A., Tessier, A. and Carignan, R. (1998). Geochemistry of trace metals associated with reduced sulfur in freshwater sediments. *Applied Geochemistry*, **13**, 213–233.

Huet, M., Paulet, Y. M. and Clavier, J. (2004). Imposex in *Nucella lapillus*: a ten-year survey in NW Brittany. *Marine Ecology Progress Series*, **270**, 153–161.

Huggett, R. J., Unger, M. A., Seligman, P. F. and Valkirs, A. O. (1992). The marine biocide tributyltin: assessing and managing environmental risks. *Environmental Science and Technology*, **26**, 232–237.

Hummel, H., Amiard-Triquet, C., Bachelet, G., *et al.* (1996). Sensitivity to stress of the estuarine bivalve *Macoma balthica* from two areas between the Netherlands and its southern limits (Gironde). *Netherlands Journal of Sea Research*, **35**, 315–321.

Hummel, H., Modderman, R., Amiard-Triquet, C., *et al.* (1997). A comparative study on the relation between copper and condition in marine bivalves and the relation with copper in the sediment. *Aquatic Toxicology*, **38**, 165–181.

Hunt, J. W., Anderson, B. S., B. M., Phillips, *et al.* (2001). A large-scale categorization of sites in San Francisco Bay, USA, based on the sediment quality triad, toxicity identification evaluations, and gradient studies. *Environmental Toxicology and Chemistry*, **20**, 1252–1265.

Hutchins, D. A., Wang, W.-X. and Fisher, N. S. (1995). Copepods grazing and the biogeochemical fate of diatom iron. *Limnology and Oceanography*, **40**, 989–994.

Hutchinson, T. C. and Whitby, L. M. (1976). The effects of acid rainfall and heavy metal particulates on a boreal forest ecosystem near the Sudbury smelting region of Canada. *Water, Air and Soil Pollution*, **7**, 421–435.

Ide, I., Whitten, E. P., Fischer, J., *et al.* (1997). Accumulation of organotin compounds in the common whelk *Buccinum undatum* and the red whelk *Neptunea antiqua* in association with imposex. *Marine Ecology Progress Series*, **152**, 197–203.

Ingersoll, C. G. (2001). Predictions of sediment toxicity using consensus-based freshwater sediment quality guidelines. *Archives of Environmental Contamination and Toxicology*, **41**, 8–21.

Ingersoll, C. G. and Nelson, M. K. (1990). Testing sediment toxicity with *Hyalella azteca* (Amphipoda) and *Chironomus riparius* (Diptera). In *Aquatic Toxicology and Risk Assessment*, vol. **13**, ed. W. G. Landis and W. H. van der Schalie, Philadelphia, PA: American Society for Testing and Materials, pp. 93–109.

Ingersoll, C. G., Brunson, E. L., Dwyer, F. J., Hardesty, D. K. and Kemble, N. E. (1998). Use of sublethal endpoints in sediment toxicity tests with the amphipod *Hyalella azteca*. *Environmental Toxicology and Chemistry*, **17**, 1508–1523.

Ingersoll, C. G., Dillon, T. and Biddinger, G., eds. (1997). *Ecological Risk Assessments of Contaminated Sediments*. Pensacola, FL: SETAC Press.

Ingersoll, C. G., Haverland, P. S., Brunson, E. L., *et al.* (1996). Calculation and evaluation of sediment effect concentrations for the amphipod *Hyalella azteca* and the midge *Chironomus riparius*. *Journal of Great Lakes Research*, **22**, 602–623.

Irving, E. C., Baird, D. J. and Culp, J. M. (2003). Ecotoxicological responses of the mayfly *Baetis tricaudatus* to dietary and waterborne cadmium: implications for toxicity testing. *Environmental Toxicology and Chemistry*, **22**, 1058–1064.

Isani, G., Andreani, G., Kindt, M. and Carpenè, E. (2000). Metallothioneins (MTs) in marine molluscs. *Cellular and Molecular Biology*, **46**, 311–330.

Jarman, W. M., Hobson, K. A., Sydeman, W. J., Bacon, C. E. and McLaren, E. B. (1996). Influence of trophic position and feeding location on contaminant levels in the Gulf of the Farallones food web revealed by stable isotope analysis. *Environmental Science and Technology*, **30**, 654–660.

Jarvinen, A. W. and Ankley, G. T. (1999). *Linkage of Effects to Tissue Residues: Development of a Comprehensive Database for Aquatic Organisms Exposed to Inorganic and Organic Chemicals*. Pensacola, FL: SETAC Press.

Jenkins, K. D. and Mason, A. Z. (1988). Relationships between subcellular distributions of cadmium and perturbations in reproduction in the polychaete *Neanthes arenaceodentata*. *Aquatic Toxicology*, **12**, 229–244.

Jenne, E. A. (1968). Controls on Mn, Fe, Co, Ni, Cu, and Zn concentration in soils and water: significant role of hydrous Mn and Fe oxides. *Advances in Chemistry Series*, **73**, 337–387.

Jernelov, A. and Lann, H. (1971). Mercury accumulation in food chains. *Oikos*, **22**, 403–406.

Jiang, G.-B., Shi, J.-B. and Feng, X.-B. (2006). Mercury pollution in China. *Environmental Science and Technology*, **40**, 3673–3677.

Jickells, T. D., An, Z. S., Andersen, K. K., *et al.* (2005). Global iron connections between desert dust, ocean biogeochemistry, and climate. *Science*, **308**, 67–71.

Johansson, C., Cain, D. J. and Luoma, S. N. (1986). Variability in the fractionation of Cu, Ag, and Zn among cytosolic proteins in the bivalve *Macoma balthica*. *Marine Ecology Progress Series*, **28**, 87–97.

Johnels, A. G., Westermark, T., Berg, W., Persson, P. I. and Sjostrand, B. (1967). Pike (*Esox lucius* L.) and some other aquatic organisms in Sweden as indicators of mercury contamination in the environment. *Oikos*, **18**, 323–333.

Johns, C. and Moore, J. N. (1985). Metal concentrations in fine grained sediments from reservoirs on the Clark Fork River, MT. In *Proceedings of the Clark Fork River Symposium*, ed. C. E. Cadsoxi and L. L. Bahls. Butte, MT: Montana Academy of Science, pp. 74–85.

Johnson, H. E. and Schmidt, C. L. (1988). *Clark Fork Basin Project: Status Report and Action Plan*. Helena, MT: Office of the Governor.

Johnson, J. (2003). Delay proposed for mercury cuts. *Chemical and Engineering News*, **Dec 15**, 20–22.

Johnson, J. (2004). Grappling with mercury. *Chemical and Engineering News*, **July 12**, 19–20.

Johnson, J. (2005). Long time cutting. *Chemical and Engineering News*, **Feb 28**, 44–45.

Johnston, E. L. and Keough, M. J. (2000). Field assessment of effects of timing and frequency of copper pulses on settlement of sessile marine invertebrates. *Marine Biology*, **137**, 1017–1029.

Johnston, E. L. and Keough, M. J. (2003). Competition modifies the response of organisms to toxic disturbance. *Marine Ecology Progress Series*, **251**, 15–26.

Johnston, T. A., Bodaly, R. A. and Mathias, J. A. (1991). Predicting fish mercury levels from physical characteristics of boreal reservoirs. *Canadian Journal of Fisheries and Aquatic Sciences*, **48**, 1468–1475.

Jones, C. A., Nimmick, D. A. and McCleskey, R. B. (2004). Relative effect of temperature and pH on diel cycling of dissolved trace elements in Prickly Pear Creek, Montana. *Water, Air and Soil Pollution*, **153**, 95–113.

Jumars, P. A. (2000). Animal guts as ideal chemical reactors: maximizing absorption rates. *American Naturalist*, **155**, 527–543.

Kalk, M. (1963). Absorption of vanadium by tunicates. *Nature*, **198**, 1010–1011.

Kammenga, J. E., Dallinger, R., Donker, M. H., *et al.* (2000). Biomarkers in terrestrial invertebrates. *Reviews in Environmental Contamination and Toxicology*, **164**, 93–147.

Kaputska, L. A., Clements, W. H., Ziccardi, L., *et al.* (2003). *Issue Paper on the Ecological Effects of Metals*, US Environmental Protection Agency Risk Assessment Forum Contract Report No.68-C-98-148. Washington, DC: Environmental Protection Agency.

Kater, B. J., Postma, J. F., Dubbeldam, M. and Prins, J. T. H. J. (2001). Comparison of laboratory and *in situ* sediment bioassays using *Corophium volutator*. *Environmental Toxicology and Chemistry*, **20**, 1291–1295.

Ke, C. and Wang, W.-X. (2001). Bioaccumulation of Cd, Se and Zn in an estuarine oyster (*Crassostrea rivularis*) and a coastal oyster (*Saccostrea glomerata*). *Aquatic Toxicology*, **56**, 33–51.

Ke, C. and Wang, W.-X. (2002). Trace metal ingestion and assimilation by the green mussel *Perna viridis* in a phytoplankton and sediment mixture. *Marine Biology*, **140**, 327–335.

Keating, J. (2003). USA ambient water quality criteria. In *Reevaluation of the State of the Science for Water-Quality Criteria Development*, ed. M. C. Reiley, W. A. Stubblefield, W. J. Adams, D. M. Di Toro, P. V. Hodson, R. J. Erickson and F. J. Keating. Pensacola, FL: SETAC Press, pp. 162–170.

Kelly, C. A., Rudd, J. W. M., Bodaly, R. A., *et al.* (1997). Increases in fluxes of greenhouse gases and methyl mercury following flooding of an experimental reservoir. *Environmental Science and Technology*, **31**, 1334–1344.

Kelly, M. (1988) *Mining and the Freshwater Environment*. Oxford,UK: Elsevier Applied Science.

Kemp, P. F. and Swartz, R. C. (1988). Acute toxicity of interstitial and particle-bound cadmium to a marine infaunal amphipod. *Marine Environmental Research*, **26**, 135–153.

Kenaga, E. E. (1982). Predictability of chronic toxicity from acute toxicity of chemicals in fish and aquatic invertebrates. *Environmental Toxicology and Chemistry*, **1**, 347–358.

Kennedy, C. J., McDonald, L. E., Loveridge, R. and Strosher, M. M. (2000). The effect of bioaccumulated selenium on mortalities and deformities in the eggs, larvae, and fry of a wild population of cutthroat trout (*Oncorhynchus clarki lewisi*). *Archives of Environmental Contamination and Toxicology*, **39**, 46–52.

Kennedy, J. H., La Point, T. W., Balci, P., Stanley, J. and Johnson, Z. B. (2003). Model aquatic ecosystems in ecotoxicological research: considerations of design, implementation, and analysis. In *Handbook of Ecotoxicology*, ed. D. J. Hoffman, B. A. Rattner, G. A. Burton Jr and J. Cairns Jr. Boca Raton, FL: CRC Press, pp. 45–74.

Kernsten, M., Leipe, T. and Tauber, F. (2005). Storm disturbance of sediment contaminants at a hot-spot in the Baltic Sea assessed by ^{234}Th radionuclide tracer profiles. *Environmental Science and Technology*, **39**, 984–990.

Kidd, K. A., Hesslein, R. H., Fudge, R. J. P. and Hallard, K. A. (1995). The influence of trophic level as measured by 15-N on mercury concentrations in freshwater organisms. *Water, Air and Soil Pollution*, **80**, 1011–1015.

Kiffney, P. M. and Clements, W. H. (1994). Effects of heavy metals on a macroinvertebrate assemblage from a Rocky Mountain stream in experimental microcosms. *Journal of the North American Benthological Society*, **13**, 511–523.

Kim, E. Y., Saeiki, K., Tanabe, S., Tanaka, H. and Tatsukawa, R. (1996). Specific accumulation of mercury and selenium in seabirds. *Environmental Pollution*, **94**, 261–265.

Kim, J. P., Kim, M. R. and Reid, R. G. (1996). Aqueous chemistry of major ions and trace metals in the Clutha River, New Zealand. *Marine and Freshwater Research*, **47**, 19–28.

King, C. K. and Riddle, M. J. (2001). Effects of metal contaminants on the development of the common Antarctic sea urchin *Sterechinus neumayeri* and comparisons of sensitivity with tropical and temperate echinoids. *Marine Ecology Progress Series*, **215**, 143–154.

King, C. K., Simpson, S. L., Smith, S. V., Stauber, J. L. and Batley, G. E. (2005). Short-term accumulation of Cd and Cu from water, sediment and algae by the amphipod *Melita plumulosa* and the bivalve *Tellina deltoidalis*. *Marive Ecology Progress Series*, **287**, 177–188.

Klerks, P. L. and Bartholomew, P. R. (1991). Cadmium accumulation and detoxification in a Cd-resistant population of the oligochaete *Limnodrilus hoffmeisteri*. *Aquatic Toxicology*, **19**, 97–112.

Klerks, P. L. and Levinton, J. S. (1989). Rapid evolution of metal resistance in a benthic oligochaete inhabiting a metal-polluted site. *Biological Bulletin Woods Hole Massachusetts*, **176**, 135–141.

Klerks, P.L. and Weis, J.S. (1987). Genetic adaptation to heavy metals in aquatic organisms: a review. *Environmental Pollution*, **45**, 173–205.

Kobayashi, N. (1984). Marine ecotoxicological testing with echinoderms. In *Ecotoxicological Testing for the Marine Environment*, ed. G. Persoone, E. Jaspers and C. Claus. Bredene, Belgium: State University Ghent and Institute of Marine Scientific Research, pp. 341–405.

Koepp, S.J., Santoro, E.D., Zimmer, R. and Nadeau, J. (1987). Bioaccumulation of Hg, Cd, and Pb in *Mytilus edulis* transplanted to a dredged-material dumpsite. In *Oceanic Processes in Marine Pollution*, vol. **1**, ed. J.M. Capuzzo and D.R. Kester. Malabar, FL: Robert E. Kreiger, pp. 51–58.

Krabbenhoft, D.K., Wiener, J.G., Brumbaugh, W.G., *et al.* (1999). *A National Pilot Study of Mercury Contamination of Aquatic Ecosystems along Multiple Gradients,* Report No. WRI 1999-4018. Reston, VA: US Geological Survey. Available online at http://toxics.usgs.gov/pubs/wri99-4018/Volume2/sectionB/2301_Krabbenhoft/pdf/2301_Krabbenhoft.pdf

Kraepiel, A.M.L., Keller, K., Chin, H.B., Malcolm, E.G. and Morel, F.M.M. (2003). Sources and variations of mercury in tuna. *Environmental Science and Technology*, **37**, 5551–5558.

Krång, A.-S. and Rosenqvist, G. (2006). Effects of manganese on chemically induced food search behaviour of the Norway lobster, *Nephrops norvegicus* (L.). *Aquatic Toxicology*, **78**, 284–291.

Krauskopf, K.B. (1956). Factors controlling concentrations of thirteen rare metals in water. *Geochimica et Cosmochimica Acta*, **9**, 1–32.

Kuykendall, J.R., Miller, K.L., Mellinger, K.N. and Cain, A.V. (2006). Waterborne and dietary hexavalent chromium exposure causes DNA–protein crosslink (DPX) formation in erythrocytes of largemouth bass (*Micropterus salmoides*). *Aquatic Toxicology*, **78**, 27–31.

La Point, T.W. and Perry, J.A. (1989). Use of experimental ecosystems in regulatory decision making. *Environmental Management*, **13**, 539–544.

Lackey, R.T. (2001). Values, policy and ecosystem health. *BioScience*, **51**, 437–443.

Lamberson, J.C., DeWitt, T.H. and Swartz, R.C. (1992). Assessment of sediment toxicity to marine benthos. In *Sediment Toxicity Assessment*, ed. I.R. Hill, P. Matthiessen and F. Heimbach. Boca Raton, FL: Lewis Publishers, pp. 183–211.

Lande, E. (1977). Heavy metal pollution in Trondheimsfjorden, Norway, and the recorded effects on the fauna and flora. *Environmental Pollution*, **12**, 187–198.

Landis, W.G. and Yu, M-H. (2004). *Introduction to Environmental Toxicology: Impacts of Chemicals upon Ecological Systems*. Boca Raton, FL: Lewis Publishers.

Landner, L. and Reuther, R. (2004). *Metals in Society and in the Environment*. Dordrecht, the Netherlands: Kluwer.

Landrum, P.F., Lee, H. II and Lydy, M.J. (1992). Toxicokinetics in aquatic systems: model comparisons and use in hazard assessment. *Environmental Toxicology and Chemistry*, **11**, 1709–1725.

Lane, T.W., Saito, M.A., George, G.N., *et al.* (2005). A cadmium enzyme from a marine diatom. *Nature*, **435**, 42.

Langston, W.J. (1980). Arsenic in UK estuarine sediments and its availability to benthic organisms. *Journal of the Marine Biological Association of the United Kingdom*, **60**, 869–881.

Langston, W.J. (1982). The distribution of mercury in British estuarine sediments and its availability to deposit-feeding bivalves. *Journal of the Marine Biological Association of the United Kingdom*, **62**, 667–684.

Langston, W.J. (1985). Assessment of the distribution and availability of arsenic and mercury in estuaries. In *Estuarine Management and Quality Assessment*, ed. J.G. Wilson and W. Halcrow. New York: Plenum Press, pp. 131–146.

Langston, W.J. (1990). Toxic effects of metals and the incidence of metal pollution in marine ecosystems. In *Heavy Metals in the Marine Environment*, ed. R.W. Furness and P.S. Rainbow. Boca Raton, FL: CRC Press, pp. 101–122.

Langston, W.J. (1995). Tributyl tin in the marine environment: a review of past and present risks. *Pesticide Outlook*, **Dec 1995**, 18–24.

Langston, W.J. (1996). Recent developments in TBT ecotoxicology. *Toxicology and Ecotoxicology News*, **3**, 179–187.

Langston, W.J. and Burt, G.R. (1991). Bioavailability and effects of sediment-bound TBT in deposit-feeding clams, *Scrobicularia plana*. *Marine Environmental Research*, **32**, 61–77.

Langston, W.J., Bebianno, M.J. and Burt, G.R. (1998). Metal handling strategies in molluscs. In *Metal Metabolism in Aquatic Environments*, ed. W.J. Langston and M.J. Bebianno. London: Chapman and Hall, pp. 219–283.

Langston, W.J., Bebianno, M.J. and Zhou, M. (1989). A comparison of metal-binding proteins and cadmium metabolism in the marine molluscs *Littorina littorea* (Gastropoda), *Mytilus edulis* and *Macoma balthica* (Bivalvia). *Marine Environmental Research*, **28**, 195–200.

Langston, W. J., Bryan, G. W., Burt, G. R. and Gibbs, P. E. (1990). Assessing the impact of tin and TBT in estuaries and coastal regions. *Functional Ecology*, **4**, 433–443.

Larison, J. R., Likens, G. E., Fitzpatrick, J. W. and Crock, J. G. (2000). Cadmium toxicity among wildlife in the Colorado Rocky Mountains. *Nature*, **406**, 181–183.

Lauenstein, G. G., Robertson, A. and O'Connor, T. P. (1990). Comparison of trace metal data in mussels and oysters from a Mussel Watch programme of the 1970s with those from a 1980s programme. *Marine Pollution Bulletin*, **21**, 440–447.

Laughlin, R. B. Jr and Linden, O. (1987). Tributyltin: contemporary environmental issues. *Ambio*, **16**, 252–256.

Lavie, B. and Nevo, E. (1986). Genetic selection of homozygote allozyme genotypes in marine gastropods exposed to cadmium pollution. *Science of the Total Environment*, **57**, 91–98.

Law, R. J., Waldock, M. J., Allchin, C. R., Laslett, R. E. and Bailey, K. J. (1994). Contaminants in seawater around England and Wales: results from monitoring surveys, 1990–1992. *Marine Pollution Bulletin*, **28**, 668–675.

Lawrence, S. G. and Holoka, M. H. (1987). Effects of low concentrations of cadmium on the crustacean zooplankton community of an artificially acidified lake. *Canadian Journal of Fisheries and Aquatic Sciences*, **44**, 163–172.

Lawrence, S. G. and Holoka, M. H. (1991). Response of crustacean zooplankton impounded in situ to cadmium at low environmental concentrations. *Verhandlungen internationale Vereinigung für theoretische und angewandte Limnologie*, **24**, 2254–2259.

Lawson, N. and Mason, R. (1998). Accumulation of mercury in estuarine food chains. *Biogeochemistry*, **40**, 235–247.

Lee, B. G. and Fisher, N. S. (1994). Effects of sinking and zooplankton grazing on the release of elements from planktonic debris. *Marine Ecology Progress Series*, **110**, 271–281.

Lee, B.-G. and Luoma, S. N. (1998). Influence of microalgal biomass on absorption efficiency of Cd, Cr, and Zn by two bivalves from San Francisco Bay. *Limnology and Oceanography*, **43**, 1455–1466.

Lee, B.-G., Griscom, S. B., Lee, J.-S., *et al.* (2000a). Influences of dietary uptake and reactive sulfides on metal bioavailability from sediments. *Science*, **287**, 282–284.

Lee, B.-G., Lee, J.-S. and Luoma, S. N. (2006). Comparison of selenium bioaccumulation in the clams *Corbicula fluminea* and *Potamocorbula amurensis*: a bioenergetic modeling approach. *Environmental Toxicology and Chemistry*, **25**, 1933–1940.

Lee, B.-G., Lee, J.-S., Luoma, S. N., Choi, H.-J. and Koh, C.-H. (2000b). Influence of acid volatile sulfide on the metal concentrations on metal bioavailability to marine invertebrates in contaminated sediments. *Environmental Science and Technology*, **34**, 4517–4523.

Lee, B.-G., Wallace, W. G. and Luoma, S. N. (1998). Uptake and loss kinetics of Cd, Cr and Zn in the bivalves *Potamocorbula amurensis* and *Macoma balthica*: effects of size and salinity. *Marine Ecology Progress Series*, **175**, 177–189.

Lee, J.-S., Lee, B.-G., Luoma, S. N., *et al.* (2000c). Influence of acid volatile sulfides and metal concentrations on metal partitioning in contaminated sediments. *Environmental Science and Technology*, **34**, 4511–4516.

Lee, J.-S., Lee, B.-G., Luoma, S. N. and Yoo, H. (2004). Importance of equilibration time in the partitioning and toxicity of zinc in spiked-sediment bioassays. *Environmental Toxicology and Chemistry*, **23**, 65–71.

Lee, J.-S., Lee, B.-G., Yoo, H., Koh, C.-H. and Luoma, S. N. (2001). Influence of reactive sulfide (AVS) and supplementary food on Ag, Cd and Zn bioaccumulation in the marine polychaete, *Neanthes arenaceodentata*. *Marine Ecology Progress Series*, **216**, 129–140.

Lee, K.-M. and Johnston, E. L. (2007). Low levels of copper reduce the reproductive success of a mobile invertebrate predator. *Marine Environmental Research*, **64**, 336–346.

Lee, K. N. (1999). Appraising adaptive management. *Conservation Ecology*, **3** (2), 3. Available online at www.consecol.org/vol3/iss2/art3/

Lee, M. R. and Correia, J. A. (2005). Effects of copper mine tailings disposal on littoral meiofaunal assemblages in the Atacama region of northern Chile. *Marine Environmental Research*, **59**, 1–18.

Lee, R. F. (1991). Metabolism of tributyltin by marine animals and possible linkages to effects. *Marine Environmental Research*, **32**, 29–35.

Lemly, A. D. (1985). Toxicology of selenium in a freshwater reservoir: implications for environmental hazard evaluation and safety. *Ecotoxicology and Environmental Safety*, **10**, 314–338.

Lemly, A. D. (1993a). Teratogenic effects of selenium in natural populations of freshwater fish. *Ecotoxicology and Environmental Safety*, **26**, 181–204.

Lemly, A. D. (1993b). Metabolic stress during winter increases the toxicity of selenium to fish. *Ecotoxicology and Environmental Safety*, **28**, 83–100.

Lemly, A. D. (1996). Assessing the toxic threat of selenium to fish and aquatic birds. *Environmental Monitoring and Assessment*, **43**, 19–35.

Lemly, A. D. (1997). Ecosystem recovery following selenium contamination in a freshwater reservoir. *Ecotoxicology and Environmental Safety*, **36**, 275–281.

Lemly, A. D. (1998). A position paper on selenium in ecotoxicology: a procedure for deriving site-specific water quality criteria. *Ecotoxicology and Environmental Safety*, **39**, 1–9.

Lemly, A. D. (1999). Selenium transport and bioaccumulation in aquatic ecosystems: a proposal for water quality criteria based on hydrologic units. *Ecotoxicology and Environmental Safety*, **42**, 150–156.

Lemly, A. D. (2002). *Selenium Assessment in Aquatic Ecosystems: A Guide for Hazard Evaluation and Water Quality Criteria*. Amsterdam, the Netherlands: Springer-Verlag.

Leonard, E. N., Mattson, V. R., Benoit, D. A., Hoke, R. A. and Ankley, G. T. (1993). Seasonal variation of acid volatile sulfide concentration in sediment cores from three northeastern Minnesota lakes. *Hydrobiologia*, **271**, 87–95.

Leonzio, C., Focardi, S. and Fossi, C. (1992). Heavy metals and selenium in stranded dolphins of the Northern Tyrrhenian (NW Mediterranean). *Science of the Total Environment*, **119**, 77–84.

Levinton, J. S., Suatoni, E., Wallace, W., *et al.* (2003). Rapid loss of genetically based resistance to metals after the cleanup of a Superfund site. *Proceedings of the National Academy of Sciences of the USA*, **100**, 9889–9891.

Lin, J. and Hines, A. H. (1994). Effects of suspended food availability on the feeding mode and burial depth of the Baltic clam, *Macoma balthica*. *Oikos*, **69**, 28–36.

Lindberg, S., Bullock, R., Ebinghaus, R., *et al.* (2007). A synthesis of progress and uncertainties in attributing the sources of mercury in deposition. *Ambio*, **36**, 19–33.

Linville, R. G. (2006). Effects of selenium on sturgeon. Unpublished Ph.D. thesis, University of California–Davis, Davis, CA.

Linville, R. G., Luoma, S. N., Cutter, L. S. and Cutter, G. A. (2002). Increased selenium threat as a result of invasion of the exotic bivalve *Potamocorbula amurensis* into the San Francisco Bay Delta. *Aquatic Toxicology*, **57**, 51–64.

Lippard, S. J. and Berg, J. M. (1994). *Principles of Bioinorganic Chemistry*. Mill Valley, CA: University Science Books.

Livingstone, D. R. (2001). Contaminant-stimulated reactive oxygen species production and oxidative damage in aquatic organisms. *Marine Pollution Bulletin*, **42**, 656–666.

Lobel, P. B. (1987). Intersite, intrasite and inherent variability of the whole soft tissue zinc concentrations of individual mussels *Mytilus edulis*: importance of the kidney. *Marine Environmental Research*, **21**, 59–71.

Lobel, P. B. and Wright, D. A. (1983). Frequency distribution of zinc in *Mytilus edulis* (L.). *Estuaries*, **6**, 154–159.

Lobel, P. B., Belkholde, S. P., Jackson, S. E. and Longerich, H. P. (1990). Recent taxonomic discoveries concerning the mussel *Mytilus*: implications for biomonitoring. *Archives of Environmental Contamination and Toxicology*, **19**, 508–512.

Löfstedt, R. E. (2005). *Risk Management in Post-Trust Societies*. New York: Palgrave Macmillan.

Löfstedt, R. E. and Renn, O. (1997). The Brent Spar controversy: an example of risk communication gone wrong. *Risk Analysis*, **17**, 131–136.

Long, E. R. (2000). Degraded sediment quality in the US estuaries: a review of magnitude and ecological implications. *Ecological Applications*, **10**, 338–349.

Long, E. R. and Chapman, P. M. (1985). A Sediment Quality Triad: measures of sediment contamination, toxicity and infaunal community composition in Puget Sound. *Marine Pollution Bulletin*, **16**, 405–415.

Long, E. R., Buchman, M. F., Bay, S. M., *et al.* (1990). Comparative evaluation of five toxicity tests with sediments from San Francisco Bay and Tomales Bay, California. *Environmental Toxicology and Chemistry*, **9**, 1193–1214.

Long, E. R., Field, L. J. and MacDonald, D. D. (1998a). Predicting toxicity in marine sediments with numerical sediment quality guidelines. *Environmental Toxicology and Chemistry*, **17**, 714–727.

Long, E. R., Ingersoll, C. G. and MacDonald, D. D. (2006). Calculation and uses of mean sediment quality guideline quotients: a critical review. *Environmental Science and Technology*, **40**, 1726–1736.

Long, E. R., MacDonald, D. D., Cubbage, J. C. and Ingersoll, C. G. (1998b). Predicting the toxicity of sediment-associated trace metals with simultaneously extracted trace metal : acid-volatile sulfide concentrations and dry weight-normalized concentrations: a critical comparison. *Environmental Toxicology and Chemistry*, **17**, 972–974.

Long, E. R., Macdonald, D. D., Smith, S. L. and Calder, F. D. (1995). Incidence of adverse biological effects within ranges of chemical concentrations in marine and estuarine sediments. *Environmental Management*, **19**, 81–97.

Lopes, I., Baird, D. J. and Ribeiro, R. (2006). Genetic adaptation to metal stress by natural populations of *Daphnia longispina*. *Ecotoxicology and Environmental Safety*, **63**, 275–285.

Lopez, J. L., Marina, A., Vazquez, J. and Alvarez, G. (2002). A proteomic approach to the study of the marine mussels *Mytilus edulis* and *Mytilus galloprovincialis*. *Marine Biology*, **141**, 217–223.

Loring, D. H. (1991). Normalization of heavy-metal data from estuarine and coastal sediments. *ICES Journal of Marine Science*, **48**, 101–115.

Lottermoser, B. (2003). *Mine Wastes: Characterization, Treatment and Impacts*. Amsterdam, the Netherlands: Springer-Verlag.

Loveley, D. R. and Phillips, E. J. P. (1986). Organic matter mineralization with reduction of ferric iron in anaerobic sediments. *Applied and Environmental Microbiology*, **51**, 683–689.

Lowe, D. M., Moore, M. N. and Clarke, K. R. (1981). Effects of oil on digestive cells in mussels: quantitative alterations in cellular and lysosomal structure. *Aquatic Toxicology*, **1**, 213–226.

Luoma, S. N. (1977). Detection of trace contaminant effects in aquatic ecosystems. *Journal of the Fisheries Research Board of Canada*, **34**, 436–439.

Luoma, S. N. (1983). Bioavailability of trace metals to aquatic organisms: a review. *Science of the Total Environment*, **28**, 1–22.

Luoma, S. N. (1989). Can we determine the biological availability of sediment-bound trace elements? *Hydrobiologia*, **176/7**, 379–396.

Luoma, S. N. (1990). Processes affecting metal concentrations in estuarine and coastal marine sediments. In *Heavy Metals in Marine Environments*, ed. R. W. Furness and P. S. Rainbow. Boca Raton, FL: CRC Press, pp. 51–66.

Luoma, S. N. (1996). The developing framework of marine ecotoxicology: pollutants as a variable in marine ecosystems? *Journal of Experimental Marine Biology and Ecology*, **200**, 29–55.

Luoma, S. N. and Bryan, G. W. (1978). Factors controlling availability of sediment-bound lead to the estuarine bivalve *Scrobicularia plana*. *Journal of the Marine Biological Association of the United Kingdom*, **58**, 793–802.

Luoma, S. N. and Bryan, G. W. (1979). Trace metal bioavailability: modelling chemical and biological interactions of sediment-bound zinc. In *Chemical Modelling in Aqueous Systems*, ed. E. A. Jenne. Washington, DC: American Chemical Society, pp. 577–609.

Luoma, S. N. and Bryan, G. W. (1981). Statistical assessment of the form of trace metals in oxidized estuarine sediments employing chemical extractants. *Science of the Total Environment*, **17**, 166–196.

Luoma, S. N. and Bryan, G. W. (1982). A statistical study of environmental factors controlling concentrations of heavy metals in the burrowing bivalve *Scrobicularia plana* and the polychaete *Nereis diversicolor*. *Estuarine, Coastal and Shelf Science*, **15**, 95–108.

Luoma, S. N. and Carter, J. L. (1990). Effects of trace metals on aquatic benthos. In *Metal Ecotoxicology: Concepts and Applications*, ed. M. Newman, and A. McIntosh. Chelsea, MI: Lewis Publishers, pp. 261–287.

Luoma, S. N. and Cloern, J. E. (1982). The impact of waste-water discharge on biological communities in San Francisco Bay. In *San Francisco Bay: Use and Protection*. ed. W. J. Kockleman, T. J. Conomos and A. E. Leviton. San Francisco, CA: Pacific Division, American Association for the Advancement of Science, pp. 137–160.

Luoma, S. N. and Davis, J. A. (1983). Requirements for modeling trace metal partitioning in oxidized estuarine sediments. *Marine Chemistry*, **12**, 159–181.

Luoma, S. N. and Fisher, N. (1997). Uncertainties in assessing contaminant exposure from sediments: In *Ecological Risk Assessments of Contaminated Sediments*, ed. T. Dillon and G. Biddinger. Pensacola, FL: SETAC Press, pp. 211–237.

Luoma, S. N. and Jenne, E. A. (1976). Factors affecting the availability of sediment-bound cadmium to the estuarine deposit-feeding clam, *Macoma balthica*. In *Radioecology and Energy Resources, Proceedings of the 4th National Symposium in Radioecology*, ed. E. Cushing. Stroudsburg, PA: Hutchinson and Ross, Inc., pp. 283–291.

Luoma, S. N. and Jenne, E. A. (1977). The availability of sediment-bound cobalt, silver, and zinc to a deposit-feeding clam. In *Biological Implications of Metals in the Environment*, ed. R. W. Wildung and H. Drucker. Springfield, VA: National Technical Information Service, pp. 213–230.

Luoma, S. N. and Presser, T. S. (2000). Forecasting selenium discharges to the San Francisco Bay-Delta estuary: ecological effects of a proposed San Luis Drain extension. *US Geological Survey Open File Report* **2000–416**. Available online at http://pubs.water.usgs.gov/ofr00-416/

Luoma, S. N. and Rainbow, P. S. (2005). Why is metal bioaccumulation so variable? Biodynamics as a unifying concept. *Environmental Science and Technology*, **39**, 1921–1931.

Luoma, S. N., Bryan, G. W. and Langston, W. J. (1982). Scavenging of heavy metals from particulates by brown seaweed. *Marine Pollution Bulletin*, **13**, 394–396.

Luoma, S. N., Cain, D. and Johansson, C. (1985). Temporal fluctuations of silver, copper and zinc in the bivalve *Macoma balthica* at five stations in South San Francisco Bay. *Hydrobiologia*, **129**, 109–120.

Luoma, S. N., Ho, Y. B. and Bryan, G. W. (1995). Fate, bioavailability and toxicity of silver in estuarine environments. *Marine Pollution Bulletin*, **31**, 44–54.

Luoma, S. N., Johns, C., Fisher, N. S., *et al.* (1992). Determination of selenium bioavailability to a benthic bivalve from particulate and solute pathways. *Environmental Science and Technology*, **26**, 485–491.

Luoma, S. N., Moore, J. N., Farag, A., *et al.* (2007). Mining impacts on fish in the Clark Fork River, Montana: a field ecotoxicology case study. In *The Toxicology of Fishes*, ed. R. Di Giulio and D. Hinton. London: Taylor and Francis, pp. 775–800.

Luoma, S. N., van Geen, A., Lee, B.-G. and Cloern, J. E. (1998). Metal uptake by phytoplankton during a bloom in South San Francisco Bay: implications for metal cycling in estuaries. *Limnology and Oceanography*, **43**, 1007–1016.

Luoma, S. N., Wellise, C., Cain, D. J., *et al.* (1997). Near field receiving water monitoring of trace metals in clams (*Macoma balthica*) and sediments near the Palo Alto and San Jose/Sunnyvale Water Quality Control Plants in South San Francisco Bay, California, 1977. *US Geological Survey Open File Report* **1998-563**.

Lytle, T. F. and Lytle, J. S. (1990). Heavy metals in the eastern oyster, *Crassostrea virginica*, of the Mississippi Sound. *Bulletin of Environmental Contamination and Toxicology*, **44**, 142–148.

MacDonald, B. G. and Chapman, P. M. (2007). Selenium effects: a weight-of-evidence approach. *Integrated Environmental Assessment and Management*, **3**, 129–136.

MacDonald, D. D., Carr, R. S., Calder, F. D. and Long, E. R. (1996). Development and evaluation of sediment quality guidelines for Florida coastal waters. *Ecotoxicology*, **5**, 253–278.

Maciorowski, A. F. (1988). Populations and communities: linking toxicology and ecology in a new synthesis. *Environmental Toxicology and Chemistry*, **7**, 677–678.

Mackay, D. (2001). *Multimedia Environmental Models: The Fugacity Approach*, 2nd edn. Boca Raton, FL: Lewis Publishers.

Mackay, D., Webster, E., Woodfine, D., *et al.* (2003). Toward consistent evaluation of the persistence of organic, inorganic and metallic substances. *Human and Ecological Risk Assessment*, **9**, 1445–1474.

Mackey, A. P. and Mackay, S. (1996). Spatial distribution of acid-volatile sulphide concentration and metal bioavailability in mangrove sediments from the Brisbane River, Australia. *Environmental Pollution*, **93**, 205–209.

Magos, L. and Webb, M. (1980). The interaction of selenium with cadmium and mercury. *CRC Critical Reviews in Toxicology*, **8**, 1–42.

Maguire, R. J. (1991). Aquatic environmental aspects of non-pesticidal organotin compounds. *Water Quality Research Journal of Canada*, **26**, 243–360.

Malm, O. (1998). Gold mining as a source of mercury exposure in the Brazilian Amazon. *Environmental Research*, **77A**, 73–78.

Malo, B. (1977). Partial extraction of metals from aquatic environments. *Environmental Science and Technology*, **11**, 277–282.

Malouf, C. (1974). Economy and land use by the Indians of western Montana. In *Interior Salish and Eastern Washington Indians*, ed. E. O. Fuller. New York: Garland Publishing, pp. 110–121.

Maltby, L. (1992). The use of the physiological energetics of *Gammarus pulex* to assess toxicity: a study using artificial streams. *Environmental Toxicology and Chemistry*, **11**, 79–85.

Maltby, L. (1999). Studying stress: the importance of organism-level responses. *Ecological Applications*, **9**, 431–440.

Maltby, L. and Naylor, C. (1990). Preliminary observations on the ecological relevance of the *Gammarus* 'scope for growth' assay: effect of zinc on reproduction. *Functional Ecology*, **4**, 393–397.

Maltby, L., Clayton, S. A., Wood, R. M. and McLoughlin, N. (2002). Evaluation of the *Gammarus pulex* in situ feeding assay as a biomonitor of water quality: robustness, responsiveness, and relevance. *Environmental Toxicology and Chemistry*, **21**, 361–368.

Maltby, L., Naylor, C. and Calow, P. (1990a). Effect of stress on a freshwater benthic detritivore: scope for growth in *Gammarus pulex*. *Ecotoxicology and Environmental Safety*, **19**, 285–291.

Maltby, L., Naylor, C. and Calow, P. (1990b). Field deployment of a scope for growth assay involving *Gammarus pulex*, a freshwater benthic invertebrate. *Ecotoxicology and Environmental Safety*, **19**, 292–300.

Manahan, S. E. (1991). *Environmental Chemistry*. Boca Raton, FL: Lewis Publishers.

Mantel, L. H. and Farmer, L. L. (1983). Osmotic and ionic regulation. In *The Biology of Crustacea, vol.* **5**, Internal Anatomy and Physiological Regulation, ed. D. E. Bliss and L. H. Mantel. New York: Academic Press, pp. 53–161.

Mantoura, R. F. C., Dickson, A. and Riley, J. P. (1978). The complexation of metals with humic materials in natural waters. *Estuarine, Coastal and Marine Science*, **6**, 387–408.

Marcus, W. A. (1987). Copper dispersion in ephemeral stream sediments. *Earth Surface Processes and Landforms*, **12**, 217–228.

Maret, T. R., Cain, D. J., MacCoy, D. E. and Short, T. M. (2003). Response of benthic invertebrate assemblages to metal exposure and bioaccumulation associated with hard-rock mining in northwestern streams, USA. *Journal of the North American Benthological Society*, **22**, 598–620.

Marigómez, I., Orbea, A., Olabarrieta, I., Etxeberria, M. and Cajaraville, M. P. (1996). Structural changes in the digestive lysosomal system of sentinel mussels as biomarkers of environmental stress in mussel-watch programmes. *Comparative Biochemistry and Physiology C*, **113**, 291–297.

Marigómez, I., Soto, M., Carajaville, M. P., Angulo, E. and Giamberini, L. (2002). Cellular and subcellular distribution of metals in molluscs. *Microscopy Research Techniques*, **56**, 358–392.

Marr, J. C. A., Bergman, H. L., Parker, M., *et al.* (1995). Relative sensitivity of brown and rainbow trout to pulsed exposures of an acutely lethal mixture of metals typical of the Clark Fork River, Montana. *Canadian Journal of Fisheries and Aquatic Sciences*, **52**, 2005–2015.

Marshall, J. S. (1978). Population dynamics of *Daphnia galeata mendotae* as modified by chronic cadmium stress. *Canadian Journal of Fisheries and Aquatic Sciences*, **35**, 461–469.

Marshall, J. S. and Mellinger, D. C. (1980). Dynamics of cadmium-stressed plankton communities. *Canadian Journal of Fisheries and Aquatic Sciences*, **37**, 403–414.

Martens, D. A. and Suarez, D. L. (1997). Selenium speciation of soil/sediment determined with sequential extractions and hydride generation atomic absorption spectrophotometry. *Environmental Science and Technology*, **31**, 133–139.

Martin, C. A., Luoma, S. N., Cain, D. J. and Buchwalter, D. B. (2007). Cadmium ecophysiology in seven stonefly (Plecoptera) species: delineating sources and estimating susceptibility. *Environmental Science and Technology*, **41**, 7171–7177.

Martin, J. L. and Gerlach, M. L. (1972). Selenium metabolism in animals. *Annals of the New York Academy of Sciences*, **192**, 193–199.

Martoja, R., Ballan-Dufrançais, C., Jeantet, A. Y., *et al.* (1988). Effets chimiques et cytologiques de la contamination expérimentale de l'huître *Crassostrea gigas* Thunberg par l'argent administré sous forme dissoute et par voie alimentaire. *Canadian Journal of Fisheries and Aquatic Sciences*, **45**, 1827–1841.

Marvin-DiPasquale, M., Agee, J., McGowan, C., *et al.* (2000). Methyl-mercury degradation pathways: a comparison among three mercury-impacted ecosystems. *Environmental Science and Technology*, **34**, 4908–4916.

Mason, A. Z. and Jenkins, K. D. (1995). Metal detoxification in aquatic organisms. In *Metal Speciation and Bioavailability in Aquatic Systems*, ed. A. Tessier and D. R. Turner. Chichester, UK: John Wiley, pp. 479–608.

Mason, R. P. and Gill, G. A. (2005). Mercury in the marine environment. In *Mercury Sources, Measurements, Cycles and Effects*, ed. M. B. Parsons and J. B. Percival. Halifax, Nova Scotia: Mineralogical Association of Canada, pp. 179–216.

Mason, R. P., Fitzgerald, W. F. and Morel, F. M. M. (1994). The biogeochemical cycling of elemental mercury: anthropogenic influences. *Geochimica et Cosmochimica Acta*, **58**, 3191–3198.

Mason, R. P., Lawson, N. M., Lawrence, A. L., Leaner, J. J., Lee, J. G. and Sheu, G.-R. (1999). Mercury in Chesapeake Bay. *Marine Chemistry*, **65**, 77–96.

Mason, R. P., Reinfelder, J. R. and Morel, M. M. (1995). Bioaccumulation of mercury and methylmercury. *Water, Air and Soil Pollution*, **80**, 915–921.

Matthiessen, P. and Gibbs, P. E. (1998). Critical appraisal of the evidence for tributyltin-mediated endocrine disruption in mollusks. *Environmental Toxicology and Chemistry*, **17**, 37–43.

Maund, S. J., Taylor, E. J. and Pascoe, D. (1992). Population responses of the freshwater amphipod crustacean *Gammarus pulex* (L.) to copper. *Freshwater Biology*, **28**, 29–36.

Mayer, L. M., Chen, Z., Findlay, R. H., *et al.* (1996). Bioavailability of sedimentary contaminants subject to deposit-feeder digestion. *Environmental Science and Technology*, **30**, 2641–2645.

Mayer, L. M., Schick, L. L., Self, R. F. L., *et al.* (1997). Digestive environments of benthic macroinvertebrate guts: enzymes, surfactants and dissolved organic matter. *Journal of Marine Research*, **55**, 785–812.

Mayr, E. (1982). *The Growth of Biological Thought*. Cambridge, MA: Harvard University Press.

McCarthy, J. F. and Shugart, L. R. (1990). *Biomarkers of Environmental Contamination*. Chelsea, MI: Lewis Publishers.

McDaniels, T. L. and Gregory, R. (2004). Learning as an objective with a structured risk management decision process. *Environmental Science and Technology*, **38**, 1921–1926.

McGee, B. L., Wright, D. A. and Fisher, D. J. (1998). Biotic factors modifying acute toxicity of aqueous cadmium to estuarine amphipod *Leptocheirus plumulosus*. *Archives of Environmental Contamination and Toxicology*, **34**, 34–40.

McGeer, J. C., Brix, K. V., Skeaff, J. M., *et al.* (2003). Inverse relationship between bioconcentration factor and exposure concentration for metals: implications for hazard assessment of metals in the aquatic environment. *Environmental Toxicology and Chemistry*, **22**, 1017–1037.

McKim, J. M. and Nichols, J. W. (1994). Use of physiologically based toxicokinetic models in a mechanistic approach to aquatic toxicology. In *Aquatic Toxicology: Molecular, Biochemical, and Cellular Perspectives*, ed. D. C. Malins and G. K. Ostrander. Boca Raton, FL: Lewis Publishers, pp. 469–521.

McRae, G., Camp, D. K., Lyons, W. G. and Dix, T. L. (1998). Relating benthic infaunal community structure to environmental variables in estuaries using nonmetric multidimensional scaling and similarity analysis. *Environmental Monitoring and Assessment*, **51**, 233–246.

Meador, J. P. (2000). Predicting the fate and effects of tributyltin in marine systems. *Reviews in Environmental Contamination and Toxicology*, **166**, 1–48.

Meador, J. P. (2002). Determination of a tissue and sediment threshold for tributyltin to protect prey species of juvenile salmonids listed under the US Endangered Species Act. *Aquatic Conservation: Marine and Freshwater Ecosystems*, **12**, 539–551.

Meador, J. P., Krone, C. A., Dyer, D. W. and Varanasi, U. (1997). Toxicity of sediment-associated tributyltin to infaunal invertebrates: species comparison and the role of organic carbon. *Marine Environmental Research*, **43**, 219–241.

Mearns, A. J., Swartz, R. C., Cummins, J. M., *et al.* (1986). Inter-laboratory comparison of a sediment toxicity test using the marine amphipod *Rhepoxynius abronius*. *Marine Environmental Research*, **19**, 13–37.

Medina, M., Andrade, S., Faugeron, S., *et al.* (2005). Biodiversity of rocky intertidal benthic communities associated with copper mine tailing discharges in northern Chile. *Marine Pollution Bulletin*, **50**, 396–409.

Medvedev, Z. (1969). *The Rise and Fall of Trofim D. Lysenko*. New York: Columbia University Press.

Meili, M., Bishop, K., Bringmark, L., *et al.* (2003). Critical levels of atmospheric pollution: criteria and concepts for operational modelling of mercury in forest and lake ecosystems. *Science of the Total Environment*, **304**, 83–106.

Melzian, B. D. (1990). Toxicity assessment of dredged materials: acute and chronic toxicity as determined by bioassays and bioaccumulation tests. In *Proceedings of the International Seminar on the Environmental Aspects of Dredging Activities*, ed. C. Alzieu and B. Gallenne. Nantes, France: Goubault, pp. 49–59.

Mergler, D., Anderson, H. A., Chan, L. H. M., *et al.* (2007). Methylmercury exposure and health effects in humans: a worldwide concern. *Ambio*, **36**, 3–11.

Mersch, J., Wagner, P. and Pihan, J. C. (1996). Copper in indigenous and transplanted zebra mussels in

relation to changing water concentrations and body weight. *Environmental Toxicology and Chemistry*, **15**, 886–893.

Mesek, S. L. and Cutter, G. A. (2006). Evaluating the biogeochemical cycle of selenium in San Francisco Bay through modeling. *Limnology and Oceanography*, **51**, 2018–2032.

Meyer, J. S., Adams, W. J., Brix, K. V., *et al.* (2005). Workshop summary and conclusions. In *Toxicity of Dietborne Metals to Aquatic Organisms*, ed. J. S. Meyer, W. J. Adams, K. V. Brix, S. N. Luoma, D. R. Mount, W. A. Stubblefield and C. M. Wood. Pensacola, FL: SETAC Press, pp. 191–200.

Meyer, J. S., Santore, R. C., Bobbitt, J. P., *et al.* (1999). Binding of nickel and copper to fish gills predicts toxicity when water hardness varies, but free-ion activity does not. *Environmental Science and Technology*, **33**, 913–916.

Meyer, J. S., Sudekamp, M. J., Morris, J. M. and Farag, A. M. (2006). Leachability of protein and metals incorporated into aquatic invertebrates: are species and metals-exposure history important? *Archives of Environmental Contamination and Toxicology*, **50**, 79–87.

Milani, D., Reynoldson, T. B., Borgmann, U. and Kolasa, J. (2003). The relative sensitivity of four benthic invertebrates to metals in spiked-sediment exposures and application to contaminated field sediment. *Environmental Toxicology and Chemistry*, **22**, 845–854.

Mileikovsky, S. A. (1971). Types of larval development in marine bottom invertebrates, their distribution and ecological significance: a re-evaluation. *Marine Biology*, **10**, 193–213.

Miller, G. E., Grant, P. M., Kishore, R., *et al.* (1972). Mercury concentrations in museum specimens of tuna and swordfish. *Science*, **175**, 1121–1122.

Millward, R. N. and Grant, A. (1995). Assessing the impact of copper on nematode communities from a chronically metal-enriched estuary using Pollution-Induced Community Tolerance. *Marine Pollution Bulletin*, **30**, 701–706.

Minchin, D. and Gollasch, S. (2003). Fouling and ships' hulls: how changing circumstances and spawning events may result in the spread of exotic species. *Biofouling*, **19**, 111–122.

Moffett, J. W., Brand, L. E., Croot, P. L. and Barbeau, K. A. (1997). Cu speciation and cyanobacterial distribution in harbors subject to anthropogenic Cu inputs. *Limnology and Oceanography*, **42**, 789–799.

Moon, E., Shouse, M. K., Parchaso, F., *et al.* (2005). Near-field receiving water monitoring of trace metals and a benthic community near the Palo Alto Regional Water Quality Control Plant in South San Francisco

Bay, California. *US Geological Survey Open File Report* 2005–1279.

Moore, J.N. and Landrigan, E.M. (1999). Mobilization of metal-contaminated sediment by ice-jam floods. *Environmental Geology*, **37**, 96–101.

Moore, J.N. and Luoma, S.N. (1990). Hazardous wastes from large-scale metal extraction. *Environmental Science and Technology*, **24**, 1279–1285.

Moore, J.N., Brook, E.J. and Johns, C. (1989). Grain size partitioning of metals in contaminated, coarse-grained river floodplain sediment: Clark Fork River, Montana, USA. *Environmental Geology*, **14**, 107–114.

Moore, M.N. (1985). Cellular responses to pollutants. *Marine Pollution Bulletin*, **16**, 134–139.

Moore, M.N. (1988). Cytochemical responses of the lysosomal system and NADPH-ferrihemoprotein reductase in molluscan digestive cells to environmental and experimental exposure to xenobiotics. *Marine Ecology Progress Series*, **46**, 81–89.

Moore, M.N. (2002). Biocomplexity: the post-genome challenge in ecotoxicology. *Aquatic Toxicology*, **59**, 1–15.

Moore, M.N., Allen, J.I. and McVeigh, A. (2006). Environmental prognostics: an integrated model supporting lysosomal stress responses as predictive biomarkers of animal health status. *Marine Environmental Research*, **61**, 278–304.

Moore, M.N., Depledge, M.H., Readman, J.W. and Leonard, P. (2004). An integrated biomarker-based strategy for ecotoxicological evaluation of risk in environmental management. *Mutation Research*, **552**, 247–268.

Moore, P.G. and Rainbow, P.S. (1984). Ferritin crystals in the gut caeca of *Stegocephaloides christianiensis* Boeck and other Stegocephalidae (Amphipoda: Gammaridea): a functional interpretation. *Philosophical Transactions of the Royal Society of London B*, **306**, 219–245.

Moore, P.G. and Rainbow, P.S. (1987). Copper and zinc in an ecological series of talitroidean Amphipoda (Crustacea). *Oecologia*, **73**, 120–126.

Moore, P.G., Rainbow, P.S. and Hayes, E. (1991). The beach-hopper *Orchestia gammarellus* (Crustacea: Amphipoda) as a biomonitor for copper and zinc: North Sea trials. *Science of the Total Environment*, **106**, 221–238.

Morel, F.M.M, Kraepiel, A.M. and Amyot, M. (1998). The chemical cycle and bioaccumulation of mercury. *Annual Review of Ecology and Systematics*, **29**, 543–566.

Morel, F.M.M. and Hering, J.G. (1993). *Principles and Applications of Aquatic Chemistry*. New York: John Wiley.

Moriarty, F. (1999). *Ecotoxicology: The Study of Pollutants in Ecosystems*. London: Academic Press.

Morse, J.W. (2002). Sedimentary geochemistry of the carbonate and sulphide systems and their potential influence on toxic metal bioavailability. In *Chemistry of Marine Water and Sediments*, ed. A. Gianguzza, E. Pelizzetti and S. Sammartano. Berlin, Germany: Springer-Verlag, pp. 164–189.

Morse, J.W. and Rickard, D. (2004). Chemical dynamics of acid volatile sulfide. *Environmental Science and Technology*, **38A**, 132–139.

Motelica-Heino, M., Naylor, C., Zhang, H. and Davison, W. (2003). Simultaneous release of metals and sulfide in lacustrine sediment. *Environmental Science and Technology*, **37**, 4374–4381.

Mouneyrac, C., Amiard, J.-C., Amiard-Triquet, C., *et al.* (2002). Partitioning of accumulated trace metals in the talitrid amphipod crustacean *Orchestia gammarellus*: a cautionary tale on the use of metallothionein-like proteins as biomarkers. *Aquatic Toxicology*, **57**, 225–242.

Mouneyrac, C., Mastain, O., Amiard, J.-C., *et al.* (2003). Trace-metal detoxification and tolerance of the estuarine worm *Hediste diversicolor* chronically exposed in their environment. *Marine Biology*, **143**, 731–744.

Mount, D.R. (2005). Introduction. In *Toxicity of Dietborne Metals to Aquatic Organisms*, ed. J.S. Meyer, W.J. Adams, K.V. Brix, S.N. Luoma, D.R. Mount, W.A. Stubblefield and C.M. Wood. Pensacola, FL: SETAC Press, pp. 7–13.

Mount, D.R., Barth, A.K., Garrison, T.D., Barten, K.A. and Hockett, J.R. (1994). Dietary and waterborne exposure of rainbow trout (*Oncorhynchus mykiss*) to copper, cadmium, lead and zinc using a live diet. *Environmental Toxicology and Chemistry*, **13**, 2031–2041.

Mourgaud, Y., Martinez, E., Geffard, A., *et al.* (2002). Metallothionein concentration in the mussel *Mytilus galloprovincialis* as a biomarker of response to metal contamination: validation in the field. *Biomarkers*, **7**, 479–490.

Munger, C. and Hare, L. (1997). Relative importance of water and food as cadmium sources to an aquatic insect (*Chaoborus punctipennis*): implications for predicting Cd bioaccumulation in nature. *Environmental Science and Technology*, **31**, 891–895.

Munkittrick, K.R. and Dixon, D.G. (1989). A holistic approach to ecosystem health assessment using fish population characteristics. *Hydrobiologia*, **188/9**, 123–135.

Munthe, J., Bodaly, R.A., Branfireun, B.A., *et al.* (2007). Recovery of mercury-contaminated fisheries. *Ambio*, **36**, 33–44.

Murray, C. and Marmorek, D. (2003). Adaptive management and ecological restoration. In *Ecological Restoration of Southwestern Ponderosa Pine Forests*, ed. P. Freiderici. Washington, DC: Island Press, pp. 417–428.

Myers, C. R. and Nealson, K. H. (1988). Bacterial manganese reduction and growth with manganese oxide as the sole electron acceptor. *Science*, **240**, 1319–1321.

Nagorski, S. A., Moore, J. N., McKinnon, T. E. and Smith, D. B. (2003). Scale-dependent temporal variations in stream water geochemistry. *Environmental Science and Technology*, **37**, 859–864.

NAS (1980). *The International Mussel Watch*. Washington, DC: National Academy of Sciences.

Nasci, C., Da Ros, L., Campesan, G., *et al.* (1999). Clam transportation and stress-related biomarkers as useful tools for assessing water quality in coastal environments. *Marine Pollution Bulletin*, **39**, 255–260.

Nassiri, Y., Rainbow, P. S., Amiard-Triquet, C., Rainglet, F. and Smith, B. D. (2000). Trace metal detoxification in the ventral caeca of *Orchestia gammarellus* (Crustacea: Amphipoda). *Marine Biology*, **136**, 477–484.

Nehring, S. (2000). Long-term changes in Prosobranchia (Gastropoda) abundances on the German North Sea coast: the role of the anti-fouling biocide tributyltin. *Journal of Sea Research*, **43**, 151–165.

Nehring, S. (2001). After the TBT era: alternative antifouling paints and their ecological risks. *Sneckenbergiana Maritima*, **31**, 341–351.

Nehring, S. (2005). International shipping: a risk for biodiversity in Germany. *Neobiota*, **6**, 125–143.

Nevo, E., Noy, R., Lavie, B., Beiles, A. and Muchtar, S. (1986). Genetic diversity and resistance to marine pollution. *Biological Journal of the Linnean Society*, **29**, 139–144.

Newman, J. R. and Schreiber, R. K. (1988). Air pollution and wildlife toxicology: an overlooked problem. *Environmental Toxicology and Chemistry*, **7**, 181–190.

Newman, M. C. and Unger, M. A. (2003). *Fundamentals of Ectoxicology*. Boca Raton, FL: Lewis Publishers.

Ng, T. Y.-T. and Wang, W.-X. (2004). Detoxification and effects of Ag, Cd, and Zn pre-exposure on metal uptake kinetics in the clam *Ruditapes philippinarum*. *Marine Ecology Progress Series*, **268**, 161–172.

Ng, T. Y.-T, Amiard-Triquet, C., Rainbow, P. S., Amiard, J.-C. and Wang, W.-X. (2005). Physico-chemical form of trace metals accumulated by phytoplankton and their assimilation by filter-feeding invertebrates. *Marine Ecology Progress Series*, **299**, 179–191.

Nichols, F. H., Cloern, J. E., Luoma, S. N. and Peterson, D. H. (1986). The modification of an estuary. *Science*, **231**, 567–673.

Nicholson, S. (2003a). Tachycardia in the mussel, *Perna viridis* (L.) on exposure to tributyltin antifouling paint. *Australasian Journal of Ecotoxicology*, **9**, 137–140.

Nicholson, S. (2003b). Cardiac and branchial physiology associated with copper accumulation and detoxication in the mytilid mussel *Perna viridis* (L.). *Journal of Experimental Marine Biology and Ecology*, **295**, 157–171.

Nicholson, S. and Lam, P. K. S. (2005). Pollution monitoring in Southeast Asia using biomarkers in the mytilid mussel *Perna viridis* (Mytilidae: Bivalvia). *Environment International*, **31**, 121–132.

Nichson, R., MacArthur, J., Burgess, W., *et al.* (1998). Arsenic poisoning of Bangladesh groundwater. *Nature*, **395**, 338.

Nieboer, E. and Richardson, D. H. S. (1980). The replacement of the nondescript term 'heavy metals' by a biologically and chemically significant classification of metal ions. *Environmental Pollution B*, **1**, 3–26.

Nigro, M. and Leonzio, C. (1996). Intracellular storage of mercury and selenium in different marine invertebrates. *Marine Ecology Progress Series*, **135**, 137–143.

Nigro, M., Falleni, A., Del Barga, I., *et al.* (2006). Cellular biomarkers for monitoring estuarine environments: transplanted versus native mussels. *Aquatic Toxicology*, **77**, 339–347.

Nimick, D. A. and Moore, J. N. (1991). Prediction of water-soluble metal concentrations in fluvially deposited tailings sediments, Upper Clark Fork Valley, Montana, USA. *Applied Geochemistry*, **6**, 635–646.

Niyogi, S. and Wood, C. M. (2003). Effects of chronic waterborne and dietary metal exposure on gill metal-binding: implications for the biotic ligand model. *Human and Ecological Risk Assessment*, **9**, 813–846.

Niyogi, S. and Wood, C. M. (2004). Biotic Ligand Model, a flexible tool for developing site-specific water quality guidelines for metals. *Environmental Science and Technology*, **38**, 6177–6192.

Nordstrom, D. K. and Alpers, C. N. (1999). Negative pH, efflorescent mineralogy, and consequences for environmental restoration at the Iron Mountain Superfund site, California. *Proceedings of the National Academy of Sciences of the USA*, **96**, 3455–3462.

Norton, S. B., van der Schalie, W. H., *et al.* (2003). Ecological risk assessment: US EPA's current guidelines and future directions. In *Handbook of Ecotoxicology*, ed. D. J. Hoffman, B. A. Raffner, G. A. Burton Jr and J. Cairns Jr Boca Raton, FL: CRC Press, pp. 951–983.

Nott, J. A. and Nicolaidou, A. (1990). Transfer of metal detoxification along marine food chains. *Journal of the Marine Biological Association of the United Kingdom*, **70**, 905–912.

Nott, J. A. and Nicolaidou, A. (1994). Variable transfer of detoxified metals from snails to hermit crabs in marine food chains. *Marine Biology*, **120**, 369–377.

Novelli, A. A., Losso, C., Ghetti, P. F. and Ghirardini, A. V. (2003). Toxicity of heavy metals using sperm cell and embryo toxicity bioassays with *Paracentrotus lividus* (Echinodermata: Echinoidea): comparisons with exposure concentrations in the lagoon of Venice, Italy. *Environmental Toxicology and Chemistry*, **22**, 1295–1301.

NRC (2000). *Toxicologic Effects of Methylmercury*. Washington, DC: National Research Council.

NRC (2001). *Assessing the TMDL Approach to Water Quality Management*. Washington, DC: National Research Council.

NRC (2003). *Bioavailability of Contaminants in Soils and Sediments: Processes, Tools, and Applications*. Washington, DC: National Research Council.

Nriagu, J. O. (1979). Global inventory of natural and anthropogenic emissions of trace metals to the atmosphere. *Nature*, **279**, 409–411.

Nriagu, J. O. (1988). A silent epidemic of environmental metal poisoning? *Environmental Pollution*, **50**, 139–161.

Nriagu, J. O. (1990). Global metal pollution: poisoning the biosphere? *Environment*, **32**, 7–11.

Nriagu, J. O. (1994). Mercury pollution from the past mining of gold and silver in the Americas. *Science of the Total Environment*, **149**, 167–181.

Nriagu, J. O. and Pacyna, J. M. (1988). Quantitative assessment of worldwide contamination of air, water and soils by trace metals. *Nature*, **333**, 134–139.

Nriagu, J. O. and Sprague, J. B. (1987). *Cadmium in the Aquatic Environment*. New York: John Wiley.

Nriagu, J. O., Lawson, G., Wong, H. K. T. and Cheam, V. (1996). Dissolved trace metals in Lakes Superior, Erie and Ontario. *Environmental Science and Technology*, **30**, 178–187.

Nriagu, J. O., Wong, H. K. T. and Coker, R. D. (1982). Deposition and chemistry of pollutant metals in lakes around the smelters at Sudbury, Ontario. *Environmental Science and Technology*, **16**, 551–559.

Nugegoda, D. and Rainbow, P. S. (1989). Effects of salinity changes on zinc uptake and regulation by the decapod crustaceans *Palaemon elegans* and *Palaemonetes varians*. *Marine Ecology Progress Series*, **51**, 57–75.

Nugegoda, D. and Rainbow, P. S. (1988a). Effect of a chelating agent (EDTA) on zinc uptake and regulation by *Palaemon elegans* (Crustacea: Decapoda). *Journal of the Marine Biological Association of the United Kingdom*, **68**, 25–40.

Nugegoda, D. and Rainbow, P. S. (1988b). Zinc uptake and regulation by the sublittoral prawn *Pandalus montagui*

(Crustacea: Decapoda). *Estuarine, Coastal and Shelf Science*, **26**, 619–632.

Nunez-Nogueira, G., Rainbow, P. S. and Smith, B. D. (2006). Assimilation efficiency of zinc and cadmium in the decapod crustacean *Penaeus indicus*. *Journal of Experimental Marine Biology and Ecology*, **332**, 75–83.

O'Brien, P., Rainbow, P. S. and Nugegoda, D. (1990). The effect of the chelating agent EDTA on the rate of uptake of zinc by *Palaemon elegans* (Crustacea: Decapoda). *Marine Environmental Research*, **30**, 155–159.

O'Connor, T. P. (1996). Trends in chemical concentrations in mussels and oysters collected along the US coast from 1986 to 1993. *Marine Environmental Research*, **41**, 183–200.

Odum, H. T. (1986). *Ecosystem Theory and Application*. New York: John Wiley.

Ohlendorf, H. M. (2003). The ecotoxicology of selenium. In *Handbook of Ecotoxicology*, ed. D. J. Hoffman, B. A. Rattner, G. A. Burton Jr and J. Cairns Jr. Boca Raton, FL: CRC Press, pp. 465–500.

Ohlendorf, H. M., Hoffman, D. J., Saiki, M. K. and Aldrich, T. W. (1986). Embryonic mortality and abnormalities of aquatic birds: apparent impacts by selenium from irrigation drainwater. *Science of the Total Environment*, **52**, 49–63.

Oken, E., Kleinman, K. P., Berland, W. E., *et al.* (2003). Decline in fish consumption among pregnant women after a national mercury advisory. *Obstetrics and Gynecology*, **102**, 346–351.

Olsen, T., Ellerbeck, L., Fisher, T., Callaghan, A. and Crane, M. (2001). Variability in acetylcholinesterase and glutathione s-transferase activities in *Chironomus riparius* Meigen deployed *in situ* at uncontaminated field sites. *Environmental Toxicology and Chemistry*, **20**, 1725–1732.

Oremland, R. S., and Stolz, J. F. (2003). The ecology of arsenic. *Science*, **300**, 939–944.

Oremland, R. S., Hollibaugh, J. T., Maest, A. S., *et al.* (1989). Selenate reduction to elemental selenium by anaerobic bacteria in sediments and culture: biogeochemical significance of a novel, sulfate-independent respiration. *Applied and Environmental Microbiology*, **55**, 2333–2343.

Oremland, R. S., Steinberg, N. A., Maest, A. S., Miller, L. C. and Holllbaugh, J. T. (1990). Measurement of in situ rates of selenate removal by dissimilatory bacterial reduction in sediments. *Environmental Science and Technology*, **24**, 1157–1164.

Orihel, D. M., Paterson, M. J., Blanchfield, P. J., Bodaly, R. A. and Hintelmann, H. (2007). Experimental evidence of a linear relationship between inorganic mercury loading and methylmercury accumulation by aquatic biota. *Environmental Science and Technology*, **41**, 4952–4958.

Orr, P. L., Guiger, K. R. and Russel, C. K. (2005). Food chain transfer of selenium in lentic and lotic habitats of a western Canadian watershed. *Ecotoxicology and Environmental Safety*, **63**, 175–188.

Outridge, P. M., Scheuhammer, A. M., Fox, G. A., *et al.* (1999). An assessment of the potential hazards of environmental selenium for Canadian water birds. *Environmental Reviews*, **7**, 81–96.

Owens, S. (2006). Risk and precaution: changing perspectives from the Royal Commission on Environmental Pollution. *Science in Parliament*, **63**, 16–17.

Ozawa, T., Ueda, J. and Shimazu, Y. (1993). DNA single strand breakage by copper (II) complexes and hydrogen peroxide at physiological conditions. *Biochemistry and Molecular Biology International*, **31**, 455–461.

Pacyna, J. M. (1986). Atmospheric trace elements from natural and anthropogenic sources. In *Toxic Metals in the Atmosphere*, ed. J. O. Nriagu and C. I. Davidson. New York: John Wiley, pp. 33–52.

Páez-Osuna, F., Frías-Espericueta M. G. and Osuna-López, J. I. (1995). Trace metal concentrations in relation to season and gonadal maturation in the oyster *Crassostrea iridescens*. *Marine Environmental Research*, **40**, 19–31.

Pagenkopf, G. K. (1983). Gill surface interaction model for trace-metal toxicity to fishes: role of complexation, pH, and water hardness. *Environmental Science and Technology*, **17**, 342–347.

Pain, D. J., Sánchez, A. and Meharg, A. A. (1998) The Doñana ecological disaster: contamination of a world heritage estuarine marsh ecosystem with acidified pyrite mine waste. *Science of the Total Environment*, **222**, 45–54.

Palheta, D. and Taylor, A. (1995). Mercury in environmental and biological samples from a gold mining area in the Amazon region of Brazil. *Science of the Total Environment*, **168**, 63–69.

Paquin, P. R., Gorsuch, J. W., Apte, S., *et al.* (2002). The biotic ligand model: a historical overview. *Comparative Biochemistry and Physiology* C, **133**, 3–35.

Pascoe, D., Kedwards, T. J., Maund, S. J., Muthi, E. and Taylor, E. J. (1994). Laboratory and field evaluation of a behavioural bioassay: the *Gammarus pulex* (L.) precopula separation (GaPPS) test. *Water Research*, **28**, 369–372.

Patarnello, T., Guinez, R. and Battaglia, B. (1991). Effects of pollution on heterozygosity in the barnacle *Balanus amphitrite* (Cirripedia: Thoracica). *Marine Ecology Progress Series*, **70**, 237–243.

Patterson, C. C. (1994). Delineation of separate brain regions used for scientific versus engineering modes of thinking. *Geochimica et Cosmochimica Acta*, **58**, 3321–3327.

Paulsen, A. J. (1995). Tracing water and suspended matter in Raritan and Lower New York Bays using dissolved and particulate elemental concentrations. *Marine Chemistry*, **97**, 60–77.

Paulsson, K. and Lundbergh, K. (1991). Treatment of mercury contaminated fish by selenium addition. *Water, Air and Soil Pollution*, **56**, 833–841.

Peakall, D. B. (1992). *Animal Biomarkers as Pollution Indicators*. London: Chapman and Hall.

Peeters, E. T. H. M., Dewitte, A., Koelmans, A. A., van der Velden, J. A. and den Besten, P. J. (2001). Evaluation of bioassays versus contaminant concentrations in explaining the macroinvertebrate community structure in the Rhine-Meuse delta, The Netherlands. *Environmental Toxicology and Chemistry*, **20**, 2883–2891.

Pellacani, C., Buschini, A., Furlini, M., Poli, P. and Rossi, C. (2006). A battery of in vivo and in vitro tests useful for genotoxic pollutant detection in surface waters. *Aquatic Toxicology*, **77**, 1–10.

Pelletier, E. (1986). Mercury–selenium interactions in aquatic organisms: a review. *Marine Environmental Research*, **18**, 111–132.

Pelletier, E. (1988). Acute toxicity of some methylmercury complexes to *Mytilus edulis* and lack of selenium protection. *Marine Pollution Bulletin*, **19**, 213–219.

Pelletier, E. (1995). Environmental organometallic chemistry of mercury, tin, and lead: present status and perspectives. In *Metal Speciation and Bioavailability in Aquatic Systems*, ed. A. Tessier and D. R. Turner. Chichester, UK: John Wiley, pp. 103–148.

Pentreath, R. J. (1973). The accumulation from water of Zn-65, Mn-54, Co-58, and Fe-59 by the mussel, *Mytilus edulis*. *Journal of the Marine Biological Association of the United Kingdom*, **53**, 127–143.

Perceval, O., Couillard, Y., Pinel-Alloul, B., Giguère, A. and Campbell, P. G. C. (2004). Metal-induced stress in bivalves living along a gradient of Cd contamination: relating sub-cellular metal distribution to population-level responses. *Aquatic Toxicology*, **69**, 327–345.

Perez, M. H. and Wallace, W. G. (2004). Differences in prey capture in grass shrimp, *Palaemonetes pugio*, collected along an environmental gradient. *Archives of Environmental Contamination and Toxicology*, **46**, 81–89.

Pesch, C. E. and Stewart, N. E. (1980). Cadmium toxicity to three species of estuarine invertebrates. *Marine Environmental Research*, **3**, 145–156.

Pesch, C. E., Munns, W. R., and Gutjahr-Gobell, F. (1991). Effects of a contaminated sediment on life history traits and population growth rate of *Neanthes arenaceodentata* (Polychaeta: Nereidae) in the laboratory. *Environmental Toxicology and Chemistry*, **10**, 805–817.

Peters, G. M., Maher, W. A., Krikowa, F., *et al.* (1999). Selenium in sediment, pore waters and benthic infauna of Lake Macquarie, New South Wales, Australia. *Marine Environmental Research*, **47**, 491–508.

Phillips, D. J. H. (1977). The common mussel *Mytilus edulis* as an indicator of trace metals in Scandinavian waters. I. Zinc and cadmium. *Marine Biology*, **43**, 283–291.

Phillips, D. J. H. (1978). The common mussel *Mytilus edulis* as an indicator of trace metals in Scandinavian waters. II. Lead, iron and manganese. *Marine Biology*, **46**, 147–156.

Phillips, D. J. H. (1985). Organochlorines and trace metals in green-lipped mussels, *Perna viridis*, from Hong Kong waters: a test of indicator ability. *Marine Ecology Progress Series*, **21**, 251–258.

Phillips, D. J. H. and Muttarasin, K. (1985). Trace metals in bivalve molluscs from Thailand. *Marine Environmental Research*, **15**, 215–234.

Phillips, D. J. H. and Rainbow, P. S. (1988). Barnacles and mussels as biomonitors of trace elements: a comparative study. *Marine Ecology Progress Series*, **49**, 83–93.

Phillips, D. J. H. and Rainbow, P. S. (1994). *Biomonitoring of Trace Aquatic Contaminants*, 2nd edn. London: Chapman and Hall.

Phillips, D. J. H. and Yim, W. W.-S. (1981). A comparative evaluation of oysters, mussels and sediments as indicators of trace metals in Hong Kong waters. *Marine Ecology Progress Series*, **6**, 285–293.

Phillips, G. and Lipton, J. (1995). Injury to aquatic resources caused by metals in Montana's Clark Fork River basin: historic perspective and overview. *Canadian Journal of Fisheries and Aquatic Sciences*, **52**, 1990–1993.

Phinney, J. T. and Bruland, K. W. (1994). Uptake of lipophilic organic Cu, Cd and Pb complexes in the coastal diatom *Thalassiosira weissflogii*. *Environmental Science and Technology*, **28**, 1781–1790.

Piper, D. Z., Skorupa, J. P., Presser, T. S., *et al.* (2000). The Phosphoria Formation at the Hot Springs Mine in southeast Idaho: a source of selenium and other trace elements to surface water, groundwater, vegetation, and biota. *US Geological Survey Open File Report* 2000-020.

Playle, R. C., Dixon, D. G. and Burnison, K. (1993). Copper and cadmium binding to fish gills: estimates of metal–gill stability constants and modelling of metal accumulation. *Canadian Journal of Fisheries and Aquatic Sciences*, **50**, 2678–2687.

Plénet, S. (1995). Freshwater amphipods as biomonitors of metal pollution in surface and interstitial aquatic systems. *Freshwater Biology*, **33**, 127–137.

Plénet, S. (1999). Metal accumulation by an epigean and a hypogean freshwater amphipod: considerations for water quality assessment. *Water Environment Research*, **71**, 1298–1309.

Popham, J. D. and D'Auria, J. M. (1982). Effects of season and seawater concentrations on trace metal concentrations in organs of *Mytilus edulis*. *Archives of Environmental Contamination and Toxicology*, **11**, 273–282.

Presley, B. J., Taylor, R. J. and Boothe, P. N. (1990). Trace metals in Gulf of Mexico oysters. *Science of the Total Environment*, **97/8**, 551–593.

Presser, T. S. (1994). The Kesterson effect. *Environmental Management*, **18**, 437–454.

Presser, T. S. (1999). Selenium pollution. In *Encyclopedia of Environmental Science*, ed. D. E. Alexander and R. W. Fairbridge. Boston, MA: Kluwer, pp. 554–556.

Presser, T. S. and Luoma, S. N. (2006). Forecasting selenium discharges to the San Francisco Bay-Delta Estuary: ecological effects of a proposed San Luis Drain extension. *US Geological Survey Professional Paper* 1646. Available online at http://pubs.usgs.gov/pp/p1646/

Presser, T. S. and Ohlendorf, H. M. (1987). Biogeochemical cycling of selenium in the San Joaquin Valley, California, USA. *Environmental Management*, **11**, 805–821.

Presser, T. S., Piper, D. Z., Bird, K. J., *et al.* (2004). The Phosphoria Formation: a model for forecasting global selenium sources to the environment. In *Life Cycle of the Phosphoria Formation: From Deposition to the Post-Mining Environment*, ed. J. R. Hein. Amsterdam: Elsevier, pp. 110–125.

Preston, A. (1979). Standards and environmental criteria: the practical application of the results of laboratory experiments and field trials to pollution control. *Philosophical Transactions of the Royal Society of London B*, **286**, 611–624.

Prospero, J. (2001). African dust in America. *Geotimes*, Nov 01. Availabel online at www.geotimes.org/nov01/feature_dust.html

Prusha, B. A. and Clements, W. H. (2004). Landscape attributes, dissolved organic C, and metal bioaccumulation in aquatic macroinvertebrates (Arkansas River Basin, Colorado). *Journal of the North American Benthological Society*, **23**, 327–339.

Pullen, J. S. H. and Rainbow, P. S. (1991). The composition of pyrophosphate heavy metal detoxification granules in barnacles. *Journal of Experimental Marine Biology and Ecology*, **150**, 249–266.

Purkerson, D. G., Dobin, M. A., Bollens, S. M., Luoma, S. N. and Cutter, G. A. (2003). Selenium in San Francisco Bay zooplankton: potential effects of hydrodynamics and food web interactions. *Estuaries*, **26**, 956–969.

Pyle, G. G., Rajotte, J. W. and Couture, P. (2005). Effects of industrial metals on wild fish populations along a metal contamination gradient. *Ecotoxicology and Environmental Safety*, **61**, 287–312.

Quinnell, S., Hulsman, K. and Davie, P. J. F. (2004). Protein model for pollutant uptake and elimination by living organisms and its implications for ecotoxicology. *Marine Ecology Progress Series*, **274**, 1–16.

Rada, R. G., Weiner, J. G., Winfrey, M. R. and Powell, D. E. (1989). Recent increases in atmospheric deposition of mercury to North-Central Wisconsin lakes inferred from sediment analyses. *Archives of Environmental Contamination and Toxicology*, **18**, 175–181.

Rainbow, P. S. (1987). Heavy metals in barnacles. In *Barnacle Biology*, ed. A. J. Southward. Rotterdam, the Netherlands: A. A. Balkema, pp. 405–417.

Rainbow, P. S. (1993). The significance of trace metal concentrations in marine invertebrates. In *Ecotoxicology of Metals in Invertebrates*, ed. R. Dallinger and P. S. Rainbow. Chelsea, MI: Lewis Publishers, pp. 3–23.

Rainbow, P. S. (1995a). Physiology, physicochemistry and metal uptake: a crustacean perspective. *Marine Pollution Bulletin*, **31**, 55–59.

Rainbow, P. S. (1995b). Biomonitoring of heavy metal availability in the marine environment. *Marine Pollution Bulletin*, **31**, 183–192.

Rainbow, P. S. (1998). Phylogeny of trace metal accumulation in crustaceans. In *Metal Metabolism in Aquatic Environments*, ed. W. J. Langston and M. J. Bebianno. London: Chapman and Hall, pp. 285–319.

Rainbow, P. S. (2002). Trace metal concentrations in aquatic invertebrates: why and so what? *Environmental Pollution*, **120**, 497–507.

Rainbow, P. S. (2007). Trace metal bioaccumulation: models, metabolic availability and toxicity. *Environment International*, **33**, 576–582.

Rainbow, P. S. and Black, W. H. (2001). Effects of changes in salinity on the apparent water permeability of three crab species: *Carcinus maenas, Eriocheir sinensis* and *Necora puber*. *Journal of Experimental Marine Biology and Ecology*, **264**, 1–13.

Rainbow, P. S. and Black, W. H. (2002). Effects of changes in salinity and osmolality on the rate of uptake of zinc by three crabs of different ecologies. *Marine Ecology Progress Series*, **244**, 205–217.

Rainbow, P. S. and Black, W. H. (2005a). Physicochemistry or physiology: cadmium uptake and the effects of salinity and osmolality in three crabs of different ecologies. *Marine Ecology Progress Series*, **268**, 217–229.

Rainbow, P. S. and Black, W. H. (2005b). Cadmium, zinc and the uptake of calcium by two crabs, *Carcinus maenas* and *Eriocheir sinensis*. *Aquatic Toxicology*, **72**, 45–65.

Rainbow, P. S. and Blackmore, G. (2001). Barnacles as biomonitors of trace metal availabilities in Hong Kong coastal waters: changes in space and time. *Marine Environmental Research*, **51**, 441–463.

Rainbow, P. S. and Kwan, M. K. H. (1995). Physiological responses and the uptake of cadmium and zinc by the amphipod crustacean *Orchestia gammarellus*. *Marine Ecology Progress Series*, **127**, 87–102.

Rainbow, P. S. and Moore, P. G. (1986). Comparative metal analyses in amphipod crustaceans. *Hydrobiologia*, **141**, 273–289.

Rainbow, P. S. and Moore, P. G. (1990). Seasonal variation in copper and zinc concentrations in three talitrid amphipods (Crustacea). *Hydrobiologia*, **196**, 65–72.

Rainbow, P. S. and Phillips, D. J. H. (1993). Cosmopolitan biomonitors of trace metals. *Marine Pollution Bulletin*, **26**, 593–601.

Rainbow, P. S. and Wang, W.-X. (2001). Comparative assimilation of Cr, Cr, Se, and Zn by the barnacle *Elminius modestus* from phytoplankton and zooplankton diets. *Marine Ecology Progress Series*, **218**, 239–248.

Rainbow, P. S. and White, S. L. (1989). Comparative strategies of heavy metal accumulation by crustaceans: zinc, copper and cadmium in a decapod, an amphipod and a barnacle. *Hydrobiologia*, **174**, 245–262.

Rainbow, P. S., Amiard, J.-C., Amiard-Triquet, C., (2007). Trophic transfer of trace metals: subcellular compartmentalization in bivalve prey, assimilation by a gastropod predator and *in vitro* digestion simulations. *Marine Ecology Progress Series*, **348**, 125–138.

Rainbow, P. S., Amiard-Triquet, C., Amiard, J.-C., *et al.* (1999). Trace metal uptake rates in crustaceans (amphipods and crabs) from coastal sites in NW Europe differentially enriched with trace metals. *Marine Ecology Progress Series*, **183**, 189–203.

Rainbow, P. S., Blackmore, G. and Wang, W.-X. (2003). Effects of previous field exposure history on the uptake of trace metals from water and food by the barnacle *Balanus amphitrite*. *Marine Ecology Progress Series*, **259**, 201–213.

Rainbow, P. S., Emson, R. H., Smith, B. D., Moore, P. G. and Mladenov, P. V. (1993c). Talitrid amphipods as biomonitors of trace metals near Dunedin, New Zealand. *New Zealand Journal of Marine and Freshwater Research*, **27**, 201–207.

Rainbow, P. S., Fialkowski, W., Wolowicz, M., Smith, B. D. and Sokolowski, A. (2004a). Geographical and seasonal variation of trace metal bioavailabilities in the Gulf of Gdansk, Poland using mussels (*Mytilus trossulus*) and barnacles (*Balanus improvisus*) as biomonitors. *Marine Biology*, **144**, 271–286.

Rainbow, P. S., Geffard A., Jeantet A.-Y., *et al.* (2004b). Enhanced food-chain transfer of copper from a diet of copper-tolerant estuarine worms. *Marine Ecology Progress Series*, **271**, 183–191.

Rainbow, P. S., Huang, Z. G., Yan, S.-K. and Smith, B. (1993a). Barnacles as biomonitors of trace metals in the coastal waters near Xiamen, P. R. China. *Asian Marine Biology*, **10**, 109–121.

Rainbow, P. S., Malik, I. and O'Brien, P. (1993b). Physicochemical and physiological effects on the uptake of dissolved zinc and cadmium by the amphipod crustacean *Orchestia gammarellus*. *Aquatic Toxicology*, **25**, 15–30.

Rainbow, P. S., Moore, P. G. and Watson, D. (1989). Talitrid amphipods as biomonitors for copper and zinc. *Estuarine, Coastal and Shelf Science*, **28**, 567–582.

Rainbow, P. S., Ng, T. Y.-T., Shi, D. and Wang, W.-X. (2004c). Acute dietary pre-exposure and trace metal bioavailability to the barnacle *Balanus amphitrite*. *Journal of Experimental Marine Biology and Ecology*, **311**, 315–337.

Rainbow, P. S., Poirier, L., Smith, B. D., Brix, K. V. and Luoma, S. N. (2006). Trophic transfer of trace metals: subcellular compartmentalization in a polychaete and assimilation by a decapod crustacean. *Marine Ecology Progress Series*, **308**, 91–100.

Rainbow, P. S., Scott, A. G., Wiggins, E. A. and Jackson, R. W. (1980). Effect of chelating agents on the accumulation of cadmium by the barnacle *Semibalanus balanoides*, and complexation of soluble Cd, Zn and Cu. *Marine Ecology Progress Series*, **2**, 143–152.

Rainbow, P. S., Smith, B. D. and Lau, S. S. (2002). Biomonitoring of trace metal availabilities in the Thames estuary using a suite of littoral biomonitors. *Journal of the Marine Biological Association of the United Kingdom*, **82**, 793–799.

Ralston, N. V. C., Lloyd Blackwell, J. III and Raymond, L. J. (2007). Importance of molar ratios in selenium-dependent protection against methylmercury toxicity. *Biological Trace Element Research*, **119**, 255–268.

Ramsey, P. W., Rillig, M. C., Feris, K. P., Moore, J. N. and Gannon, J. E. (2005). Mine waste contamination limits soil respiration rates: a case study using quantile regression. *Soil Biology and Biochemistry*, **37**, 1177–1183.

Rand, G. M., Wells, P. G. and McCarty, L. S. (1995). Introduction to aquatic toxicology. In *Fundamentals of Aquatic Toxicology*, ed. G. M. Rand. Washington, DC: Taylor and Francis, pp. 3–67.

Ranville, M. A. and Flegal, A. R. (2005). Silver in the North Pacific Ocean. *Geochemistry, Geophysics, Geosystems*, **6**, 1–12.

Ratcliffe, D. A. (1970). Changes attributable to pesticides in egg breakage frequency and eggshell thickness in some British birds. *Journal of Applied Ecology*, **7**, 67–115.

Reboucas do Amaral, M. C., Rebelo, M. F., Torres, J. P. M. and Pfeiffer, W. C. (2005). Bioaccumulation and depuration of Zn and Cd in mangrove oysters (*Crassostrea rhizophorae*, Guilding, 1828) transplanted to and from a contaminated tropical coastal lagoon. *Marine Environmental Research*, **59**, 277–285.

Regoli, F. (1992). Lysosomal responses as a sensitive stress index in biomonitoring heavy metal pollution. *Marine Ecology Progress Series*, **84**, 63–69.

Regoli, F. (2000). Total oxyradical scavenging capacity (TOSC) in polluted and translocated mussels: a predictive biomarker of oxidative stress. *Aquatic Toxicology*, **50**, 351–361.

Regoli, F., Gorbi, S., Frenzilli, G., *et al.* (2002). Oxidative stress in ecotoxicology: from the analysis of individual antioxidants to a more integrated approach. *Marine Environmental Research*, **54**, 419–423.

Rehfeldt, G. and Söchtig, W. (1991). Heavy metal accumulation by *Baetis rhodani* and macrobenthic community structure in running waters of the N' Harz mountains (Lower Saxony/FRG) (Ephemeroptera: Baetidae). *Entomologia Generalis*, **16**, 31–37.

Reilly, M. C., Stubblefield, W. A., Adams, W. J., *et al.*, eds., (2003). *Reevaluation of the State of the Science for Water-Quality Criteria Development*. Pensacola, FL: SETAC Press.

Reinfelder, J. R. and Fisher, N. S. (1991). The assimilation of elements ingested by marine copepods. *Science*, **251**, 794–796.

Reinfelder, J. R. and Fisher, N. S. (1994). Retention of elements absorbed by juvenile fish (*Menidia menidia, Menidia beryllina*) from zooplankton prey. *Limnology and Oceanography*, **39**, 1783–1789.

Reinfelder, J. R., Fisher, N. S., Luoma, S. N., Nichols, J. W. and Wang, W.-X. (1998). Trace element trophic transfer in aquatic organisms: a critique of the kinetic model approach. *Science of the Total Environment*, **219**, 117–135.

Reinfelder, J. R., Wang, W.-X., Luoma, S. N. and Fisher, N. S. (1997). Assimilation efficiencies and turnover rates of trace elements in marine bivalves: a comparison of oysters, clams, and mussels. *Marine Biology* **129**, 443–452.

Reish, D. J. (1978). The effects of heavy metals on polychaetous annelids. *Revue Internationale d'Océanographie Médicale*, **49**, 99–104.

Reitsema, T. J., Field, S. and Spickett, J. T. (2003). Surveying imposex in the coastal waters of Perth, Western Australia, to monitor trends in TBT contamination. *Australasian Journal of Ecotoxicology*, **9**, 87–92.

Rember, R. and Trefrey, J. (2004). Increased concentrations of dissolved trace metals and organic carbon during snowmelt in rivers of the Alaskan arctic. *Geochimica et Cosmochimica Acta*, **68**, 477–489.

Renner, R. (2005). Asia pumps out more mercury than previously thought. *Environmental Science and Technology*, **39**. Available online at http://pubs.acs.org/subscribe/journals/esthag-w/2005/jan/science/rr_asia.html

Rhoads, D. C. and Boyer, L. F. (1983). The effects of marine benthos on physical properties of sediments: a successional perspective. In *Animal–Sediment Relations: The Biogenic Alteration of Sediments*, ed. P. L. McCall and M. J. S. Tevesz. New York: Plenum Press, pp. 3–52.

Riba, I., Casado-Martínez, C., Forja, J. M. and DelValls, A. (2004). Sediment quality in the atlantic coast of Spain. *Environmental Toxicology and Chemistry*, **23**, 271–282.

Rice, J. (2003). Environmental health indicators. *Ocean and Coastal Management*, **46**, 235–259.

Rickard, D. G. and Morse, J. W. (2005). Acid volatile sulfide. *Marine Chemistry*, **97**, 141–197.

Ridout, P. S., Rainbow, P. S., Roe, H. S. J. and Jones, H. R. (1989). Concentrations of V, Cr, Mn, Fe, Ni, Co, Cu, Zn, As, Cd in mesopelagic crustaceans from the north east Atlantic Ocean. *Marine Biology*, **100**, 465–471.

Riedel, G. F. and Sanders, J. G. (1998). Trace element speciation and behavior in the tidal Delaware River. *Estuaries*, **21**, 78–90.

Ringwood, A. H. and Keppler, C. J. (2002). Comparative *in situ* and laboratory sediment bioassays with juvenile *Mercenaria mercenaria*. *Environmental Toxicology and Chemistry*, **21**, 1651–1657.

Risebrough, R. W., Menzel, D. B., Jun, D. J. M. and Olcott, H. S. (1967). DDT residues in Pacific sea birds: a persistent insecticide in marine food chains. *Nature*, **216**, 589–591.

Ritson, P. I., Bouse, R. M., Flegal, A. R. and Luoma, S. N. (1999). Stable lead isotopic analyses of historic and contemporary lead contamination of San Francisco Bay estuary. *Marine Chemistry*, **64**, 71–83.

Rivera-Duarte, I. and Flegal, A. R. (1997). Pore-water silver concentration gradients and benthic fluxes from contaminated sediments of San Francisco Bay, California, USA. *Marine Chemistry*, **56**, 15–26.

Robinson, W. E., Ryan, D. K. and Wallace, G. T. (1993). Gut contents: a significant contaminant of *Mytilus edulis* whole body metal concentrations. *Archives of Environmental Contamination and Toxicology*, **25**, 415–421.

Roditi, H. A. and Fisher, N. S. (1999). Rates and routes of trace element uptake in zebra mussels. *Limnology and Oceanography*, **44**, 1418–1429.

Roditi, H. A., Fisher, N. S. and Sanudo-Wilhelmy, S. (2000). Field testing a bioaccumulation model for zebra mussels. *Environmental Science and Technology*, **34**, 2817–2825.

Roesijadi, G. (1986). Mercury-binding proteins from the marine mussel, *Mytilus edulis*. *Environmental Health Perspectives*, **65**, 45–48.

Roesijadi, G. (1992). Metallothioneins in metal regulation and toxicity in aquatic animals. *Aquatic Toxicology*, **22**, 81–114.

Roesijadi, G. (1996). Metallothionein and its role in toxic metal regulation. *Comparative Biochemistry and Physiology C*, **113**, 117–123.

Roesijadi, M. E. and Unger, G. (1993). Cadmium uptake in gills of the mollusc *Crassostrea virginica* and inhibition by calcium channel blockers. *Aquatic Toxicology*, **24**, 195–206.

Roper, D. S., Nipper, M. G., Hickey, C. W., Martin, M. L. and Weatherhead, M. A. (1995). Burial, crawling and drifting behaviour of the bivalve *Macomona liliana* in response to common sediment contaminants. *Marine Pollution Bulletin*, **31**, 471–478.

Rouleau, C., Gobeil, C. and Tjalve, H. (2000). Accumulation of silver from the diet in two marine benthic predators: the snow crab (*Chironectes opilio*) and American plaice (*Hippoglossoides platessoides*). *Environmental Toxicology and Chemistry*, **19**, 631–637.

Rouleau, C., Pelletier, E. and Pellerin-Massicotte, J. (1992). Uptake of organic mercury and selenium from food by nordic shrimp *Pandalus borealis*. *Chemical Speciation and Bioavailability*, **4**, 75–81.

Roy, I. and Hare, L. (1999). Relative importance of water and food as cadmium sources to the predatory insect *Sialis velata*. *Canadian Journal of Fisheries and Aquatic Sciences*, **56**, 1143–1149.

Rudd, J. W. M., Turner, M. A., Townsend, B. E., Swick, A. and Furutani, A. (1980). Dynamics of selenium in mercury-contaminated freshwater ecosystems. *Canadian Journal of Fisheries and Aquatic Sciences*, **37**, 848–857.

Ruiz, J. M. (2004). Oil spills versus shifting baselines. *Marine Ecology Progress Series*, **282**, 307–309.

Rule, J. H. (1986). Assessment of trace element geochemistry of Hampton Roads harbor and lower Chesapeake Bay area sediments. *Environmental Geology*, **8**, 209–219.

Ryan, A. C., Van Genderen, E. J., Tomasso, J. R. and Klaine, S. J. (2004). Influence of natural organic matter source on copper toxicity to larval fathead minnows (*Pimephales promelas*): implications for the biotic ligand model. *Environmental Toxicology and Chemistry*, **23**, 1567–1574.

Rygg, B. (1985). Effects of sediment copper on benthic fauna. *Marine Ecology Progress Series*, **25**, 83–89.

Saiki, M. K. and Lowe, T. P. (1987). Selenium in aquatic organisms from subsurface agricultural drainage water, San Joaquin Valley, California. *Archives of Environmental Contamination and Toxicology*, **16**, 657–670.

Salazar, M. H. and Salazar, S. M. (1995). *In-situ* bioassays using transplanted mussels. I. Estimating chemical exposure and bioeffects with bioaccumulation and growth. In *Environmental Toxicology and Risk Assessment, vol. 3*, ed. J. S. Hughes, G. R. Biddinger and E. Mones. Philadelphia, PA: American Society for Testing and Materials, pp. 216–241.

Salice, C. J. and Miller, T. J. (2003). Population-level responses to long-term cadmium exposure in two strains of the freshwater gastropod *Biomphalaria glabrata*: results from a life-table experiment. *Environmental Toxicology and Chemistry*, **22**, 678–688.

Salomons, W. and Eagle, A. M. (1990). Hydrology, sedimentology and the fate and distribution of copper in mine-related discharges in the Fly River system, Papua New Guinea. *Science of the Total Environment*, **97/8**, 315–334.

Salomons, W. and Förstner, U. (1984). *Metals in the Hydrocycle*. Berlin, Germany: Springer-Verlag.

Sanders, B. M. and Jenkins, K. D. (1984). Relationships between free cupric ion concentrations in sea water and copper metabolism and growth in crab larvae. *Biological Bulletin*, **167**, 704–712.

Santillo, D., Johnston, P. and Langston, W. J. (2001). *Tributyltin (TBT) Antifoulants: A Tale of Ships, Snails and Imposex*, Environment Issue Report No. 22. Copenhagen: European Environment Agency.

Santore, R. C., Di Toro, D. M., Paquin, P. R., Allen, H. E. and Meyer, J. S. (2001). Biotic ligand model of the acute toxicity of metals. II. Application to acute copper toxicity in freshwater fish and daphnia. *Environmental Toxicology and Chemistry*, **20**, 2397–2402.

Sanudo-Wilhelmy, S. A. and Gill, G. A. (1999). Impact of the Clean Water Act on the levels of toxic metals in urban estuaries: the Hudson River Estuary revisited. *Environmental Science and Technology*, **33**, 3477–3481.

Sanudo-Wilhelmy, S. A., Tovar-Sanchez, A., Fisher, N. S. and Flegal, A. R. (2004). Examining dissolved toxic metals in US estuaries. *Environmental Science and Technology*, **38**, 34A–38A.

Schaule, B. and Patterson, C. C. (1981). The occurrence of lead in the Northeast Pacific and the effects of anthropogenic inputs. In *Lead in the Marine Environment: Proceedings of an International Experts' Discussion*, ed. M. Branica and Z. Konrad. New York, Pergamon, pp. 31–45.

Schiff, K. C. and Weisberg, S. B. (1999). Iron as a reference element for determining trace metal enrichment in Southern California coastal shelf sediments. *Marine Environmental Research*, **48**, 161–176.

Schiller, A., Hunsaker, C. T., Kane, M. A., *et al.* (2001). Communicating ecological indicators to decision makers and the public. *Conservation Ecology*, **5**, 19. Available online at www.consecol.org/vol5/iss1/art19/

Schindler, D. W., Mills, K. H, Malley, D. F., *et al.* (1985). Long-term ecosystem stress: the effects of years of experimental acidification on a small lake. *Science*, **228**, 1395–1401.

Schlekat, C. E., Decho, A. W. and Chandler, G. T. (2000a). Bioavailability of particle-associated silver, cadmium, and zinc to the estuarine amphipod *Leptocheirus plumulosus* through dietary ingestion. *Limnology and Oceanography*, **45**, 11–21.

Schlekat, C. E., Dowdle, P. R., Lee, B.-G., Luoma, S. N. and Oremland, R. S. (2000b). Bioavailability of particle-associated Se to the bivalve *Potamocorbula amurensis*. *Environmental Science and Technology*, **34**, 4504–4510.

Schlekat, C. E., Purkerson, D. G. and Luoma, S. N. (2004). Modeling selenium bioaccumulation through an arthropod food webs in San Francisco Bay. *Environmental Toxicology and Chemistry*, **23**, 3003–3010.

Schlenk, D., Perkins, E. J., Hamilton, G., Zhang, Y. S. and Layher, W. (1996). Correlation of hepatic biomarkers with whole animal and population-community metrics. *Canadian Journal of Fisheries and Aquatic Sciences*, **53**, 2299–2309.

Schneider, A. L. and Ingram, H. (1997). *Policy Design for Democracy*. Lawrence, KS: University of Kansas Press.

Schneider, S. H. (2002). Keeping out of the box. *American Scientist*, **90**, 496–498.

Schnoor, J. (2004). Top ten environmental success stories. *Environmental Science and Technology*, **38**, 319A.

Schrenk, M. O., Edwards, K. J., Goodman, R. M., Hamers, R. J. and Banfield, J. F. (1998). Distribution of *Thiobacillus ferrooxidans* and *Leptospirillum ferrooxidans*: implications for generation of acid mine drainage. *Science*, **279**, 1519–1522.

Schuster, P. F., Krabbenhoft, D. P., Naftz, D. L., *et al.* (2002). Atmospheric mercury deposition during the last 270 years: a glacial ice core record of natural and anthropogenic sources. *Environmental Science and Technology*, **36**, 2303–2310.

Schutz, D. F. and Turekian, K. K. (1965). The investigation of the geographical distribution of several trace elements in sea water using neutron activation analysis. *Geochimica et Cosmochimica Acta*, **29**, 259–313.

Schwarzbach, S. E., Albertson, J. D. and Thomas, C. M. (2006). Effects of predation, flooding and contamination on reproductive success of California clapper rails (*Rallus longirostris obsoletus*) in San Francisco Bay. *The Auk*, **123**: 45–60.

Scott, A. and Clarke, R. (2000). Multivariate techniques. In *Statistics in Ecotoxicology*, ed. T. Sparks. Chichester, UK: John Wiley, pp. 149–178.

Scott, D. M. and Major, C. W. (1972). The effect of copper (II) on survival, respiration and heart rate in the common blue mussel, *Mytilus edulis*. *Biological Bulletin*, **143**, 679–688.

Scott, J. K. and Redmond, M. S. (1989). The effects of a contaminated dredged material on laboratory populations of the tubicolous amphipod *Ampelisca abdita*. In *Aquatic Toxicology and Hazard Assessment*, vol. 12, ed. U. M. Cowgill and L. R. Williams. Philadelphia, PA: American Society for Testing of Materials, pp. 289–299.

Seas at Risk (2007). *The TBT Ban*. Available online at www.seas-at-risk.org/n3.php?Page=81

Seed, R. (1992). Systematics, evolution and distribution of mussels belonging to the genus *Mytilus*: an overview. *American Malacological Bulletin*, **9**, 123–137.

Seiler, R. L., Skorupa, J. P. and Peltz, L. A. (1999). Areas susceptible to irrigation-induced selenium contamination of water and biota in the western United States. *US Geological Survey Circular* **1180**.

Shafer, M. M., Hoffmann, S. R., Overdier, J. and Armstrong, D. E. (2004). Physical and kinetic speciation of copper and zinc in three geochemically contrasting marine estuaries. *Environmental Science and Technology*, **38**, 3810–3819.

Shen, G. T. and Boyle, E. A. (1987). Lead in corals: reconstruction of historical industrial fluxes to the surface ocean. *Earth and Planetary Science Letters*, **82**, 289–304.

Shi, D., Blackmore, G. and Wang, W.-X. (2003). Effects of aqueous and dietary preexposure and resulting body burden on silver biokinetics in the green mussel *Perna viridis*. *Environmental Science and Technology*, **37**, 936–943.

Sholkovitz, E. R. (1976). Flucculation of dissolved organic and inorganic matter during the mixing of river water and sea water. *Geochimica et Cosmochimica Acta*, **40**, 831–845.

Shouse, M. K. (2002). The effects of decreasing trace metal contamination on benthic community structure. Unpublished M. A. thesis, San Francisco State University, San Francisco, CA.

Sigg, L., Xue, H., Kistler, D. and Shönenberger, R. (2000). Size fractionation (dissolved, colloidal and particulate) of trace metals in the Thur River, Switzerland. *Aquatic Geochemistry*, **6**, 413–434.

Silva, C. A. R., Rainbow, P. S. and Smith, B. D. (2003). Biomonitoring of trace metal contamination in mangrove-lined Brazilian coastal systems using the oyster *Crassostrea rhizophorae*: comparative study of regions affected by oil, salt pond and shrimp farming activities. *Hydrobiologia*, **501**, 199–206.

Silva, C. A. R., Rainbow, P. S., Smith, B. D. and Santos, Z. L. (2001). Biomonitoring of trace metal contamination in the Potengi estuary, Natal (Brazil), using the oyster *Crassostrea rhizophorae*, a local food source. *Water Research*, **35**, 4072–4078.

Silva, C. A. R., Smith, B. D. and Rainbow P. S. (2006). Comparative biomonitors of coastal trace metal contamination in tropical South America (N. Brazil). *Marine Environmental Research*, **61**, 439–455.

Simpson, S. L. (2005). Exposure-effect model for calculating copper effect concentrations in sediments with varying copper binding properties: a synthesis. *Environmental Science and Technology*, **39**, 7089–7096.

Simpson, S. L. and Batley, G. E. (2003). Disturbances to metal partitioning during toxicity testing of iron(II)-rich estuarine pore waters and whole sediments. *Environmental Toxicology and Chemistry*, **22**, 424–432.

Simpson, S. L. and King, C. K. (2005). Exposure-pathway models explain causality in whole-sediment toxicity tests. *Environmental Science and Technology*, **39**, 837–843.

Simpson, S. L., Apte, S. C. and Batley, G. E. (1998). Effect of short-term resuspension events on trace metal speciation in polluted anoxic sediments. *Environmental Science and Technology*, **32**, 620–625.

Sindermann, C. J. (1996). *Ocean Pollution: Effects on Living Resources and Humans*. New York: CRC Press.

Singh, N. P., McCoy, M. T., Tice, R. R. and Schneider, E. L. (1988). A simple technique for the quantitation of lowlevels of DNA damage in individual cells. *Experimental Cell Research*, **175**, 184–191.

Skorupa, J. P. (1998). Selenium poisoning of fish and wildlife in nature: lessons from twelve real-world examples. In *Environmental Chemistry of Selenium*, ed. W. T. Frankenberger Jr and R. A. Engberg. New York: Marcel Dekker, pp. 315–354.

Skorupa, J. P. (1999). Beware missing data and undernourished statistical models: comment on Fairbrother *et al.*'s critical evaluation. *Human and Ecological Risk Assessment*, **5**, 1255–1262.

Skorupa, J. P. and Ohlendorf, H. M. (1991). Contaminants in drainage water and avian risk threshold. In *The Economics and Management of Water and Drainage in Agriculture*, ed. A. Dinar and D. Zilberman. Boston, MA: Kluwer, pp. 345–368.

Smith, B. S. (1971). Sexuality in the American mudsnail, *Nassarius obsoletus* Say. *Proceedings of the Malacological Society of London*, **39**, 377–378.

Smith, B. S. (1980). The estuarine mud snail, *Nassarius obsoletus*, abnormalities in the reproductive system. *Journal of Molluscan Studies*, **46**, 247–256.

Smith, B. S. (1981). Male characteristics in female mud snails caused by antifouling bottom paints. *Journal of Applied Toxicology*, **1**, 22–25.

Smith, D. R., Stephenson, M. D. and Flegal, A. R. (1986). Trace metals in mussels transplanted to San Francisco Bay. *Environmental Toxicology and Chemistry*, **5**, 129–138.

Smith, G. J. and Flegal, A. R. (1993). Silver in San Francisco Bay estuarine waters. *Estuaries*, **16**, 547–558.

Smith, R. A., Alexander, R. B. and Wolman, M. G. (1987). Water quality trends in the nation's rivers. *Science*, **235**, 1607–1615.

Smith, R. E. W., Ahsanullah, M. and Batley, G. E. (1990). Investigations of the impact of effluent from the OK Tedi copper mine on the fisheries resource in the Fly River, Papua New Guinea. *Environmental Monitoring and Assessment*, **14**, 315–331.

Smolders, R., Bervoets, L., De Coen, W. and Blust, R. (2004). Cellular energy allocation in zebra mussels exposed along a pollution gradient: linking cellular effects to higher levels of biological organization. *Environmental Pollution*, **129**, 99–112.

Soga, T., Ohashi, Y., *et al.* (2003). Quantitative metabolome analysis using capillary electrophoresis mass spectrometry. *Journal of Proteome Research*, **2**, 488–494.

Sokolowski, A., Wolowicz, M., Hummel, H. and Bogaards, R. (1999). Physiological responses of *Macoma balthica* to copper pollution in the Baltic. *Oceanologica Acta*, **22**, 431–439.

Solà, C. and Prat, N. (2006). Monitoring metal and metalloid bioaccumulation in *Hydropsyche* (Trichoptera, Hydropsychidae) to evaluate metal pollution in a mining river: whole body versus tissue content. *Science of the Total Environment*, **259**, 221–231.

Solà, C., Burgos, M., Plazuelo, A., *et al.* (2004). Heavy metal bioaccumulation and macroinvertebrate community changes in a Mediterranean stream affected by acid mine drainage and an accidental spill (Guadiamar River, SW Spain). *Science of the Total Environment*, **333**, 109–126.

Somerfield, P. J., Gee, J. M. and Warwick, R. M. (1994). Soft sediment meiofaunal community structure in relation to a long-term heavy metal gradient in the Fal estuary system. *Marine Ecology Progress Series*, **105**, 79–88.

Song, K.-H. and Breslin, V. T. (1998). Accumulation of contaminant metals in the amphipod *Diporeia* spp. in western Lake Ontario. *Journal of Great Lakes Research*, **24**, 949–961.

Spallholz, J. E. (1997). Free radical generation by selenium compounds and their prooxidant toxicity. *Biomedical Environmental Science*, **10**, 260–270.

Spallholz, J. E. and Hoffman, D. J. (2002). Selenium toxicity: cause and effect in aquatic birds. *Aquatic Toxicology*, **57**, 27–37.

Spence, S. K., Bryan, G. W., Gibbs, P. E., *et al.* (1990). Effects of TBT contamination on *Nucella* populations. *Functional Ecology* **4**, 425–432.

Spies, R. B. (1989). Sediment bioassays, chemical contaminants and benthic ecology: new insights or just muddy water? *Marine Environmental Research*, **27**, 73–75.

Spry, D. J. and Wiener, J. G. (1991). Metal bioavailability and toxicity to fish in low-alkalinity lakes. *Environmental Pollution*, **71**, 243–304.

Squire, S., Scelfo, G. M., Revenaugh, J. and Flegal, A. R. (2002). Decadal trends of silver and lead contamination in San Francisco Bay surface waters. *Environmental Science and Technology*, **36**, 2379–2386.

St Louis, V. L., Rudd, J. W. M., Kelly, C. A., *et al.* (1994). Importance of wetlands as sources of methyl mercury to boreal forest ecosystems. *Canadian Journal of Fisheries and Aquatic Sciences*, **51**, 1065–1076.

Stebbing, A. R. D. (1976). The effects of low metal levels on a clonal hydroid. *Journal of the Marine Biological Association of the United Kingdom*, **56**, 977–994.

Stebbing, A. R. D. (1979). An experimental approach to the determinants of biological water quality. *Philosophical Transactions of the Royal Society of London B*, **286**, 465–481.

Stewart, A. R., Luoma, S. N., Schlekat, C. E., Dobin, M. A. and Hieb, K. A. (2004). Food web pathway determines how selenium affects aquatic ecosystems: a San Francisco Bay case study. *Environmental Science and Technology*, **38**, 4519–4526.

Stokes, P. M., Hutchinson, T. C. and Krauter, K. (1973). Heavy metal tolerance in algae isolated from polluted lakes near the Sudbury, Ontario smelters. *Water Pollution Research in Canada*, **8**, 178–201.

Stokstad, E. (2005). Europe, U. S. differ on mercury. *Science*, **307**, 655.

Strong, C. R. and Luoma, S. N. (1981). Variations in correlation of body size with concentrations of Cu and Ag in the bivalve *Macoma balthica*. *Canadian Journal of Fisheries and Aquatic Sciences*, **38**, 1059–1064.

Stumm, W. and Morgan, J. J. (1996). *Aquatic Geochemistry*. New York: John Wiley.

Suchanek, T. H., Richerson, P. J., Nelson, D. C., *et al.* (2004). Evaluating and managing a multiply-stressed ecosystem at Clear Lake, California: a holistic ecosystem approach. In *Managing for Healthy Ecosystems: Case Studies*, ed. D. Rapport, W. Lasley, D. Rolston, O. Nielson, C. Qualset and A. Damania. Boca Raton, FL: CRC Press, pp. 1233–1265.

Sunda, W. G. and Guillard, R. R. L. (1976). The relationship between cupric ion activity and the toxicity of copper to phytoplankton. *Journal of Marine Research*, **34**, 511–529.

Sunda, W. G., Tester, P. A. and Huntsman, S. A. (1987). Effects of cupric and zinc ion activities on the survival and reproduction of marine copepods. *Marine Biology*, **94**, 203–210.

Sunda, W. P., Tester, P. A., and Huntsman, S. A. (1990). Toxicity of trace metals to *Acartia tonsa* in the Elizabeth River and Southern Chesapeake Bay. *Estuarine, Coastal and Shelf Science*, **30**, 207–221.

Sundby, B., (1990). Geochemical aspects of metal bioavailability: an overview of sediment geochemistry. In *Proceedings of the 17th Annual Aquatic Toxicity Workshop*, Vancouver, BC.

Sundby, B., Anderson, L. G., Hall, P. O. J., *et al.* (1986). The effect of oxygen on release and uptake of cobalt, manganese, iron and phosphate at the sediment–water interface. *Geochimica et Cosmochimica Acta*, **50**, 1281–1288.

Sundelin, B. (1983). Effect of cadmium on *Pontoporeia affinis* (Crustacea: Amphipoda) in laboratory soft-bottom microcosms. *Marine Biology*, **74**, 203–215.

Suter, G. W. (1990). Endpoints for regional ecological risk assessments. *Environmental Management*, **14**, 9–23.

Suter, G. W. (1993). A critique of ecosystem health concepts and indexes. *Environmental Toxicology and Chemistry*, **12**, 1533–1539.

Suter, G. W. (2006). *Ecological Risk Assessment*. Boca Raton, FL: Lewis Publishers.

Swales, S., Storey, A. W., Roderick, I. D., *et al.* (1998). Biological monitoring of the impacts of the OK Tedi copper mine on fish populations in the Fly River system, Papua New Guinea. *Science of the Total Environment*, **214**, 99–111.

Swartz, R. C. (1987). Toxicological methods for determining the effects of contaminated sediment on marine organisms. In *Fate and Effects of Sediment-Bound Chemicals in Aquatic Systems*, ed. K. L. Dickson, A. W. Maki and W. A. Brung. New York: Pergamon, pp. 101–114.

Swartz, R. C., DeBen, W. A., Jones, J. K. P., Lamberson, J. O. and Cole, F. A. (1985). Phoxocephalid amphipod bioassay for marine sediment toxicity. In *Aquatic Toxicology and Hazard Assessment: 7th Symposium*, ASTM Special Technical Publication No. 854. Philadelphia, PA: American Society for Testing and Materials, pp. 284–307.

Swartz, R. C., DeBen, W. A., Sercu, K. A. and Lamberson, J. O. (1982). Sediment toxicity and the distribution of amphipods in Commencement Bay, Washington, USA. *Marine Pollution Bulletin*, **13**, 359–364.

Swartz, R. C., Kemp, P. F., Schults, D. W. and Lamberson, J. O. (1988). Effects of mixtures of sediment contaminants on the marine infaunal amphipod, *Rhepoxynius abronius*. *Environmental Toxicology and Chemistry*, **7**, 1013–1020.

Szefer, P. (2002) *Metals, Metalloids and Radionuclides in the Baltic Sea Ecosystem*. Amsterdam, the Netherlands: Elsevier.

Szefer, P., Fowler, S. W., Ikuta, K., *et al.* (2006). A comparative assessment of heavy metal accumulation in soft parts and byssus of mussels from subarctic, temperate, subtropical and tropical marine environments. *Environmental Pollution*, **139**, 70–78.

Szefer, P., Ikuta, K., Kushiyama, S., Frelek, K. and Geldon, J. (1997). Distribution of trace metals in the Pacific oyster, *Crassostrea gigas*, and crabs from the east coast of Kyushu Island, Japan. *Bulletin of Environmental Contamination and Toxicology*, **58**, 108–114.

Talbot, V. (1986). Seasonal variation of copper and zinc concentrations in the oyster *Saccostrea cuccullata* from the Dampier archipelago, Western Australia: implications for pollution monitoring. *Science of the Total Environment*, **57**, 217–230.

Tannenbaum, L. V. (2005). A critical assessment of the ecological risk assessment process: a review of misapplied concepts. *Integrated Environmental Assessment and Management*, **1**, 66–72.

Taylor, E. J., Jones, D. P. W., Maund, S. J. and Pascoe, D. (1993). A new method for measuring the feeding activity of *Gammarus pulex* (L.). *Chemosphere*, **26**, 1375–1381.

Taylor, G., Baird, D.J. and Soares, A.M.V.M. (1998). Surface binding of contaminants by algae: consequences for lethal toxicity and feeding to *Daphnia magna* Straus. *Environmental Toxicology and Chemistry*, **17**, 412–419.

Teh, S.J., Deng, X., Deng, D.-F., *et al.* (2004). Chronic effects of dietary selenium on juvenile Sacramento splittail (*Pogonichthys macrolepidotus*). *Environmental Science and Technology*, **38**, 6085–6093.

Temara, A., Warnau, M., Ledent, G., Jangoux, M. and Dubois, P. (1996). Allometric variations in heavy metal bioconcentration in the asteroid *Asterias rubens* (Echinodermata). *Bulletin of Environmental Contamination and Toxicology*, **56**, 98–105.

Tessier, A. and Campbell, P.G.C. (1987). Partitioning of trace metals in sediments: relationships with bioavailability. *Hydrobiologia*, **149**, 43–52.

Tessier, A., Campbell, P.G.C., Auclair, J.C. and Bisson, M. (1984). Relationships between the partitioning of trace metals in sediments and their accumulation in the tissues of the freshwater mollusc *Elliptio complanata* in a mining area. *Canadian Journal of Fisheries and Aquatic Sciences*, **41**, 1463–1471.

Tessier, A., Campbell, P.G.C. and Bisson, M. (1979). Sequential extraction procedure for the speciation of particulate trace metals. *Analytical Chemistry*, **51**, 844–854.

Thompson, B.E., Bay, S.M., Anderson, J.W., *et al.* (1989). Chronic effects of contaminated sediments on the urchin *Lytechinus pictus*. *Environmental Toxicology and Chemistry*, **8**, 629–637.

Thompson, D.R., Furness, R.W. and Walsh, P.M. (1992). Historical changes in mercury concentrations in the marine ecosystem of the North and North-East Atlantic Ocean as indicated by seabird feathers. *Journal of Applied Ecology*, **29**, 79–84.

Thompson-Eagle, E.T. and Frankenburger, W.T. (1992). Bioremediation of soils contaminated with selenium. *Advances in Soil Science*, **17**, 261–310.

Thomson, E.A., Luoma, S.N., Cain, D.J. and Johansson, C.E. (1980). The effect of sample storage on the extraction of Cu, Zn, Fe, Mn and organic material from oxidized estuarine sediments. *Water, Air and Soil Pollution*, **14**, 215–233.

Thomson, E.A., Luoma, S.N., Johansson, C.E. and Cain, D.J. (1984). Comparison of sediments and organisms in identifying sources of biologically available trace metal contamination. *Water Research*, **18**, 756–765.

Thomson-Becker, E.A. and Luoma, S.N. (1985). Temporal fluctuations in grain size, organic materials and iron concentrations in intertidal surface sediment of San Francisco Bay. *Hydrobiologia*, **129**, 91–107.

Thorndycraft, V.R., Pirrie, D. and Brown, A.G. (2004). Alluvial records of medieval and prehistoric tin mining on Dartmoor, southwest England. *Geoarchaeology*, **19**, 219–236.

Timmermans, K.R., Spijkerman, E. and Tonkes, M. (1992). Cadmium and zinc uptake by two species of aquatic invertebrate predators from dietary and aqueous sources. *Canadian Journal of Fisheries and Aquatic Sciences*, **49**, 655–662.

Timmermans, K.R., Van Hattum, B., Kraak, M.H. and Davids, C. (1989). Trace metals in a littoral food web: concentrations in organisms, sediment and water. *Science of the Total Environment*, **878**, 477–494.

Tipping, E. (1994). WHAM: a chemical equilibrium model and computer code for waters, sediments, and soils incorporating a discrete site/electrostatic model of ion-binding by humic substances. *Computers and Geoscience*, **20**, 973–1023.

Tipping, E., Lofts, S. and Lawlor, A.J. (1998). Modelling the chemical speciation of trace metals in the surface waters of the Humber system. *Science of the Total Environment*, **210**, 63–77.

Tohtz, J. (1992). *Survey and Inventory of Cold Water Streams, Oct. 21, 1991 through June 30, 1992*, Progress Report for Statewide Fisheries Investigation, Project No. F-46-R-5. Helena, MT: Montana Department of Fish, Wildlife and Parks.

Tokunaga, T.K., Pickering, I.J. and Brown, G.E. (1996). Selenium transformations in ponded sediments. *Soil Science Society of America Journal*, **60**, 12–24.

Tokunaga, T.K., Sutton, S.R., Bajt, S., Nuessle, P. and Shea-McCarthy, G. (1998). Selenium diffusion and reduction at the water-sediment boundary: micro-XANES spectroscopy of reactive transport. *Environmental Science and Technology*, **32**, 1092–1098.

Tollefson, L. and Cordle, F. (1986). Methylmercury in fish: a review of residue levels, fish consumption and regulatory action in the United States. *Environmental Health Perspectives*, **68**, 203–208.

Tong, S.L., Yap, S.Y., Ishak, I. and Devi, S. (1999). ASEAN marine water quality criteria for copper. In *ASEAN Marine Water Quality Criteria: Contextual Framework, Principles, Methodology and Criteria for 18 Parameters*, ed. C.A. McPherson, P.M. Chapman, Q.A. Vigers and K.S. Ong. Vancouver, BC: ASEAN-Canada CPMS-II Cooperative Programme on Marine Science, pp. IX-1–IX-40.

Topping, B.R. and Kuwabara, J.S. (2003). Dissolved nickel and benthic flux in South San Francisco Bay. *Bulletin of Environmental Contamination and Toxicology*, **71**, 46–51.

Trannum, H. C., Olsgard, F., Skei, J. M., et al. (2004). Effects of copper, cadmium and contaminated harbour sediments on recolonisation of soft-bottom communities. *Journal of Experimental Marine Biology and Ecology*, **310**, 87–114.

Trefry, J. H., Metz, S., Trocine, R. P., and Nelson, T. A. (1985). A decline in lead transport by the Mississippi River. *Science*, **230**, 439–441.

Tremblay, G. A. (2004). Sustainable development in the mining industry. In *Life-Cycle Assessment of Metals: Issues and Research Directions*, ed. A. Dubreuil. Pensacola, FL: SETAC Press, pp. 143–146.

Troup, B. N., Bricker, O. P. and Bray, J. T. (1974). Oxidation effect on the analysis of iron in the interstitial water of recent anoxic sediments. *Nature*, **249**, 237–239.

Trudel, M. and Rasmussen, J. B. (1997). Modeling the elimination of mercury by fish. *Environmental Science and Technology*, **31**, 1716–1722.

Truhaut R. (1977). Ecotoxicology: objectives, principles and perspectives. *Ecotoxicology and Environmental Safety*, **1**, 151–173.

Tsui, M. K. and Wang, W.-X. (2004). Uptake and elimination routes of inorganic mercury and methylmercury in *Daphnia magna*. *Environmental Science and Technology*, **38**, 808–816.

Turner, A., Martinoa, M. A. and Leroux, S. M. (2002). Trace metal distribution coefficients in the Mersey Estuary, UK: evidence for salting out of metal complexes. *Environmental Science and Technology*, **36**, 4578–4584.

Turner, D. R., Whitfield, M. and Dickson, A. G. (1981). The equilibrium speciation of dissolved components in freshwater and seawater at 25 °C and 1 atm pressure. *Geochimica et Cosmochimica Acta*, **45**, 855–881.

Udo de Hias, H. A. (2004). Land-use impacts of mining in the life-cycle initiative. In *Life-Cycle Assessment of Metals: Issues and Research Directions*, ed. A. Dubreuil. Pensacola, FL: SETAC Press, pp. 159–163.

Underwood, A. J. and Peterson, C. H. (1988). Towards an ecological framework for investigating pollution. *Marine Ecology Progress Series*, **46**, 227–234.

USEPA (1980). *EPA Ambient Water Quality Criteria for Chromium*, EPA No.440/5-80-035. Washington, DC: US Environmental Protection Agency.

USEPA (1984). *Ambient Water Quality Criteria for Mercury: 1984*, EPA No.440-5-84-026. Washington, DC: US Environmental Protection Agency.

USEPA (1985). *Wastes from the Extraction and Beneficiation of Metallic Ores, Phosphate Rock, Asbestos, Overburden from Uranium Mining, and Oil Shale*, EPA/No.530-SW-85-033.

Washington, DC: US Environmental Protection Agency. Available online atwww.epa.gov/epaoswer/other/mineral/execsum.pdf

USEPA (1987). *Ambient Water Quality Criteria for Selenium: 1987*, EPA No.440/5-87-006. Washington, DC: US Environmental Protection Agency.

USEPA (1992) *Sediment Classification Methods Compendium*. EPA No.823-R-92-006. Washington, DC: US Environmental Protection Agency.

USEPA (1993). *Wildlife Exposure Factors Handbook*, EPA No.600-R-93-187a. Washington, DC: US Environmental Protection Agency.

USEPA (1995a). *An SAB Report: Review of the Agency's Approach for Developing Sediment Criteria for Five Metals*, EPA-SAB-EPC No.95-020.Washington, DC: US Environmental Protection Agency.

USEPA (1995b). Great Lakes Water Quality Initiative methodology for development of wildlife criteria. *Federal Register*, **60**(56), 15410–15412.

USEPA (1997). *Ambient Aquatic Life Water Quality Criteria for Tributyltin – Draft*, EPA No.822-D-97-001. Washington, DC: US Environmental Protection Agency.

USEPA (1998). *Guidelines for Ecological Risk Assessment*, EPA No.630/R-95/002F. Washington, DC: US Environmental Protection Agency.

USEPA (1999a) *Chemical Safety in Your Community: EPA's New Risk Management Program*, EPA No.55-B-99-010. Washington, DC: US Environmental Protection Agency.

USEPA (1999b). *Clark Fork River Ecological Risk Assessment – Draft*. Helena, MT: US Environmental Protection Agency. Available online athttp://epa.gov/region8/superfund/sites/mt/milltowncfr/millterap.html

USEPA (2000). *Update of Ambient Water Quality Criteria for Cadmium*, EPA No.822-R-01-001, 1–166. 2001. Washington, DC: US Environmental Protection Agency.

USEPA (2001a). *Draft Guidance for Implementing the Methylmercury Water Quality Criterion*. Washington, DC: US Environmental Protection Agency. Available online at www.epa.gov/waterscience/criteria/methylmercury/guidance-draft.html

USEPA (2001b). *Methods for Collection, Storage, and Manipulation of Sediments for Chemical and Toxicological Analyses: Technical Manual*, EPA No.823-B-01-002. Washington, DC: US Environmental Protection Agency.

USEPA (2001c). *Update of Ambient Water Quality Criteria for Cadmium*, EPA No.822-R-01-001. Washington, DC: US Environmental Protection Agency.

USEPA (2002a). *Ambient Aquatic Life Water Quality Criteria for Tributyltin (TBT) – Draft*. EPA No.822-B-02-001. Washington, DC: US Environmental Protection Agency.

USEPA (2002b). *AQUIRE (Aquatic Toxicity Information Retrieval Database)*. Duluth, MN: National Health and Environmental Effects Research Laboratory. Available online at http://cfpub.epa.gov/ecotox/

USEPA (2002c). *National Recommended Water Quality Criteria: 2002*, EPA No.822-R-02-047. Washington, DC: US Environmental Protection Agency. Available online at www.epa.gov/waterscience/criteria/wqcriteria.html

USEPA (2002d). *Emissions Estimates for Lead and Cadmium in U.S. for Year 2002*. Available online at www.epa.gov/ttn/chief/net/2002inventory.html

USEPA (2003). *Strategy for Water Quality Standards and Criteria: Setting Priorities to Strengthen the Foundation for Protecting and Restoring the Nation's Waters*, EPA No.823-R-03-010. Washington, DC: US Environmental Protection Agency.

USEPA (2004). *Aquatic Life Water Quality Criterion for Selenium: 2004*, EPA No.82-D-04-001. Washington, DC: US Environmental Protection Agency.

USEPA (2006). *Sediment Quality Guidelines*. Washington, DC: US Environmental Protection Agency. Available online at www.epa.gov/waterscience/cs/guidelines.htm

Väinölä, R. (2003). Repeated trans-arctic invasions in littoral bivalves: molecular zoogeography of the *Macoma balthica* complex. *Marine Biology*, **57**, 51–60.

van den Berg, G.A., Loch, J.P.G., van der Heijdt, L.M. and Zwolsman, J.J.G. (1998). Vertical distribution of acid-volatile sulfide and simultaneously extracted metals in a recent sedimentation area of the River Meuse in The Netherlands. *Environmental Toxicology and Chemistry*, **17**, 758–763.

van der Sloot, H.A., Hoede, D., Wijkstra, J., Duinker, J.C. and Nolting, R.F. (1985). Anionic species of V, As, Se, Mo, Sb, Te and W in the Scheldt and Rhine estuaries and the Southern Bight (North Sea). *Estuarine, Coastal and Shelf Science*, **21**, 633–651.

van Geen, A., Adkins, J.F., Boyle, E.A., Nelson, C.H. and Palanques, A. (1997). A 120-year record of widespread contamination from the Iberian mining belt. *Geology*, **25**, 291–294.

van Geen, A., Luoma, S.N., Fuller, C.C., *et al.* (1992). Evidence from Cd/Ca ratios in foraminifera for greater upwelling off California 4000 years ago. *Nature*, **358**, 54–56.

van Geen, A., Takesue, R. and Chase, Z. (1999). Acid mine tailings in southern Spain. *Science of the Total Environment*, **242**, 221–229.

van Geen, A., Thoral, S., Rose, J., *et al.* (2004). Decoupling of As and Fe release to Bangladesh groundwater under reducing conditions. II. Evidence from sediment incubations. *Geochimica et Cosmochimica Acta*, **68**, 3475–3486.

Van Kirk, R.W. and Hill, S.L. (2007). Demographic model predicts trout population response to selenium based on individual-level toxicity. *Ecological Modelling*, **206**, 407–420.

van Zyl, D.J.A. (2004). Towards improved environmental indicators for mining using life-Cycle thinking. In *Life-Cycle Assessment of Metals: Issues and Research Directions*, ed. A. Dubreuil. Pensacola, FL: SETAC Press, pp. 117–122.

Vasak, M. (1991). Metal removal and substitution in vertebrate and invertebrate metallothioneins. In *Methods in Enzymology*, **205**, 452–458.

Velinsky, D.J. and Cutter, G.A. (1991). Geochemistry of selenium in a coastal salt marsh. *Geochimica et Cosmochimica Acta*, **55**, 179–191.

Veron, A.J., Church, T.M., Flegal, A.R., Patterson, C.C. and Erel, Y. (1993). Responses of lead cycling in the surface Sargasso Sea to changes in tropospheric inputs. *Journal of Geophysical Research*, **98**, 18269–18276.

Verslycke, T., Vercauteren, J., Devos, C., *et al.* (2003). Cellular energy allocation in the estuarine mysid shrimp *Neomysis integer* (Crustacea: Mysidacea) following tributyltin exposure. *Journal of Experimental Marine Biology and Ecology*, **288**, 167–179.

Vethaak, A.D., Bucke, D. Lang, T., *et al.* (1992). Fish disease monitoring along a pollution transect: a case study using dab *Limanda limanda* in the German Bight. *Marine Ecology Progress Series*, **91**, 173–192.

Viant, M.R. (2003). Improved methods for the acquisition and interpretation of NMR metabolomic data. *Biochemical and Biophysical Research Communications*, **310**, 943–948.

Viarengo, A., Moore, M.N., Mancinelli, G., *et al.* (1987). Metallothioneins and lysosomes in metal toxicity and homeostasis in marine mussels: the effects of cadmium in the presence and absence of phenanthrene. *Marine Biology*, **94**, 251–257.

Viarengo, A., Moore, M.N., Pertica, M., *et al.* (1985). Detoxification of copper in the cells of the digestive gland of mussels: the role of lysosomes and thioneins. *Science of the Total Environment*, **44**, 135–145.

Viarengo, A., Pertica, M., Mancinelli, G., *et al.* (1982). Evaluation of general and specific stress indices in mussels collected from populations subjected to different levels of heavy metal pollution. *Marine Environmental Research*, **6**, 235–243.

Viarengo, A., Pertica, M., Mancinelli, G., et al. (1984). Biochemical characterization of Cu-thioneins isolated from the tissues of mussels exposed to the metal. *Molecular Physiology*, **5**, 41–52.

Vijver, M. G., Van Gestel, C. A. M., Lanno, R. P., Van Straalen, N. M. and Peijnenburg, W. J. G. M. (2004). Internal metal sequestration and its ecotoxicological relevance: a review. *Environmental Science and Technology*, **38**, 4705–4712.

Villavicencio, G., Urrestarazu, P., Carvajal, C., et al. (2005). Biotic ligand model prediction of copper toxicity to daphnids in a range of natural waters in Chile. *Environmental Toxicology and Chemistry*, **24**, 1287–1299.

von der Ohe, P. C. and Liess, M. (2004). Relative sensitivity distribution of aquatic invertebrates to organic and metal compounds. *Environmental Toxicology and Chemistry*, **33**, 150–156.

Wageman, R., Trebacz, E., Boila, G. and Lockhart, W. L. (1998). Methylmercury and total mercury in tissues of arctic marine mammals. *Science of the Total Environment*, **218**, 19–31.

Waldichuck, M. (1979). The assessment of sublethal effects of pollutants in the sea. *Philosophical Transactions of the Royal Society of London B*, **286**, 483–505.

Walker, C. H., Hopkin, S. P., Sibly, R. M. and Peakall, D. B. (1996). *Principles of Ecotoxicology*. London: Taylor and Francis.

Walker, G. (1977). 'Copper' granules in the barnacle *Balanus balanoides*. *Marine Biology*, **39**, 343–349.

Walker, G., Rainbow, P. S., Foster, P. and Crisp, D. J. (1975a). Barnacles: possible indicators of zinc pollution? *Marine Biology*, **30**, 57–65.

Walker, G., Rainbow, P. S., Foster, P. and Holland, D. L. (1975b). Zinc phosphate granules in tissue surrounding the midgut of the barnacle *Balanus balanoides*. *Marine Biology*, **33**, 161–166.

Wallace, W. G. and Lopez, G. R. (1996). Relationship between the subcellular cadmium distribution in prey and cadmium transfer to a predator. *Estuaries*, **19**, 923–930.

Wallace, W. G. and Lopez, G. R. (1997). Bioavailability of biologically sequestered cadmium and the implications of metal detoxification. *Marine Ecology Progress Series*, **147**, 149–157.

Wallace, W. G. and Luoma, S. N. (2003). Subcellular compartmentalization of Cd and Zn in two bivalves. II. The significance of trophically available metal (TAM). *Marine Ecology Progress Series*, **257**, 125–137.

Wallace, W. G., Hoexum Brouwer, T. M., Brouwer, M. and Lopez, G. R. (2000). Alterations in prey capture and induction of metallothioneins in grass shrimp fed cadmium-contaminated prey. *Environmental Toxicology and Chemistry*, **19**, 962–971.

Wallace, W. G., Lee, B. G. and Luoma, S. N. (2003). Subcellular compartmentalization of Cd and Zn in two bivalves. I. Significance of metal-sensitive fractions (MSF) and biologically detoxified metal (BDM). *Marine Ecology Progress Series*, **249**, 183–197.

Wallace, W. G., Lopez, G. R. and Levinton, J. S. (1998). Cadmium resistance in an oligochaete and its effect on cadmium trophic transfer to an omnivorous shrimp. *Marine Ecology Progress Series*, **172**, 225–237.

Walling, D. E. and Moorehead, P. W. (1989). The particle size characteristics of fluvial suspended sediment: an overview. *Hydrobiologia*, **1767**, 125–149.

Wallner-Kersanach, M., Theede, H., Eversberg, U. and Lobo, S. (2000). Accumulation and elimination of trace metals in a transplantation experiment with *Crassostrea rhizophorae*. *Archives of Environmental Contamination and Toxicology*, **38**, 40–45.

Walters, C. (1997). Challenges in adaptive management of riparian and coastal ecosystems. *Conservation Ecology*, **1**, 1–16. Available online at www.consecol.org/vol1/iss2/art1

Wang, Q., Shen, W. and Ma, Z. (2000). Estimation of mercury emission from coal combustion in China. *Environmental Science and Technology*, **34**, 2711–2713.

Wang, W.-X. (2001). Comparison of metal uptake rate and absorption efficiency in marine bivalves. *Environmental Toxicology and Chemistry*, **20**, 1367–1373.

Wang, W.-X. (2002). Interactions of trace metals and different marine food chains. *Marine Ecology Progress Series*, **243**, 295–309.

Wang, W.-X. and Fisher, N. S. (1996). Assimilation of trace elements and carbon by the mussel *Mytilus edulis*: effects of food composition. *Limnology and Oceanography*, **41**, 197–207.

Wang, W.-X. and Fisher, N. S. (1997). Modeling the influence of body size on trace element accumulation in the mussel *Mytilus edulis*. *Marine Ecology Progress Series*, **161**, 103–115.

Wang, W.-X. and Fisher, N. S. (1999). Assimilation efficiencies of chemical contaminants in aquatic invertebrates: a synthesis. *Environmental Toxicology and Chemistry*, **18**, 2034–2045.

Wang, W.-X. and Ke, C. (2002). Dominance of dietary uptake of cadmium and zinc by two marine predatory gastropods. *Aquatic Toxicology*, **56**, 153–165.

Wang, W.-X. and Rainbow, P. S. (2000). Dietary uptake of Cd, Cr, and Zn in the barnacle *Balanus trigonus*: influence of diet composition. *Marine Ecology Progress Series*, **204**, 159–168.

Wang, W.-X. and Rainbow, P.S. (2005). Influence of pre-exposure on trace metal uptake in marine invertebrates. *Ecotoxicology and Environmental Safety*, **61**, 145–159.

Wang, W.-X., Fisher, N.S. and Luoma, S.N. (1996). Kinetic determinations of trace element bioaccumulation in the mussel *Mytilus edulis*. *Marine Ecology Progress Series*, **140**, 91–113.

Wang, W.-X., Fisher, N.S. and Stupakoff, I. (1999). Bioavailability of dissolved and sedimentbound metals to a marine deposit-feeding polychaete. *Marine Ecology Progress Series*, **178**, 281–293.

Wang, W.-X., Griscom, S.B. and Fisher, N.S. (1997). Bioavailability of Cr(III) and Cr(VI) to marine mussels from solute and particulate pathways. *Environmental Science and Technology*, **31**, 603–611.

Wang, W.-X., Stupakoff, I., Gagnon, C. and Fisher, N.S. (1998). Bioavailability of inorganic and methylmercury to a marine deposit-feeding polychaete. *Environmental Science and Technology*, **32**, 2564–2571.

Wang, W.-X., Yan, Q.-L., Fan, W. and Xu, Y. (2002). Bioavailability of sedimentary metals from a contaminated bay. *Marine Ecology Progress Series*, **240**, 27–38.

Wangersky, P.J. (1986). Biological control of trace metal residence time and speciation: a review and synthesis. *Marine Chemistry*, **18**, 42–59.

Warren, L.A., Tessier, A. and Hare, L. (1998). Modelling cadmium accumulation by benthic invertebrates *in situ*: the relative contributions of sediment and overlying water reservoirs to organism cadmium concentrations. *Limnology and Oceanography*, **43**, 1442–1454.

Warwick, R.M. and Clarke, K.E. (1991). A comparison of some methods for analysing changes in benthic community structure. *Journal of the Marine Biological Association of the United Kingdom*, **71**, 225–244.

Watanabe, H. and Iguchi, T. (2006). Using ecotoxicogenomics to evaluate the impact of chemicals on aquatic organisms. *Marine Biology*, **149**, 107–115.

Wedderburn, J., McFadzen, I., Sanger, R.C., *et al.* (2000). The field application of cellular and physiological biomarkers, in the mussel *Mytilus edulis*, in conjunction with early life stage bioassays and adult histopathology. *Marine Pollution Bulletin*, **40**, 257–267.

Weed, W.H. (1912). Geology and ore deposits of Butte District, Montana. *US Geological Survey Professional Paper* **74**.

Weeks, J.M. (1992). The talitrid amphipod (Crustacea) *Platorchestia platensis* as a biomonitor of trace metals (Cu and Zn) in Danish waters. In *Proceedings of the 12th Baltic Marine Biologists Symposium*, ed. E. Bjørnestad, L. Hagerman and K. Jensen. Fredensborg, Denmark: Olsen and Olsen, pp. 173–178.

Weeks, J.M. (1993). Effects of dietary copper and zinc concentrations on feeding rates of two species of talitrid amphipods (Crustacea). *Bulletin of Environmental Contamination and Toxicology*, **50**, 883–890.

Weeks, J.M. and Moore, P.G. (1991). The effect of synchronous moulting on body copper and zinc concentrations in four species of talitrid amphipods (Crustacea). *Journal of the Marine Biological Association of the United Kingdom*, **71**, 481–488.

Weeks, J.M. and Rainbow, P.S. (1991). The uptake and accumulation of zinc and copper from solution by two species of talitrid amphipods (Crustacea). *Journal of the Marine Biological Association of the United Kingdom*, **71**, 811–826.

Weeks, J.M. and Rainbow, P.S. (1993). The relative importance of food and seawater as sources of copper and zinc to talitrid amphipods (Crustacea; Amphipoda; Talitridae). *Journal of Applied Ecology*, **30**, 722–735.

Weeks, J.M., Rainbow, P.S. and Moore, P.G. (1992). The loss, uptake and tissue distribution of copper and zinc during the moult cycle in an ecological series of talitrid amphipods (Crustacea: Amphipoda). *Hydrobiologia*, **245**, 15–25.

Weis, J.S., Smith, S., Zhou, T., Santiago-Bass, C. and Weis, P. (2001). Effects of contaminants on behavior: biochemical mechanisms and ecological consequences. *BioScience*, **51**, 209–217.

Wellise, C.J. (1999). Effect of carbamate pesticides on Cd and Zn bioaccumulation in a bivalve. Unpublished M.Sc. thesis, San José State University, San José, CA.

Welsh, P.G., Lipton, J., Chapman, G.A. and Podrabsky, T.L. (2000). Relative importance of calcium and magnesium in hardness-based modification of copper toxicity. *Environmental Toxicology and Chemistry*, **19**, 1624–1631.

Wenning, R.J., Adams, W.J., Gatley, G.E., *et al.* (2005). Executive summary. In *Use of Sediment Quality Guidelines and Related Tools for the Assessment of Contaminated Sediments*, ed. R.J. Wenning, G.E. Batley, C.G. Ingersoll and D.W. Moore. Pensacola, FL: SETAC Press, pp. 11–39.

Westerlund, S.F.G., Anderson, L.G., Hall, P.O.J., *et al.* (1986). Benthic fluxes of cadmium, copper, nickel, zinc and lead in the coastal environment. *Geochimica et Cosmochimica Acta*, **50**, 1289–1296.

White, S. L. and Rainbow, P. S. (1982). Regulation and accumulation of copper, zinc and cadmium by the shrimp *Palaemon elegans*. *Marine Ecology Progress Series*, **8**, 95–101.

White, S. L. and Rainbow, P. S. (1984a). Regulation of zinc concentration by *Palaemon elegans* (Crustacea: Decapoda): zinc flux and effects of temperature, zinc concentration and moulting. *Marine Ecology Progress Series*, **16**, 135–147.

White, S. L. and Rainbow, P. S. (1984b). Zinc flux in *Palaemon elegans* (Crustacea: Decapoda): moulting, individual variation and tissue distribution. *Marine Ecology Progress Series*, **19**, 153–166.

White, S. L. and Rainbow, P. S. (1985). On the metabolic requirements for copper and zinc in molluscs and crustaceans. *Marine Environmental Research*, **16**, 215–229.

White, S. L. and Rainbow, P. S. (1986). A preliminary study of Cu-, Cd- and Zn-binding components in the hepatopancreas of *Palaemon elegans* (Crustacea: Decapoda). *Comparative Biochemistry and Physiology C*, **83**, 111–116.

Whitfield, M. and Turner, D. R. (1987). The role of particles in regulating the composition of seawater. In *Aquatic Surface Chemistry*, ed. W. Stumm. New York: John Wiley, pp. 457–493.

WHO (1989). *DDT and Its Derivatives: Environmental Aspects*, WHO Task Group on Environmental Health Criteria for DDT and its derivatives – environmental aspects, International Program on Chemical Safety: Environmental Health Criteria No.83. Geneva, Switzerland: World Health Organization.

Widdows, J. (1985). Physiological responses to pollution. *Marine Pollution Bulletin*, **16**, 129–134.

Widdows, J. and Johnson, D. (1988). Physiological energetics for *Mytilus edulis*: scope for growth. *Marine Ecology Progress Series*, **46**, 113–121.

Widdows, J. and Page, D. S. (1993). Effects of tributyltin and dibutyltin on the physiological energetics of the mussel, *Mytilus edulis*. *Marine Environmental Research*, **35**, 233–249.

Widdows, J., Donkin, P., Brinsley, M. D., *et al.* (1995). Scope for growth and contaminant levels in North Sea mussels. *Marine Ecology Progress Series*, **127**, 131–148.

Widdows, J., Donkin, P., Salkeld, P. N., *et al.* (1984). Relative importance of environmental factors in determining physiological differences between two populations of mussels (*Mytilus edulis*). *Marine Ecology Progress Series*, **17**, 33–47.

Widdows, J., Donkin, P., Staff, F. J., *et al.* (2002). Measurement of stress effects (scope for growth) and contaminant levels in mussels (*Mytilus edulis*) collected from the Irish Sea. *Marine Environmental Research*, **53**, 327–356.

Widdows, J., Nasci, C. and Fossato, V. U. (1997). Effects of pollution on the scope for growth of mussels (*Mytilus galloprovincialis*) from the Venice Lagoon, Italy. *Marine Environmental Research*, **43**, 69–79.

Wiederholm, T. (1984). Incidence of deformed chironomid larvae (Diptera: Chironomidae) in Swedish lakes. *Hydrobiologia*, **109**, 243–249.

Wielinga, B., Lucy, J. K., Moore, J. N., Seastone, O. F. and Gannon, J. E. (1999). Microbiological and geochemical characterization of fluvially deposited sulfidic mine tailings. *Applied and Environmental Microbiology*, **65**, 1548–1555.

Wiener, J. G. and Giesy, J. P. (1979). Concentrations of Cd, Cu, Mn, Pb, and Zn in fishes in a highly organic softwater pond. *Journal of the Fisheries Research Board of Canada*, **36**, 270–279.

Wiener, J. G. and Shields, P. J. (2000). Mercury in the Sudbury River (Massachusetts, USA): pollution history and a synthesis of recent research. *Canadian Journal of Fisheries and Aquatic Sciences*, **57**, 1053–1061.

Wiener, J. G. and Spry, D. J. (1996). Toxicological significance of mercury in freshwater fish. In *Environmental Contaminants in Wildlife: Interpreting Tissue Concentrations*, ed. W. N. Beyer, G. H Heinz and A. W. Redmon-Norwood. Boca Raton, FL: Lewis Publishers, pp. 297–339.

Wiener, J. G., Bodaly, R. A., Brown, S. S., *et al.* (2007). Monitoring and evaluating trends in methylmercury accumulation in aquatic biota. In *Ecosystem Responses to Mercury Contamination*, ed. R. Harris, D. P. Krabbenhoft, R. Mason, M. W. Murray, R. Reasch, and T. Saltman. Pensacola, FL: SETAC Press, pp. 887–122.

Williams, R. J. P. and Fraústo da Silva (1996). *The Natural Selection of Chemical Elements*. Oxford: Clarendon Press.

Willis, J. N. and Jones, N. Y. (1977). The use of uniform labeling with zinc-65 to measure stable zinc turnover in the mosquito fish, *Gambusia affinis*. I. Retention. *Health Physics*, **32**, 381–387.

Windom, H. L. (1990). Flux of particulate metals between East Coast North American rivers and the North Atlantic Ocean. *Science of the Total Environment*, **97/8**, 115–124.

Windom, H. L., Byrd, J. T., Smith, R. G. Jr and Huan, F. (1991). Inadequacy of NASQAN data for assessing metal trends in the nation's rivers. *Environmental Science and Technology*, **25**, 1137–1142.

Winge, D., Krasno, J. and Colucci, A.V. (1974). Cadmium accumulation in rat liver: correlation between bound metals and pathology. In *Trace Element Metabolism in Animals*, ed. G. Hoekstra, J.W. Suttie, H.E. Ganther and W. Mertz. Baltimore, MD: University Park, pp. 500–502.

Winner, R.W., Boesel, M.W. and Farrel, M.P. (1980). Insect community structure as an index of heavy-metal pollution in lotic ecosystems. *Canadian Journal of Fisheries and Aquatic Sciences*, **37**, 647–655.

Wood, L.B. (1982). *Restoration of the Tidal Thames*. Philadelphia, PA: Hayden.

Woodward, D.F., Brumbaugh, W.G., DeLonay, A.J. and Smith, C.E. (1994). Effects on rainbow trout of a metals-contaminated diet of benthic invertebrates from the Clark Fork River, Montana. *Transactions of the American Fisheries Society*, **123**, 51–62.

Woodward, D.F., Farag, A.M., Bergman, H.L., *et al.* (1995). Metals-contaminated benthic invertebrates in the Clark Fork River, Montana: effects on age-0 brown trout and rainbow trout. *Canadian Journal of Fisheries and Aquatic Sciences*, **52**, 1994–2004.

Woodward, D.F., Goldstein, J.N., Farag, A.M. and Brumbaugh, W.G. (1997). Cutthroat trout avoidance of metals and conditions characteristic of a mining waste site: Coeur d'Alene River, Idaho. *Transactions of the American Fisheries Society*, **126**, 699–706.

Woodwell, G.M. (1970). Effects of pollution on the structure and physiology of ecosystems. *Science*, **168**, 429–433.

Woodwell, G.M., Craig, P.P. and Johnson, H.A. (1971). DDT in the biosphere: where does it go? *Science*, **174**, 1101–1107.

Wrench, J.J. (1978). Selenium metabolism in the marine phyoplankters *Tetraselmis tetrathele* and *Dunaliella minuta*. *Marine Biology*, **49**, 231–236.

Wrench, J.J. and Measures, C.I. (1982). Temporal variations in dissolved selenium in a coastal ecosystem: *Nature*, **299**, 431–433.

Wright, D.A. and Welbourne, P. (2002). *Environmental Toxicology*, Cambridge, UK: Cambridge University Press.

Wright, D.A., Mihursky, J.A. and Phelps, H.L. (1985). Trace metals in Chesapeake Bay oysters: intra-sample variability and its implications for biomonitoring. *Marine Environmental Research*, **16**, 191–197.

Wu, J.F. and Boyle, E.A. (1997). Lead in the western North Atlantic Ocean: completed response to leaded gasoline phaseout. *Geochimica et Cosmochimica Acta*, **61**, 3279–3283.

Wu, Y., Wang, S., Streets, D.G., *et al.* (2006). Trends in anthropogenic mercury emissions in China from 1995 to 2003. *Environmental Science and Technology*, **40**, 5312–5318.

Xu, Y. and Wang, W.-X. (2001). Individual responses of trace-element assimilation and physiological turnover by the marine copepod *Calanus sinicus* to changes in food quality. *Marine Ecology Progress Series*, **218**, 227–238.

Xu, Y. and Wang, W.-X. (2002a). Exposure and potential food chain transfer factor of Cd, Se and Zn in marine fish *Lutjanus argentimaculatus*. *Marine Ecology Progress Series*, **238**, 173–186.

Xu, Y. and Wang, W.-X. (2002b). The assimilation of detritus-bound metals by the marine copepod *Acartia spinicauda*. *Limnology and Oceanography*, **47**, 604–610.

Xu, Y. and Wang, W.-X. (2004). Silver uptake by a marine diatom and its transfer to the coastal copepod *Acartia spinicauda*. *Environmental Toxicology and Chemistry*, **23**, 682–690.

Xu, Y., Wang, W.-X. and Hsieh, D.P.H. (2001). Influences of metal concentration in phytoplankton and seawater on metal assimilation and elimination in marine copepods. *Environmental Toxicology and Chemistry*, **20**, 1067–1077.

Yan, N.D. (1983). The effects of changes in pH on transparency and on thermal regimes of Lohi Lake, near Sudbury, Ontario. *Canadian Journal of Fisheries and Aquatic Sciences*, **40**, 621–626.

Yan, N.D. and Miller, G.E. (1984). Effects of deposition of acids and metals on chemistry and biology of lakes near Sudbury, Ontario. In *Environmental Impacts of Smelters*, ed. J. Nriagu. New York: John Wiley, pp. 244–282.

Yan, N.D. and Strus, R. (1980). Crustacean zooplankton communities of acidic, metal-contaminated lakes near Sudbury, Ontario. *Canadian Journal of Fisheries and Aquatic Sciences*, **37**, 2282–2293.

Yan, N.D., Girard, R., Heneberry, J.H., *et al.* (2004). Recovery of copepod, but not cladoceran, zooplankton from severe and chronic effects of multiple stressors. *Ecology Letters*, **7**, 452–460.

Yan, N.D., Keller, W., MacIsaac, H.J. and McEachern, L.J. (1991). Regulation of zooplankton community structure of an acidified lake by *Chaoborus*. *Ecological Applications*, **1**, 52–65.

Yan, N.D., Keller, W., Scully, N.M., Lean, D.R.S. and Dillon, P.J. (1996). Increased UV-B penetration in a lake owing to drought induced acidification. *Nature*, **381**, 141–143.

Yan, N.D., Leung, B., Keller, W., *et al* (2003). Developing conceptual frameworks for the recovery of aquatic biota from acidification: a zooplankton example. *Ambio*, **32**, 165–169.

Yan, N. D., Pérez-Fuentetaja, A., Ramcharan, C. W., *et al.* (2001). Changes in the crustacean zooplankton communities of Mouse and Ranger Lakes: Part 6 of the Dorset food web piscivore manipulation project. *Archiv für Hydrobiologie, Special Issue, Advances in Limnology*, **56**, 127–150.

Yan, Q. L. and Wang, W.-X. (2002). Metal exposure and bioavailability to a marine deposit-feeding Sipuncula, *Sipunculus nudus*. *Environmental Science and Technology*, **36**, 40–47.

Yeats, P. A. and Bewers, J. M. (1987). Evidence for anthropogenic modification of global transport of cadmium. In *Cadmium in the Aquatic Environment*, ed. J. O. Nriagu and J. B. Sprague. New York: John Wiley, pp. 19–30.

Yohn, S. S., Long, D. T., Fett, J. D., *et al.* (2002). Assessing environmental change through chemical-sediment chronologies from inland lakes. *Lakes and Reservoirs: Research and Management*, **7**, 217–230.

Zhang, H. and Davison, W. (2000). Direct in situ measurements of labile inorganic and organically bound metal species in synthetic solutions and natural waters using diffusive gradients in thin films. *Analytical Chemistry*, **72**, 4447–4457.

Zhang, H., Davison, W., Mortimer, R. J. G., *et al.* (2002). Localised remobilization of metals in a marine sediment. *Science of the Total Environment*, **296**, 175–187.

Zhang, Y.-Q. and Moore, J. N. (1996). Selenium fractionation and speciation in a wetland sediment. *Environmental Science and Technology*, **30**, 2613–2619.

Zhang, Y.-Q. and Moore, J. N. (1997a). Reduction potential of selenate in wetland sediment. *Journal of Environmental Quality*, **26**, 910–916.

Zhang, Y.-Q. and Moore, J. N. (1997b). Controls on sediment distribution in wetland sediment, Benton Lake, Montana. *Water, Air and Soil Pollution*, **97**, 323–340.

Zhang, Y.-Q. and Moore, J. N. (1997c). Environmental conditions controlling selenium volatilization from a wetland system. *Environmental Science and Technology*, **31**, 511–517.

Zirino, A. and Yamamoto, S. (1972). A pH-dependent model for the chemical speciation of copper, zinc, cadmium and lead in seawater. *Limnology and Oceanography*, **17**, 661–671.

Zorita, I., Ortiz-Zaragoitia, M., Soto, M. and Cajaraville, M. P. (2006). Biomarkers in mussels from a copper site gradient (Visnes, Norway): an integrated biochemical, histochemical and histological study. *Aquatic Toxicology*, **78**, 109–116.

Index

Page numbers of definitions are in **bold**.

abalone 249–250
  see also Haliotis
absorption **126**
  see uptake
abundance *see* univariate measures
Acartia 150, 254, 342
accumulation *see* bioaccumulation
acidification 279, 288, 302, 311–312
acid mine drainage (AMD) 48, 62,
    72, 79, 279, 297, 301–302,
    305, 312–313
acid volatile sulphide (AVS) 96–97,
    105, 107, 117–119, 122–123, 150,
    426, 442–447, 451, 453, 456
Acipenser 265, 344, 397, 471, 475, 486
actinides 9
Actitis 352
active biomonitoring
  see biomonitoring translocation
acute-to-chronic ratio (ACR) **394**,
    402, 407, 409, 412, 414, 417,
    421–422, 483
adaptive management **14**, 24–25, 31,
    459, 503
adduct 210–211
Aden 186
adsorption
  biota **126**, 126–127, 129, 150, 172,
    190–192, 262
  sediment 100, 115, 121, 337, 339
advisory (food consumption) 388, 390
Afon Goch 72
Africa 49, 186, 294, 369
AFS Convention 364, 367
Agusan River 379
Alaska 58, 122
Alberta 329–330, 352
alderfly 165, 190
  see also Sialis
Alice in Wonderland 369
alkali disease 346
alkali metals 7
alkaline earth metals 7

alkyl mercury **354**, 369
  see also methyl mercury
Allorchestes 135
Almadén 369, 376, 378
aluminium 9, 48, 55, 73, 86, 312
  sediment 96–98, 101–102, 103–104,
    122, 193
Amazon 79, 379
ambient criterion **394**
ambient water quality criteria 15, 61,
    362, 364, 390–391, 393–394, 399,
    404, 408–409, 416, 418, 421,
    460–461, 463, 472, 479, 486, 495
American dipper 352
americium 74, 144, 146
amino acid 10, 130, 133, 157, 212,
    338, 341, 382–383, 439
  membrane transport 131–132, 146,
    382–383, 439
  see also cysteine, histidine,
    methionine, glutamic acid,
    glycine
aminopeptidase 10
ammonia 429, 432
Ampelisca 252, 443, 447, 467
ampharetid polychaete 184, 476
amphibian 412
amphipod
  assimilation efficiency 146, 150, 163
  bioaccumulation 154, 158–159, 198,
    343, 359, 381, 438, 447, 471
  biomarker 221
  biomonitor 185, 187–188,
    193–194, 201
  prey 188
  uptake 135, 137, 139, 141
  toxicity 220–221, 224, 226–227,
    244, 247, 252, 262, 285, 317,
    412, 421, 428–429, 435, 441,
    443, 455, 471
  *see also Allorchestes, Ampelisca,
    Corophium, Diporeia,
    Echinogammarus, Eohaustorius,*

  Gammarus, Hyalella, Niphargus,
    phoxocephalid, *Pontoporeia,*
    stegocephalid, talitrid
Anaconda 299–301, 307, 311
analysis of covariance (ANCOVA)
    196–197
analysis of variance (ANOVA) 197, 284
Anas 388, 474–475
Andes Mountains 280
Anglesey 72
anion 22, 84–85, 119, 132, 136
Ankeveen Lake 444
Antarctic 54, 251, 371, 474
anthropogenic input 49, 50–66, 76,
    79, 88, 106, 269, 280, 370–371
antifouling 5, 41, **354**, 362, 364,
    367–368
antimony 7, 9, 57, 88, 97, 297,
    301, 310
antioxidant 213–214
application factor 36, 258, 400–401,
    409, 424, 483
applied ecotoxicology **33**, 43, 46, 268,
    270–272
  applied metal ecotoxicology 32, 34
Aquatic Toxicity Information
    Retrieval (AQUIRE) database 235
Arabian Sea 49
Arcachon Bay 362
Arctic 54, 76, 79, 122, 371, 474
Arctopsyche 283, 320
Arenicola 146, 150, 184
Argopecten 251, 261
Arkansas River 284, 286
Armandia 359, 386
arsenic 4, 7, 9, 18, 120
  bioaccumulation 198
  biogeochemical cycle 57, 63, 310
  biomonitor 176, 179–182, 184–185,
    200, 447
  organic forms 84, 88
  oxidation state 84, 119–120, 253
  mining 63, 297, 301

rock 97
sediment 63, 105, 107, 119,
    310, 447, 453
speciation 84, 86
toxicity 253, 256, 311–312, 315, 350
uptake 132, 136
water 73, 79, 112, 303
Artemia 224–226, 229, 256, 261
ascidian 290
    see also Ascidiella, Pyura
Ascidiella 290
Ascophyllum 175, 190
Asellus 165
Asia 55, 65, 108, 330, 365, 369,
    372–374
southeast 186, 221
assimilation efficiency **126**, 142–143,
    147–148, 165, 167, 438, 448,
    464–466
measurement 142
Association of Southeast Asian
    Nations (ASEAN) 417
Astragalus 346
Atlantic Ocean 48–49, 78, 356
    north Atlantic 38, 52, 75, 78, 203
    south Atlantic 186
atmosphere
    biogeochemical cycle 47–54, 62,
        107, 370
    smelter waste 301, 309, 311
ATP 132
ATPase 132, 134, 207, 214
Australia 15–17, 61, 203, 228, 290,
    294, 323, 327–329, 343, 356,
    363–365, 367, 380, 418, 451, 453,
    455–456, 461
Australia–New Zealand Environment
    Council (ANZEC) 418
Austria 363
Austrocochlea 343
avocet 348–349, 474
    see also Recurvirostris
Avon 105
Axinopsida **317**
Aznalcollar 309, 313

background concentrations
    (sediments) 96, 99, 306–307
bacteria *see* microbes, *see also* iron
    sulphate oxidizing bacteria,
    nitrate-reducing bacteria,
    Pseudomonas, sulphate-reducing
    bacteria
Baetis 192, 199, 201, 245, 261, 320
Balanus 143–144, 147, 154, 162–163,
    181, 186–187, 198–201, 203, 225

bald eagle 2–3
ballast water 366
Baltic Sea 106–107, 163, 180, 185,
    200–201, 450
Baltimore Harbor 377
Bangladesh 4, 120
Barbados Islands 49
barium 7
barnacle
    assimilation efficiency 142–144,
        147, 151
    bioaccumulation 8, 149, 152–154,
        157–160, 163, 173, 198–200, 468
    biodynamics 162
    biomonitor 177, 181, 186–187, 197,
        200–201
    uptake 139
    toxicity 132, 225
    see also Balanus, Capitulum, Elminius,
        Fistulobalanus, Semibalanus,
        Tetraclita
Bay of Brest 367
Bay of Papua New Guinea 305, 309, 322
Beaufort Sea 58, 122
bed sediment **93**
Bedford Basin 338
Beethoven 369
beetle *see* Coleoptera
behaviour 220, 223, 261, 321, 472
Belews Lake 341, 347–348, 351, 461,
    474, 482
Belgium 416, 451–453
    see also Flanders, Walloon
beluga whale 387
    see also Delphinapterus
Bembicium 343
Bengal, West 4
benthos **426**
Benton Lake 335–340
Bermuda 62, 78, 373
beryllium 7
Biala Przemsza River 192–193, 199
binding site 89, 98, 110, 115–116,
    121, 232
bioaccumulation 28, **33**, **126**–168,
    172, 396, 467, 491
    bioaccumulation factor (BAF) **328**,
        340, 359, 390, 396–397, 436, 477
    see also BSAF
    bioaccumulated metal guideline
        (BMG) 30, **460**, 463,
        467–476, 488
    dietary **33**
bioaccumulation pattern 126, 152,
    155, 156–159, 173
    crustaceans 156–159, 160

regulation 156, 158, 160, 173,
    184–185, 189, 464, 467, 470
strong net accumulation 158, 160,
    172–173, 177, 187, 201, 467
weak net accumulation 160, 173,
    192, 467
    see also efflux, uptake
bioassay **33**, 227, 234
    see also toxicity test
bioavailability **14**, 26, 30, 39, 41,
    126–**127**, 129, 131, 135, 138, 142,
    167, **169**–173, 205, 252–253, 255,
    258, 269, 287–288, 291, 400,
    404–405, 491
    see also uptake
bioconcentration **33**
    bioconcentration factor (BCF) **328**,
        340–341, 357–359, 385, 390,
        396–398
biodiversity *see* community structure
biodynamic modelling **127**, 134, 159,
    161–162, 164, 167, 174, 438–439,
    441, 491
    see also assimilation efficiency,
        efflux rate, ingestion rate,
        uptake rate
bioindicator
biokinetic modelling 126–**127**, 129,
    150, 172, 190–192, 262
    see biodynamic modelling
biogeochemical cycle 25, 37, **47**–66,
    49–50, 53–54, 94–95, 108, 490
    flux 47–49, 52–53
biomagnification 3, **33**, 164, 166,
    254, **327**, 344–345, 364, 369,
    382, 386, 391
biomarker 28, 43, **204**, 209–231, 244,
    277, 292, 488, 492
    behaviour 223–224
    biochemical 211–213, 214, 278, 430
    community 221, 228–230, 292
    cytological 213–217
    histological 217, 256, 263, 315, 320
    molecular 210
    morphological 217
    physiological 218–223
    population 226–228
    suite 221
    tolerance 226, 229
biomics 212
biominification 164, 166
biomonitor 28, 41, 43, **169**–203,
    230, 284, 291, 319, 464, 471,
    473, 491
    coastal waters 169, 174, 175–187,
        183–185, 200

biomonitor (cont.)
   cosmopolitan **169**, 173, 175–193,
     203
   gut contents 190, 192, 193–194
   interspecific comparison 200
   intraspecific comparison 199
   lakes 187–189
   moult cycle 185, 197
   sampling 197
   season 196–197
   stream/river 189–193
   systematics 198–199
   translocation 201
   required characteristics 173, 188
   suite **169**, 174, 175–187, 183–185,
     200
Biomphalaria 227
biosphere **47**, 301
Biotic Ligand Model (BLM) 30, 135,
   137, 146, 207, 253, 407–410,
   415, 472, 491
biotransformation **328**, 337–338
bioturbation **93**, 112–113, 229–230,
   240, 251, 279–281, 310, 318,
   426–428, 433
bismuth 7, 9, 58, 297
bivalve mollusc
   assimilation efficiency 142–145, 341
   bioaccumulation 149, 166, 195,
     358, 379
   biomonitor 177, 184, 193, 318
   clearance/filtration rate 140, 142,
     150, 220
   digestion 145–146, 215
   efflux 152–153, 166, 342
   toxicity 217–218, 228–229, 246,
     248, 317
   uptake 135–136, 140
   see also Axinopsida, clam, cockle,
     lucinid, mussel, oyster,
     Petricola, Pyganodon, scallop,
     tellinid
Blackfoot River 299, 304, 320–321
Blackwater Estuary 466
bladder wrack 185, 224, 241, 256, 291
   see also Fucus
blind staggers 346, 350
blood cell 211, 214, 216
bluegill 241, 267, 348, 474–475
   see also Lepomis
blue-green algae *see* cyanobacteria
boron 7
Bosmina 288, 290
bottom-up effect 291
Bourgneuf Bay 213
bradycardia 220

Brazil 109, 186, 379
Brent Spar 23
brine shrimp 249, 256
   see also Artemia
Brisbane River 456
Bristol Channel 82, 105
British Columbia 16, 217, 317–318,
   329, 352, 427
   see also Vancouver Island
Brittany 367
brittle star *see* ophiuroid
   echinoderm
brown alga 175, 183, 185–186, 225
   see also fucoid, kelp, *Sargassum*
brown pelican 2–3
brown trout *see* trout, *see also Salmo*
BSAF 358–359
bug *see* Heteroptera
Bunker Hill 62, 309–310
Butte 59, 293, 297–300, 302, 307–309,
   312, 321

caddisfly
   bioaccumulation 138, 166,
     191–193, 283–284, 438
   biomonitor 175, 190–192, 194, 203,
     284, 313, 319, 473, 476
   toxicity 282
   see also hydropsychid, *Mystacides*,
     Plectrocnemia, Rhyacophila
cadmium 4, 7, 9–11, 18, 37, 41–42
   assimilation efficiency 143–149,
     151, 165, 438, 448, 464–466
   bioaccumulation 74, 154, 158–159,
     166, 189, 192, 198, 208, 213,
     227, 261, 283–284, 288, 313,
     315, 322, 355, 398, 437–439,
     446–447, 466, 471–472
   biodynamics 164, 180, 283, 472
   biogeochemical cycle 50–51, 53–54,
     58, 74, 88, 310
   biomonitoring 176, 178–183, 185,
     188, 190, 192–193, 199–200,
     319, 473
   colloid 69
   detoxification 149, 157, 212
   efflux 152–153
   mining 297, 301
   rock 97
   sediment 54, 95, 100, 105, 107,
     117, 120, 123, 190, 225, 308,
     310, 313, 430, 437, 440,
     443–447, 453
   speciation 86–87, 90, 121, 264, 422
   stable isotopes 166
   tolerance 149, 225

   toxicity 150, 166, 209, 215, 224,
     227, 240, 244–245, 247, 249,
     251–253, 255, 257, 261–263,
     282, 311–312, 350, 398–399,
     401, 410–412, 418–421, 430,
     443, 464, 472
   uptake 130, 132–141, 163, 165,
     170, 438
   water 54, 72–73, 75–76, 79, 87, 95,
     121, 257, 288–289, 303–304
caesium 4, 7
calanoid copepod 240, 341
   see also Acartia, Calanus, Temora
Calanus 147
calcium 3, 7, 9, 42, 73, 86–87, 115,
   121, 141, 155, 166
   carbonate 148, 404, 420
   channel 130, 132–134, 141
California 52, 82, 106, 121, 298, 302,
   330–331, 334, 339, 345, 349, 379,
   389, 461, 495
California Bight 99, 107, 294,
   304, 306
California clapper rail 389
   see also Rallus
California Gold Rush 379
California Regional Water Quality
   Control Board 461, 479
Camel River 105
Campanularia 218–219, 227
Canada 4, 16, 18, 58, 63, 76, 91, 108,
   123, 188, 217, 240, 263, 265, 279,
   287–288, 294, 311, 314, 317, 324,
   327, 329–331, 338, 350, 352, 356,
   369, 371, 374–375, 383, 385–386,
   401, 417–418, 445, 461–462
   see also Alberta, British Columbia,
     Newfoundland, Nova Scotia,
     Ontario, Quebec,
     Saskatchewan
Canadian Environmental Quality
   Guidelines 400
Canadian Shield 240, 287, 311
Cancer 343, 467, 471
cap and trade scheme 60, 373, 393
Cape Cod 266
Cape Fear 121
Capitulum 200
carbonate 38, 79, 85, 97, 103, 112,
   116, 297
carbonic anhydrase 10, 153
carcinogenesis 210, 217
Carcinus 141, 214, 216, 222, 225
Cardiff 82
Cardigan Bay 82
Caribbean Sea 48–49

caridean decapod 139, 153–156, 159–160
  see also grass shrimp, *Macrobrachium*, *Palaemon, Palaemonetes, Pandalus*, prawn, shrimp
Carnon River 72, 90, 112
carp 189, 348
  see also Cyprinus
Carquinez Strait 469
carrier proteins **127**, 129–130
  see uptake transporters
Carson, Rachel 3
catalase 212, 214, 221
catfish 322
  see also Neosilurus
cation 38, 85–87, 111, 405
Catostomus 265
cattle 4, 311, 346
Cecina 211, 214–216, 221
cell
  membrane 18, 129–133, 145
    see also uptake
  organelles 149, 154–155, 160, 215, 230, 464, 472
  solution *see* cytosol
  see also lysosome
cellular energy allocation (CEA) **204**, 219
Central Valley 330, 339, 379, 461
  Central Valley Project (CVP) 331
centrarchid fish 267
  see also bluegill
Cerastoderma 193, 214, 222, 358
Ceriodaphnia 247–248, 322, 400, 405, 410
chalcopyrite 301
Chanaral 280
Chaoborus 138, 175, 188–189, 203, 288, 290, 440
Charleston Harbor 430–431
chelation *see* complexation
chelating agent
  lipid-soluble 135
  water-soluble 135
  see also EDTA, NTA
chemoautotrophy 184
Chesapeake Bay 83, 198, 377–378
Chevron Marsh 344
chicken 3
Chile 280–282, 286, 316, 318, 408
China 48–49, 187, 199, 295, 372–373, 378–379
  see also Xiaoquinling
chinook salmon 411, 474
  see also Oncorhynchus
Chipps Island 469

chironomid midge 188–189, 192, 218, 410, 444
  see also Chironomus, Micropsectra, Tanytarsus
Chironomus 193–194, 218, 244, 246, 248, 428, 430–431, 435, 444
Chlorella 143
chloride 80, 85
  cells 139, 153–156, 159–160
  complexation 39, 86–87, 121, 131, 136, 138, 163, 253, 370, 382, 422–423
chromium 9, 11, 18
  assimilation efficiency 143–147, 149, 165, 438
  bioaccumulation 159, 439
  biodynamics 180
  biogeochemical cycle 51, 57–58, 310
  biomonitoring 176, 178, 180–182, 200, 447
  efflux 140, 152–153, 187, 199, 203
  mining 55
  oxidation state 84, 119, 253
  rock 97–98
  sediment 96–97, 105, 107, 117–119, 122–123, 150, 426, 442–447, 451, 453, 456
  speciation 84, 86, 115, 121
  toxicity 150, 211, 217, 253, 398–400, 409, 412
  uptake 132, 134–136, 140, 159, 163, 165
  water 73
ciliate protozoan 264
Cinclus 352
cirratulid polychaete 198
  see also Tharyx
cladoceran
  toxicity 211, 240, 246–248, 287–289, 314, 322, 408, 413–414, 421–422, 435
  see also Bosmina, daphnid, *Halopedium*
Cladophora 187
clam
  assimilation efficiency 143–144, 146–148, 151, 341, 448
  bioaccumulation 45, 207–208, 275, 278, 343, 358, 378, 438–439, 446, 448, 465, 469–470
  biodynamics 180, 341–342, 439
  biomarker 208, 217
  biomonitor 175, 182–183, 197, 437, 447
  efflux 152–153, 165–166, 342–343
  toxicity 208, 219, 428, 431, 468

uptake 136, 138, 140, 177
  see also Corbicula, Corbula, Macoma, Mercenaria, Mya, Ruditapes, Scobicularia, tellinid
Clark Fork River 99–100, 108, 189–193, 198, 216, 256, 270–271, 283–284, 300, 305–307, 309, 312, 315, 318–323, 485
  Clark Fork Ecological Risk Assessment (USEPA) 315, 484–485, 487
clay 98, 101–103, 111, 330–331
Clean Air Act 373
Clean Air Mercury Rule 373
Clean Water Act 13, 15, 45, 55, 59, 109, 233
Clear Lake mercury mine 298
Clibanarius 148
Clutha River 69
Cnidaria 217, 228
  see also scyphozoan
coal 295, 329–330, 336, 348, 352, 370, 372–373
Coast Range 331
coastal waters
  dissolved concentrations 82, 83, 266, 280
  bimonitoring 169, 174, 175–187, 183–185, 200
  sediment 97–98, 104–106, 413
  speciation 87, 122
cobalt 9, 297
  assimilation efficiency 146
  biodynamics 180
  biomonitoring 176, 178, 180–182, 184
  efflux 153
  mining 311
  particles 57
  rock 97
  sediment 105, 311
  speciation 86, 115
  uptake 134, 140, 439
  water 73
coccolithophore 263
cockle 193, 214, 222
Coeur d'Alene 62, 99, 190, 284, 309–310, 314, 318–319, 323
Coleoptera 313
colloid **67**, 81, 123, 312
Colombia 379
Colorado 42, 284–285, 287, 316, 318
Colorado River 330, 332
Columbia River 306
Colville River 122
comet assay 211, 221

community 5–6, 28, 33, **232**, 241, 249, 263, 277
  community structure 228, 232, 268, 276, 278–281, 285–288, 290–291, 313, 316–320, 435, 458, 485
  community function 285, 291, 316
compensation 209, 265
complexation 38, 89, 115, 121, 131, 135, 252, 264, 370, 404–405
  inorganic 83, 85–86, 138
  organic 81, 83, 86–87, 91, 115, 122–123, 135, 145, 264, 266, 406–407, 419
concentration units
  volume **33**
  weight 33, 94, **127–128**
condition index 219, 272–273
conservative distribution **67**, 80, 85, 88
containment pond *see* tailings pond
contamination
  dissolved 71–72
  primary 62, **294**, 298–299
  secondary 62, **294**, 299, 303
  tertiary 62, **294**, 299, 303
  see also mining
content **127–128**
Continuous Concentration Criteria 15, 61, 362, 364, 390–391, 393–394, 399, 404, 408–409, 416, 418, 421, 460–461, 463, 472, 479, 486, 495
coot 331, 351
copepod 145, 147–148, 150, 153, 155, 157, 159, 205, 212, 215–216, 221–222, 225, 240, 254, 264, 288–289, 342, 380–381, 420, 422, 465
  see also calanoid, cyclopoid, harpacticoid
copper 4, 7, 9–11, 18, 25, 37, 153
  assimilation efficiency 145, 150, 165, 448
  bioaccumulation 45, 154–156, 158, 160, 172, 191–192, 195, 197–199, 275, 283, 313, 315, 322, 437, 439, 448, 476
  biogeochemical cycle 50–51, 56–59, 88, 122, 310
  biomonitoring 176, 178–185, 187, 189–190, 192–193, 199–200, 473
  colloid 121
  detoxification 148, 156–157, 160, 212

environmental management 480, 495
  mining 51, 55, 59, 294, 296–298, 300–302, 304–305, 311, 322, 324, 329
  oxidation state 84
  rock 97–98, 309
  sediment 45, 95, 97, 102, 104–108, 112, 117, 120–121, 172, 190, 228–229, 278, 306, 308, 310–311, 313, 317, 437, 448, 453, 480
  speciation 86–87, 90–91, 115, 121, 264, 266, 406, 419
  stable isotopes
  tolerance 224–226, 229, 256, 261
  toxicity 211–212, 214–215, 217–221, 224, 226, 228–229, 246–251, 253–255, 257, 263–264, 266, 275, 278, 281–282, 290, 312, 314–317, 321, 355, 398–400, 408–409, 413–414, 417–418, 423, 472, 480, 485, 495
  uptake 132, 134, 146, 163, 165, 225, 280
  water 45, 69, 72–73, 76, 79, 87, 95, 112, 257, 264, 266, 280–281, 288, 303–304, 313, 315
coral 78
Corbicula 207–208, 342, 344
Corbula 136, 138, 140, 142, 144, 147, 149, 151–153, 197, 342–343, 446, 465, 467, 469–470
Coregonus 385
corixid bug 190
cormorant 349
Cornwall 72, 90, 105, 225, 228–229, 287–288, 294, 298–299, 308
Corophium 225, 467
Cossura 317
County Adit 298
crab 132, 139, 141, 214, 216, 222, 225, 255, 358, 472
  see also Carcinus, Dungeness crab
Crassostrea 140, 152–153, 177, 187, 198, 202–203, 213, 217, 250, 341, 356, 361–362, 417
crayfish 279
criterion continuous concentration **394**, 412, 482
criterion maximum concentration **394**
critical body concentration 467–468, 471, 474
critical load model 62, 64

crustacean
  bioaccumulation 129, 153–154, 379
  bioaccumulation patterns 156–159, 160
  see amphipod, barnacle, brine shrimp, cladoceran, copepod, decapod, malacostracan
cumulative criterion unit (CCU) 283, 286, 288
cut bank **294**
cyanobacteria 263–264, 266
cycle **47**
cyclopoid copepod 240
Cyprinus 348–349, 474
cysteine 130, 146, 157, 212, 377, 382–383
cytoplasm 380, 464, 465
cytosol 74, 192, 227, 380, 472

dab 217
  see also Limanda
Dangerous Substance Directive 327
Danube River 69
Daphnia 150, 165, 209, 240, 246–248, 261, 288, 290, 381, 402, 408–409, 415, 421, 435, 491
daphnid 211, 220, 243, 246, 408, 410
  see also Ceriodaphnia, Daphnia
Dart River 105
Darwin 6
DDD *see* DDT
DDE *see* DDT
DDT 9, 367
  metabolites (DDD, DDE)
decapod crustacean
  assimilation efficiency 142, 149
  bioaccumulation 153–156, 158, 195
  toxicity 224, 428, 472
  uptake 135–136, 139, 141, 156
  see also caridean, crab, crayfish, grass shrimp, hermit crab, *Macrobrachium*, Norway lobster, *Palaemon, Palaemonetes, Pandalus*, penaeid, prawn, shrimp
Dee 72, 82
deep sea clay 97–98
deer 4
Deer Lodge Valley 4, 321
deficiency 10–11, 25, 28, 39, 132, 153, 157, 204–208, 233, 253, 396, 405, 413, 467, 491
Delaware 83, 104
Deleatidium 411
Delphinapterus 387
demethylation 374, 383, 387

deposit feeding 114, 140, 143, 150, 171, 174, 177, 180, 184, 193, 198, 269, 317, 378, 426–427, 438–439, 447
depuration 193–194
desorption **127**, 150
DET 123
detoxification 10–11, 28, **127**, 133, 148, 153, 205–208, 212, 215, 225, 265, 283, 347, 349, 396, 405, 466, 468, 473, 491
  detoxified fraction 153, 155, 157–158, 160, 225, 283, 466
detritivore 185, 191
detritus 74, **93**–94, 111, 122, 174, 177, 337, 341, 426–427
development stage 248–251, 428, 435, 472
Devon 105, 230, 299, 308
DGT (diffusive gradient in thin film) 91, 123, 280, 490
diatom 74, 146–148, 177, 230, 245, 254, 263, 266, 341, 433, 465
  see also Phaeodactylum, Skeletonema, Thalassiosira
dibutyl tin 356, 358
  see also tin organic form
dietary metal guideline (DMG) **460**, 463, 465–466, 475, 488
digestion 132, 145–146, 147, 150, 261, 439
digestive gland (mollusc) 213, 215, 217, 219, 222–224
dinoflagellate 147, 150, 263–264, 288
  see also Prorocentrum
Diporeia 193
discrimination analysis 201
dispersion train 99, 107, **294**, 304, 306
dissimilatory reduction **328**, 339–340
dissolved metal **67**, 94
dissolved metal concentrations 36, 70
  determination 78, 91
  coastal waters 83
  estuaries 83, 122
  lakes 80
  oceans 73–77
  rivers 77–80
dissolved organic carbon (DOC) 58, 122, 135, 240, 289, 408, 413
dissolved organic matter (DOM) 39, 111, 121, 123, 163, 253, 257, 264, 370, 384, 406–407, 495
distribution coefficient 121
diversity index *see* univariate measures
DNA 210–211, 214

dogwhelk 360, 362–363, 365
  see also Nucella
Domestic Substances List 18
Donana Park 309
dose **232**
  dose–response curve 235–237, 239, 274, 349, 402, 428, 443
Dreissena 193, 215, 342
Drunella 134, 282, 319
duck 331
  see also mallard duck
Dulas Bay 72, 154, 160
Dunedin 154
Dungeness crab 343, 467, 471
  see also Cancer
dust 48–49, 55, 62, 78
dynamic stability 270–271, 273
dytiscid beetle 190

eared grebe 351
Earth Day 5, 13
earthworm 256, 315
  see also Lumbriculus
East Fork Poplar Creek 384
Ebro River 76
echinoderm 142, 227, 379, 400
  see also brittle star, sea urchin, starfish
EC_{25} 244, 263
EC_{50} **232**, 236, 246, 249, 251, 349
Echinogammarus 139, 159
ecoepidemiology **261**, 267
ecological risk assessment 402
ecology 6, 25, 41
ecosystem 5, 39, 46, 260
  ecosystem services 285, 287
  ecosystem simplification 276, 279, 289
ECOTOX database (USEPA) 235
ecotoxicology **1**, 5–7, 26, 29, 32, 34, 170, 230, 413
  field evidence 260–292
Ecuador 379
EDTA 84, 135
efflux 150–153, 206–207, 283
  efflux rate constant **127**, 151–153, 161, 174, 183, 283
egg 262
eggshell 3
El Salvador mine 280
Elbe River 108
Elizabeth River 97, 121, 264
Elk River 352
Elminius 132, 139, 157, 159, 200
Elodea 165
elutriate test 452

endocrine disruptor 360, 364, 368
endpoint **33**, **232**, 236, 241–243, 255, 262, 428
  assessment endpoint 232, 241–242
  measurement endpoint 232, 241–242
energy reserves 196, 219, 273, 275
England 81–82, 154, 180, 183, 185, 222, 228–230, 285, 365, 430, 447
  southwest 41, 107, 109, 117, 183–184, 287, 294, 298–299, 318, 323, 356, 363, 365, 367, 380, 437, 448
  see also Cornwall, Devon
English Channel 82
enrichment factor 103
Enteromorpha 175, 186, 203
Environmental Defense 397–398
environmental geochemistry 32, 36–39, 42, 108, 432
environmental goals 23
Environmental Monitoring and Assessment Program (EMAP-E) 230
Environmental Protection Agency (USEPA) 18–19, 44, 51, 59–60, 230, 235, 246, 271, 283, 302, 315, 347, 349, 355, 361–362, 364, 383, 388, 390, 397–398, 401–402, 404, 409, 411, 416–417, 420–421, 451, 474, 482
environmental quality objective (EQO) **14**, 27–28, 30
  see also water quality objective
environmental quality standard (EQS) **14**–15, 18, 26, 28–29, 59
  see also water quality standard
environmental standard **394**, 416, 463
environmental toxicology
  see ecotoxicology
Environment Canada 18, 419
enzyme inhibition 214
Eohaustorius 247, 359, 429, 435
Epeorus 282, 319–320
Ephemerella 283
ephemerellid mayfly 283
  see also Serratella, Timpanoga
Ephemeroptera *see* mayfly
equilibrium constant 83, 89, 115, 408
equilibrium model 89
equilibrium partitioning 357, 359, 436
Esox 350, 380, 382, 385, 389
essentiality 8–9, 11, 48, 153, 204, 206, 346, 397, 413, 415
estuary 80
  biogeochemical cycles 48, 53, 88

estuary (cont.)
  biomonitoring 175, 180, 183–185,
    437
  dissolved concentrations 71, 76, 82,
    83, 112, 170, 264, 356, 365
  mining 309
  sediment 97–98, 101, 105–108, 112,
    117, 298, 437, 447
  speciation 87–88
European Commission 19, 327, 398, 461
European Court of Justice 22
European Food Safety Authority 20
European Union (EU) 16–17, 19, 22, 29,
    46, 59, 61, 78, 364, 373, 393, 415
excretion *see* efflux
Exe River 230
experimentalist approach 40
exposure **233**, 269, 276
  exposure assessment **14**, 402

facilitated diffusion **127**, 130,
    132–133, 250
faecal pellet 74
Fal 105, 108, 363
Far East 186
fathead minnow 211, 289, 400,
    405–406, 408–409, 474
  see also Pimephales
fecundity *see* reproductive rate
feeding rate *see* ingestion rate
fertilisation 227, 400
Ficopomatus 290
final acute value (FAV) 402, 409
final chronic value (FCV) 402
Finland 385
Firth of Clyde 149, 154–155, 160,
    215, 230, 464, 472
fish 5, 192, 215
  assimilation efficiency 142–143,
    166, 341
  bioaccumulation 64, 166–167, 189,
    207, 262, 321, 341, 358, 369,
    386, 470
  biomarker 217, 241, 256, 315,
    320, 348
  biomonitor 187, 189
  efflux 342
  fish kill 2, 302, 311–312, 319
  liver 256, 262, 320, 322, 348,
    464, 470, 474
  muscle 470, 474
  toxicity 216–218, 224, 230, 241,
    246, 255–256, 262–264, 267,
    279, 319, 332, 341, 347–348,
    412, 415, 461, 474
  uptake 132, 141

Fistulobalanus 186
flagellate 177
flagfish 411
  see also Jordanella
Flambeau Mine 324
Flanders 416, 451–453
flatworm 227, 280
  see also Stylochus, turbellarian
floodplain 79, 109, 122, 300, 305,
    307–308, 310, 312, 319, 322
Florida 230, 383
fluoride 85
flux 85
  see also biogeochemical cycle
Fly River 305, 309, 322
food chain 4, 345, 385
food web 3–4, 41–42, 48, 62, 94, 166,
    328, 344–345, 350–351, 380,
    382–383, 386, 486
footprint **294**, 296, 303, 306
foraminiferan 280
Fording River 352
Foundry Cove 149, 225
Fowey River 105
Framework for Ecological Risk
    Assessment (USEPA) 409
France 76, 79, 88, 177, 180, 202, 208,
    213, 219, 356, 361–362, 364–365,
    367, 416, 452–453
Free Ion Activity Model (FIAM) 131,
    134–137, 146, 167, 170, 253,
    404–405
free metal ion 39, 83, 85, 87, 90–91,
    131, 133, 135–138, 141, 165, 189,
    252–254, 263–264, 266, 405
Fremont Glacier 371
fucoid 175
  see also Ascophyllum, bladder wrack,
    Fucus
Fucus 175–176, 180, 183, 185, 196,
    225–226, 358
fugacity **1**, 11
Fujian Province 199
fulvic acid **67**, 91, 116
functional ecology **261**, 277–278,
    285, 426, 441, 444, 457
Fundulus 398

gallium 9, 439
Galveston Bay 83
Gambusia 225, 251, 331, 348, 382, 388
Gammarus 193, 201, 220–221, 224,
    226–227, 285, 412
Ganges 4
Gannel River 105
Garroch Head 230

gasoline *see* petrol
Gasterosteus 398
gastropod mollusc
  assimilation efficiency 142, 144,
    148–149
  bioaccumulation 132, 148, 152,
    166, 185, 343, 379
  biomarker 213
  biomonitor 185
  toxicity 218, 227, 268, 360, 367
  see also abalone, *Austrocochlea*,
    Bembicium, Biomphalaria,
    limpet, littorinid winkle,
    neogastropod, *Nerita*, snail
gene frequency 225–226
genome **204**, 212
genomics **205**
genotoxin 210, 217, 221
genotype 225
geochemistry 6, 25–26, 32–**33**, 36–37,
    39, 41, 370, 403
  see also environmental
    geochemistry
geophysics 26
germanium 7, 297
Germany 4, 78, 108, 192, 363–364,
    401, 416, 450, 452–453
giant petrel 474
  see also Macronectes
Gironde 213
Global Mining Initiative 2002 295
glutamic acid 383
glutathione 156, 214, 382–383
glutathione peroxidase 10, 346
Glycera 228
glycine 383
glycogen 196, 219, 273
Gobi Desert 49
gold 7, 9–10, 73, 295, 297, 300, 302,
    322, 324, 329, 371, 379, 495
grain size *see* granulometry
Grand Coulee Dam 310
Grand Junction 332
granule 10–11, 148–149, 155, 157,
    160, 225, 265, 465
  see also detoxification, lysosome
    residual body
granulometry 98–103, 171, 280,
    426, 429
grass shrimp 224, 251
  see also Palaemonetes
Great Lakes 56, 76, 369, 419,
    421, 476
  see also Lake Erie, Lake Huron,
    Lake Michigan, Lake Ontario,
    Lake Superior

Great Marsh, Delaware 335, 339
grebe 331
    see also eared grebe
Greece 230
green alga 175, 186–187, 402
    see also Cladophora, Enteromorpha,
        Scenedesmus, Ulva
green lipped mussel see Perna
Grizzly Bay 195, 345, 469
groundwater 119–120, 305
growth dilution 195
growth rate 161, 219, 244–245, 256,
    263, 315, 431, 472, 474
Guadiamar River 190, 309, 313–314
Guangdong 187
Gulf of Cadiz 304
Gulf of California 332
Gulf of Gdansk 163, 177, 185,
    200–201
Gulf of Papua 305, 309, 322
Gulf of Trieste 378
Gulf Stream 54
gut passage time (GPT) **127**, 147
gut residence time (GRT) **127**,
    146, 439

haematite 297
Haematopus 475
haemocyanin 10, 153–156, 160
haemoglobin 10, 401
Haliotis 249–250
Halopedium 240, 288, 290
Hamble Estuary 365
Hamilton Harbour 38
hardness 404–405, 408, 420, 492
harpacticoid copepod 280, 318
Hayle River 105
hazard **14**, 18, 35, 59, **233**, 327,
    393–394
    hazard characterization 402
    hazard identification **394**–395
    hazard quotient 402, 485
    hazard ranking 397–399, 483
Hazardous Waste Minimization
    Prioritization Program
    (WMPT) 18
health status 210, 221–224
heart rate 220
    see also bradycardia, tachycardia
heat shock protein see stress protein
heavy metal 1, 7, 9, 25
Helford River 105
Heliaster 281
Helisoma 443
hepatopancreas (decapod crustacean)
    160

heptageniid mayfly 282–284, 286,
    473, 476
    see also Drunella, Epeorus, Rhythrogena
herbivore 174, 185, 285
hermit crab 148
    see also Clibanarius, Pagurus
Heteromastus 446
Heteroptera 313
heterozygosity 225
Hexagenia 194, 244
hierarchy of toxic effects,
    208–209, 212, 219, 221, 226–227,
    230, 243
    see also biomarker
Himantopus 274, 344, 349,
    474–475, 481
Hippocrates 368
histidine 130, 146
holistic management 32, 40, 44,
    420, 422–423, **460**, 477–478,
    482, 505
Hong Kong 154, 162–163, 186–187,
    197, 200–201, 203, 453
horses 4, 311, 346
hotspot 107–109, 124, 269, 278, 367,
    450, 490
Hudson River 72, 82, 102, 149,
    225, 261
Hudson-Raritan Bay 81–82, 109
Humber estuary 72, 82, 222
humic substance (acid) **67**, 86, 91,
    111, 447, 461
Hyalella 193, 244, 262, 421, 428–429,
    438, 440, 471
hybrid distribution 74
hydrochloric acid 96, 117, 183, 437,
    447, 449, 456
hydrodynamic modelling 64
hydrofluoric acid 95–96
hydroid 217–219, 227
    see also Campanularia
Hydroides 290
hydrophobic **354**
Hydropsyche 152, 175, 190–194,
    283–284, 313, 319–320,
    473, 476
hydropsychid caddisfly 203, 473
    see also Arctopsyche, Hydropsyche
hydrothermal source 74
hydroxide 38, 85, 382
hyporheic zone 109, 271, 305, 307

Iberian pyrite belt 304
ice core 371, 376
Idaho 62, 284, 299, 309–310, 318
immunotoxicity 217, 364

imposex 218, 268, **354**, 359–360
    see also TBT
India 295, 372
Indian Ocean 203
indium 9, 367
Indrija Mine 371
IndoPacific 186–187, 198
Industrial Revolution 369
infauna 114, 171, 184, 269, 278, 426
ingestion rate 142, 150, 220, 223–224,
    244–245, 256, 261, 263, 315, 438
inherent variability 197
integrative criteria 460, 463,
    477–478, 483
International Maritime Organisation
    (IMO) 364
interstitial water see sediment
    pore water
invasive species 366–368
iodine 7
ionic regulation 139, 214
Iraq 369
iridium 7, 9
Irish Sea 82, 221
iron 7, 9–10, 48
    biogeochemical cycle 49, 58, 122
    biomonitoring 185, 192–193, 199–200
    colloid 121
    detoxification 156, 158
    mining 302, 311
    oxidation state 84, 113
    oxide 39, 81, 98, 101, 110–117, 119,
        121–123, 132, 180, 191, 302,
        304, 314, 337, 426–427,
        432–434, 447
    rock 97, 121
    sediment 96–97, 101–103, 105, 107,
        111–112, 120, 183
    speciation 86–87, 115
    sulphide 113, 117, 297, 301, 314
    water 72–73, 112, 303
Iron Mountain Mine 302
iron sulphate oxidising bacteria 301
    see also Leptospirillium, Thiobacillus
irrigation 330–332, 335
Irving–Williams series 115
Island Copper Mine 317
Isle of Cumbrae 365
    see also Millport
Isochrysis 448
itai-itai disease 4
Italy 211, 214–216, 221, 225, 444, 452

Japan 4–5, 65–66, 369
Jervis Bay 343
Jordanella 411

Kara Sea 74, 76
kelp 150, 281, 318
    see also Laminaria, Lessonia
Kesterson Reservoir/National Wildlife
    Refuge 331, 333, 335, 338–339,
    344, 351, 461, 481
keystone species 279, 281, 360, 363
killifish 398
    see also Fundulus
Koch's postulates *see* ecoepidemiology
Korea 372
Kuparak River 58, 122

labile metal 90–91
Lagopus 42
Lahonten Reservoir 378
lake
    biogeochemical cycle 47, 51, 64
    biomonitoring 187–189
    contamination 188, 247, 263, 288,
        309, 311, 350, 380
    dissolved concentration 76
    ecotoxicological effects 287–290
    experimental lakes 240, 279, 350, 375
    sediment 97–98, 109, 116, 123,
        311, 376
    speciation 72, 90, 112
Lake Erie 2, 378
lake herring 385
    see also Coregonus
Lake Huron 38
Lake Michigan 56, 209
Lake Moira 63
Lake Oltertjarn 346, 350
Lake Ontario 38, 378, 388
Lake Superior 80
Laminaria 150
lanthanides 7, 9
largemouth bass 211, 348
    see also Micropterus
larval stage *see* development stage
lateral management 2, 24, 28, 43,
    259, 292, 355, 392, 425, 457,
    476, 478, 489, 505
lateral thinking **1**–2, 5–6, 26, 327,
    414, 424, 460, 481
Lavaca Bay 377
LC_{10} 236–237, 246
LC_{50} **233**, 236–237, 241, 246, 248–251,
    400, 417, 471
lead 5–7, 9–11, 18, 20, 22, 25, 30,
    36–38, 51–52, 108
    bioaccumulation 158, 165, 172,
        284, 315, 322, 437–438
    biogeochemical cycle 50–59, 62, 65,
        74, 78, 88, 122, 310

biomonitoring 176, 178–185,
    199–200, 319, 447
organic forms 88
    see also tetra-ethyl Pb
mining 192, 296–297, 301, 311
rock 97
sediment 56, 59, 95, 97, 102,
    104–106, 117, 120, 172, 183,
    228–229, 310–311, 437, 447,
    450, 452–453
speciation 86–87, 115, 406
stable isotopes 52, 305, 450
toxicity 250, 253, 255, 311, 355,
    398–399, 417–418
water 56, 62, 72–73, 76, 78, 95, 112
Lemanea 187
Lepomis 267, 348, 474–475
Leptocheirus 441
Leptophlebia 261
Leptospirillium 301–302
lethal **33**, 35, 232, 235, 243–244
Lewis acid 8–9
    see also metal ions
life cycle analysis 20, 26, 51
ligand **1**, 8, 10–11, 39, **68**, 83, 85, 89,
    91, 95, 111, 115–116, 131, 135,
    252, 408
Limanda 217
lime 280, 302–303, 324
limestone 97
Limnesia 438
Limnodrilus 149, 225, 465
limpet 185
    see also Patella
Linnaeus 6, 41
lipid peroxidation 214–216, 256,
    263, 320
lipofuscin **205**, 215–216, 221–222, 268
lithium 7
Littorina 185, 213, 358
littorinid winkle 185, 192–193
Liverpool Bay 82
load **47**
Loch Ewe 338
Long Island Sound 106, 341
Looe, East 105
Looe, West 105
Loligo 464
Loripes 185
Los Angeles 107
Los Frailes 309, 313–314, 323
Lot River 208
Loughor River 105
lowest observed effect concentration
    (LOEC) 227, **233**, 236–237, **394**,
    409, 417

lucinid bivalve 184
    see also Loripes, Lucinoma
Lucinoma 185
lugworm *see Arenicola*
Lumbriculus 194, 256, 315, 443
Lumbrinereis 317
Luscar Creek 352
Lutjanus 341–342
Lymington 365
Lynher River 190
lysosome 145, 147–148, 150, 153,
    155, 157, 159, **205**, 212, 215–216,
    221–222, 225, 240, 254, 264,
    288–289, 342, 380–381, 420,
    422, 465
    intracellular digestion 145
    membrane stability 215, 221–222,
        224
    residual body 159–160, 216
    see also lipofuscin
Lytechinus 428

Macoma 45, 138, 140–141, 143–144,
    146–149, 152–153, 177, 180, 182,
    195–196, 202, 217, 219, 224, 273,
    275, 278, 341, 358, 437–439, 441,
    446–448, 469–470
Macomona 224
Macrobrachium 322
Macronectes 474
macrophytic alga 86, 174–176, 180,
    184–187, 196, 203, 226, 314, 464
    see also brown alga, green alga, kelp,
        red alga
mad cow disease 497
mad hatter 369
magnesium 7, 9, 73, 87, 115, 121,
    148, 155
major ion 9, 132, 134, 139, 141,
    253, 405–406
    channel **127**, 147
Makaira 389
malacostracan crustacean
    153–154, 197
    see also amphipod, decapod, mysid
malaria 3, 23, 367
mallard duck 351, 388–389, 474–475
    see also Anas
malondialdehyde 214
mammal 349, 387, 477
manganese 9–10, 48
    bioaccumulation 127, 129–130
    biomonitoring 200
    detoxification 148, 155
    mining 316
    oxidation state 84, 113

oxide 39, 98, 101, 110–115, 117, 122–123, 312, 426–427, 432, 434
rock 97–98, 109, 116, 123, 311, 376
sediment 84, 97, 105, 224
speciation 86, 115, 224
toxicity 224, 312
water 72–73, 79
mangrove 186–187
mangrove snapper 341–342
see also Lutjanus
mariculture 187, 199, 203
marlin 389
see also Makaira
Martin's Lake 461, 482
Maryland 377
Massachusetts 266, 375
mass balance (budget) **47**, 94, 330, 334, 495
mass media models 59, 62–64
Maximum Contaminant Levels 15
maximum acceptable toxic concentration (MATC) **233**, 236
maximum achievable control technology (MACT) 373
mayfly 134, 138–139, 165, 188, 199, 201, 283, 476
biomonitor 192, 194
toxicity 244–245, 247, 261, 263, 282–284, 286–287, 319–320, 411, 472–473
see also Baetis, Deleatidium, ephemerellid mayfly, heptageniid mayfly, Hexagenia, Leptophlebia, Siphlonuris
McLeod River 352
Mecklenberg Bay 450
Mediterranean Sea 48–49, 249, 304
meiofauna 93, 112–113, 229–230, 240, 251, 279–281, 310, 318, 426–428, 433
Melanitta 486
Menai Strait 154
Menidia 143
Mercenaria 341, 358, 428, 431
Mersey 72, 82, 121
mercury 4–5, 7, 9–11, 18, 50, 53, 164, 368
atmosphere 369–373, 377
bioaccumulation 64, 369, 371, 374, 376–377, 379–380
biodynamics 381
biogeochemical cycle 49, 51, 53–54, 64, 108, 369, 370–371, 373, 375, 377, 391

biomonitoring 179–180, 184–185, 200
detoxification 122, 157, 212, 374, 387
environmental management 6, 61, 393, 476, 495
mining 271, 297–298, 371–372, 377–379, 495
organic forms 132, 374
see also methyl mercury
rock 97
sediment 64, 97, 105, 111, 271, 310, 376, 378–379, 384, 452–453
selenium 349, 387
sources 369–374
speciation 86–87, 370, 384
toxicity 132, 250, 253–254, 262, 389, 398–399, 417–418, 485
water 72–73, 376–378
mesocosm 238, 240, 247, 258, 260, 383, 410, 420, 430, 484, 491
metabolically available metal fraction (MAF) **127**, 153, 155, 157–158, 160, 206–207, 468, 470, 472–473
metabolic requirement 155–156, 160, 205, 346, 415
metabolism 8, 11
metabolite **205**, 212
metabolome **205**, 212
metabolomics **205**
metal **2**, 7
metal-activated protein 10
metal ions 8–9
class A 8–9
class B 8–9
borderline 8–9, 11
metalloid 2, 7, 9, 327
metalloprotein 10
metallothionein (MT) 10, **127**, 141, 151, 156–157, 159–160, **205**, 208, 212–213, 215, 221–222, 225, 227, 244, 265, 321, 472
induction 157, 212–213, 227, 278
turnover 157, 212, 216
metallothionein-like protein (MTLP) **127**, 151, 208, 213, 472
metal rich granules (MRG) 148–149, 151, 465–466
see also granules
metal sensitive fraction (MSF) 472
metal-sensitive species 277, 291, 316, 318, 411, 473
see also cladoceran toxicity, ephemerellid mayfly, heptageniid mayfly

methionine 338, 382–383
methylation **354**
methyl mercury 2, 4, 41, 88, 132, 164, 253, **354**, 368–392
bioaccumulation 377–382, 383, 385–386
biodynamics 381
concentrations 375–378, 387
cysteine 382–383, 387–388, 391
formation 374–375, 384
risk management 388
uptake 377, 380–383
toxicity 132, 350, 377, 386–388, 389–390, 426, 473
trophic transfer 378, 382–385, 386, 391, 465
methyl mercury criterion 390
Meuse River 104, 118
Mexico 331–332
Meyenaster 281
Miami 49
microalgae 448
see also phytoplankton
microarray 212
microbes 48, 85, 88, 112–113, 120, 151, 184, 301, 339, 342, 357, 369–370, 374, 402, 426, 433
see also bacteria
microbial biogeochemistry **328**
micronucleus **205**, 214, 221
Micropsectra 218
Micropterus 211, 348
Middle East 49, 330
Middle Lake 288, 311
midge 188, 192, 246
biomonitor 190, 194
toxicity 217–218, 227–228, 244, 248, 428, 431, 435
see also chironomid
Millport 154
Milltown reservoir 299, 309
Minamata disease 4, 34, 73, 217, 369
Mindanao 379
mining 4–5, 51, 55–56, 72, 79, 105, 107, **294**
ecological risk 296, 311–321
estuary contamination 309
lake contamination 309
marine contamination 280, 316–317
primary contamination 298–302, 306
river contamination 108, 303–308, 305–306, 318, **318**–321

mining (cont.)
  stream contamination 190–192,
    271, 318–321
  sustainable mining 295, 496
Minnesota 457
Mississippi River 38, 76, 95
mite 438
  see also Limnesia
molybdenum 9, 132, 198, 297, 301
  biogeochemical cycle 58
  rock 97
  speciation 84
  water 73
monitoring 169
  biota see biomonitor
  sediment 171
  water 170
Monodon 387
Monodonta 149, 471
Montana 4, 59, 79, 189, 191, 193, 198,
    216, 270–271, 283–284, 287, 293,
    297, 299–301, 304, 307–308, 311,
    318–319, 336, 339–340
Montreal 58
Morone 143, 342, 389, 421, 471, 474
Morula 343
mosquito 3
mosquitofish 225, 331, 348, 382, 388
  see also Gambusia
motor vehicle 51–52, 55–56, 78,
    88, 108
mudsnail 362
  see also Nassarius
multidimensional scaling (MDS) 201,
    222–223, 230
multivariate measures 223, 229,
    280, 286, 435
mussel
  assimilation efficiency 143–144,
    146–148, 151, 341
  bioaccumulation 153, 159, 161,
    163, 177, 197–198, 201,
    212–213, 381, 447, 468
  biodynamics 162, 177, 341–342
  biomarker 211–217, 221–222, 224
  biomonitor 175, 177–178, 180,
    183, 186–187, 193, 196,
    199–202
  efflux 152–153
  toxicity 211–212, 214, 219–221,
    249–250, 361, 405, 413
  uptake 132, 140, 341, 350
  see also Dreissena, Mytilidae, Mytilus,
    Perna, Septifer
Mussel Watch 174, 177, 180, 199–200
Mya 358

mysid 220, 246, 342, 385, 409, 412,
    417, 421
  see also Mysidopsis, Mysis, Neomysis
Mysidopsis 409, 417
Mysis 385
Mystacides 438
Mytilus 132, 140, 143–144, 146–147,
    153, 161, 163, 175, 177–178,
    193, 196–203, 211–217, 219–223,
    225, 249–250, 341–342, 350,
    361, 381, 405, 447
Mytilidae 177, 287

nanoparticles 68
nanotechnology 5, 18, 20
Narragansett Bay 83, 221
narwhal 387
  see also Monodon
NASQUAN Program (US Geological
    Survey) 38
Nassarius 362
National Academies of Sciences (USA)
    61, 390, 441
National Emissions Inventory NEI
    (USEPA) 47, 51, 64
National Irrigation Drainage
    Study 345
National Marine Fisheries Service 482
National Oceanic and Atmospheric
    Administration (NOAA) 230, 451
National Pollutant Discharge
    Elimination System (NPDES) 14
National Research Council
    (National Academies of
    Sciences, USA) 61, 121, 264,
    377, 390
National Status and Trends
    Monitoring Program 104, 118
National Water Quality Assessment
    Program (US Geological Survey)
    101, 271
natural history 39–40, 42–43, 278
Neanthes 409, 428
Neath 105
nematode 229–230
neogastropod mollusc 358, 361, 465
  see also dogwhelk, mudsnail, whelk
Neomysis 342
neoplasia 217
Neosilurus 322
Nephrops 224
nephtyid polychaete 184
  see also Nephtys
Nephtys 225
nereid polychaete 318
  see also Neanthes, Nereis

Nereis
  bioaccumulation 149, 184,
    225–226, 229, 342, 358, 437,
    439, 447–448, 466
  biomonitor 183–184, 194, 318, 437
Nerita 343
Netherlands 118, 219, 330, 336, 401,
    435, 444, 451, 453, 475
neutral red 216
New Chemicals Pre-manufacture
    Notification Program 18
Newfoundland 198
Newport Bay 461
New York 102, 106, 149, 225
  see also Hudson, Long Island Sound,
    Raritan Bay
New Zealand 15–16, 69, 76, 154,
    160, 282, 356, 401, 411, 418,
    451, 453, 455
nickel 9, 11
  bioaccumulation 148, 165, 446
  biogeochemical cycle 50, 56, 58, 310
  biomonitoring 176, 178, 180–182,
    200
  colloid 69
  mining 55, 296, 311
  rock 96–98, 101–102, 103–104,
    122, 193
  sediment 58, 95–97, 105, 107, 117,
    271, 311, 443, 453
  speciation 86, 90, 115
  toxicity 253, 314, 398–399, 413,
    417–418, 443
  water 58, 72–73, 76, 95, 121, 288
Niphargus 193
nitrate-reducing bacteria 339
nitric acid 95–96, 105
nitrogen 8–10, 83, 130, 206
  oxides 55
  stable isotope ratio 166, 345,
    382, 386
noble metals 7
Norfolk 97, 121
normalization 101–102, 107, 122,
    171, 180, 449
no observed effect concentration
    (NOEC) **233**, 236–237, 246, 250,
    394, 409, 412, 415–416, 477
North Carolina 348, 461
northern pikeminnow 391
  see also Ptychocheirus
North Sea 217, 221, 368
Norway 212, 217, 219, 221, 228–229
Norway lobster 224
  see also Nephrops
Nova Scotia 338

NTA (nitrilotriacetic acid) 135, 254
Nucella 358–363, 365, 367
nuclear magnetic resonance (NMR) 212
nutrient 74

Ob River 88
oceans
    biogeochemical cycling 48, 50, 53,
        74, 370
    dissolved concentrations 73, 108,
        264, 350
    vertical distributions 74
octanol water partitioning 357
oil 2, 311, 329
    refinery effluent 333–336
OK Tedi 108, 305, 309, 321–322
oligochaete 149, 194, 224–225, 246,
    261, 443, 465
    see also earthworm, *Limnodrilus*
Oncorhynchus 141, 246, 256, 289, 315,
    319–320, 346, 351–352, 382, 411,
    421, 474
Ontario 63, 188, 263, 288–289, 311, 386
    see also Lake Ontario
Oonopsis 346
Ophiura 317
ophiuroid echinoderm 250, 317
    see also Ophiura
Ophryotrocha 249
Orchestia 135–137, 141, 150, 154, 185,
    198–199
ore 37, 55, 99, **294**, 296–298, 371
Oregon 391
organic carbon/matter 84, 98, 101,
    103, 110, 112, 115–117, 180
    see also dissolved organic carbon,
        particulate organic matter,
        total organic carbon
organic contaminants 5–6, 11, 19, 215,
    217, 356, 383, 396, 443, 460, 467
organochlorine 2–3, 11, 41, 164,
    218, 368
    see also PCB
organometal **2**, 7, 87, 131–132, **354**, 473
    see also methyl mercury, TBT,
        tetra-ethyl lead
organo-selenide *see* selenium
    organic form
Organotin Paint Control Act 356
Osemerus 385
Oslofjord 228
osmoregulation 139, 214
OSPAR Convention 22, 84–85, 119,
    132, 136
osprey 3
Ostrea 343

Ostreidae 177, 179–180, 183, 187,
    196, 198–200, 202
overlying water **93**, 174, 184
over-protection 403, 412, 416, 419,
    421–422, 460, 473, 476
oxygen 8, 84, 111, 113
oxyradical **205**, 211, 213–216
oyster
    assimilation efficiency 341
    bioaccumulation 8, 148, 152–153,
        177, 212, 343
    biomarker 213
    biomonitor 177, 179–180, 183, 187,
        196, 198–200, 202
    efflux 152–153
    toxicity 250, 356, 361, 365, 417
    uptake 134
    see also Crassostrea, Ostrea, Ostreidae,
        Saccostrea
oystercatcher 475
    see also Haematopus

Pacific Flyway 331
Pacific Ocean 48, 87, 180, 203, 331,
    345, 461
    north Pacific 75
    western Pacific 65
Pagurus 251
Palaemon 135–136, 139, 149, 156,
    158–160, 465
Palaemonetes 136, 139, 149, 156, 224,
    251, 428, 465–466, 472
palladium 7, 9–10
Pallanza 444
Palo Alto 45, 273, 275
Panama 364
Pandalus 139, 155–156, 160, 350
Papua New Guinea 108, 305, 322–323
Paracentrotus 249
Parrett River 105
particulate metal **68**, 81, 93–94, 112,
    121–122
particulate organic matter (POM) 111,
    122
partitioning *see* sediment
    partitioning, subcellular metal
    distribution
Patella 185, 224, 241, 256, 291
Pautaxant 377
PCB (polychlorinated biphenyl) 3, 383
penaeid 464
    see also Penaeus
Penaeus 464
Pend Oreille Lake 307, 309
Pennsylvania 405
peptide 328

perchloric acid 95
Percuil Estuary 229
peregrine falcon 2–3
Periodic Table 1–2, 7–9
Perca 187, 200, 289–290, 314, 350
perch 179–180, 203, 350
    see also yellow perch, *Perca*
periphyton 189, 245, 261, 284–285, 410
permeability 139–140, 195
    see also water permeability
Perna 135, 140, 143–144, 147–148,
    151–153, 162–163, 175, 177–178,
    186–187, 203, 214, 220–221, 447
persistence, bioaccumulation,
    toxicity (PBT) 18–19, 394–395
Perth 367
pesticide 2–3
    carbamate 'Ziram' 135–136
    see also organochlorine
Petricola 358
petrol 22, 52, 56, 62, 78, 88, 108
petroleum 330
pH 79, 87, 95, 112, 137, 146, 189,
    253, 279, 288–289, 301–302,
    311–313, 405
Phaeodactylum 143
phantom midge 175, 188, 190, 203,
    288, 290
    see also Chaoborus
Phasianus 388
pheasant 389
    see also ring necked pheasant
Philippines 379
    see also Mindanao
phosphate 51, 132, 295, 329
    dissolved 75
    granule 11, 148, 155, 158
phosphorus 74
phoxocephalid amphipod 247
    see also Rhepoxynius
Physa 410
physicochemistry *see* uptake
    physicochemistry, *see also*
    speciation
physiological stress indicator
    see biomarker
phytoplankton 4, 10, 48, 74, 81,
    85–86, 88, 143, 147, 150–151,
    153, 174, 177, 246, 251, 254,
    263–264, 266, 279, 288–289,
    337–338, 341, 380, 385,
    420–422, 448, 464–465, 480
    see also coccolithophore,
        cyanobacteria, diatom,
        dinoflagellate, flagellate,
        Isochrysis

pike 350, 380, 382, 385, 389
    see also Esox
Pileolaria 290
Pimephales 211, 289, 407, 409, 474
Pista 317
plant 337–338, 346, 413
    see also Astragalus, Oonopsis, Stanleya,
        willow, Xylorhiza
platinum 7, 9–10
Platorchestia 187–188
Plecoptera see stonefly
Plectrocnemia 194, 284, 287
pluteus larva (echinoderm) 251
Plymouth Harbour 298
Pogonichthys 265, 397, 471, 486
Poland 177, 192–193, 199, 201
    see also Gulf of Gdansk, Upper Silesia
policy 1, 489
policy community 32, 44, 489–490,
    499–500
policy–science interaction
    see science–policy interaction
pollution induced community
    tolerance (PICT) 229, 284
polyaromatic hydrocarbon (PAH) 11,
    222
polychaete 145–146, 149, 163,
    183–184, 193, 198, 225, 227–229,
    249–250, 290, 317–318, 342, 359,
    379, 381, 409, 412, 428, 433, 437,
    439, 446–448, 466
    see also ampharetid, arenicolid,
        Armandia, cirratulid, Cossura,
        Glycera, Heteromastus,
        Lumbrinereis, nephtyid, nereid,
        Ophyrotrocha, Pista, serpulid,
        spionid, terebellid
Polydora 228
polyp 217–218, 227–228, 244, 248,
    428, 431, 435
Pontoporeia 252
population 5–6, 28, **33**, 41, 226, **233**,
    243, 245, 248–249, 251–252, 263,
    265, 351
    age structure 226–227, 251, 265
    growth rate 226–227, 251
pore water see sediment pore water
porphyry 296, 298, 301, 322
Port Philip Bay 290
Portugal 304
Potamocorbula see Corbula
potassium 7, 9, 20, 86, 130, 132
power model 195
prawn 139, 322, 464
    see also caridean, decapod crustacean,
        Macrobrachium, penaeid

precautionary **394**, 400, 404, 453,
    462, 473
Precautionary Principle **14**, 21–22,
    493
precious metals 7
predicted environmental
    concentration (PEC) 394,
    402–403, 415, **460**, 464–465,
    470, 477, 485, 488
predicted no effect concentration
    (PNEC) **394**, 402–403, 411, 415,
    460, 464, 466, 477, 485, 488
Prickly Pear Creek 79
principal components analysis (PCA)
    222–223
probit 236
productivity 48–50, 53–54, 63, 94–95,
    110, 114, 336
prokaryotes 10, 86
    see also bacteria
Prorocentrum 143, 150
proteome **205**, 212
proteomics **205**
protozoan 240
    see also ciliate
Pseudomonas 402
ptarmigan 41–42, 166, 319
    white-tailed 42
    see also Lagopus
Ptychocheirus 391
Pyganodon 227
Purdue University 398–399
Pusa 290
pyrite 295–296, 301–302, 304, 329
pyrophosphate 158, 465
Pyura 290

quantitative structure activity
    relationship (QSAR) 11
Quebec 58, 116, 123, 188–189, 311,
    444–445
Queensland 456

radical see oxyradical
radium 7
rainbow smelt 385
    see also Osemerus
rainbow trout see trout
    see also Oncorhynchus
Rallus 389
rare earth metals 7
    see also lanthanides
Raritan Bay 106
    see also Hudson
razorback sucker 332
    see also Xyrauchen

REACH 18–19, 30, 59, 415
recovery 278, 284, 289, 291, 309,
    313, 317–318, 323, 325,
    348, 365
Recurvirostris 349, 474
recruitment 226, 228, 249, 263, 265,
    351, 363, 431
recycled (nutrient-type) distribution 74
red alga 187
    see also Lemanea
Red List 22
redox 10, 84–85, 103, 111–113, 119
    redox interface 112–113, 122–123
    redox potential 83, 113
reductionist approach 24, 40
reference site 270
regeneration 74
relative penis size (RPS) index 360
remediation 297, 309, 323–324
reproduction 227, 243, 245, 248–249,
    261–263, 265, 272, 274, 329, 351,
    361, 363–364, 388–389, 422, 461,
    469–471, 473–474
reservoir 47
residence time 81
respiration rate 220
Restronguet Creek 72, 105, 112, 154,
    225–226, 228–229, 466
Rhepoxynius 247, 359, 429, 443
Rhine–Meuse delta 435
Rhine River 108
Rhode Island 221
rhodium 7, 9
Rhone River 76, 79, 87–88, 95
Rhyacophila 152, 191, 319
Rhythrogena 152, 282–283, 319
ring necked pheasant 388
    see also Phasianus
Rio Declaration 14, 21
Rio Odiel 304, 309
Rio Tinto 297, 304, 309
risk **14**, 124, **233**
    risk assessment **14**, 18–21, 25, 28,
        91, 208, 241, 258, 292, 311,
        316, 393, 402–403, 408,
        414–415, 460, 466, 477,
        483–484, 489
    risk management 1–2, 5, 12, 20, 43,
        56, 61, 258, 292, 327, 354, 362,
        388, 393, 400, 460, 466
    risk quotient (RQ) 402, 464
river
    biogeochemical cycles 47–48, 53
    colloidal metal 69
    contamination gradient 108,
        303–308, 305–306, 318–321

dissolved concentrations 72, 76, 79, 95, 122, 304
  mining 191, 303–306
  sediment 95, 100, 108–109, 191, 305–306, 310, 322
Rocky Mountains 4, 42, 282, 286–287, 319
Roe Island 469
Romans 294, 297, 299, 355
Rörvik 373
rotifer 288
Rouyn-Noranda 72, 90, 105, 225, 228–229, 287–288, 294, 298–299, 308
Royal Commission on Environmental Pollution 22
rubidium 7
Ruditapes 135, 140, 143–144, 147, 151–153, 447
Rupert Inlet 317
Russia 76, 88, 330

Saccostrea 140, 152–153, 187, 199, 203
Sacramento River 190, 302, 333–336
Sacramento splittail 265, 397, 471, 486
  see also Pogonichthys
Sagavanirtok River 58, 122
Sahara Desert 48–49
salinity
  gradient 88, 121
  metal uptake 136–139, 141, 163, 170, 421
  speciation 86
Salix 42
Salmo 189, 246, 256, 315, 319–320, 421
salmonid fish 246
  see also chinook salmon, trout
Salvelinus 224, 279, 316, 319, 352, 385, 388–389
sand 98, 101, 103
Sander 388–389
San Diego Bay 82, 121
sandstone 97
San Francisco Bay 2, 45, 58, 81–83, 91, 96–97, 102, 106–107, 109–110, 119, 166, 180, 196–197, 202, 262, 265, 269, 271–273, 275–276, 278, 330–331, 333–337, 339, 341, 343–345, 376–379, 389, 397, 405, 423, 433, 435, 447, 450, 461–462, 465, 467, 469, 471, 475, 480, 484, 487, 495
  see also Palo Alto, Suisun Bay
San Joaquin River/Valley 331–333, 335, 344, 461, 484

San Pablo Bay 469
Santa Monica Bay 107
Sargasso Sea 54, 62, 78
Sargassum 175, 186
Saskatchewan 330
scallop 251
  see also Argopecten
Scandinavia 177, 364, 384
  see also Finland, Norway, Sweden
scavenging 81, 85, 111, 113, 119, 123, 132, 301, 303, 312, 337, 339, 462
  scavenged distribution 74
Scenedesmus 402
Scheldt Estuary 330, 336
science 1, 489
science community 489–490, 499
science–policy interaction 489, 492, 494, 501
  advocacy 498
  dialogue 494–495, 500
  goals 494, 496, 499
  mechanics 499
  science agenda 482, 500–505
  trust 496
scope for growth (SFG) **205**, 220–222, 226
Scotland 154–155, 160, 230, 338, 365
Scrobicularia 177, 182–184, 193, 202, 225, 358, 367, 378, 380, 437, 447–448
scyphozoan 228
seasonal variability 272
sea urchin 249–251, 435
  see also Paracentrotus, Serechinus, Lytechinus
seal
  ringed seal 387
  see also Pusa
seaweed *see* macrophytic alga
sediment
  analysis 95–96, 98–104
  bioaccumulation 436–437, 436–440, 439–440, 455
  bioavailability 425, 436, 441–442, 446–447, 455, 459
  biogeochemical cycles 48–50, 53–54, 63, 94–95, 110, 114, 336
  biomonitoring 177–185, 183–184, 437
  coastal waters 98
  concentrations 94
  contamination 45, 96, 104–110, 124, 322, 425
  core 93, 96, 98, 124, 304, 310–311, 340, 376, 450

decomposition/extraction 95–96, 100, 116
distribution coefficient 95–96, 100, 116
estuaries 98, 122, 309
lakes 97–98, 309
management 425, 450–459
partitioning **93**–94, 111–114, 116, 121, 122, 238, 356, 375, 425, 433, 436, 438, 440, 443, 456–457, 490
pore water 93, 110, 113, 117, 122–123, 174, 184, 430, 433–436, 440, 442–443, 445–446
organometals 88
redox 84, 110–113, 119, 123, 425–426, 432–433, 442, 444, 446, 449
suspended sediment 57, 70, 96, 102, 120, 148, 177, 180, 305
toxicity 224, 227–229, 241, 247, 251–252, 278, 427–428, 431, 435, 443, 451
sediment quality 6, 44, 364, 425–459
  apparent effects threshold (AET) **426**, 454–455
  effects range low (ERL) **426**, 451, 454–455
  effects range median (ERM) **426**, 454–455
  probable effects level (PEL) **426**, 455
  sediment quality guideline 15, 451–454, 457
  sediment quality guideline quotient (SQGQ) 454–455
  sediment quality triad 175, 238, 276, 436, 457, 478, 484
  sediment quality value 455
  threshold effects level (TEL) **426**, 451, 454–455
selenate 337–341, 344, 462, 482
  see also selenium oxidation state, speciation
selenite 337–338, 341, 344, 348, 350, 482
  see also selenium oxidation state, speciation
selenium 4–5, 7, 9–10, 18, 29, 41, 107, 262
  assimilation efficiency 145–148, 341–342
  bioaccumulation 142, 152, 166, 247, 340–343, 341–348, 439, 467, 471–472, 474–475, 483, 486–487

selenium (cont.)
  bioavailability 337–338, 341
  biodynamics 162, 164, 341–343,
      486
  biogeochemical cycle 57, 74, 81,
      85, 310, 327, 329–333, 336,
      340, 487
  bird 344–345, 347, 349–350, 352
  biomonitoring 179–180, 203, 350
  detoxification 347, 349–350
  efflux 152–153, 165–166, 342–343
  egg 262, 344, 347–349, 351–352,
      473–475, 481
  environmental management 6,
      460, 462, 471, 475, 477, 479,
      482, 484, 487, 495
  fish 332, 341–342, 344–348, 350
  organic form 81, 85, 88, 336–341,
      482
  oxidation state 84, 253, 336–337
  mercury 349, 387
  mining 297
  partitioning 339–340
  sediment 335, 339–340, 348, 483
  speciation 81, 85–86, 336–337,
      339, 341
  toxicity 150, 218, 243–244,
      253–254, 262, 265, 267, 274,
      328–329, 331–332, 340–341,
      346–351, 352, 397–399,
      417–419, 461, 472–474, 479,
      481, 483
  trophic transfer 247, 341, 343–346,
      351, 397, 399, 465, 483, 487
  uptake 132, 134–136, 140, 163,
      165, 337–338, 340, 341–342,
      350, 397
  volatilisation 338, 340
  water 73, 330, 332–333, 333–336,
      345, 348, 461, 483
Selsey Bill 82
Se-methionine 341
Semibalanus 154
semimetal 7
Septifer 135, 140
sequential extraction 116
Serechinus 251
serpentine 97–98, 104–106, 413
serpulid polychaete 290
  see also Ficopomatus, Hydroides,
      Pileolaria
Serratella 283, 320
sewage 58, 102
  sludge 47–48, 51, 53, 230
shale 97, 329–330
Shasta Lake 302

sheep 311
shrimp 139, 350, 466, 472
  see also caridean, decapod
      crustacean, grass shrimp
Sialis 165
Sierra Nevada 379
Siphlonuris 134
silicate 75
silicon 7, 74
silt 98, 101, 481
silver 4, 7, 9–11, 18, 108
  assimilation efficiency 143–145,
      147–149, 151, 165, 438,
      448, 466
  bioaccumulation 45, 159, 198, 262,
      275, 437–439, 468–470, 473
  biodynamics 180
  biogeochemical cycle 58, 65, 74
  biomonitoring 176, 178–182,
      184–185, 200, 447
  detoxification 157, 212
  mining 297, 300, 302, 324, 329,
      371, 379
  rock 97
  sediment 45, 97, 102, 105–107, 110,
      117, 120, 272, 278, 310, 437,
      447, 453
  speciation 86–87, 121, 253, 423
  trophic transfer 166, 399
  toxicity 224, 227, 253, 255, 261,
      275, 278, 350, 399, 417–418,
      423, 468, 473
  uptake 131–132, 134, 136, 163,
      165, 342
  water 45, 73, 112, 333
Silver Bow Creek 270, 299, 306, 308,
      312, 319
simultaneously extracted metal (SEM)
      117–118, 122, 426, 442–446,
      451, 456
sipunculid 142–145, 163, 439
Sipunculus 439
site-specific assessment 16, 19,
      390–391, 400, 404, 423, 479, 495
site-specific objective (SSO) 479–480
Skeletonema 266
slickens 308
Slovenia 371
smelter 51, 59, 62, 90, 107, 188,
      263, 287–288, 295–297, 301,
      309, 311, 329, 450
snail 41, 149, 358, 410, 443
  see also Biomphalaria, Helisoma,
      littorinid winkles,
      neogastropods, Physa
socio-economic consequences 321

sodium 7, 9, 73, 86, 121
  channel 130, 132
  chloride 80
soil 122, 301, 311, 315, 370, 374, 413
  biogeochemical cycles 47–50,
      52–54, 62, 64
solid phase 93, 116
South Africa 17, 61, 363
South America 49, 187, 294,
      330, 371
  see also Amazon, Brazil, Chile,
      Colombia, Ecuador
Southampton Water 222, 364
South Carolina 431
Southend 154
Soviet Union 4, 373
  see also Russia
Spain 76, 190, 294, 297, 304, 309, 313,
      355, 369, 373, 376, 378
spawning 273, 275
speciation 38–39, 42, 83–91, 95, 121,
      238, 252–254, 263, 404–405, 408,
      419, 490
  modelling 89, 116
  see also WHAM
species diversity see univariate
      measures
species number see univariate
      measures
species richness see univariate
      measures
species sensitivity distribution (SSD)
      411–415, 421
spillover model 208
spionid polychaete 228
  see also Polydora
Spokane River 310
sponge 228
spotted sandpiper 352
squid 464
  see also Loligo
stability constant see equilibrium
      constant
standing stock 47
Stanleya 346
starfish 281, 318
  see also Heliaster, Meyenaster
statistical power 274, 470
stegocephalid amphipod 156, 158
stickleback 398
  see also Gasterosteus
stilt 274, 331, 344, 349, 351,
      474–475
  see also Himantopus
St Lawrence River 58, 76, 104, 385
stonefly 138, 282

STORET database (USEPA) 33, 44, 232, 235, 243, 246, 250, 361, 401, 405, 407, 409, 411–412, 421, 492
Straits of Gibraltar 304
stratosphere **47**–48
stream
  benthos 280–282, 286, 316, 318, 408
  biogeochemical cycles 47–48, 51, 53, 230
  biomonitoring 190, 199
  dissolved concentrations 79, 271
  ecotoxicological effects 282, 286, 410–411
  mining 190–192, 271, 318–321
stress 204, 209, 212–213, 215
  see also biomarker
stress protein **205**, 213–214
Strickland River 309, 322
striped bass 143, 342, 345, 389, 421, 471, 474
  see also Morone
strontium 7, 292
sturgeon 265, 272, 345, 397, 471, 475, 486
  see also Acipenser, white sturgeon
Stylochus 227
subcellular metal distribution 146–150, 208, 227, 254, 380, 465, 491
subchronic toxicity **233**
sublethal toxicity 33, 233, 235, 243, 265
sucker 189
Sudbury 107, 263, 287, 289, 311, 314, 323, 350
Suisun Bay 81, 195, 334, 467, 469, 471
sulphate 81, 85, 111, 113, 119, 123, 132, 301, 303, 312, 337, 339, 462
sulphate reducing bacteria 374
sulphide 11, 39, 113, 116, 146, 296–297, 302–304, 312, 374, 382, 426
  sediment 110–111, 117, 122–123, 184, 318, 429, 432, 442–443, 447, 456
  see also acid volatile sulphide (AVS)
sulphur 7–11, 130, 148, 156–157, 206, 262, 294, 304, 346, 349
  dioxide 55, 287–288, 311
Superfund 301, 485
superoxide dismutase 214
surf scoter 486
  see also Melanitta
suspended particulate material (SPM) 57, 93–94, 269

suspension feeding 114, 148, 174–175, 180, 184, 269, 278, 426
Swan Lake 288
Sweden 64, 350–351, 369, 373, 380, 387
Switzerland 363
swordfish 369, 380, 389
  see also Xiphias
Sydney 367

tachycardia 220
Taf River 105
tailings 51, 62, 166, 217, 280–281, **294**, 297–299, 308–309, 312–314, 316–317
  tailings pond 62, **294**, 299, 308–310
Talitrus 185
talitrid amphipod 160, 185, 187–188, 194, 198
  see also Orchestia, Platorchestia, Talitrus, Talorchestia
Talorchestia 160
Tamar River 105, 226, 298–299
Tanytarsus 218
Tavy River 105
taxonomy 6, 41, 143–144, 177, 186, 193, 198–199
TBT (tributyl tin) 2, 5–6, 20, 30, 41, 43, 88, 218–221, 268, **354**, 368, 473
  bioaccumulation 357–359, 361
  concentration 356, 361–365
  environmental management 362, 366, 495
  imposex 358–363, 365, 367
  metabolic breakdown 358, 361, 368
  partitioning 356
  recovery 365, 367
  toxicity 356, 358–363, 368
Tees 72, 82
Teign River 105
Tellina 439, 448
tellinid bivalve 175, 177, 180, 182–184, 318, 447–448
  see also Macoma, Macomona, Scrobicularia, Tellina
tellurium 7, 297
Temora 342
Tennessee 384
  University of 398
teratogenesis 218, 262, 268, 347–348, 351
terebellid polychaete 184
Tetraclita 154, 186, 200
tetra-ethyl lead 52, 78, 88
Tetraselmis 143
Texas 377, 461

Thais 143, 148, 166, 358, 360, 367
Thalassiosira 143, 146, 254, 338, 341, 448, 465
Thames 2, 82, 177
Thiobacillus 301
Tharyx 198
Thunnus 389
Thur River 69
Timpanoga 283
tin 9, 11
  organic form 88, 355, 363
    see also TBT
  rock 97
  sediment 105
  toxicity 357
  water 73
tissue residue 467
tissue residue criteria 467
titanium 4, 34, 73, 217, 369
tolerance
  ecotoxicological monitoring 224–226, 229, 277, 291, 312
  physiology 141, 149, 225
Tomales Bay 97, 121, 264
top-down effect 290
Torridge River 105
total maximum daily load (TMDL) 16, 26, 44, **47**, 59–62, 65, 68, 393–394, 419, 423
total organic carbon (TOC) 357, 359, 378
total oxyradical scavenging capacity (TOSC) **205**, 214, 221–222
toxic inventory *see* Toxics Release Inventory
toxicity 10–11, 25, 28, 39, 132, 153, 157, 204–208, **233**, 253, 396, 405, 413, 467, 491
toxicity testing 6, 14, 28, 32, 34–35, 41, 44, 170, 204, 207, 259, 276, 315, 361, 402, 419, 424
  acute 33, 44, 232, 235, 243, 246, 250, 361, 401, 405, 407, 409, 411–412, 421, 492
  chronic 33, **232**, 235, 246, 250, 361, 401, 405, 409, 411, 421
  dietary 243, 245, 254–256, 261–263, 315, 398–399, 420, 422, 461, 464, 470, 476
  in situ 429
  multi-species 239–240
  ranking 397
  sediment 428, 431–432, 434, 436, 451, 458
  single species 238, 400
  test species 244, 248

toxicology 6
Toxics Release Inventory (TRI) 18,
    44, 397
Toxic Substances Control Act 13
transporter proteins *see* carrier
    proteins
tributyl tin *see* TBT
trace metal **2**, 7, 9
transcriptome **205**, 212
transition metals 7, 9
Trichoptera *see* caddisfly
Trondheimsfjorden 219, 228
trophically available metal fraction
    (TAM) **127**, 149, 464
trophic level 3, 164, 166, 277, 291,
    345, 380, 382, 385–386, 466
trophic transfer **33**, **127**, 148–149,
    164, 166–167, 192, 229, 247, 254,
    285, 398–399, 466, 491
    trophic transfer factor (TTF) 165
troposphere **47**–48
trout 185, 224, 241, 256, 291
    brook trout 319, 352, 388
    brown trout 189, 256, 315–316,
        319–320, 421
    bull trout 224, 241, 316, 319
    cutthroat trout 241, 319, 351–352
    lake trout 279, 385, 389
    rainbow trout 141, 246, 256,
        289–290, 315–316, 319–320,
        346, 352, 381, 408, 421, 474
    toxicity 150, 256, 263, 270, 312,
        314–315, 319, 321, 347,
        351–352, 422, 485
    westslope cutthroat trout 319
    see also Oncorhynchus, Salmo,
        salmonid, *Salvelinus*
Tulare Lake 349
tuna 350, 369, 380, 389–390
    see also Thunnus
tunicate 132, 290
    see also ascidian
turbellarian flatworm 280, 287
Tuscany 211, 221
Tweed 72, 82
Tyne 72, 82
Type I error 274, 454
Type II error 274, 279, 454

ultra-clean method 38, 170, 376
Ulva 175, 186, 203
uncertainty 20, 24, **33**, 35, 257, 268,
    270, 352, 403, 407, 415, 432–433,
    462, 497
under-protection 403, 405, 412, 416,
    421, 460–461, 493

unit world model 62, 64
univariate measures 229, 276,
    284–285, 291, 317
Upper Silesia 192–193, 199, 201
uptake 129–150, 130–132, 135–137,
    162, 165, 172, 206–208, 283, 405
    channel 130–134, 131–132
    diet 141–150
    organometals
    physicochemistry 134–138
    physiology 138–141
    pinocytosis/endocytosis 132–133,
        145
    transporter 127, 130–132,
        130–134, 139
    uptake rate constant **127**, 133–141,
        161, 163, 165, 341–342
    see also Biotic Ligand Model, Free
        Metal Ion Activity Model
upwelling 76
uranium 198, 297, 329, 352
urban runoff 51, 57–59
US Army Corps of Engineers 451
US Bureau of Reclamation 331, 461
US Department of Interior 331
USEPA *see* Environmental Protection
    Agency
USEPA Science Advisory Board 400
US Fish and Wildlife Service (USFWS)
    461, 471
US Food and Drug Administration
    388, 390
US Geological Survey 38, 101, 333

vas deferens sequence (VDS) 360
vanadium 9–10, 73, 84, 86, 96–97,
    132, 271, 398
Vancouver Island 317
variability **33**, 43, 46, 268, 270–272
Venice Lagoon 221, 225
ventral caecum (amphipod) 156,
    158–160
Victoria, Australia 228, 290
Virginia 61, 121, 264, 377, 390
Visnesfjord 212, 217, 221
Vistula River 177
vitellogenin 207, 262, 347, 349,
    464, 473

Wadden Sea 453, 475
Wales 105, 154, 183, 185, 447
    see also Anglesey
walleye 350, 388–389
    see also Sander
Walloon 416
Warm Springs Creek 299, 306, 321

Wash 82
Washington 100
water effects ratio (WER) **394**,
    404–405, 408, 480, 485, 492
Water Framework Directive 16–17,
    29, 46, 59, 61, 65, 292, 393,
    401, 423
water permeability 140–141
water quality 2, 6, 15, 44, 56, 227,
    271, 327, 391, 393, 401
    water quality criteria (WQC) 15, 61,
        362, 364, 390–391, 393–**394**,
        399, 404, 408–409, 416, 418,
        421, 460–461, 463, 472, 479,
        486, 495
    water quality guidelines 16, 59,
        240, 261, 327, **394**, 400, 419
    water quality objectives 16, **394**, 479
    water quality standards 6, 16, 20,
        45, 61, 240, **394**
    see also ambient water quality
        criteria
Water Quality Standards for
    Protection of Aquatic Life 327
Wear 72, 82
weight of evidence (WOE) 276, **460**,
    476, 478, 480–484, 486
West Cleddau 105
West Virginia 329
WHAM 91, 406
whelk 143–147, 149, 165, 438
    see also dogwhelk, *Morula, Nucella,
        Thais*
whitefish 189
white sturgeon 344
    see also Acipenser
white sucker 265, 279
    see also Catostomus
white urchin 428
    see also Lytechinus
wildlife criteria 476–477
    reference dose 476–477
Willamette River 391
Williamstown 228
willow 42, 166, 308
    see also Salix
window of essentiality 415
window of tolerance 277, 397
window of vulnerability 277
winkle *see* littorinid winkle
winter stress syndrome 267, 475
Wisconsin 324, 376
World Bank Guidance (mine
    reclamation) 297
World Health Organisation 3–4,
    321, 367

Wye River 105
Wyoming 332, 371

xenobiotic *see* organic contaminant
Xiamen 199
Xiaoquinling 379
Xiphias 389
Xylorhiza 346
Xyrauchen 332

yellow perch 187, 289–290, 314
    see also Perca

zebra mussel *see Dreissena*
zinc 4, 7–11, 18, 25, 37, 54–55, 153
    assimilation efficiency 143–149,
        151, 165, 438, 466
    bioaccumulation 74, 152–159, 161,
        163, 184, 192, 201, 208, 251,
        284, 313, 322, 438–439,
        446–447, 466, 468

biodynamics 162, 180, 283
biogeochemical cycle 50–51, 54–58,
    74, 122, 310
biomonitoring 176, 178–182,
    184–185, 187, 190, 192–193,
    199–200, 319
colloid 69
detoxification 148, 155, 157, 212
efflux 140, 152–153, 177, 187, 198,
    202–203, 213, 217, 250, 341,
    356, 361–362, 417
mining 55, 59, 192, 296–297,
    302, 309
rock 97–98, 101, 105–108, 112, 117,
    298, 437, 447
sediment 97, 100, 102, 105–106,
    117, 120–121, 172, 190,
    228–229, 308, 310–311, 313,
    433, 444–445, 447, 450,
    452–453
speciation 86–87, 90, 115, 406

tolerance 225–226, 229
toxicity 215, 219, 221, 224,
    227–228, 246–247, 250,
    253–255, 257, 262, 282, 312,
    316, 399, 411, 415, 418,
    444–445
uptake 130, 132, 134–136,
    138–140, 146, 156, 159, 163,
    165, 170, 438
water 69, 72–73, 75–76, 79, 121,
    288, 304, 309, 313
zooplankton 174, 177, 180, 188, 209,
    211, 227, 243, 342, 345, 358, 383
    bioaccumulation 86, 142, 152, 166,
        288, 332, 343, 385, 464–465,
        467, 470
    feeding 74
    toxicity 240, 246–248, 250–251,
        254–256, 261, 264, 266,
        287–288, 421–422
    see also cladoceran, copepod, rotifer